Fundamentals of
Cytogenetics and Genetics

MAHABAL RAM
Formerly Project Coordinator
Indian Council of Agricultural Research (ICAR)
New Delhi

PHI Learning Private Limited
New Delhi-110001
2010

Best Seller
Reference Book

FUNDAMENTALS OF CYTOGENETICS AND GENETICS
Mahabal Ram

ISBN-978-81-203-4033-6

Published by Asoke K. Ghosh, PHI Learning Private Limited, M-97, Connaught Circus, New Delhi-110001 and Printed by Mohan Makhijani at Rekha Printers Private Limited, New Delhi-110020.

Dedicated to

my beloved wife ***Smt. Neelam Ram*** (***Malini***)
for her sacrifice and support extended to me during
my most difficult days of life
and
my three lovely children
Rahul, Rubal and Nupur

Contents

PART III MOLECULAR GENETICS

PART IV QUANTITATIVE GENETICS

Preface

Genetics is the science which deals with how a character is transmitted from parents to their progeny. This science has grown explosively during the last one hundred years, as a result, a lot of developments have taken place througout the world in this field from Mendelian (classical) genetics to the finest structure of gene discovered in viruses and microbes. Not the least, fingerprinting of complete genomic complex of *Drosophila*, rice and human beings has been done and man is trying to amalgamate the evolutionary process by transferring alien genes not only from unrelated species and genera but also from unrelated life forms like human gene in mice, pigs and bulls, chicken gene and fish gene in tomato, potato, corn and banana, as well as bacterial gene (*Bacillus thuringiensis*) in cotton and rice by developing modern genetic engineering techniques, though DNA structure and genetic code prepared by Watson and Crick (1953) are still hypothetical and unsolved problems. Because on the basis of Chargaff's chemical data, Wilkins and Franklin's X-ray diffraction photographs and inferences drawn from the model building, Watson and Crick (1953) proposed a hypothetical DNA structure stating that DNA exists as a double helix in which two polynucleotide chains are roped together in a spiral fashion which is doubtful for universal application because chromosomes which are known as packaging material for DNA are reported to exist in relational coiling manner like "spring" with their homologue in the nucleus. In addition, a polynucleotide chain consisting of sequence of nucleotides linked together by phosphate-diester bond is joined together adjacent to deoxyribose moieties. The two polynucleotide strands are held together in their helical configuration by hydrogen bonds between bases in opposite strands, the resulting base pair, being stacked between two chains perpendicular to the axis of the molecule like in steps of a spiral staircase, is also hypothetical. Therefore, readers are advised to be more clear about the fundamentals of cytogenetics and genetics instead of getting bogged down in the jungle of molecular genetic theories which are still in fluid state and will take time to clear many unsolved mysteries.

But definitely the science of genetics has emerged today as one of the most miraculous branches of biological sciences to be called an "oracle" on earth and Mendel, incarnation of God, because there is hardly any area connected to human life where it has not made home ranging from food security to health, hygiene as well as security from destructive war weapons like biological war weapons developed through "Recombinant DNA technology".

In the middle of 20th century, some books on the fundamentals of genetics such as *General Genetics* by A.M. Srb and R.D. Oven (1952), *Principles of Genetics* by E.W. Sinnot, L.C. Dunn and T. Dobzansky (1958), *Quantitative Genetics* by Falconer (1960), *Elements of Genetics* by W.R. Singleton (1967) and *Molecular Genetics* by G.S. Stent (1968) were published. But these are now outdated books. Some new and more exhaustive books have been published in the late 20th century such as *Principles of Genetics* by Eldan John Gardener, Michael J. Simmons and D. Peter Snutad (1978), *Cytogenetics* by C.P. Swanson (1981), *Science of Genetics* by P.W. Brun and P.J. Brunitro (1989), and *Basic Genetics* by D.N. Hartl (1991). These are standard books written by learned authors of repute, very costly and rarely available in the libraries of the developing countries. On the other hand, this book is written in a very simple, cohesive language so that students can easily grasp the subject matter.

The author has tried his best to provide the students with up-to-date information on cytogenetics and genetics by dividing the book into four parts. Part I deals with principles of cytogenetics. Part II covers principles of genetics. Part III discusses molecular genetics, and the last Part IV is devoted to quantitative genetics.

The book is intended as a textbook for undergraduate and postgraduate students of Botany, Agriculture, Zoology and Biotechnology for their courses in Genetics and Cytogenetics. In addition, the book will also be useful to students sitting different competitive examinations.

Mahabal Ram

Introduction

Science of genetics is a new branch of biological sciences. It deals with heredity and variation among the related organisms. Heredity is the cause of resemblance between the parents and the progeny as well as between the siblings. Variation is the reason of differences between the individuals. The science of genetics deals with the mechanisms and the basis for the resemblance and variation between the individuals of a species.

In the fifth century (B.C.) Hippocrates taught in the medical school the theory of heredity. He was of the opinion that a child possesses the qualities of his parents because semen contains within it small representative elements from all parts of the parental body. However, Aristotle argued that the filial embryo could not have been reconstituted from representative elements collected from parental bodies because:

(i) Parents produce offsprings endowed with parental traits (for example, grey hair) that are manifested only in the post-reproductive stage of life.
(ii) Children of crippled parents do not always show the defects of their progenitors.
(iii) Body is only the wrapping of the embryo; acquired characters cannot send their representatives in the semen.

However, variation and the evolution of new types and species were first studied by Lamark (1744–1829). There are variations from individual to individuals in a species. Lamark opined that causes of variations in nature are the conscious efforts, changed environmental conditions and the use and disuse of organs. According to Lamark, variations acquired in the lifetime of an organism are transmitted to the progenies which give the new species to form in due course of time. The main acquired characters are transmitted from parents to the progenies.

However, Weisman (1834–1914) disproved the theory of Lamark by contemplating that acquired characters are not inherited. Weisman conceptualized the theory that all organisms are made up of somatoplasm and germplasm. The somatoplasm builds up the body and gives energy. The germplasm are the source of transmitting the 'characters' from the parents to the progenies. Germplasms are thus immortal and continuously flowing from generation to generation. Thus concept of Weisman is of fundamental values.

In spite of some interesting experimental attempts to know the reality behind the inheritance, the predominant concept was blending inheritance, which means hereditary contributions of two parents are blended to produce an intermediate type.

The most scientific (realistic) explanation, which caught the attention of the biologists all over the world, came from Charles Darwin who was a naturalist and published his book "Origin of Species" in 1859. Darwin gave a fantastic and believable theory of 'Pangenesis'. According to the theory of pangenesis, every organ of the body gives rise pangenes, which together form the germ cells. Darwin believed that variation is a universal phenomenon occurring in nature. In all organisms, overproduction is a rule and this leads to competition, i.e. struggle for existence and survival of the fittest. Through natural selection, the unfits and the weak are eliminated in the competition. The quality which helps to the organism to the success, are inherited from generation to generation through pangenes.

The resemblance between the species could be explained that 'like begets like'. For example, wheat plant gives rise to wheat, barley to barley, corn to corn and man to man. The phenomenon of 'like begets like' is responsible for conserving the existing types. But yet there is a phenomenon of variation, which may alter the structure of the progeny.

The period from pangenes to present could be divided into the following four era:

(i) Mendelian era of genetics (qualitative genetics).
(ii) Quantitative (biometrical) genetics era.
(iii) Molecular genetics era.
(iv) Genetic engineering era (an era of alien-gene transfer).

Mendelian genetics begins in the middle of nineteenth century with the discovery of laws of inheritance in the garden pea by Greger Mendel (1822–1884), an Austrian monk. He started his experiments in 1857, and published his results in 1865. He made crosses between the different strains of garden pea, with contrasting characters and studied the segregation patterns of the different characters instead of the parents or the progenies as units. Mendel proposed the concept of hereditary units which are transmitted from the parents to the progeny, which determine the characters in the hybrid/progeny. The characters itself are not transmitted, but particles, units or factors determining the characters are transmitted from the parents to the offsprings. The appearance of the characters in the offspring is determined by gene, a term was given by Danish geneticist W. Johannsen (1909). This was a great challenge to the prevailing concept of the blending inheritance of the parental characters in their progeny, hence it did not get the support of the biologists of that era like Negelie because Mendel gave a new concept which was against the blending inheritance.

Mendel did not use the word gene. He used the word 'something' for the hereditary determinants. Some geneticists called them 'factor'. The significance of Mendel's principles of heredity was not appreciated and approved by the renowned biologist Negelie, who was chairman of Brunn Society of the Natural Sciences and Mendel's discovery remained almost a forgotten subject till 1900, when three biologists, Hugo DeVries, Carl Correns and Erikvon Tschermak independently rediscovered the reality of the Mendel's laws of inheritance. Thus, the science of genetics dawned in the twentieth century as a new and most powerful field of biological sciences.

Following the rediscovery of Mendel's laws, many theories and principles of inheritance of characters were discovered in 1902. Walter S. Sutton proposed the chromosome theory of heredity in which he postulated that behaviour of Mendelian factor of heredity and that of chromosomes are parallel. Walter S. Sutton proposed that Mendel's heredity units (factors) are located on the chromosomes, as a result of meiotic division pairs of factors located on the chromosomes are separated out along with the factors determining the characters. Sutton's discovery brought here the terms 'cytology' and 'genetics' under the terminology 'cytogenetics'.

In 1906, William Bateson discovered the phenomenon of linkage, when genetic factors remain associated with each other on the chromosomes. He coined the term 'genetics', a new field of biological sciences, though the term 'gene' was first used by W. Johnnsen in 1909. Bateson was the first who translated Mendel's paper in English and first showed that Mendel's laws of inheritance are also applicable to animals.

Around 1912, the concept of genetics was migrated from Europe to United States to the laboratory of Thomas Hunt Morgan at Columbia University. Initially, Morgan was skeptical of chromosome theory of heredity postulated by S. Sutton. But on the basis of elegant works of his three Ph.D. students—C.B. Bridges, H.I. Muller and A.H. Sturtevant, Morgan not only believed, but also upgraded the chromosome theory of heredity into the concept of genes being located in the linear array on each chromosome. Morgan (1912) enunciated his linkage theory and showed that genes located on the same chromosome are linked. Only the process of 'crossing over' could break down this linkage which assist in the recombination of genes coming from two parents and origin of a new type having characters of both parents. Johannsen (1909) by studying meiosis pointed out chiasmata, which was formed later on and considered the basis of explaining the phenomenon of crossing over. These researches proved that chromosomes are physical basis of heredity.

Mutation was another important genetic phenomenon discovered by De Vries in 1901, which was considered an important source of variation in nature and played important role in the evolution of new species. However, discovery of Muller in 1927 demonstrated that x-rays can induce mutations and these gene mutations can alter the developmental processes in the fruit fly (*Drosophila melanogaster*), opened new era in the science of genetics to create new genetic variability. Muller and Stadler (1927) demonstrated that x-ray treatment could be equally effective in inducing mutations in corn and other crops. Muller and Dobzhansky reported that genes when translocated to new position behaved differently. It was termed 'position effect'. In the study of nature and function of gene, 'position effect' occupy important place. Ram and Zaman (1972) reported that application of gamma rays can induce economic mutation in the rice and new induced characters (genes) could be utilized in conventional breeding programmes for development of better varieties of crop plants.

The theoretical basis of quantitative genetics was established around 1920 by work of Fisher (1928), Haldane (1928–1932) and Wright (1921). The theory consists of deduction of the consequences of Mendelian inheritance when extended to study the properties of population and simultaneous segregations of the genes at many loci. However, genetic properties of a population are influenced in the process of transmission of genes from one generation to the next by a number of factors in which environment is one of them.

Nineteen forties witnessed two significant discoveries leading to the chemical nature of gene. Oswald and Avery established that genes are composed of a specific type of nucleic acid called deoxyribonucleic acid (DNA) and not protein. Avery, Macleod and McCarty (1944) collectively demonstrated that when DNA from heat-killed, pneumonia-causing (virulent) bacteria *Pneumococcus*(s) were C mixed with avirulent bacteria (R), the avirulent bacteria were transformed into virulent form. This was beginning of the molecular genetics. Secondly, Beadle and Tatum (1941) proposed one-gene one-enzyme hypothesis, through which they indicated that action of each gene is carried out through the synthesis of protein (enzyme) which in turn catalyses a single chemical reaction. They proved this hypothesis using a multitude of biochemical mutants in *Neurospora crassa*, a fungus.

Molecular genetics took upward lead with the discovery of the molecular structure of DNA by James Watson and Francis Crick (1953). They made it possible to understand the chemical nature of gene and how genetic information is stored and released from a cell as well as transmitted from one generation to next.

In 1961, Francis Jacob and Jacques Monod gave the evidence of mechanism of gene regulation in bacteria called 'operon'. Operon consists of a series of genetic subunits (regions) interacting in such a way as to regulate transcription of a messenger RNA (mRNA) for forming copy of a gene. If formed, whether this mRNA directs (specifies) the synthesis of protein?

The concept of genetic code was cleared by M.W Nirenberg and Mathai (1961), M.W. Nirenberg and P. Leder (1964). They synthesized small RNA molecule of known composition and identified the exact code word for each amino acid by testing sixty-four possibilities for four DNA bases taken three (triplet) at a time.

It is believed that a molecule discovered in 1869 by Kossel was probably DNA. But the timing of this discovery was not ripe as a result like suppression of Mendel's laws of inheritance, this discovery as an important 'event' second time in the history of genetics was overlooked. Mendel's laws of inheritance were not believed upon probably because ironically the genes were discrete 'objects' was not conceived. The idea that DNA is the genetic material was overlooked because the molecular structure of gene was not understood, because then the consensus was that genes were made of proteins.

Nineteen seventies went with a revolution in the field of genetics with the development of 'recombinant DNA' technology called 'genetic engineering'. Applying this technology, the molecular geneticists can transfer the gene at will from one organism (species) to another organism (species). The molecular scientists have succeeded to decipher the complete DNA sequences of the genomes of bacteria, yeast and nematode, *Drosophila* and rice. Now they are trying to decipher the human genome and prepare a complete 'blue print' of gene composition of man.

The significance of science of genetics could be understood from the fact that applications of genetic principles have brought green revolution (increased food grain production), white revolution (increased milk production), yellow revolution (increased oil seeds production) and blue revolution (increased fish production) in the 20th century and this way, the human population was saved from devastation due to huger specially in Asia (China and India) where population growth have been like eruption of the volcano. The growth in the food grains production in India rose from 50 million tons in 1950 to 215 million tons in 2008, which has been beyond the projection of Malthus 1798. This happened by breeding genetically superior varieties of wheat and rice, as well as hybrids of corn, sorghum and pearl millet, etc. responsive to high input management and also tolerant to the major diseases and pests. Genetically, superior varieties of wheat, barley (hull-less and hulled both) and rice have been evolved tolerant to non-biotic environments also such as drought, saline, alkaline soils, heat and cold atmospheric conditions.

Today, genetic engineering with advancement of molecular genetics has entered in such an advance stage that the desirable genes could be identified, cloned and transferred not only from one plant species to the other but from one animal species to the other. The bacterial and viral genes could be transferred to the plant and animal species too. Such plants and animals are called transgenic plants and animals. Transgenic animals, transgenic crop plants—tobacco, cotton, Bt-cotton and corn have been developed by transferring genes from alien sources. Bt-cotton is the best example of transgenic crop evolved by Monsanto Seed Company. Similarly, in USA molecular geneticists have succeeded to transfer fish genes, chicken genes in tomato and human genes in pigs and mouse.

Not the least, 'Recombinant DNA Technology' is going to play havoc to the human race. The scientists can develop virulent races of not only plants, animals and birds diseases, but also of human diseases too, which could be called Biological War Weapon of Mass Destruction (WMD). The scientists have estimated that 4 mg of smallpox virulent virus culture can cause death of 0.1 million people. This is adverse impact of genetics for the survival of human race.

However, we can solve many legal issues related to paternity, crime and corruption through DNA fingerprinting. At the same time, we can produce noble human race, if the genetic principles are taken into account in settling planned marriages.

Science of genetics is now hundred years old. There are some good books available in this field written by foreign writers. But the students and teachers of the underdeveloped countries like Asia and Africa are having medium of instruction either in their mother tongues or in regional languages where English is the link language as a result they are not able to grasp too well, the fundamentals of genetics, cytogenetics, etc. from these books written in English by the Western writers. Secondly, the prices of these books are so high (thousands rupees) that majority of students coming from poor and middle-class families feel inability to purchase. Even the libraries of the rural agriculture colleges do not dare to buy the high-priced books written by foreign writers. Moreover, there are separate books for cytogenetics, quantitative genetics and molecular genetics and the students of undergraduate and postgraduate classes find very less time to consult a number of books written on different subjects due to limited time in trimester/semester system of education in the agriculture universities and colleges. Moreover, good books are rare and these books are generally kept by the teachers. Author, therefore, has tried to write this book in very simple language and the price of this book is also well within the reach of the students, teachers and the research workers of the agriculture universities and colleges in the Third World countries. In addition, this book covers the major portion of the courses on the principles of genetics, cytogenetics, quantitative genetics, molecular genetics and biotechnology taught in the colleges and the universities. Therefore, students have not to run to consult a number of books to prepare them for examinations and interviews etc.

For the above said reasons, author has consulted almost recent books available in our libraries on genetics, cytogenetics, quantitative genetics, molecular genetics and also on genetic engineering, etc. Conclusively, this book has been written to help the students coming from economically poor to middle-class families and also to assist fund starved rural agriculture colleges who due to financial crunch are not able to buy costly books. This is a textbook written according to the syllabus prescribed by the Indian Council of Agricultural Research (ICAR) and UGC for undergraduate and postgraduate classes.

FACTS: HUNDRED YEARS JOURNEY OF GENE

1865 Mendel read his paper in Brunn Society of Natural History.

1866 Mendel's paper was published in Proceedings of the Brunn Society of Natural History.

1900 Mendel's work was discovered by DeVries, Correns and von Tschermak.

1902 Boveri and Sutton demonstrated the presence of paired chromosomes (homologous) in diploid species.

1905 Bateson named the Science of Genetics.

1908 Hardy and Weinberg formulated Hardy–Weinberg Law, related to gene frequencies in a randomly-mating population.

1909 Johannsen gave the word 'gene'.

1909 Garrod published the book the 'Inborn Errors of Metabolism'.

1910 Morgan (Nobel Prize, 1933) discovered sex linked inheritance.

1911 Morgan postulated chromosomal basis of inheritance.

1927 Muller (Nobel Prize, 1946) reported the use of CLB technique to demonstrate that x-rays possess mutagenic effects.

1928 Griffith discovered transformation in *Diplococcus pneumoniae.*

1931 Creighton and McClintock in maize, and Stern in *Drosophila* demonstrated that genetic recombination can be correlated with the exchange of chromosomal marker.

1940 Oliver demonstrated recombination within gene—the lozenge functional unit in *Drosophila.*

1941 Beadle and Tatum's work published (Nobel Prize, 1958) on *Neurospora* giving one gene-one enzyme concept.

1944 Avery, Macleod and McCarty demonstrated that in *Pneumococcus*, transforming principle is DNA.

1946 Lederberg (Nobel Prize, 1958) and Tatum discovered conjugation in bacteria.

1950 McClintock's (Nobel Prize, 1983) first paper was on transposable element in maize.

1952 Hershey (Nobel Prize, 1969) and Chase reported that genetic material of bacteriophage is DNA.

1952 Zinder and Lederberg reported phase-mediated transduction in bacteria.

1953 Watson and Crick (1953) (Nobel Prize, 1962) worked out the double-helix structure of DNA, using x-ray diffraction data of Wilkinson (Nobel Prize, 1962) and base composition data of Chargaff.

1953 Benzer's first paper was on the *fine structure of gene*—the phase T4 rII locus.

1956 Tjio and Levan established that a normal human possesses 46 chromosomes.

1957 Fraenkel-Conrat and Singer reported that genetic information of Tobacco Mosaic Virus is stored in RNA.

1958 Meselson and Stahl reported that DNA replication is semi-conservative.

1958 Kornberg (Nobel Prize, 1959) isolated DNA polymerase from *E. coli.*

1959 Ochoa (Nobel Prize, 1959) discovered RNA polymerase.

1961 Jacob and Monod (Nobel Prize, 1965) proposed 'Operon model' for regulating gene expression in bacteria.

1964 Colinearity between gene and polypeptide was shown by Yanofsky and colleagues.

1965 Holley worked out nucleotide sequences of RNA—yeast alanine tRNA.

1966 Cracking of complete genetic code was reported by Narenberg and Khurana (Nobel Prize, 1968) and coworkers.

1970 Nathan and Smith (Nobel Prize, 1978) isolated first restriction endonucleases.

1970 *Reverse transcriptase* of RNA tumour viruses were identified by Baltimore (Nobel Prize, 1975).

1972 First recombinant DNA was produced *in vitro* in Berg's laboratory (Nobel Prize, 1980).

1976 Bishop and Varmus (Nobel Prize, 1989) reported relationship between *proto-oncogene* and *oncogene*.

1977 Publication of DNA sequencing technique of Maxam and Gilbert and of Sanger, Nickden and Coulson (Nobel Prize, 1980).

1977 Publication of complete 538' nucleotide sequence of phase '$\Phi \times 174$' by Sanger and colleagues.

1978 Discovery of splicing of adenovirus RNA's.

1982 Publication of complete 48502 nucleotide pairs of phase Lambada by Sanger and colleagues.

1987 Start of Human Genome Project.

1988 Watson became coordinator of Human Genome Project.

1989 Tsui, Collins and colleagues cloned the *cystic fibrosis* gene, which is responsible for dreaded disease among children.

1995 *Haemophilus influenzae* genome was sequenced.

1996 *Saccharomyces cerevisiae* genome was sequenced.

1998 *Caenorhabditis elegans* genome was sequenced.

1999 Human chromosome 22 was sequenced.

2000 *Drosophila melanogaster* fruit fly genome was sequenced.

2005 A draft sequence of the entire human genome was published.

PART I

PRINCIPLES OF CYTOGENETICS

CHAPTER

1 Evolution of Cell

The evolution of cell was the most crucial event in the origin of life. The molecular biologists say that the first cell was 'obcell', a semiorganism, halfway between open population of molecules and true cellular organism since its genes were not yet enclosed by the membrane and in this condition, obcells could have been more susceptible than cells to the invasion by the parasitic nucleic acid. Only when obcells invaginated to form true cells, a complete individual metabolism and full organic individuality could have been evolved. The mechanism of evolution of 'obcell' which is known as the obcell theory of the origin of first cell (obcell) is described below (Cavalier-Smith, 1991).

Life is a product of a systematic evolutionary process. It is believed that evolution of efficient DNA; its replication, transcription, protein synthesis, membrane growth and division may have taken place on the outside of liposome like lipoprotein vesicle—the 'obcell'. The different stages involved in the origin of obcell as shown in Figure 1.1 (Cavalier-Smith, 1991) are:

Stage I: The obcell got its energy by photophosphorelation and its organic food ready-made from the prebiotic soup existing on the newly formed earth. It evolved porters to conserve food molecules temporarily inside the obcell and had an internal murien skeleton.

Stage II: After it becomes flattened, it invaginated possibly by the asymmetrical growth of its skeletal murien peptidoglycan.

Stage III: It formed a double-membraned vesicle that fully enclosed the formerly exposed chromosome and ribosome.

Stage IV: The porter could have modified to import food molecules into newly created cytoplasm, plasma membrane (PM) and outer membrane (OM). The obcell proteins were either embedded in the membrane by hydrophobic region or bound to ribosomal RNA or DNA which were attached to the obcell surface by amphipathetic proteins. Evolution of CO_2 and fixation would have greatly stimulated the evolution of intracellular metabolism to create the first autotrophic negibacteria (probably an anaerobic bacteria) from which all other cells could have evolved.

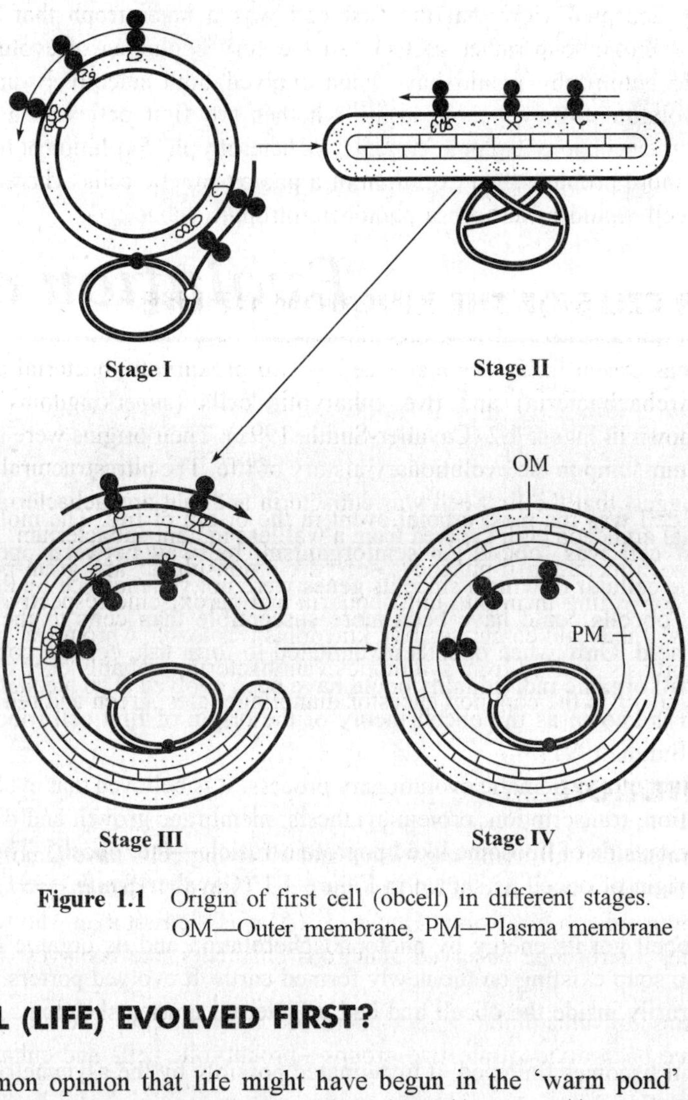

Figure 1.1 Origin of first cell (obcell) in different stages.
OM—Outer membrane, PM—Plasma membrane

WHERE CELL (LIFE) EVOLVED FIRST?

There is a common opinion that life might have begun in the 'warm pond' instead of a sea because of higher concentration of salts like K^+, Na^+, Mg^{++}, Cl^-, Ca^{++} in the sea than cytoplasm of the microorganisms. It is believed that small isolated pockets of wet soil perhaps in a subarctic, subtemperate climate subject to volcanism might have provided an ideal condition for the origin of obcell. The synthesis of activated molecules would have been favoured by drying of the organic solution and their heating by sun or volcanism. After being dissolved in rain, any degree of concentration could have occurred by partial drying or cooling. Light covering of certain mineral could have protected from ultraviolet damage while transmitting sufficient visible light for photosynthesis to evolve.

RNA is considered as the first molecule arose in the prebiotic soup. It is believed that early RNA world might have involved in the evolution of cooperative population of different RNA molecules, which would have been very sensitive to the invasion of selfish RNA molecules (Cavalier-Smith, 1987).

It is generally accepted view that the first cell was a heterotroph that fed on organic molecules in the prebiotic soup rather on CO_2. At the very beginning of evolution, numerous enzymes to enable autotrophy would have been evolved both selectively unnecessary and mutationally improbable. Many people argue whether the first cell was a photosynthetic heterotroph like a green eubacterium or a fermentative heterotroph? Evolution of fermentative cell would have posed more problem than evolution of a photosynthetic cells. Therefore, it is better to think that first cell would have been a photoheterotrophic eubacteria.

FOUR MAJOR CELLS OF THE KINGDOM OF LIFE

The living world has arisen from four major cells—two prokaryotic bacterial cells (kingdoms Eubacteria and Archaebacteria) and two eukaryotic cells (superkingdoms Archezoa and Metakaryota) as shown in Figure 1.2 (Cavalier-Smith, 1991). Their origins were the most significant event of quantum jump in the evolutionary history of life. The ultrastructural, molecular and fossil evidences suggest that the first cell was eubacteria and that archaebacteria and the primitive, a mitochondrial archezoa both evolved from a wallless mutant eubacterium. Subsequently, a tetraciliated archezoan metamonad phagocytosed purple nonsulphur and gram-positive bacteria (both eubacteria), converting them into mitochondria and peroxisome respectively, creating the first metakaryote. Metakaryote comprises five kingdoms: Protozoa, Chromista, Fungi, Animalia and Plantae. Chloroplasts evolved from symbiotic cyanobacteria, probably once only either in the ancestral metakaryote or in the common ancestor dianoflagellates, green and red algae.

CELL CONSTITUTION

Cell is the structural unit of all living organisms and all living cells have almost structural and functional similarities. However, size of cells varies from 2 μm to 10 μm in prokaryotes and from 2 μm to 10,000 μm in eukaryotes. Robert Hooke (1665) was the first man who while examining the dead cork under microscope observed numerous chambers like beehives. He called these units as cell.

Many organisms are unicellular while others are multicellular. Therefore, for the study purposes, cells have been divided into two groups—prokaryotic cells and eukaryotic cells.

Prokaryotic Cells

These are found in unicellular organisms like bacteria, protozoans and blue-green algae. These cells possess the following structures:

Cell membrane

Prokaryotes have only one membrane, called cell membrane or plasma membrane in which all the components of cell are enclosed.

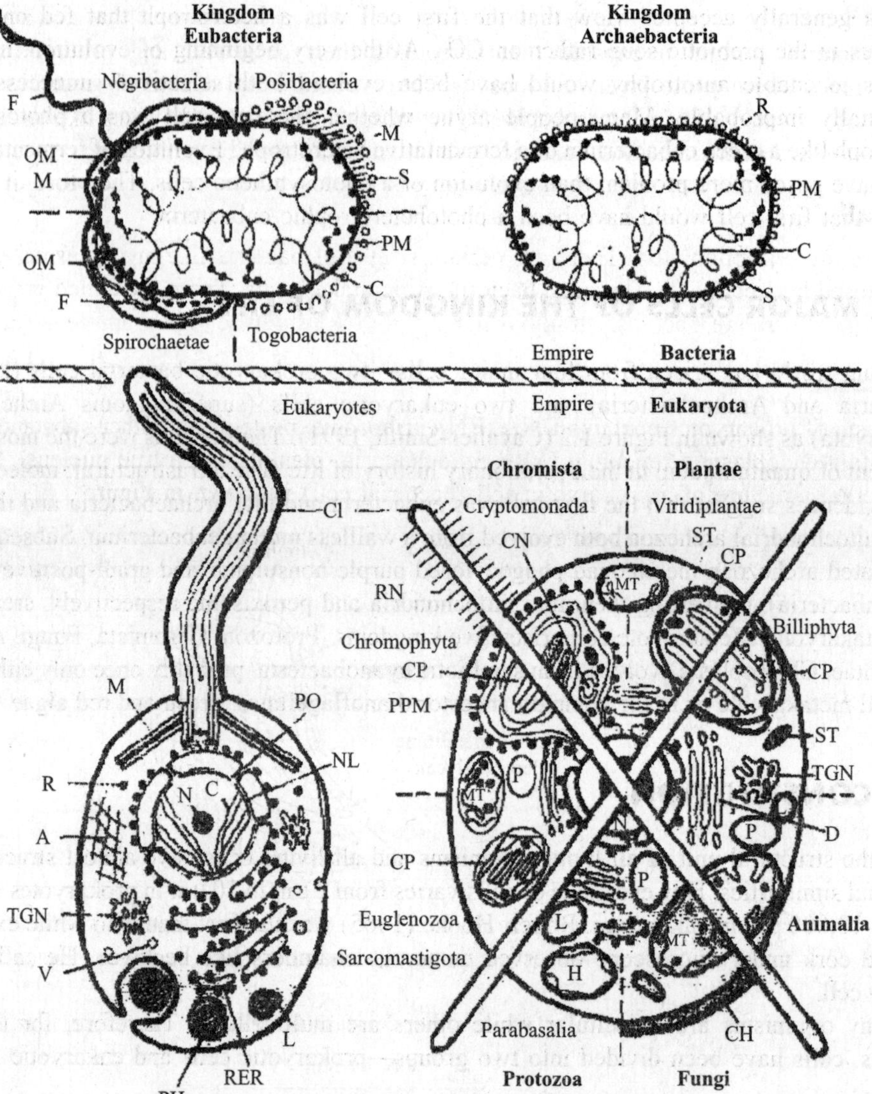

Figure 1.2 Four major cells formed the origin of life. Because eubacterial and metakaryote cells are substantially diverse than those of archaebacteria and archeozoa, their major subtypes are indicated by subdividing cells into separate sectors.

PM-plasma membrane, OM-outer membrane, M-murein wall, F-flagellum, S-proteinaceous, S-layer of cell wall, C-chromosome, R-ribosome, CI-cilium, RN-retroneme, M-microtubules, A-actin filament, NL-nucleus lamila, PO-nuclear pore complex, N-Nucleolus, RER-rough endoplasmic reticulum, D-Golgi dictyosome, TGN-trans-Golgi network, PH-phagosome, V-cytotic transport vehicle, L-lysosome, MT-mitochondrion, P-peroxisome, CP-chloroplast, PPM-periplastid membrane, CN-cryptonucleus, ST-starch, H-hydrogenosome, CH-chitin.

Capsule

Some bacteria contain a capsule that surrounds their cell wall and protect them from phagocytosis. The capsule layer is a loose gel-like structure and can be observed by negative staining. It is composed of polysaccharides. The encapsulated bacterial strains like *Streptococcus pneumoniae* are pathogenic and bacteria without capsules are non-pathogenic.

Cell wall

But there are exceptions too. There are certain strains of bacteria like mycoplasma, which are not enclosed by a cell wall. There are bacteria, which take stains. These are called gram-positive bacteria and there are bacterial strains which do not take stains, called gram-negative bacteria.

Cytoplasm

Cytoplasm of prokaryotes contains DNA, RNA, ribosome, proteins and other substances but do not contain endoplasmic reticulum (ER), mitochondria, plastids and a true nucleus.

Internal structure of a prokaryotic cell (bacterial cell) is shown in Figure 1.3.

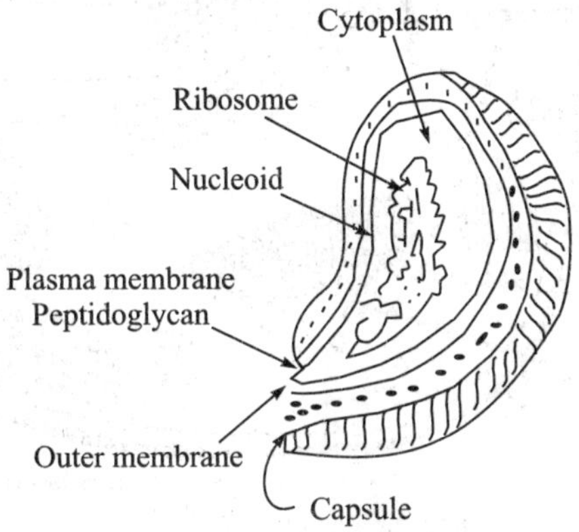

(a) Longitudinal section of a bacterial cell.

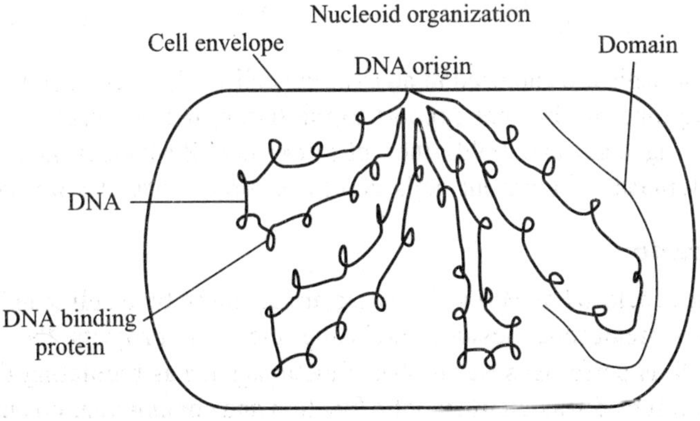

(b) Nucleoid organization of a bacterial cell.

Figure 1.3 Internal structure of a prokaryotic cell (bacterial cell).

Eukaryotic Cells

The eukaryotic cells on the other hand possess internal membrane in addition to plasma membrane. Plant cells possess cell wall. Cytoplasm possesses several distinct structural bodies such as nucleus, endoplasmic reticulum, Golgi complex, mitochondria, (chloroplast in plant cells) lysosomes and ribosomes, etc. (Figure 1.4).

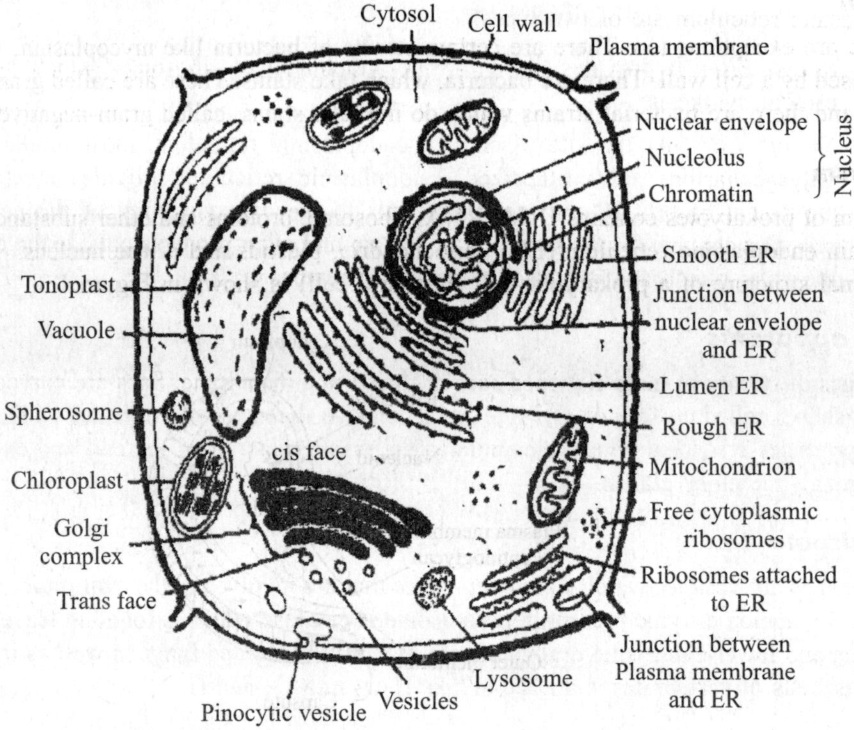

Figure 1.4 Structure of eukaryotic cell (plant cell).

Cell wall

The eukaryotic plant cells are enclosed in an external cell wall which encloses plasma membrane and provide rigidity and mechanical support to the tissue and organism. A middle lamella lies between neighbouring cells and holds the cells together. Plasmodesmata are small openings through which cytoplasm of adjacent cells comes in direct contact with each other.

Plasma membrane

Animal cells are not enclosed by cell wall. These are enclosed by a cell membrane called plasma membrane. Plasma membrane besides enclosing the cell, protects the cell from external environment as well as performs several other functions such as regulating the movement of the material inside the cell by diffusion, metabolic function and communication and adhesion between the neighbouring cells.

Cytoplasm

Eukaryotic cytoplasm contains several organelles such as endoplasmic reticulum, Golgi apparatus, peroxisome, vacuoles, centrosome, ribosomes, chloroplast (plant cell), mitochondrion, nucleus, nucleolus and cytoskeleton, etc.

Endoplasmic reticulum (ER)

Endoplasmic reticulum are of two types:

(a) *Rough ER:* It provides place for the attachment of ribosomes which are sites for protein synthesis.

(b) *Smooth ER:* At the site of smooth endoplasmic reticulum, most of the lipids and polysaccharides are synthesized. Endoplasmic reticulum provides channel for the movement of protein and other materials to the different parts of the cell and also provides channel for the secretion of protein and other products outside the cell through the golgi membrane → vesicle → plasma membrane.

Golgi apparatus

It consists of twenty or more flattened sacs enclosed in a membrane. Sacs are curved and attain convex shape, called *cis* face on one side and a concave shape called *trans* face on the other side.

It produces lysosome which accumulates the waste material of the cell and exudes out of cell through secretory glands.

Peroxisome

Peroxisomes are vesicles which contain enzymes that are involved in the respiratory mechanism. The most common enzyme present in peroxisome is catalase which is found in leaves and seeds of plants and lower eukaryotic organism such as yeast, algae and fungi as well as in the kidney and liver cells of vertebrates. Catalase breaks H_2O_2 into O_2 and H_2.

$$H_2O_2 + \text{Catalase} \rightarrow O_2 + H_2$$

Vacuoles

Vacuoles of different sizes are found in the plant cells. These are enclosed in a membrane called tonoplast and function as water reservoirs and also store cellular waste materials.

Centrosome

Centrosome is located adjacent to the nucleus in the animal cells but not in the plant cells. Each centrosome contains two cylindrical centrioles in the animal cells. They are always found in pair; one member is oriented vertical to other. Replication of centriole takes place in S stage and each centrosome moves to the pole to form spindle fibre at metaphase. Thus, centrioles are involved in the formation of mitotic apparatus during cell division.

Ribosome

Ribosomes are found free in the cytoplasm of prokaryotes and are attached to the endoplasmic reticulum and the outer membrane of nuclear envelope in eukaryotes. Ribosome consists of two subunits: 50s and 30s, jointly 70s in prokaryotes and 80s (two subunits 60s and 40s) in eukaryotes. Ribosome consisting of protein and RNA is called ribosomal RNA (rRNA). Ribosomes are called 'workshop' for protein synthesis.

Chloroplast

Chloroplasts are the seat of photosynthesis in the plants and produce carbohydrates in presence of light, CO_2 and water.

$$6CO_2 + 6H_2O \xrightarrow{\text{Light}} C_6H_{12}O_6 + 6O_2$$

Chloroplast is a symbiont and it is hypothesized that cynobacteria were engulfed by metakaryotes which would have been a non-photosynthetic phagotrophs. If this theory is correct, there would have been three major symbionts—organelle conversion to produce phototrophs; the origin of chloroplasts, chromista and photosynthetic euglenoids.

There are different views about origin of chloroplast. Mereschkowsky (1910) favoured five separate symbiotic origins for chloroplast; Margulis (1970, 1981) suggested 20. But phylogenetic data do not favour more than one or two; chloroplast rRNA sequences favour one and nuclear rRNA sequence not more than two. But Cavalier-Smith (1982, 1987) favours a single ancestral cynobacterial symbiont origin.

There are two different ways of producing a chloroplast with a three-membrane envelope. It is even possible that those of dinoflagellates and euglenoids originated by a different mechanism. In fact, a primary monophyletic origin of chloroplasts in an early dinoflagellate host and a secondary transfer of a chlorophyll b, containing eukaryotic symbiont to euglenoid would be very neat way of explaining all the apparently contradicting phylogenetic data.

Mitochondria

Kolliker in 1850 discovered an oval particles in the cytoplasm and Berrda in 1897 called it mitochondria. Mitochondria are associated with energy generation, therefore, they are called 'Power House' of the cell. They are involved in electron transfer and oxidative phosphorylation in Krebs cycle. They contain protein, lipids, RNA and DNA. The interior of mitochondria is called matrix. Mitochondria are involved in glucose oxidation.

$$C_6H_{12}O_6 + 6O_2 \rightarrow 6CO_2 + 6H_2O + 686 \text{ kcal}$$

Mitochondria originate by growth and division of pre-existing mitochondria. Mitochondria also have symbiotic origin from purple respiring bacteria. Schnepf in 1964 proposed that outer mitochondrial membrane is evolved from food vacuole membrane of the eukaryotic host that engulf the bacterium.

Nucleus

Nucleus was first discovered by Robert Brown in the year 1833. It is a key part of eukaryotic cell. Nucleus is usually spherical or oval in shape. There is usually a single nucleus in the cell, but multinucleate condition is also observed in some protozoa and fungi, which results due to repeated nuclear division without division of cytoplasm.

Nucleus controls and commands essential genetic and biosynthetic activities related to growth and reproduction of the cell. It contains chromosomes (chromatin-singular) and genes. The cell's DNA is confined, replicated and transcribed in the nucleus. Thus, it is a vault of genetic information encoding the past history and future prospect of the plant and animal life.

Nucleus of a plant cell is shown in Figure 1.5.

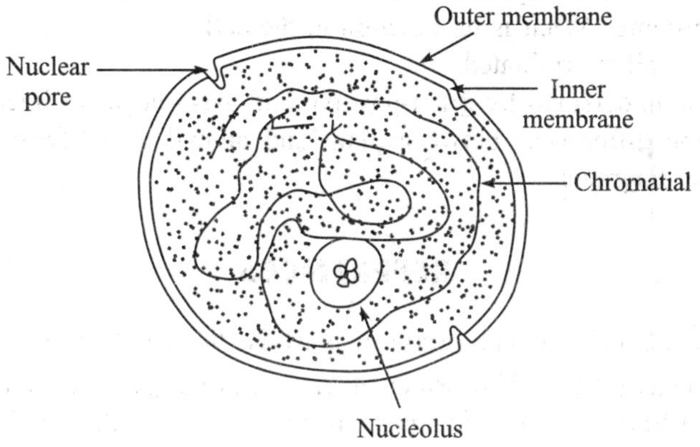

Figure 1.5 Nucleus of a plant cell.

Nucleolus

A spherical body found in the nucleus is called nucleolus. It is found in the higher organisms and is attached with specific region of a particular chromosome. It disappears during prophase of mitosis and meiosis and reappears during telophase. Usually, there is only one nucleolus per cell but in polyploids, there may be more than one nucleoli per cell. The nucleoli sometimes fuse and form one large nucleolus. The nucleolus consists of granules, fibrils and matrix.

The important function of nucleolus is the formation of ribosomes and synthesis of RNA. It can synthesize 70–90 per cent of the cellular RNA in many cells.

Chromosomes

Chromosomes form the main components of the nucleus. Each chromosome consists of two chromatins which are composed of proteins (55%), DNA (40%) and RNA (4–5%). Chromatin forms the basic unit of chromosomes and play important role in governing the inheritance of characters from parent to progeny. Chromosomes are darkly stained rod-shaped bodies visible under the light microscope in a cell during metaphase stage of mitosis. Strausburger discovered chromosomes in 1875 and Waloleyer in 1888 named chromosome. Shape, size and number of chromosomes differ from species to species. In plants and animals, chromosomes have prepotency of self-duplication.

REVIEW QUESTIONS

1. Draw typical structures of plant and animal cells and label their parts.
2. Draw structures of prokaryotic (bacterial) and eukaryotic (higher plants) cells and label their constituents.

3. Write brief notes on plasma membrane, endoplasmic reticulum and cell wall.
4. Draw the structure of a mitochondrion and state its function. Does mitochondrion possess DNA?
5. What is the function of chloroplast? Does chloroplast possess genetic material, the DNA?
6. What do you mean by cytoskeleton? What is the function of lysosome?
7. What is ribosome? What is its function in the cell?
8. Explain how cell is originated.
9. What do you understand by gram-positive and gram-negative bacteria?
10. What are the differences between plant and animal cells? Draw their well-labelled comparative diagrams.

REFERENCES

Cavalier-Smith, Thomas (1982): The origin of plastids, *Biol. J. Linn Soc.*, **17,** 289–306.

Cavalier-Smith, Thomas (1987): The origin of cell: A symbiosis between genes, catalyst and membrane, Cold Spring Harbor Symposium, *Quant. Biol.*, **52,** 805–824.

Cavalier-Smith, Thomas (1991): The evolution of cell, in: *Evolution of Life,* Osawa, S. and Honjo, T. (Eds.), Springer-Verlag, Tokyo, 271–304.

Margulis, L. (1970): *The Origin of Eukaryotic Cells,* Yale University Press, New Haven.

Margulis, L.(1981): *Symbiosis in Cell Evolution*, Freeman, San Francisco.

Mereschkowsky, C. (1910): Theoric der Zwie Plasmaarten als Grundlage der Symbiogenesis, ciner veven Lehre von der Entstehung der orhanismen, *Biol. Zentralbl.*, **30,** 278–303, 321–347, 353–367.

Ram, M. (2005): *The Unfolding Reality of Life: The DNA world*, Dominant Publishers, New Delhi.

CHAPTER

2 Cell Division

Schleiden and Schwann in 1838 defined cell as the unit of structure and function. The great Naturalists and Philosophers Aristotle in antiquity and Paracelsus in the renaissance concluded that "all animals and plants, however, complicated, are constituted by a few elements which are repeated in each of them". Many centuries later, it was found that a single cell can constitute an entire organism as in protozoa or it can be made of many cells that are grouped and differentiated into different tissue systems and organs forming of a multicultural organism. The cell is thus a fundamental structural and functional unit of living organism, just as the atom is in the chemical structure of matter. The cell can be considered as the organism itself, often very specialized and composed of many elements, the sum of which not only constitutes the cellular unit, but also has particular significance in organism as a whole.

The cell consists of a small mass of protoplasm, the cytoplasm containing a nucleus, and surrounded by plasma membrane. The cytoplasm appears as an amorphous homogeneous substance, the ground cytoplasm containing refractive particles of various sizes are mitochondria, the spherical bodies termed plastids (granules), chloroplasts (green pigments, chlorophyll), xanthophyll and carotenoids. The function of chlorophyll is photosynthesis, a process of immense importance in the biological world. The mitochondria and chloroplasts are considered cell organelles. Some other cell organelles such as Golgi substance and centrioles are also present in the plant cell. Figure 2.1 shows cellular components of bacterial, yeast, animal and plant cells.

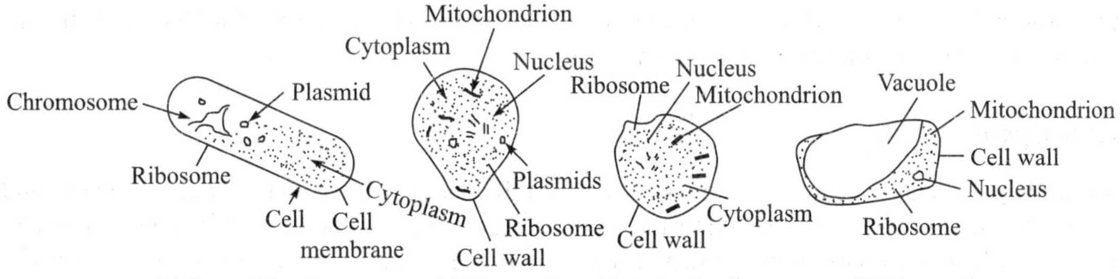

Figure 2.1 Cellular components of bacterial, yeast, animal and plant cells.

The prokaryotes lack true nuclei. Bacteria and blue-green algae are prokaryotes. In prokaryotes, the genetic material is a simple circle of DNA although ancillary circles of DNA called plasmids are found frequently in the cell.

On the other hand, eukaryotic cells are characterized by a true nucleus enveloped in a nuclear membrane, which divides the cell into two main compartments—the nucleus and cytoplasm. The cytoplasm is restricted by the plasma membrane. In plant cell, the plasma membrane is covered and protected on the outside by a thick cell wall through which there are tunnels, the plasmodesmata, by which the cell intercommunicates with neighbouring cells by means of various cell processes.

The non-dividing (interphase) nucleus is bound by a nuclear membrane, which is not clearly visible except, by electron microscope. The nuclear membrane is continuous with a cytoplasmic double membrane system. The endoplasmic reticulum to which the dense granular structures, the ribosomes are seen attached. The ribosomes are rich in ribonucleic acid (RNA) and play an important role in the protein synthesis.

The nucleus consists of nuclear sap, karyolymph—a clear non-staining proteinaceous colloidal material. The second is generally spherical, densely staining body called nucleolus. Many nuclei contain two or more nucleoli. The nucleoli contain ribonucleic acid (RNA), some deoxyribonucleic acid (DNA) and proteins. The third, the chromatin fine thread-like strands of deoxyribonucleic acid (DNA) which consist of histones (five kinds of basic, positively charged, low-molecular weight proteins) and non-histone proteins (all other proteins of chromatin), may be hundred or even more in higher organisms.

During nuclear division, the chromatin strands contract and appear thicker, which become visible and known as chromosome. The DNA carries the coded genetic information of most of the cellular activities. However, many of them activate through coded information by DNA found in cytoplasm.

MITOSIS

Mitosis is a continuous process. If the chromosomes of a cell are not visible, that means the cell is not ready for mitotic division. This stage is called *Interphase.* There are four major stages of mitosis excluding interphase. These are called *Prophase, Metaphase, Anaphase* and *Telophase.* Before mitosis takes place, the DNA in the chromosome is synthesized during the period of late interphase called S stage. When DNA synthesis does not take place, this period is called G1 and G2 respectively. Watson in 1946 referred G as a sequence of preprogrammed operation in preparation for DNA replication. This may include cytoplasmic synthesis of a DNA polymerase (an enzyme system catalyzing DNA replication).

Cell Cycle

The cell cycle means life cycle of an individual cell. It consists of two phases, i.e. *interphase* and *mitotic* phase. *Interphase* is very important phase because it prepares the cell for undergoing cell division, which is important phenomenon for growth and repair of the torn tissues of the body. Therefore, interphase takes longer period because it has to perform three important functions which can be grouped into three phases—G1 phase (pre-DNA synthesis phase), S phase (DNA synthesis phase), G2 phase (post-DNA synthesis phase) (Figure 2.2).

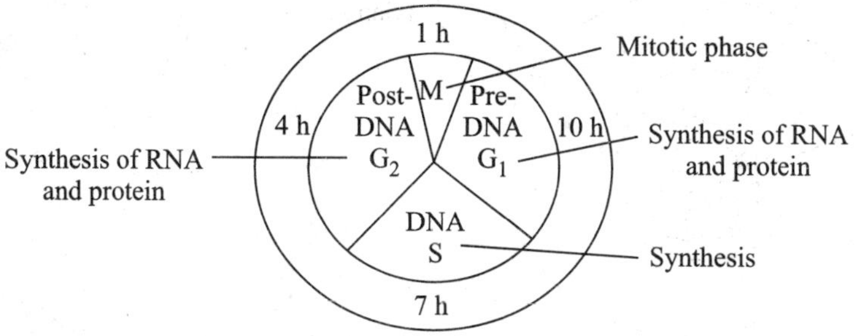

Figure 2.2 Cell cycle of a mammalian cell.

The cell cycle period differs from species to species (Table 2.1). Notwithstanding, it varies from cell to cell of different tissues within the same organism. Cell cycle is also influenced by environmental factors such as sunlight, temperature and humidity, etc.

TABLE 2.1 Time taken by different phases of cell cycle in different organisms

Organism	*Duration* (h)				*Total*
	Interphase			*Mitosis*	
	Gl	S	G2		
Root tip of *Vicia faba*	12	6	12	1	31
Tissue culture of human HeLa cell	12	10	3	1	26
Tissue culture of *Mouse* L. cells	12	6–8	3–4	1	22–25

Various stages of mitosis (Figure 2.3) are described below.

Interphase

The nucleus is separated from the cytoplasm by a nuclear membrane. The chromosomes are not visible under ordinary microscope. They are being replicated during *interphase* for coming cell division. In the cytoplasm, near nuclear membrane, there are two 'parent' centrioles, each paired with a small daughter centriole. Centrioles are found in animal cells only (Figure 2.3(a)).

Prophase

The beginning of *prophase* is marked with the condensation of chromosomes into compact, visible structures. Constriction of the chromosomes by shortening and thickening is the most conspicuous feature of the *prophase.* The two sister chromosomes having arisen from the replication of a parent chromosome during the preceding *interphase* are longitudinally held together at centromere as double structure, which are randomly arranged in the nucleus. The sister chromatids of each chromosome are closely aligned and somewhat coiled on themselves. The tightening of these coils contributes in a large measure to the shortening and thickening of the chromosomes.

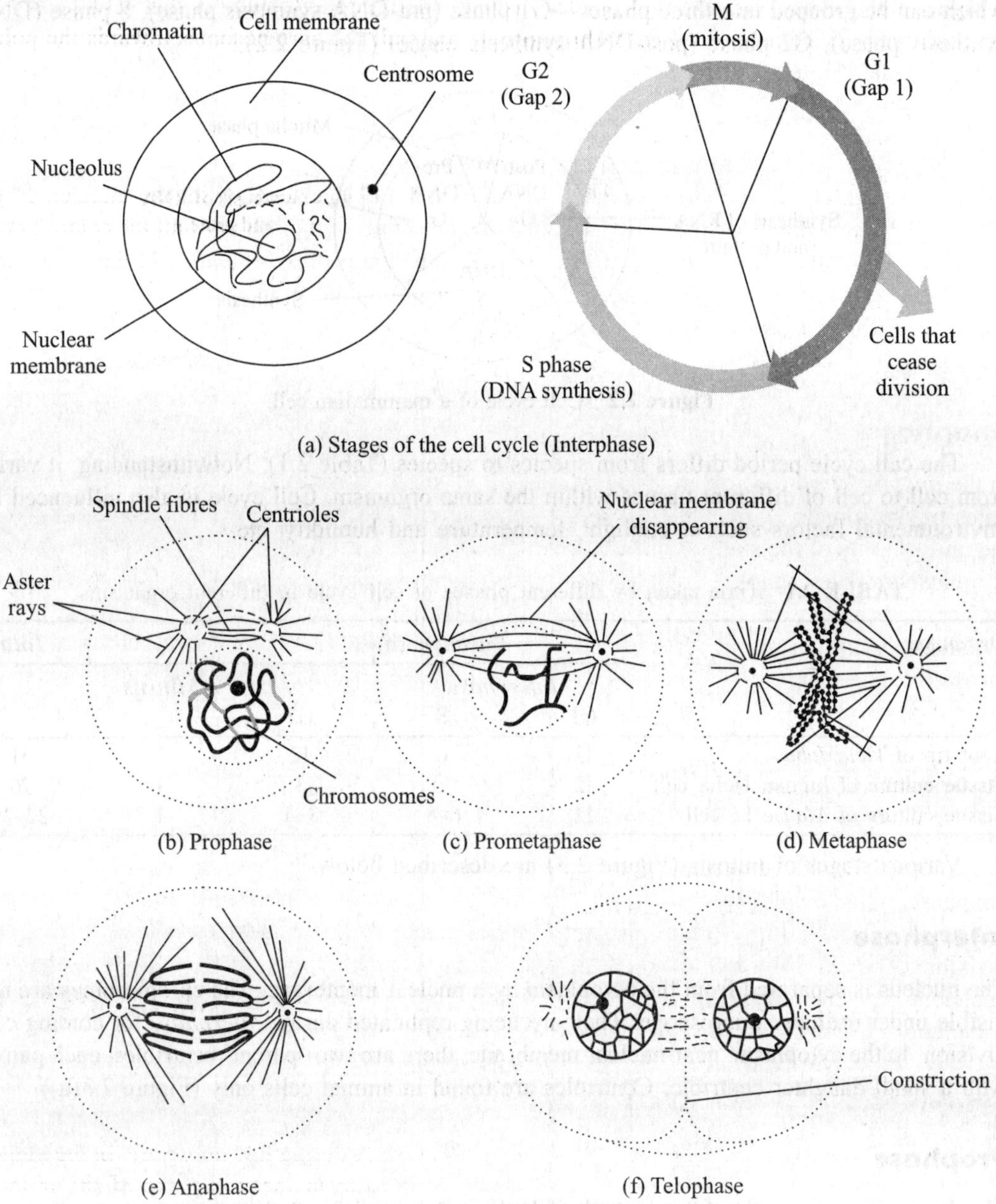

Figure 2.3 Various stages of mitosis (in animal cell).

The *centrioles* (in animal cell), which are duplicate, migrate to the opposite poles of the cell and the spindle apparatus is formed. During movement, *centrioles* radiate aster of fibres in all direction. These fibres make bridge between two *centriole* pairs located on two poles are called spindle apparatus. The centrioles are cylindrical organelles found in every eukaryotic cell except in higher plants. They organize the spindle—a conglomerate of microtubules that get attached to the centromere (kinetochore) of each chromosome and pull the chromosomes towards the poles of the spindle (Figure 2.3(b)).

Metaphase

During the metaphase, the chromosomes are shortened and thickened. With the attachment of the spindle fibres, the chromosomes move towards the centre of the cell until all the centromeres lie on a plane of the *equator* of the mitotic apparatus called *metaphase plate.* At this stage, the sister chromatids are still held together by connecting chromatid fibre at the centromere region. Alignment of chromosome on the *metaphase plate* is marked with the end of metaphase (Figure 2.3(c),(d)).

Anaphase

In the *anaphase,* the centromere divides and the two sister chromatids move toward the opposite poles of the spindle fibres attached to the centromere. The movement of chromatids occur at the uniform rate of the speed. Once the centromere divides, each chromatid becomes a separate chromosome. The anaphase, therefore, accomplishes quantitatively equal. The position of metacentric chromosome looks like V-shaped, sub-telocentric like J-shaped and telo-centric like rod-shaped. At this stage process of cell cleavage begins which in the animal cell begins like a furrow from periphery to inwards and in the plant cell, the growth of a new cell wall begins from inside (Figure 2.3(e)).

Telophase

The sister chromosomes reach to opposite poles on their migration. The arrival of the longitudinally single daughter chromosome at the spindle pole marks the beginning of the telophase. A nuclear membrane is formed around each group of chromosomes; nuclei are formed and spindle fibres disappear and cytokinesis takes place. The chromosomes become distant and invisible. The chromosomes uncoil and begin to carry out the physiological function and two nuclei slowly enter into the *interphase* (Figure 2.3(f)). The cell-cleavage process goes on completion, which generates two daughter cells in their *interphase.*

Cytokinesis

The cytokinesis occurs during the *telophase* though it initiates during late *anaphase.* In the higher plants, *cytokinesis* occurs with the formation of a cell plate in the mid-plane of the mitotic apparatus which is later converted to a middle lamella with a new cross-wall between the daughter cells. Cytokinesis initiates with the formation of vesicle at the mid-plane of the mitotic cell, which coalesce at the centre of spindle to form a cell plate.

In the animal cell, the cytokinesis occurs by cytoplasm, simply pinching cell in half. Thus, cleavage first begins as a furrowing process, around the cell in the plane of metaphase plate. It grows as a contractile ring closing in a way similar to purse string (Figure 2.4).

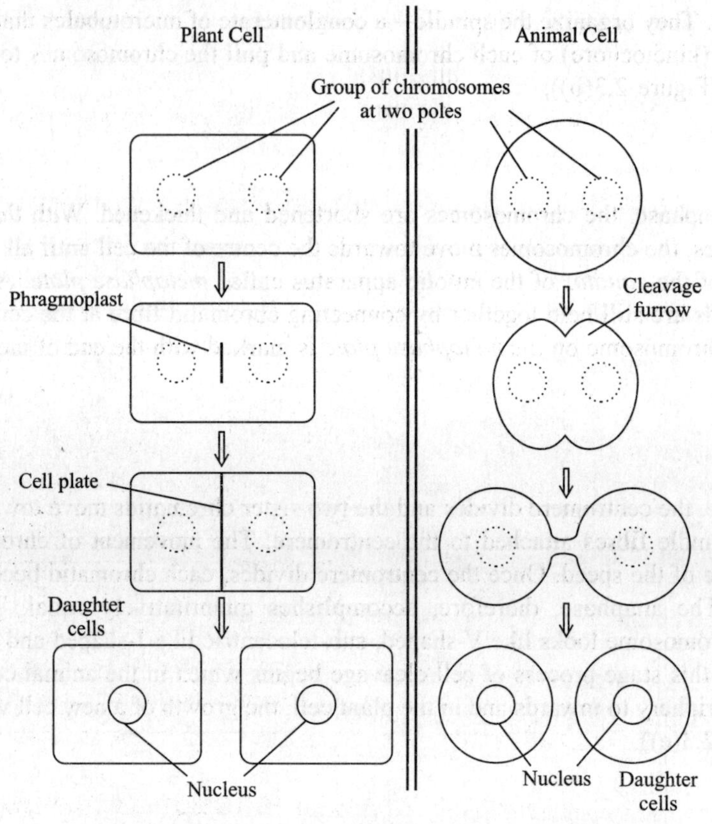

Figure 2.4 Cytokinesis in plant and animal cells.

Mechanism of Chromosome Movement

Chromosome movement during cell division primarily happens due to spindle apparatus. This apparatus consists of three kinds of microtubules.

(a) Astral radiating from centrioles (animals).

(b) Continuous or interpolar microtubules passing from one pole to other.

(c) Kinetochore microtubules originating from poles and are attached to the centromere.

The force applied for movement of chromosome acts at kinetochore and draws the chromosomes at poles (Figure 2.5).

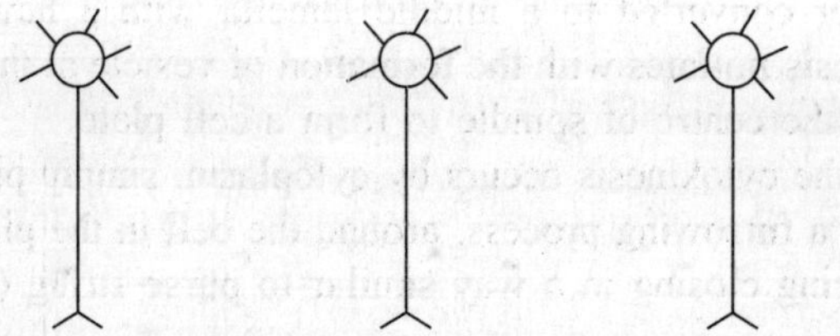

Figure 2.5 Mechanism of chromosome movement.

Regulation of Mitosis

It is believed that an enzyme M-phase-kinase (protein) is the cause of mitotic division. It regulates the mitotic division by activating and inactivating chromosomes at the different phases.

Significance of Mitosis

The cytokinesis in mitosis result into equal distribution of genetic materials in the form of chromosomes provided cell makes no mistake. The mechanism of mitosis proves that Mendel's theory and other later theories about location of genes are correct. Mitosis is also called somatic division of cell, which is responsible for growth and differentiation of the different parts of the body of plants and animals both. Mitosis is also responsible for the healing of the wounds and fractures in the body.

MEIOSIS

Syngamy, i.e. union of male and female gametes to form the zygote is one of the fundamental cytological and genetic events in the life cycle of sexually reproducing plants and animals. In *syngamy,* all the chromosomes of male (n) and female (n) gametes come together to form a diploid zygote (2n). The gametes are haploid (n), i.e. they possess half the number of chromosomes of the somatic cell (2n). Such a natural phenomenon requires a counterbalancing event where by doubling in the number of chromosomes at *syngamy* is offset by a nuclear division that halves the amount of chromatin per nuclei again prior to gamete formation. Such a reduction division is fundamental rule in all the sexually reproducing plants and animals. This is called *meiosis.*

Meiosis consists of two successive divisions. In the *first* division too, the members of each pair of chromosomes, which are known as homologous to each other, become closely associated along their length. Each homologous chromosome consists of two sister *chromatids,* joined at the centromere. The pairing of homologous chromosomes makes four-stranded structure. In the first meiotic division, the homologous chromosomes are separated from one another and distributed equally into two nuclei, each containing half the number, i.e. haploid (n) of duplex chromosomes.

Second division resembles a mitotic division in which chromatids of each chromosome are separated and distributed into daughter nuclei. These two successive nuclear divisions result into formation of four haploid daughter nuclei (tetrad), each containing n number of chromosomes, i.e. half of the somatic cell. Thus, the two successive nuclear divisions in meiosis are called *first* meiotic division and the *second* meiotic division. *First* meiotic division is also called *reduction division.*

First Meiotic Division

The first meiotic division consists of four *stages—Prophase* I, *Metaphase* I, *Anaphase* I and *Telophase* I. Before entering in prophase I, like mitosis, replication of DNA takes place at interphase and each chromosome becomes double but they are not visible. Part of DNA replication is completed at zygotene stage also.

Various stages of first meiotic division (Figure 2.6) are described below.

Prophase I

This is a longest stage, losting several days. Prophase I consists of five substages—leptotene, zygotene, pachytene, diplotene and diakinesis. (In Greek, *lepto* means "thin"; *zygo* means "yoke-shaped"; *pachy* means "thick", *diplo* means "double" and *dia* means "across".)

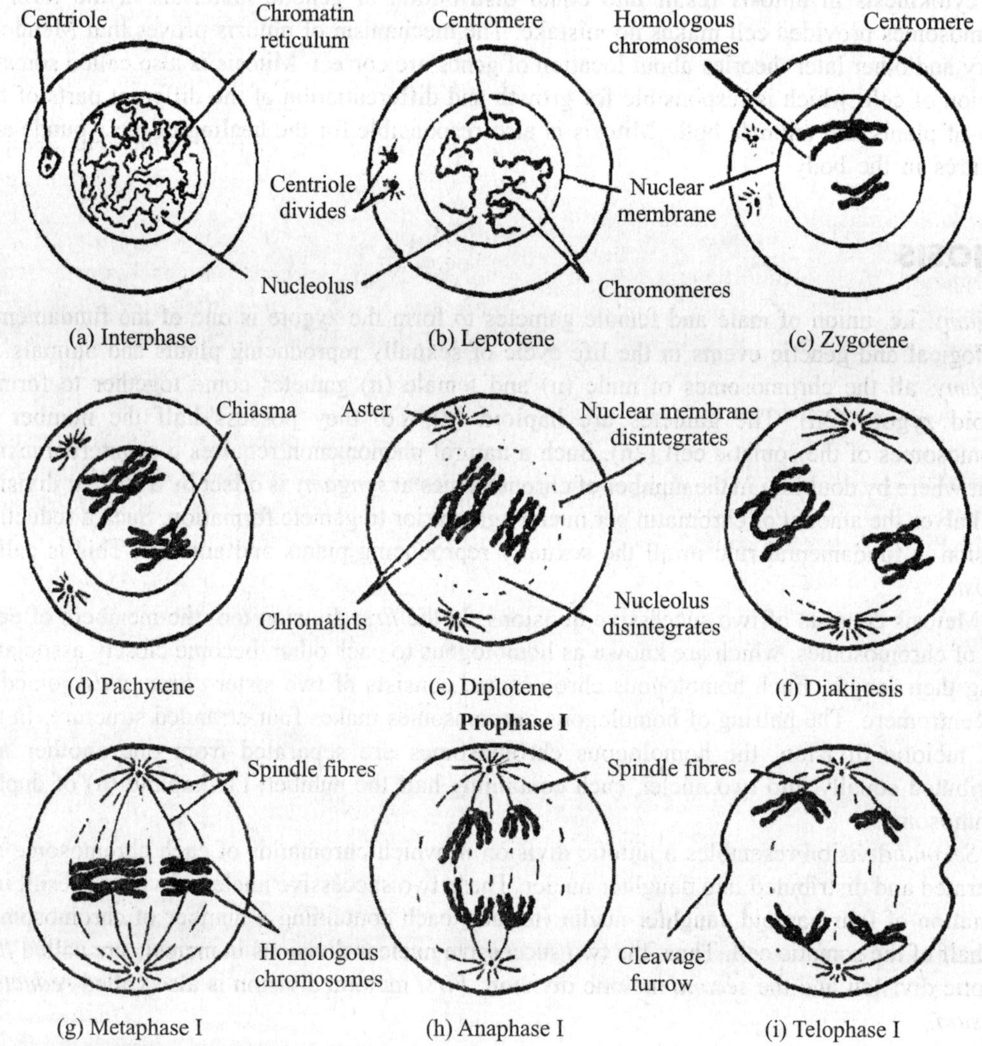

Figure 2.6 Various stages of first meiotic division.

Leptotene: In this stage, the chromosomes become visible as a long slender, thread-like structures. Condensation of chromosome, i.e. shortening and thickening begins at this stage. Numerous dense granules appear at irregular interval along their length. These localized contractions are called *chromomeres* (Figure 2.6(b)). In some plants, chromosomes are clumped to one side of nucleus. This is called *synizesis*. In the animals, chromosomes are polarized and chromosome ends are drawn to the nuclear membrane close to centrioles. Such polarization is called *bouquet* and where chromosomes are attached at both ends of the nuclear membrane is called *attachment plaque*.

Zygotene: While contraction of the chromosome is underway, the homologous chromosomes begin to pair or *synapse*. The *synapsis* (pairing) is remarkably exact and specific and takes place point by point (chromomere by chromomere association) along the length of two homologous chromosomes; somewhat twined about each other called *bivalent*. In other word, each pair of *synapsed* homologous chromosomes is called a *bivalent*. The *synapsis* is mediated in an unknown way, by a proteinaceous complex appearing between the homologous chromosomes is referred as a *synaptonemal complex*. At this point, chromosome figures are referred as *bivalent*, one per homologous pair (Figure 2.6(c)). If there are more homologous chromosomes then, *trivalent*, *quadrivalent* and *multivalents* may be formed.

Pachytene: As chromosome continue to shorten and thickened, at pachynema, each chromosome could be seen made of two sister chromatids; each *bivalent* can be seen as a *tetrad* because they are made up of four chromatids. While the homologous chromosomes are held together at *synapsis*, an important genetical event called *crossing over* takes place in *pachytene*. But it becomes apparent only at transition *diplotene* stage.

The point at which the exchange of genetic materials takes place between two non-sister chromatids is evidenced by X-shaped configuration called *chiasmata* (singular chiasma) (Figure 2.6(d)). The longer the chromosome, greater is the likelihood of more chiasma formation, though one chiasma formation interferes the formation of another chiasma in a close vicinity of other chiasma on the same side of centromere.

Diplotene: At the onset of *diplotene* the synapsed chromosomes begin to separate except at point where *chiasmata* occur. *Chiasma* is the result of breakage and joining between two non-sisters chromatids, i.e. a physical exchange between chromatids of two homologous chromosomes (Figure 2.6(e)). The number of chiasma per *bivalent* depends upon the species and the length of chromosome also.

Normally, each *bivalent* has at least one chiasma and there may be more, if chromosomes are longer. The chromosomes continue to contract and nucleolus begins to disappear. *Synaptonemal complex* dissolves but two homologous chromosomes remain joined together by one or more chiasmata. Due to forces of repulsion, *chiasmata* start terminalization.

Diakinesis: This is the last stage of prophase I. The chromosomes attain their maximum condensation, the homologous chromosomes in a bivalent remain connected by one or two terminal chiasmata which persist until first meiotic anaphase. The chiasmata gradually terminalize, i.e. move towards the end of the arms and finally slip-off due to the continued shortening of the chromosomes. The nucleolus disappears, the nuclear membrane degenerates and spindle fibre attachment is formed or initiated (Figure 2.6(f)).

Hypothesis of Chiasma Terminalization

There are three main hypothesis proposed by different workers.

1. *Electrostatic hypothesis:* This hypothesis was given by Darlington and his associates. They presumed that movement of chiasma towards chromosome ends may be caused by two separate repulsion forces: (a) the localized repulsion force which repels the centromere, and (b) the generalized repulsion force which repels entire homologous chromosomes apart.
2. *Coiling hypothesis:* This hypothesis was proposed by Swanson. According to him, movement of chiasma occurs due to mechanical tension developed by coiling and shortening of the homologous chromosomes.
3. *Elastic chromosome repulsion hypothesis:* This hypothesis was proposed by Ostergren, who said, chiasmata are pushed towards the chromosome end by the repulsion force generated between the homologous chromosomes.

Metaphase I

The bivalents are pulled to a metaphase plate by spindle fibre attachment. The orientation of the undivided centromere of each bivalent relative of the two poles of the spindle is random, i.e. for a given pair; it is likely that the paternal member be directed towards north pole as for maternal member to be south pole. In contrast to the metaphase of mitosis, here the centromere does not split (Figure 2.6(g)).

The random alignment of non-homologous chromosomes on the metaphase plate is the physical basis of Mendel's law of independent assortment.

Anaphase I

In anaphase I, the disjunction (separation) of the *synapsed* homologous chromosomes occurs, as a result one longitudinally double chromosome (two sister chromatids attached with undivided centromere,) moves to the opposite pole of the spindle (Figure 2.6(h)), thereby complete the process of terminalization. It differs from mitosis, because in mitosis two sister chromatids separate out as a result of division of centromere and move to opposite poles. Whereas in meiosis, whole chromosome of each homologous pair separates so that each pole receives either a paternal or maternal double chromosome of each pair. This results changes in the chromosome number from diploid (2n) to haploid (n) in the resultant reorganized daughter nuclei. This meiotic division is, therefore, called reduction division.

The physical separation of homologous chromosomes during anaphase I is the physical basis of Mendel's law of segregation.

Telophase I

The haploid set of chromosomes consisting of one homologous from each *bivalent* move to each pole of the spindle. During telophase, the chromosomes are in the condensed state, and spindle breaks down. The nucleolus and nuclear membrane are reconstituted and cytokinesis may occur (Figure 2.6(i)) depending upon the species. In some plant species like *Trillium*, meiocytes progress directly to form anaphase I to prophase II or even metaphase II without loss of condensation. In some species, either short or fairly long interphase between first and second meiotic divisions may occur. Chromosome duplication (DNA replication) never occurs between two divisions.

Interkinesis

This stage falls between telophase I and prophase II. During this stage DNA replication does not take place as we have seen in pachytene stage where chromosomes are already replicated during pre-meiotic interphase. In many cases interkinesis is absent and the two nuclei of telophase I directly enter into prophase II of meiotic division.

Second Meiotic Division

Basically, the second meiotic division is a mitotic division in which sister chromatids of each chromosome are pulled to the opposite poles. Like mitotic cell division, there are four stages in second meiotic division (Figure 2.7) which are described below.

Prophase II

Prophase II is short and superficially resembles mitotic prophase except that sister chromatids of each chromosome are genetically different and the chromatids possess no relational coiling. After short prophase II, the formation of second division of spindles takes place and the centromere of each chromosome in each nucleus begins to align on the central plane of the spindle as metaphase II (Figure 2.7(a)).

Metaphase II

On the two spindles, normally oriented at right angle to the first division spindle, the haploid number of chromosomes, each consisting of two chromatids, attached at centromere are arranged in the *equatorial plane* (Figure 2.7(b)).

Anaphase II

In this stage, centromere divides and the two sister chromatids of each chromosome at anaphase II separate and move to the opposite poles as daughter chromosomes, similar to as their movements in mitosis (Figure 2.7(c)).

Telophase II

On the arrival of haploid number of daughter chromosomes at poles, the chromosomes return to a long attenuated reticulate configuration. The nuclear membranes are reconstituted, nuclei reappear and cytokinesis results into formation of four haploid nuclei accompanied by division of cytoplasm (Figure 2.7(d)). Figure 2.7(e) shows gamete formation at the end. This may be also called interphase.

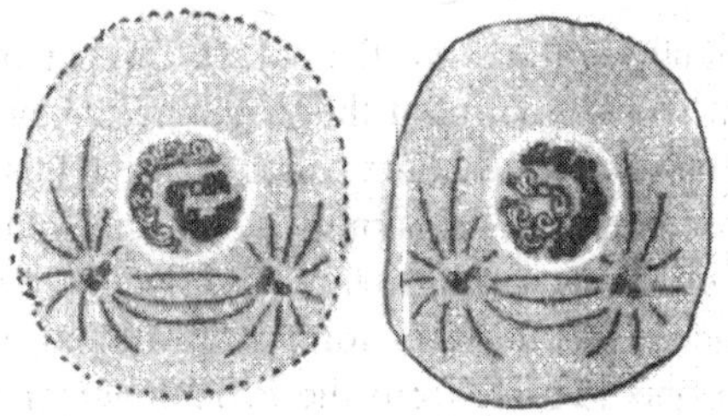

(a) **Prophase II**

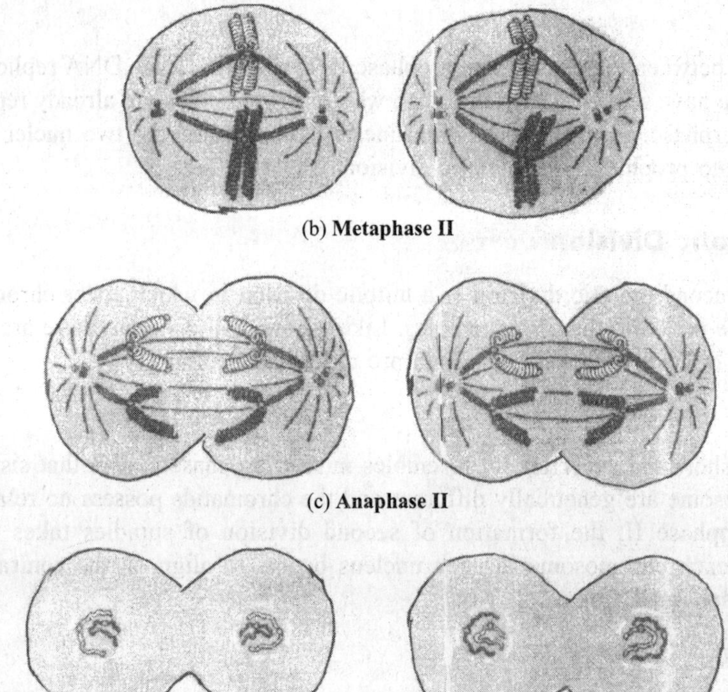

(b) **Metaphase II**

(c) **Anaphase II**

(d) **Telophase II**

(e) **Interphase** (gamete formation)

Figure 2.7 Various stages of second meiotic division.

The mechanisms of gamete formation in the animal cell is almost the same as in plants (Figure 2.8).

Synaptonemal Complex

Synaptonemal complex is found in almost all the cells of higher organisms (eukaryotes) during meiosis. The pairing starts at the beginning of the zygotene; when synaptonemal complex starts formation of the rope-like proteinaceous structure called axis and this ribbon-like structure proceeds along the homologous chromosomes. As synapsis advances, the axes of the homologous chromosomes get adhered to each other to form lateral elements of *synaptonemal complex.* These lateral elements of two chromosomes together form a third band, i.e, *central element.* This way three bands, two lateral and one central, form the synaptonemal complex (Figure 2.9).

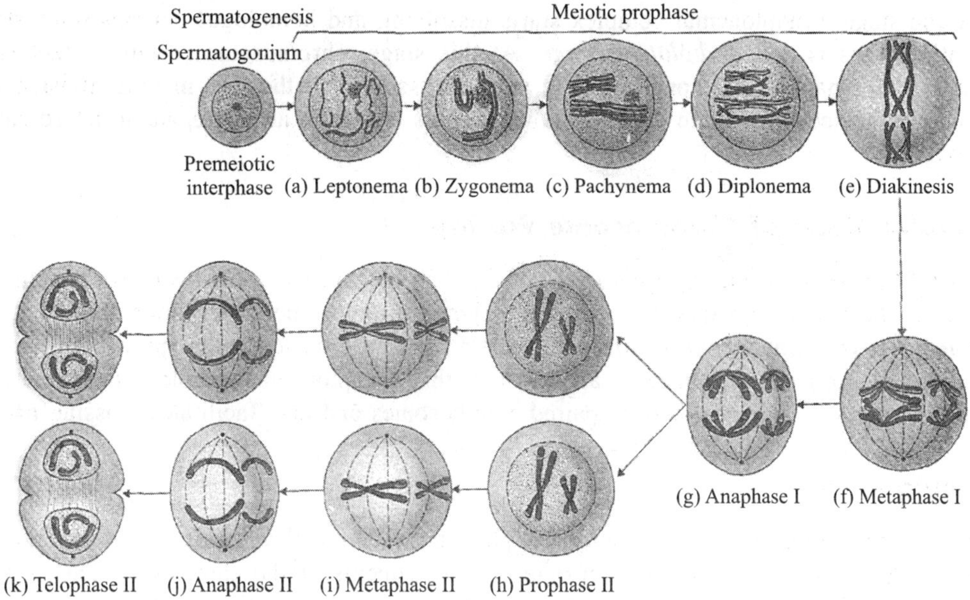

Figure 2.8 Different stages of meiosis in gamete formation in animals.

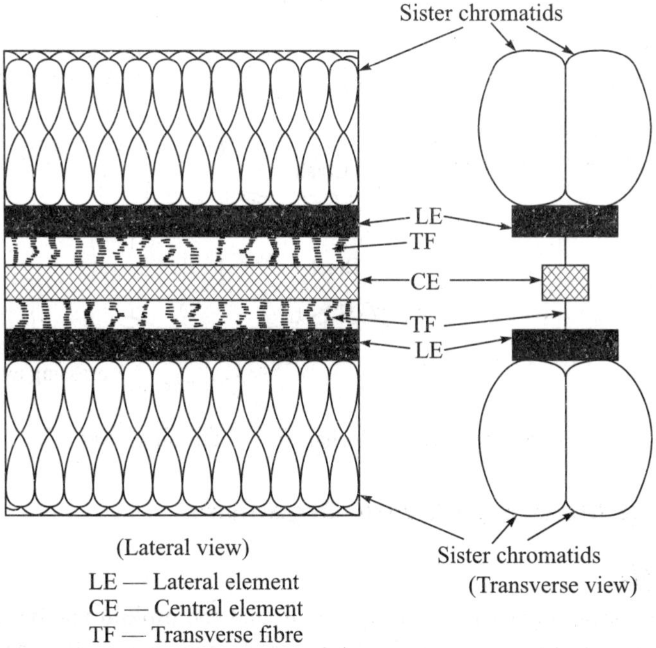

Figure 2.9 Synaptonemal complex.

Synaptonemal complex consists of protein and lateral elements of DNA. DNA is also present in transverse fibre that crosses the space between central and lateral elements. At the end of pachytene stage, synaptonemal complex starts dissolving and homologous chromosome starts separating. This is called *diplotene stage*. At this stage, chromosome remains attached to chiasmata. If synaptonemal complexes did not get dissolved in the first meiotic division and continue to the second meiotic division in some cases and form an aggregate structure called *polycomplex*.

Molecular Basis of Chromosome Pairing

Moses (1956) working with spermatocytes of crayfish observed that a ribbon-like structure is formed at the site of synapsis of two homologous chromosomes. He called this structure *synaptonemal* or *synaptinemal complex*. Since this structure occupies the space between two paired homologous chromosomes, it appears that this synaptonemal complex is the cause of aligning of the DNA molecules from paired chromosomes and also facilitates crossing over.

Crossing Over

Crossing over is a physical step, which takes place between the chromatids of two homologous chromosomes and results into the exchange of the segments of the chromatids. This is very important step for creation of new variability through the recombination of genes where genes are linked. The Mendelian assortment of chromosomes does not help in the recombination of genes, if they are located on the same chromosomes (Figure 2.10).

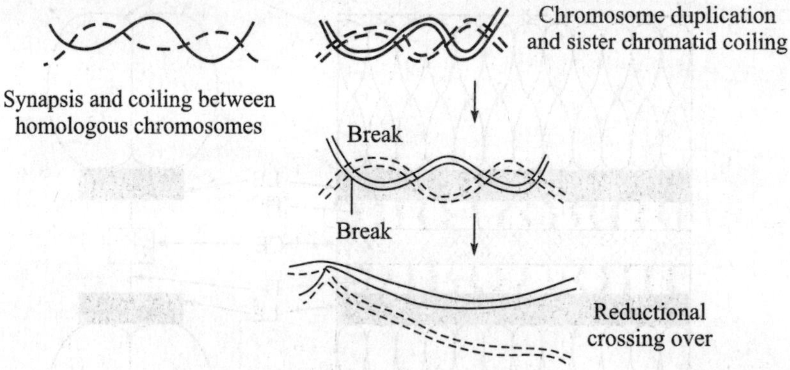

Figure 2.10 Mechanism of crossing over.

Significance of Meiosis in Living Organisms

Gametogenesis

Cytologically, the basic significance of meiosis is that the diploid number of chromosomes are reduced in such a way that each of the four tetrad (daughter cells) possesses one complete haploid chromosome set (n) which on syngamy (fertilization) forms the zygote, restoring 2n chromosomes set again in the somatic cell. Second, because of crossing over, gametes with new allelic gene combinations arise which on syngamy give rise to new genotype. Third, if the genes are located on the chromosomes, the process of meiosis creates new genetic viability in two ways:

(a) Random assortment of paternal and maternal chromosomes at anaphase I.

(b) Crossing over of linked genes, which is collectively called recombination. Any mistake, in the meiotic cell division results into chromosome aberration.

Suppose there are two pairs of contrasting genes *RRWW* (red flower and smooth seed coat) and *rrww* (white flower and wrinkled seed coat) located on two pairs of chromosomes. On their crossing, dihybrid (F_1) with red flower and smooth seed coat (*RrWw*) is obtained wherein gene alleles *RW* are derived from their paternal parents and gene alleles *rw* are derived from their maternal parents. If these two pairs of gene alleles are located on two different homologous chromosome pairs which may be designated as pair 1 and pair 2, then in prophase I, the paternal and maternal chromosomes, carrying gene *R*, will synapse with chromosome carrying allele *r*, and chromosome carrying allele *W*, will synapse with chromosome carrying gene allele *w*.

The arrangement of each pair of synapse homologous chromosomes on metaphase I spindle is random, i.e. the paternal member of each pair of chromosomes has equal chance of getting oriented towards either of the pole as the maternal chromosomes. In anaphase I, each chromosome moves towards the nearer pole soon after it separates from its homologue. This way, each telophase I nucleus has an equal chance of receiving a paternal or a maternal homologue of each chromosome. Hence with two pairs of genes, four possible gametic combinations may be formed in both male and female parents at telophase I.

Male (♂) gametes	**Female (♀) gametes**
RW	*RW*
Rw	*Rw*
rW	*rW*
rw	*rw*

Second meiotic division will simply increase the number of each gametic gene arrangement from one to two, i.e. the number of possible gamete types in a dihybrid cross would be 2n, i.e. a given gamete has (1/2)n chance of receiving any particular arrangement of paternal and maternal chromosome/gene. Thus, segregation and independent assortment of genes located on chromosomes form four types of gametes in F_1 and four types of individuals in F_2 in the ratio of 9: 3: 3:1 wherein two additional types of phenotypes white—smooth and red—wrinkled appeared, not earlier available in the lines intercrossed. Thus recombination, through independent assortment of genes is the major cause of variability in nature. However, if the two pairs of genes are located on a single pair of homologous chromosome, i.e. linked as shown here

*R*______*W*	*r*______*w*
*R*______*W*	*r*______*w*
(Male)	**(Female)**

If crossing over and chiasmata formation do not take place, then only two kinds of gametes will be formed, viz. *RW* and *rw*, in equal proportion. However, if crossing over takes place between the chromatids of two pairs of homologous chromosomes and chiasmata are formed between gene alleles *R*, and *W*, then four types of gametes will be formed: *RW*, *Rw*, *rW* and *rw*, but proportion of each type produced will depend upon the frequency with which crossing over takes place between two gene pairs *RR* and *WW*. But there would be four types of phenotypes in F_2 generation, irrespective of their proportion.

Meiosis

In the normal sexual reproduction, the fusion of male and female gametes is essential for the seed production in plants and child in human beings. But formation of gamete (n) through the process of meiosis is more or less the same. Figure 2.11 is a diagrammatic longitudinal section of a typical flower representing both female and male organs (Hermaphrodite) as well as gamete (egg) formation in the embryo sac.

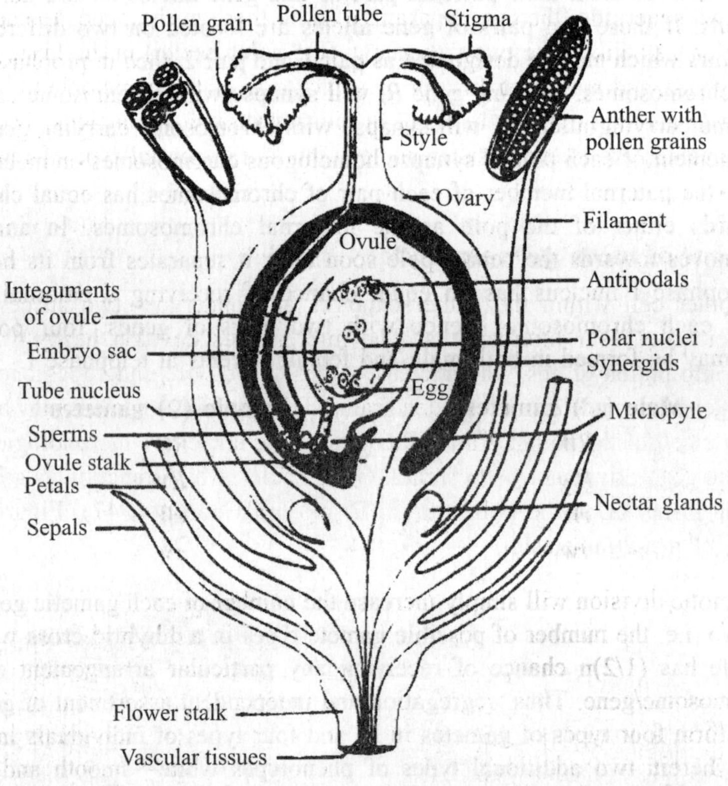

Figure 2.11 Diagram of a complete flower in longitudinal section at fertilization as well as female gamete (egg) formation in the embryo sac.

Development of female gametophyte

Each ovule is born at the apex of a funicules or stalk, which arises from the placenta, a portion of the inner surface of the ovary. The ovule is consisting of a central portion, the megasporangium or nucellus, surrounded by one or two integuments with an opening, the micropyle, opposite the apex of the nucellus. A sub-epidermal cell of the nucellus immediately opposite the micropyle, either at once or after division, enlarges and differentiates as a megasporocyte. This diploid cell increases in volume until it becomes several times of its original size and then, by two successive divisions (meiosis) gives rise to an axial row of a four haploid megaspores. The other cell may not divide again after the first division, so that the quartet is incomplete. Usually, the innermost megaspore develops into the female gametophyte and remaining spores disintegrate. Thus functional spore enlarges and elongates and as a result of a series of divisions of its nucleus, eight nuclei are formed which are arranged in two groups of four, one at each end of the developing gametophyte. During the end of mitosis, cell plates are formed across the spindles in such a manner as to cut off three cells at each end of the gametophyte and leave one nucleus from each group in the central cell so that a seven-celled, eight-nucleate structure is formed. The three cells at its micropylar end become differential into an egg and two synergids (the egg apparatus), while those at the opposite end (the antipodals) remain more or less alike. The two free nuclei (the polar nuclei) in the large central cell come to lie side by side and may even fuse prior to maturity of the gametophyte. The two synergid nuclei are sister nuclei, i.e. they are result of division of a single nucleus. The egg nucleus and upper polar nucleus are likewise sister nuclei. This is the normal type of female gametophyte development.

Development of male gametophyte

A pollen mother cell within the anther enlarges and undergoes two successive divisions to produce a tetrad of four microspores. The microspores develop heavy walls and become transformed into pollen grains. The nucleus within microspore undergoes one division to form the tube and generative nuclei of the pollen grains. The generative nucleus may be walled off early into a separate cell of the pollen grains. Frequently, the nuclei are morphologically distinct. The division of the generative nucleus to form the two nuclei, which function as gametes, may occur in the pollen grains or in the pollen tube (Brink and Cooper, 1947). Figure 2.12 shows the development of male gametes.

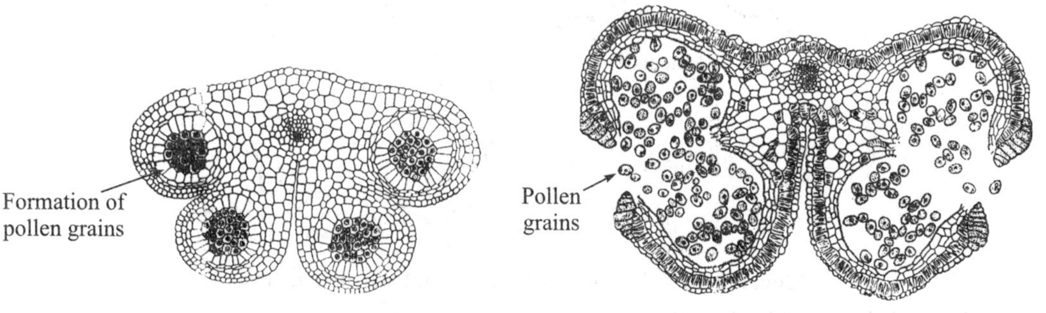

(a) Cross-section of a young anther

(b) Cross-section of a mature anther

Figure 2.12 Development of male gametes.

DIFFERENCES BETWEEN MITOSIS AND MEIOSIS

Following are the differences between mitosis and meiosis (Figure 2.13):

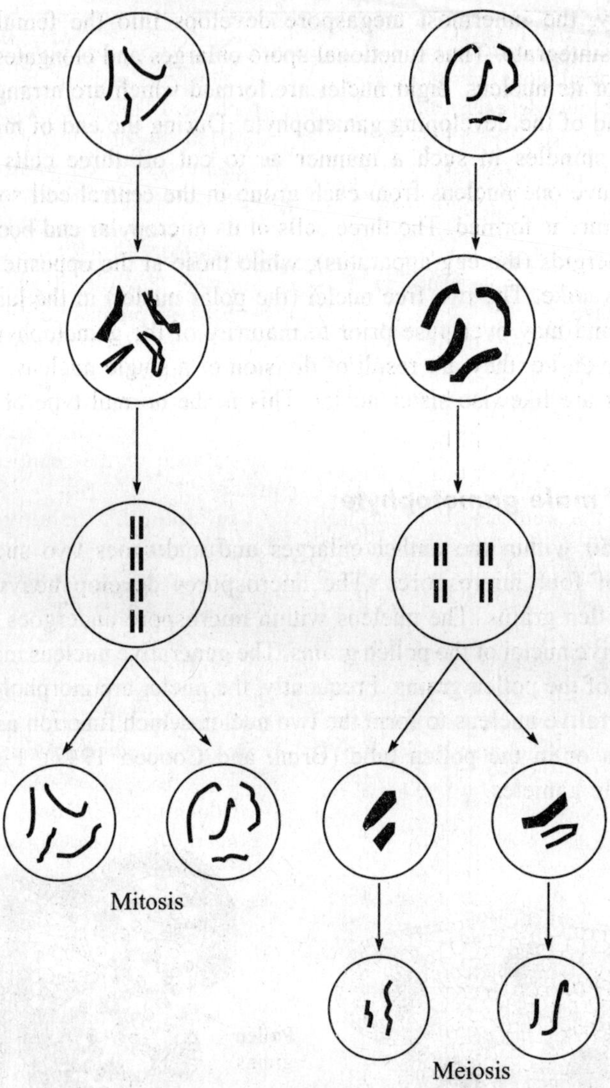

Figure 2.13 Differences between mitosis and meiosis.

Mitosis	Meiosis
i. **Interphase** (resting stage) proceeding mitosis is long.	**Interphase** (resting stage) proceeding meiosis is short.
ii. **Prophase** is short without substages.	**Prophase** I is long with five substages: leptotene, zygotene, pachytene, diplotene and diakinesis.
iii. The chromosomes are longitudinally split into two sister chromatids, as a result of early prophase and look like double thread.	The chromosomes are not longitudinally split into chromatids and look like a single thread.

	Mitosis	Meiosis
iv.	The split up homologous chromosomes are not attracted towards each other.	At zygotene and in the succeeding stages in prophase I, the homologous chromosomes are attracted towards each other and they form *bivalent*. Each chromosome splits into two sister chromatids and four strands are formed. Crossing over takes place between chromatids of paternal and maternal chromosomes.
v.	Complete DNA replication occurs in S phase of interphase.	Complete DNA replication does not occur in S phase of premeiotic interphase only. A small amount of DNA replication remains to be completed during prophase I (pachytene) with exchange of segments of chromatids. The chromosomes then fall apart with characteristic appearance of the number and types of chiasmata.
vi.	At **metaphase**, the centromere of each chromosome divides and the sister chromatids move towards opposite poles and the somatic number (2n) is maintained.	At **metaphase** I, the centromere does not divide but the homologous chromosomes move toward opposite poles as *bivalent*. Thus, the number of chromosomes moving towards the opposite pole is reduced to half the number of somatic cell (n).
vii.	Each chromosome of paternal or maternal origin is longitudinally divided and their movements to the opposite poles result into the formation of identical daughter cells.	At metaphase I, orientation of chromosomes on the equatorial plate (metaphase plate) is such that either paternal or maternal chromosomes only go to the opposite poles. The orientation in one pair is independent of the other pair. The two daughter cells born at the end of the first division are non-identical.
viii.	The movement of the chromosomes at the late metaphase is due to repulsion between the parts of divided centromere (Figure 2.14(a)).	The movement of the chromosomes at the late metaphase I is due to repulsion between the centremeres of the homologous chromosomes (Figure 2.14(b)).
ix.	At **anaphase**, further separation of chromosome is due to the formation of stem body. At telophase, each of the two daughter nuclei are reconstituted with diploid (2n) number of chromosomes.	At **anaphase** I, the changes in the spindle are the same as in mitosis. At telophase I, two daughter nuclei are formed. Each possesses haploid (n) number chromosomes.
x.	No synatonemal complex is formed. There is no second division. The two daughter cells after a period of resting stage again undergo under repeated mitosis. Thus, at the end of mitosis, there is no change in the quality or quantity of the chromosomes in the cell.	Synaptonemal complex is formed. There are two cell divisions. First division is called *reduction* division as the two daughter cells possess n number of chromosomes. Second division is mitotic division. Thus, at the end each daughter cell of the tetrad contain n number of chromosomes with changes in quality due to crossing over.
xi.	Crossing over and chiasmata formation do not takes place.	Crossing over between chromatids of homologous chromosomes takes place and chiasmata are formed.

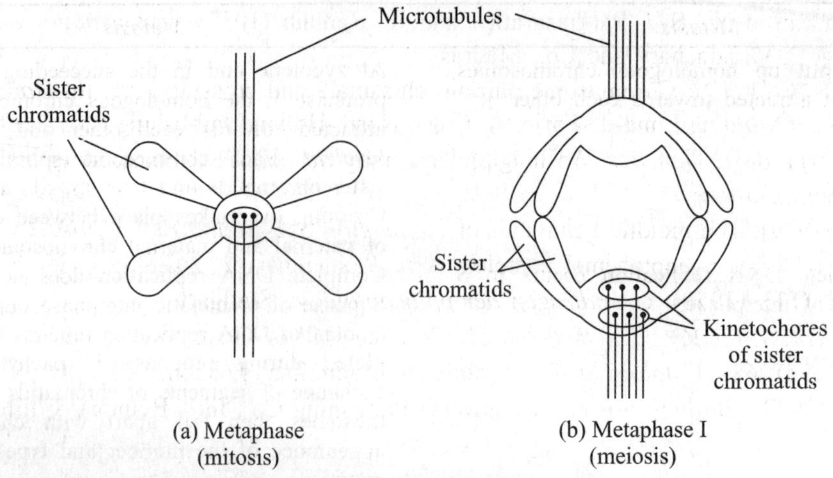

Figure 2.14 Arrangement of microtubules in mitosis and meiosis.

REVIEW QUESTIONS

1. What is cell cycle? Draw a typical cell cycle and discuss its different substages.
2. What is mitosis? Discuss different stages of mitosis with the help of suitable diagrams.
3. Write short notes on the importance of mitosis in the grown and development of life.
4. Define fission and cytokinesis.
5. What is meiosis? Describe with the help of suitable diagrams the substages of prophase I, taking two or four pairs of chromosome numbers in a cell.
6. What are the differences between mitosis and meiosis?
7. What is the significance of meiosis in the life cycle of plants.
8. What do you mean by equational and reductional divisions?
9. Describe briefly the different stages and substages of meiosis.
10. Write brief notes on:
    (a) Synaptonemal complex
    (b) Interphase
    (c) Relational or plectonemic coiling
    (d) Chiasma terminalization
    (e) Differences between pachytene and diptotene stages
11. If a cell contains four pairs of chromosomes, after completion of the mitotic telophase, how many chromosomes remain in the daughter cell. Explain giving suitable diagrams.

REFERENCES

Brink, R.A. and D.C. Cooper (1947): *Botany Review,* 423–541.

Chandrasekharan, S.N., S.V. Parthasarathy, and K. Ramaih (1953): *Cytogenetics and Plant Breeding*, P. Varadachary and Co., Madras.

Handerson, S.A. (1969): Chromosome pairing, chiasmata and crossing over, in: *Handbook of Molecular Cytology*, Lima-de-Fario, A. (Ed.), North-Halland Publishing Co., Amsterdam.

Hartl, D. (1991): *Basic Genetics,* Washington University School of Medicine, Jones and Bartlett Publication, Boston.

King, R.C. (1970): The meiotic behaviour of *Drosophila* oocyte, *Int. Rev. Cytol.,* **2**, 215–218.

Moses, M.J. (1968): Synaptonemal complex, *Ann. Rev. Genet.,* **2**, 263–412.

Schindewolf, O.H. (1950): *Grundfragen der palacontologie*, Stuttgart.

Strickberger, M.W. (2006): *Genetics,* 3rd ed., Prentice-Hall of India, New Delhi.

Swanson, C.P. (1968): *Cytology and Cytogenetics,* Macmillan, London.

Wolfe, S.L. (1972): *Biology of Cell,* Wadsworth Publishing Co., Inc., Belmont, California.

CHAPTER

3

Organization of Chromosome and Theory of Heredity

In 1903, Walter Sutton, a cytologist gave the idea of chromosome, as a basis of heredity. The behaviour of chromosome at first meiotic division (–1) explains about Mendel's principle of *independent assortment of genes.* Therefore, gene must be located on the chromosome. The location of gene(s) can be determined by monosomic (2n – 1) and nullisomic (2n – 2); trisomic (2n + 1) and tetrasomic (2n + 2) analyses. Normally, chromosomes remain in the pair in somatic cells, hence called diploid (2n); at gametic stage, they become haploid (n) and at polyploid level, chromosome numbers are doubled called triploids (3n), tetraploid (4n), pentaploid (5n), hexaploid (6n), octaploid (8n), etc.

Chromosomes were discovered by C. Nageli in 1842. The term 'chromosome' was coined by W. Waldeyer in 1888, means coloured body and is used because chromosome stains a bright colour, with certain techniques of histology, the nucleoprotein materials of chromosome is called as *chromatin*. Each chromosome with very few exceptions has a distinct attachment point for fibres that make up the mitotic and meiotic spindle apparatus. The attachment point is called *kinetochore* and the constriction in the chromosome where this occur is called *centromere.* The kinetochore seems to be composed of protein and RNA. On the basis of the position of centromere on the chromosome, the chromosomes could be classified as metacentric (if the centromere is located in the centre); telocentric (if the centromere is located at the end of chromosome) and acrocentric (very near the end of the chromosome) and sub-telocentric or sub-metacentric (somewhere in between). In the diploid state, members of same chromosome pairs are called *homologous.* The homologus pairs are made up of identical partners called *homomorphic* chromosomes. A potential exception is sex chromosomes, which in some species are of unequal size, called *heteromorphic* chromosome.

Earlier, it was believed that entire chromosomes of the genome carry the genes and it was estimated that in human beings, there are about 60,000–100,000 genes, which determine the different characters. But now, molecular biologists involved in deciphering the human chromosomes feel that hardly 3.3% of DNA of the chromosomes are involved in coding of the genetic information and rest of the DNA of chromosomes are non-coding or inert matters because of repetitive DNA forming heterochromatic region of chromosomes.

GENOME SIZE OF PROKARYOTES AND EUKARYOTES

The DNA content analysis of the genomes of viruses, bacteria and lower and higher eukaryotes indicate that genome size increases roughly with evolutionary complexity, i.e. the single nucleic molecule of a typical virus is smaller than the DNA molecule in bacterial chromosome. The unicellular eukaryote such as yeast possesses more DNA than a typical bacterium and the DNA content of multicelluar eukaryotes is much higher in amount and organized in several chromosomes. But on the whole, DNA content is not proportionately correlated with the number of genes present (Table 3.1) (Hartl, Daniel L; 1991). For example, bacteriophage MS2 is one of the smallest viruses and possesses only four genes in a single thread of RNA molecule consisting of 3569 nucleotides. SV_{40} virus which infects monkey as well as human beings possesses only five genes in a circular double-stranded DNA molecule consisting of 5 kb (5000 nucleotides pairs). Large nucleotides are measured in kilobase (thousand nucleotide base pair). The complex phases and animal viruses consist of 50–300 kb. Bacterial genomes are little larger. The chromosome of *E. coli* contains 1500–3000 genes.

The eukaryotic genomes are complex genetic structures because their genomes are packed in chromosomes. The chromosome composition is a species characteristic. But number of chromosomes do not form any specific pattern. For instance, nematode (*Caenorhabditis elegans*) contains 20 times higher DNA content than *E. coli* genomes. *Drosophila* possesses 40 times and human being 700 times more DNA than *E. coli* respectively.

TABLE 3.1 DNA composition of some representatives of prokaryotes (virus, bacteria) and eukaryotes

Organisms	*Genome Approximate number of nucleotide pairs* (kb)	*Form*
Prokaryotes		
Virus		
ΦX_{174}	5	Circular single-stranded
SV_{40}	5	Circular single-stranded;
M_{13}	6	double-stranded replicate form
λ	50	
Herpes simplex	152	Linear, double-stranded
T_2, T_4, T_6	165	
Smallpox	267	
Bacteria		
		Circular, double-stranded
Escherichia coli (*E. coli*)	4,700	Chromosome pairs
Eukaryotes		
Yeast (*Saccharomyces cerevisiae*)	14,000 → n number	16
Fruit fly (*Drosophila melanogaster*)	165,000	4
Nematodes (*Caenorhabditis elegans*)	100,000	6
Maize (*Zea mays*)	4,500,000	10
Human being (*Homo sapiens*)	3,000,000	23

But there are exception too. Among amphibia and fishes there are species whose genome sizes are many time bigger than mammalian genomes. At the same time, there are higher living forms whose genome sizes are much longer. For example, genome size of humans contains 30,000–100,000 genes and each gene may be made up of 500 to 5000 nucleotides or even more, which are located on a linear DNA molecule of 1 metre length. The largest of the 23 chromosomes in the human genome contains a DNA molecule that is 82 mm (8.2×1000 μm) long. But at metaphase, it get condensed to 10 μm and 1 μm in diameter (Hartl, D, 1991).

BACTERIAL CHROMOSOME

The chromosome of *E. coli* is also called *nucleoid* which are seen in a folded form when condensed. The DNA is arranged in form of loop. This is also the characteristic of eukaryotic chromosomes. Figure 3.1 indicates that *E. coli* chromosome is supercoiled but some are not due to deoxyribonuclease (DNase) activities. The supercoiling is get eliminated on the break of a single thread of DNA molecule. The supercoiling could also be eliminated by treating the DNA by topoisomerase.

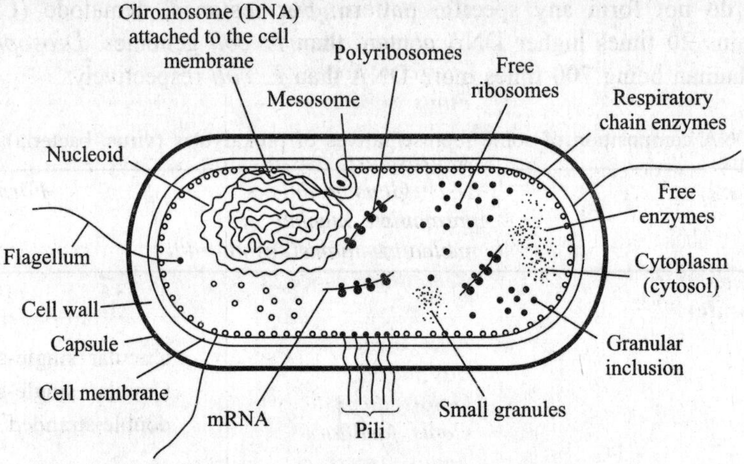

Figure 3.1 Supercoiled structure of chromosome in *E. coli.*

STRUCTURE OF EUKARYOTIC CHROMOSOME

The eukaryotic chromosome contains a long, single double DNA helix molecule. The largest chromosome in *Drosophila* genome possesses a DNA molecule of about 65,000 kb (6.5×10^7) nucleotide pair, which is equivalent to 22 mm in length.

STABILITY OF CHROMOSOMES

There are three properties of chromosomes:

1. Each somatic cell (a cell of body) contains a fixed number of chromosomes. This is most typical in a particular species (Table 3.2).

TABLE 3.2 Chromosome number of some plant and animal species

Common name	*Species*	*Haploid number* (n)
Higher plant		
Barley	*Hordeum vulgare*	7
Wheat	*Triticum aestivum*	21
Rice	*Oryza sative*	12
Corn	*Zea mays*	10
Rye	*Secale cereale*	7
Oat	*Avena sativa*	21
Broad bean	*Vicia faba*	6
Cotton	*Gossypium hirsutum,*	52
	Gossypium arborium	26
Onion	*Allium cepa*	8
Pea	*Pisum sativum*	7
Potato	*Solanum tuberosum*	24
Soya bean	*Glycine max*	20
Squash	*Cucurbita pepo*	20
Tomato	*Lycopersicon esculentum*	12
Watermelon	*Citrullus vulgaris*	11
Tobacco	*Nicotiana tabacum*	24
Oak	*Quercus alba*	12
Pine	*Pinus* species	12
Lower plant		
Bread mold (pink)	*Neurospora crassa*	7
Bread mold (black)	*Aspergillus nidulans*	8
Pencillin mold	*Penicillin* species	4
Green algae	*Chlamydomonas reinhardtii*	16
Penicillium (bacteria)	*Penicillium* species	2, 4, 5
Animal (vertebrate)		
Cat	*Felis domestica*	19
Cattle	*Bos taurus*	30
Chicken	*Gallus domesticus*	39
Dog	*Canis familiaris*	39
Donkey	*Equus asinus*	31
Horse	*Equus calibus*	32
Frog	*Rana pipiens*	13
Rat	*Rattus norvegicus*	21
Mouse	*Mus musculurs*	20
Guinea pig	*Cavia cobaya*	32
Chimpanzee	*Pan troglodytes*	24
Man	*Homo sapiens*	23
Orthopode (invertebrate)		
Fruit fly	*Drosophila melanogaster*	4
Grasshopper	*Melanoplus differentialis*	23
		24
House fly	*Musca domestica*	6
Honeybee	*Apis mellifora*	16
Mosquito	*Culex pipiens*	3

2. The chromosomes in the nuclei of somatic cells are usually present in pairs. One chromosome of each pair is derived from the maternal parent (female) and the other from the paternal parent (male) of the organism. For example, the chromosomes of pea may be represented by 7 pairs, and so on. The cells with nuclei of this sort containing two similar sets of chromosomes are called *diploid.*
3. The germ cells (gametes) contain only half of the total number of chromosomes of the somatic cells are called *haploid.* For example, the chromosomes in pea may be reduced to 7 chromosomes in the each gamete (germ cell).

The chromosomes of each pair called two sister chromatids, share a common centromere.

CONTINUITY OF THE CHROMOSOMES

A particular chromosome at the metaphase of a cell division is identical to that of the metaphase of a later cell division. This indicates that chromosomes do not lose their identity during *interphase* period. Baveri was the first who provided this evidence with the eggs of nematodes (*Ascaris megalocephala*).

Morphology of Chromosomes

The morphological characteristics of chromosomes could be best studied at metaphase and anaphase of the cell division. The chromosomes look like cylindrical, intensely-stained bodies, with basic dyes like acetocarmine and feulgen. Chromosomes can be studied in tissue sections, sex-organs, plant meristems, pollen mother cells and other tissues. Human chromosomes can be easily studied in the smears of bone marrows, cultures of leucoplasts or other tissues. Based on the position of centromere, the chromosomes can be classified into following four types (Figure 3.2).

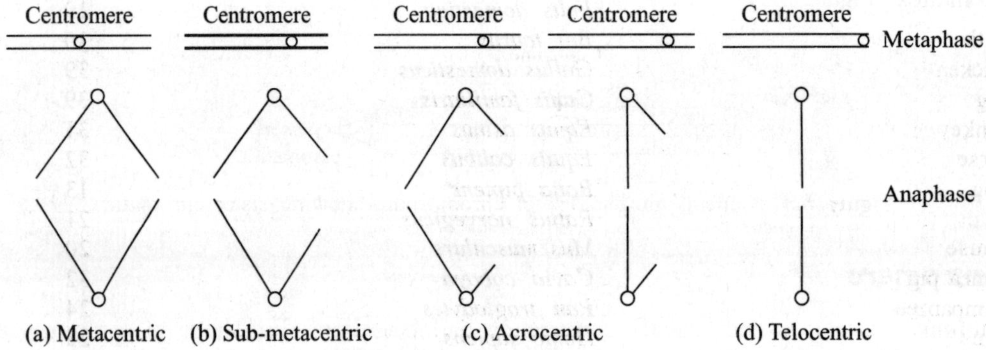

Figure 3.2 Shapes of chromosomes according to the position of centromere.

(a) *Metacentric:* Chromosomes have equal or almost equal arms and thus look like V-shaped.
(b) *Submetacentric:* Chromosomes have unequal arms and are thus L-shaped.

(c) *Acrocentric:* Chromosomes are rod-like structures and have a small or even imperceptible position of centromere near end. One arm is very long, other arm is short; ratio is 1:7.

(d) *Telocentric:* Chromosomes are rod-like structures with centromeres located on the proximal end.

The different chromosomal types are constant for each homologous chromosome. The chromosome numbers are also constant throughout a species as given in Table 3.1.

There are other two parameters to characterize a chromosome in a karyotype:

1. *Centromeric index* (*C.I.*): It is expressed as a ratio of short arm length to total chromosome length
2. *Proportion of arms*: It is a ratio between the long and short arms of the chromosome. In the metacentric chromosome, this ratio is 1:1.

It is important to note that morphology of chromosomes can be changed under certain chemicals like by *Colchicine* treatment, with that chromosome becomes straight and two chromatids separate.

Chromatid

It is structural and functional unit of a chromosome. At metaphase, each chromosome consists of two longitudinal parts called chromatids. Chromatids of a chromosome are held together at a point called centromere.

During telophase and G_1 stage of interphase, each chromosome consists of a single chromatid. During S phase of interphase, DNA replicates, as a result the single chromatid also replicates and becomes double structure. Each chromosome thereafter consists of two chromatids at prophase and metaphase. Two chromatids of same chromosome are called sister chromatids (Figure 3.3).

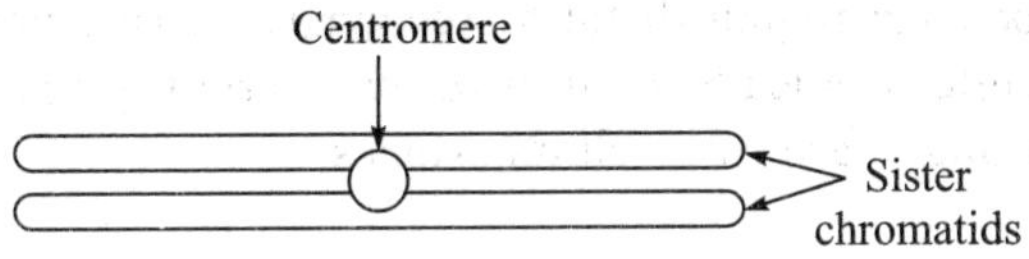

Figure 3.3 General morphology of chromosome showing sister chromatids.

Centromere

Centromere is also called primary constriction or kinetochore. The shape of chromosome is determined by the primary constriction located at the point where the two arms of a chromosome meet. Within the constriction, a clear zone containing a small granules is located. This clear region is called *centromere* or *kinetochore*. It is functionally related to the chromosome movement that occurs during mitosis. Cytologically, this indicates that the centromere contains DNA (Figure 3.4).

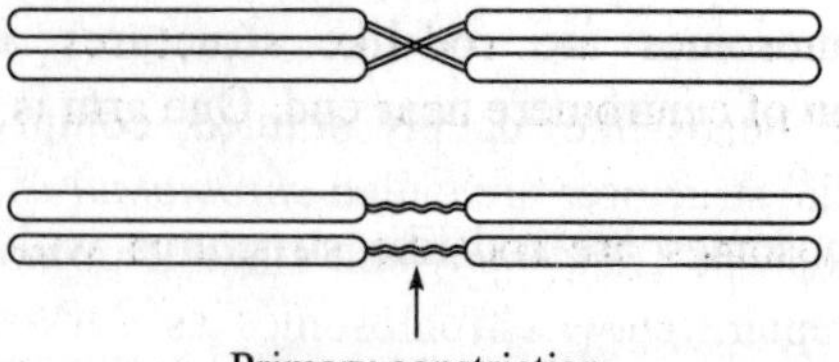

Figure 3.4 Pattern of centromere arrangement.

Secondary Constriction

Secondary constrictions are constant in their position and extent. These constrictions are useful in identifying any particular chromosome in a set. Secondary constrictions may be short or long. They are distributed along the chromosomes and are distinguished from the primary constrictions by their absence of marked angular deviations of the chromosomal segments (Figure 3.5).

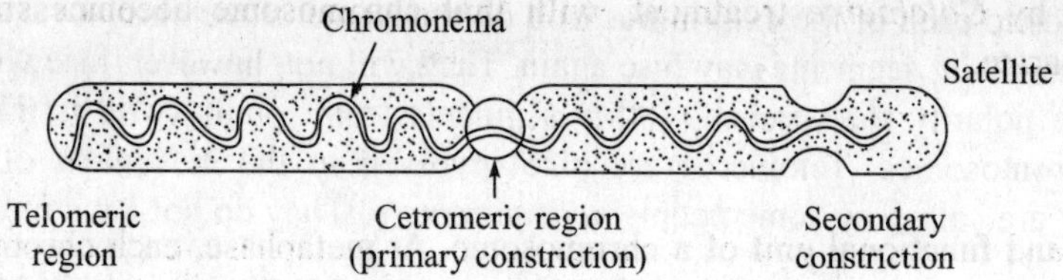

Figure 3.5 Model of a chromosome showing secondary constriction.

Some chromosomes have one or more non-centromeric secondary constriction in addition to their primary constriction. It may be nucleolar or non-nucleolar secondary constriction. Nucleolar constriction is also called *nucleolar organizers region* (NOR) because nucleolus is formed and attached to this region.

Nucleolus is site of ribosomal RNA synthesis. NOR of chromosome contains genes for coding RNA. The chromosome region distal to secondary constriction is called *satellite* and chromosome-bearing satellite is called *SAT. chromosome*. Certain genes are located in this region. In barley chromosome 6 and 7 are SAT. chromosomes.

Knob

In some species, chromosome possesses spherical *heterochromatic* bodies called *knobs*. They are located at terminal position on the chromosome, but sub-terminal and interstitial knobs are also found. They are best known in maize. The knobs are used as marker to identify the chromosome. The knobs are constant in a species but size is variable at different stages of development (Figure 3.6).

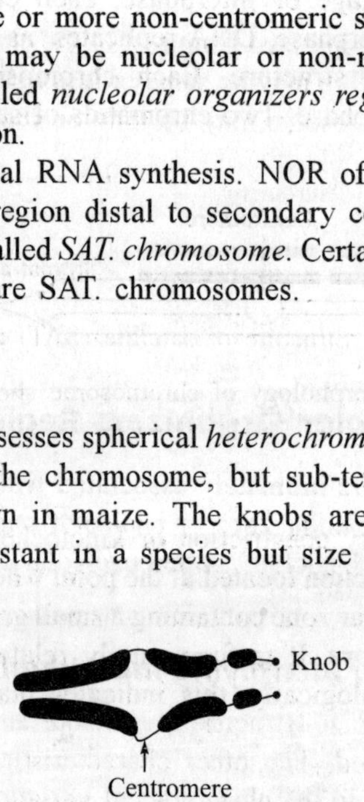

Figure 3.6 Chromosome showing structure of knob at anaphase stage of mitosis.

Chromomere

The chromomeres are small, beads-like, deeply stained, compact structures located on the chromosomes. These beads-like structures are called *chromomeres.* In mitosis, chromosomes are not visible but in meiotic prophase I, at *pachytene* stage, chromomeres are visible. The giant polytene chromosomes in Deptera show chromosomes as dark-stained bonds. Chromomeres were once considered genes but now, the earlier view has been changed and chromomeres are being considered as locally coiled structure of DNA and they are considered to involve in DNA replication and mRNA synthesis (transcription) (Figure 3.7).

Figure 3.7 Structure of chromomere.

Telomere

This term applies to each of the extremities of a chromosome. If the chromosomes are fractured by X-rays, the resulting segments may fuse again. They will not, however, fuse with the telomere. Telomere has a polarity that prevents other segments from joining with it. Thus, it provides stability to chromosomes. Telomeres are also involved in the movement of chromosomes. Therefore, they are called by some people *neocentromere.* They do not have any morphological feature.

Satellite

There is a round, elongated body separated from rest of the chromosome by a delicate chromatin filament. The diameter of the satellite may be same or less than chromosome. Similarly, filament joining the satellite and the chromosome may be long or short (Figure 3.8).

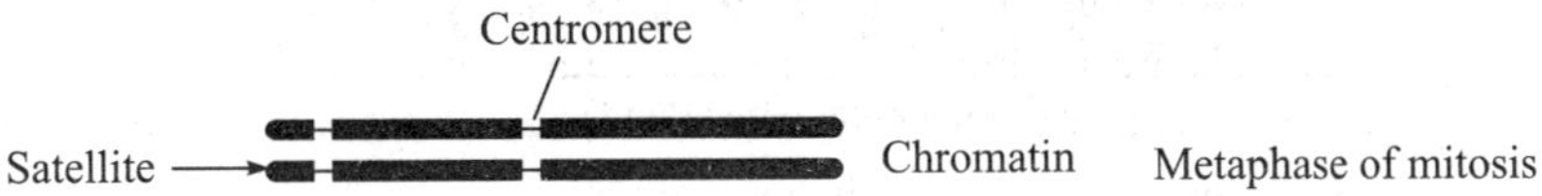

Figure 3.8 Structure of satellite (SAT) chromosome.

Nucleolar Zone or Nucleolar Organizers Region (NOR)

Certain secondary constrictions are intimately associated with the formation of nucleoli. These are specialized regions called nucleolar zone or nucleolar organizers region (NOR). Generally, there are two chromosomes in each nucleus, called nucleolar chromosomes that have this special characteristics (Figure 3.9).

Important characteristics for identifying individual chromosome

In mitosis their number, relative size, structure, behaviour and the internal organization can be ascertained by karyotype analyzed. The other characteristics such as linear constriction and degree of coiling may be subject to the physiological variation.

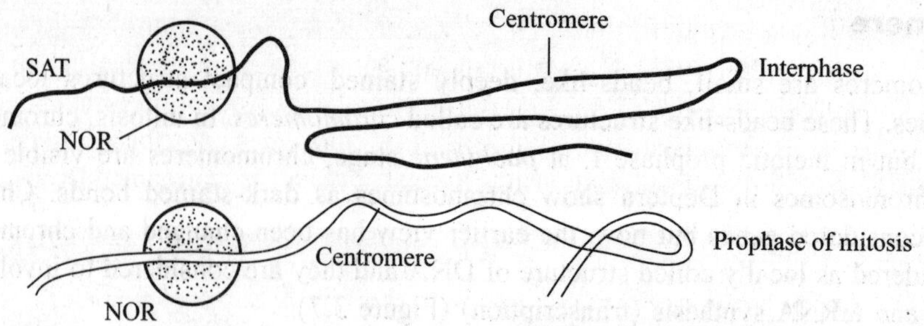

Figure 3.9 Nucleolar organizers region (NOR).

Karyotype: Plant and animal species are characterized by a set of chromosomes which have constant features. These features include—chromosome number, size and shape of individual chromosome.

The term 'karyotype' is used to the group of characteristics that identifies a particular chromosome set and usually presented by a diagram called *idiogram* where chromosomes of haploid set of an organism are ordered in a series of decreasing size. The karyotypes of different groups are compared to find evolutionary relationship.

A karyotype, showing large differences between the smallest and the largest chromosomes of the set and having fewer metacentric chromosomes is called *asymmetric karyotype,* which is considered to be a relatively advance feature when compared with *symmetrical karyotype.* Those having large number of metacentric chromosomes are called symmetrical (Figure 3.10).

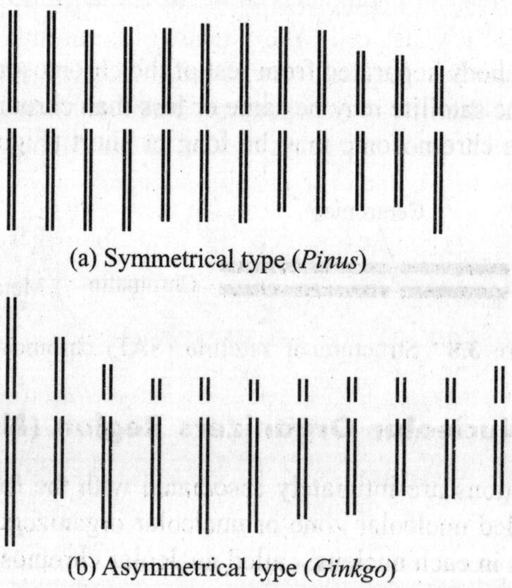

(a) Symmetrical type (*Pinus*)

(b) Asymmetrical type (*Ginkgo*)

Figure 3.10 Different types of karyotypes.

The number of chromosome is one of the essential constants to determine the phylogeny and taxonomic position of the plant and animal species. The organism with the lowest number of chromosomes is the nematode, *Ascaris megalocephala.* It has two chromosomes in each somatic cell (n = 1).

Chromonema

Each chromosome is a double-coiled structure. Each thread-like structure is called chromonema (plural chromatids). The chromonema may be single during one stage of development and two- or four-stranded during another.

CHROMOSOME MODELS

Since a very long DNA molecule is packed in a short dimension of a chromosome, for this purpose, several models have been proposed by different workers. They are:

1. Molecular model
2. Multistranded model
3. General chromosome model
4. Folded fibre model
5. Nucleosome–solenoid model

Molecular Model

This model was proposed by Taylor and coworkers in 1957 based on semi-conservative replication of DNA. In this model, it is believed that chromatid consists of only one DNA chain wherein several DNA double helices are attached end to end to form a chain with the support of protein molecule. Protein is responsible for coiling and uncoiling of the chromosome (Figure 3.11).

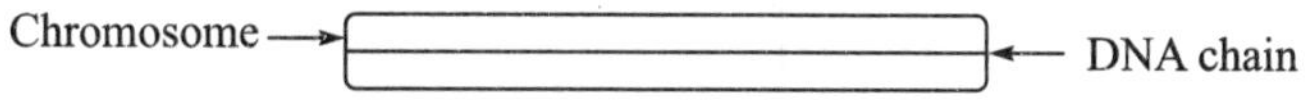

Figure 3.11 Molecular model.

Multistranded Model

Ris (1961) and Ris and Chandler (1963) expressed that chromosome is multistranded, i.e. it contains several DNA double helices arranged parallel to each other. In fact, the proponent of this theory say that each chromosome is divided into two chromatids, each chromatid is again divided into two half chromatid and each half chromatid composed of twoquarter chromatid is further divided into four chromatid fibres which possess two DNA double helices, Two DNA molecules in association with protein form the chromatid fibre. The diameter of DNA helices is 2 nm while diameter of chromatid is 10 nm; quarter chromatid is 40 nm; half chromatid is 80 nm and the last chromatid is 160 nm. This way each chromatid possesses 32 DNA molecules at *interphase* and each chromosome possesses 64 DNA molecules at *metaphase*. This is not a valid proposition, i.e. chromosomes are not multistranded (Figure 3.12).

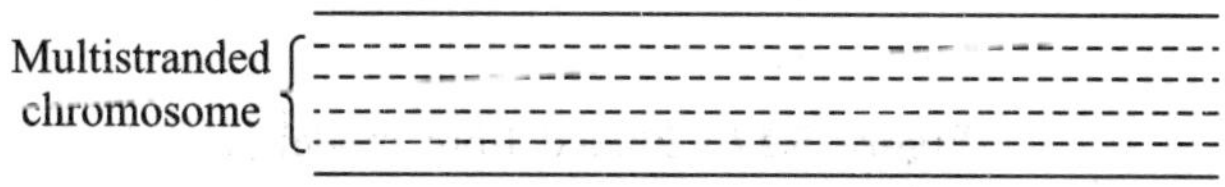

Figure 3.12 Multistranded model.

General Chromosome Model

This chromosome model was proposed by Crick in 1971, which suggests that DNA in a chromatin is monomer which is stretched from one end to other end. The bands and inter-bands have special function (Figure 3·13).

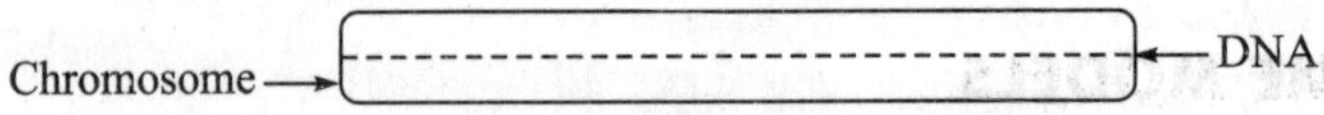

Figure 3.13 General chromosome model.

Folded Fibre Model

Chromatin fibres are the basic unit of chromosomes which are about 230 Å in diameter. A single chromatin fibre is found in each Chromatid which consists of single coiled double DNA helix. The folding of chromatin fibre in different ways results in the development of chromatin structure which is seen at metaphase II.

Du Praw (1965) proposed this model on the basis of observations recorded in the electron microscope of human chromosome. According to him, each chromatid possesses a single, long nucleoprotein complex in which a single DNA double helix forms the main structure of the axis (Figure 3.14). This model is correct and has wide support. The main features of this model are:

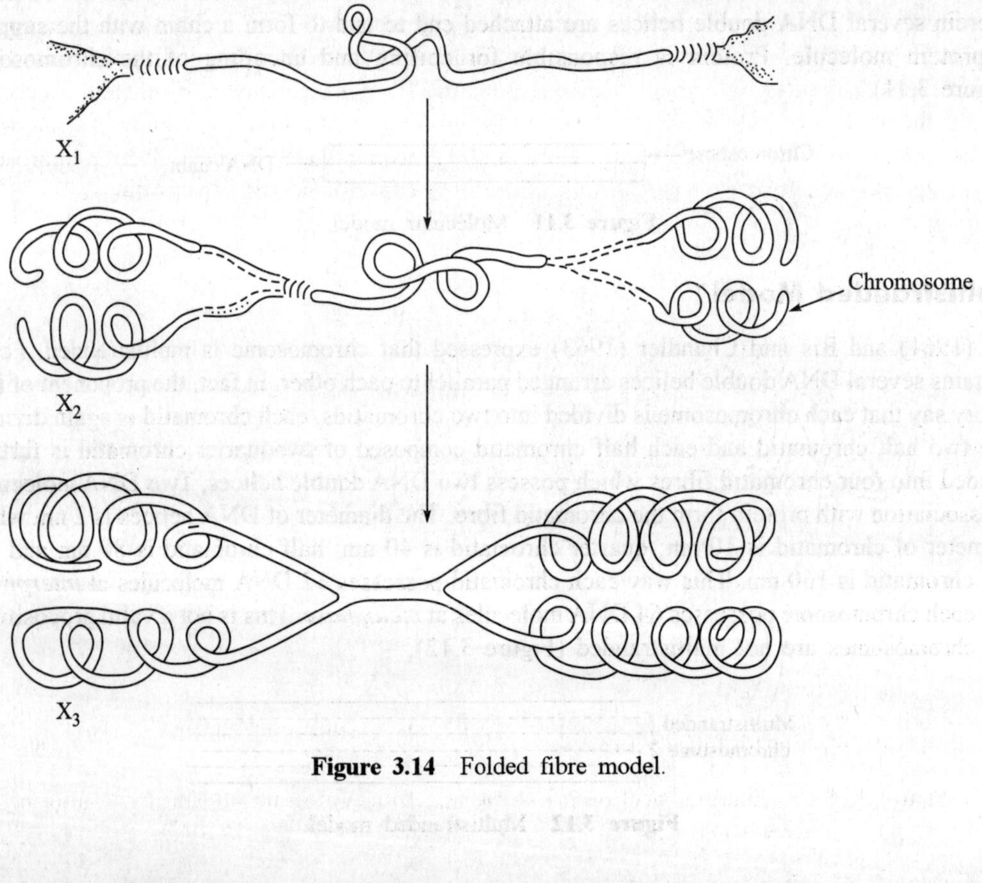

Figure 3.14 Folded fibre model.

1. In the resting phase (interphase) the chromosome consists of a single chromatid and each chromatin fibre possesses a single DNA double helix along with protein and RNA.
2. The chromatin fibre replicates on the consequence of DNA replication in S phase of cell cycle but at axis, it does not divide as a result the pairs of chromatin remain held together and become visible at *prophase* of mitosis.
3. The two sister chromatin fibres remain held together with the centromere. They are separated during division of centromere at anaphase of mitosis.

Coiling of DNA occurs around the three types of histone proteins: (a) lysine-rich, (b) arginine-rich, and (c) DNA-dependent enzyme, DNA polymerase.

Nucleosome—Solenoid Model

This model of chromatin organization was proposed by Kornburg and Thomas in the year 1974. Chromatin is composed of DNA, RNA, Histones and other proteins. Chromatin fibres are 300 Å in diameter. The nucleosomes are subunits of chromatin and have bead like appearance. Each nucleosome is composed of a histone octamer and 16 base pairs (bp) of DNA. Each nucleosome consists of one core particle, one linker (spacer) DNA. The core particle consists of two copies, each made of H2B, H3 and H4 histone molecules, i.e. it has histone octomer. The core particle is 60 Å in height and 100 Å in diameter. A double stranded DNA is tightly wound up around this core particle.

Figure 3.15 shows structure of nucleosome solenoid model of chromosome. The spacer (linker) DNA consists of 4 base pairs. A single molecule of histone (H1) protein is connected with linker DNA. The supercoiled nucleosome (chromatin) fibre is also known as solenoid. According to this theory, a very long molecule of DNA (146bp) is packed into a single units of nucleosome make a chromatin fibre. The chromatin fibre of 300 Å which is visible under electron microscope at metaphase, develops from nucleosome fibre as a consequence of supercoiling.

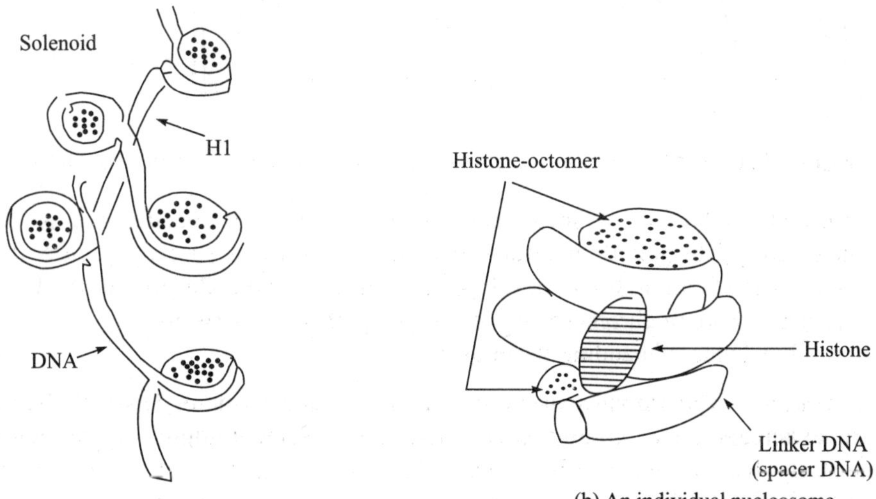

(a) Nucleosome chain

(b) An individual nucleosome

Figure 3.15 Structure of nucleosome chain, and structure of an individual nucleosome.

BANDING

Bands are localized areas in the chromosome that stain deeply from rest of the chromosome. Bands are GC or AT-rich regions (Guanine–Cytosine or Adenine–Thymine) of constitutive chromatin. It consists of repetitive DNA sequences. There are a varieties of DNA bands (e.g., a, c, G or R bands), which have been studied in animals, human, mouse, pig and mosquitoes. Bands have been used as marker to identify the chromosomes (Figure 3.16(a), (b)).

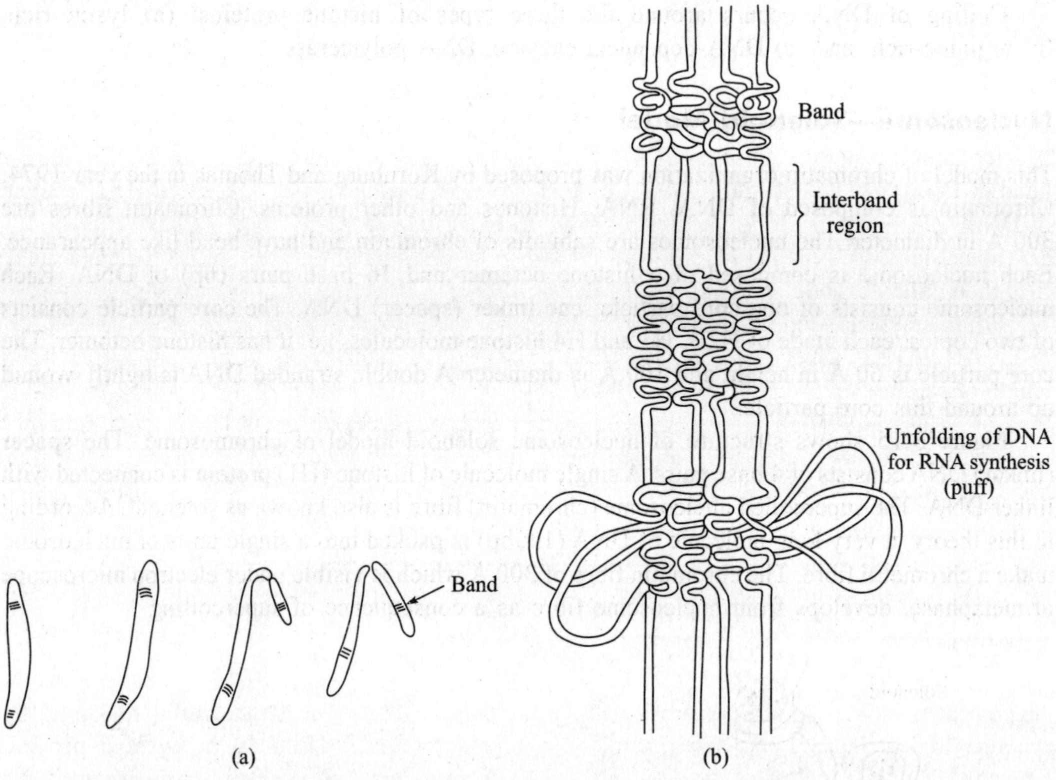

Figure 3.16 (a) Banding in the chromosome, (b) formation of bands in DNA.

The artificial production of such bands by treatment with specific dye is known as chromosome banding. The pattern of chromosome banding is highly species-specific. It is useful for chromosome identification. It was developed by Caperson and colleagues in 1971 in Sweden for identification of human chromosome with the help of quinacrine due.

There are four kinds of banding techniques:

1. *Q-banding:* The chromosomes are treated with quinacrine or Hoechst-3325 dye and are examined under fluorescence microscope. Some regions of chromosome get intensely dyed farming bands because of repetitive sequences of adenine and thymine, whereas, lightly fluorescent bands are richer in guanine and cytosine sequences.
2. *R-bands:* In this technique, chromosomes are treated with *Giesma* strain at a high temperature. R-bands are reverse of G-bands.

3. *G-bands:* Chromosomes are treated with saline deterants and urea may be also used. The chromosomes are visible under light microscope.
4. *C-bands:* Chromosomes are treated first with moderately strong alkali and then by warm saline solution. Chromosomes are then stained with *Giesma* stain. The chromosome around centromere which contains satellite DNA exhibits dark stain under microscope.

SATELLITE DNA

The term satellite DNA was applied to the repetitive sequences. Satellite DNA is highly repetitive. DNA sequences may differ from other classes of DNA regarding buoyant density so that they usually form a separate minor band under Cscl density gradient ultracentrifugation. This minor band of DNA is called a satellite DNA or SAT DNA. Buoyant density depends on G:C content of DNA.

Sometimes, the repetitive DNA does not form a minor band, but it lies within the main band. It is then called *cryptic satellite.*

Prominent base sequences of satellite bands I, II and III are 5′ A CAA ACT3′, 5′ ATAAACT3′ and 5′ ACAATT3′. Most of the satellite DNA consists of tandem repeats of these sequences. Total length of satellite DNA sequences of band I is 1.1×10^7 bp, II and III is 3.6×10^6 bp. A sudden type of DNA replication that produces families of similar base sequences is called *saltatory replication.* Certain tandem repeats of short DNA sequences occur less frequently. They may range from 5 to 50 repeats in mammalian genomes. Such repetitive sequences are called *minisatellete* or *Variable Number Tandem Repeats* (VNTRs), as the size of repeats are variable. Minisatelletes are used in *DNA fingerprinting* in man.

REPETITIVE DNA IN CHROMOSOME

The organization of DNA in chromosomes may be studied by denaturation followed by renaturation technique. The DNA strand could be separated by heating or by high pH salt concentration. This phenomenon is called *denaturation* and on cooling or by treating the denatured DNA/RNA, by low pH solution (acidic), the single DNA strands could be reassociated by complementary base pairing rule. This is called *renaturation.* During cooling, the complementary strands collide and reassembled. This technique can be used to form DNA–RNA hybrid too. Some people called it nucleic acid hybrid. The procedures of denaturation and renaturation of DNA have revealed that the eukaryotic genomes possess two types of DNA base sequences. They are:

1. Non-repetitive DNA sequences (unique or single copy DNA sequences)
2. Repetitive DNA sequences

Non-repetitive DNA Sequences

The base sequences of unique DNA are not repeated in genome, rather DNA base sequences are present in a single copy in each chromosome. A large number of genes—structural genes are present in a single copy in genome. Bacterial and viral genomes are considered as unique DNAs, which consist of very negligible quantity of repetitive DNA.

Repetitive DNA Sequences

In contrast, the repetitive DNA sequences constitute 30 per cent of total genome of some sea urchin, 9 per cent in human and 70 per cent in *Drosophila.* Again the repetitive DNA are of following two types.

Moderately repetitive DNA sequences

In this case, respective sequences are non-tandemly dispersed throughout entire genome; the number varies from 2 to 10^5 copies per genome, but the figure varies from species to species. In *Drosophila melanogaster* 12 per cent genome DNA form moderately repetitive DNA; in man 13 per cent DNA and in *Nicotiana* 65 per cent. In the repetitive DNA, genes are found for ribosomal RNA and for ribosomal proteins. The short sequences consist CACACA------CA; GATA or GATA-500 to several thousand.

Highly repetitive DNA sequences

Highly repetitive DNA constitutes 5 to 45 per cent of genome. Some of these sequences are satellite DNA. The sequences in this class are from 5 to 30 base pairs per repeat and duplicated as many as 10^5 times per genome. Many highly repetitive sequences are localized in blocks of tandem repeats while others are dispersed throughout the genome. In human genome, the sequence called 'Alu' sequence which is restriction site for enzyme 'Alu' contains 300 base pairs in length and are present in about 500,000 copies in human genome. This repetitive DNA family alone accounts for about 5 per cent of human DNA. These highly localized repetitive sequences are very short, which make 6 per cent of human and 45 per cent of *Drosophila viridis.* In *Drosophila* genome, the major components of the highly repetitive class are three related sequences of seven base pairs such as:

(i) 5′-ACAAACT 3′
(ii) 5′-ATAAACT 3′
(iii) 5′-ACAAATI 3′

Blocks of satellite (highly repetitive) sequences in the genomes of several organisms have been located by *in situ* hybridization with *metaphase* chromosome. The satellite sequences have been found in the heterochromatin region of the chromosome. These are the regions that condense early in *prophase* than rest of the chromosomes and stain darkly. Sometime heterochromatin remains highly condensed throughout the cell cycle. Numbers of genes located in heterochromatin are lesser than euchromatin. The present view is that no genes are found in the heterochromatic region.

IMPORTANCE OF CHROMOSOMES

Chromosomes, apart from being little packaging medium of DNA, perform some important functions, without chromosomes, it is difficult to think of genes. Thus,

1. Chromosomes provide raw materials for the formation of new genes. The large number of genes, we find in the most organisms are the result of evolutionary process, of course, mutation and natural selection have played important role in the evolution of genes.
2. The genetic materials in the form of chromosomes are well recognized. Integration of thousands of genes into chromosome provides an obvious mechanism for their balanced distribution into newly arising cells following mitotic division.
3. Apart from ensuring this accuracy and maintaining the genetic stability of a cell line, the chromosomes play important role to hold together the favourable gene combination through their linkage. The chromosomes provide mechanism for regulating the amount of genetic recombination.
4. Chromosomes have also been important in the evolution of new species. Most of cultivated plants like bread wheat (*Triticum aestivum*), rice (*Oryza sativa*), maize (*Zea mays*) cotton (*Gossypium hirsutum*) arose as polyploids.
5. Chromosomal aberrations like translocation and inversion have been used as important tools by natural selection in evolving new genomes and to suppress crossing over, if this breaks favours to get gene complexes.

SEMICONSERVATIVE REPRODUCTION OF CHROMOSOMES

In *Escherichia coli* (Prokaryotes), the chromosomal material takes the form of a single DNA molecule. In higher organisms, the organization of DNA in the chromosome was not properly understood. Taylor, Wood and Huge in 1957 made an important discovery. According to them, DNA in the chromosomes of higher plants show a semiconservative mode of replication, i.e. DNA is spread throughout the entire length of chromosomes and when it reproduces, the old and the new DNA are divided equally between two daughter chromosomes (chromatids). They had used 'tritium', a radioactive isotope of hydrogen for labelling the newly synthesized DNA in the chromosome. However, it is not possible to distinguish the old and newly synthesized DNA strand except that old DNA is in the form of conserved template.

NEW CONCEPT OF CHROMOSOME STRUCTURE

The semiconservative replication of chromosome in respect of their DNA content agrees completely with the Watson and Crick scheme who proposed that two strands of duplex molecule act as templates for the synthesis of the complementary strands, i.e. a chromatid contains a single DNA double helix and a single molecule (duplex) extends from one end to other end of the chromosome. Thus, earlier concept that chromosome may have a multistranded structure is no longer tenable.

Subsequently, that single chromomeres on the chromosome correspond to single gene had no validity. Chromomeres were found to have far more DNA than the amount most genes can be expected to have, i.e. DNA amount in single chromomere tends to be 10 to 100 times more.

EUCHROMATIN AND HETEROCHROMATIN

In 1928, Heitz defined heterochromatin as, the chromosomal regions that remain condensed during *interphase*, forming so-called chromocenters or false nuclei. In contrast, the rest of the chromosome substance uncoils and swells during same period are called *euchromatin.* Euchromatin does not take stain. This phenomenon is called *negative heteropycnosis.* Heterochromatic regions of *interphase* and early *prophase* are the chromosomal portions that do not take stain. This phenomenon is called *heteropycnosis*, which forms massive or condensed blocks of chromatin. Heterochromatin should be considered as a state of the chromosome and not a special substance. During mitosis, the heterochromatic regions may stain strongly or more weakly than the euchromatic regions. These phenomena are called respectively, *positive* and *negative heteropycnosis* (differential staining). Heteropycnosis is the characteristic of the sex chromosomes of many species. But may be observed in other chromosomes also. Morphologically, these heterochromatic regions are intercepted as segments in which the chromonema has a different degree of coiling or packing. Heterochromatic regions consist of chromomeres, chromocentres and knob. Salivary gland chromosomes of Deptera contain one large chromocentre on account of fusion of chromocentres of other chromosomes present in the cell.

Heterochromatins are of two types:

Constitutive Heterochromatin

This contains repetitive DNA and non-repetitive DNA both, and A–T rich main bands of DNA. Repetitive DNA contains millions of copies of base sequences.

Facultative Heterochromatin

This type of heterochromatin can revert back to euchromatin. It is expressed under certain circumstances. It is presumed that this takes part in gene regulation. The euchromatic region of chromosomes possesses genes, while genes located in the heterochromatic region are repressed. The cause of repression is the highly condensed nature of heterochromatin due to highly repetitive nature of DNA.

The Y chromosome of *Drosophila* is heterochromatic but possesses babbed bristle gene *bb*.

Heterochromatic regions contain more DNA, means more genes than euchromatin that is not correct. They are almost genetically inert.

SPECIALIZED CHROMOSOMES

In certain cells of living organisms—plants and animals species, there exist some functionally specialized chromosomes. These chromosomes permanently reside in the cell; occasionally, extra chromosomes are noted in some individuals of a species. These are known as *specialized chromosomes.* Following are specialized chromosomes:

1. B-chromosomes
2. Nucleolar chromosomes
3. Sex chromosomes

4. Holocentric/holokinetic chromosomes
5. Giant chromosomes

B-chromosomes

These chromosomes are found in certain animals and plants in addition to the basic numbers constituting the genome. These chromosomes are called supernumerary or accessory chromosomes. These chromosomes were discovered first in 1905 but in 1928 Randolph gave the term B-chromosome while normal chromosomes are autosomes (A) which are involved in performing normal metabolic functions.

B-chromosomes are polymorphic in character. In insects (heteroptera, lepidoptera, coleoptera and mollusk) B-chromosomes have been found, but not in fishes and primates. In plants, B-chromosomes have been found in 1000 species from angiosperm, gymnosperm to fern and mosses. But they have been predominantly found in cross-pollinated species specially in forage grasses, maize, rye, verul grass and *Poa* species (Bosemark 1954a, 1954b, Muntzing, 1948, Rhoades, 1953 and Darlington, 1956). Therefore, origin in plants appears in wide crosses like *T. aestium* × *Aegilops* inter-specific crosses and inter-generic crosses wheat × rye due to non-disjunction of chromosomes. In the animals, its origin may be sex chromosome.

In general, B-chromosomes are telocentric or acrocentric and are highly heterochromatic (Figure 3.17). Heterochromatin is present but B-chromosomes neither show pairing with A-chromosomes nor show any genetic consequence though, they contain more A–T, than A-chromosomes in grasshopper.

Figure 3.17 B-chromosome.

B-chromosomes are genetically inert but it has been observed that B-chromosomes reduce pairing, chiasma formation and pollen fertility, decrease vigour, defective seed, scarred endosperm, increased nucleus and cell size, variation in pollen size wherever they are present.

Translocation of B-chromosome with normal chromosome were induced in maize through x-ray radiation (Roman, 1947 and Rhoades, 1965). Randolph (1941) noticed that in the presence of B-heterochromatin in nucleus. B-chromosomes undergo mitotic non-disjunction in the second division of microspore.

The other characteristics of B-chromosomes are:

(a) They are not found in all the species and also may not be found in all the cells of an individual and hence are dispensable.
(b) They are smaller than A-chromosomes (autosomes).
(c) B-chromosomes have non-Mendelian mode of inheritance because they are not homologous to any of the basic A-chromosome.

Nucleolar Chromosomes

The chromosomes, which are involved in the formation of nucleolus are known as *nucleolar chromosomes* and the region which is actively involved in the nucleolus formation, is known as *nucleolar organizers region* (NOR). These chromosomes possess ribosomal DNA (rDNA). Figure 3.9 indicates the structure of nucleolar chromosomes.

The portion of chromosome separated by secondary constriction from the main chromosome is called *satellite*. The RNA is synthesized in the core region of the nucleolus in which ribosomal RNA genes are located. In the eukaryotes, the ribosomal subunits occur within the nucleolus but ribosomal proteins are synthesized in the cytoplasm.

Sex Chromosomes

In general, genes determine the sex in the organism, but in some plants like *Melandrium* and man there are some specific chromosomes, which determine the sex. These specific chromosomes carry sex-determining genes, hence these are called *sex chromosomes* and the remaining chromosomes are called *autosomes* (A). In such cases, one of the pairs (sex) is homogametic which produces one type of gametes and the other is heterogametic which produces two types of gametes (Figure 3.18).

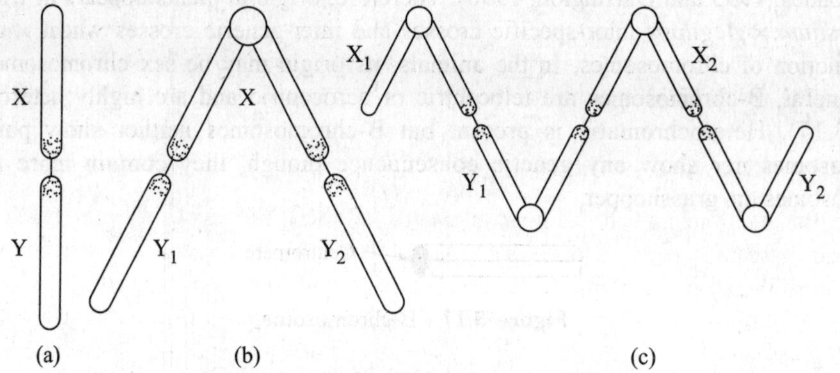

Figure 3.18 Orientation of X and Y chromosomes during meiosis I. X and Y chromosomes oriented towards opposite poles.

In human beings, melandrium and *Drosophila melanogaster*, X-Y system determines the sex. Females are homogametic and consist of two XX chromosomes and produce one type of gametes X and X, while males are heterogametic XY and produce two types of gametes X and Y. In birds and reptiles, males are homogametic and possess ZZ (XX) chromosomes while females are heterogametic and possess ZW (XY) chromosomes (silkworm), and produce two types of gametes. On the other hand, in moth XX determines male and XO determines female. Details are given in Chapter 11 dealing with sex determination.

Holocentric/Holokinetic Chromosomes

In some animals and plants species, centromeric activities are spread over a considerable part or even on the entire chromosome. Such chromosomes are called *holocentric chromosomes*. They are also found in some monocotyledenous plants like juncales and cyperales (Figure 3.19).

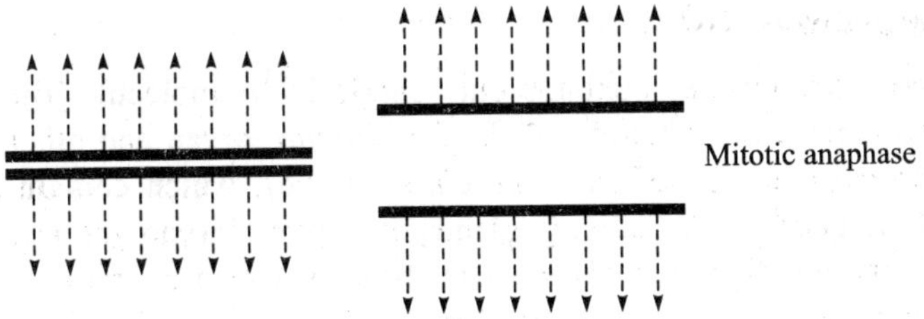

Figure 3.19 Holocentric chromosomes.

During cell division, entire chromosomes get attached with the spindle fibres which form sheet-like appearance towards pole (Figure 3.20).

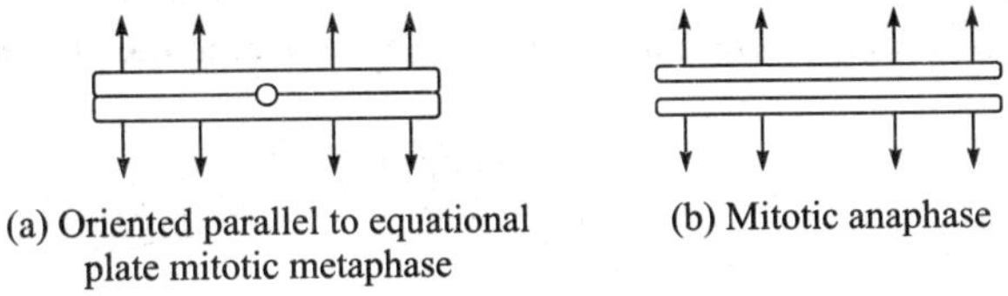

Figure 3.20 Structure of holocentric chromosome at metaphase (a) anaphase of mitosis (b).

Giant Chromosomes

In some cells and particularly at certain stages of their life cycle special types of giant chromosomes are observed. These are characterized by their enormous size and by a corresponding increase in volume of the nucleus and the cell (Figure 3.21). *Polytene chromosomes* are found in salivary glands of Dipteran larvae. *Lampbrush chromosomes* also belong to this category.

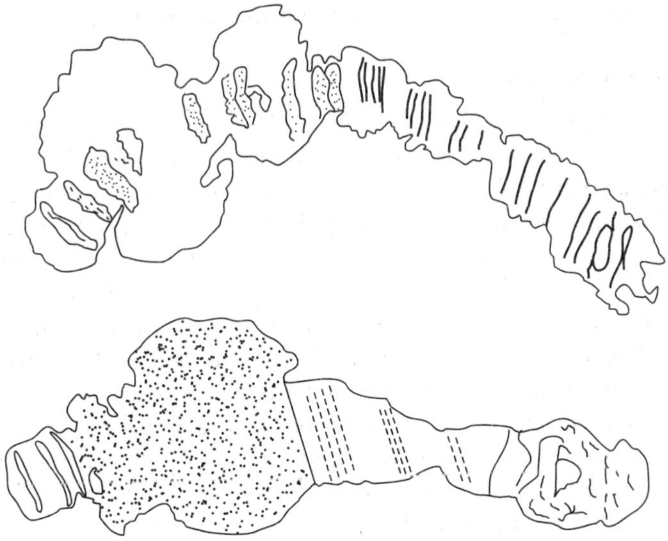

Figure 3.21 Puff on the giant chromosomes in the salivary gland chromosome of *Diptera larvae*.

Polytene chromosomes

A typical eukaryotic chromosome contains only single DNA molecule. But the nuclei of the salivary glands and certain other tissues of the *Drosophila* larvae and other two-winged flies, there are giant chromosomes called *polytene chromosomes*, which contain about 1000 DNA molecules, laterally aligned. In the salivary glands of Diptera larvae gut, trachea, fat body cell, ovarian nurse cells and malpighian tubules, some chromosomes are strikingly different from the somatic chromosomes of the same organism. This enormous size is reached by a series of 9 to 10 consequent duplication cycles of chromosomes which do not separate from each other as a result they increase in number and also in the DNA content about 1000 times. These are generally found in the intestinal cells of mosquitoes. Another characteristic of polytene chromosome is that the homologous pairs are closely associated as in meiotic prophase. This phenomena is called *somatic pairing.* The process of reduplication of the strands is called *endomitosis.* The length of a polytene chromosome is almost same as a mitotic chromosome during prophase (Figure 3.22).

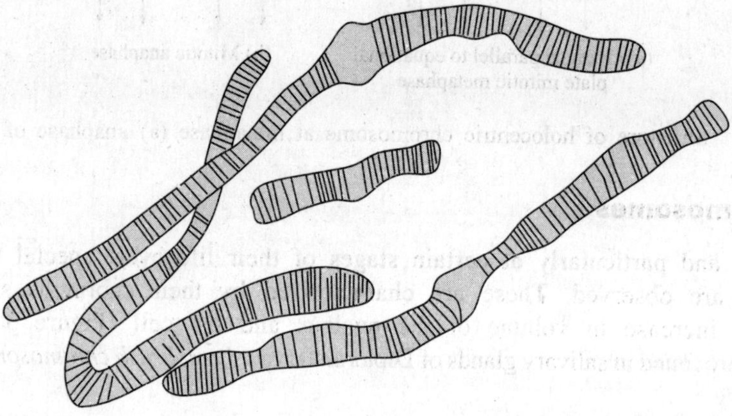

Figure 3.22 Polytene chromosomes found in salivary gland of *Drosophila melanogaster.*

Lampbrush chromosomes

They are longer than polytene chromosomes and are found in oocytes during extended diplotene phase of the first meiotic division, which lasts for several years/months/days. At this stage, a large amount of protein and RNA are synthesized, which increases the cellular mass. The length may reach three times of polytene chromosome. The growth of lampbrush chromosome is the result of an increase in the size of the chromonemata. When diplotene is blocked and bivalent decondense and take brush-like structure, the bivalent chromosomes remain attached by chiasmata. The lampbrush chromosomes have many fine lateral projections giving them the appearance of a test tube brush or lampbrush. The central axis of the lampbrush chromosome is composed of four chromatids to which fine lateral projections are attached. The loops are the extended portion of the chromatid (half chromosome). The loops are coated with protein and DNA materials. In urodelic amphibians, the length of lampbrush chromosomes reaches up to 1 mm (1000 μm). Chiasma is located in between loop, not within loop. The loop possesses core DNA in highly condensed state (Figure 3.23).

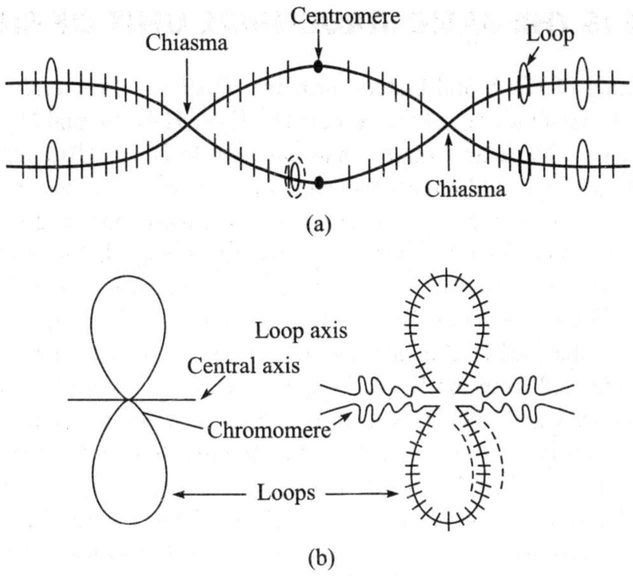

Figure 3.23 Lampbrush chromosome (bivalent).

BIOCHEMICAL STRUCTURE OF AN EUKARYOTIC CHROMOSOME

Eukaryotic chromosome consists of protein (50–60%), DNA (30%–40%) and RNA (5%–10%). However, this composition varies from species to species as well as in different cells within the same species. It also varies in the different stages of the cell cycle (G1–S–G2–M). The proportion of DNA in metaphase is lower than in interphase. However, the proportion of protein is higher in metaphase than in interphase. Eukaryotic chromosome contains a single DNA molecule of enormous length. The largest chromosome of *Drosophila melanogaster* genome has a DNA content of about 6500 kb or 65×10^5 nucleotide pairs (1 kilobase = 1000 nucleotides) that is equivalent to a continuous linear duplex of about 22 mm long. DNA is measured in picogram (pg) and 1 pg = 10^{-12} nucleotides in DNA helix.

The eukaryotic chromosomes constitute a linear array of genes of which DNA is a major constituent. DNA is also the genetic material in bacteria and eukaryotes. That is to say, genes are embodied in a stretch of double-helical, polynucleotide chains into which the amino acid sequence of a polypeptide chain is inscribed via universal genetic code. In the eukaryotic nucleic acid, the negatively charged amino acid and DNA exist in association with an approximately equal weight of positively charged or basic proteins (histones). The association of chromosomal DNA and chromosomal histones appears to proceed by formation of an ionic bond between the phosphate group of polynucleotide chains and the side chain; amino groups of polypeptide chain. The DNA of all the organisms is associated with numerous protein molecules in a stable, orderly aggregated form called *chromatin.* Some of these proteins are present in the chromatins that determine the chromosome structure and also changes take place in the division of the cell; some chromatin proteins regulate the chromosome functions.

NUCLEOSOME IS THE BASIC STRUCTURAL UNIT OF CHROMATIN

Chromatin is aggregate of DNA and histone proteins. Histones are mainly responsible for the structure of chromatin. Histones are of five types: H_1, H_2A, H_2B, H_3 and H_4 in the chromatin of almost all the eukaryotes. These are equal in mass amount to DNA. Histones are small proteins, each molecule contains 100 to 200 amino acids and differ from most other proteins, in that from 20 to 30 per cent of the amino acids are lysine and arganine while only a few per cent of proteins, both lysine and arganine have positive charge. The positive charge helps histone proteins to bind to DNA, primarily due to electrostatic attractions to the negatively charged phosphate bond of the sugar–phosphate, the backbone of DNA. If the chromatin is placed in a solution of high salt concentration (e.g. 2 molar NaCl), the electrostatic attraction of histone proteins will disappear and this will get dissociated from the DNA. Histones also bind each other, i.e. histone–histone in chromatin as stated by Hartl (1991), is doubtful when these are positively charged; of course DNA–histone association is possible because of the negatively charged DNA phosphate bond.

The study indicates that histone proteins of different organisms are almost similar with exceptions of H_1. The amino acid sequences of H_3 molecules of different species are also identical. The histone proteins of type H_3 of cow chromatin for example and amino sequences of pea chromatin differ by only 4 amino acids out of 135 amino acids. The histone proteins of type H_4 of almost all the living forms are similar. The histone proteins (H_4) of cow and pea differ by only 2 amino acids out of 120 amino acids.

When the variation in the protein between the organisms is very small, the amino acid sequence is considered highly conserved. The extraordinary conservation in histone protein composition for hundreds of millions of years of evolutionary divergence is consistent with the importance of these proteins in the structural organization of eukaryotic chromosome.

Under electron microscope, the chromatin looks like a regularly beaded thread (Figure. 3.24).

If the chromatin is treated with DNases that gives rise to a collection of small particles of quite uniform size consisting of histones and DNA (Figure 3.21), and when histones are removed from these particles, the DNA fragments are found to the length of about 200 nucleotide pairs or small multiples of that unit size; the specific size differs from species to species and tissue. The bead-like units on the chromatin is called *nucleosome.*

Each nucleosome unit is composed of one molecule of H_1, two molecules of each of H_2A, H_2B, H_3 and H_4 and one fragment of DNA containing about 200 nucleotide pairs. If these

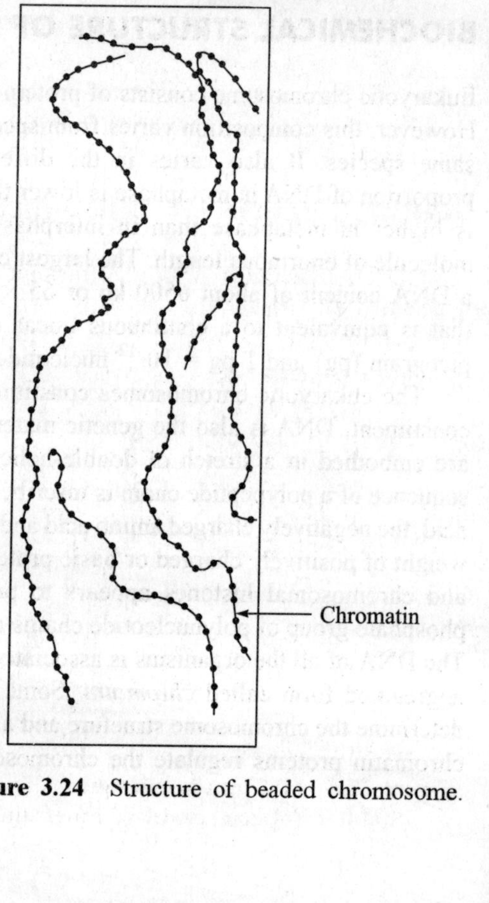

Figure 3.24 Structure of beaded chromosome.

units are digested with nuclease, it removes some of the DNA and also causes loss of OH. The resulting structure is called 'core particle', which consists of an octamer of pairs of H_2A, H_2B, H_3, H_4 around which the remaining 145 nucleotide pairs of length of DNA are wounded in about one and three-fourth turns (Figure 3.25). Thus, a nucleosome is composed of a core particle (additional DNA) that links adjacent core particle (the DNA that is removed by nuclease digestion) and one molecule of H_1; the H_1 bonds to the histone octamer and to the linker DNA causing the linkers extending from both sides of the core particles to cross and draw nearer to the octamer, though some of the linker DNA do not come in contact with any histones. The sizes of the linker DNA vary from 20 to 200 nucleotide pairs, which vary from species to species and also in the different cell types within same organism. (200–155 nucleotide pairs are considered an average size). Nothing is known about structure and function of the linker DNA.

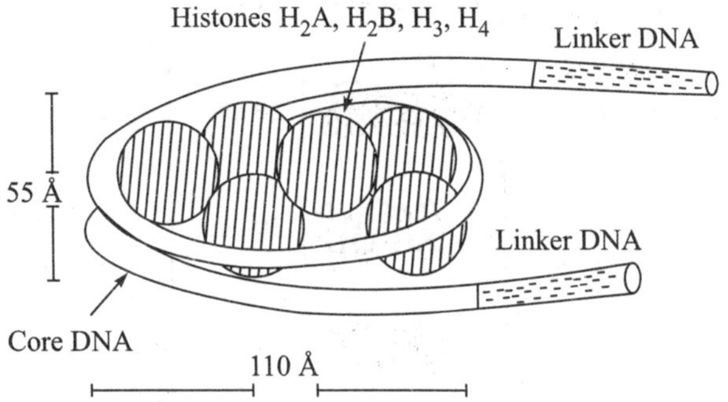

Figure 3.25 Diagram of a nucleosome core particle. The DNA molecule has wounded in one and three-fourth turns around a histone octamer.

ARRANGEMENT OF CHROMATIN FIBRE IN A CHROMOSOME

The DNA molecules of a chromosome are folded and refolded in such a way that long chromosomal strands also get shortened. The assembly of DNA and histones cause first- and seven-fold reduction in the length of DNA and form a beaded fibre of 110 Å (11 nm) wide, roughly five times width of free DNA (Figure 3.26).

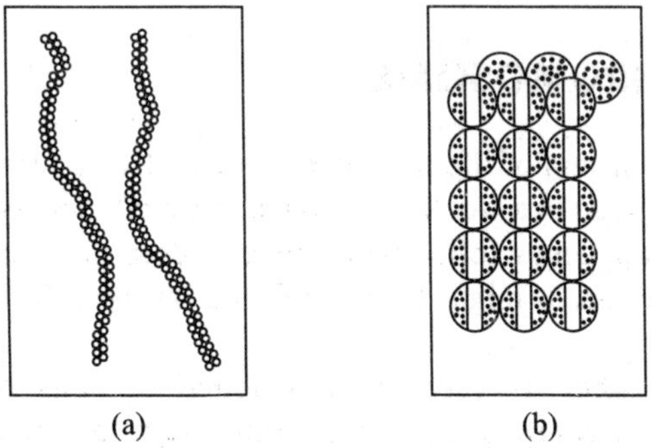

Figure 3.26 (a) Electron mictographic view of metaphase of mouse chromosome, (b) A proposed solenoid model of chromatin.

The structures of chromatin vary with the concentration of salt and 110 Å fibre is present when salt concentration is low. If the salt concentration is increased slightly, the fibre becomes shortened somewhat by forming a gigas arrangement of closely spaced bands between which the linking DNA is no longer visible under electron micrograph. However, if the salt concentration is further increased to that present in the cell, a second level compaction is reached that makes 110 Å level nucleosome fibre into a shorter and thicker fibre with an average diametre of 300–350 Å (30 nm fibre). In the final level of organization, 30 nm fibre condenses into a chromatid of compact metaphase chromosome (Figure 3.27).

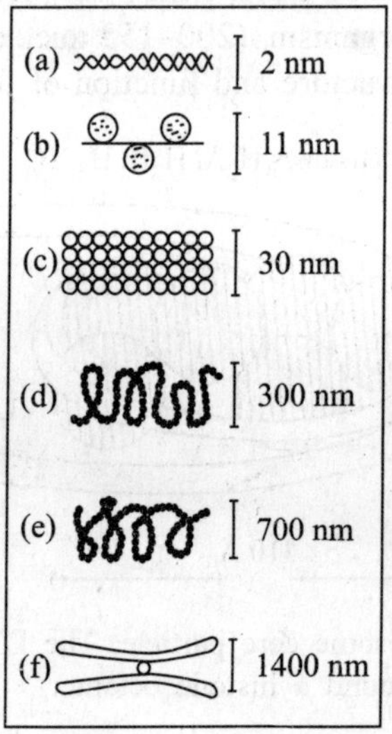

Figure 3.27 Different stages in condensation of DNA(a) and chromatin formation(b–c). A metaphase chromosome(f).

The genetic significance of compaction of DNA and protein into chromatin and ultimately into chromosome is that it greatly facilitates the movement of genetic material during nuclear division. Without chromosome condensation, there could have been many more abnormalities in the distribution of genetic material into daughter cells.

EVOLUTION OF CHROMOSOME

Most of the theories on the origin of life postulate that first living forms might have appeared as a molecule in the prebiotic soup in a wet land. Cells of more efficient type could have arisen from duplication of some of their genes followed by independent mutation of the duplicated copy, which performed new function. It might have conferred on the cell the capacity to adopt itself to adverse environments.

The characteristic number of chromosomes found in cells of different species is a consequence of the evolutionary process. Starting from a single chromosome in most prokaryotes, the basic chromosome number has been considerably increased in higher organism. Basic number of chromosome has been acquired by a species in the course of evolution.

Asexual reproduction appears to be the rule in some of the unicellular blue-green algae and fungi imperfecti as well as flagellate protozoa. But most significant event in the history of evolution of living organism has been the emergence of sexual reproduction which paved the way for origin of diploids. The corresponding number of chromosomes contributed equally by male and female gametes give rise to a diploid zygote and this balance is maintained by meiosis which gives rise to haploid gametes.

ARTIFICIAL CHROMOSOME

A new technique has been developed through which an artificial chromosome can be constructed and then placed into a yeast cell (*Saccharomyces cerevisiae*), where it functions like real yeast chromosome. The importance of this technique is that any gene or small block of genes can be added to this artificial chromosome and its expression in the yeast cell could be studied.

The *technique* is used to isolate the various components of the yeast chromosomes that are essential for the chromosomal activities.

Components

The components include the *centromere* responsible for chromosome movement; so called the *autonomous replicating sequence* (ARC), which is responsible for duplicating of the chromosome during cell division, and the *telomere*—the chromosome end that is essential for chromosome duplication.

After isolation of centromere, autonomous replicating sequence and telomere, these are fused together and reintroduced into a yeast cell. They behave like normal chromosome during cell division. This is *artificial chromosome construction technique.*

The importance of this technique is that non-yeast genes (alien genes) could be added to artificial chromosome to learn their expression such as human genes. This way a large topics of alien gene/human gene can be isolated for genetical studies.

THEORIES OF HEREDITY

Chromosomes as a Carrier of Hereditary Material

Continuity of life is the rule of nature. This happens because of the continuity of chromosomes. How a particular chromosome at metaphase of one cell division is identical to that at the metaphase of a later cell division? This indicates that chromosomes do not lose their identity during interphase period. Baveri was the first who provided this evidence in his experiment with the eggs of nematode, *Ascaris megalocephala.* Ram (2005) has given ample evidences to prove that DNA is the source of life and this DNA is packed in the chromosomes, a thread-like coiled structure.

Parallelism between Behaviours of Chromosomes and Genes

W.S. Suttan in 1902 in Columbia University and in the same year, German biologist (cytologist) Baveri found that behaviours of chromosomes in thc formation of gametes and inheritance of genes are almost similar because of the principles of segregation and independent assortment of genes and chromosomes in the anaphase I of the meiotic division.

There are certain specific genetic laws which draw the parallelism between the behaviours of genes and chromosomes. They are:

1. Both, chromosomes and genes behave in their inheritance as individual units. The individuality of the chromosomes could be seen under microscope. Each pair of chromosome is different from other pair. Likewise, each gene has individuality, which is inferred from its visibility in inheritance and its emergence intact and unaltered after a cross (hybridization) between the parents having contrasting characters.
2. Genes occur in pairs (allelic pairs) in every individual. One member of each comes from the male and the other from the female parent. This situation resembles the chromosomes, which are also seen in pairs, each member has been derived from one of the two parents (Figure 3.28).

Figure 3.28 Genes are located on chromosomes in allelic form.

3. Each gamete is seen to contain only one member of each pair of chromosome. Likewise, each gamete contains only one allele of a gene pair (*AA*, *Aa*, *aa*), that is each gamete contains only haploid number of chromosome (n) and each gamete contains only one allele of a gene pair, is evidenced from Mendel's first law of inheritance as well as by breeding experiments. In fact, important point in the Mendel's first law is the segregation by which each factor separates sharply from its alternative form or allele. This way two alternative members of a gene pair enter independently which are genetically pure.
4. Such separation takes place between the two members of a pair of chromosome during reduction division. Thus, both gene and chromosome undergo segregation at anaphase I, as a result each gamete is pure containing only member of a pair of gene/chromosome.
5. *Independent assortment of genes and chromosomes:* The pairing and disjunction of homologous chromosomes at meiosis provide a mechanism of segregation of allelic genes (first law of Mendel). Likewise, the independent assortment of two gene pairs (second law of Mendel) could be deduced from the chromosome behaviour at anaphase I of meiotic division (prophase I).

   Diploid cell (2n) contains two sets of chromosomes—one set is derived from mother and the other from father. At meiosis different pairs of chromosomes are associated and distributed to the different gametes independently. It is a matter of chance that either two paternal or two maternal chromosomes go to the same pole during reduction division or maternal or paternal chromosomes go together to one pole. The gamete may contain a combination of both paternal and maternal chromosomes. This is actually the biological principle and genes behave according to this principle. However, it is not possible to distinguish maternal and paternal chromosomes under microscope.
6. *Chromosomal determination of sex:* Sex chromosomes are best example to support that they carry the genes responsible for determination of characters or there is a parallelism between the genes and chromosomes. Sex chromosomes are an exception to the rule that all the chromosomes of a diploid organisms are morphologically homologous.

CYTOLOGICAL EVIDENCE TO PROVE THAT CHROMOSOMES ARE CARRIERS OF HEREDITARY PARTICLES—GENES

In 1931 Curl Stern discovered two X chromosomes in *Drosophila* that had undergone structural changes, that made them distinguishable from each other and from normal X chromosome. He used these X chromosomes in an experiment that provided one of the classical proofs on the physical basis of crossing over (Figure 3.29).

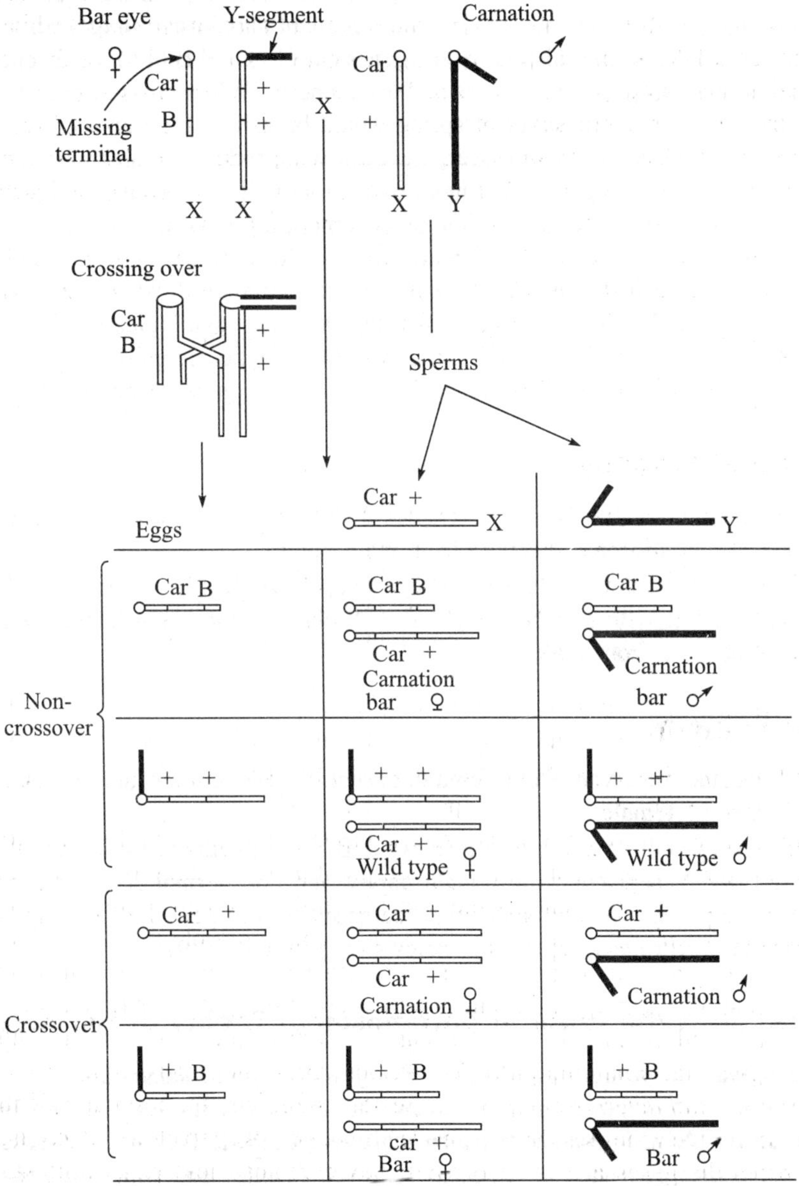

Figure 3.29 Cytological evidence for crossing over.

Curl Stern in 1931 crossed *Drosophila* stock carrying structurally-altered chromosome (one of the X chromosomes was separated into two parts by an exchange of fragments between the X chromosome and the chromosome 4) whose terminal at centromere was missing. The second of X chromosome was attached with a segment of Y chromosome in the second arm that provided one of the classical proofs of the physical basis of crossing over. The other X chromosome was composed of two approximately equal fragments, each with its own centromere. Each X chromosome was thus recognizable microscopically.

In addition, one of the X chromosome fragments carried the mutant allele for 'carnation' (car, recessive light eye colour) and Bar (B-semi-dominant narrow eye shape) while the long XY chromosome carried the normal alleles for the genes car (++) and B (++). In absence of crossing over, the female may produce only car B and ++ gametes. When crossed by a male with car-eye colour, the female non-crossover offspring would be as car and bar-eyed (car B/car +) or wild type (++/car +). Genetic crossing over, however, will produce females with carnation (car+/car+) and bar-eye shape (+B/car +), but more interesting will also produce uniquely recognized chromosome, if the chromosomal exchange accompanies genetic crossing over. The two crossover products, produced by the female are one long X chromosome and the other a short X fragment attached with Y segment. When compared with the normal long X chromosome of the male, the crossover offsprings are thus cytologically distinct from non-crossover (Strickberger, 1991). Figure 3.29 shows that genetic recombination between marker genes is associated with physical exchange between homologous chromosomes.

Non-crossover Progeny

1. Carnation bar eye (car B/ car +) females with broken X chromosomes terminus and one normal X chromosome received from male parent.
2. Wild-type eye colour with normal eye shape female (++ / car +) had one normal X chromosome with attached segment of Y chromosome from female and one normal X chromosome from male.

Crossover Progeny

1. Chromosome received from female carnation eye colour and normal eye shape (car +/car +) female.
2. Wild-type eye colour and bar-shaped eye (+B/ car +) with one broken chromosome with translocated Y segment from female parent and one normal X chromosome received from male parent. This indicates that recombination of gene was accompanied with an exchange of chromosome during meiosis (see Figure 3.29).

Nondisjunction is the Proof of Chromosomal Theory of Heredity

Calvin Bridges was the student of Morgan. Bridges discovered some rare exceptions in the expected pattern of the inheritance between the crosses with several sex-linked genes (characters). When he made crosses between white-eyed female flies and red-eyed male flies, most of the offsprings consisted of expected red-eyed females and white-eyed males. But he observed one in every 2000 F_1 flies was an exception: either a white-eyed female or a red-eyed male. Bridges reported that these were exceptional offsprings which had resulted due to occasional failure of two XX chromosomes in the female flies to separate from each other during meiosis. This he called *nondisjunction* of homologous chromosome. The consequence of nondisjunction of two XX chromosomes during meiosis was that some eggs had two XX chromosomes and some eggs had no X chromosome. The eggs without X chromosomes were non-viable (Figure 3.30).

Four types of phenotypes were expected from fertilization of these abnormal eggs.

1. The offsprings without X chromosomes were not detected, because zygote lacking an X chromosome was not viable.
2. The offsprings with XXX chromosomes also died during early development.
3. The exceptional white-eyed females with XX chromosomes and Y chromosome survived.
4. The exceptional red-eyed males had a single X chromosome but lacking Y chromosome were sterile (XO) males.

This experiment proved that genes are located on the chromosomes because chromosomal behaviour is parallel to the exceptional inheritance of genes.

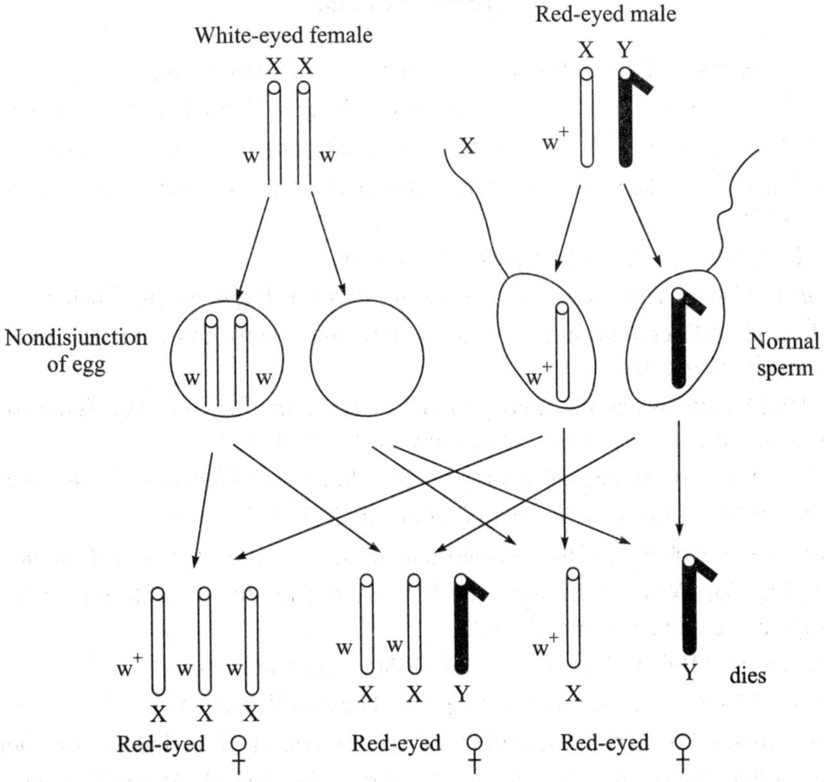

Figure 3.30 Meiotic nondisjunction of X chromosome in *Drosophila melanogaster.*

REVIEW QUESTIONS

1. What is chromosome and its importance in living organisms? On the basis of shape, give the different types of chromosomes found in the cell.
2. What do you mean by stability of chromosome? Give three properties of chromosomes.
3. What do you understand by secondary constriction? Give brief description of chromomere telomere, satellites and nucleolar region zone (NOR).
4. What is karyotype? Give different types of karyotypes and their evolutionary significance.
5. Write short notes on euchromatin, heterochromatin and chromosome banding.
6. What are the different models of chromosomes? Write short notes on B chromosomes, nucleolar chromosome, sex chromosomes and holocentric chromosomes.
7. Give differences between giant chromosomes and lampbrus chromosomes.
8. Write in brief about biochemical composition of eukaryotic chromosomes.
9. How will you prove that chromosomes are carriers of hereditary materials?
10. Give cytological evidences to prove that chromosomes are carriers of hereditary particles.

REFERENCES

Bridges, C.B. (1935): Salivary chromosomes map, *J. of Heredity,* **26,** 60–64.

Burnham, C.R. (1962): *Discussion in Cytogenetics*, Burgess Publishing Co., Mineapolis, USA.

Callan, H.G. (1963): Nature of lampbrush chromosome, *Inter. Review of Cytology,* **15,** 1–34.

Camy, S.D.E. and T.A. Okadu, (1971): Triple chromosome pairing in triploid, Chicken, *Nature*, **231,** 119–121.

Darlington, C.D. (1965): *Cytology*, Churchill, London.

Hartl, Daniel L. (1991): *Basic Genetics*, Jones and Bertlett Publications, Boston.

Jones, R.M. (1975): B chromosome system in flowering plants and animals species, *Intern. Review of Cytology*, **40,** 1–100.

Mazia, D. (1961): Mitosis and physiology of cell division, in: *The Cell,* Vol. **III,** Brachet, L. and Mirsky, A.E. (Eds.), Academic Press, New York, 77–412.

Moses, M.J. (1968): Synaptonemal complex, *Ann. Review of Genetics*, **2,** 363–4/2.

Nicklas, R-8. (1971): Mitosis, *Adv. in Cell Biol.*, **2,** 225–297.

Rish, H. and D.F.D. Kubol, (1970): Chromosome structure, *Ann. Review of Genetics*, 263–294.

Robertis, E., D.P. De, W.W. Nowinski, and F.A. Saez, *Cell Biology,* 5th ed., W.B. Saunders, Philadelphia, London, Toronto, 25–42.

Scientific American (1961): *The Living Cell*, **205**(3), separate issue.

Sharp, L.W. (1934): *Introduction to Cytology*, McGraw-Hill, New York.

Strickberger, Monroe W. (2006): *Genetics,* 3rd ed., Prentice-Hall of India, New Delhi.

Wilson, E.B. (1995): *The Cell in Development and Heredity*, 3rd ed., Macmillan Inc., New York.

Woll, S.L. (1972): *Biology of the Cell*, Wadsworth Publishing Co., Belmont, California.

CHAPTER

4 Chromosomal Aberrations (Structural)

According to the chromosomal theory of heredity, each species and race possess constant chromosome number (Chandrashekhar et al., 1953). Each chromosome has a definite gene complex, in form of DNA (deoxyribonucleic acid) and their linear order is fixed in form of polynucleotides chain. These stable structural features of the nucleus constitute the *karyotype* of the organism. If the chromosomal structure undergoes a change on account of natural forces or by human intervention by way of physical or chemical treatment, etc., it may reflect in the phenotypic behaviour of the organism. These intrachromosomal changes have played important roles in creating variability and also in evolution of new species.

Under the normal condition, during the resting phase (interphase), the chromosomes inside the nucleus look like a mass of tangled threads. During the prophase, chromosomes are the longest but, slowly and slowly they disentangle them and enter into the second phase of cell division. While getting disentangle and shortening, sometimes the chromosomes may break and the broken ends of the chromosomes reunite causing rearrangement of the genes. The centromere is an important part of the chromosome. When chromosomes break and reunite, there are three possibilities:

1. A fragment with centromere may unite with another fragment without centromere.
2. A segment with centromere unites with another fragment with centromere.
3. Two fragments without centromere may unite.

Under such a chromosomal breakage and reunion, only first arrangement is functional, the other two, second and third are lost in ontogeny. There are four types of structural changes that have been reported so far.

1. Deletion
2. Duplication
3. Translocation
4. Inversion

DELETION

When a segment of a chromosome is lost from the intercalary region, it is termed *deletion,* and if the lost segment is from the terminal region, it is called *deficiency.* In both the situations, the effect of loss is almost the same as changes do arise because total gene content is affected. If *A B C D E F G H* is the arrangement of genes on the chromosome; the deletion of chromosome segment containing genes *G* and *H* is called deficiency and if loss consists of *EF,* it is called deletion. These conditions are shown below:

A B C D E F G H	Normal
A B C D E F	Deficiency
A B C D G H	Deletion

If *deletion* is small, cytologically, it may not be visible; homozygous deletion (deletion of chromosome) may prove lethal but heterozygous deletion (deletion in one of the chromatin) may survive. Very large deletions are usually lethal even when heterozygous with a normal chromosome. Small deletions are viable when they are heterozygous with a normal chromosome, but lethal when homozygous because deletion eliminate one or two genes essential for survival.

Detection

Deletion can be identified and its size can be assessed genetically by test crosses in *Drosophila* by crossing a normal red-eyed fly with a white-eyed fly (recessive genes and a mutant (deletion of region C *D E*) fly (Figure 4.1).

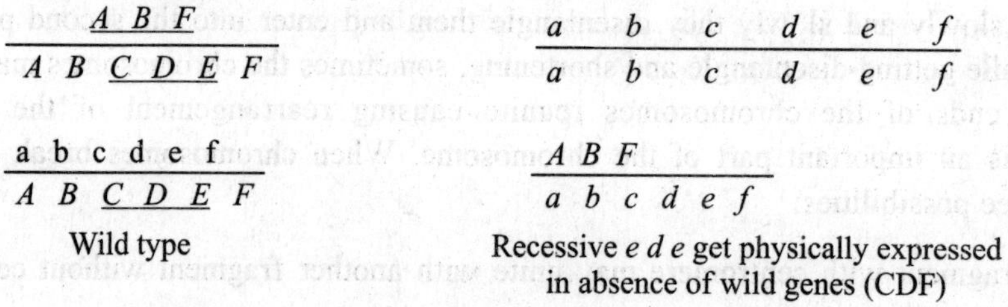

Figure 4.1 Mapping of a deletion by test cross.

This is because chromosome with deletion does not carry wild alleles of the genes that have been eliminated. In *Drosophila,* notch deletions are large enough to remove nearby wild-type allele of white also.

The comparison of cytological—genetical map also confirms the linear order of genes which are uniformly distributed all over the chromosome (Figure 4.1). The deletion in the chromosomes shows loop formation during pairing with homologue (Figure 4.2).

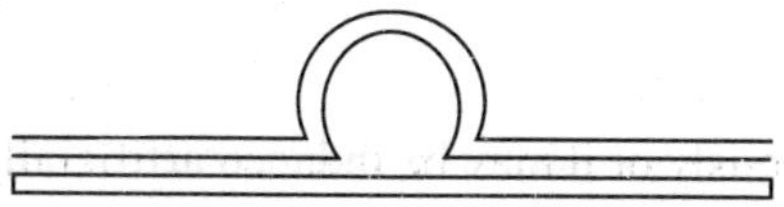

Figure 4.2 Loop formation in deleted chromosome.

An intercalary deficiency in the salivary gland chromosome of *Drosophila* demonstrates bucking effect. It gives notch effect at the tips of wings (Figure 4.3).

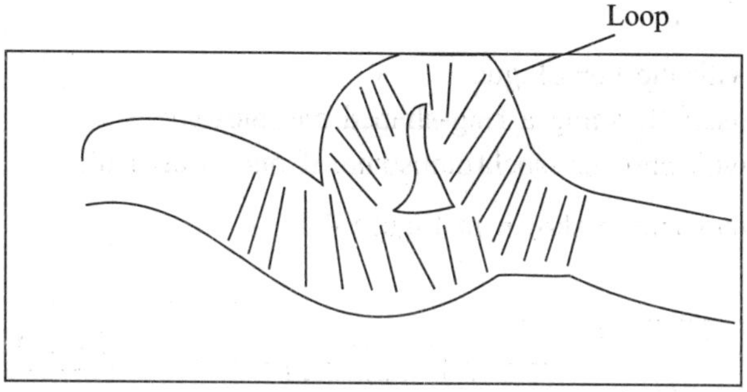

Figure 4.3 Salivary gland of *Drosophila* showing buckling effect.

Probably the best-known disorder associated with the chromosome deletion in man is *cri-du-chat* or *cat-cry syndrome*, (Figure 4.4) as reported by Lejeume and colleagues in 1963, and based on three cases, it was seen that 'plaintive' (sufferer) cry resembling kitten cry in distress. This symptom is associated with a malformation of larynx due to deletion. Other symptoms of malformation involve, the head and face; severe mental retardation with reduced IQ, below 20 per cent brain, heart, eyes, kidney and skeleton disorder may also occur.

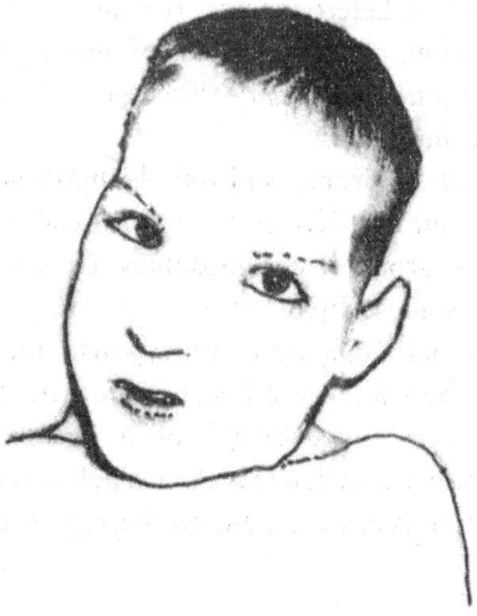

Figure 4.4 Cri-du-chat syndrome in human beings due to deletion in chromosomes.

DEFICIENCY

Deficiency originates spontaneously or it may be induced artificially. Terminal deficiency requires a single break; intercalary deficiency requires two breaks. Terminal deficiency may also arise through a misdivision of the centromere, which also produces isochromosomes. The chromatid arms of one of the telocentrics become oriented at 180° configuration and an isochromosome with identical arms may be formed. Rhoades (1940, 1955) and Sears (1952(a), (b)) indicated that isochromosome may not be stable. At meiosis this new chromosome may act in one of several ways:

(i) It may pair with the homologue.
(ii) It may pair itself, forming a ring-shaped chromosome
(iii) It may pair with another isochromosome, if one is present.

Origin of an isochromosome is shown in Figure 4.5.

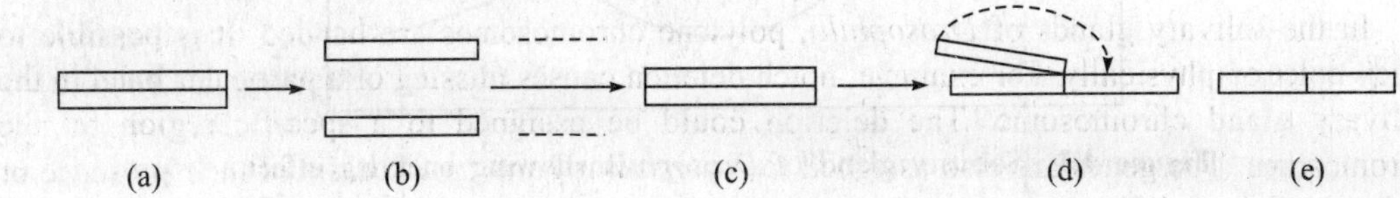

Figure 4.5 Possible origin of an *isochromosome* from a univalent.
(a) Univalent with break developing transversely through centromere.
(b) Telecentrics formed from break in the centromere.
(c) (d) and (e) show origin of *isochromosome* with identical arm.

Phenotypic Effect

Large deficiencies are lethal even in heterozygotes. Because, heterozygotes are homozygous for missing loci. Under such a situation, recessive alleles of missing genes express their phenotypic appearance. This is also called *pseudodominance.* Recessive alleles can also act as lethal. Small deficiencies resemble recessive mutations.

In human being, *cri-du-chat* syndrome and *anti-Mongolism* are due to deletion of small segment. The *cri-du-chat* syndrome (46 XX or YY – 5p–) and *anti-Mongolism* (46 XX or XY, Gq–) are due to deficiency in G^- group of chromosomes. *Cri-du-chat* is caused due to deletion in the short-arm of the chromosomes (Figure 4.6).

When chromosomes having deletions are heterozygous with a normal chromosome having recessive gene *w* in the fly will be white-eyed because the wild-type '*W*' allele is absent in the deleted Notch chromosome. This 'uncovering' of a recessive allele indicates that the corresponding wild-type allele of white is deleted. Once a deletion has been identified, its size can be assessed by determining which recessive gene in the region is uncovered by deletion. This method is shown in Figure 4.7.

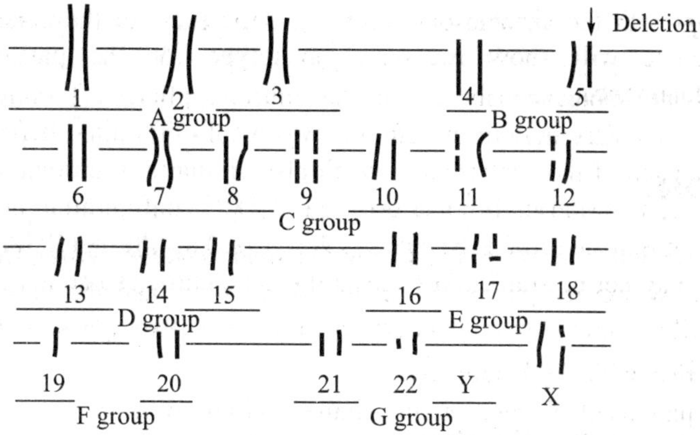

Figure 4.6 Karyotype of human being where deletion had occurred in one of the short arm of the chromosomes 5.

In the salivary glands of *Drosophila*, polytene chromosomes are banded. It is possible to study deletion physically. For example, notch deletion causes missing of a particular band in the salivary gland chromosome. The deletion could be assigned to a specific region on the chromosome. The genes are also assigned to a particular band on the basis of their presence or absence (Figure 4.7).

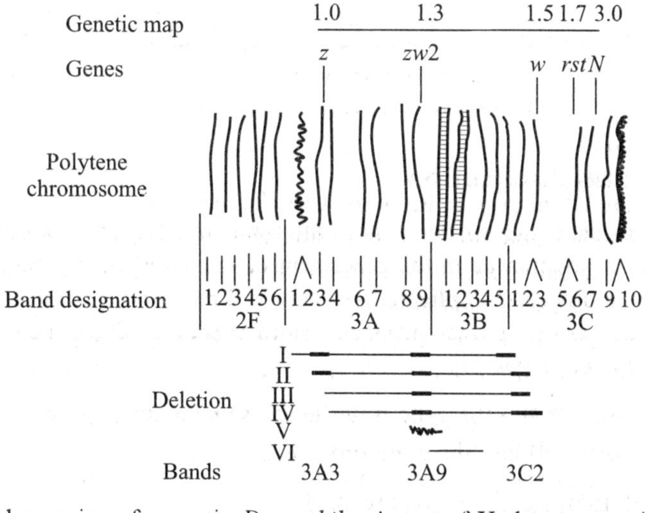

Figure 4.7 Physical mapping of genes in *Drosophila:* A part of X chromosome in the salivary gland chromosome. Any recessive uncovered by a deletion is considered to be located on a particular band in the chromosome.

Uses

1. To study chromosome pairing and its behaviour during cell division.
2. It may be used to locate the gene on a particular chromosome.

3. During growth the chromosome carrying deficiency, in the concerned region of the chromosome will show recessive phenotype for the particular gene due to pseudodominance.

DUPLICATION

Bridges (1919) was first to locate duplication in *Drosophila melanogaster.* Doubling of a segment of a chromosome within the genome is called *duplication*, and the individual carrying such duplication is called *hyperploids.* Duplication may be of two types—interchromosomal and intrachromosomal.

Interchromosomal Duplication

If the duplicated segment of a chromosome is inserted in another chromosome of the same genome, it is called *interchromosomal duplication.* Figure 4.8 shows interchromosomal duplication. It is of two types:

(i) The duplicated segment can be inserted in a nonhomologous chromosome.
(ii) The duplicated segment present separately survives, if it has centromere.

1 2 3 4 5 6 7 8 5 6 7 8 Homologous chromosome

1 2 3 4 5 6 7 8 Normal chromosome

9 10 11 12 13 14 5 6 7 8 Nonhomologous chromosome

Figure 4.8 Interchromosomal duplication

Intrachromosomal Duplication

In this case, the duplicated segment remains in the same arm, distant from the original position or next to the original position or in the other arm of the same chromosome. It is also called *tandem duplication.* In this case, duplicated segment is directly adjacent to the normal region of chromosome. Unequal crossing over produces more copies of duplicated segments. Tandem duplication is of following types:

(i) *Direct tandem:* When the gene order is the same as in original.
(ii) *Reverse tandem:* When the gene order is reversed.

Both these conditions are shown in Figure 4.9.

1 2 3 4 5 6 7 8 9 Original order

1 2 3 4 5 6 4 5 6 7 8 9 Direct tandem

1 2 3 4 5 5 4 3 6 7 8 9 Reverse tandem

Figure 4.9 Intrachromosomal duplication.

Origin

Unequal crossing over is the main cause of duplication, but crossing over in inversion heterozygote and translocation heterozygote are also responsible for duplication. Ac–Ds (Activator–Dissociation) factor was known as *Jumping transposon* discovered in maize by McClintock (1950). It has been also reported that jumping transposon can cause chromosomal aberrations such as deletion, deficiency, translocations and duplication etc. Ac–Ds can cause transposition of a segment of any chromosome or within the same chromosome. The Ds causes breakage in the chromosome at the place where it was inserted earlier. The various mutagenic agents both chemical and physical also cause chromosomal duplication. The manner in which unequal crossing over take place in the 16 A region of X chromosome of *Drosophila melanogaster* resulting bar-eye is shown in Figure 4.10.

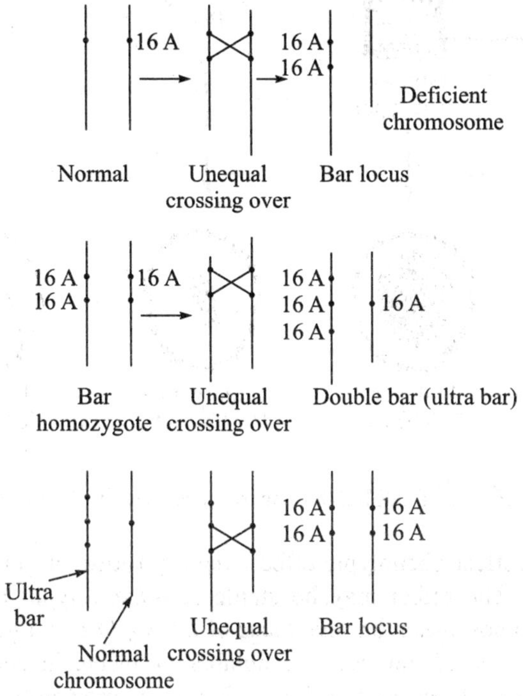

Figure 4.10 Unequal crossing over resulting duplication of 16 A region in X chromosome of *Drosophila* resulting bar-eye.

The duplicated segment form a loop during pachytene of meiosis in duplicated region of the chromosome of heterozygote. Unequal crossing over in the duplicated region of chromosome of heterozygote produces further duplication of the segment, as shown in Figure 4.10. Bar eye locus gives rise to ultrabar (double bar) eye in *Drosophila melanogaster.* Bridges and Muller discovered independently that the bar phenotype was the result of a duplication involving a part of X chromosome already present in the wild type *Drosophila* that reduces the size of eye, which was called *double bar.* The polytene chromosome section 16A is present in the flies with bar-type eyes. When this section is duplicated, it produces the phenotypes of bar type and when it is present in three times in a single chromosome, it produces the 'double bar' phenotypes.

Reverse tandem duplication may also form loop, while pairing with normal chromosome at pachytene stage. A crossover within the loop may produce a dicentric fragment, which may form bridge at anaphase I of meiosis (Figure 4.11).

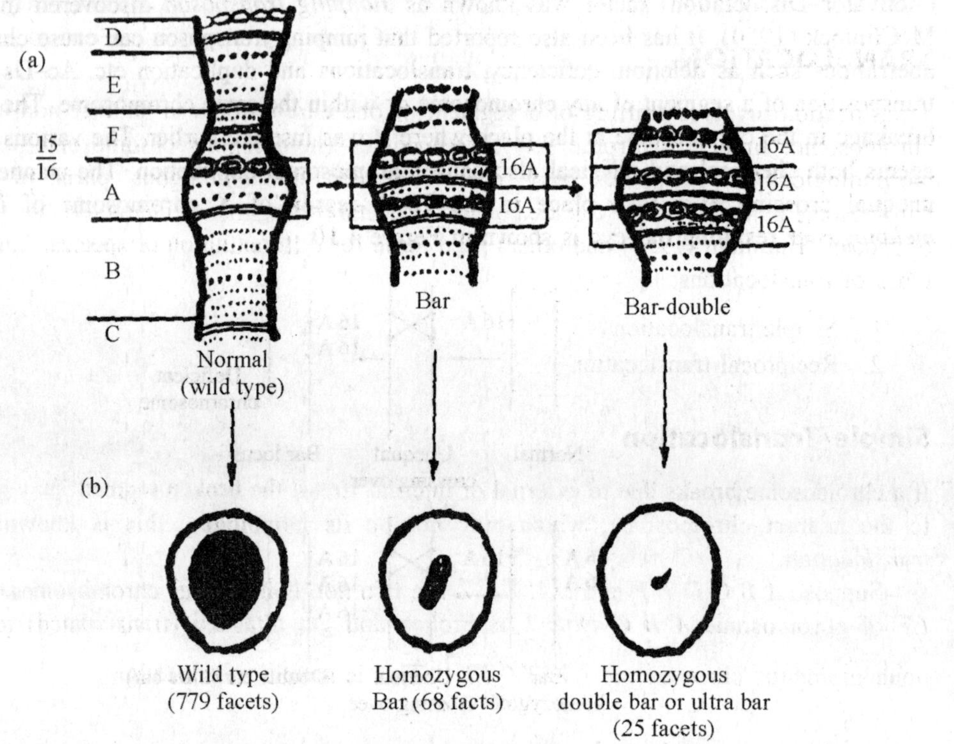

Figure 4.11 Effect of duplication on bar-eye size in *Drosophila melanogaster*.

Duplication may reflect phenotypic effect where position of gene is changed. This is also called *position effect.* The effect may be stable (*cis–trans* type) or mosaic. Duplication of segment l6A of X chromosome in bar eye locus of *Drosophila* is a good example of stable type of position effect. In case of bar eye, the number of facets in compound eye of developed *Drosophila* fly is reduced from normal 779 to 358 in heterozygotic (*BB*+) fly, but in case of homozygotic (*BB*) fly the number is reduced to 68. If segment 16A is duplicated three times, that produces 'double bar' or ultrabar fly.

Uses of Duplication

There are a number of uses of duplication in biological sciences:

1. Duplication can be used to study the dosage effect of nuclear organizer.
2. It can be used to study the position effect of a gene/gene complex.
3. Development of true breeding heterosis in self-pollinated crops can be achieved by duplication breeding.

4. Duplication can be used to study the chromosome behaviour during meiosis, e.g. in pairing and in crossing over.
5. It can be used to increase the dosage of some desirable gene(s) for increasing the durability of disease resistance; rust resistance or enzymatic activities, etc.

TRANSLOCATION

Translocation involves shifting of a segment of one chromosome to another non-homologous chromosome. In heteromorphic sex chromosomes (homologous chromosomes that differs morphologically), exchanges may occur between nonhomologous parts of the same chromosome pair. Translocations are common varietal distinguishing characteristics in a number of species. Burnham (1956) and others used these for differentiation of species. There are two types of translocations:

1. Simple translocation
2. Reciprocal translocation

Simple Translocation

If a chromosome breaks due to external or internal stress, the broken segment may get attached to the nearest chromosome, which may not be its homologue, this is known as *simple translocation*.

Suppose *A B C D E F* and *U V X Y Z* are two non-homologous chromosomes. If segment *EF* of chromosome *A B C D E F* is broken and get attached (translocated) to the other nonhomologous chromosome $UVXYZ\boxed{EF}$. This is simple translocation.

Interchromosomal translocation

It is of following two types:

(i) *Intrachromosomal:* Shift occurs in the same arm, like

A B C D E F	*A B E C D F*
A B C D E F	*A C D B E F*

(ii) *Extraradical:* Shift occurs from one arm to other arm, like

A B C D E F	*A B C D E C D F*
A B C D E F	*A B C D E D C F*

Interchromosomal translocation

There are three types of interchromosomal translocation:

(i) *Transposition:* Transfer of a chromosome segment from one chromosome to other.
(ii) *Insertional:* Transposition occurs in an intercalary position.
(iii) *Terminal:* Segment is attached to the chromosome end.

Reciprocal Translocation

However, when breaking occurs at a point where two nonhomologous chromosomes touch each other and get attached each other reciprocally in such a way that the segments of nonhomologous chromosomes are interchanged, this is called *reciprocal translocation* or a *segmental interchange*. It can be explained as:

A B C D E F	*U V X Y Z*	Normal chromosomes
A B C D ¦*Y Z*¦	*U V X* ¦*E F*¦	Reciprocal chromosomes

Though this type of structural change, i.e. reciprocal translocation has been seen in irradiated materials (*X*-rays, alpha rays, gamma rays), but it has also been reported in the natural population. It can be identified by chromosome ring formation in *Oenothera*, *Rhoeo discolor*, *Chrysanthemum* (Figure 4.12, after H.K.Jain, 2000) and hybrids of Himalian × European peas, which indicates that reciprocal translocation has also been one of the reasons of differentiation of new species. Detailed studies have been conducted is *Datura* and *Drosophila*.

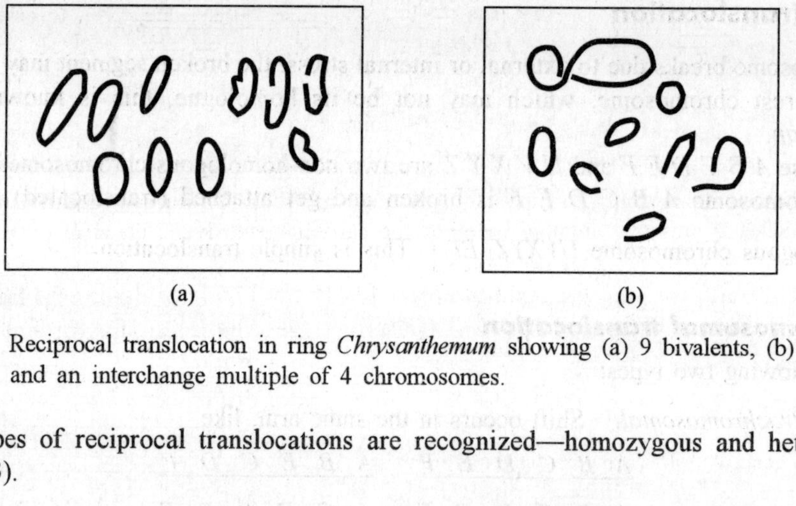

(a) (b)

Figure 4.12 Reciprocal translocation in ring *Chrysanthemum* showing (a) 9 bivalents, (b) 7 bivalents and an interchange multiple of 4 chromosomes.

Two types of reciprocal translocations are recognized—homozygous and heterozygous (Figure 4.13).

A B C D E F   *A B C D* ¦*Y Z*¦

V U X Y Z   *A B C D* ¦*Y Z*¦

Normal pair   *V U X Y* ¦*E F*¦

V U X Y ¦*E F*¦

(a)

A B C D E F   *V U X* ¦*Y Z*¦

A B C D X Y   *V U X* ¦*E F*¦

(b)

Figure 4.13 (a) Homozygous translocation and (b) heterozygous translocation.

On selfing the translocation heterozygotes, two viable combinations form offsprings in a ratio of 1 normal : 2 interchanged heterozygote : 1 homozygote. Figures 4.14 (after Riley, 1948) and 4.15 show the results of F_2 generation.

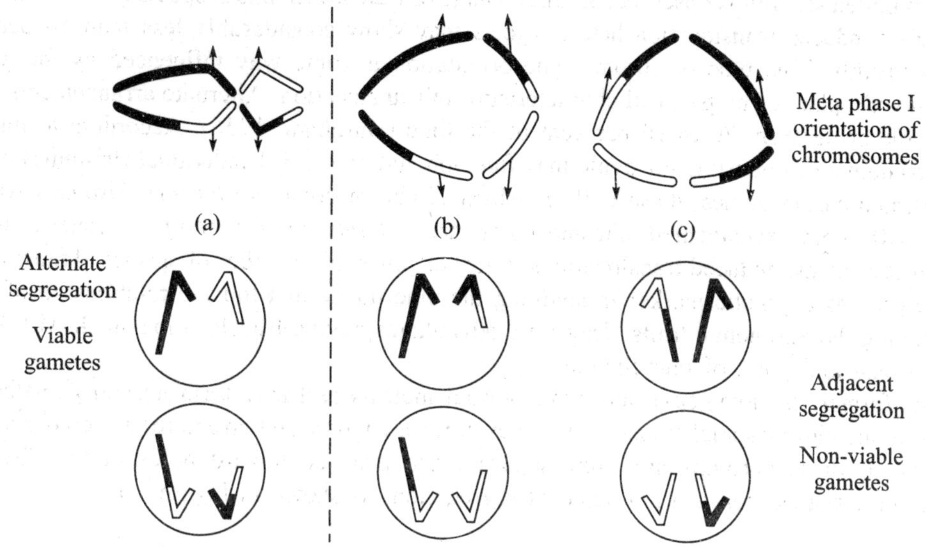

Figure 4.14 Metaphase orientation and gamete formation of a ring-shaped reciprocal translocation (a) viable gametes having normal and translocated chromosomes; (b) and (c) non-viable due to deficient gamete combinations.

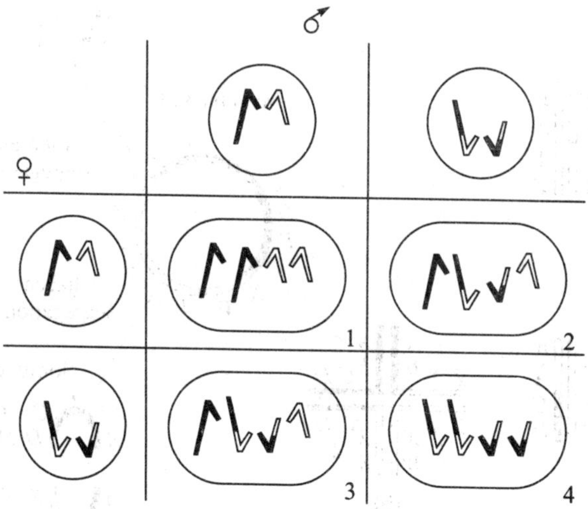

Figure 4.15 F_2 results of viable gametes combination in reciprocal translocation heterozygotes crosses. (1) homozygous normal; (2) and (3) heterozygous translocation; and (4) homozygous translocations.

Translocation heterozygotes can be detected in the breeding nursery by pollen examination using a drop of stain on the crushed anthers on a slide, (covered with glass-slide) under microscope. Empty pollen grains are assumed to arise from inviable combinations. If about 50 per cent pollen sterility is observed, an interchange or translocation is suspected. Often, however, natural or induced translocation heterozygotes may show considerably less than 50 per cent, pollen sterility. This type of chromosome orientation is some way influenced by the genetic make-up of a plant or by physical characteristics of translocation. Alternate arrangements which are viable occur from 70 to 90 per cent of the time (Burnham, 1956). According to him, the n – 1 combinations abort; n + 1 ovule may function and give 2 + 1 individual (trisomic) or both translocated chromosomes; those with translocated chromosome are *tertiary trisomics* whereas those with both translocated chromosomes are *interchanged trisomics*. Geneticists and cytogenetecists use induced translocations to locate genes in a number of species. Translocation can be used as a genetic marker in studying the inheritance and linkage relationship of genes determining the agronomic traits. These are advantages having translocation stock with known chromosome and point of interchange.

The former (homozygous) may have normal meiosis and may form normal gametes. But heterozygous chromosomal translocation may form a ring-shaped structure at metaphase I. At anaphase I, these chromosomes may separate and proceed toward poles in two fashions: (1) alternate segregation and (2) adjacent segregation, as shown in Figure 4.16.

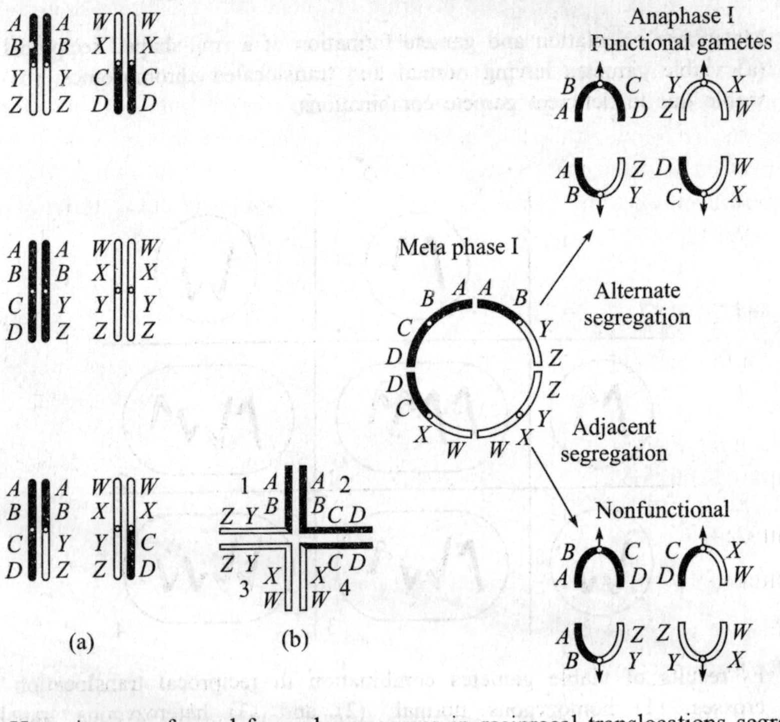

Figure 4.16 How rings are formed from chromosomes in reciprocal translocations segregates during gametes formations.

In alternate segregation, the normal chromatids move to one pole and the translocated chromatids to the opposite pole. Since none of the two daughter cells possesses deletion or duplication of the genetic materials, hence both gametes will be normal (Hartl, 1991).

If one each of translocated and normal chromatids moves to the opposite poles, deletions and duplications will occur in the gametes. Such meiospores and gametes will be nonfunctional in each case, where such adjacent segregation will take place. Semisterility resulting from reciprocal translocation have been reported in corn.

In human beings

Translocations in addition to the 14/21 associated with Down syndrome have been reported in human beings also. In most cases, there are many abnormalities and spontaneous abortions or deaths within a few months after birth. Those who live longer life exhibit mental retardation.

INVERSION

If there is reversal in the linear order of genes in a segment without change in the total gene content; it is termed *inversion.* If in a chromosome, gene order is *A B C D E F G* and if inversion of the segment containing *C D E* takes place, the order of gene will be changed from *A B C D E F G* to *A B C E D F G.* Thus, an inversion requires two breaks. In such inverted segment the centromere may not be present. The inverted segment may either include the centromere (peracentric inversion) or not (paracentric inversion). The homologous chromosomes with identical inversions in each member pair undergo normal distribution.

Both spontaneous and induced inversion occur in nature but they are best known in *Drosophila* and in *Zea mays.* When the linear order of a chromosome segment becomes inverted at 180°, the inverted segment must form a loop in early stages of meiosis in order to synapse fully with a normal homologue. This cytological configuration is characteristics of inversion heterozygotes (Pattern and Stone, 1952).

Types of Inversion

Depending upon the number of inverted segments within a chromosome, inversions are of two types:

1. Single inversion
2. Complex inversion

Single inversion

Single inversion can be further subdivided into:

(i) Pericentric inversion
(ii) Paracentric inversion

***Pericentric inversion*:** The inverted segment possesses centromere, i.e. two breaks are located in the two arms of the chromosome, e.g.

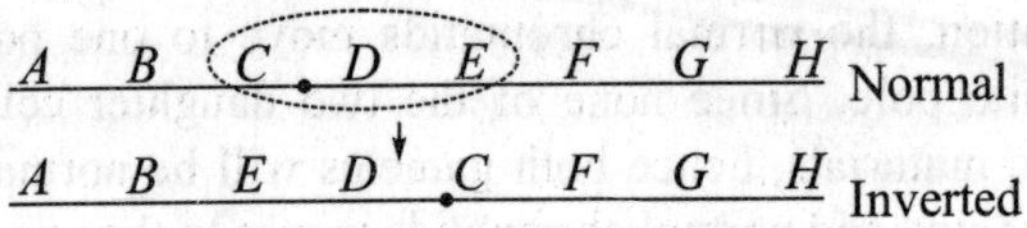

Paracentric inversion: It is located in a single arm of the chromosome, i.e. both inversion breakpoints are located in the same arm, centromere is not involved.

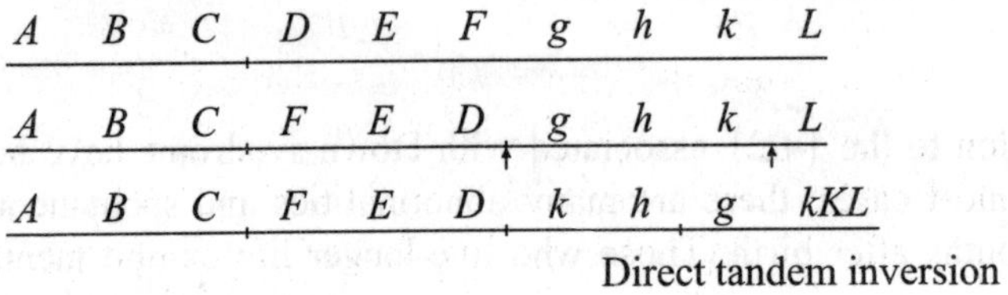

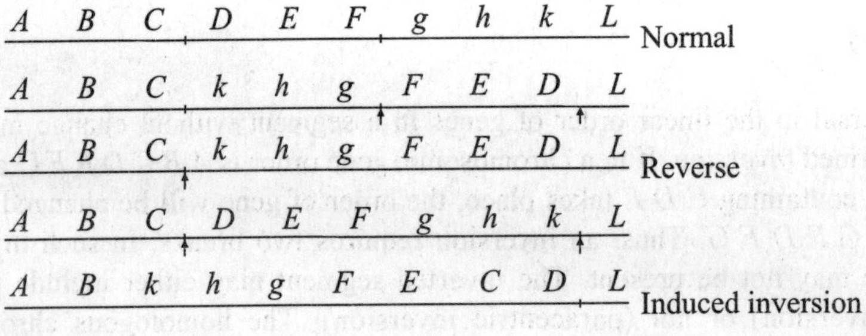

Complex inversion

When inversion occurs more than once in a chromosome is called *complex inversion.* The complex inversions are of following types:

(i) Independent inversion
(ii) Included inversion
(iii) Overlapping inversion

Independent inversion: When inversion occurs in different regions of the chromosome, which is separated, from each other by normal chromosal segments.

(a) *Direct tandem:* When two or three or more inverted segments are linked to each other.
(b) *Reverse tandem:* When two inverted segments are linked to each other but in a reverse direction.

Included inversion: When one inversion is confined within another inversion, i.e. when a segment within inverted segment is inverted.

Overlapping inversion: When inversions have a common segment, i.e. part of inverted segment is inverted again together with an adjacent inverted segment.

Pairing configuration

1. If inverted segment is very small, pairing may not take place in this region while rest of the chromosome parts pair, as a result, crossing over does not take place within the inverted segment, therefore, genetic exchange will not occur.

2. If inverted segment is large, chromomere to chromomere (gene to gene) pairing will take place and a loop will be formed. Such type of pairing will result crossing over within the inverted loop.
3. When most of the Parts of chromosomes are included in inversion, chromomere to chromomere (gene to gene) pairing may occur in the inverted segments.

Chromatid bridge may not be formed, if crossing over does not take place in the inverted segment. Formation of chromatid bridge and a fragment at meiosis indicates that the organism is inversion heterozygote (Figure 4.17). However, non-appearance of chromatid bridge is not proof for the absence of inversion. Formation of inversion loop by chromosome pairing is shown below:

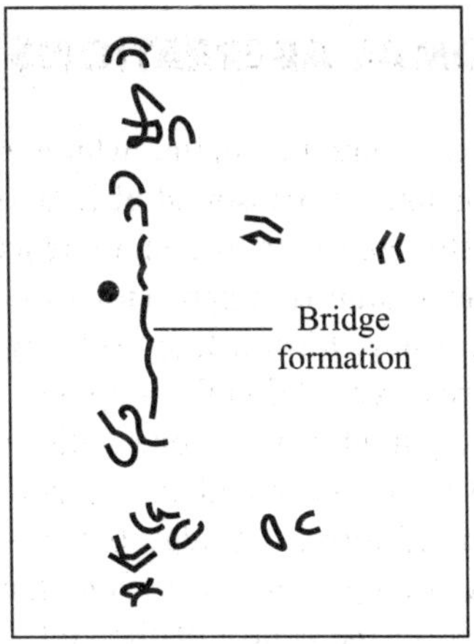

Figure 4.17 An inversion showing bridge formation during anaphase due to dicentric chromosome.

A B C D E F G	Normal
A E D C B F G	Inverted segment

The products of inversion are shown in Figure 4.18

(a)

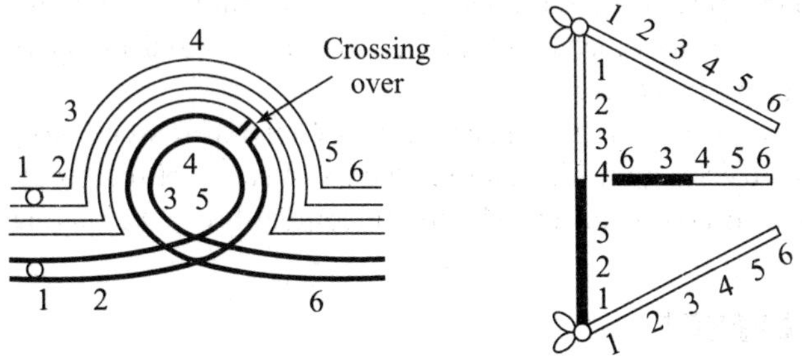

Paracentric inversion

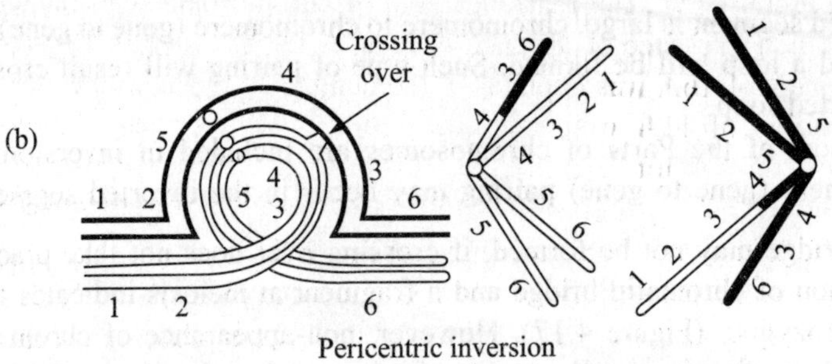

Figure 4.18 (a) Paracentric inversion and (b) Pericentric inversion.

CAUSES OF CHROMOSOMAL ABERRATIONS

Among natural causes, cosmic rays are the main, which induce chromosomal aberrations. However, many substances also cause chromosomal aberration. These are marijuana, nicotine, cyclamates, DDT, caffeine, asbestos and hair dye ingredients.

The genetic consequence of inversion is suppression of crossing over. In the heterozygotes for inversion, the inverted segment may form a loop to bring the homologous part opposite to each other for pairing. A simple crossover within the loop will lead to the formation of four types of chromatids; two resembling the parental types; one with two centromere (*dicentric*) and one without a centromere. The latter two dicentric and one without centromere are abnormal and do not function. The dicentric chromatid forms a bridge between two poles and hence finally get deleted; the chromatid without centromere (*acentric*) lose the power of movement during cell division and hence lags a fragment in the centre of the cell.

EVOLUTIONARY SIGNIFICANCE OF CHROMOSOMAL ABERRATIONS

The structural changes bring about a new genetic interaction within the genotype and the consequence interaction with the environment may bring changes in the genotype to be look like a new species or have to acquire better tolerance to the environmental charges. Many inversions are known to differentiate several species. In *Drosophila, D. pseudoobscura* and *D. persimilis* which are morphologically alike produce sterile male but fertile female hybrids and differ in four major inversions. On the other hand, *D. pseudoobscura* and *D. miranda* produced completely sterile offsprings when crossed.

Translocations in the evening primose and *Oenothera* were largely responsible for differences, on which DeVries gave his mutation theory. A series of reciprocal translocations in isolated population may lead complete reproductive incompatibility, belonging to the same species.

POSITION EFFECT

During rearrangement of chromosome, genetic material is not lost, only order of gene(s) is changed; it is called *position effect.* The first example of this type of change was reported by Sturtevant Beadle (1939) on the bar-eye duplication in *Drosophila.* This also indicates the *dosage effect* of gene (Figure 4.18). The rearrangement of chromosome segment with respect to each other as well as their presence or absence influences the eye size in *Drosophila.*

When a segment of 16A was duplicated, the number of facets in the eyes averaged 68. But, when three segments of 16A were duplicated side by side in one homologous chromosome and one segment in another homologous chromosome, the eyes averaged 45 facets. The variegated position effect in *Drosophila* is associated with structural changes in the chromosome (Figure 4.19).

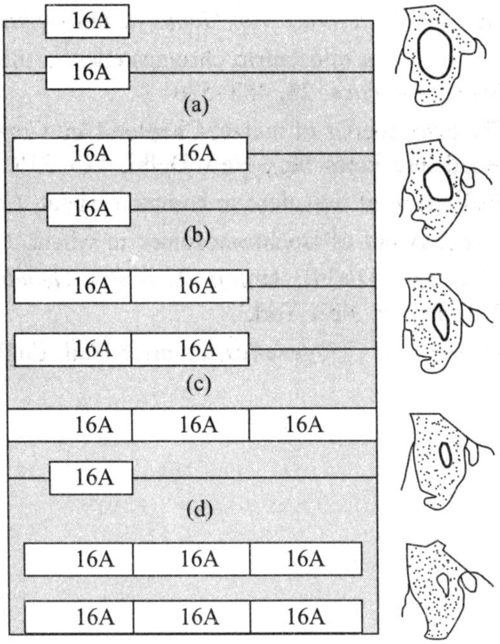

Figure 4.19 Dosage effect of 16A governing eye size.

REVIEW QUESTIONS

1. What is deficiency deletion? Give its evolutionary significance.
2. What is duplication? Discuss its role in the creation of genetic variability in nature.
3. Explain briefly the different types of translocations.
4. What are inversion and inversion heterozygous?
5. What is evolutionary significance of chromosomal aberration in crop plants?
6. What are paracentric and pericentric inversion? Differentiate between both diagrammatically.

REFERENCES

Burnham, C.R. (1956): Chromosomal and interchange in plants, *Bot. Rev.*, **22,** 414–552.

Chandrashekhar, S.N., S.V. Parthasarthy, and K. Ramiah (1953): *Cytogenetics and Plant Breeding*, Vardhanachary and Co., Madras.

Darlington, C.D (1965): *Recent Advances in Cytology*, Churchill, London.

Hartl, D.L. (1991): *Basic Genetic*, Jones Bartlett, Boston.

Jain, H.K. (2000): *Genetics, Concept, Principles and Implications,* Oxford and IBH, New Delhi.

McClintock, B. (1950): The origin and behaviour of mutable loci in maize, *Proc. Nati. Acad. Sci.*, USA, **36,** 344–355.

Pattern, J.T. and W.S. Stone (1952): *Evolution of Genus in Drosophila,* Macmillan, New York.

Rhoades, M.M. (1940): Studies of a telocentric chromosomes in maize with reference to the stability of its centromere, *Genetics*, **25,** 483–520.

Rhoades, M.M. (1955): The cytogenetics of maize, Chapter 4 in: *Corn and Corn Improvement*, G.F. Sprague (Ed.), Academic Press Inc., New Delhi, 123–219.

Sears, E.R. (1952(a)): Misdivision of univalent in common wheat, *Chromosome,* **4,** 535–550.

Sears, E.R. (1952(b)): The behaviour of isochromosomes in wheat, *Chromosome*, **4,** 551–562.

Sturtuevant, A.H. and G.W. Beadle, (1939): *An Introduction to Genetics*, Saunder, Philadelphia (Republished in 1962 by Dover, New York).

Swanson, C.P. (1968): *Cytology and Cytogenetics,* Prentice Hall, Englewood Cliffs, New York.

CHAPTER

5

Chromosomal Variation (Numerical): Polyploidy and Its Evolutionary Significance

A normal plant or animal possesses in its each cell twice the number of chromosomes found in the gametes. This doubling is on account of the union of male and female gametes during fertilization. Such plants and animals are called *diploids*. When each gamete possesses half the numbers of somatic cells is called haploid. Thus, an increase in the number of chromosomes in multiples of the haploid number is termed *polyploidy*. If n represents the haploid number of chromosomes in the cells, then the cells possessing chromosome numbers (Figure 5.1) are called:

- 2n—Diploid
- 3n—Triploid
- 4n—Tetraploid
- 5n—Pentaploid
- 6n—Hexaploid
- 7n—Heptaploid
- 8n—Octaploid
- 9n—Nonaploid
- 10n—Decaploid, etc.

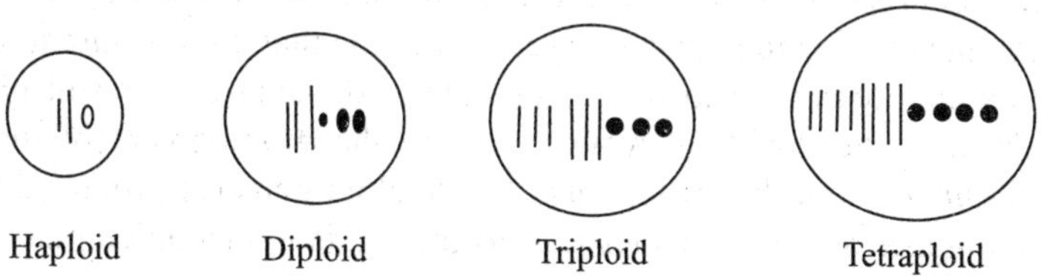

Figure 5.1 Different types of numerical chromosomal variations.

A general survey of plant species indicates that there are many related plant species whose numbers form multiples of the basic number. For examplc, in the family Solanaceae, there are a number of species which are multiples of their basic numbers:

Species of family solanaceae	Multiples of basic chromosome number (n = 12)
1. *Solanum nigrum*	12
2. *Capsicum annum*	24
3. *Atropa belladona*	72
4. *Datura fastuosa*	24
5. *D. ferox*	24
6. *D. metel*	24
7. *D. quercifolia*	24
8. *D. stramonium*	24
9. *Hyocyamus albus*	36
10. *H. canadensie*	72
11. *H. niger*	36
12. *Nicotiana digluta*	72
13. *N. rusbyi*	24
14. *N. sylvestris*	24
15. *N. tabacum*	48
16. *Solanum nigrum*	72
17. *S. muriccatum*	24
18. *S. nigrum var gagas*	144
19. *S. tuberosum*	48
20. *Physalis philadelphica*	24

Thus, species of solanaceae mentioned above are multiples of the basic number 12. The rose genus *Rosa* includes species with 80 metric numbers 14, 21, 28, 35, 42 and 56 which are multiples of basic number 7 (haploid).

ORIGIN OF POLYPLOIDY

All kinds of polyploidy are the result of cytological accidents. The monosomics (2n – 1), nullisomics (2n – 2) and trisomics (2n + 1) arise due to nondisjunction of one or another pair of chromosome during gamete formation. Monoploids and triploids are another cause of aneuploid. Aneuploids arise due to troubles at meiosis, that are caused by asynaptic genes.

Euploidy arises mostly due to misadventures in meiosis that causes unreduced (2n) gametes formation. Euploids also arise due to failure of mitotic division as a result chromosomes are doubled. If the chromosome doubling occurs early in the development of the plants, it may produce a sectoral *chimera* that bears flowers. The haploid gametes produced by such flowers will have the same number of chromosomes (2n) as the original undoubled plant. Fertilization of unreduced gametes by normal gametes of the same species produces triploids with 3n chromosomes. Triploids are unstable cytologically, therefore, they may often turn to produce a great array of aneuploid progeny.

If the original plant is a diploid species, the result of a self-fertilization in the polyploid sector is the production of an autotetraploid. However, if the original plant is an F_1 (interspecific hybrid), the doubling of chromosome numbers of two species will produce a new amphidiploid species. The formation of unreduced gametes appears to be particularly frequent in wide crosses as result of disturbances in the meiosis because of the lack of pairing of the nonhomologous chromosomes.

Importance of Polyploidy in Cultivated Plants

The importance of polyploidy among crop plants can be obtained from the range of cultivated crop species as given below:

Crop species	Common name	Chromosome number(n)
Hordeum vulgare	Cultivated barley	7
H. jubatum	Squirrel tail barley	14
H. nodosum	Fox tail barley	21
Gossypium thurberi	Wild American cotton	13
G. arborium	Cultivated Asiatic cotton	13
G. hirsutum	American upland cotton	26
G. barbadense	Island cotton	26
Sorghum versicolar	Wild sorghum	5
S. vulgare	Cultivated sorghum	10
S. helepense	Johanson grass	20
Avena strigosa	Sand oat	7
A. brevis	Short wild oat	7
A. barbata	Slender wild oat	14
A. sativa	Cultivated oat	21
A. fatua	Wild oat	21
A. byzantina	Red cultivated oat	21
Medicago hispida	California burclover	7
M. lupulina	Black medic	8, 16
M. falcata	Yellow alfalfa	8, 16
M. sativa	Common alfalfa	16
Nicotiana sylvestris	Wild tobacco	12
N. tomentosa	Wild tobacco	12
N. tabacum	Cultivated tobacco	24
Prunus americana	American plum	8
P. avium	Sweat cherry	8
P. persica	Peach	8
P. cerasus	Sour cherry	16
P. domestica	European plum	24
Trifolium agrarium	Hop clover	7
T. incarnatum	Crimson clover	7
T. pratense	Red clover	7
T. alexandrium	Berseem clover	8
T. fragiferum	Strawberry clover	8

T. hybridum	Alsike clover	8
T. subterranean	Subterranean clover	8
T. dubium	Small hop clover	16
T. resoens	White clover	16
T. medium	Zigzag clover	40, 42, 48, 49

The original monoploid (x) sets of chromosomes are known in only few polyploid species of crop plants such as in brassica, wheat, tobacco and cotton.

The changes in the number of whole chromosome are called *heteroploidy.* It may be of two types:

1. Euploidy, and
2. Aneuploidy

Euploidy

When changes involve in entire chromosome number, such as 2n, 4n, 6n. This position is known as euploidy.

Aneuploidy

When loss (n – 1) or addition (n + 1) of a single (whole) chromosome takes place, this condition is known as *aneuploidy.* Each may produce phenotypic changes, modification of phenotypic ratio or alteration of linkage groups.

Euploids

Euploids are characterized by possession of entire set of chromosomes, haploid (n), diploid (2n), triploid (3n) and so on. Haploidy is rare in animals (male honeybee is an exception) but most common in plants.

Polyploids

Euploids with three or more complete sets of chromosomes are called *polyploids*. They are of two types:

1. Autopolyploids
2. Allopolyploids

Autopolyploids

Monoploids (haploids) are typically smaller and less vigorous plants, if viable, since chromosomes (n) have no regular partners with which to pair during meiosis. At telophase some chromosomes go to one pole and some to the other pole. Thus, deficient for one or more chromosomes, results nonfunctional gametes. But haploids are useful to the plant breeders to produce diploid pure-lines by doubling the chromosomes as a short-cut method. The increase in the number of identical chromosomes as a result of doubling is termed as *autopolyploids.*

Thus, if *AA* represents the chromosome of a diploid species, *AAA* constitute the chromosome complement of an autoploids; *AAAA* represent that of autotetraploids and so on.

Triploids are highly sterile

Triploids are fairly common among crop plants in which sterile barrier can be overcome by vegetative propagation, viz. banana and other ornamental plants. Among other fruits, apple (Baldwin, wine sap staymax) and many European varieties of pears. Triploids can be grown with diploids as pollen producing plants. Triploids are vigorous growers and produce large seedless fruits.

Autotetraploidy

Autopolyploids result in a complex meiosis. It is explained in Chapter 2 that during meiotic division similar chromosome pairs and crossing over result between the chromatids of homologous chromosomes. When they have more than two homologous chromosomes such as three in triploids and four is tetraploids, pairing is more complex. At any one point, two homologous chromosomes can pair. In autopolyploids, homologous chromosomes complement, come together and pairing may take place between the two homologues. In triploids, two chromosomes may pair leaving third unpaired. Pairing between two chromosomes of same parental origin is called *autosyndesis*. Multivalent appear at heterotic metaphase and separation of chromosomes is not half and half as in diploid. Unequal separation may cause improper reduction division where some gametes may carry more than 2n and some less than 2n chromosomes.

Formation of multivalent at metaphase interferes normal disjunction of chromosomes at anaphase I of meiosis. The nondisjunction of the paired homologous chromosomes at anaphase I of meiosis cause abnormal ties in the gamete formation and because of this reason, some sterility is found in autopolyploids.

Cytological behaviour of autotetraploids

In the four homologous chromosomes of each type coming out of resting stage, if they become associated in pairs; pairing begins at a number of points along the chromosomes so that different chromosomes pair with different partner at different places.

In *pachytene*, pairing get completed, a chromosome may thus have been paired first with one chromosome and then with another (Figure 5.2). At the end of *pachytene*, each chromosome is divided into two chromatids and *chiasmata* is formed as a result of crossing over between paired chromosomes. This, of course, causes portion of different chromatids to become attached to the same kinetochore and leads to equational separation at anaphase. Whenever a number of chiasmata are formed between a locus and the kinetochore at first anaphase, division of kinetochore (centromere) always leads reduction division. Variations in the number and position of chiasmata lead to the formation of different configurations such as *bivalents*, chains, rings at *diplotene*.

At metaphase, a ring may disjoin and the adjacent chromosomes go to either same or to opposite poles. Similar possibilities are for the 'chains'. But when two chiasmata are not involved in the exchange of the segments between the partner chromatids, the two bivalents are formed thus two chromosomes of each go to opposite poles at anaphase. The disjunction of quadrivalents at anaphase I is normally two-by-two, but numerical aberrations sometimes may

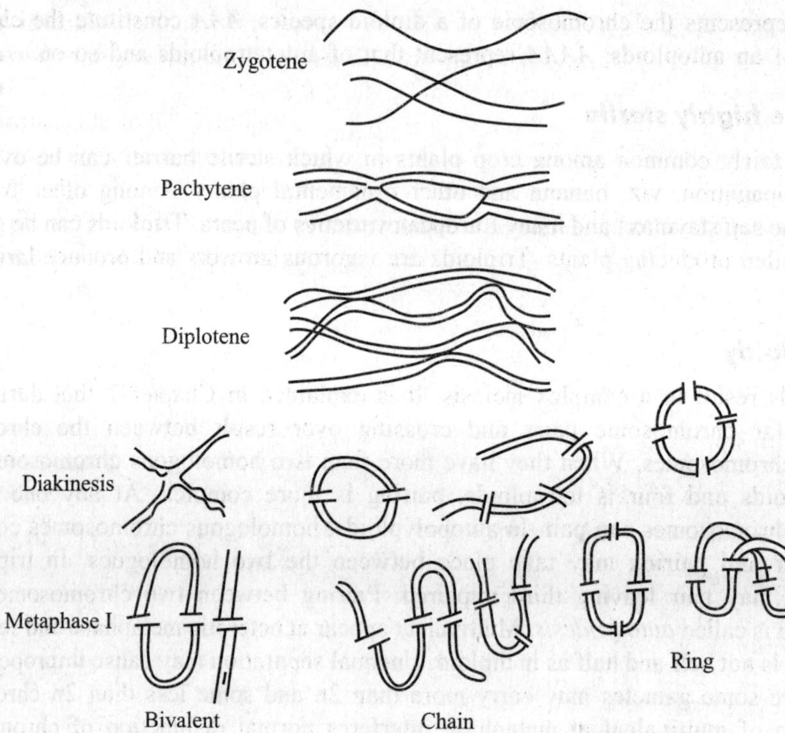

Figure 5.2 First meiotic division in a quadrivalent.

occur. The numerical aberrations at the level of disjunction of the chromosomes at anaphase I may lead to unbalanced gametes production.

At the second division of meiosis, the two chromatids attached to the same centromere at the first division always go to the opposite poles and also into different gametes.

The essential feature of the cytology of tetraploids is the partition of eight chromatids of the four pairs of chromosomes at first meiotic division; each pair corresponding to one of the four gametes produced by each sporocyte. This is much more complicated than diploids which involve partition of four chromatids, one to each of the four gametes.

In autotetraploids, five genotypes—*AAAA* (quadruplex), *AAAa* (triplex) *AAaa* (duplex), *Aaaa* (simplex) and *aaaa* (nulliplex) are possible at each locus. These zygotic combinations arise from the fusion of three different types of gametes, *AA*, *Aa* and *aa*, whose relative frequency for a given genotype depends on the cytological events of meiosis.

Random chromosomal assortment in autotetraploid

The factors affecting the gamete formation in the autotetroploids are:

1. Regularity with which quadrivalents are formed.
2. Randomness of disjunction of the chromosomes from quadrivalents, which depends upon the distance between the centromere and the locus (chiasma formation).

Let us see, how gametes are formed in simplex (*Aaaa*) individual in which bivalents are formed. The two chromatids carrying dominant alleles (*AA*) are always joined to the same centromere. Therefore, in the anaphase, they always travel to the same pole; but in the anaphase II, they never go to the same pole (gamete) because of division of the centromere. The three pairs of chromatids bearing recessive alleles (*aa*) behave in the similar way, i.e. the sister chromatids derived from same chromosome never go to the same pole (gametes), therefore in autotetraploid the eight chromatids (A_1A_2), (a_3a_4), (a_5a_6) and (a_7a_8); A_1A_2 never appear in the same gamete because of the division of centromere at anaphase II, but each has equal chance of appearing in the same gamete carrying gene alleles a_3, a_4, a_5, a_6, a_7 and a_8.

Similarly, a_3 and a_4 have no chance of appearing in the same gamete; but they have equal chance of appearing with A_1, A_2, a_5, a_6, a_7, a_8 and so on. This type of division and distribution of sister chromatids during gamete formations is called random chromosomal assortment (Figure 5.3).

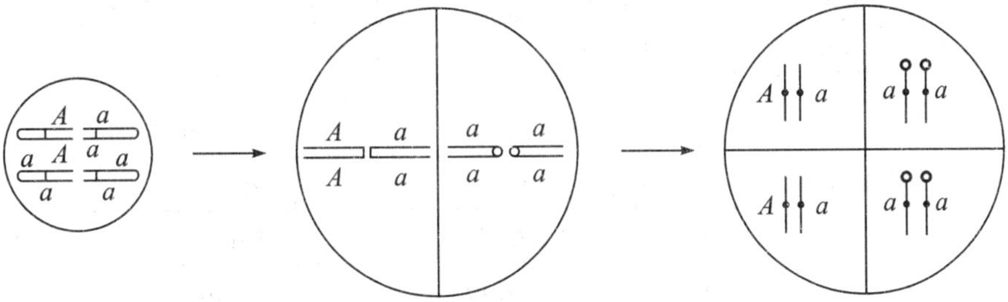

Figure 5.3 Gamete formation in a simplex (*Aaaa*) quadrivalent. First division is reductional and the second division is equational.

The gametic output in the simplex would be as follows:

♀ \ ♂	A_1	A_2	a_3	a_4	a_5	a_6	a_7	a_8
A_1			A_1a_3	A_1a_4	A_1a_5	A_1a_6	A_1a_7	A_1a_8
A_2			A_2a_3	A_2a_4	A_2a_5	A_2a_6	A_2a_7	A_2a_8
a_3					a_3a_5	a_3a_6	a_3a_7	a_3a_8
a_4					a_4a_5	a_4a_6	a_4a_7	a_4a_8
a_5							a_5a_7	a_5a_8
a_6							a_6a_7	a_6a_8
a_7								
a_8								

The gametic ratio would be 1 *AA* : 1 *aa* (simplex), 1 *AA* : 4*Aa* : 1*aa* (duplex), and 1 AA : 1*aa* (triplex) respectively. Similarly, the zygotic ratio on selfing would be as follows:

Simplex (*Aaaa*)

Gametes: ♀ \ ♂	1*Aa*	1*aa*
1*Aa*	1*AAaa*	1*Aaaa*
1*aa*	1*Aaaa*	1*aaaa*

Duplex (*AAaa*)

Gametes: ♀ \ ♂	1*AA*	4*Aa*	1*aa*
1*AA*	1*AAAA*	4*AAAa*	1*AAaa*
4*Aa*	4*AAAa*	16*AAaa*	4*Aaaa*
1*aa*	1*AAaa*	4*Aaaa*	1*aaaa*

Triplex (*AAAa*)

Gametes: ♀ \ ♂	1*AA*	1*aa*
1*AA*	1*AAAA*	*AAaa*
1*aa*	1*AAaa*	1*aaaa*

If *A* denotes the dominant effect of gene in the phenotype, then expected phenotypic ratio in F_2 would be 3*A*: l*a* provided there is a random assortment of the chromosome. Similar expected phenotypic ratio may be obtained in other cases. If assortment of chromosomes is random, quadrivalent formation does not take place in the first meiotic division. However, exactly same result would be obtained even if quadrivalents are formed, when locus in question is inseparably linked to the kinetochore. The linkage causes the locus to separate reductionally at the anaphase I and equationally at the anaphase II exactly in the same way as when quadrivalent formation fails in the first meiotic division.

Random chromatid assortment

When quadrivalents are formed and when locus is far enough away from the centromere to allow the formation of chiasmata between the centromere and the locus, the sister chromatids can end up their attachment at this locus, as a result sister alleles may be included in the same gamete or in different gametes depending upon the distribution of chromatids at the two meiotic anaphases (Figure 5.4).

Figure 5.4 shows gametic formation in a simplex quadrivalent in which crossing over takes place between chromosomes carrying dominant *A* and recessive alleles at *a* locus. The two *AA* chromosomes go to the same pole in the first division in one third time (Top) and to the opposite poles in two third time (Bottom). When, they pass to the same pole, in the first division, and also in the second division, some homozygous gametes are formed. This is called 'double reduction'. During gamete formation in duplex, quadrivalent is much more difficult because many different combinations are possible.

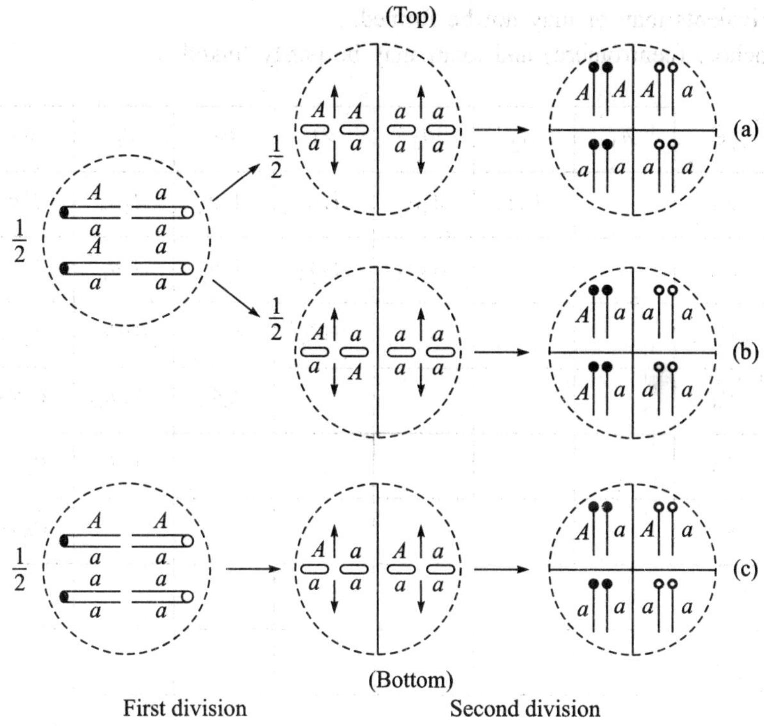

Figure 5.4 Gamete formation in simplex quadrivalent in which chromosomes carrying dominant alleles *A* crossover another chromosome carrying recessive allele *a*.

The gametic ratio for simplex plant is thus 1*AA*: 12*Aa*: 15*aa*; for duplex 3*AA*: 8*AA*: 3*aa*, and for triplex quadrivalent 15*AA*: 12*Aa*: 1*aa*. On selfing, the zygotic ratio for simplex individual may be as follows:

• Quadriplex	1
• Triplex	24
• Duplex	174
• Simplex	360
• Nulliplex	225
Total	784

The gametic output for simplex (*Aaaa*) individual would be as shown on p. 92.

Random chromosome assortment with its characteristic gametic output occurs when separation is reductional at first meiotic division owing to the failure of quadrivalent formation or due to complete linkage between the kinetochore and locus. On the other hand, random chromatid assortment with its different gametic output occurs when quadrivalents form but 50 per cent crossing over takes place between kinetochore (centromere) and the locus. Thus, random chromosome assortment and random chromatid assortment represent two extreme situations as regards gametic output in tetrasomic inheritance. But a third situation (intermediate) may also occur in which:

1. quadrivalents may or may not be formed.
2. kinetochore (centromere) and locus may be partly linked.

Gametes:

♂ / ♀	A_1	A_2	a_3	a_4	a_5	a_6	a_7	a_8
A_1		A_1A_2	A_1a_3	A_1a_4	A_1a_5	A_1a_6	A_1a_7	A_1a_8
A_2			A_2a_3	A_2a_4	A_2a_5	A_2a_6	A_2a_7	A_2a_8
a_3				a_3a_4	a_3a_5	a_3a_6	a_3a_7	a_3a_8
a_4					a_4a_5	a_4a_6	a_4a_7	a_4a_8
a_5						a_5a_6	a_5a_7	a_5a_8
a_6							a_6a_7	a_6a_8
a_7								a_7a_8
a_8								

Autotriploids (AAA)

Autotriploids have been found in a number of crops. Triploids may arise in the following ways:

1. Due to failure of reduction division at megasporogenesis, diploid gametes (2n) are formed, which when fertilized by haploid pollen grains give rise to triploid zygotes.
2. Due to somatic doubling in germ tract, cells with unreduced chromosomes undergo reduction revision and give eggs that are fertilized by haploid pollen grains and triploids are formed.
3. When tetraploid is crossed with diploid, triploid progenies arise.
4. When a haploid egg is fertilized twice by haploid male gametes, this is called *dispermy*.
5. Sometimes diploid eggs are fertilized by haploid pollen grains.

No irregularities are noticed in the somatic division because of equal halving of the chromosomes. In *prophase*, the three chromosomes are attached with the nucleolus. However, at meiosis, in few cells, trivalents are seen at diakinesis. In majority of cells, trivalents, bivalents and univalents are seen. Trivalents are y or rod-shaped at metaphase; bivalents and trivalents are arranged in haphazard manner on the equatorial plate and the univalents lie outside the spindle fibres.

The disjunction of bivalents is normal but when trivalent separates two chromosomes move to one pole and the third to the opposite pole, causing numerical abnormalities in the daughter cells. The lagging univalent splits, but split halves may go to the same pole or may be lost in the cytoplasm.

In the second meiotic phase, the division of the chromosomes is regular with exception a few laggards. The divided univalents do not divide again but pass on unequally to the two poles. In Figure 5.5 pairing process of triploid at various phases in meiotic I division is shown.

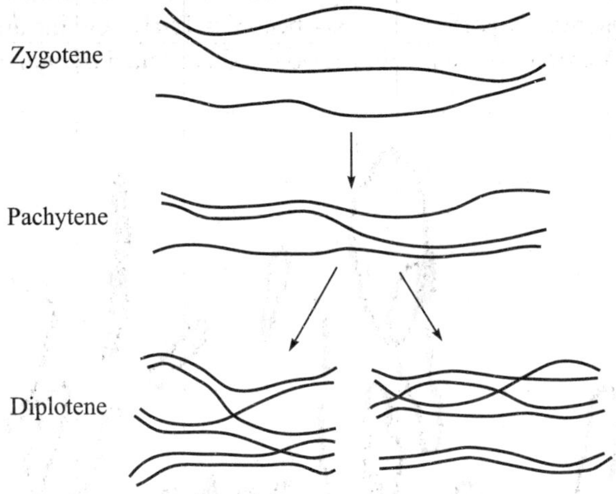

Figure 5.5 Pairing behaviour of chromosomes in triploid.

On account of chromosomal abnormalities in the first meiotic division, no functional gametes are formed or very rarely formed, they are mostly sterile, therefore, seeds are not formed. In Japan, Sweden seedless triploid watermelon was developed. Banana is also a seedless natural triploid.

Tetraploids exhibit gigas characters such as thicker and broader leaves; bigger flowers. They are partially sterile. Because quadrivalents are formed in meiosis, as a result paired chromosome do not regularly disjoin and pass on to the opposite poles as in normal diploid. Non-disjunction at anaphase affects the genic as well as the chromosomal constitution of the gametes because gametes possess varying number of chromosomes.

Principles of polyploid genetics

A number of crop plants are polyploids, therefore, plant breeders face problems in understanding the genetics of polyploids. B.P. Gaplen (California) selected some resistant plants of alfalfa from a commercial field to the root knot nematode (*Meloidogyne hapla*), a soil-borne nematode disease. When he selfed a clonal material, the resulting 51 populations scored 314 resistant and 119 susceptible plants. At the same time when a resistant clone was test-crossed to a susceptible plant, there were 46 resistant and 15 susceptible plants. This indicated that the clone which was selfed as well as test-crossed to a susceptible plant is heterozygote for a single gene (*Aa*), segregating in a disomic manner or alternatively a tetrasomic simplex (*Aaaa*). However, a simplex plant giving 3:1 ratio in S_1 generation is expected to have quite different pattern of segregation in S_2 generation that is a disomic heterozygote.

Allopolyploids

When a diploid species is crossed with another diploid species, the hybrid possesses two sets of unrelated haploid chromosome complements from each parental species which during meiosis form univalents, their separation at anaphase I is irregular, therefore the resulting gametes are non-functional and the net result is the hybrid, which is sterile. The sterility is caused due to non-homology between the chromosomes, which do not pair during meiosis. However, if the chromosomes of the hybrids are doubled, each chromosome is duplicated. The pairing between the duplicated chromosomes is perfect, meiosis becomes fairly regular and fertility is restored. Such a hybrid in which chromosome number is doubled is termed *allotetraploid* or *amphidiploid* (Figure 5.6).

Figure 5.6 An allopolyploid (2n = 42) from a cross between tetraploid × diploid species of *Hordeum.*

Thus, if *A* and *B* represent the cross (*AA* × *BB*), then the amphidiploid is represented by *AA BB*; pairing is strictly between *A–A* and *B–B*, this is called *auto-syndesis.* However, if pairing takes place between *A–B* due to presence of some homology such pairing is called *allosyndesis.* In case of autosyndesis bivalents are formed, whereas in case of allosyndesis quadrivalents are formed.

In case of autotetraploids, sterility increases due to irregular meiosis on account of nondisjunction of multivalents. In case of allopolyploid, F_1 (hybrid) is sterile due to non-pairing of dissimilarity between two sets of chromosomes (species × species crosses); but sterility is overcome by doubling the chromosome complements of F_1. Sterility is more in allotriploids (3X) because of dissimilarity in the chromosomes coming from different species (Figure 5.7).

Raphanobrassica is an example of experimental production of new species by doubling the chromosome numbers in F_1 hybrid.

Radish (*Raphanus sativus*) × Cabbage (*Brassica oleracea*)

↑ (2n = 18) ↓ (2n = 18)

F_1 hybrid sterile

(n = q + n = q)

Figure 5.7 PMC from an allotriploid (3X) plant of *Hordeum* showing bivalents, univalent and a trivalent (arrow).

In F_1, some seeds were produced by chance. Their cytological analyses indicated that they possess 36 chromosomes instead of 18. Plants having such types of chromosomes showed foliage like radish and root like cabbage; the allotetraploid bred true. This cross was made by Karpechenko whose dream was to produce a new plant *species—Raphanobrassica,* whose root should be like radish and foliage like cabbage, but it happened just reverse.

Brassica juncea (n = 18) is an amphidiploid.

The F_1 of the cross *Brassica compestris* (n = 10) × *Brassica nigra* (n = 8) shows 18 chromosomes in the somatic cell. Reciprocal cross completely failed. The hybrid was intermediate and sterile.

The amphidiploid of this cross was produced by colchicine treatment of one bud, the seeds from anphidiploid bred true and progenies were fertile. They were closely resembling naturally existing species of *Brassica juncea.* This indicates that *Brassica juncea* is an amphidiploid. Figure 5.8 indicates how different species of *Brassica* have arose in nature.

Other examples of allopolyploid (amphidiploid) originated crops are *Nicotiana, Triticum Gosspium* and *Primula kewenensis* species.

Allosyndesis indicates a common origin in a progenitor, which may be extant or extinct. The cytological analysis establishes the degree of relationship between different species, based on the pairing behaviour of the chromosomes during meiosis in the hybrids.

Hexaploid wheat consists of three sets of chromosome complements *A, B, D*; each consists of 7 chromosomes.

Emmer group is an allotetraploid (*AABB*), a hybrid between an einkorn species (*A* genome) and another species (with *B* genome). Cross between Emmer (*AB* genome) and a species of *Aegilops squarosa* (genomc *D*) has given rise to hexaploid wheat (*Triticum aestivum*).

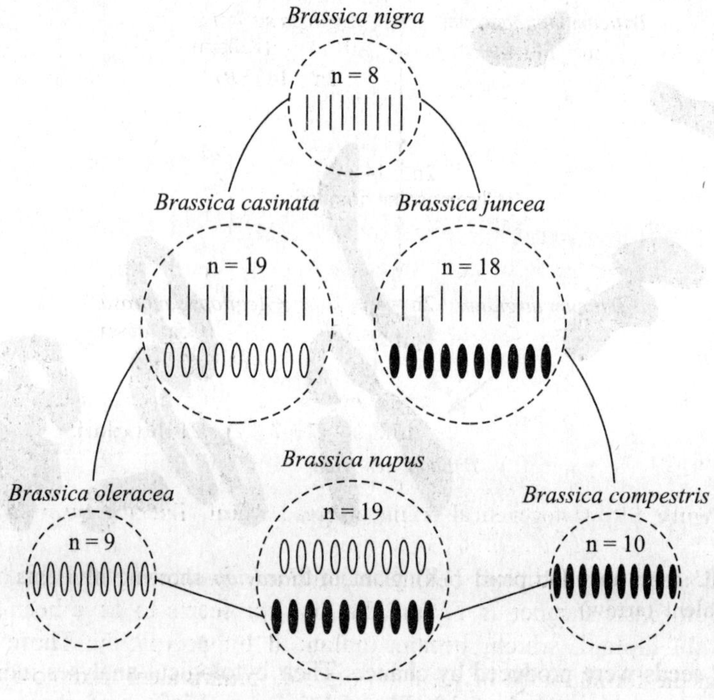

Figure 5.8 Origin of different *Brassica* species.

The three groups of wheat species are:

Einkorn group (n = 7)	*Emmer group* (n = 14)	*Vulgare group* (n = 21)
Triticum aegiloploides	*T. dicoccoides*	*T. spelta*
T. monococcum	*T. dicoccum*	*T. aestivum*
	T. durum	*T. compactum*
	T. turgidum	*T. spherococcum*
	T. polonicum	
	T. persicum	
	T. timopheevii	
Genome (*A*)	Genome (*A, B*)	Genome (*D*)

Thus source of evolution of various species has been allopolyploids and differentiation of genomes themselves has been mutation of structural changes.

Cytogenetical origin of bread wheat (*Triticum aestivum*) has been explained in Figure 5.9 which indicate how nature has produced this great grain for man.

Secondary Polyploids

Plants with extra unbalanced number of chromosomes are termed polysomics (2n + 1). When the polysomies (2n + 1) form the basis of polyploidy, the polyploids derived from them are termed

Triticum monococcum × *Aegilops speltoids* (Kihara)
2n = 14 (*AA*) 2n = 14 (*BB*)
↓
2n = 14
Chromosome doubled
↓
Triticum turgidum 2n = 4x × *Aegilops squorosa* (Goat grass)
AABB *DD*
↓
ABD (7 + 7 + 7) = 21 Hexaploid.
Triticum aestivum

Figure 5.9 Cytogenetical origin of bread wheat (*Triticum aestivum*).

secondary polyploids. Apple and pears belong to the family Rosaceae with basic number 7. In the pomoideac section, basic number is 17 and this number seems to have been derived from a trisomic (2n + 1) diploid, which, became balanced by polyploidy. These are secondary polyploids. From the formation of bivalent in, diploid, quadrivalent in diploid and trivalent in triploid of *Oryza sativa,* it is concluded that *Oryza sativa* has arose by secondary polyploids of an unknown species with n – 5. The 2n + 2 of this unknown species has formed the basic number.

The species of Zizineae are hexaploid with basic number 5, by secondary polyploidy. While Zizineae has retained the basic number 5, and formed polyploid species. The Oryzineae has developed from a secondary basic numbers and has formed polyploid species. The non-pairing of *Oryza sativa* × *O. officinalis* indicates that latter species has been developed by structural changes in this chromosome. Secondary polyploids may also arise by loss of chromosomes. *Crepes artificialis* is an example. The F_1 of C. *bienis* (n – 20) × C. *setosa* (n – 4) shows 10 pairs of *bienis* complements by autosyndesis and 4 *setosa* chromosomes are univalent. By selfing, a progeny with 10 pairs of *bienis* and two pairs of *setosa* was obtained. The latter pair arose by loss of *setosa* chromosomes and duplication of remaining two. The new species is called *Crepis artificialis*.

Induction of Polyploidy

Polyploids arise in nature, but polyploidy can be induced artificially. Diploidy is more primitive than polyploidy. Polyploids have arose in nature due to interference in the somatic cell division which give rise to tetraploid branches and also due to interference of anaphase I in meiotic division which produced unreduced gametes.

Application of alkaloids

Colchicine derived from *Autumn crocus* (*Colchicine autumnale*) either in liquid or in a lanolin paste, induces polyploidy. Chromosome replication is unaffected but spindle formation is prevented. Thus, chromosomes are doubled within a common nuclear membrane. Subsequent nuclear division is normal. Polyploidy can also be induced by some other chemicals like acenaphthene and veratrine or by exposure to heat and cold.

Methods tested for colchicine treatment

1. Twigs could be immersed in solutions of various concentrations for different periods.
2. Seeds were immersed in colchicine solution for different periods.
3. Seeds were germinated on blotting paper soaked in different concentrations of colchicine solution.
4. Young flower buds could be immersed in test tubes filled with different concentrations of colchicine for different periods.
5. Vegetative buds could be immersed in 0.5 per cent and 0.25 per cent aqueous solutions of colchicine for different periods.
6. Colchicine could be applied as a paste in anhydrous lanolin with 0.4–0.5 per cent solution.
7. Colchicine solution could be sprayed on the growing buds.
8. Growing buds could be wetted with aqueous solution.
9. Solution of colchicine in agar may be applied to growing buds on interval.

In all the cases, materials were washed with distilled water. Best result is obtained by seed treatment.

Methods of treatment with colchicine

Crop	*Treatment*	*Method*
Monocots		
1. Small grain crops	Seedlings with coleoptiles 2–4 mm are inverted through a net in a petridish containing 0.25 per cent acqueous colchicine solution for 30 minutes.	Svalof
2. Einkorn (*T. monococcum*) oats, barley (2 rows, 6 rows)	Seedlings wilts shoots $\frac{1}{8}-\frac{1}{4}$ inch are immersed in 0.25 per cent aqueous solutions for 20-30 minutes.	Dorsey (1939)
3. Interspecific and intergeneric hybrids.	Tillers of the young plants are cut back and capped wilts vial containing 6.3 per cent colchicine solution for 72 hours.	Bell (1950)
4. Rice	Seeds germinated in petridishes and shoots of 3–5 mm are split at base wilts razor blade. Cotton ball or blotting paper soaked is 0.05–0.1 per cent aqueous colchicine solution is inserted in split.	Luoug (1951)
Dicots		
1. Rape etc.	Seed sown in flats when primary leaves open, a drop of 0.5–1.0% aqueous solutions is placed.	Sval of method.

	on growing points with a soft brush. Repeat daily per 10 days.	
2. Cotton	Seed coat is removed after soaking in water per 24 hours and after 4 days of germination, seedlings are inverted and shoots treated with 0.5 per cent colchicine solution per 6 hours at 23–26°C.	Douwer (1952)
3. Beet	Seedlings with coleoptiles and roots, a few mm in length should immersed in a thin layer of 0.1 per cent acqueous solution for 3 hours at 27°C.	Brewer Reinders and Bremer (1952)
4. Gram (*Cicer arietinum*)	Seed treatment into 0.25 per cent aqueous colchicine solutions for 30 minutes.	Ramanujam and Joshi (1941)
5. Muskmelon	Seedlings in cotyledon stage treated with 0.4 per cent emulsion of colchicine to shoot apices in two treatments at 4 days intervals	Bafra (1952)
6. Watermelon	One drop of 0.2–0.4 per cent aqueous solution of colchicine should be applied to the shoot apices of young seedlings for two successive days.	Lcihara (1951)
7. Potato	Seeds are germinated in sterilized petridishes containing a colchicine jelly made up of equal parts of 0.5 per cent colchicine and 2 per cent agar. Soon after germinations the seeds are washed, then planted.	Swaninalch (1950)

Adaptability of plants to colchicine treatment

The different crop plants show different range of adaptability to the induction of polyploidy by colchicine treatment. Not the least, even different genotypes within a species too show different response to double the chromosomes. According to loue (1953) and Levan (1948):

1. Plants low in chromosome numbers are more likely to respond well in doubling the chromosome than those higher in chromosome number.
2. Cross-fertilizing plants are more likely to be successful for doubling the chromosome numbers than self-fertilizing species.
3. Vegetatively propagated plants may be more successful as autopolyploids than seed grown plants.
4. Fertility of autotetraploids may be reduced to a great extent.

Effects of treatment

The changes brought by chromosome doubling are of great significance. Both individually as well as doubling of chromosomes bring morphological, physiological and genetic changes.

Morphological: Leaves become large, darker in colour, leathery in texture; leaf lobes gct overlapped and twisted. Stomata, bracts, pollen grains, bolls and seeds become bigger than control. Maturity is delayed.

Immediate effect of seed treatment causes swelling of radicles and retardations of their growth. Seedlings show slow growth with thicker stem, darker and broader leaves.

Size of pollen grains and guard cells are bigger than control plants.

The stems of polyploids are stouter and thicker; leaves are larger and broader; hair on the vegetative parts are coarse and thicker. The floral parts, fruits and seeds are larger than diploid.

Normally, polyploids form gigas character because individual cell size becomes larger. Gigas character occurs in both naturally as well as artificially-induced polyploids.

Physiological: Becker in 1931 reported that osmotic concentration in autopolyploids is proportional to the chromosome number.

Autotetraploid maize takes longer time to mature than diploid. In fruits and vegetables, vitamin content is increased. In tobacco, nicotine content is increased and also nitrogen, calcium, potash and magnesium but carbohydrate, sulphur and phosphorus decreases. In triploid beet, sugar content is increased.

The decrease in growth rate slowly leads to the formation of perennial from annual types in cotton. Sharp in 1934 reported that autotetraploid *Zea* race is perennial, while diploid is annual. *Sorghum sudanense* and *Sorghum halepense* look alike but former is annual (n – 10) and latter is perennial (n – 20). Similarly, two species of *Enchlaena, E. mexicana* is diploid and annual, and *E. perenis* is an autotetraploid and perennial.

Muntzing (1936) concluded based on large collection of data that a large number of perennial species might have originated from annual diploid with lower chromosome number.

Aneupolyploidy

Change in the one or a few chromosomes of a genome is called aneupolyploidy. An ordinary hexploid wheat plant with two pairs of chromosomes is disomic at meiosis. However, when one member of a chromosome pair is missing, the plant is monosomic (2n – 1). When both the members of a pair are missing, the plant is nullisomic (2n – 2). When an extra chromosome is present, the plant is trisomic (2n + 1) and when an extra chromosome pair is present (2n + 2), the plant is tetrasomic. The cytological behaviour of different types of trisomies are given in Figure 5.10(a, b, c).

Primary trisomics: Where the extra chromosome is an exact replica of any of the monoploid set, so that the homologous chromosomes are present, the plant is called a primary trisome. During meiosis, these three homologous chromosomes may become multivalent of three chromosomes in chain (Figure 5.10(a)).

Secondary trisomics: Where the extra chromosome is an isochromosome with two identical arms, the carrier is a secondary trisome. It may from rings of three chromosomes at meiosis (Figure 5.10(b)).

Tertiary trisomics: If parts of two nonhomologous chromosomes make up the extra chromosome, the plant is tertiary trisome. Chains of five chromosomes are possible at meiosis in such individuals. Tertiary trisome may arise as a results of nondisjunction of the ring of four chromosomes at anaphase I in translocation heterozygote (Figure 5.10(c)). In general, trisomes are obtained from segregation of unbalance chromosome complements of triploids.

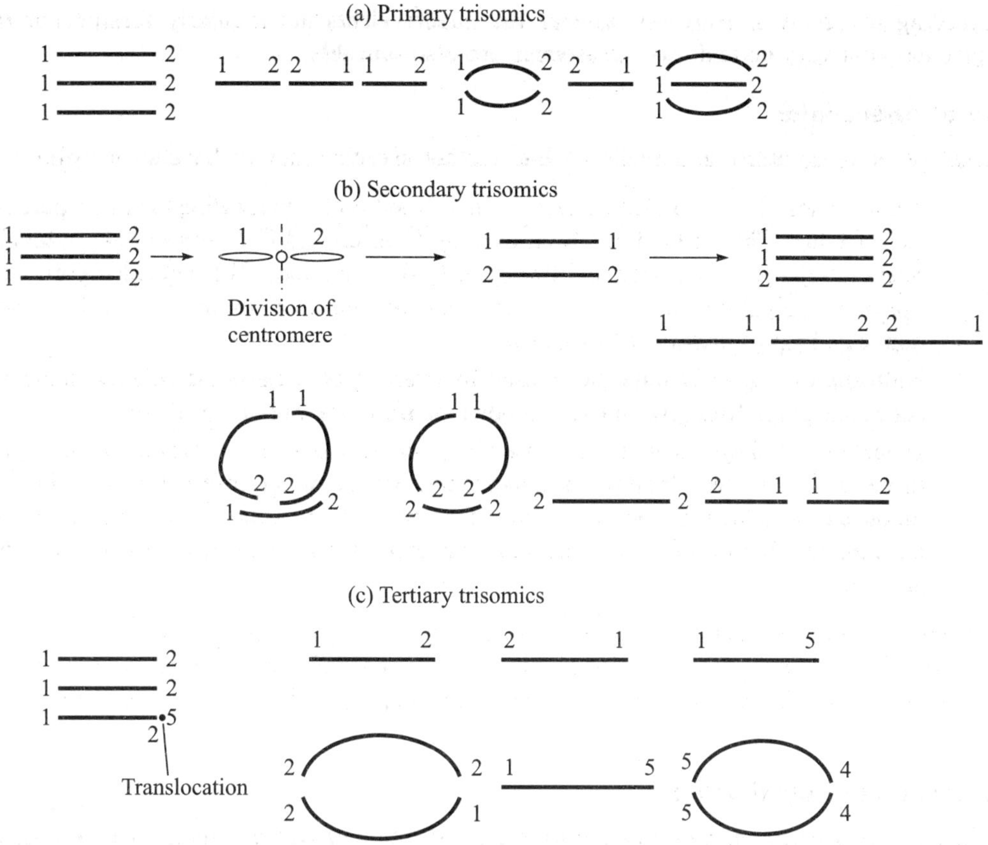

Figure 5.10 Metaphase configuration of chromosomes in (a) Primary trisomics, (b) secondary trisomics, (c) Tertiary trisomics involving translocation.

Cytogenetics

In aneuploids, with an uneven number of some particular chromosome (monosomics, trisomics), there is clearly no possibility for an equal number of each chromosome to end at the two poles in the division. Monosomics (2n – 1) for instance produce two types of gametes n, and n – 1. The odd chromosome passes at random to either pole in meiosis but frequently it will lag at anaphase I between two poles and may not get included in either daughter cell nucleus at reduction division. The gametes with n – 1 chromosomes often do not function and further more, zygotes with 2n – 2 (nullisomics) are inviable except in few polyploid species. Thus, most of the progenies of monosomics are either normal diploid or monosomics.

***Trisomics* (2n + 1):** Trisomics are more stable genetically than monosomics (2n –1). However, there is no assurance that two chromosomes will always go to one pole and one to other pole at meiosis.

***Tetrasomics* (2n + 2):** Tetrasomics often behave more regularly than trisomics. The four homologous chromosomes tend to form a quadrivalent at meiosis and disjunction is often proceeding two-by-two. However, neither the quadrivalents are regularly formed nor their disjunction is always normal. So, tetrasomics are also unstable.

Use of aneuploids

Aneuploids have no utility in agriculture, but in plant breeding they do have some utility.

1. Monosomics (2n – 1) and nullisomics (2n – 2) are useful in locating genes on particular chromosome. This method has been used by Clausen and Cameron (1944) in tobacco; Sears (1953) in wheat; Rick and Barton (1954) in tomatoes. The method is particularly useful in polyploids in which monosomics and nullisomics are viable because of duplicated genes present in polyploids.
2. Nullisomics (2n – 2) have been used in transfer of a particular chromosome with desirable genes from one variety to other or from one species to other.
3. Detection of linkage. Sears in 1953 used latter method in locating certain dominant genes in wheat for red seed located on chromosome xvi by direct observation that the seeds produced by nullisomics xvi are white. Similarly, chromosomes viii and x, each carry a dominant inhibitor of awn as shown by the increased awn development of nullisomics viii and x.

Genes in varieties other than Chinese spring wheat can be located by crossing it to each of 21 nullisomics and observing the ratios obtained in F_2. In actual practice, monosomics are used instead of nullisomics because of weak plant and greatly reduced vigour and fertility of nullisomics.

Chromosome substitution

The use of nullisomics and monosomics for the substitution of whole chromosome and the part of chromosomes so that they can be studied for gene content, has been described by Unrau (1950). Later Kuspira and Unrau (1957) found that the substitution of a particular homologous chromosome from one variety to another caused significant changes in yield, earliness, lodging resistance and other characters.

Substitution must be from related genera and species, The germplasm involved in the substitution must be fairly closely related so that substitution or addition may be viable and not so alien to upset the normal physiological functions. Within *Triticinae* many possibilities are present.

EVOLUTIONARY SIGNIFICANCE OF POLYPLOIDY

Polyploidy is found to be of wide occurrence in nature. It has been observed that after species hybridization, doubling of chromosomes results into appearance of new types.

The wild and cultivated wheats are diploids and polyploids, from 2n = 14 to 2n = 42 with basic number 7, which was discovered around 1920, by a number of workers. McFadden and Sears (1946) in USA, Kihara in Japan and Russian workers were able to synthesize such wheats by crossing *Aegilops squarosa* with emmer wheats and then doubling the chromosome of the hybrid. The *Aegilops* species is native of the region of ancient Babylonian civilization, where present day bread wheat arose.

The new world cotton (2n = 52) arose through allopolyploidy from a hybrid or hybrids of an Asiatic (2n = 26) and American (2n = 26) parents. In 1940 Harland had independently synthesized a new world cotton (2n = 52) by doubling the chromosome number of F_1 hybrid of *Gossypium arboreum* (Asiatic cotton) × *Gossypium thurberi*(*American* cotton). Although these partially sterile, polyploids were crossed readily with new world cotton, offsprings obtained from these hybrids were fully fertile.

In tobacco, *Nicotiana tabacum* (2n = 48) is polyploid of the cross *Nicotiana sylvestris* (2n = 2n) × *N. tomentosa* group (2n = 24).

The mustard provides another interesting example of the role of polyploid in the evolution of certain crops (Figure 5.8).

ARTIFICIALLY-INDUCED POLYPLOID

Triticale is an amphidiploid crop, a cross of rye and wheat (Muntzing, 1948). But this rye and wheat crossing programme was not very successful because of the sterility and its unproductive nature. Another classical example of artificially-induced polyploidy is *Raphanobrassica* (a cross of radish and cabbage), an allotetraploid, which unfortunately combined noncommercial features of both the parents (the above ground part of radish and the root of cabbage). Third is the synthesis of an amphidiploid of *Brassica*, spp., *B. juncea*, *B. napus* and *B. corinata*.

Importance of Induced Allopolyploids

To Transfer Plant Characters

The new species developed by allopolyploidy may serve as intermediary in the transfer of disease resistance and other characters from related species and genera.

1. Necrotic lesion type of mosaic resistance was transferred from *Nicotiana glutinosa* to *N. tabacum*, which was made possible by means of amphidiploid, *N. digluta* (Gerstel 1945).
2. Transfer of resistance, to phyllody disease of *Sesamum prostratum* to cultivated. *S. orientale* by means of amphidiploid *S. indicatum.* (Ramunujan, 1924, 1944).
3. Transfer of large, berry-size and earliness from high bush berry (*Vacciniun australe*) (4X) to the rabbit eye blue berry (*V. ashei*) (6X).
4. Pearson, et al. (1951) combined insect resistance from *Cucurbita moschata* with fruit flesh character from *C. maxima* by colchicine treatment of F_1 hybrid, which was sterile.
5. Extrachromosomal complements in polyploids confer certain advantages to the species lacking in the diploid form. For example, two extra sets of genes are available in tetraploid which buffer the deleterious effect of mutation; however, if mutation is favourable, it adds the selection value, therefore polyploids are important from evolutionary point of view.
6. Where breeder is dealing with asexually propagated plants such as sugarcane, potato, zinger, etc. and garden flower and where vegetative parts, fruits and seeds are of economic importance: gigas forms and large variation can be induced by polyploidy.

7. Polyploidy in parental species enables wide crosses to be attempted. Crosses of sugar cane, *Saccharum officinarum* with *S. spontaneum* and *Sorghum durra* were possible due to high degree of polyploidy in sugar cane.

REVIEW QUESTIONS

1. What is polyploidy? Give its evolutionary significance.
2. Give brief account of different types of polyploidy.
3. What are auto and allopolyploids? Explain briefly their role in evolution of crop plants.
4. How can you produce seedless watermelon and oranges? Why banana does not bear seeds?
5. Give cytological configuration of pairing difficulty in autopolyploid.
6. Describe two methods through which polyploids may occur.
7. How autopolyploids and allopolyploids originate in nature? Explain the significance that each might have in evolution.
8. Why do tetraploids behave more regularly in meiosis than triploids and perpetuate in the population?
9. Why is chromosome irregularity causes low fertility in plants?
10. A plant species × which has 7 chromosomes in its gametes was crossed with a related species B, which had 9. F_1 were sterile, but a section of hybrid grew through vegetative propagation and a plant was found fertile with 32 pairs of chromosomes in its somatic cells. What step might have been involved. Show diagrammatically.
11. A plant species of (n = 5) was crossed with a related species B, ($n = 7$), only a few pollen grains were produced by F_1 hybrids. These were used to fertilize the ovules of species B. A few plants were produced with 19 chromosome numbers. They were highly sterile, but self-fertilization produced few plants with 24 chromosome numbers. These plants were quite different from species A and B. Phenotypically and their progenies were fertile. Explain how it happened so.
12. How does colchicine treatment result in chromosome doubling?
13. What practical and theoretical significance may be associated with colchicine-induced polyploidy?
14. How could the polyploidy be a significant event in the evolution of cultivated plants like wheat and cotton?
15. Give plausible explanation for origin of *Triticum aestivum*, *Raphanobrassica* (rubbage) and new world cotton.
16. What following chromosome arrangement is symbolized for:
(a) n, (b) 2n, (c) 2n + 1, (d) 2n – 1, (e) 2n + 2, (f) 3n and (g) 4n.

REFERENCES

Allard, R.W. (1960): *Principles of Plant Breeding*, John Wiley & Sons, New York.

Chandrasekharan, S.N., S.V. Parthsarthy, and K. Ramiah (1953): *Cytogenetics and Plant Breeding*, Vardhanacharya Co., Madras.

Clausen, R.E. and Cameron (1944): Inheritance in *Nicotiana tobacum* XVIII: monosomic analysis, *Genetics*, **29,** 447–477.

Dawson, G.W.P. (1962): *An Introduction to the Cytogenetics of Polyploids*, Blackwell, Oxford.

Gerstel, D.U. (1945): Inheritance in *Nicotiana tobacum* XIX: identification of *tobacum* chromosomes replaced by one from *N. glutinos* in mosaic resistance, Holmes Samsjun. *Tobacco Genetics*, **30,** 448–454.

Kihara, H. (1951): Triploid watermelon, *Proc. Am. Soc. Hort Sci.*, **58,** 217–230.

McFadden, E.S. and E.R. Sears (1946): The origin *of Triticum speltu* and its free-threshing hexaploid relatives, *J. Heredity,* **37,** 81–89, 107–116.

Muntzing, A. (1948): Experience from work with induced polyploidy in cereals, 324–337 Svalof (1888–1946), edited by A. Aderman and others, Carl bloms Boktry ckeri. A.B. Lund. (1948).

Pearson, O.H., H. Hopp, and G.W. Bohn (1951): Notes on species crosses in cucurbits, *Proc. Am. Soc. Hart. Sci.*, **37,** 10–322.

Ramanujan, S. (1944): Cytogenetics of an amphidiploid, *Sesamum orientale* × *S. prostratum*, *Current Science* (India), **13,** 40–41.

Ric, C.M. and D.W. Barton (1954): Cytological and genetical identification of the primary trisomics of tomato, *Genetics*, **39,** 640–666.

Sears, E.R. (1953): Nullisomic analysis in common wheat, *Am. Nat.*, **87,** 245–252.

Stebbin, G.W. (1950): *Variation and Evolution in Plants*, Columbia University Press, New York.

Strickberger, M.W. (1990): *Evolution*, Jones and Barttett, Inc., USA.

Unrau, J. (1950): The use of monosomics and nullisomics in cytogenetical studies in common wheat, *Sci. Agri.*, **30,** 66–89.

Unrau, J. (1958): Cytogenetics and wheat breeding, *Proc. of Tenth Intern. Congress on Gene*, **1,** 129–141.

Part II

Principles of Genetics

CHAPTER

6 Mendel and Principles of Heredity

Mendel was an Austrian monk of Brunswick and a teacher of mathematics in a local school. However, he was a keen observer of the natural phenomenon, specially how characters are transferred from the parents to their progenies. He started his experiment in mid-nineteenth century with garden peas. He chose pea for his studies because of easy pollination of the flowers, plants being annual; every year one crop could be raised and easy in the handling of the materials. Before initiating hybridization work in pea, Mendel obtained 34 varieties of garden peas. He grew these varieties for about two years to purify them for genetic homozygosity of the varieties for the characters to be studied. He did so to see that parents are true breeding and finally he selected 22 varieties for his experimental use. For the genetical studies, constancy in the character is must.

Mendel chose garden pea because he had to access the varieties that differed in observable alternative characters and because he had studied the floral biology of garden pea. Mendel effected crosses between the types differing in some contrasting characters. He chose seven pairs of characters in the garden pea for his studies. Those characters were seed shape (round vs wrinkled), cotyledon colour (yellow vs green), flower colour (purple vs white), pod shape (inflated vs constricted), pod colour (green vs yellow), pod position (axial vs terminal) and plant height (tall vs dwarf). He made seven single crosses between plants having contrasting characters and all gave fruitful results.

1. Out of the two alternative parental characters, only one appeared in the first generation (F_1).
2. The character that did not appear in the first filial generation; appeared in the second generation (F_2) in one fourth of the total population.

The results of Mendel's experiment on garden peas are summarized below:

S. No.	*Parental character*	F_1	*No. of* F_2 *progeney*	F_2 *ratio*
1.	Round × Wrinkled	Round	3474 Round, 1850 Wrinkled	2.90:1
2.	Yellow × Green	Yellow	6022 Yellow, 2001 Green	3.01:1
3.	Purple × White	Purple	750 Purple, 224 White	3.15:1
4.	Axial × Terminal	Axial	651 Axial, 207 Terminal	3.14:1

5.	Inflicted × Constricted	Inflicted	882 Inflicted, 299 Constricted	2.15:1
6.	Green × Yellow	Green	428 Green, 152 Yellow	2.82:1
7.	Long × Short	Long	787 Long, 277 Short	2.84:1

Mendel had a feeling that the plants as a composite structure are composed of a large number of unit characters in the hybrids and further generations. He considered one character pair at a time from his above said experiments. Mendel derived two conclusions, which he called laws of inheritance.

1. Law of segregation, and
2. Law of independent assortment.

Let us examine, how carefully Mendel, handled his experiments. He made the crosses of the pea varieties having round vs wrinkled seed shape; the seed of first filial generation (F_1 hybrid) was round. He made reciprocal crosses also. Mendel picked up the F_1 hybrid plants and threshed them separately. He sowed the seeds in a plant progeny row to raise F_2 generation. Mendel again counted number of plants having round and wrinkled seeds in F_2 generation and harvested them separately. He raised F_3 (third filial generation) by sowing the seeds of F_2 in the plant progeny rows and studied the pattern of segregation.

♀ Round seed × Wrinkled seed ♂ Parents

F_1 Round seed

F_2 3 round seed:1 wrinkled seed

Similar results were obtained in other six crosses made by Mendel.

Mendel maintained pedigree records for every cross. He used simple mathematics, i.e. 'ratio' to class his observations in F_2 and F_3 generations. The study of the progenies of different crosses, character-wise maintenance of simple mathematical ratios to class the different types of plants, grain shape, pod shape and size, etc. enabled Mendel to conclude the law of segregation of the hereditary units (gene).

Before Mendel, the prevailing concept of the heredity was the blending of the parental traits in their hybrids (F_1). Mendel's experimental results indicated that one of the parental characters such as wrinkled seed shape was absent in the F_1 hybrid but reappeared unchanged in pure form in the second generation (F_2) which was exception to the idea of the blending heritance.

From his experimental results, Mendel made brilliant conclusion that each pea plant possesses a homologous pair of such elements (factors) of which, it has received one from the male (pollen parent) and the other from the female parent (ovum). Of the two homologous units, one which appeared in F_1 hybrid is called 'dominant' and the other which remained absent in F_1 but reappeared in F_2 generation is called 'recessive'.

LAW OF SEGREGATION

In Mendel's monohybrid crosses round seed which appeared in F_1 was dominant and symbolized as *WW* and wrinkled seed which did not appear in F_1 was symbolized as ww. However, F_1 hybrid plants with round seeds were symbolized *Ww*, where *W* was dominant over *w*. When F_1 was self-crossed, it formed two types of gametes *W*:*w* in the male (pollen) parent and *W*:*w* in the female (ovule) parent. On their random mating (fertilization), they formed four types of genotypes as shown in checkerboard, where *WW* and *ww* were true breeding for round and wrinkled seeds. But *Ww* were hybrid (F_1) types with round seed. On clubbing all the genotypes with round seed shape, three genotypes with round seed shape fell into one class, and wrinkled in one class and thus they appeared in 3:1 ratio. Mendel called it law of segregation or monohybrid ratio. Thus, law of segration could be defined as, *in the formation of gametes, the pair of hereditary units* (*Ww*) *separate* (*segregate*) *out in such a manner that each gamete is equally likely contains either of one allele*.

	♀	*WW* (Round seed)	×	*ww* (Wrinkled seed) ♂	Parents
F_1		*Ww* (Round seed)			

Gametes: ♀ \ ♂	*W*	*w*
W	*WW* (Round)	*Ww* (Round)
w	*Ww* (Round)	*Ww* (Wrinkled)

Note:

The alternative form of gene is called allele. The alleles are hereditary determinants which are always present in pairs (*WW*/*Ww*) *in both parents and in the progeny, but individually* (*only one*) *are present in the reproductive cells* (*gametes*). *The plants possessing WW and ww gene pairs are called homozygous and those possess gene pairs Ww are called heterozygous. In other words, when two hereditary determinants* (*genes*) *are identical* (*WW or ww*), *the plant is homozygous, but when two factor pairs differ* (*Ww*), *the genotypes are called heterozygous.*

In the sexually reproducing plants, the progeny arises with the fusion of male (♂) and female (♀) gametes. Therefore, all the characters of the two parents are transmitted to the progeny through the gametes only. The gametes thus form the physical link between the parents and the progenies. The gametes are microscopic and unicellular, each carrying a prominent nucleus. Mendel presumed, there was something in gamete, which was later on determined by Johannsen in 1909 as *gene*.

Like hereditary determinant to which Mendel called 'something', other points clearly explained by Mendel were the external and internal constitution of different segregants. Johannsen (1909) defined that phenotypes denote the observable external appearance of a character of an organism. Similarly, he also defined genotypes which denote the internal genetic constitution of the somatic cell. The two different genotypes *WW* and *Ww* have the same phenotypic appearance round, but genetically they are different. The phenotypic appearance of a character is not always the index for its breeding behaviour, therefore, latter must be judged by genotypic constitution of the organism.

Reciprocal Crosses

The crosses are attempted between two distinct types differing in the characters where any one type can be used as male or female parent. Thus, if type *AA* is crossed with type *BB*, i.e. *AA* × *BB*, the cross *BB* × *AA* is called reciprocal. Normally, which way the cross is made, the F_1 behaviour and the mode of genetic segregation in F_2 generation are the same provided no plasma genes are interacting.

	♀ *WW* (Round seed)	×	*ww* ♂ (Wrinkled seed)	Parents
F_1		*Ww* (Round seed)		
F_2		3 round seed:1 wrinkled seed		
	ww (Wrinkled seed)	×	*WW* (Round seed)	Parents
F_1		*Ww* (Round seed)		
F_2		3 round seed:1 wrinkled seed		

Back Cross

If F_1 hybrid is crossed with any one of the two parents say round or wrinkled seed shape, it is called *back cross*. The F_1 will be all round seed shape. If F_1 is back-crossed with the parent having dominant seed shape (round), all will be round, means no segration. However, if F_1 is back-crossed with the parents having *recessive* character say wrinkled seed shape, the back cross progeny will appear in 1:1 ratio for round: wrinkled seed shape, because recessive parent gives one type of gamete. The back-crossing of F_1 hybrid with recessive character parent is also called *test cross*. It is useful in testing the purity of gametes.

Let me explain the two situations:

(i) ***Back cross with parent possessing dominant (WW) character:***

	♀ *WW* (Round seed)	×	*ww* ♂ (Wrinkled seed)	Parents
F_1		*Ww* (Round seed)		
F_1	back-crossed: *Ww* (Round seed)	×	*WW* (Dominant character)	
		WW (Round seed)	*WW* (Round seed)	

Gametes of dominant parents:

♀ \ ♂	*W*	*W*
W	*WW* (Round seed)	*WW* (Round seed)
w	*Ww* (Round seed)	*Ww* (Round seed)

All progenies with round seed.

(ii) ***Back cross with parent possessing recessive (ww) character:***

♀ *WW* × *ww* ♂ Parents
(Round seed) (Wrinkled seed)

F_1 *Ww*
(Round seed)

Back cross to recessive character *ww* (Wrinkled seed) × *Ww* (F_1 round seed)

Back cross 1: *W* : *w*
(1 round:1 wrinkled)
Gametes of F_1

Gametes produced by recessive parent:

♀ \ ♂	*W*	*w*	
			F_1 gametes
w	*Ww* 1	*ww* 1	1:1 ratio

Modifications of Monohybrid Ratio

Incomplete dominance

Mendel in his paper, which he read before the Natural History Society at Brunn on February 8, 1965, designated round seed shape as complete dominant to wrinkled seed shape. But when cotyledons of three genotypes *WW*: *Ww*: *ww* were examined microscopically; they differed in the starch grain contents. The genotypes with *WW* plants were found with abundance of filled starch grains in seeds, very few in *ww* and the heterozygote (*Ww*) showed an intermediate number of starch grains. Therefore, on microscopic and biochemical (molecular) level allele *W* must be considered only incomplete dominant to the *w* allele.

In the *Snapdragon*, where a red flower plant was crossed with white flower plant, the F_1 hybrid was pink. In the F_2 generation, 1/4 plants had red flowers, 1/4 plants had white flowers and 1/2 plants had pink flowers: the red; pink and white flower plants were in the ratio of 1:2:1 in F_2 generation.

♀ *rr* × *RR* ♂ Parents
(White flower) (Red flower)

F_1 *Rr*
(Pink flower)

F_2 *rr* *Rr* *RR*
(White flower) (Pink flower) (Red flower)
1 : 2 : 1

Incomplete dominance was also observed in horses, chestnut horse (*dd*) with full reddish colour when crossed with cremello horse (*DD*), the progeny was palmino horse (*Dd*) with beautiful golden colour. This is a case of cumulative genetic effect of *DD* gene in determining in coat colour in horse. When both dominant gene alleles (*DD*) are present, the coat colour of horse was ceramist white; when only one dominant gene was present, the coat colour of animal was golden (palmino) and when only recessive (*dd*) gene allele was present in homozygous condition, the animal was reddish in colour.

Incomplete dominance in 4 O'clock *Mirabilis jalapa* has been also reported.

Codominance

In short-horned cattle, alleles for red and white coat colours occur. The cross between red ($r^1 r^1$) and white ($r^2 r^2$) produced F_1 offspring whose coat colour from distance appear reddish-grey or roan, which looks like a case of incomplete dominance. But on close examination as, it would be seen that coat is composed of mixture of red and white hairs rather than hairs are intermediate between red and white. Thus, where the heterozygote F_1 exhibits a mixture of the phenotypic characters of both the homozygotes is called *codominance.*

Another example of codominance is *A, B, AB* and *O* blood groups found in human beings. The blood type is determined by the types of polysaccharides (polymers of sugar) found on the surface of red blood cells. *A* type of blood is synthesized by an enzyme coded by I^AI^A allele and *B* type, by the enzyme coded by the blood group $I^B I^B$ allele. The people of genotype I^AI^A are called *A* blood group and genotype $I^B I^B$ are called blood group *B*. The heterozygous I^AI^B are called *AB* blood group. The genotype I^AI^B of blood group *AB* is codominant because both the alleles are expressed.

Codominance genetic phenomenon is also found in human beings with blood group *M* and *N*. The persons with blood groups *M* and *N* produce antigenic substance *M* and *N* respectively on the surface of red blood cells. The person of genotype *MM* produces antigen *M* and *NN* individual produces antigen *N*, and heterozygous *MN* produces both *M* and *N* antigens instead of an intermediate substance.

Lethal alleles

Sickle-cell anaemia is most prevalent in blacks in United States and in Africa. The haemoglobin of most persons is chemically designated as haemoglobin *A*. The person showing sickle-cell anaemia is classed as haemoglobins Hb^S. The gene alleles responsible for haemoglobin types are Hb^A and Hb^S. The persons with sickle-cell anaemia have $Hb^S\,Hb^S$ genotypes. In the blood of such persons, the blood circulation in the capillaries is unable to carry oxygen and CO_2 to and fro from the tissues and die. The persons with genotypes Hb^AHb^A and Hb^AHb^S survive. But microscopic examination of blood $Hb^A\,Hb^S$ under low oxygen tension shows both normal and sickled erythrocytes, which is a case of codominance, but externally $Hb^A\,Hb^S$ individuals do not show symptoms of Hb^SHb^S individuals.

The genotypic and phenotypic characteristics of such individuals are:

- Hb^AHb^A—normal haemoglobin *A* carries oxygen and carbon dioxide (CO_2) to and fro from the tissues.
- Hb^AHb^S—sickle-cell trait (haemoglobins *A* and *S*), externally does not show symptoms of sickle-cell anaemia but shows sickling under reduced oxygen tension.
- Hb^SHb^S—sickle-cell anaemia (haemoglobin), shows sickling under normal oxygen tension.

The marriages of Hb^AHb^A (male) × Hb^AHb^S (female) and vice versa give rise to 1 Hb^AHb^A: 2 $Hb^A Hb^S$:1 Hb^SHb^S. The offspring (baby) with Hb^SHb^S genotypic constitution either die or get aborted. Therefore, the surviving babies would be in the proportion of 1:2 only.

In mouse, L. Cuenot early in the 20th century reported that yellow and black when mated, produced offsprings in 1:1 ratio. On this basis, he concluded that yellow-coloured mice were heterozygous because when yellow × yellow were crossed, they produced two types of offsprings—yellow and black in the ratio of 2:1 (Figure 6.1). The experimental results of Cuenot can be explained as:

Cross I		♀ A^ya (Yellow mouse)	×	aa (Black mouse) ♂		Parents (A)
	F_1	A^ya (1/2 yellow)	:	aa (1/2 black)		
			1:1			
Cross II		♀ Yellow mouse (A^ya)	×	Yellow mouse (A^ya)		Parents (B)
	F_1	1/4 (A^ya^y) Die	: 2:1	1/2 (A^ya)	1/4 (aa)	

Genotypic explanation	Yy (Yellow)	×	Yy (Yellow)		
	YY (Dies)	Yy (Yellow)	Yy (Yellow)	yy (Black)	2:1 (yellow : black)

Phenotypic explanation

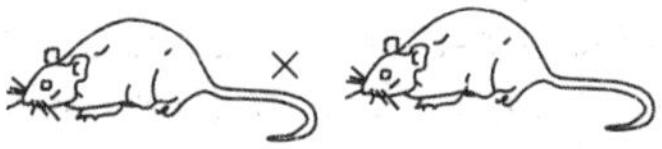

(Yellow) Yy × (Yellow) Yy

YY Dies	Yy (Yellow)	Yy (Yellow)	yy (Black)	2:1 (2 yellow:1 black)

Figure 6.1 Lethal factor in mouse. Homozygous YY dies.

Gene yellow is dominant over aa for coat colour, but lethal when present in homozygous state, and hence the carrier offspring die.

Lethal gene has also been reported in fowl. The creeper birds possess shortest and deformed legs and wings, which give them squatty appearance and creeping habit. The crosses between creeper × creeper produced two creeper and one normal birds. The homozygous creeper bird dies in the embryonic stage.

	♀ C^1C^2 (Creeper)	×	C^1C^2 (Creeper)	♂		Parent
F_1	$1C^1C$ (Normal)	:	$2C^1C^2$ (Creeper)		:	$1C^2C^2$ (Die)

The creeper gene allele retards the growth of the embryo with severe effects during limb bud formation.

In crop plants such as wheat, barley, and maize, seedling lethality have been observed. The green and 'albino' seedlings appear in 3 (green):1 (yellow) ratio. The albino seedlings die within 10 to 15 days after germination begin. They do not possess chlorophyll to manufacture carbohydrates. Behaviour of some lethal genes in the organisms are given in Table 6.1.

TABLE 6.1 Behaviours of lethal genes

Organism	*Phenotype*	*Dominant phenotype*	*Lethal*	F_1 *phenotypic ratio of heterozygote × heterozygote*		
				Before lethality	*After lethality*	*Age at death*
Human	Sickle-cell	Codominance	Recessive	1 normal 2 sickle trait 1 sickle anaemia	2 normal 1 sickle-cell trait	Adolescence or later
Corn	Albino	Recessive	Recessive	3 green, 1 albino	All green	10–14 days
Mouse	Yellow	Incomplete dominance	Recessive	Unknown	2 yellow, 1 black	Post-zygote
Fowl	Creeper	Incomplete	Recessive	Unknown	2 creeper, 1 normal	Early embryonic

To explain Mendel's findings of law of segregation, R.C. Punnett used a checkerboard called *Punnett Square*. The main philosophy behind Punnett's checkerboard was that the two pure lines, red and white flowers with genetic constitution *AA* and *aa* when crossed produced only one kind of germ cell (gamete) either *A* or *a*. The hybrid (*Aa*) produces two types of gametes, viz. *A* and *a*. Mendel felt that gametes would be equal in number and *A* and *a* would be pure gametes because they had come from pure lines. The F_1 hybrid (*Aa*) if self-fertilized, produces following genotypes and phenotypes, which are explained through the checkerboard, showing gametes and progeny produced.

Gametes:

♀ \ ♂	*A*	*a*
A	*AA*	*Aa*
a	*Aa*	*aa*

Phenotypic ratio—3:1 (red:white)
Genotypic ratio—1:2:1

The monogenic segregation could also be explained, using elementary algebra $(a + b)^2$, if substituted *a* for *AA* and *b* for *aa*. The expension of binominal $(a + b)^2$ would give $a^2 + 2ab + b^2$ genotype. If we substitute the probability 1/2 for *A* and 1/2 for *a*, we get $(1/2)^2 + 2(1/2 + 1/2) + (1/2)^2$, which gives us the 1/4 + 1/2 + 1/4 or 1:2:1 genotypic ratio for one pair of genic difference.

The original studies on inheritance were conducted in plants, but it has been intensively studied in *Drosophila melanogaster* (fruit fly), which is also called *veneger fly*. Thomas Morgan and his associates utilized *Drosophila* for detailed genetical studies at Columbia University. They separated brown (*bwbw*), scarlet (*stst*), dark red eye colour with black dark near centre. These flies are mutants of wild type (++) and brown eye colour (*bwbw*). *Drosophila* produces red-coloured F_1 offspring (+/*bw*). F_1 hybrid produces two types of gametes (+) and (*bw*). When two F_1 (+/*bw*) offspring male and female are inter-crossed, they produce following offspring in the ratio of 3 red:1 brown: eye flies in F_2 generation.

Gametes: ♀ \ ♂	+	*bw*
+	++ (Wild red-eyed)	+*bw* (Wild red-eyed)
bw	+*bw* (Wild red-eyed)	*bw* (Brown-eyed)

Thus, inheritance pattern is the same whether mutant is used as male or female.

LAW OF INDEPENDENT ASSORTMENT

In the first experiment on pea, Mendel studied the inheritance pattern of a single pair of allele. To determine whether the same pattern of inheritance can be observed in each pair of alleles, two and more pairs were taken at a time. So in the second experiment, Mendel crossed two true breeding pea lines, one line having round seed with yellow cotyledon and the other line having wrinkled seed with green cotyledon.

The F_1 seeds of this cross were round and yellow. They were hybrid for both the characters, in other words, they were dihybrid. It should be noted that in a single pair cross, round seed (*WW*) was dominant over wrinkled seed (*ww*), and yellow cotyledon (*YY*) was dominant over green cotyledon (*yy*). A cross between round yellow and wrinkled green pea seeds produced following four types of phenotypes in F_2:

Round yellow : 315
Wrinkled yellow : 101
Round green : 108
Wrinkled green : 32

Mendel observed that when the pairs of alternative characters were considered independently, the phenotypic ratios 423 (315 + 108) round: 133 (101 + 32) wrinkled seed (shape) and 416 (315 + 101) yellow: 140 (108 + 32) green seed (cotyledon) were in each case close to 3:1 ratio, as observed in the F_2 progenies of the monohybrid crosses, while studying the segregation pattern of a single pair of gene.

Applying the simple mathematical ratio, Mendel observed that four phenotypes in the F_2 of dihybrid cross were approximately in proportion of 9/16 round yellow, 3/16 wrinkled yellow, 3/16 round green and 1/16 wrinkled green. Similar results (9:3:3:1) were also obtained in the F_2 progeny of other dihybrid crosses.

The results of F_2 generations could be presented by construction of Punnett's square (checkerboard). Each parent produced four types of gametes in equal proportions. Hence, there would be 4 × 4 = 16 boxes in the square. Mendel's dihybrid results are shown in Punnett's (checkerboard) square as:

Gametes: ♀ \ ♂	*WY*	*Wy*	*wY*	*wy*
WY	*WWYY* (Round yellow)	*WWYy* (Round yellow)	*WwYY* (Round yellow)	*WwYy* (Round yellow)
Wy	*WWYy* (Round yellow)	*WWyy* (Round green)	*WwYy* (Round yellow)	*Wwyy* (Round green)
wY	*WwYY* (Round yellow)	*WwYy* (Round yellow)	*wwYY* (Wrinkled yellow)	*wwYy* (Wrinkled yellow)
wy	*WwYy* (Round yellow)	*Wwyy* (Round green)	*wwYy* (Wrinkled yellow)	*wwyy* (Wrinkled yellow)

The rule of independent assortment came into force when allelic pair for seed colour (yellow/green) was assorting independently of the allelic pair for the round/wrinkled seed shape. The F_1 formed four types of gametes as shown in checkerboard (*WY* : *Wy*:*wY* : *wy*). The gametes occur in equal frequencies. Regardless of which allele for seed shape, a gamete ends up with, it has 50:50 chance of getting either of allele for colour, i.e. the two genes are segregating and assorting independently. Mendel concluded that members of different pairs of alleles assort independently into gametes. This is Mendel's second law. The principle of independent assortment or the law of independent assortment could be defined as, *'segregation of the members of a pair of alleles is independent of the segregation of other pairs of alleles in the gametes formation.'*

Mendel's hypothesis for transmission of inherited characters that alleles of a gene segregate and undergo independent assortment has been explained in Punnett's square.

Test Cross

The purity of gametes could be ascertained by test-cross of the dihybrid (F_1) with recessive parent. In the present case, the dihybrid pea plant with round and yellow seed having genetic constitution *WwYy* will produce four types of gametes *WY*, *Wy*, *wY* and *wy* in equal frequencies whereas wrinkled green with genetic constitution *wwyy* will produce only one kind of gamete *wy*. If this hybrid (*WwYy*) is crossed with recessive parents (*wwyy*), four types of phenotypes consisting of round yellow, round green, wrinkled yellow and wrinkled green would appear in the ratio of 1:1:1:1. This ratio is the directly reflection of the kinds of gametes produced by the double heterozygote because no dominant alleles are recessive produced by the parent (*wwyy*) to obscure the results. In actual study, Mendel obtained 55 round yellow, 51 round green, 49 wrinkled yellow and 53 wrinkled green which were in the goodness of fit to Mendel's predicted ratio 1:1:1:1. The results were the same as in the reciprocal crosses. This result confirmed Mendel's hypothesis.

♀ *WwYy* × *wwyy* ♂ Parents
(Round yellow seeds) (Wrinkled green seeds)
Gametes—*Wy Wy, wY wy* *wy*

Test cross progeny—*WwYy* (55), *Wwyy* (51), *wwYy* (49), *wwyy* (52).

TRIHYBRID CROSS

All crosses of fertilizing plants and animals differ from each other in one or two or three or more pairs of gene alleles. Their mating in the natural population produces new recombinations of many genes. Genetic analysis of such crosses become cumbursome. But such complication could be simplified by resolving monohybrid crosses. For example, when two lines differing in three contrasting characters, viz. *AA*, *BB*, *CC* and *aa*, *bb*, *cc* are crossed, the F_1 of trihybrid obtained is actually combination of three single pair of genes operating together. Thus, (*AA* × *aa*) (*BB* × *bb*) and (*CC* × *cc*) could be combined together in the same cross as *AA*, *BB*, *CC* × *aa*, *bb*, *cc*, the F_1 of such crosses would be *Aa*, *Bb*, *Cc* which may produce 8 types of gametes $(2)^3$. When F_1 hybrid is selfed, in F_2 generation, 27 types of genotypes $(3)^3$ and 8 types of phenotypes $(2)^3$ are produced in a ratio of 27:9:9:9:3:3:3:1 = 64. This could be explained in Punnett's square (checkerboard) as well as in forked line methods.

Forked Line Method

The forked line method could be used to explain the results of a hybrid cross. In this method, visualize a trihybrid cross as three monohybrid crosses, i.e. *AA* × *aa*; *BB* × *bb* and *CC* × *cc*, operating together. If one member of each gene pair is dominant a 3:1 ratio would be predicted from each monohybrid cross. Since each pair of gene is independent of other, each monohybrid segregation is possible from three pairs of gene alleles. Sixty-four pairs of combination may be worked out by forked line system given as

3 round
- 3 yellow
  - 3 tall → 27 round, yellow, tall
  - 1 dwarf → 9 round, yellow, dwarf
- 1 green
  - 3 tall → 9 round, green, tall
  - 1 dwarf → 3 round, yellow, dwarf

1 wrinkled
- 3 yellow
  - 3 tall → 9 wrinkled yellow, tall
  - 1 dwarf → 3 wrinkled, yellow, dwarf
- 1 green
  - 3 tall → 3 wrinkled, yellow, tall
  - 1 dwarf → 1 wrinkled, green, dwarf

Bracket Method

Dominance is assumed in using the forked line method for determining the genotypic and phenotypic proportions of the individuals to be obtained when two or more pairs of genes are segregating. The genotypes can also be ascertained by the use of bracket or by the zygotic checkerboard, as given below.

Bracket method for combining segregation of individual genes

Gene A/a	*Gene B/b*	*Combination of A/a and B/b*	*Aa, Bb, Cc*
3/4 *A/-*	3/4 *B/-*	9/16 *A-B*	27/64 *A-B-C*
			9/64 *A-B-cc*
	1/4 *b/b*	3/16 *A-bb*	9/64 *A-bb-C*
			3/64 *A-bb-cc*
1/4 *a/a*	3/4 *A*	3/16 *aa-B*	9/64 *aa-B-C*
			3/64 *aa-B-cc*
	1/4 *bb*	1/16 *aa-bb*	3/64 *aa-bb-C*
			1/64 *aa-bb-cc*

Number of Genotypes and Phenotypes with n-gene Segregation

Sometimes, a researcher try to know many genotypes and phenotypes he/she can obtain from the segregation of any known number of genes and the minimum population needed to obtain all the possible recombination of all the genes present. This information for one to six segregating genes is given in Table 6.2. If there is complete dominance, the number of phenotypes (2n) is always lower than the number of genotypes (3n), i.e. the number of genotypes is greater than (3n) while the population is greatest of all (4n). However, if there is no dominance, there will be as many phenotypes as genotypes. This continues to increase.

TABLE 6.2 Number of gametes, phenotypes, genotypes and minimum population size for 1 – n genes segregating (dominance assumed)

Number of gene	*Kind of gamete*	*Number of phenotype*	*Number of genotype*	*Population size*
1	2	2	3	4
2	4	4	9	16
3	8	8	27	64
4	16	16	81	264
5	32	32	243	1024
6	64	64	729	4096
n	2^n	2^n	3^n	4^n

This will be noticed that all kinds of gametes are a multiple of 2, i.e. 2^n; the number of F_2 genotype is a multiple of 3, i.e. 3^n and the number of phenotypes is multiple of 2^n, where there is a complete dominance. This method could be used for predicting the results of any number of independent gene pair.

The population of homozygous recessive offspring, the number of phenotypes with complete dominance and the number of different genotypes or phenotypes, if there is no dominance from a multiple hybrid self-fertilizing F_2 population, could be predicted as given below:

Multihybrid with self-fertilization

S.No.		*Monohybrid* (n = 1)	*Dihybrid* (n = 2)	*Trihybrid* (n = 3)	*General rule*
1.	F_1 gametes	2	4	8	2^n
2.	Population: homozygous, recessive among F_2 individuals	1/4	1/16	1/64	$1/(2^n)^2$
3.	Number of different F_2 phenotypes with complete dominance	2	4	8	2^n
4.	Number of different genotypes or phenotypes, if there is no dominance	3	9	27	3^n

PURITY OF MENDEL'S FACTORS

One of the most fundamental principles about genes is that these are considered pure and constant. One of the convincing evidences is that no recessive genes show modification even when kept for many generations in the heterozygous organism of dominant phenotype. In *Drosophila* certain stock (CLB) have been kept heterozygous for both recessive and dominant genes since 1919. During this time period, there is no evidence of 'dilution' or blending of genes.

Another direct observation of gene purity was in an experiment carried out in guinea pigs by Castle and Phillips in the year 1909 who transplanted an ovary from a black guinea pig to an albino female whose ovary had been removed and thus female mated with an albino male. The resultant offsprings were all black. In absence of mutation, the genotype is therefore constant. Albino is a recessive character in the mammals. An albino guinea pig if mated with an albino male produces albino offsprings only. Thus this experiment cleared that genes are not affected by the genotypic or phenotypic background.

REASONS OF MENDEL'S SUCCESS

The main reasons of Mendel's success were his careful selection of garden pea, which is a self-fertilizing plant. He was well conversant with its floral biology and his intelligent ways of manning his experience. Before making the crosses, he purified the pea cultures raising two years to make them true breeding. He made crosses between lines differing in contrasting characters. It should be noted that every character may have a contrasting character in other type/line. This character contrast is important in the study of mode of inheritance in the genetic studies to classify the progenies.

Mendel selected seven characters in his study, which were fortunately located on seven pairs of chromosomes in garden pea as a result showed independent assortment. In the hybrid and in the F_2 and F_3 generations, he studied one character at one time.

He considered the plant as a composite structure made up of a large number of unit characters. There were few other characters in which the parents and progenies could not show resemblance. Mendel did not study them all conjointly with main character pair he studied. Because of this reason some people believe that Mendel did not report the inheritance of trait that did not show independent assortment.

He counted the number of plants showing contrasting characters such as tall and dwarf in F_2 generation, classed them, harvested and threshed each F_2 plant separately and raised plant progeny rows in F_3 generation to confirm the results of F_2. He maintained pedigree records of each plant, counted the progenies falling into different character groups such as round and wrinkled. He introduced simple mathematics such as ratio, which was responsible for his success.

He believed that characters are transmitted from parents to the progeny through factors (elements) or genes which are located in linear orders on the chromosomes found in the nucleus (Figure 6.2).

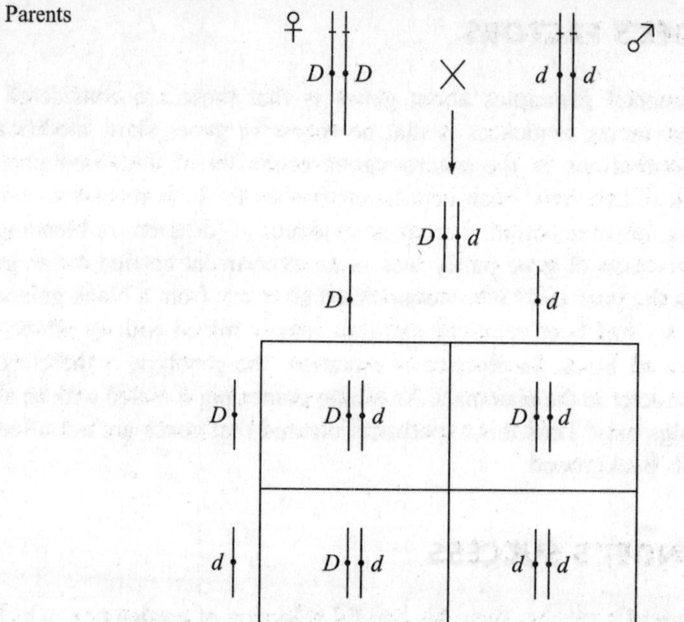

Figure 6.2 Location of genes on the chromosome.

HYPOTHESIS TESTING

The statistical parameters are useful to the geneticists in the testing of hypothesis. To determine whether to reject or support a hypothesis is based on how close the data are to the prediction of the hypothesis. For example, the observed data, 307: 101 are closed to the ratio 3: 1. But we cannot answer with an absolute 'yes'.

To know what kind of variations exist, this could be determined by calculating a sampling distribution, the frequencies with which various possible events could occur in an experiment. For instance, in a cross between and tall and dwarf plants, tall and dwarf plants segregated in a ratio of 3 :1 is expected in F_2 population. Our hypothesis of 3 :1 ratio is based on assumption that plant height is controlled by one pair of genes. The probability of getting three tall and one dwarf plants could be calculated by using the following formula for the term of the bionominal expansion.

$$p = \frac{4!}{3!!!} = (3/4)^3\ (1/4)^1$$

$$= 108/256$$

$$= 0.42$$

The hypothesis against which the data are tested is referred as *null hypothesis*. Null hypothesis involves testing the assumption that there is no difference between the observed and expected values. If the *null hypothesis* does not reject it means, data are consistent with it. But it does not mean that hypothesis is proved. However, if the hypothesis is to be rejected as 1:1 ratio the other choice is not to reject the alternative hypothesis that there is difference between observed and expected values. The data should be tested against some other hypothesis, e.g. 3:1.

It is a convention to use the probability 0.05 level as a cut-off limit for rejecting a hypothesis. It is called the level of significance. The other level of significance is 0.01 probability. If a χ^2 value is greater than the critical value in the probability table at 0.01 level, it means the data depart in a highly significant manner from the null hypothesis.

The χ^2 (chi-square) test is a very useful tool for obtaining an objective approximation of *goodness of fit*, but it is reliable only when the observed value in any class is more than five numbers of individuals.

Goodness of Fit and Chi-square (χ^2) Ratio

In the genetical experiments one does not expect always the results would be exactly to his/her expectations. The deviations could not be ruled out. But how much deviations from the calculated probabilities can be accepted is likely due to purely a chance. Even Mendel's results were not exactly to his expected ratio though they too were very close to 3:1 ratio. Mathematical tools are needed to determine the *goodness of fit*. Such mathematical parameter is called *chi-square* (χ^2) *test*. The formula for calculating *chi-square* is given below:

$$\chi^2 = \sum \frac{(O - E)^2}{E}$$

where

Σ is the Greek letter for summation of values of all the classes
O is the observed values
E is the expected values

Both O and E should be calculated in actual numbers, but not in percentage. Each class possesses at least five individuals.

Let us say whether the data given in Table 6.1 is fit in the 1:2:1 expectation or not.

TABLE 6.1 Hypothetical data to calculate χ^2 value

Class	*Observed* (O)	*Expected* (E)	*Deviation* ($O - E$)	*Square deviation* $(O - E)^2$	$\frac{(O-E)^2}{E}$
HH	12.0	12.5	−0.5	0.25	0.02
HT	27.0	25.0	−12.0	4.0	0.16
CT	11.0	12.5	−1.50	2.25	0.18
	50.0	50.0			$\chi^2 = 0.36$

With χ^2 value 0.36, one can question as 'how often' by chance alone will a deviation of the order of 0.36 may arise, where a 1:2:1 ratio is expected for segregation of a character or how often by chance one can expect a value of $\chi^2 = 0.36$. If the probability is high only then validity of the assumption could be acceptable under which expectancy was determined as well as the hypothesis that the observed deviation was due to chance for testing the *goodness of fit* to proposed hypothesis. The chi-square (χ^2) table could be consulted.

To use this table, it is necessary to know the 'degree of freedom' operating in a particular case. This is one less than the number of phenotypic classes involved in calculations of χ^2 values. For example, in Table 6.1, the number of phenotypic classes are three, therefore, the degree of freedom would be two. It will be seen that in Chi-square (χ^2) table, the χ^2 value = 0.36 lies between 0.03 and 0.446 at two degree in the probability column, it is seen that χ^2 = 0.36 corresponds to a p value of between 0.95 and 0.80.

EXAMPLE 1: Chi-square analysis of Mendel's experiment (3:1)

	Tall plant	*Dwarf plant*	*Total*
Observed number (O) ratio	787	277	1064
Expected ratio	3/4	1/4	
Expected number	798	266	1064
$O - E$	−11	+11	
$(O - E)^2$	121	121	
$\frac{(O-E)^2}{E}$	0.15	0.45	

Therefore, χ^2 = 0.60

EXAMPLE 2: Chi-square analysis of Mendel's experiment (1:1)

	Tall plant	*Dwarf plant*	*Total*
Observed number (O)	787	277	1064
Expected ratio	1/2	1/2	
Expected number (E)	582	532	1064

$O - E$	255	−255
$(O - E)^2$	65.025	05.025
$\frac{(O - E)^2}{E}$	122.23	122.23

Therefore, $\chi^2 = 244.45$
Suppose Mendel had thought of 1:1 ratio, he could have arrived at a wrong conclusion.

REVIEW QUESTIONS

1. On the basis of Mendelian laws of inheritance and observation, find the results coming out of the following crosses:
   (a) Tall × dwarf
   (b) Progeny of selfed F_1
   (c) Progeny of F_1 × tall variety
   (d) Progeny of F_1 × original dwarf parent
2. Mendel crossed pea plants having round and wrinkled seeds. Out of total 7324 F_2 seeds, 5474 were round and 1850 were wrinkled. Using symbols ensure for genes governing round and wrinkle seeds, symbolize
   (a) The original parental cross
   (b) The gametes
   (c) The F_1 progeny
   (d) Write F_1 × F_1 (selfed and symbolize the gametes)

   Give F_2 result after selfing of F_1, in the terms of phenotypes, genotypes, genotypic frequency and phenotypic ratio.
3. A gray-coloured mouse was crossed with a white-coloured (albino) mouse. In the first generation, (F_1) all were gray. By crossing F_1 gray-coloured male and female mice, 198 gray and 72 white (albino) mice were obtained. Propose a hypothesis to explain this result. On the basis of hypothesis, draw a diagram and compare the observed and expected results.
4. In pigeon, the checkered pattern is governed by a dominant gene *C* and a plain exterior by recessive allele *c*. Red colour is governed by dominant gene *B* and brown by a recessive gene *b*. Make a cross between homozygous, checked red pigeon and plain brown pigeon and summarize the F_2 results.
5. If in a rabbit, black fur is governed by dominant gene *B* and brown by recessive *b*. Normal fur length is determined by a dominant gene *R* and short (rex) by a recessive allele *r*. Make a diagram and summarize the F_1 and F_2 results of cross between homozygous black rabbit with normal length fur and a brown short (rex) fur rabbit. What proportion of normal black F_2 rabbits from the preceding cross may be expected to be homozygous for both gene pairs. Diagrammatically, summarize the test cross between the F_1 and fully recessive brown rex parents.

6. In a F_2 generation of a certain tomato cross, 3672 fruits were fed and 1175 were yellow
   (a) What is expected ratio?
   (b) Are discrepancies between observed and expected ratios are significant?

   In the same cross, 671 plants with green foliage and 569 with yellow foliage were observed. This was back-crossed and hypothetical ratio was 1:1. Test with χ^2 and explain the result.
7. Following are the Mendel's results with the hypotheses they are fitted. Test goodness of fit and indicate whether each is significantly different from others.

Cross	*Results*	*Hypothesis*
Round × Wrinkled seeds (F_2)	572:185	3:1
Violet × White flowers (F_2)	705:224	3:1
Green × Yellow pods (F_2)	428:152	3:1
Round yellow (F_1) × Wrinkled green seeds	31:26:26:27	1:1:1:1
Round yellow (F_1) × Wrinkled green seed	24:25:22:27	1:1:1:1

REFERENCES

Bateson, W. (1909): *Mendel's Principles of Heredity*, Cambridge University Press.

Burn, G.W. and P.J. Brunito (1989): *Science of Genetics*, Macmillan Publishing Company, New York.

Chandrasekharan, S.N. and S.V. Parthasarthy (1953): *Cytogenetics and Plant Breeding*, Vardhanocharya and Co., Madras.

Correns, C. (1909): *Birth of Genetics* (English translation), Published as a supplement to 1950 Volume of Genetics.

Dunn, L.S. (1951): *Genetics in 20th Century,* Macmillan, New York.

Dunn, L.S. (1965): *A Short History of Genetics,* McGraw-Hill, New York.

Ford, E.B. (1938): *The Study of Heredity*, Thomton Butterworth, London.

Gardner, E.J., M.J. Simmons and D.P. Shustad (1991): *Principles of Genetics*, John Wiley & Sons, New York.

Hartl, D.L. (1991): *Basic Genetics*, Jones and Bartlett, Boston.

Olvy, R.C. (1966): *Origin of Mendelism*, Shocken Books, New York.

Sturtevant, A.H. (1965): *A History of Genetics*, Harper and Row, New York.

CHAPTER

7 Exceptions to Mendelian Principles of Inheritance

In Chapter 6, we have discussed that seven pairs of genes, controlling the seven characters in the garden pea, studied by Mendel, segregated independently. This was mere a chance that seven pairs of genes studied by Mendel were located on seven pairs of chromosomes in the pea, which assorted independently in F_2 generation (second filial generation). But genes do not function always in isolation. The phenotype of an organism is the result of action of the individual gene and interaction of the entire gene complex as well as interaction of the environmental factors in which organism grows. Thus, there are several exceptions to the independent assortment of the genes (9:3:3:1), which modify the 9:3:3:1 ratio. Some of them identified so far are discussed in this Chapter. These are:

1. Incomplete dominance
2. Complementary genes (9:7)
3. Supplementary genes (9:3:4)
4. Dominant epistatic genes (12:3:1)
5. Recessive epistatic genes (9:3:4)
6. Inhibitory genes (13:3)
7. Duplicate genes (15:1)
8. Polymeric genes (9:6:1)
9. Xenia
10. Cytoplasmic behaviour
11. Pleiotropic genes

INCOMPLETE DOMINANCE

One of the important reasons of Mendel's success was that in all the seven pairs of characters he studied, there was clear-cut dominance of one gene–allele over the other, as a result, there were independent assortment of genes giving 9:3:3:1 ratio. However, this is not always true. There are exceptions to this rule also.

In tomato, two pairs of alleles located on different pairs of chromosomes are:

- *DD* (tall plant): h^1h^1 (hairless stem)

- *dd* (dwarf plant): h^1h^2 (scattered short hairs)
- h^2h^2 (very hairy stem)

Crossing of $D_1D_1\, h^1h^1$ and $d_1d_1\, h^2h^2$ individuals produces $Dd\, h^1h^2$ hybrid in F_1 generation (tall plant with scattered short hairs), which when self-crossed produced the following progenies:

- 3/16 tall hairy
- 6/16 scattered hairs
- 3/16 tall very hairy
- 1/16 dwarf hairless
- 2/16 dwarf scattered hairs
- 1/16 dwarf very hairy

In this case, tall and dwarf plants segregated into 3:1 ratio, where tall was dominant over dwarf plant height. But hairless, scattered hairs and very hairy showed incomplete dominance. As a result, in case of monohybrid cross, they segregated in 1:2:1 ratio. In a dihybrid cross, they segregated in 3:6:3:1:2:1 phenotypic proportion. This is because F_1 hybrid ($D_1d_1 \times Dd$) produces two phenotypic classes in F_2 generation. The $h^1h^2 \times h^1h^2$ F_1 hybrid in absence of dominance produces three phenotypic classes of offspring, and the cross between $D_1d_1\, h^1h^2$ and $D^1d^1\, h^1h^2$ produces $2 \times 3 = 6$ classes of phenotypes. This is one of the exceptions to the Mendelian theory of dominance and independent assortment of genes in a dihybrid cross.

Novel Gene Interaction: (Interallelic Gene Interaction)

A classical example of nonallelic gene interaction was reported by Willium Bateson and R.C. Punnett by crossing two different breeds of chicken. The domestic breeds of chicken possess, rose, pea, and single type of combs (Figure 7.1). The rose type comb is dominant over single type. Similarly, the comb type pea is monogenic dominant over single type, however, when rose and pea comb types of chicken were crossed, the F_1 hybrid was of walnut type comb. If genic constitution of rose is symbolized as *RRpp*, pea as *PPrr* and single as *pprr*, in F_2 generation chicken with walnut comb appeared, where two nonallelic dominant gene alleles *R* and *P* were brought together. Individually, two nonallelic genes *RR* and *PP* acted independently and produced two different types of combs. This is similar to the ways in which subdominant alleles act.

	♀	*ppRR* (Rose)	×	*pprr* (Single) ♂	Parents
F_1			*Rrpp* (Rose)		
F_2			3 rose: 1 single		
	♀	*PPrr* (Pea)	×	*pprr* (Single) ♂	Parents
F_1			*Pprr* (Pea)		
F_2			3 pea: 1 single		
	♀	*RRpp* (Rose)	×	*PPrr* (Pea)	Parents
F_1			*RrPp* F_1 (Walnut)		

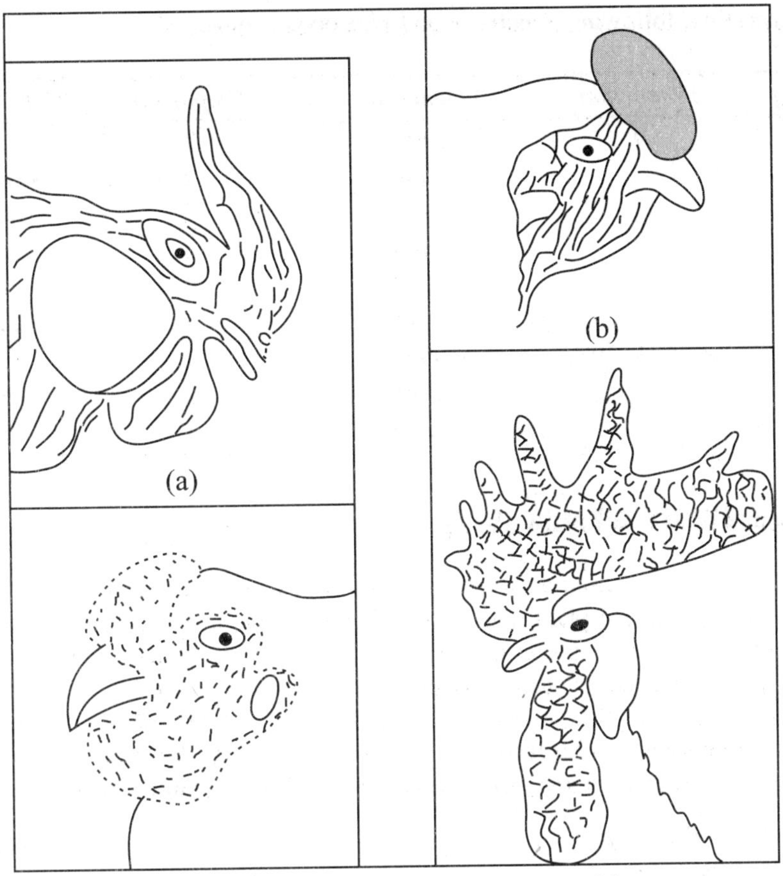

(c) (d)

Figure 7.1* Nonallelic gene interaction in chicken—(a) rose, (b) pea, (c) walnut, and (d) single.

F_2 behaviour of F_1 hybrid is detailed below:

Gametes: ♀ \ ♂	*RP*	*Rp*	*rP*	*rp*
RP	*RRPP*	*RRPp*	*RrPP*	*RrPp*
Rp	*RRPp*	*RRpp*	*RrPp*	*Rrpp*
rP	*RrPP*	*RrPp*	*rrPP*	*rrPp*
rp	*RrPp*	*Rrpp*	*rrPp*	*rrpp*

* Figure 7.1 is based on Bateson and Punnett's work in the first decade of 20th century, i.e. about 100 years ago. They might have died about 50 years ago.

In F_2 generation, following genotypes and phenotypes appeared:

Phenotype	*Proportion*	*Genotype*	*Proportion*	*F_3 behaviour*
Walnut	9	*PPRR*	1	True breed
		PPRr	2	3 walnut: 1 pea
		PpRR	2	3 walnut: 1 rose
		PpRr	4	Segregated like F_1
Rose	3	*ppRR*	1	True breed
		ppRr	1	3 rose:1 single
Pea	3	*PPrr*	1	True breed
		Pprr	2	3 pea:1 single
Single		*pprr*	1	True breed

COMPLEMENTARY GENES (9:7)

Bateson and Punnett came across a case in sweat pea where two lines having white flowers when crossed together gave hybrid plant with purple flower colour in F_1 generation. On selfing, F_1 hybrid plants segregated in the proportion of 9 purple and 7 white flowers in F_2 generation. This hypothesis could be explained that two pairs of nonallelic dominant genes interacted in complementary manner in the same way as mixing of two colourless chemicals—solution phenolphthalein (colourless) when added in KOH (colourless) produces pink-coloured solution. The mechanism of complementary gene interaction could be explained as follows:

♀ $P^1P^1\,p^2p^2$ (White flower)	×	$p^1p^1\,P^2P^2$ (White flower) ♂	Parents
F_1		$P^1p^1\,P^2p^2$ (Purple flower)		

F_2 behaviour of F_1 hybrid is detailed below.

Gametes: ♀ \ ♂	P^1P^2	P^1p^2	P^2p^1	p^1p^2
P^1P^2	$P^1P^1P^2P^2$	$P^1P^1P^2p^2$	$P^1p^1P^2p^2$	$P^1p^1P^2p^2$
P^1p^2	$P^1P^1P^2p^2$	$P^1P^1p^2p^2$	$P^1p^1P^2p^2$	$P^1p^1p^2p^2$
p^1p^2	$P^1p^1P^2p^2$	$P^1p^1P^2p^2$	$p^1p^1P^2P^2$	$p^1p^1P^2p^2$
p^1p^2	$P^1p^1P^2p^2$	$p^1p^1p^2p^2$	$p^1p^1P^2p^2$	$p^1p^1p^2p^2$

In F_2 generation, following phenotypes and genotypes appeared:

Phenotype	*Proportion*	*Genotype*	*Proportion*	F_3 *behaviour*
Purple	9	$p^1p^1 p^2p^2$	1	True breed
		$P^1p^1 P^2P^2$	2	3 purple
		$P^1P^1 P^2p^2$	2	3 purple:1 white
		$P^1p^1 P^2p^2$	4	9 purple:7 white
White	7	$P^1P^1 p^2p^2$	1	True breed—white
		$p^1p^1 P^2p^2$	2	3 purple:1 white
		$p^1p^1 P^2P^2$	1	True breed—white
		$p^1p^1 P^2p^2$	2	3 purple:1 white
		$p^1p^1 p^2p^2$	1	True breed—white

Similar results were obtained in many other plants. For example, in rice, crosses of two lines with grey grain types resulted F_1 hybrid with red grain rice plant which on selfing in F_2 generation gave rise to rice plants with red and grey grain types in 9:7 proportion. This is another example of nonallelic gene interaction with complementary gene effect. The 9:7 ratio is the modification of 9:3:3:1 ratio because in case of complementary gene action, last three classes of phenotypes become indistinguishable from one and other. This is also called *Dominant epistasis*, because dominance is due to interaction between different alleles of same gene.

Biochemical Basis of Dominant Epistasis

In clover (*Trifolium repens*), some strains test high in hydrocyanin (HCN) content whereas others test negatively for this substance. The strains positive to HCN are more vigorous in growth but do not harm the cattle eating the fodder. The crosses of positive and negative strains result F_1 with positive test for HCN. In F_2, 342 positive and 265 negative individuals appeared which were near to 9:7 ratio. This is biochemical basis of dominant epistasis.

The formation of HCN involves biochemical reaction in the plant cell that may be represented as

$$\text{Precursor} \rightarrow \text{Cyanogenic glucoside} \rightarrow \text{Enzyme} \rightarrow \text{Gene} \rightarrow \text{HCN}$$

Suppose two pairs of genes *AA* and *BB* are responsible for production of HCN in white clover.

	♀ *AABB* (HCN positive)	×	*aabb* (HCN negative)	♂	Parents
F_1		*AaBb* (HCN positive)			

In F_2 generation following phenotypes and genotypes appeared:

Phenotype		*Proportion*	*Genotype*	*Proportion*
9	HCN positive	9	*AABB*	1 ⎤
			AaBB	2 ⎥ 9
			AABb	2 ⎥
			AaBb	4 ⎦

(Contd.)

Phenotype		Proportion	Genotype	Proportion	
7	HCN negative	3	*aaBB*	1	7
			aaBb	2	
	HCN negative	3	*AAbb*	1	
			Aabb	2	
	HCN negative	1	*aabb*	1	

Two Pairs Dominant Gene Interaction

Complementary gene interaction has also been reported in maize by East and Hayes for aleurome colour. But instead of two, three gene pairs complementary gene effect was shown by Emerson. Genes *CCrr* and *RRcc* independently show white flowers but when they were crossed, F_1 (*CcRr*) showed coloured flowers. In F_2 generation, coloured and white flower plants appeared in 9:7 ratio. An additional gene *AA* was responsible for anthocyanin production along with gene *CCrr*, *Rrcc* and AA_1, as they interact together in a complementary fashion to produce colour when all the three are present.

Two Pairs Recessive Gene Interaction

In the above said example, two nonallelic dominant genes P^1P^1 and P^2P^2 interacted and gave a new type of phenotype, purple colour flower. However, in *Drosophila melanogaster*, interaction of two recessive gene alleles *br* (brown eye) and *st* (scarlet eye) produces a white-eyed fly (*brbrstst*). These genes are located on chromosomes 2 and 3 of *Drosophila*. The fly with *br*/*br*, genic constitution is brown-eyed, but if it has one dominant allele at *br* locus *st*+/– shows pinkish eye. Similarly, a scarlet fly homozygous for *st st* gene, if possesses any one allele of two dominant gene alleles *br*+/– possess bright scarlet eye. The two recessive gene alleles *br*/*br*, *st*/*st* possess white eye which is distinguishable from fly containing *br* gene in chromosomic 1. Thus, this is a good case of the interaction of two pairs of recessive genes/alleles located at two different loci to produce quite distinct phenotypic effect.

SUPPLEMENTARY GENES (9:3:4)

In this case, it was observed that two gene pairs *P* and *Q* interact and give rise to a third phenotype different from the parental stocks, but *Q* has no effect of its own if dominant allele *P* is absent. The factorial scheme is as under:

♀	*PPqq*		*ppQQ* ♂	Parents
	(Blackish-purple)	×	(Brown)	
F_1		*PpQq*		
		(Reddish-purple)		

F_1 produces four types of gametes *PQ*, *Pq*, *pQ*, *pq* on both male and female plants. In F_2 generation, the segregation pattern of genes *P* and *Q* is given in the checkerboard.

Gametes: ♀ \ ♂	*PQ*	*Pq*	*pQ*	*pq*
PQ	*PPQQ*	*PPQq*	*PpQQ*	*PpQq*
Pq	*PPQq*	*PPqq*	*PpQq*	*Ppqq*
pQ	*PpQQ*	*PpQq*	*ppQQ*	*ppQq*
pq	*PpQq*	*Ppqq*	*ppQq*	*ppqq*

In F_2 generation, following phenotypes and genotypes appeared:

Phenotype	*Proportion*	*Genotype*	*Proportion*	*F_3 behaviour*
Reddish-purple	9	*PPQQ*	1	True breed
		PPQq	2	3 reddish-purple: 1 blackish-purple
		PpQQ	2	3 reddish-purple 1 brown
		PpQq	4	Behaved like F_1 brown
Blackish-purple	3	*PPqq*	1	1 True breed (blackish-purple)
		Ppqq	2	3 blackish-purple: 1 brown
Brown	4	*ppQQ*	1	1 True breed (brown)
		ppQq	2	*Q–q* do segregates but all brown
		ppQQ	1	True breed (brown)

Thus, F_2 showed 9:3:4 phenotypic ratio. Though brown-glumed parents do possess dominant gene *QQ* but in absence of gene *PP*, they do not show any effect as a result genotypes *ppQQ* and *ppQq* were phenotypically brown-glumed. However, genotypes *ppQQ* and *ppQq* carry gene *Q* which interact with dominant gene *PP*, when it is present and produces reddish-purple flower.

The phenotypic ratio 9:3:4 is modification of 9:3:3:1 where last two genotypes were phenotypically indistinguishable and thus they were with brown glume. Here two pairs of genes produce the same character; where gene pair *PP* has visible effect but *QQ* does not. But when two factor pairs (*P* – *Q*) come together, give rise to a new phenotype. Here *QQ* supplement the action of gene *PP*, and hence called *supplementary gene interaction*. This is also called **recessive epistasis**.

RECESSIVE EPISTATIC GENES (9:3:4)

Thc laboratory mice are of different colour patterns, viz. *agouti*, is characterized by colour bounded hairs in which the part nearest the skin is grey, then a yellow band, and finally, the distal part is either black or brown. There are two other colour mice—albino and solid black.

The cross between black and albino gives rise uniformally *agouti* in F_1. When *agouti* female is crossed with *agouti* male (inbreed), the F_2 population consists of three classes of individuals, *agouti*, *black* and *albino* in 9:3:4 ratio. This is a modification of 9:3:3:1 ratio in which 1/16 class and one of the 3/16 classes are indistinguishable. The cross in mouse could be factorized as:

♀ *aaCC* (Black)	×	*AAcc* (Albino) ♂		Parents
F_1	*AaCc* (Agouti)			

On inbreeding, F_1 produces four types of gametes—*AC*, *Ac*, *aC*, *ac*. Checkerboard shows F_2 behaviour:

Gametes: ♀ \ ♂	*AC*	*Ac*	*aC*	*ac*
AC	*AACC*	*AACc*	*AaCC*	*AaCc*
Ac	*AACc*	*AAcc*	*AaCc*	*Aacc*
aC	*AaCC*	*AaCc*	*aaCC*	*aaCc*
ac	*AaCc*	*Aacc*	*aaCc*	*aacc*

In F_2 generation, following phenotypes and genotypes appeared:

Phenotype	*Proportion*	*Genotype*	*Proportion*
Agouti	9	*AACC*	1
		AACC	2
		AaCC	2
		AaCc	4
Black	3	*aaCC*	1
		aaCc	2
Albino	3	*AAcc*	1 *c* masks the expression of gene *A*
		Aacc	2
	1	*aacc*	1

In this case allele *c*, which is recessive to its own allele *C* masks the effect of gene *A* or *aa* so that *AAcc*, *Aacc*, *aacc* individuals are albino. Such a gene, which masks the effect of one or both members of another pair of alleles is said to be *Epistatic*, and the masked gene may be termed *hypostatic*. Here *c* is epistatic to *AA* and *aa*. This is the case of **recessive epistasis** because recessive gene is epistatic to another gene. Epistasis is quite different from dominance. In epistasis, both alleles of one pair of gene mask the gene alleles of another pair. But in dominance, one allele of a pair of gene masks the effect of other allele of the same gene.

DOMINANT EPISTATIC GENES (12:3:1)

In epistasis, expression of one dominant gene is masked by another dominant gene. In *Sorghum*, red, white pearly and chalky seed coat colour of grains are found. Red (*RR*) is dominant over white pearly (*rr*) and white pearly (*WW*) is dominant over chalky (*ww*). When red grain type is crossed with white pearly grain type, F_1 is red. But in F_2 generation, three classes of phenotypes, red, white pearly and white chalky appeared in 12:3:1 ratio. The factorial representation of this cross is:

	♀ *RRww* (Red)	×	*WWrr* ♂ (White pearly)	Parents
F_1		*RrWw* (Red)		

On selfing, F_1 formed four types of gametes: *RW*, *Rw*, *rW*, *rw*
The segregation pattern of gene is shown in checkerboard.

Gametes: ♀ \ ♂	*RW*	*Rw*	*rW*	*rw*
RW	*RRWW*	*RRWw*	*RrWW*	*RrWw*
Rw	*RRWw*	*RRww*	*RrWw*	*Rrww*
rW	*RrWW*	*RrWw*	*rrWW*	*rrWw*
rw	*RrWw*	*Rrww*	*rrWw*	*rrww*

In F_2 generation, following phenotypes and genotypes appeared:

Phenotype	*Proportion*	*Genotype*	*Proportion*
Red	12	*RRWW*	1
		RrWW	2
		RRWw	2
		RrWw	4
		RRww	1
		Rrww	2
White pearly	3	*rrWW*	1
		rrWw	2
White chalky	1	*rrww*	1

In this case, gene *RR* masks the expression of gene *WW*. However, in absence of *RR* gene, *WW* shows its phenotypic expression. Phenotypic chalky grain appears, when both dominant gene pairs red (*RR*) and white pearly (*WW*) are absent.

INHIBITORY GENES (13:3)

This is also a case of two pairs of nonallelic gene interaction, when one of the dominant gene pairs suppresses or inhibits the expression of other dominant gene. This may be explained, taking pigmentation in rice plant. In a cross between purple and green plant, F_1 was green. In F_2 generation, there appeared 13 green: 3 purple plants. This may be explained as under:

	♀ *IIpp* (Green)	×	*iiPP* ♂ (Purple)	Parents
F_1		*IiPp* (Green)		

F_2 behaviour of F_1 hybrids:

Gametes: ♀ \ ♂	*IP*	*Ip*	*iP*	*ip*
IP	*IIPP*	*IIPp*	*IiPP*	*IiPp*
Ip	*IIPp*	*IIpp*	*IiPp*	*Iipp*
iP	*IiPP*	*IiPp*	*iiPP*	*iiPp*
ip	*IiPp*	*Iipp*	*iiPp*	*iipp*

In F_2 generation, following phenotypes and genotypes appeared:

Phenotype	*Proportion*	*Genotype*	*Proportion*	
Green	13	*IIPP*	1	13
		IiPP	2	
		IIPp	2	
		IiPp	4	
		IIpp	1	
		Iipp	2	
		iipp	1	
Purple	3	*iiPP*	1	3
		iiPp	2	

In this case, visible effect of gene *PP* is obliterated by inhibitory effect of factor II. But II itself does not show its own effect in absence of *PP*. This is called *inhibitory nonallelic gene interaction*. This is also called dominant and recessive epistasis where *I* suppresses the expression of *P* and *p* and *P* suppresses the expression of *I* and *i*.

DUPLICATE GENES (15:1)

In this type of nonallelic gene interaction, both the dominant gene pairs show identical effect for the expression of character. And presence of any one of the dominant alleles gives full type effect.

This is generally the case with tetra or hexploid types. An example could be presented taking awned and awnless characters in barley and rice. In a cross between awned and awnless type, F_1 was awned and in F_2 generation, awned and awnless appeared in 15:1 ratio. If genes for awned and awnless could be symbolized as A^1A^1, A^2A^2 awned and a^1a^1, a^2a^2 awnless following factorial design could be represented for the segregation pattern in F_2 generation.

	♀ $A^1A^1A^2A^2$		$a^1a^1a^2a^2$ ♂	Parents
	↑		↑	
	(Awned)	×	(Awnless)	
F_1		$A^1a^1A^2a^2$ (Awned)		

In F_2 generation the proportion of phenotypes and genotypes would be as:

Gametes: ♀ \ ♂	A^1A^2	A^1a^2	a^1A^2	a^1a^2
A^1A^2	$A^1A^1A^2A^2$	$A^1A^1A^2a^2$	$A^1a^1A^2A^2$	$A^1a^1A^2a^2$
A^1a^2	$A^1A^1A^2a^2$	$A^1A^1a^2a^2$	$A^1a^1A^2a^2$	$A^1a^1a^2a^2$
a^1A^2	$A^1a^1A^2A^2$	$A^1a^1A^2a^2$	$a^1a^1A^2A^2$	$a^1a^1A^2a^2$
a^1a^2	$A^1a^1A^2a^2$	$A^1a^1a^2a^2$	$a^1a^1A^2a^2$	$a^1a^1a^2a^2$

Phenotype	*Proportion*	*Genotype*	*Proportion*
Awned	15	$A^1A^1A^2A^2$	1
		$A^1A^1A^2a^2$	2
		$A^1a^1A^2A^2$	2
		$A^1a^1A^2a^2$	4
		$A^1A^1a^2a^2$	1
		$a^1a^1A^2A^2$	1
		$A^1a^1a^2a^2$	2
		$a^1a^1A^2a^2$	2
Awnless	1	$a^1a^1a^2a^2$	1

Ratio 15:1 arise when two dominant gene alleles affect the same character and in this case presence of any one can give the same type effect. In absence of both dominant gene alleles, recessive character, awnless appears. This is also called *Duplicate recessive* epistasis where $a'a'$ is epistatic to A_2 and a^2 and a^2a^2 epistatic to gene A^1 and a^1.

POLYMERIC GENES (9:6:1)

Polymeric or polygenes show additive or cumulative effect to express a character. Individually, they are not able to show full type effect to express a phenotype. A large number of genes are involved, which show graded effect on the phenotype. With the gradual increase in the number of gene alleles, the character expression is proportionately increased. Such characters are called *quantitative characters*.

In hull-less barley author made crosses between purple and amber colour grains. In F_1 generation, purple was dominant over amber. In F_2 generation, purple, light purple and amber grains appeared in 9:6:1 ratio.

	♀			♂	
	$P^1P^1P^2P^2$		$p^1p^1p^2p^2$	♂	Parents
	(Purple)	×	(Amber)		
F_1	$P^1p^1P^2p^2$				
	(Purple)				

In F_2 generation, three classes of phenotypes appeared as given in the checkerboard.

Gametes: ♀ \ ♂	P^1P^2	P^1p^2	p^1P^2	p^1p^2
P^1P^2	$P^1P^1P^2P^2$	$P^1P^1P^2p^2$	$P^1p^1P^2P^2$	$P^1p^1P^2p^2$
P^1p^2	$P^1P^1P^2p^2$	$P^1P^1p^2p^2$	$P^1p^1P^2p^2$	$P^1p^1p^2p^2$
p^1P^2	$P^1p^1P^2P^2$	$P^1p^1P^2p^2$	$p^1p^1P^2P^2$	$p^1p^1P^2p^2$
p^1p^2	$P^1p^1P^2p^2$	$P^1p^1p^2P^2$	$p^1p^1P^2p^2$	$p^1p^1p^2p^2$

Phenotype	*Proportion*	*Genotype*	*Proportion*	
Purple	9	$P^1P^1P^2P^2$	1	9
		$p^1P^1P^2p^2$	2	
		$P^1p^1P^2P^2$	2	
		$P^1p^1P^2p^2$	4	
Light purple	6	$P^1P^1P^2P^2$	1	6
		$P^1p^1p^2p^2$	2	
		$P^1P^1p^2p^2$	1	
		$p^1p^1P^2p^2$	2	
Amber	1	$p^1p^1p^2p^2$	1	

The purple-coloured grains appeared when both dominant gene alleles p^1–p^2 in homozygous or heterozygous ($P^1P^1P^2p^2$ or $P^1p^1P^2P^2$, $P^1p^1P^2p^2$) were present. However, where only one dominant (P^1P^1/P^2P^2) pair of alleles in homozygous ($P^1P^1p^2p^2$ or $p^1p^1P^2P^2$) or heterozygous ($P^1p^1p^2p^2$ or $p^1p^1P^2p^2$) were present, the grains were light purple. When both dominant alleles

(p^1p^1 either of p^2p^2) were absent, amber grains appeared. Thus, the effects of gene alleles P^1P^1 and P^2P^2 are cumulative or additive in nature, the contribution of either of the alleles of a gene P^1–P^1 or P^2–P^2 are identical to each other, i.e. p^1 is identical to p^2. But, when p^1 and p^2 come together in homozygous ($P^1P^1P^2P^2$) conditions or in heterozygous ($P^1p^1P^2p^2$) conditions, the nonallelic gene interaction is complete and purple-coloured grain appeared. It is also known as Duplicate gene interaction. As a summary of exception to the dihybrid ratio in F_2 in a hypothetical cross *AABB* × *aabb* is given in Table 7.1.

TABLE 7.1 Summary of exceptions to the dihybrid ratio in F_2 in a hypothetical cross *AABB* × *aabb*

Phenotypic classes	*Behaviour of genes in F_2*	*Genotypic classes*								
		AABB	*AABb*	*AaBB*	*AaBb*	*AAbb*	*Aabb*	*aaBB*	*aaBb*	*aabb*
More than four phenotypic class	*A* and *B*, both incompletely dominant	1	2	3	4	1	2	1	2	1
	B-completely dominant *A*-incompletely dominant	3		6		1	2	3		1
Four phenotypic class	*A* and *B* completely dominant (classical ratio)	9				3		3		1
Fewer than four phenotypic class	Recessive epistasis: *aa* epistatic to *B* and *b*	9				3		4		
	Dominant epistasis: *A* epistatic to *B* and *b*	12						3		1
	Dominant and recessive epistasis: *A* epistatic to *B* and *b*, *B* epistatic to *A* and *a*	13						3		
	Duplicate recessive epistasis: *aa* epistatic to *B* and *b*, *bb* epistatic to *A* and *a*	9			7					
	A epistatic to *B* and *b*, *B* epistatic to *A* and *a*	15								1
	Duplicate interaction	9				6				1

Mosaic Expression

There are some cases where F_1 hybrids have shown mosaic expression of both dominant and recessive characters. In the animals, if a cross between red and white breeds of short horn cattle is made, F_1 hybrid is always roan in colour, which is intermediate between the two parental characters. The roan colour is the mixture of red and white hairs which gives mosaic appearance to the animals.

Similarly, Andalusian fowls are blue in colour. This is a heterozygous character, as a result Andalusian never breed true and when they are intercrossed (Andalusian × Andalusian), they give rise to 1 black: 2 blue and 1 white splashed with blue. The blue-feathered Andalusian fowls are the hybrids of black-feathered and white splashed with blue type. Both the black and white splashed with blue types are true breeding types, but with the pigmentation of feather colour, they differ. In black, feather type pigmentation is uniform. In the white splashed blue types pigment bodies are blue round, but barbules of the feather are not pigmented. In the hybrid, all feathers are pigmented like black point, but in the distribution of pigment, it is identical to other parent (white splashed blue). This hybrid gives rise mosaic appearances of both the parents.

CYTOPLASMIC BEHAVIOUR

The details of cytoplasmic behaviour will be discussed in Chapter 12 where, it is briefly stated that in Mendel's experimentation, there was no reciprocal differences which have been reported in many cases, such as snail coiling and moth pigmentation. Reciprocal differences were also observed in Combodian cotton of Coimbatore. The reciprocal crosses between *Epilobium hirsutum* and *Jena* also showed cytoplasmic differences. Thus, cytoplasmic behaviour is also a cause of modification in Mendelian laws of inheritance. This is due to the effect of plasma genes, present in the cytoplasm of certain organisms.

XENIA

Normally, the segregation of contrasting characters are observed in F_2 generation, F_1 may show complete or incomplete dominance. But in maize plant, the crossed seeds developing on the plant after crossing may exhibit changes in form or colour. If the segregation of character is evidenced in one generation ahead, i.e. in F_1, this phenomenon is called *xenia*. In maize, *Xenia* has been reported due to jumping gene effect (*Ac–Dc*) or moving transposans (DNA-fragments) within the genome.

PLEIOTROPIC GENES

A single gene may have more than one effect, i.e. simultaneously, it may be influencing size, shape, colour, etc. Pleiotrophic genes control expression of more than one character. The gene may have only one function, say production of one specific enzyme, but it may affect the expression of several related traits in the plants. The *uzu* gene in barley in recessive condition may shorten stem, rachis-internodes, reduces seed size and produces an erect coleoptile leaf.

- *Uz* shows normal appearance
- *uz* shows semidwarf, dense spike, small seed and short erect leaves

A semidwarf mutant barley called RDB-1 developed in Jaipur (Rajasthan) by mutation has similar appearance as *uz* gene to which author himself has examined. It is poor combiner, therefore, it was not useful to evolve semidwarf, high-yielding barley varieties in India.

PENETRANCE AND EXPRESSIVITY

Mendel was lucky to get clear-cut phenotypic segregation in his attempted crosses made in sweet pea. But it has been amply clear that a gene may fail to produce the character associated with it, if one or more modifying genes are present in its background. Thus, the term expressivity refers to the variation in phenotypic expression of a gene, as a result of gene or environment or both. The percentage of individuals showing expression of gene is called *penetrance*. Suppose out of 100 individuals, 65 exhibited expression of a character, then the allele is said to have exhibited 65 per cent penetrance. *Expressivity* is the degree of expression of a gene that varies, whereas penetrance refers to the inability of a given allele to express itself at all in the same individual. The above said cases can be factorized as follows:

Penetrance means number of individuals showing expected phenotypic expression. If 80 out of 100 genotypes carrying a particular gene show expected phenotypic expression, then penetrance of gene is 80 per cent. Penetrance provides a measure of gene expressivity in a population.

Incomplete penetrance is common for some lethal genes known to be present in human. Huntington's chorea is a serious disease of nervous system, which generally results in death. The disease is caused by a dominant gene mutation; the person afflicted by this disease belonged to different age groups. It shows that in some persons, disease starts very early; while in others quite late, in old age.

The variable expressivity and incomplete penetrance are the functions of gene modifiers. Environmental factors are also responsible for it.

REVIEW QUESTIONS

1. In a short-horned cattle, the gene R, for red coat is not dominant over white (R^1). The heterozygous (RR^1) produces roan. A breeder has white, red and roan cow and bulls. What phenotypes might be expected from the following crosses:

(a) Red × Red	(b) Red × White
(c) Roan × White	(d) Red × Roan
(d) Roan × Roan	(e) White × White

2. From the above said genetic interactions, would it be possible to establish a true breeding (homogeneous for colour) herd of red or a true breeding herd of roan short-horned cattle? Explain.
3. The shape and colour of rish are controlled by two independent pairs of alleles that do not show dominance; each genotype is distinguished by phenotypic appearance. The

colour may be red (RR), pink (R_1R_1) and white (R^1R^1) and shape may be long (LL) and round (L^1L^1). Using the Punnett square, draw a cross between red long ($RRLL$) and white round ($R^1R^1L^1L^1$), and summarize the F_2 results, showing phenotypes, genotypes, genotypic frequencies and phenotypic ratio.

4. In poultry, genes for rose comb (R) and pea comb (P) together produce walnut comb, the alternative alleles of both in a homozygous condition ($rrpp$) produce a single comb. According to the information given in the chapter 7 determine the phenotypes and their proportions expected from the following crosses:

   (a) $RRPP \times rrpp$ (b) $rrpp \times RrPp$
   (c) $RrPp \times Rrpp$ (d) $Rrpp \times rrpp$

5. What phenotypic ratio would be expected from a test cross of F_1 and a full recessive ($AaBb \times aabb$), if F_2 results from $F_1 \times F_1$ ($AaBb \times AaBb$) were:

   (a) 13:3 (b) 15:1
   (c) 9:3:4 (d) 12:3:1

6. A white leghorn breed of poultry is known to carry in homozygous condition a coloured allele (C) and a dominant inhibitor (I) that prevents the expression of (C). The white Wyandotte ($iicc$) has neither the inhibitor nor the colour gene. Give F_1 phenotypes and proportion expected from a cross made between white leghorn ($IICC$) and a white Wyandotte ($iicc$).

7. Three possible phenotypes in barley are 'hooded', long-awned and short-awned. When a cross was made between the two breeding lines, hooded × short-awned, F_1 was phenotypically 'hooded'. When F_1 was selfed ($F_1 \times F_1$), in F_2 450 hooded, 150 long-awned and 200 short-awned plants were obtained. How many genes are involved in determining these traits?

8. Two dwarf corn plants, dwarf-1, and dwarf-2, have different origin, but are phenotypically identical. When crossed with a variety of tall plants, they were known to be homozygous, for all the genes determining the size both dwarf plants produced in F_1 only tall plants. When any one of these F_1 was self-fertilized, the F_2 appeared in the ratios of 3:1 (tall:dwarf). In crosses between dwarf-1 and dwarf-2, F_1 was tall. However, when F_1 of this cross was selfed, the proportion of dwarf plants in F_2 was 7/16. Give the gene symbols and explain the results in terms of the genotypes involved. Also give the phenotypic ratio, if F_1 (dwarf-1 × dwarf-2) is back-crossed with dwarf-1.

9. In mice, the gene (C) for coloured fur is dominant over its allele (c) for white. The gene for normal behaviour (V) is dominant over the one for waltzing (v). Give the probable genotypes of the parent mice, (each pair was mated repeatedly and produced the following results:

   (a) Coloured normal mice mated with white normal mice and produced 29 coloured normal and 10 coloured waltzers.
   (b) Coloured normal mice mated with white normal and produced 38 coloured normal, 15 coloured waltzers, 11 white normal and 4 white waltzers.
   (c) Coloured normal mice mated with white waltzer and produced 8 coloured normal, 7 coloured waltzers, 9 white normal and 6 white waltzers.

10. A woman, in general does not have ptosis, an abnormality of eyelids which make impossible to open the eyes completely. This is governed by one dominant gene *P*. The woman's father had ptosis, but her mother had normal eyelids. Her paternal grandmother too had normal eyelids. What would be the probable genotypes of woman, her father and her mother? What proportion of her children would be expected to have ptosis, if she marries a man with normal eyelids?

11. In garden pea, the genes for tall vine (*D*) yellow seed (*G*) and Round seed (*W*) are dominant over the respective alleles for dwarf (*d*), green (*g*) and wrinkled (*w*), symbolize a cross between a homozygous tall, green, round plant and a dwarf yellow, wrinkled plant.

(a) Represent the gametes possible from each parent and the F_1.

(b) Symbolize a cross between two F_1 plants. Complete this cross by making use of the forked-line method and summarize the expected phenotypes.

REFERENCES

Bateson, W. (1909): *Mendel's Principles of Heredity*, Cambridge University Press.

Burn, G.W. and P.J. Brunitto (1989): *Science of Genetics*, New York.

Chandrasekharan, S. and S.V. Parthasarthy (1953): *Cytogenetics and Plant Breeding*, Vardhanacharya and Co., Madras.

Correns, C. (1909): *The Birth of Genetics* (English Translation), as supplement to 1950 volume of *Genetics*, 33–41.

Dunn, L.S. (1965): *A Short History of Genetics*, McGraw-Hill, New York.

—— (1951): *Genetics in 20th Century*, Macmillan, New York.

Ford, E.B. (1931): *Mendelism and Evolution*, Methuen and Co., London.

Gardener, E.J., M.J. Simmons, and D.P. Snustad (1991): *Principles of Genetics*, 8th ed., John Wiley & Sons, Inc.

Hartl, D.L. (1991): *Basic Genetics*, Jones and Bartlet, Boston.

Sinnot, E.W., L.C. Dunn, and T. Dabzansky (1958): *Principles of Genetics*, McGraw-Hill, New York.

Singleton, W.R. (1967): *Elementary Genetics*, 2nd ed., D. Van Nostrand Company, Inc., Princeton.

Srb, A.M. and R.D. Owen (1952): *General Genetics*, W.H. Freeman, San Francisco.

Stern, C. and E.R. Sherwood (Eds.) (1966): *The Origin of Genetics—A Mendel Source Book*, W.H. Freeman, San Francisco.

Sturtevant, A.H. and G.W. Beadle (1939): *An Introduction to Genetics*, Saunders, Philadelphia.

Strickberger, M.W. (1961): *Genetics*, Macmillan, New York.

CHAPTER

8 Linkage and Crossing Over

Rediscovery of Mendel's second law, i.e. law of independent assortment of genes indicated that two factor pairs (genes) assort and recombine independently in F_2 generation, giving rise 9:3:3:1 ratio. But Suttan in 1903 proposing relationship between Mendelian factors and chromosomes suggested that some genes specially on the same chromosome are adhered or linked together in a group and are transmitted as a single unit. Thereafter, genetical studies conducted on both plants and animals indicated that in a number of cases, parental characters do not recombine independently or recombination appears in a very low number than the expected frequencies. Bateson and Punnett (1906) were the first to report a departure from 9:3:3:1 ratio in sweet pea. The factor pairs involved in their studies were purple flower, long pollen, red flower and round pollen grains where purple flower and long pollen grains were dominant over red flower and round pollen grains, the recessive. In F_2 generation, parental types were more than the recombinant types (Table 8.1).

TABLE 8.1 Results of Bateson and Punnett's sweat pea crosses

Phenotype	*Number*	*Frequency*	*Expected ratio* (9:3:3:1)	
			Number	*Frequency*
Purple long	296	0.693	240	0.562 (9/16)
Purple round	19	0.044	80	0.187 (3/16)
Red long	27	0.863	80	0.187 (3/16)
Red round	85	0.199	27	0.062 (1/16)

Bateson and Punnett were unable to explain how and why 9:3:3:1 ratio was not obtained. Later genetic studies provided evidences that genes for flower colour and pollen grain shape were linked. However, within short time studies conducted on *Drosophila* by Morgan and his colleagues (1910–1915) demonstrated that law of independent assortment was not universally accepted as in certain crosses, there were certain limitations for the free segregation of two or more allelic pairs of genes. In each case, there was a marked tendency for the parental character to remain linked and thereby to produce lesser proportion of new combinations. If two genes *a* and *b* are on the same chromosomes, only following two classes of gametes will be obtained:

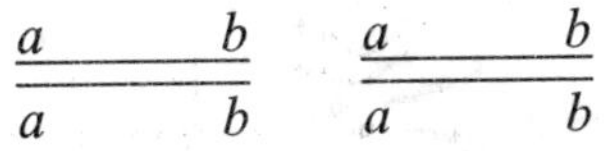

The coexistence of two or more genes in the same chromosome is called linkage.

After studying a number of crosses in *Drosophila*, Morgan reached to the conclusion that all genes in the fruit fly were clustered into four groups, each containing a single chromosome. In *Drosophila*, there are four pairs of chromosomes, each form one linkage group of genes. Further studies indicated that linkage is not absolute. It may be broken down with certain frequencies. This phenomenon is called *crossing over*. For example, Morgan crossed a hybrid *Drosophila* female having grey and long wing (double dominant) with a male having black and vestigial wing (double recessive). Four classes of progeny were obtained instead of expected two. The first two were parental types, which appeared in 83 per cent cases. The other two types—grey and vestigial wing and black and long wing were new recombinants. The new recombinants were 17 per cent of the cases (Figure 8.1(a)).

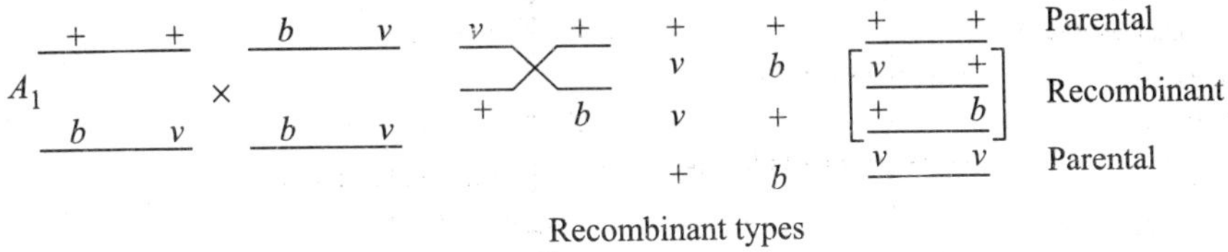

Figure 8.1(a) A cross involving two linked genes.

Linkage may be of two types: (i) coupling, and (ii) repulsion phase according to the arrangement of genes on the chromosomes. Coupling—when dominant and recessive (++/*bv*) genes are located on the same homologous chromosomes. Repulsion—when dominant and recessive (+*u*/*b*+) genes are located on the opposite homologous chromosomes. (Figure 8.1(b)).

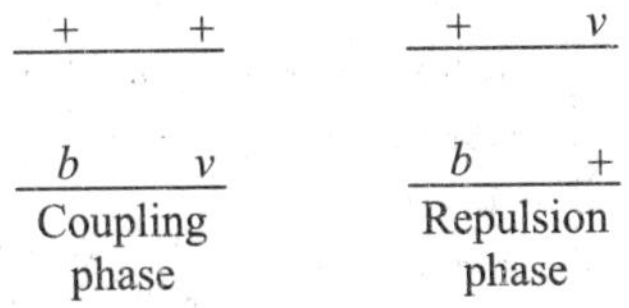

Figure 8.1(b) Linked genes in coupling and repulsion phase.

LINKAGE GROUP

Morgan and Bridges (1916) found many hereditary characters associated together in a group in *Drosophila*. For example, white eye and miniature wings, yellow body, winged bristle, etc. are sex-linked genes and tend to get inherited together. The female fly heterozygous for yellow (*y*) and white (*w*) in which these genes came from one parent and the wild types alleles (++) from another, produced male offsprings that were mostly either yellow-bodied (++) and white-eyed (*yw*) or wild type (y^+w^+). The new recombinants such as $y^-\ w^+$ and $y^+\ w^-$ were quite rare. This reflects a strong connection or linkage between *y* and *w* in one of the female *X* chromosome and *y*+ and *w*+ on the other. Bridges found that shortened wing and black (dark) body colour tend to be transmitted together. In *Drosophila*, four groups of linked genes are found (Figure 8.2), which correspond to the four pairs. The linkage groups are always equivalent to the haploid number of chromosomes such as in pea 7; in maize 10 and in barley 7, etc. The number of genes located on a chromosome are related to the physical length of the chromosome.

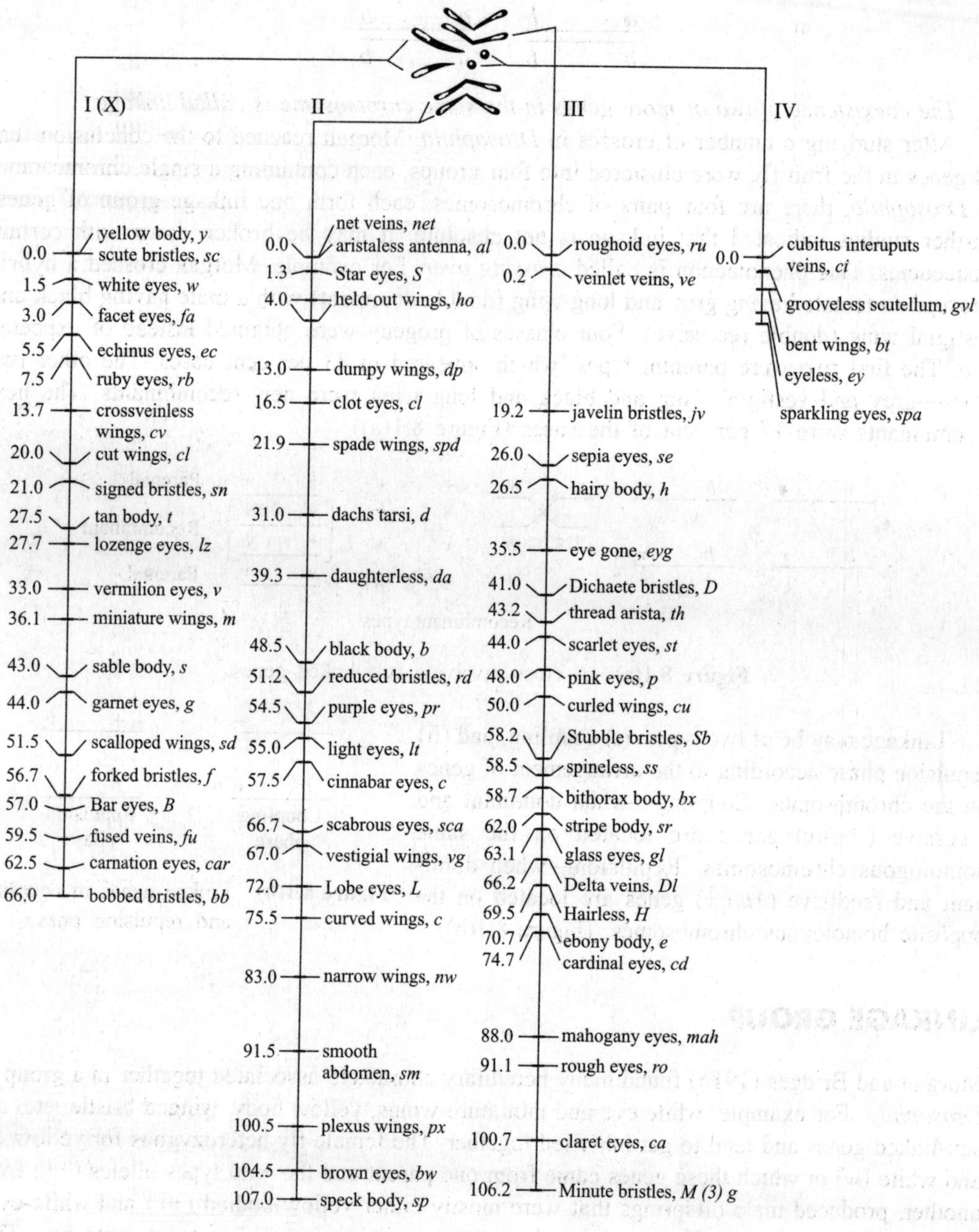

Figure 8.2 Abbreviated genetic map of *Drosophila melanogaster*.

Complete Linkage

When genes are closely linked on the same chromosome they are transmitted together from parent to the progeny. The linkage between such genes are considered complete. The mutation in the fourth chromosome which is very small in *Drosophila*, there is hardly any assortment from each other during transmission. Complete linkage is observed for all gene pairs in each of the other groups when *Drosophila* males are used as heterozygous parent in the test crosses. For example, gene for purple eye (*p*) and vestigial wing (*v*) are on the second linkage group and are recessive to those red eyes and long wing. When homozygous purple with vestigial wing fly is crossed to a normal red-eyed, long-winged fly, all F_1 offsprings are normal. A back cross of an F_1 male to a homozygous purple, vestigial-winged female fly produces only two phenotypic classes of fly—a homozygous purple, vestigial-winged fly and heterozygous normal. Complete linkage between alleles of purple and vestigial genes have been observed when only males are used as heterozygous parents in the back cross to the purple vestigial stock. The *pr* and *vg* genes remain linked on one homologue in the heterozygous male as do the pr^+ vg^+ alleles on the other homologue and their progeny inherit only one of these combination (*pr vg*) and the other (pr^+ vg^+). Absence of any assortment between all genes in a linkage group appears to be a peculiarity of *Drosophila* male and some heterogametic sex such as in silkworm (Figure 8.3).

Parents: $$♀ \frac{pr \quad vg}{pr \quad vg} \times \frac{pr^+ \quad vg^+}{pr^+ \quad vg^+} ♂$$

F_1 back cross: $$\frac{pr \quad vg}{pr^+ \quad ug^+} (♀) \times \frac{pr \quad vg}{pr \quad vg} (♂)$$

F_2 back cross progeny: $$1 \frac{pr \quad vg}{pr^+ \quad vg} : 1 \frac{pr \quad vg^+}{pr^+ \quad vg^+}$$

(Purple) (Normal)

Figure 8.3 Complete linkage between purple and vestigial genes.

However, purple gene on the second linkage group and bent gene on the fourth linkage group (pr/pr^+: bt/bt^+) produce all four possible combinations of gametes according to the law of independent assortment of genes. Such males can be back-crossed to the homozygous purple bent; female and four classes of phenotypes in equal frequencies could be obtained (Figure 8.4).

Parents: $$♀ \frac{pr \quad bt}{pr \quad bt} \times \frac{pr^+ \quad vg^+}{pr^+ \quad vg^+} ♂$$

F_1 back cross: $$\frac{pr \quad bt}{pr^+ \quad bt^+} (♂) \times \frac{pr \quad bt^+}{pr \quad bt^+} (♀)$$

F_2 back cross progeny: $$1 \frac{pr \quad bt}{pr^{\pm} \quad bt^{\pm}} : 1 \frac{pr \quad bt}{pr^{\pm} \quad bt^{\pm}} : 1 \frac{pr \quad bt}{pr^{\pm} \quad bt} : 1 \frac{pr \quad bt}{pr \quad bt}$$

(Normal) (Purple) (Bent) (Purple bent)

Figure 8.4 Complete linkage between purple and bent genes.

Incomplete Linkage

In the sexually reproducing organisms, linkage may be or may not be complete. Incomplete linkage was first reported by Morgan in crosses between white-eyed (*w*) and miniature-wing (*m*) *Drosophila*. When a white-eyed miniature-winged male (*wm*) fly was crossed to a homozygous wild type (normal) female (w^+m^+). F_1 females were wild type since one of two X chromosomes carries normal allele (wm/w^+m^+). In the second generation (F_2) males were miniature (*wm*) depending on the X chromosome, they received from F_1 female parent (Figure 8.5).

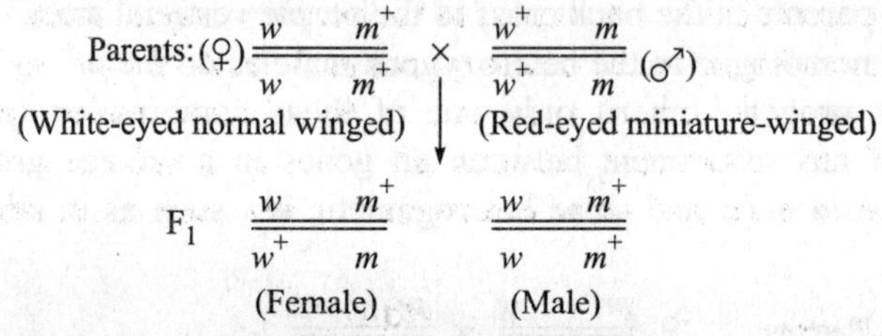

Figure 8.5 Incomplete linkage.

The resulting F_1 progeny consisted of wild-type females and white-eyed, non-miniature males. When the F_1 progeny were intercrossed, the females were in the ratio of 1:1 of red: white eyes (all non-miniature), but the males (*y*) were in the following proportions:

- White-eyed, normal-winged ($w\ m^+/y$) 226] 66.5% parental types
- Red-eyed, miniature-winged ($w^+\ m/y$) 206]
- Red-eyed, normal-winged ($w^+\ m^+/y$) 114] 33.5% recombinant types
- White-eyed, miniature-winged ($w\ m/y$) 102]

Here *y* represents males.

The result of this experiment showed a great departure from 1:1:1:1 ratio of the four male stocks expected with the independent assortment of genes. This deviation is possible if genes are located on the same chromosome, which tend to remain together but were not completely linked. This is called *couplings phase* (Figure 8.6). The 33.5 per cent recombinant types is the frequency of recombination.

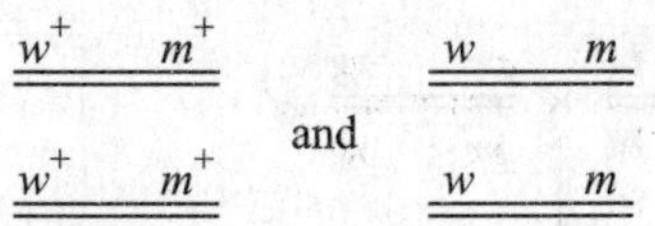

Figure 8.6 Dominant and recessive genes are located in coupling phase on the same homologous chromosomes.

The recombination are the result of crossing over during meiosis in the F_1 female. The frequency of recombination between linked genes *w* and *m* was 33.5 per cent but with other linked genes, it ranges from near 0 to 50 per cent. The genes with recombination frequencies less than 50 per cent are present in same chromosome (linked). The two genes, which undergo independent assortment with 50 per cent recombination frequency, may be located on two homologous chromosomes or even on the same chromosome located apart from each other. Figure 8.7 shows genes are located on the repulsion phase in two homologous chromosomes.

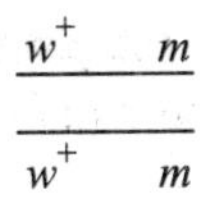

Figure 8.7 Dominant and recessive genes are located in repulsion phase on the opposite homologous chromosomes.

The genes w^+m and wm^+ can also be written as $^+m/w^+$. The location of genes on the opposite homologous chromosome in $w^+/^+m$ fashion is said to have *trans* (repulsion) configuration (a). The alternate arrangement of the mutant alleles in the same homologous chromosome, i.e. ++/*wm* or *wm*/++ configuration (b) is called *cis* (Coupling) (Figure 8.8).

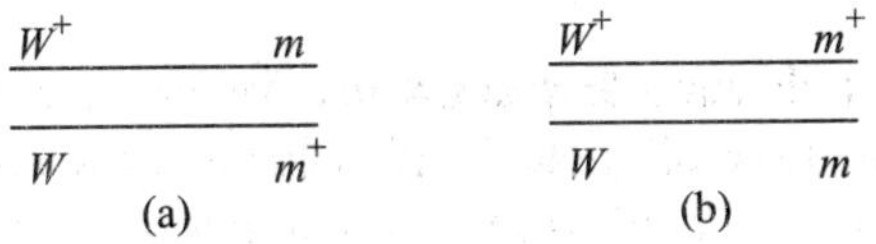

Figure 8.8 (a) *Trans* configuration, (b) *Cis* configuration.

CROSSING OVER

During meiosis, at later part of the *pachytene*, an important genetical event takes place in which each of the paired chromosome begins to divide lengthwise into the chromatin which reach at completion during *deplotene* stage. The chromatins (chromatids) belonging to the same (original) chromosome, however, are still paired along their length. The two paired chromosomes (bivalents) start separating from each other but are kept together lengthwise at some location by so called *chiasmata*. As assumed by Morgan (1926), a chiasma represents a change of partners between the chromatids of two homologous chromosomes. This phenomenon is called *crossing over*. The chromosomes are paired by two, but a chromatid *a* may be paired in length with chromatid *b* and in another segment with chromatid *c*. The number of such chiasma may be higher but rarely more than four or five. At least one chiasma is needed to keep the two homologous-paired chromosome together (Figure 8.9).

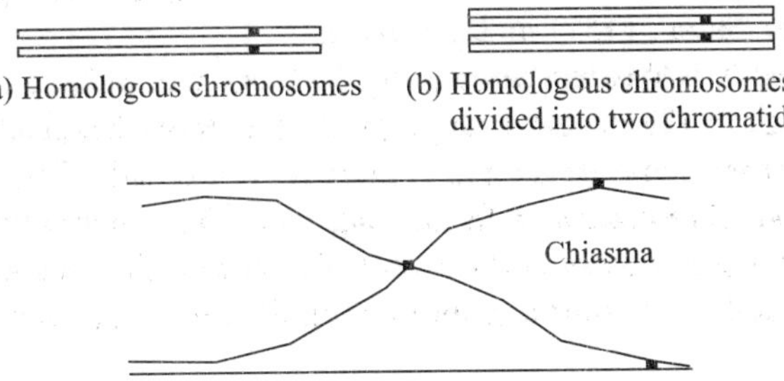

(a) Homologous chromosomes (b) Homologous chromosomes are divided into two chromatids each

(c) An exchange of partners between the chromatids occur at the transition between the *pachytene* and *diplotene* stages of meiosis

Figure 8.9 Mechanism of chiasma formation.

The chiasma is expression of the fact that crossing over, i.e. breakage and exchange of chromatids has occurred (Darlington, 1965).

When there are more than one chiasma, the pairing occurs in a way similar to Figure 8.10. This figure shows a case of three chiasmata in which all four chromatids are involved in a single or double crossovers.

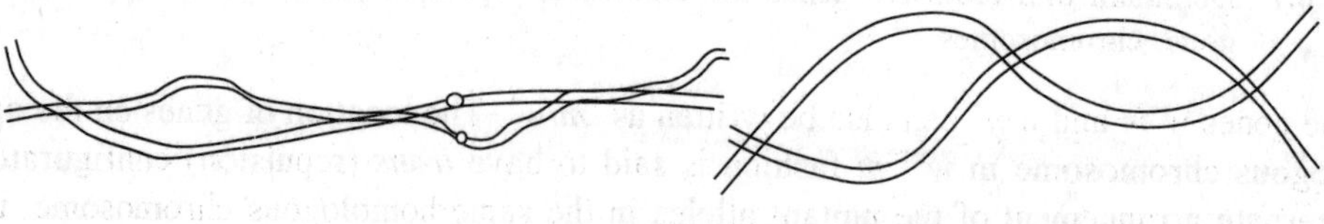

Figure 8.10 Double crossing over and three chiasma formation between two homologous chromosomes.

It could be specified that chiasma formation has double function. One is genetic crossing over and consequent recombination of 'genes'; the other is a mechanism of keeping together of the homologous chromosomes from *diplotene* to *metaphase* I stage of meiosis. Figure 8.10 shows formation of several chiasmata in a chromosome at meiosis (After Roling modified).

There are organisms where such chromosomal association occurs during part of meiosis but no crossing over, i.e. no exchange of chromatid segments take place. For example, in *Drosophila* male pairing is purely an attraction without interchange of chromosomal segments.

Relationship between Genetics and Cytological Crossing Over

In 1931, the experimental results reported by C. Stern in *Drosophila melanogaster* and by Creighton and McClintock in Corn demonstrated that genetic crossing over is parallel by an actual physical crossing over between the chromatids of two homologous chromosomes. In both experiments, a cytologically visible marker was employed on one member of the pairs of the homologous chromosomes, which also differed with respect to the alleles of certain gene pairs. The genetic crossing over resulted in an unique phenotypes which were accompanied in each case by a transfer of the cytological markers and a visible change in the chromosomal structure identified by its missing terminal (Figure 8.11).

In 1931, Creighton and McClintock demonstrated a very convincing correlation between cytological crossing over and genetical result, which was confirmed as a landmark in the experimental genetics by Stern (1931) in *Drosophila.*

In corn, chromosome 9 is shortest among the 10 pairs and lacks a small knob (satellite). Creighton and McClintock (1931) found a single plant in a strain that possesses a dissimilar 9th chromosome, i.e. one chromosome possessed a satellite at the end of its short arm and also a translocated segment from chromosome 8 to the long arm. The chromosome possesses the gene for coloured aleurome (*C*) and starchy endosperm (*Wx*) on one chromosome and the alleles for colourless aleurome (*c*) and waxy endosoperm (*wx*) on the other chromosome of the pair given below:

$$\text{Chromosome 9} \; \frac{C \qquad Wx\bullet}{c \qquad wx} \; \text{Translocated chromosome}$$

Creighton and McClintock (1931) crossed female individual (♀) having such a dissimilar character with a normal male (♂) plant having no knob (knobless) on chromosomes and no translocation of a segment on chromosome 8.

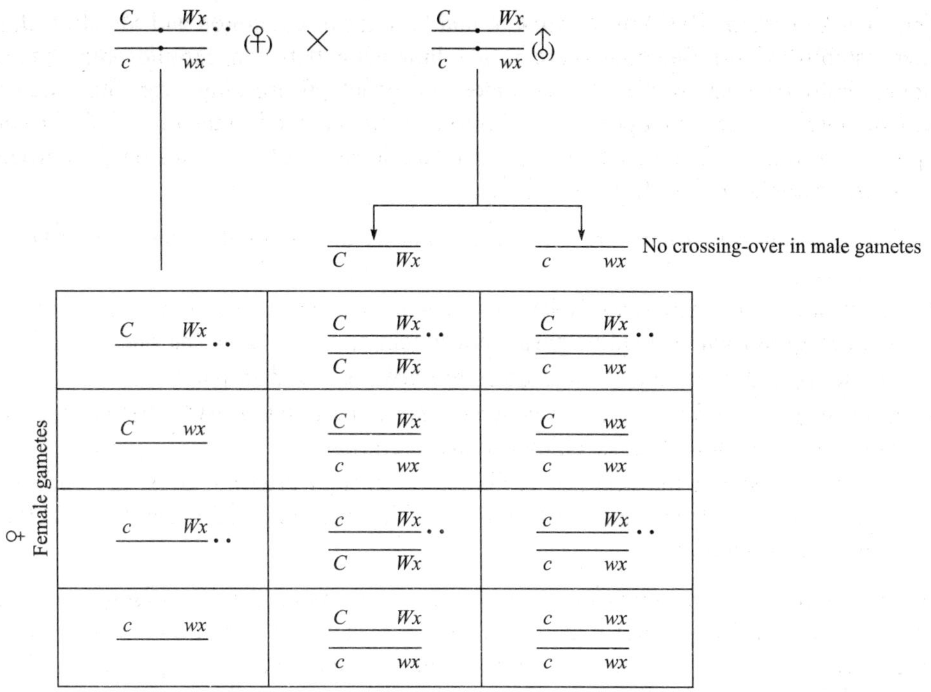

Figure 8.11 Exchange of segments between homologous chromosomes with cytological markers which helped to identify recombined chromatids in maize, as domonstrated by Creighton and McClintock in the year 1931.

The outcome of the results are demonstrated here:

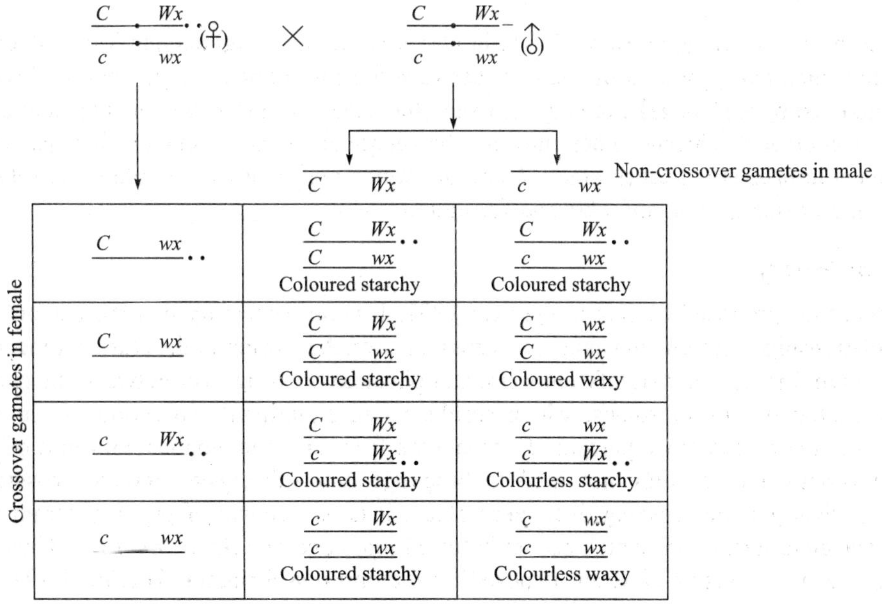

Thus, work of Stern (1931) in *Drosophila* and findings of Creighton and McClintock (1931) in maize established the fact that there is a relationship between chromosomal interchange (crossover) between homologous chromosomes and genetical crossing over. Stern had used a female *Drosophila* which had a portion of *Y* chromosome attached to the one of *X*, chromosome of its pair *XX* as marker. The genetical aspects of the location of the genes on the chromosome can be briefly summarized as follows:

1. There are genes, which are not linked, and assort randomly from each other during meiosis.
2. There are genes, which are linked on the same chromosomes, and form a group called linkage group which tend to pass from parent to the progeny as one group.
3. In the diploid cells, the chromosomes are always found in pairs.
4. Linked genes oftenly get exchanged through crossing over between non-sister chromatids of homologous chromosomes reciprocally.
5. While exchanging the segment reciprocally through crossing over, the chromosomes form knot-like structure called *chiasma*. The exchange of segment is reflected in genetical crossing over.

The old concept that *chiasma* formation and genetic crossing over (recombination of genes) may be having similar frequencies may not be correct because according to recent findings only 10–15 per cent of DNA packed in chromosome is functional (exon), the rest is either nonfunctional (intron) or the function of genes located on these DNA segments are not yet known. This needs further investigation because if present view of molecular biologist is correct, in that case, frequency of *chiasma* formation and frequency of crossing over may not be correlated.

Relationship between Crossing Over and Chiasma Formation

The experimental findings of Stern (1931) in *Drosophila* and Creighton and McClintock (1931) in corn indicated that genetic crossing over between the chromatids of homologous chromosome is accompanied by a physical exchange between the same chromosome. The physical exchange between homologous chromosomes may not be identical to the extent of chiasma formation during meiotic prophase I takes place. There are following theories to explain the relationship between the crossing over and chiasma formation.

Classical theory

This theory was proposed by Sax in the year 1932. It is not necessary that there should be one to one relationship between chiasmata and crossing over. According to the classical theory, when chiasmata are formed, they signify an accidental physical crossing over between the chromatids of the homologous chromosomes, which results to an equational separation of chromosome segment, caused by the close pairing of chromatids from one homologous with that of another. Such chiasmata may or may not lead to breakage and subsequent genetic crossing over. However, when genetic crossing over takes place, it is on account of physical strain imposed by the chiasmata formation, which occurs between *diplotene* and the *anaphase* I. Thus, genetic crossing over is the result of strain imposed by the chiasma formation (Figure 8.12).

Chiasma type theory

This theory was proposed by Janssens in 1909 and later developed more accurately by Belling and Darlington in 1935. According to this theory, crossing over occurs somewhere at *pachytene* stage when homologous chromosomes are closely paired and a chiasma is formed at a point where crossing over has already occurred. Thus, according to chiasma type theory, each chiasmata represent one genetic crossing over and there is always reductional separation on both sides of a chiasma (Figure 8.13).

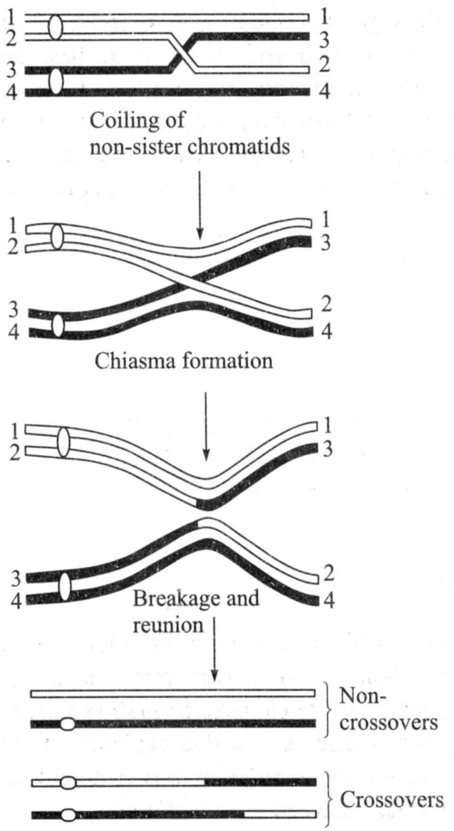

Figure 8.12 Classical theory.

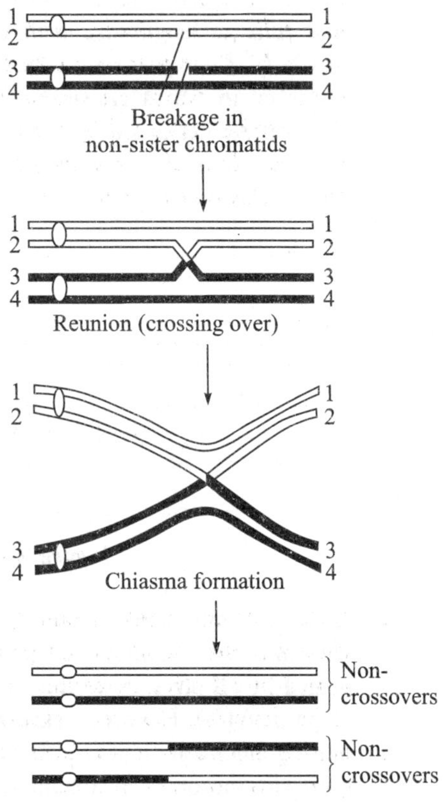

Figure 8.13 Chiasma type theory.

In explaining the origin of observed chiasmata in pairing between two homologous chromosomes, in classical theory, breakage and reunion (genetic and cytological crossing over) follow chiasma formation and one side of the chromosome is presumed to represent an equational division of chromatids. In chiasma type theory, breakage and reunion precede chiasma formation and both sides of chisama represent reduction division of chromatids.

From genetic point of views, chiasma type theory has the advantage of suggesting that crossing over at any one point occurs only between the two chromatids, one from each homologous rather than between the entire chromosome, i.e. crossovers are involved only between two out of four chromatids. This also explains recombination events are reciprocals, i.e. a crossover between the two gene loci produces both kinds of recombinant products. For example, a cross between *AB* and *ab* produces two reciprocal genotypes *Ab* and *aB*.

Darlington and others produced the evidences to support that chiasma type theory is more valid than classical theory. Reasons to support are as follows:

1. One of most important reason is the mode of pairing that occurs in an organism carrying three sets of homologous chromosomes (triploids) with only rare exception. It has been cytologically observed that no matter how many homologous are carried in an organism, a chromosome will only associate during meiosis either along its entire length or over a limited region. Thus, if there are three homologous chromosomes, *A*, *B* and *C*; *A* can pair fully in lengthwise with either *B* or *C*, but not with both at the same time and similarly, *B* can pair with either *A* or *C*. In triploids, occasionally, trivalent structures are observed in which chiasmata link together between three homologous chromosomes. This can be explained by classical theory assuming that complete synapsis has occurred between three chromosomes in the same region A_1–B_1–C_1 and A_2–B_2–C_2. Figure 8.14 shows chiasma formation in a trivalent.

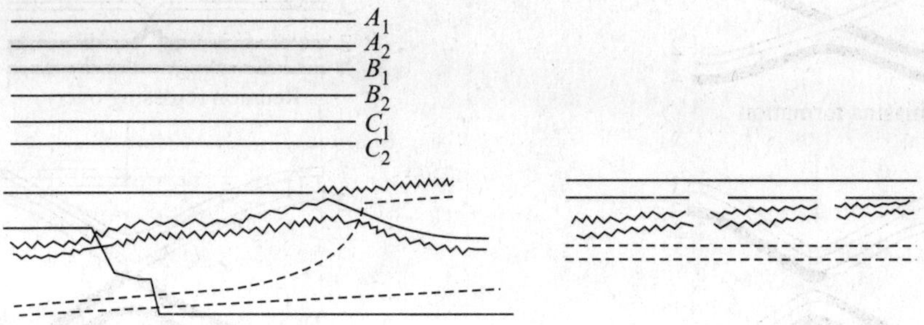

Figure 8.14 Chiasma in a trivalent.

2. Only classical types of pairing will permit equational division on both sides of each chiasma. But simultaneous pairing among three homologous chromosomes is unsupported by all other evidences. Therefore, classical theory of chiasma formation appears to be doubtful. However, chiasma type theory offers more acceptable explanation that pairing among such trivalents has occurred in a given region between only two homologous chromosomes at a time and separation at each chiasma is *reductional*.
3. The chiasma type theory should receive further support of the number of chiasmata and crossover can both be scored. For example, if chiasma type theory was correct, the presence of a single chiasma in each meiosis of a bivalent would represent a map length of 50 crossover units which means that recombination between two pairs of end genes will be observed in half time. Similarly, chiasma frequency of two and three per chromosome would represent the genetic length of 100 and 150 units respectively. There appears to be a correlation between the genetic map length of some chromosomes and their length calculated by observed chiasma frequency.

Findings of Beadle (1932) in *Zea* Euglean hybrid and Brown and Johry (1953) in *Lilium-formosanum* on relationship of chiasmata and crossing over also offer strong support for the one-to-one correspondence between chiasmata and genetic crossing over.

There are some objection to the chiasma type theory of crossing over. Kaufmann (1934) and Cooper (1949) demonstrated that chiasmata are formed in some tissues of male *Drosophila* including spermatocytes. Since genetic crossing over is absent in male *Drosophila*, by chiasmata type theory however, Slizynski (1964) reported that chiasmata observed in *Drosophila* may be surface association rather than cytological exchanges. The chiasma type theory, therefore, is most acceptable explanation for the relationship between genetic crossing over and cytologically observed chiasmata.

Theory of chromosomal crossing over

It is really very difficult to prove how genetic materials are exchanged between chromosomes. Strickberger (1976) and Darlington (1965) who proposed chiasma type theory indicated that exchange between chromatids of homologous chromosome takes place because of attractive and repulsive forces that operate between homologous chromosomes during meiotic cell division. Darlington suggested a theory known as *precocity theory* which indicates that the difference between mitosis and meiosis lies in the degree of chromosome replication, which takes place during cell division.

In mitotic cell division, chromosome replicates before *prophase*, i.e. in the *interphase* and 'pairing need' of each chromosome caused by cell division is satisfied by the newly replicated strand (chromatids). In meiosis, *prophase* is 'precocious' arising before chromosome replication has taken place. Thus, pairing need of each chromosome can only be satisfied by pairing or synapsis with its homologue.

As meiosis proceeds, the mechanical strain that results from the separation of paired homologous chromosome during *pachytene* are presumably accompanied by additional strain from the separation of the sister chromatids in each chromosome (Figure 8.15).

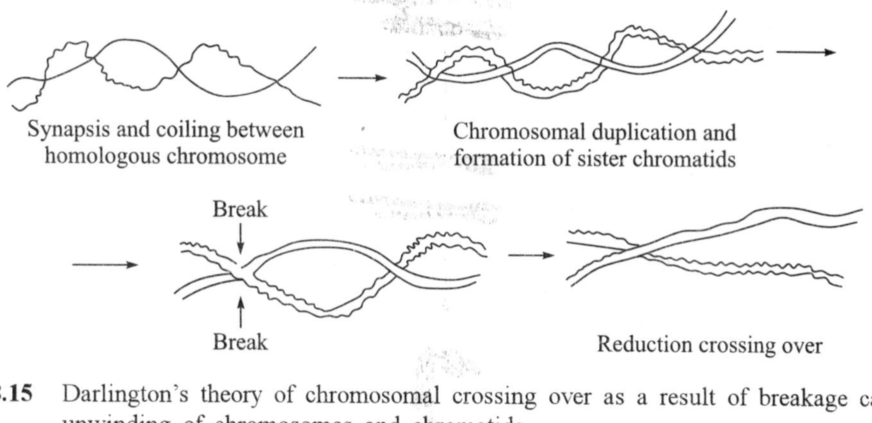

Figure 8.15 Darlington's theory of chromosomal crossing over as a result of breakage caused by unwinding of chromosomes and chromatids.

These strains and torsions lead them to break in the identical position of the homologous chromatids and to subsequent crossover when the broken ends of non-sister chromatids unite to produce chiasmata. By the end of *pachytene*, only repulsive forces are acting between the homologous chromosomes and according to Darlington, it is only the chiasmata, which then hold the homologous chromosome together till meiotic anaphase I. Darlingon's theory, therefore, correlates close pairing between the homologous chromosomes (generally observed during meiotic prophase and crossing over). Thus, when pairing between homologous chromosomes is interfered with such as in chromosomal structural changes, crossing over is restricted. According to Darlington's theory, pairing between homologous chromosomes is necessary for normal disjunction between these homologous chromosomes at meiotic anaphase I. However, if there is no pairing, the two homologous chromosomes will not align opposite to each other on the metaphase plate and may consequentally go to the same pole at anaphase causing nondisjunction.

The **partial chiasma** type theory was proposed by Brown and Zohary (1955) based on their works on *Lilium formosanum* in which one arm of the chromosome had a large deficiency, while other arm was normal. Pairing of these homologues during meiosis formed a heteromorphic bivalent as a result such heteromorphic bivalent did not form more than one chiasma (Figure 8.16).

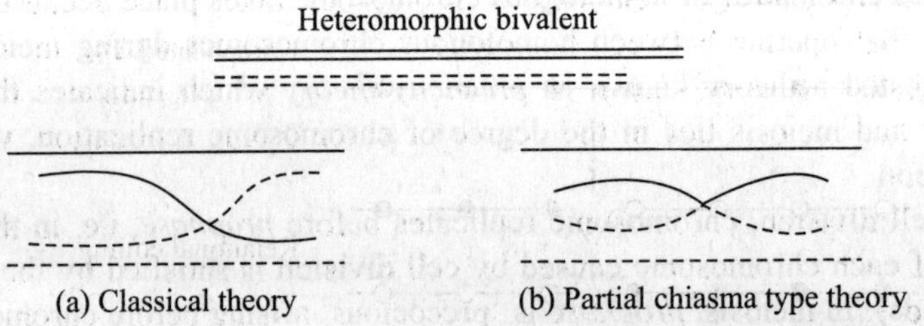

Figure 8.16 Crossing over and chiasma formation in heteromorphic bivalent.

Heteromorphic bivalent formation has been found in *Delphinium* which give rise to unequal chromatid forming a single chiasmata (Figure 8.17).

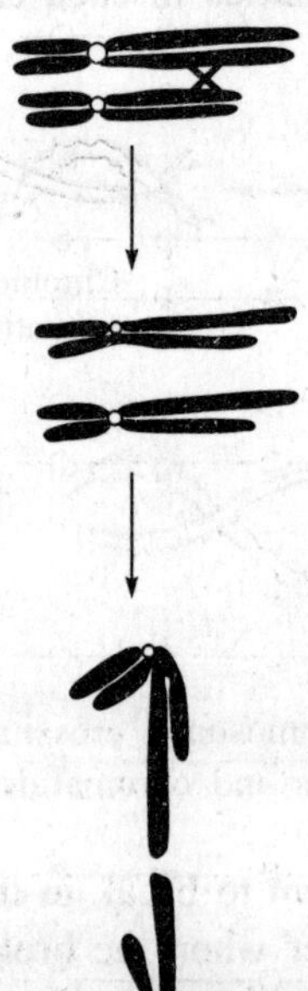

Figure 8.17 A heteromorphic bivalent of *Delphinium* gives rise unequal chromatids in two homologous chromosomes.

Belling's theory of crossing over

Belling (1931, 1935) gave another theory of chromosomal crossing over (Figure 8.18, after White). He said, breakage and reunion between chromosomes are not the causes of crossing over, but rather crossing over occurs because of novel attachments that are formed between newly duplicated genes. According to this theory, chromosome duplication takes place in following two stages:

(a) The replication of genes, and
(b) Formation of new connection between these genes

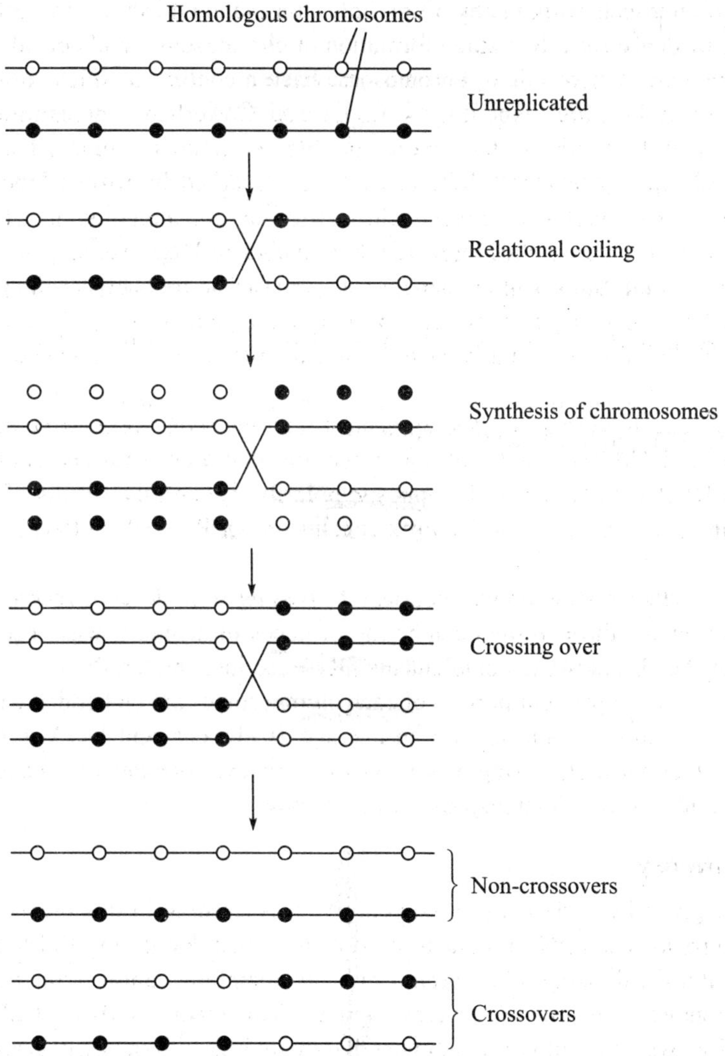

Figure 8.18 Belling's theory of crossing over as a result of new connection made between duplicated genes on two separate chromosomes that have temporarily interconnected.

In the first stage, gene duplication occurs in *pachytene* and results in a daughter gene of each pair lacking connection to either one or both of its neighbouring genes. Later, when genes interconnections are formed, the previously unconnected daughter genes are now connected to their close neighbours. If homologous chromosomes intercross, when these connections are formed, the new chains of genes will show crossover (a modified version of Belling's theory called 'copy choice' was proposed by various microbial geneticists).

Undoubtedly both Darlington's and Belling's theories are advantageous in explaining some of the observed facts; there are indications that neither of theories is completely satisfactory. Hughes Schrader raised objection on Darlington's theory that pairing among all the four chromatids exist in tetrads of certain insects without any observable chiasmata. Objections to Belling's theory are that there is no evidence for a two-stage formation of chromosomes and considerable evidences indicate that the genetic materials of chromosome have a continuous linear organization.

By chromosome labelling experiments Taylor and Cowerhen demonstrated that chromosomes and chromatids divide as they were 'double' structures composed of two subunits. Further, in grasshopper (Romalea) physical exchanges marked by tritium-labelled DNA were shown to occur during meiosis between chromatids of one homologous chromosome and chromatids of another. Later studies indicated that number of DNA exchanges were not significantly different from the number of chiasmata observed. In the root cells of the plant *Belleualia*, Taylor observed that exchanges between one chromatid and another chromatid involve the exchange of both subunits in each and each of the subunits seen to be oriented differently from each other.

These results are, therefore, in good agreement with the semiconservative replication of the chromosome at the DNA level and with the possibility that a chromosome consists of a single double helix of DNA with strand of the opposite polarity. Though the results of the Lacour and Pele were not in agreement with the Taylor's conclusion but Peacock's (1975) findings support Taylor's view.

However, in author's view above theories of crossing over deserve reconsideration in the event of recent findings through the deciphering of genes in human it was observed that about 96.6 per cent of the chromosome complements DNA are inert materials, i.e. non-coding units, called *intron* and only 3.4 per cent portion of each chromosomes are embedded with coding DNA units, called '*exon*'. In other species, on an average 10–15 per cent DNA is functional. This finding indicates that pairing/crossing over may not be always correlated with exchange of genes between chromatids of two homologous chromosomes.

Copy choice theory

This theory was given by Lederberg in the year 1955. According to this theory, recombination occurs during chromosome (DNA) replication. It is presumed that during DNA replication, new strands are synthesized using old chromosome as templates upto some point, thereafter homologous chromosome begins to serve as template. This theory is also called *complete copy choice theory*. This way, the entire recombinant chromatid arises from newly synthesized portion of chromosome (Figure 8.19).

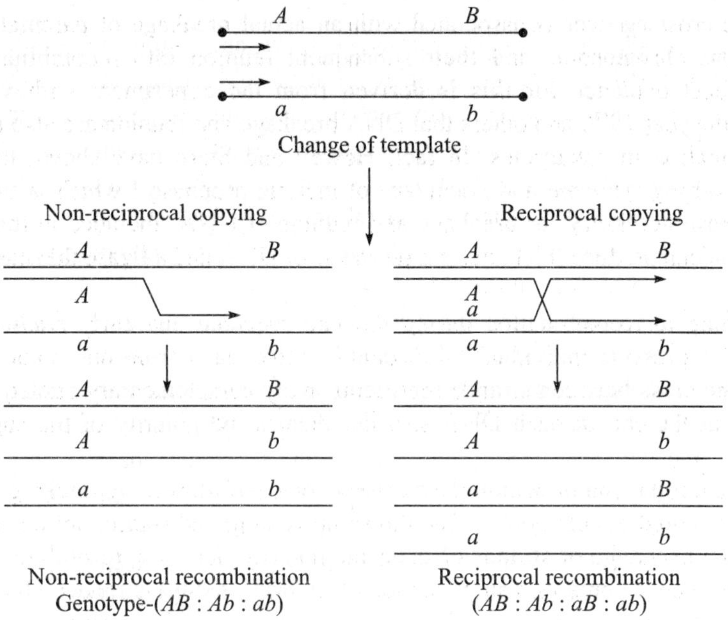

Figure 8.19 Copy choice theory of crossing over.

Molecular Mechanism of Crossing Over

The molecular mechanisms of genetic recombination was proposed on Whitehouse model which differs from the classical theory in view that breakage and reunion leading to chiasma formation are involved in the genetic recombination, the operational unit in these events is not the DNA duplex as a whole but a single DNA strand. The two parental DNA molecules form the hybrid structure for the part of their length (Figure 8.20).

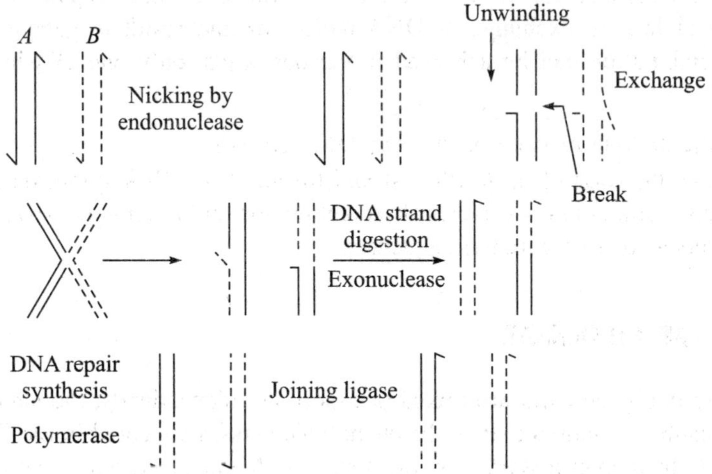

Figure 8.20 Molecular basis of crossing over.

The genetic crossing over is associated with an actual breakage of parental DNA strands from homologous chromosome and their subsequent reunion into recombination products. Although, the best evidence for this is derived from the experiment with virus given by Strickberger in the year 1976 and others that DNA breakage and reunion are also responsible for genetic recombination in eukaryotes. In fact, Howell and Stern have shown that an enzyme system appears during *zygotene* and *pachytene* of meiotic prophase I which is fully capable of performing the task necessary for breakage and reunion of DNA. Included in this system is an endonuclease that can produce 'nick, in a single strand of DNA and a ligase that can rejoin broken DNA strands.

Events leading to recombination during the late *zygotene* and early *pachytene* stages of meiosis. A or B represents individual DNA double helix, each from one homologous pair of chromosome. The cross between strands represents many complementary nucleotide base pairs and half arrow at the end of each DNA strand indicates the polarity of the sugar–phosphate backbones.

In this sequence, pairing of homologous chromosome is followed by 'nicking' of each DNA strand in the two paired double helix. This allows unwinding and rotation of each helix leading to a substitutional exchange of strands of the same polarity. The new recombinational complex formed by this exchange contains a short section of 'hybrid', double-stranded DNA perhaps 100 or more nucleotides long composed of one complementary DNA strand from each parent. *Unpaired DNA ends are digested by an exonuclease enzyme and unpaired gaps are filled in with complementary base-paired section synthesized by a polymerase enzyme.* A ligase enzyme then joins the broken strands into continuous covalently-bended nucleotide chains.

It is possible that the alignment of a single strand of DNA from different homologous chromatids into new recombinant double-stranded molecules is accomplished with the aid of the binding protein.

During *pachyatene* stage, a small burst of DNA synthesis occurs which seems to be of the type involved in breakage and reunion. That is, there is no net increase of DNA but only the repair of DNA section that may have been degraded during the breakage event that occurred upon the appearance of the endonuclease enzyme at *zygotene*. Some of these repair activities may be accompanied by or lead to exchange of DNA molecules and result in genetic crossing over. The type of recombination can be achieved in various ways, only one of which is shown in Figure 8.20.

(a) Enzymatic nicking of each of the four DNA strands.

(b) Enzymatic digestion of each DNA strand followed by DNA synthesis in the resulting gaps. This resulted into a reciprocal recombination and reunion consists of exchange of equal amount of two parental sequences.

DETECTION OF LINKAGE

To detect whether two gene pairs are linked or assorting independently, the simplest method is to compare the number of individuals in the phenotypic classes appeared in the F_2 generation of a double cross hybrid intercrossing on the basis of *Mendel's second law of independent assortment* of genes and find the goodness of fit by *chi-square* test. The second method is *test cross*, i.e. back crossing of F_1(*AaBb*) with double recessive (*aabb*). The test cross will produce four classes of genetypes (*AaBb*, *Aabb*, *aaBb* and *aabb*) in equal proportion and departure from this ratio could be detected from *chi-square* test.

Test Cross

If F_1 of a cross *AABB* × *aabb* is back-crossed with one of its recessive parents (*aabb*), four classes of phenotypes in 1:1:1:1 ratio will appear. If two pairs of genes *AA* and *BB* assort independently, the result would be:

Genotypes	*AaBb*	*Aabb*	*aaBb*	*aabb*	Total
Observed	140	38	32	150	360
Expected	90	90	90	90	360
Deviation	+50	−52	−58	+60	

In this case the *chi-square* (χ^2) value for three degree of freedom (d.f.) would be 135.19. The departure from the expected ratio is highly significant.

If χ^2 value is calculated from the expected *Aa* and *aa* phenotypes (1/2 *Aa*:1/2 *aa*), following would be the result:

Genotypes	*Aa*	*aa*	Total
Observed	178	182	360
Expected	180	180	360

The χ^2 value would be 0 = 0.25 for one degree of freedom (n – 1). This shows excellent agreement with expected ratio (1:1).

The χ^2 value for gene *Bb* can be calculated in the similar way, like

Genotypes	*Bb*	*bb*	Total
Observed	172 (140 + 32)	188 (150 + 38)	360
Expected	(1:1) 180	180	360

The χ^2 value would be 0.62 for one degree of freedom (n – 1). It is also within the limit of a normal expected 1:1 ratio.

On the other hand, χ^2 value for the random assortment of two pairs of genes (*AaBb*) gives entirely different results. Since original cross was *AABB* × *aabb*; if the assortment of the two pairs of genes had been independent of among the genes, number of new recombinant should have been equal to the number of parental types. But the observed values were just opposite to the expected ratio, as shown here:

Genotypes	Parental	Recombinant	Total
Observed	290(*AB* + *aa*)	70(*aB* + *bb*)	360
Expected (1:1)	180	180	360

The χ^2 value would be 133.22.

The departure from the random assortment is quite large. The actual χ^2 value 133.22 is highly significant at one degree of freedom, the cause of departure from 1:1:1:1 ratio could be ascribed to linkage.

Test for Linkage on the Basis of F_2 Data

If the genes are not assorting independently, the test for linkage is similar to those described above for independent assortment of genes. For example, in a dihybrid cross *AABB* × *aabb*, the F_2 population segregates, like

Genotypes	*AABB*	*AAbb*	*aaBB*	*aabb*	Total
Observed	180	30	60	10	280
Expected (9:3:3:1)	157.5	52.5	52.5	17.5	280

The χ^2 value would be 17.14.

The χ^2 value signify that probability is less than 0.001 for three degree of freedom that above said segregation is in the ratio of 9:3:3:1. The above said departure is due to linkage.

Gene Mapping

The early geneticists believed that recombination between the genes occurs by an exchange of segment between homologous chromosomes in a process, called crossing over. This idea was based on the origin of chiasmata (the cross-shaped configuration) between the homologous chromosomes in the first meiotic prophase. The idea was that a chiasma results from breakage and rejoining of the chromatids during synapsis where there is an exchange of corresponding segment between them.

The unit distance in a genetic map is called a *map unit* to describe distances between the linked genes. A map unit is equal to 1 per cent of crossover (recombination); that it represents the linear distance along the chromosome for which a recombination frequency of 1 per cent is observed. These distances can be expressed in *Morgan unit*. One Morgan unit represents 100 per cent crossing over. One per cent crossing over can also be expressed in 1 centimorgan (1c.m.); 10 per cent crossing over as 1 decimorgan and so on. The Morgan unit is obviously named in the honour of T.H. Morgan.

There is an important distinction between the distance of two genes in map units and the recombination as frequency between the genes. The map units measures how much recombination actually occurs between the genes and the recombination frequency says how much recombination is actually observed.

Figure 8.21 is an abbreviated genetic map of barley (*H. vulgare*)

Three-point Test

The most common method used in the genetic mapping of *Drosophila* chromosomes, corn and other eukaryotes (non-human) is the trihybrid (or three-point test cross). The results of a test cross in corn with three genes in a single chromosome illustrate the analysis of a three-point test-cross. Three examples have been given to explain the gene mapping.

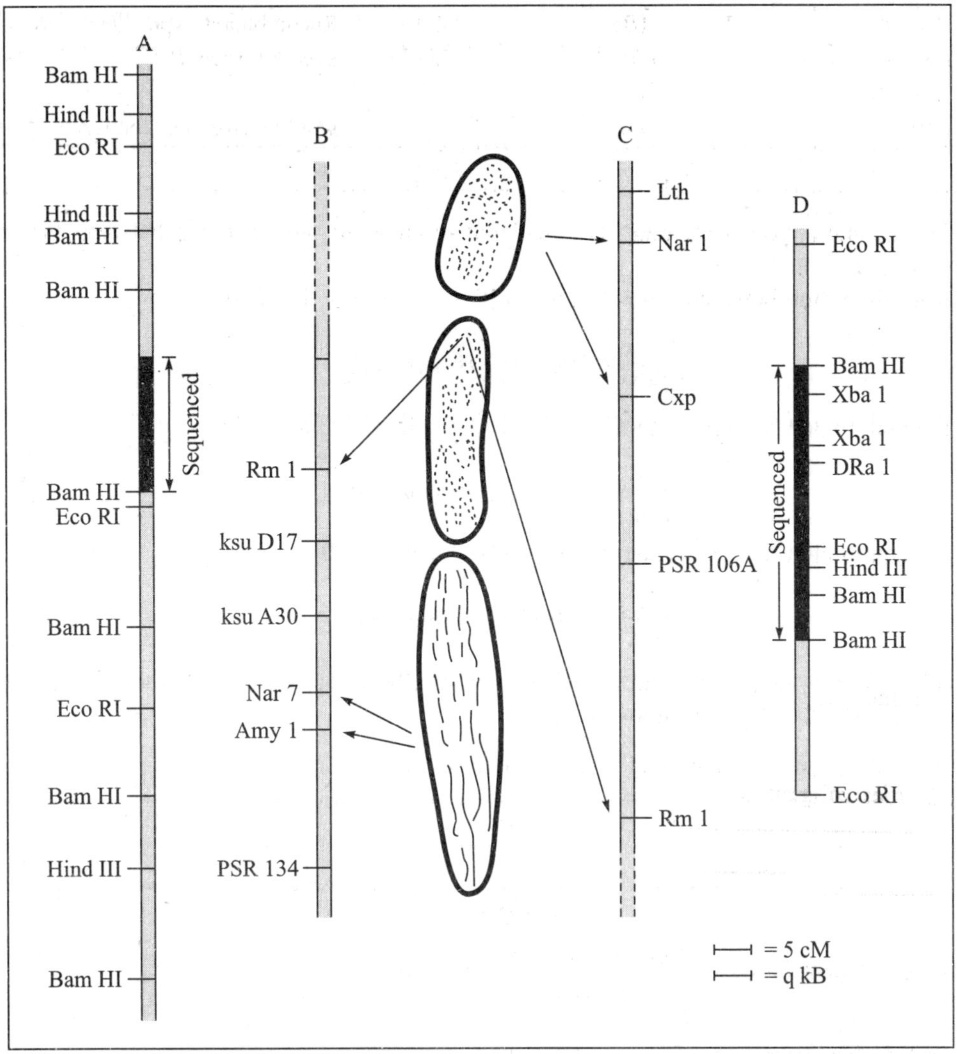

Figure 8.21 Mapping of a segment of chromosome of barley. B and C show several map genes on a barley chromosome, line A shows the details of the region B, and line D shows the details of region C.

EXAMPLE 1 Hypothetical linkage data for three-point test cross between *ABC* × *abc*—method of calculation:

Genotypic class	*Phenotypic class*		*Observed frequency*
ABC/abc	*ABC*	360	Parental type
Abc/abc	*abc*	375	
Abc/abc	*Abc*	120	Recombinant type due to crossing over between *A–B*
aBC/abc	*aBC*	115	

ABc/abc	*ABc*	130	Recombinant type due to crossing over between *B–C*
abC/abc	*abC*	125	
Abc/abc	*Abc*	7	
aBc/abc	*abc*	8	Double crossover between *A–c*
		1240	

The recombination percentage is calculated to find out unit distance between the linked genes, *A–B*, *B–c* and *A–c*.

Recombination between gene $A - B = 120 + 115 + 7 + 8 = 250$

$$\frac{250}{1240} \times 100 = 20.16 \text{ per cent}$$

Recombination between gene $B - C = 130 + 125 + 7 + 8 = 270$

$$\frac{270}{1240} \times 100 = 21.77 \text{ per cent}$$

Recombination between gene $A - C = 250 + 270 = 520$

$$\frac{520}{1240} \times 100 = 41.93 \text{ per cent}$$

This indicates that maximum crossing over (%) between gene *A–C* is 41.93 per cent. It means that gene *C* is located between *A–C*.

Hence, order of gene is *ABC* and distance between *A–C*.

A	*B*	*C*
20.16	21.77	
	41.93	

$$\text{Coefficient of coincidence} = \frac{\text{Observed double cross}}{\text{Expected}} \times 100$$

$$\text{Observed double cross} = \frac{15}{1240} \times 100 = 1.209 \text{ per cent}$$

$$\text{Expected double cross} = 20.16 \times \frac{21.77}{100} = 4.388 \text{ per cent}$$

$$\text{Coefficient of coincidence} = \frac{1.209}{4.388} \times 100 = 27.55 \text{ per cent}$$

Coefficient of interference $= 1 - 0.2755 = 72.45$ per cent

EXAMPLE 2 In *Drosophila melanogaster* three-point test was examined by a 'hypothetical cross', using three genes, symbolized as under:

Gene symbol	*Phenotype*
+	Normal wing (dominant)

cu	Curved wing (recessive)
+	Normal thorax (dominant)
sr	Stripped thorax (recessive)
+	Normal bristle (dominant)
ss	Spineless bristle (recessive)

Parents—female (♀) and male (♂) involved in this hypothetical cross were:

(++) (++) (++) *cu cu ss ss sr sr*
(Normal, normal, normal) (♀) × (Curved, spineless, stripped) (♂)

The normal individual in each hypothetical cross and the segregation pattern of three genes in F_2 generation showing linkage would be as under:

Phenotype	*Genotype*	*Number*	*Per cent*	*Type*
Normal, normal, normal	+++	430	88.2	Non-crossover
Curved, spineless, stripped	*cu ss sr*	452		(parental type)
Normal, spineless, stripped	+ *ss sr*	45	8.3	*cu-ss* single
Curved, normal, normal	*cu* + +	38		crossover
Normal, normal, stripped	+ + *sr*	16	3.3	*ss-sr* single
Curved, spineless, normal	*cu ss* +	17		crossover
Normal, spineless, normal	+ *ss* +	1	0.2	Double crossover
Curved, normal, stripped	*cu* + *sr*	1		
		1000	100	

By grouping the F_2 data as above, several important consideration arose.

1. The parental genotypes (non-crossover) received from the above cross would be largest in number and the phenotypes may disclose whether the linkage is in *cis* or *trans* phase. For example, normal, normal, normal individuals must have received +++ genes from maternal parent and *cu ss sr* genes from paternal parent. Because the non-crossover (parental) gametes would be more frequent, than crossover, the gametes, +++ and *cu ss sr* will occur in majority, if linkage is in *cis* configuration.
2. Double crossover individuals (those that would not be noted in a two-paired cross involving *cu* and *sr*) show up.
3. Double crossover could be recognized as numerically-smallest groups of individuals and may be used to determine the genes sequence.

The double crossovers are numerically the smallest group, because, the product law of probability applies. If the single crossover frequency between genes 1 and 2 is 0.1 and the crossover frequency between genes 2 and 3 is 0.2, then the probability of two crossovers in the same chromosome (chromatid) is one between genes 1 and 2 and another between genes 2 and 3 thus, a double cross should be 0.1×0.2 or 0.02—a much smaller frequency than for either single crossover.

In double crossovers + *ss* + and *cu* + *sr* support the view that gene sequence is *cis*, configuration in the parental stocks would be

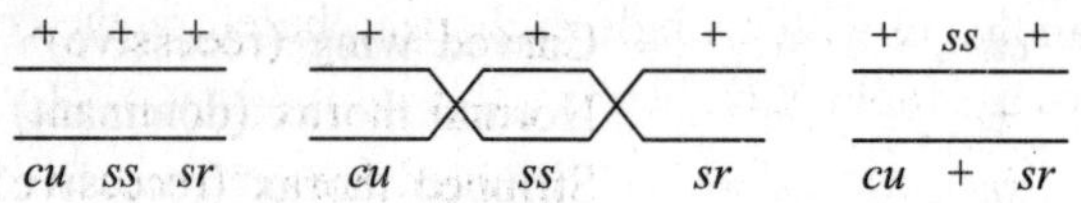

If these genes are arranged in either of the other two possible sequence (*sr cu ss* or *ss sr cu*), the outcome or double crossing over should be incompatible with the result observed in the following example:

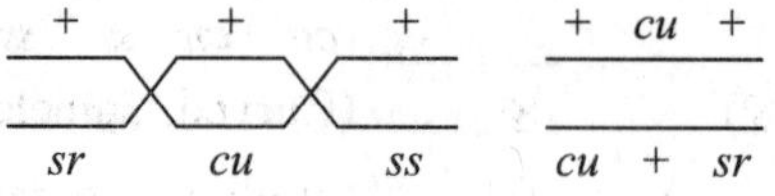

Total 8.3 in this cross

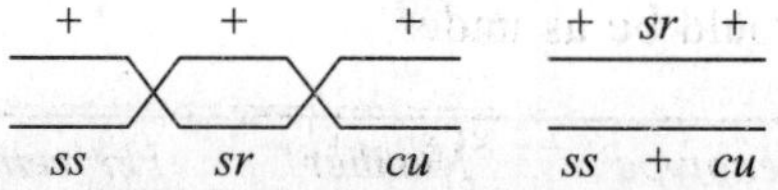

Total 3.3 in this cross

Thus, gene sequence here will be *cu ss sr*. Therefore,

- the true distance between *cu* and *ss* is 8.3 + 0.2 = 8.5 units
- the distance between *ss* and *sr* is 3.3 + 0.2 = 3.5 units.

So, the true distance between *cu* and *sr* is 8.3 + 0.2 + 3.3 + 0.2 = 12.0
The gene order along with distance could be given as under:

Cu *ss* *sr*

0.0 +8.5 12.0

EXAMPLE 3 In corn, the genetic composition of the parents of the multiple heterozygote in the cross had the genotypes *Lz Gl Su* × *lz gl su*. The recessive gene alleles symbolised as *lz* (lazy or prostrate growth) *gl* (glossy leaf) and *su* (sugary). The two class of segregates among the progeny that represents non-crossover (parental type) gametes are the normal plants. Two parental chromosomes are:

In case of linkage, parental types are larger than any of the crossover classes. Thus, if the combination of dominant and recessive alleles in the chromosomes of the heterozygous parents are unknown, one can deduce from their relative frequency in the progeny where non-crossover gametes (parental) were *Lz Gl Su*. This is important point to note as a general principle.

> *"In any genetic cross no matter how complex, the two most-frequent classes of phenotypes identify the no-recombinants; these classes provide the linkage phase (cis or trans) of the alleles in the multiple heterozygous parents."*

The gene frequency is not known in the mapping experiments, because the order in which three genes are located is entirely arbitrary in this example. But there is an easy way to find out the correct order from the *three-point data.* For example, find out first the double crossover gametes produced by the heterozygous parents (F_1) and compare them with the parental gametes. Because the possibility of the simultaneous occurrence of two exchanges is considerably smaller than that of within single exchange, therefore, the double crossover gametes will be least frequent types (Table 8.2).

TABLE 8.2 Offsprings of a trihybrid cross for three-point test cross study

Phenotype of test cross	*Genotype of gamete from hybrid* (F_1) *parents*	*Number*
Normal (wild type)	*Lz Gl Su*	286
Lazy	*lz Gl Su*	33
Glossy	*Lz gl Su*	59
Sugary	*Lz GL su*	4
Lazy, glossy	*lz gl Su*	2
Lazy, sugary	*lz GL su*	44
Glossy, sugary	*Lz gl su*	40
Lazy, glossy and sugary	*lz gl su*	272

Thus, if the parental genotypes are *Lz Lg Su* and *lz gl su*, the double crossover would be *Lz Gl su* and *lz gl Su*.

The *Su* and *su* are interchanged by double crossing over, hence should be middle pair of allele, hence, genotype of the heterozygous parent in the present cross should be:

$$\frac{Lz \quad Su \quad Gl}{lz \quad su \quad gl}$$

Then, two double crossovers should be:

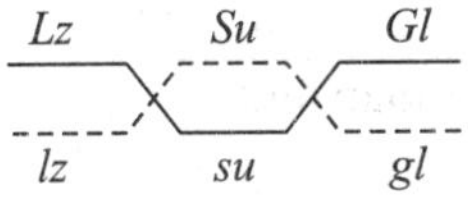

$$\frac{Lz \; Su \; Gl}{lz \; su \; gl}$$

The reciprocal product of a single crossover between *lz su* would be *Lz su gl* and *lz su Gl* and product of a single exchange between *su* and *gl* would be *Lz Su gl* and *lz su Gl*. Thus in F_2, the genotypic frequency of parental; two single crossover between *lz* and *su* as well as *su gl* and the frequency of double crossover would be as follows:

Frequency of parental types

$$Lz\ Su\ Gl = 286$$
$$lz\ su\ gl = 272$$

Single crossover between *lz* and *su*, *Lz su gl* = 40
Single crossover between *su* and *gl lz su gl* = 33
Frequency of double crossover:

$$Lz\ su\ Gl = 4$$
$$lz\ Su\ gl - 2$$

It is a point to note that two reciprocal products resulting from any crossover are expected to occur in approximately equal frequencies (280 and 272 as well as 40 and 33).

While calculating the frequency of crossover of the present data, one thing need to remember that the double recombinant resulted from two exchanges, one in each of the chromosome region defined by three genes, the crossover between *lz* and *su* could be therefore, represented by following genotypes:

Lz su gl	40
lz Su gl	33
Lz su Gl	4
lz Su gl	2
	Total = 79

That is 79

740 × 100 = 10.7%

Therefore, map distance between *lz* – *su* genes is 10.7 units.

Similarly, the crossover between *su* and *gl* may be represented as:

Lz Su gl	99
lz su Gl	44
Lz su Gl	4
lz Su gl	2
	Total = 109

The crossover frequencies between the second pair of gene = 109

The distance between gene *su* – *gl* = 109/740 × 100 = 14.8%

In other word, 14.8 is the map distance between the genes *su* – *gl*, therefore, the order of the gene on the chromosome may be

lz *su* *gl*

10.7 map unit 14.8 map unit

Construction of Linkage Map

First, determination of linkage relationship between different genes in any organism is done. The conventional procedure is to establish linkage between two genes; then rest of the genes falling in the particular linkage group are determined, using three-point cross test. The procedure is quite simple. If genes *a* and *b* are closely linked on a segment *ab*, then third gene *c*, if linked with the genes *a* and *b* on the same segment, can be determined in the sequence *a b c* or *a c b* on the basis of observed recombination values, like

a *b* *c*

$$ac = ab + bc$$

a *c* *b*

$$ac = ab - cb$$

This way, on the basis of observed recombination values the relative position of the genes located on a segment of chromosome could be fixed. A linkage map gives the relative position of genes on the chromosome, the sequence of the arrangement of the genes and distance between the genes in terms of recombination value.

One map unit is equivalent to a recombination value of 10 per cent, which is measured in Morgan unit, i.e. one map unit is equal to one Morgan map distance. However, map distance measured in terms of recombination values of two genes may be accurate, if two genes are closely related. In case of distantly located genes, map distance will be less than the true recombination value because double and multiple crossovers will be missing.

This technique was first used to map the genes located on chromosome 4 (haploid) of *Drosophila melanogaster.* Later on, 10 linkage groups were prepared in maize, 7 in pea, 7 in barley and 12 in tomato. The number of linkage groups corresponds to the haploid number of chromosomes in an organism.

How to Identify a Particular Linkage Group with a Specific Chromosome

The smallest linkage group (chromosome 4) in *Drosophila melanogaster* was found to have 12 genes while there were more than 140 genes in other groups. It appeared that these 12 genes were located on chromosome 4, which is smallest among the four chromosome pairs. This assumption was confirmed by aneuploid analysis.

An aneuploid is a trisomic (2n + 1) or a monosomic (2n – 1). The genes located on an extra chromosome (2n + 1) disturb the ratio of their segregation in crosses where other parent is normal. It is assumed that genes forming part of a particular linkage group are showing disturbed segregation. Then, these are expected to be located on the extra homologous chromosome. Using trisomics (2n + 1) for different chromosomes or monosomics (2n – 1) in which a specific chromosome is missing its homologous partner, it is possible to assign different linkage groups to specific chromosomes.

Another technique used in location of genes is *chromosomal rearrangement*, such as translocation of a segment of chromosome to other chromosome; a small segment is inverted within same chromosome or lost through process of deletion. Rearrangement of this kind is expected to change the linkage relationship of genes located on the chromosome involved in the rearrangement.

Physical Distance and Map Distance of Genes

By assigning a linkage map to a specific chromosome does not mean that we can give actual measure of physical distance between different genes located on a chromosome. The reason is that distribution of crossover point in a chromosome is not random. It has been observed that crossing over is suppressed in some parts of chromosome and double crossovers occur at a frequency less than that is expected. The expected frequency of double crossovers in a given segment can be easily calculated from the observed amount of recombination into the smaller segment. For example, in a chromosome segment *A–C*, the expected frequency of double crossovers can be derived from recombinant value of two short segments *A–B* and *B–C*, like

A *B* *C*

10% 15%

- The probability of crossover in *A–B* segment = 0.10
- The probability of crossover in *B–C* segment = 0.15
- The expected probability of double crossover in *A–C* = 0.10 × 0.15 = 0.015
- The expect frequency of double crossovers = 0.015 × 100 = 1.5 per cent

In large number of cases, crossing over in one segment has a tendency to suppress crossing over in the neighbouring segment (neighbourhood). Muller termed it *interference*.

INTERFERENCE AND COINCIDENCE

The detection of double crossover makes it possible to determine whether exchanges of chromatid segments in two different regions of a pair of chromosomes occur independent of each other. Using the information from the example of corn, we know the recombination frequencies, where probability of exchange is 0.107 between genes *lz* and *su* and 0.148 between genes *su* and *gl*.

If crossing over occurs independently in the two regions, i.e. if occurrence of one exchange does not interfere the probability of the second exchange, the probability of the occurrence of an exchange in both regions is the product of these individual probability, that is 0.107 × 0.148 = 0.0158 or 1.58 per cent. This means that in a sample of 740, the expected number of double crossovers would be 740 × 0.0158 or 12. Whereas, the numbers observed were only 6 (4 + 2). Such deficiencies in the observed values are common and has been identified a phenomenon, called *interference*. Soon a crossover occurs, the probability of occurrence of a second crossover in an adjacent region is reduced. This phenomenon is called *interference*.

Interference is unequal in different parts of the chromosome as well as among the several chromosomes of a given complement. Interference is greatest near centromere and at the end of chromosome. The degree of interference is commonly expressed as *coefficient of coincidence* or *coincidence*.

The coefficient of coincidence is the number of double recombination divided by the expected number; its value provides a simple measure of the degree of interference, defined as: Interference = 1 (coefficient coincidence).

From the data in the present example, the coefficient of coincidence is 6/12 = 0.50, means that the number of double crosses that occurred was only 50 per cent of the expected, if crossing over in the two regions were independent.

$$\text{Coincidence} = \frac{\text{Actual frequency of double crossover}}{\text{Calculated frequency of double crossover}}$$

In *Drosophila* (hypothetical cross), the coincidence

$$= \frac{0.002}{0.003}$$

$$= 0.67$$

As the interference decreases, coincidence increases. The coincidence values ordinarily vary between 0 and 1. Absence of interference gives a coincidence value of 1 whereas complete interference results in a coincidence of 0. Coincidence is quite small for short map distance.

MITOTIC RECOMBINATION

Genetic recombination has also been reported in mitosis though the frequency is about 1000 times lower than that in meiosis. Mitotic recombination was first reported in *Drosophila melanogaster*, but later, as this phenomenon was studied in fungi, such as *yeast* and *Aspergillus* in which frequencies are higher than other organisms. Genetic maps can be constructed from mitotic recombination frequencies.

RECOMBINATION WITHIN GENE

According to the earlier genetical studies (before 1940) gene was considered a smallest unit of the genetic materials capable of alternation by mutation and recombination through crossing over. Crossing over was observed in all regions of the chromosome between the genes but not within the gene and thus gene was considered indivisible unit.

The classical concept of gene changed soon as it was realized that in *Drosophila*, there are several genes where recombination takes place within the genes. The initial observation came from the investigation of *Lozenge* locus or gene *Lz* in the *X* chromosomes of fruit fly. The mutant alleles lozenge are recessive and their phenotypic effects include disturbed arrangements of facets of compound eyes and reduction in the red pigment in *Drosophila* eyes. The females heterozygous for two different *lz* alleles one in each homologous chromosome has *lozenge* eye. When such heterozygous females are crossed with males having one of the *lz* mutant alleles, the flies having normal eye are occasionally found. In one cross in which *lz bs* and *lz* were the mutant alleles of lozenge, and *ct* (cut wing) and *v* (vermillion) eye colour were two genetic markers that map 7.7 units to the left and 5.3 units to the right respectively of the *lz* locus. These rare individuals cannot be due to the mutation of the *lz* allele, because of higher frequencies of recombination than reverse mutation (Figure 8.22).

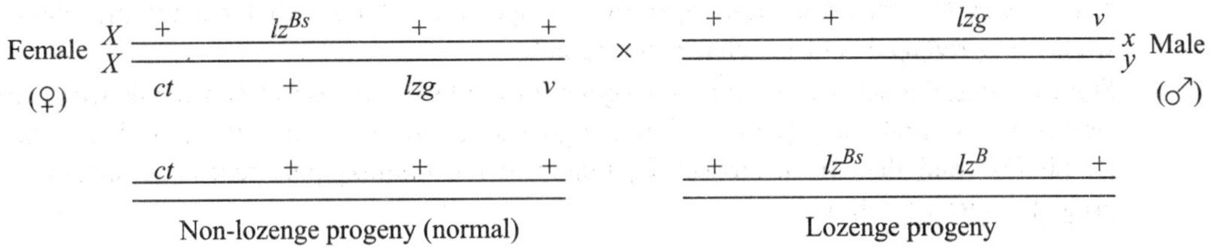

Figure 8.22 Recombination within gene.

All the rare non-lozenge progeny had *X* chromosome with constitution *ct* ++– + which could be accounted for crossover between two lozenge alleles. Proof of this conclusion came from the detection of reciprocal crossover chromosomes + lz^{Bs} Lz^{g} in five male progeny having a lozenge phenotype distinctly different from the phenotype resulting from the presence of either an lz^{Bs} or lz^{g} allele alone.

The observation of intragenic crossover indicated that genes indeed have fine structure and that the multiplicity of allelic forms of some genes might be due to mutations at different sites in the gene.

REVIEW QUESTIONS

1. If two genes are located on the same chromosome in an organism how will you determine that they are in coupling phase or repulsion? What are the advantages and disadvantages of the test cross to determine the linkage relationships?
2. From a cross between individuals with genotypes *CcDdEe* × *ccddee*, 1000 offsprings were obtained. The class appearing *CCDDee* included 351 individuals. Are the genes, *c*, *d* and *e* are linked or located on different chromosome pairs?
3. A poultry (bird) having *RrSs* genotype produced 1020 eggs of which 127 are *rst*, 121 *rST* and 130 *RST*. Are three pairs of alleles found on the same chromosome or on the two different chromosomes, or on the three different chromosomes? Explain.
4. If two loci are 10 map unit apart, what proportion of the meiotic events will contain a single crossover in the region between these two loci, assuming that no multiple crossover occurs.
5. Genes *a* and *b* are linked and are 20 map units apart. An a^+b^+/a^+b^+ female individual mated with an *abc* male.
   (a) Represent the cross on chromosome; illustrate the gamete produced by each parent and illustrate the genotypic composition of F_1.
   (b) What types of gametes will be produced by F_1 and in what proportions?
   (c) If F_1 was crossed with double recessive what offspring would be expected and in what proportion?
   (d) Is this an example of coupling or repulsions?
6. If the recombination frequency in the problems 4 and 5 were 40 per cent instead of 20 per cent, what difference should it make in the proportion of gamete and test cross progeny?
7. If problems (a) and (b) with 20 per cent recombination were carried to the F_2 ($F_1 \times F_2$) and a^+ and b^+ were dominant over their respective alleles, what phenotypic classes would be produced and in what proportion?
8. Stadler crossed a stock of corn homozygous for the linked recessive genes colourless (c), shrunken (sh), and waxy (*wx*) to a homozygous stock for dominant alleles of these genes (*C* Sh *Wx*) and then back-crossed F_1 plants to the homozygous recessive stock. The progeny were as follows:

Colourless plump waxy *c* Sh *wx*	2349
Coloured shrunken non-waxy *C* sh *Wx*	2216
Colourless plump non-waxy *c* Sh *Wx*	974
Coloured shrunken waxy *C* sh *wx*	951
Colourless non-waxy shrunken *c Wx* sh	84
Coloured waxy plump *C wx* Sh	99
Colourless waxy shrunken *c wx* sh	20
Coloured non-waxy plump *C Wx* Sh	15
	6708

(a) Draw a linkage map for three genes.
(b) Calculate the degree of coincidence and interference.
(c) With respect to double crossovers.

9. In tomato, the mutant genes *O* (oblate = flattened fruit), *P* (peach = hairy fruit) and *G* (compound inflorescence may flower is a cluster) were found to be on chromosome 2. From the undermentioned data (test cross mating of F_1 heterozygous for all three genes × homozygous recessive for all three genes) determine (a) the sequence of these three genes on chromosome 2, (b) the genotype of homozygous parents used in making F_1 heterozygote, (c) the recombination distances between genes, and (d) the coefficient of coincidences.

Phenotype of test cross progeny	*Number*
+++	73
++*s*	348
+*p*+	2
+*ps*	96
O++	110
O+*s*	2
OP+	306
Ops	63

REFERENCES

Beadle, G.W. (1932): The relation of crossing over to chromosome association in *Zea–Euglean* hybrid, *Genetics*, 17–18.

Brown and Johry, D. (1915): Relation of chiasmata and crossing over in *Lilium formosanum*, *Genetics*, 850–873.

Cooper, K.W. (1949): The cytogenetics of meiosis in *Drosophila*: mitotic and meiotic autosomal chiasmata without crossing over in the male, *J. Morph.*, **84,** 81–122.

Creighton, M.R. and B. McClintock (1931): A correlation of cytological and genetical crossing over in *Zea mays.*, *Proc. Matl. Acad. Sci.*, **17,** 485.

Darlington, C.D. (1965): *Recent Advances in Cytology*, Churchill, London.

Kaufmann, B.P. (1934): Somatic mitosis in *Drosophila melanogaster*, *J. Morph.*, **56,** 125–155.

Morgon, T.H. (1910): Sex-limited inheritance in *Drosophila*, *Science*, **32,** 120.

Morgan, T.H. (1926): *The Theory of the Gene*, Yale University Press, New Haven.

Morgan, T.H. and Bridges, C.B. (1916): Sex-linked inheritance in *Drosophila*, **237,** 1.

Slizynski, B.M. (1964): Chiasmata in spermatocytes of *Drosophila melanogaster*; *Gene. Res.*, **5,** 80–84.

Strickberger, Monroe, W. (1976): *Genetics*, Prentice-Hall of India, New Delhi.

CHAPTER

9 *Evolution of Species*

The species of sexually reproducing organisms may be defined as, *population of interbreeding organisms that are reproductively isolated from other such population.* This indicates that interbreeding population of an organism/species shares in a common gene pool. Second, that the member of different species all reproductively isolated from each other, as a result do not interbreed.

Lysenko (1954) defined species as, *a distinct qualitatively definite state of loving matter*. According to him, species arise by leap. However, Darwin in 1859 said, species arise as a result of slow and gradual modifications through descent and natural selection. According to him, nature does not leap. This is true. Darwin's theory of evolution was, however, based on competition and survival of fittest. Natural selection does not in any way direct the variations in the organisms but it operates on the variation when they arise. The changes are not in any way directive or purposive. Natural selection acts as sieve to dropout unfit to adopt and diversify.

Hurst defined species as, *a group of individuals of common descent, with certain constant specific characters in common, which are represented in the nucleus of the cell, by constant and characteristic set of chromosomes, carrying homologous specific genes, causing inter-fertility and inter-sterility.* However, this is too long definition.

The species could be defined as, *a group of interbreeding populations with common descent and common genomic constitution.* A species is a discrete and independent evolutionary unit. A Japanese biologist and thinker Imanishi has stated that unit for explaining evolution is not the individual but species. Some major qualitative changes in the history of vertebrates such as transition from water to land; the development of endothermy in mammals and origin of humans could have taken place over a vast period of time. According to palaeontology, speciation arise from the slow and steady transformation of entire population (phyletic, gradualism), under its influence. We speak unbroken fossils series linking two forms by insensible gradation as the only complete mirror of Darwinian processes.

MECHANISM OF SPECIATION

So far Stebbins (1977) and others have given three types of isolation mechanisms for the origin of species:

1. Allopatric and sympatric isolation mechanism
2. Reproductive isolation mechanism (RIM)
3. Genetic mechanism

Allopatric and Sympatric Isolation Mechanism

Allopatric populations are those, which are isolated by geographic barriers, and sympatric population are those, which inhabit in the same geographic areas, but isolated by the ecological and edaphic barriers, as a result they are unable to interbreed.

The model of speciation through allopatric divergence and sympatric reinforcement is the most common model for the evolution of reproductive isolation mechanism (RIM) among the sexually reproducing population. Peterson (1980, 1981) and others (Carson, 1989) have challenged this. According to them, isolation is misplaced and is neither necessary nor desirable in a genetic evolutionary definition of species. This is because one may come across the population, which have been isolated geographically (sea, mountain and river, etc.) for millions of years and have developed morphological distinction, but still they do not fail to exchange gene freely if brought together. For example, in a peanut (*Arachis hypogea*), there are two types—spreading and erect plant types but they are monogenically-controlled character and they interbreed.

But there are population inhabiting even in the same area but they have still maintained their distinct genomic entity and they do not interbreed. Hence, they are called *species*. In U.S.A., there are eight species of frogs, the leopard frog (*Rana pipiens*); wood frog (*Rana sylvastica*) and bull frog (*Rana cateshina*), etc., which live in the same area in other region but they do not interbreed. Similarly, song sparrow (*Melospiza melodia*) and Linclon's sparrow (*Melospiza lincolni*) live together but do not interbreed.

In plants, sugar maple (*Acer saccharam*), red maple (*Acer rubrus*), and in pines, *Pinus palustris*; *P. virginiana*, *P. ponderosa* and *P. sabiniana* grow in the same area but do not intercross with each other, so that exchange of genes does not occur.

Reproductive Isolation Mechanism

The biological properties of organisms that prevent interbreeding are known as Reproductive Isolation Mechanism (RIM). The RIM may be further classified into:

1. Prezygotic isolation
2. Postzygotic isolation

Prezygotic isolation

The prezygotic mechanisms impede the formation of hybrid zygotes because of the following reasons:

(i) Spatial and ecological isolation
(ii) Seasonal or temporal isolation
(iii) Mechanical isolation

(i) *Spatial and ecological isolation:* As stated above, geographical isolation alone would not result into speciation of species or subspecies without ecological barriers such as climatic, biotic and edaphic factors. There population inhabiting in the different ecological habitats are called *ecotypes*. Such differentiation lead to speciation. For example, oaks, *Quercus coccinea* and *Quercus velutina* grow in the same region (sympatric population) but both are identically different and do not intercross. Because, *Quercus coccinea* grow in swamps on poorly-drained soil and *Quercus velutina* grow on dry soils. Thus, due to adaption to two ecological conditions these two species maintain their identity in the sympatric population.

(ii) *Seasonal or temporal isolation:* This type of isolation is brought about due to differences in the flowering duration on account of photoperiodic sensitivity, so that no hybridization is possible in nature. But this kind of isolation would not be effective if flowering period is brought at the same period by vernalization. For example, spring wheat and winter wheat differ in their flowering durations due to photoperiodic sensitivity but crossing can be done and fertile hybrids could be obtained, if plants of winter wheat are grown under phytotron under controlled temperature and light.

(iii) *Mechanical isolation:* Mechanical isolation is caused due to differences in the genital organs in the animal species so that copulation may not be easily brought about.

Postzygotic isolation

In some cases, fertilization can take place between crosses of different species but hybrid (F_1) would be sterile or in second generation F_2 progeny would be eliminated due to following reasons:

(i) The F_1 hybrid produced may be enviable or weak.
(ii) The F_1 hybrid may be viable and vigorous, but may be sterile.
(iii) The F_2 progeny of the vigorous fertile F_1 hybrid may be weak and sterile.

However, allopatric and sympatric isolations are external factors. The history of life is more adequately represented by a picture of 'punctuated equilibria' which get disturbed rarely to produce a major events of evolution (Gould and Eldredge, 1977). The fossil history of man is an example of punctuated equilibria while formation of species is a genetic phenomenon. The theory of allopatric (geographical) speciation suggests a different interpretation of palaeontological data. If new species arise very rapidly in small peripherally isolated local population, the great expectation of insensibly-graded fossil sequence is a Chimera. A new species does not arise in an area of its ancestors. It does not arise from slow transformation of all its forebears. Many breaks in the fossil records are real and necessary too.

The question arise how new species came about or how reproductive isolation arises between the population. First, answer is that reproductive isolation mechanism (RIM) is a by-product of the accumulation of genetic differences between geographically separated diverging population and second, the condition under which the development of reproductive isolation might be accelerated. The natural selection favours genetic variants promoting mating between the individuals of the same population. This accelerates the reproductive isolation and ensures emergence of new species.

Genetic Mechanism

Now the questions arise, what are those genetic mechanism, which are responsible for reproductive isolation, and ensure emergence of new species? Let me illustrate them briefly. These are:

1. Gene mutation
2. Migration
3. Natural selection
4. Genetic drift
5. Chromosomal aberrations
6. Polyploidy

Gene mutation

Gene mutation is a potent source to create new variability in a population. Mutations are a slow but a continuous process. When a group of individuals of an organism is isolated by allopatric or sympatric barriers, barriers such as mountains and rivers, etc. natural hybridization between the groups is impossible. The new genetic variability is created by mutation (macromutation or micromutation or both) which brings proportional morphological changes in the population. The accumulation of mutation in an isolated population over the generations spread followed by shifting of gene frequencies due to *genetic drift* may proceed in a different direction to create a novel interactive system of genes. If process of genetic divergence continues for longer period of time, geographically separated population might become distinct as to be unable to interbreed, if the opportunity did eventually arise. Natural selection seive out the weak as a result in the course of time, new species originate.

Mayr, E. (1949) argued that evolution was faster in a small population than large areas because inbreeding lead to random genetic drift, accidental gene loss, and consequence increase of homozygosis and decrease of genotypic and phenotypic variability in the small population. These mechanisms caused rapid divergent evolution in the small population. The potentiality for rapid divergent evolution in small population, explains also why we have on islands so many dwarf, or races giant or races with peculiar colour characters (albinism, melanism) or with peculiar structures (long bills in birds) or other peculiar characters (loss of special male plumage in birds).

Oenothera brevistylis and *Oenothera lamarkiana*, six-rowed barley (*Hordeum vulgare*) and two-rowed barley (*Hordeum vulgare*) show monogenic differences. These species have arisen in nature due to gene mutation.

Mutations and genetic recombination: Mutations are ultimate source of genetic variability without which evolution is not possible. Had there been no mutation, all the loci in an organism would have been having a single allele; sexual reproduction would not have been effective in producing new genotypes and selection, which is a powerful force of evolutionary change, might have failed to operate on such a population that provides no alternatives to select from.

But mutations are very rare. The rates of spontaneous mutation in nature are one between 10^{-5} and 10^{-6} per generation (Falconer and Mackey, 1996). These values are almost equivalent to Muller's (1927) findings in *Drosophila melanogaster*.

In fact, major source of genetic variability in nature is recombination of genes because mutations alone in absence of selective advantage have less chance of survival. Of course recurrent mutations do bring about visible changes.

Mutational equilibrium: The effects of mutation on the population structure differ according to whether the mutational event is occurring rarely or repeatedly. Thus, according to the nature of occurrence, mutations are classified into non-recurrent and recurrent mutations.

The *non-recurrent* mutation gives rise to just one representative of the mutated gene or chromosome in the whole population. This type of mutation is of little importance as a cause of change in the gene frequency because such mutation has little chance of survival in a large population unless it has a selective advantage. Such mutation must be major gene mutation like dwarfing gene "Dee-Geo-Woo-Gen" in rice and 'Daruma' (Norin-10) in wheat. Because frequency of mutant gene A_1–A_2 is extremely low. According to the Hardy–Weinberg equilibrium, the gene frequency should not change in the subsequent generations. Therefore, a single gene mutation from A_1–A_2 without selective advantage cannot bring a permanent change in the population.

On the other hand, the *recurrent gene* mutation can bring a permanent change in the population structure because each mutational event occurs regularly with characteristic frequency and in a large population, the frequency to mutant gene is never low. For example, gene A_1 mutates to A_2 with a frequency x per generation (x is the proportion of all A_1 genes that mutate to A_2 between one generation and the next. If the frequency of A_1 in one generation is p, the frequency of newly mutated A_2 gene in the next generation may be xp, therefore, the frequency of new gene A_2 may be $p - xp$ and the change of gene frequency of new may be $-xp$.

Now suppose, gene mutation occurs in both directions $A_1 \rightleftarrows A_2$. For simplicity, let there be only two alleles A_1 and A_2 with initial frequencies of p and q, where A_1 mutates to A_2 at a rate of x per generation and A_2 mutates to A_1 at a rate of y. Then, after one generation, there may be a gain of A_2 genes equal to xp due to mutation in one direction, and a loss equal to yq due to mutation in other direction, thus, we have the situation

Mutation rate $A_1 \rightleftarrows A_2$

Initial gene frequency pq

Change in gene frequency in one generation $\Delta q = xp - yq$

This situation may lead to an equilibrium in gene frequency at which no further change may take place, because mutation is taking place in both the directions, which may lead to a point of equilibrium. The point of equilibrium can be found by equating the change of frequency Δq to zero. Thus, at equilibrium

$$px = qy$$

or

$$\frac{p}{q} = \frac{y}{x}$$

and

$$q = \frac{x}{x + y} \quad \text{(Falconer, 1960)}$$

It may take 10,000 generations and at the equilibrium, the frequency of mutant allele (a) may be higher than wild type (AA).

Migration

Migration is similar to mutation in that new alleles can be introduced into a local population although these alleles derive from another population rather than from mutation. Migration may be uni or bidirectional. If the migration of a new individual organism among the population is sufficiently rare, the local (recipient) population can undergo considerable genetic differentiation. The local population witness genetic differentiation, if there are differing frequencies of common alleles or if each populations, possesses one or more rare alleles not found earlier in other population. The genetic differences among the local populations can be swamped, if the rate of population exchange, i.e. migration is at sufficiently high rate. However, if the migration is very low and the rare gene alleles are not very active due to biological or genetic reasons, the migration may not be able to bring high levels of evolutionary genetic changes.

Natural selection

The operation of natural forces is a continuous process, which some time turns calamitous causing erratic changes in the population structure of the organisms. The driving force behind adaptive evolution is the natural selection, which is consequence of the genetic ability of the individual organism to survive and reproduce in the prevailing environment. Charles Darwin's concept of natural selection (1859) has been revised. With little revision in Darwin's theory of natural selection and survival of fittest, natural selection rests on three promises:

(i) In all organisms, more offsprings are produced than which survive and reproduce.
(ii) Those survive differ in their genetic constitution from those which are eliminated by natural forces.
(iii) The gene alleles that enhance the survival of an organism favours to build up a better and more adaptive population after a few generations because of increase in the frequency of favoured genotypes.

Concept of genetic fitness: We have enough understanding of Darwinian's classic of the survival of fittest and natural selection pressure. A Mendelian population is expected to have enough genetic variability due to mutation and recombination. But the fitness of a genotype in a population is counted not on account of its own fitness but its reproductive capacity and it is its relative reproductive capacity, which decides its gene frequencies in a population, which is collectively responsible for its fitness. If a genotype fails to reproduce or leaves fewer progeny than other competitive genotypes in the population; the proportion of gene alleles it carries will be reduced in the next generation. Darwin believed that natural selection provided forceful mechanism for the origin of the species. In my observation, Darwin's classical view of the natural selection and survival of fittest should be taken as 'genetic fitness' to survive in competitive life not only in the biotic environment but also in the socioeconomic and political world too. And if it is so, genetic fitness should not be taken limited to longer reproductive lifespan; say in human beings, lifespan has almost doubled or trebled, i.e. from 25 to 60–70 years causing population explosion the world is facing today. But genetic fitness is correlated to IQ (Intelligent Quotient) and Darwin's fitness could be equated to EQ (Emotional Quotient) because growth of three Indian billionaires—Mr. Mittal ($ 22 billions), Ambani brothers ($ 11 billions) and Azim Premji ($ 6 billions) worth wealth in a short span in the international business world, is not due to wealth competing wealth (capitalistic war) alone but they succeeded to excel other riches within country/abroad through their high managerial skill (EQ) supported with high IQ. IQ is genetically-controlled character but EQ is not. But high IQ without high EQ may prove fatal. It is high EQ, which creates conducive environment for the success of high IQ in career advancement; in business, world, trade and also in politics. In politics, even illiterate or semiliterate are very successful because of high EQ. Here, EQ could be equal to Darwinian's natural environment and IQ, governing the 'genetic fitness' of the individual. Same principle governs the fitness in biological world too, where polygenic variation play important role. Thus, many factors contribute the fitness of a genotype, not one.

Genetic load and genetic death: The new equilibrium of gene frequencies as a result of the combined effects of mutation and selection makes it clear that recurrent mutations make important contributions to the gene frequency of recessive gene. Mutation is thus an important cause of genetic load in a population. The term genetic load could be used to the impaired fitness of a population relative to the perfect fitness caused by wild genes. At this stage, selection coefficient of a genotype will be almost zero. If a beauty queen dies without bearing a child, then certain gene governing beauty spot may die forever. This may be called genetic death.

Adaptive radiation: The term adaptive radiation is often used to describe the process of radiating out of new evolutionary lines from a base population, leading to the formation of new races and species. Late blight of potato (*Phytoplethora infestans*) is a most damaging disease of potato throughout world. But this disease has spread its tentacles to halt dozen plant species by producing new races and species, viz. late blight of potato (*P. infestans*), tobacco black shank (*P. parasitica* var. *nicotianae*), colocasia rhizome rot (*P. colocasiae*), ruber pod rot (*P. fiberi*), apple collar rot (*P. cactorum*), pepper root rot (*P. palmivora* var. *peperina*) and palm bud rot (*P. palmivora*). There is another good example of adaptive radiation in 3,50,000 species of beetles which become adapted to varying habitats, habits and bionomics including rain forests, deserts and ice-covered terrains (Jain, 2000).

Rice: A wild rice *Oryza rufipogan* is considered the progenitor of the cultivated rice (*Oryza sativa*). Cytological evidences indicate that *Oryza sativa* is an aneuploid. Its 12 chromosomes (haploid) are: $a, b, c, d, e, a^1, b^1, c^1, d^1, e^1$ and a^1, b^1. It is hypothesized that a species with a haploid set of 5 chromosomes (a, b, c, d, e) hybridized with a second species with a haploid set of 5 chromosomes (a^1, b^1, c^1, d^1, e^1). Due to some meiotic irregularity two chromosomes (a^1, b^1) were duplicated in the hybrid and this was followed by doubling of the chromosomes to produce fertile progeny of *Oryza sativa*, the wild progenitors of *Oryza sativa* with 5 chromosomes appear to have disappeared (Figure 9.1).

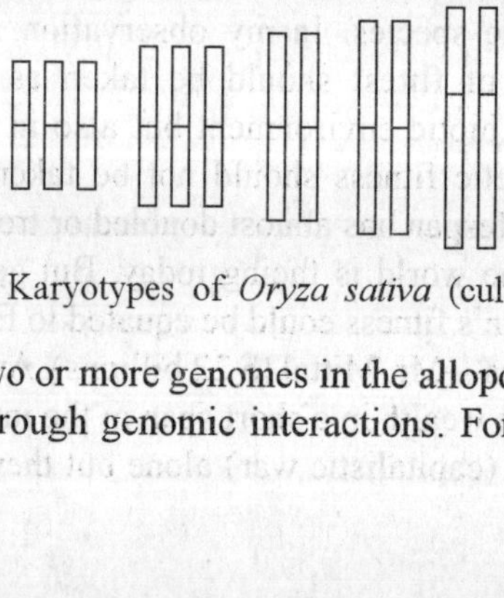

Figure 9.1 Karyotypes of *Oryza sativa* (cultivated rice).

Cotton: The combination of two or more genomes in the allopolyploids provide an opportunity for the origin of new species through genomic interactions. For example, *Gossypium hirsutum* and *Gossypium barbadense*, the two tetraploid species have evolved from hybrids between diploid species like *Gossypium herbaceum* which had their origin in Asia and new world's diploid species of cotton grow wild and they hardly produce any fibre (lint) on their seed. The Asian diploid species do produce lint, but lint yield is very poor. Combining genes from two groups of cotton species (allotetraploids) produced *G. hirsutum*, which produces lint of improved quality through genomic interaction.

$$\frac{\textit{Gossypium herbaceum}}{(2n = 26)} \times \frac{\textit{Gossypium raimondii}}{(2n = 26)}$$

$$\frac{\textit{Gossypium hirsutum}}{(2n = 52)}$$

Cultivated cotton arose on doubling of F_1 (n = 13 + n = 13) chromosomes of two crossed species.

Maize: Wild maize was a form of pod-corn, native to the low lands of South America. It is accepted view that corn was originated from *teosinte*, the nearest known relative of corn. The wild annual forms of *teosinte* have the same chromosome number as corn and cross readily, with corn to produce fertile hybrids. A simple recessive gene like corn-grass can entirely change the morphology of maize plant. This mutation hypothesis appears a possibility. But such mutations in *teosinte* or maize have not yet been produced (Figure 9.2).

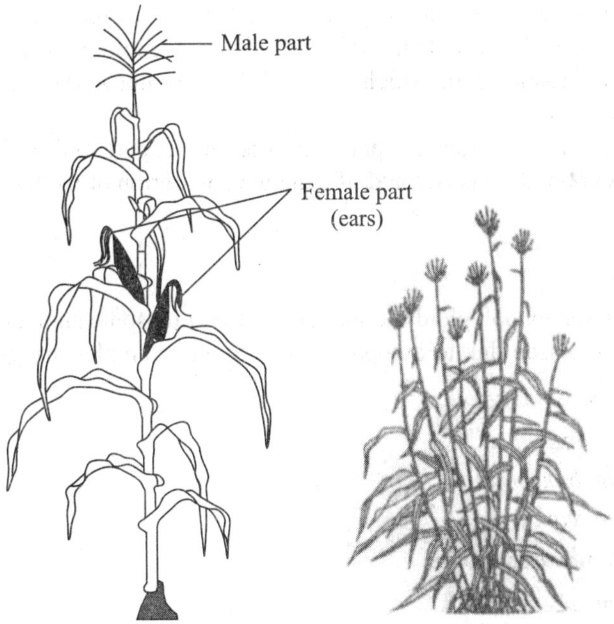

(a) Maize plant (b) *Teosinte* (ancestor of corn)

Figure 9.2 Morphology of maize and *teosinte*. The main difference in morphology is that maize produces a few large ears, while *teosinte* produces many small ones with seeds in a very hard casing.

Genetic drift

Random genetic drift is a statistical process of change in the allele frequency in a small population resulting from the inability of every individual to contribute equally to the offspring of successive generations. In a subdivided population, random genetic drift is the cause of divergence in the allele frequency. In an isolated population, barring mutation, an allele will ultimately become fixed or lost as a result of *genetic drift*. In general, random *genetic drift* causes differences in the allele frequency among sub-population and, therefore, cause genetic divergence among sub-population. The population that are small in size are particularly subject to larger effect of drift and, therefore, differentiate more. If selective conditions also change from one area to other due to adaptation to other local environments frequencies of certain genes may differ in different areas in such a way that a gradient of gene frequencies may ensue such a gradient, called a *cline*. Extreme genetic of drift occurs when some individuals break a way from main population.

Chromosomal aberrations

Drosophila nasuta and *Drosophila albomicans* are two closely related (sibling) species, widely distributed throughout the Inco-Pacific tropics. Although, morphologically-indistinguishable, these two species due to 'inversion' in the chromosome, are intersterile. Chang and Ayala (1984) have studied various reproductive isolating mechanisms and demonstrated how reproductive divergence is initiated as allopatric population which differentiated into separate species.

There are several types of aneuploids. But there are two main groups—first group is called *hypoploids*, which include all those aneuploids in which the chromosome number is less than 2n(2n – 1) or monosomics and second group is called *hyperploids* or *polysomics*, in which chromosome number exceeds more than 2n(2n + 1). Some aneuploids such as rice have received greater attention than trisomics in which, one of the chromosomes is represented by three homologues (2n + 1).

A genetic revolution in the founder population is only a potentiality, but does not need to happen every time. A population is isolated, if genetic constitution of the founder population does not favour it.

Polyploidy

Many species have arisen by polyploidy (duplication of chromosome numbers). A number of wild species of potato have arisen due to autopolyploidy. Some examples are given as under:

1. 2n = 24 *chromosome group:*
   - *Solanum ajanhuiri*
   - *Solanum bukasovil*
   - *Solanum vavilovii*
2. 2n = 36 *chromosome group:*
   - *Solanum juzepezukii*
   - *Solanum medians*
3. 2n = 48 *chromosome group:*
   - *Solanum acaule*
   - *Solanum ajuscoense*

- *Solanum andigenum*
- *Solanum edinense*
- *Solanum tuberosum*
- *Solanum antipoviezii*

4. 2n = 60 *chromosome group:*
   - *Solanum curtilobum*
   - *Solanum semidemisum*
5. 2n = 72 *chromosome group:*
   - *Solanum demissum*

Allopolyploidy: Amphidiploid: *Brassica juncea* has been originated as an amphidiploid of the cross between *Brassica compestris* and *Brassica nigra*. It is shown in the following cross:

Brassica compestris (n = 10) × *Brassica nigra* (n = 8)

↓

F_1(10 + 8)

↓

Doubling of chromosomes
2n = 36 (amphidiploid)

↓

Brassica juncea

The upland American cottons such as Cambodia are allopolyploids from a cross between 13 chromosomes of Asiatic cotton, probably *Gossypium arboreum* and new world cotton, probably *Gossypium raimondii*.

Hexaploidy: Origin of bread wheat (*Triticum aestivum*) is the best example of hexaploidy. Wheat species consist of three genomic groups, according to their chromosome numbers.

1. *Einkorn group* (2n = 14): *genome-A*
   - *Triticum aegiloploides*
   - *Triticum monococcum*
2. *Emmer group* (2n = 28): *genome-AB*
   - *Triticum dicocoides*
   - *Triticum dicoccum*
   - *Triticum durum*
   - *Triticum turgidum*
   - *Triticum polinicum*
   - *Triticum persicum*
   - *Triticum timopheevii*
3. *Vulgare group* (2n = 42): *genome ABD*
   - *Triticum spelta*
   - *Triticum vulgare*
   - *Triticum compactum*
   - *Triticum spaerococcum*

Emmer group is allotetraploid, and a hybrid between an einkorn group of species (*A* genome) and another species (*B* genome). A cross between emmer (*AB*) and species of *Aegilops* (*D* genome), viz. *Aegilops squorosa* by hexaploidy has given rise to bread wheat (*Triticum aestivum*).

Bread wheat (*Triticum aestivum*) is also a product of polyploidy (hexaploid). Its origin is shown in the following cross:

Triticum monococcum × *Aegilops speltoides* (goat grass)
$2n = 14(AA)$ ↓ $2n = 14(BB)$

F_1 $2n = 14(AB)$

↓

Doubling of chromosome

Amphidiploid ↓ *Aegilops squorosa* (goat grass)

×

(*AABB*) ↓ $2n – 14(DD)$

$4n = 28$ $(7A + 7B + 7D)$

Doubling of chromosome

↓

Hexaploid, $6n = 42$

↓

Bread wheat (*AABBDD*)

MANY CROPS EVOLVED FROM WEEDS

There are some cultivated crops which have been derived from weeds growing in the wheat, barley, maize, soy beans, flax and cotton. For example, 'Rye' was carried northward from south-western Asia as a weed in winter wheat fields. On reaching northern limits of winter wheat, rye completely replaced wheat and became an important crop due to winter hardiness. Cultivated oats also achieved the status of primary crop from weedy oats. Wulff (1950) listed a number of crops which originated from weeds, such as rye, oats, hemp, coriander and most probably kenaf (*Hibiscus cannabinus*) and Chinese jute (*Abutilon avincennae*).

Similarly, it is quite convincing that corn originated from the introgression of *teosinte* germplasm into corn since it has been under cultivation (Mangelsdorf, 1952). Since long time, *teosinte* is wild, weedy, near-relative to corn, which grows in and around corn fields in Central America. This type of introgression along with interracial hybridization, has increased enormously the range of variation in maize (*Zea mays*). Such introgressive hybridization have played important role in the evolution of many variable crop species (Anderson, 1949).

Near the distribution periphery of a plant species, there are often found unusual recessive forms, which might have resulted from selection following inbreeding, mutation or chance fixation in relatively small and partially isolated population. The seeds and fruits in some crop plants become larger in size as one proceed from centre of diversity to the periphery of the distribution. The species in the peripheral areas of distribution are often variable because of the unique ecological stresses imposed upon the more or less isolated population. The same species at the centre of distribution may be relatively stable. Carson (1953–1955) has shown this to be true for *Drosophila* species and it may be true in the plant and animal species too.

There is no relationship between the area of origin of a crop and the areas where the same crop is now extensively grown. Central North America is unsurpassed anywhere in the world in the total production of a number of crops, but this area does not contribute a single endemic major crop to this array. A few crop-plant genera, which have representatives in Asia and in New World, include *Fragaria*, *Gossypium*, *Helianthus*, *Hicoria*, *Juglans*, *Malus*, *Oxycoccus*, *Prunus*, *Rubus*, *Vaccinium* and *Vitis*. In addition, there are a number of grasses and legume species with relic distribution, particularly in the intermountain region and the Black Hill areas have close relative in Asia.

REFERENCES

Anderson, E. (1949): *Introgressive hybridization*, John Wiley & Sons, Inc., New York.

Carson, H.L. (1955): Genetic characteristics of marginal populations of *Drosophila*, *Cold Spring Harbour Symp on Quant. Biol.*, **20,** 276–287.

Carson, H.L. (1989): Genetic imbalance, religned selection and origin of species, in: *Genetics: Speciation and Founder Principles,* L.V. Giddengs, K.Y. Kaneshiro, and W.W. Anderson (Eds.), Oxford University Press, New York, 345–362.

Gould, S.J. and N. Eldredge (1977): Punctuated equilibria: The tempo and mode of hominid evolution, *Nature*, **292,** 113–22.

Hartl, D.L. (1991): *Basic Genetics*, Jones and Burtlet, Boston.

Jain, H.K. (2000): *Genetics: Principles, Concepts and Implications*, Oxford I.B.H., New Delhi.

Lysenko, T.D. (1954): *Agrobiology*, FLPH, Moscow.

Mangelsdorf, P.C. (1952): Hybridization in evolution of maize, Chapter 2, *in*: *Heterosis*, Gowery, J.W. (Ed.), Iowa State Col. Press, Iowa.

Mayr, E. (1949): Speciation and systematics, in: *Mechanism of Speciation*, Barigozzi, C. (Ed.), Alan, R. Liss, New York, 1–19.

Peterson, H.E.H. (1981): The continuing search for the unknown and the unknowable: A critique of contemporary idea on speciation, *S. Afr. J. Sci.*, **77,** 113–119.

Peterson, H.E.H. (1985): The recognition concept of species, in: *Species and Speciation*, Vrba, E.S. (Ed.), Transveel Museum Monograph No. 4, Pretoria, 21–92.

Provine, W.B. (1991): Mechanism of speciation: A review, in: *Evolution of Life*, Osawa and Honjo, T. (Eds.), Springer-Verlag, Tokyo, 201–214.

Stebbins, G.H. (1977): *Processes of Organic Evolution*, 3rd ed., Prentice Hall, Inc., New Jersey.

Wulff, E.V. (1950): *An Introduction to Historical Plant Geography*, Translated by *E*. Brissenden, The Chronica Botanica Company, Waltham Mass.

CHAPTER

10 Multiple Alleles and Blood Groups

Alleles can be defined as, *'an alternative forms of gene, affecting same character'*. A gene pairs in a diploid organism may contain, by definition, two alleles at a time, one for each member of the pair. This is because of homologous pairs of chromosomes; each of the two homologous chromosomes contains the allele of the gene pairs. If *R* determines the red colour and *r* the white of the flowers; since *R* and *r* affect the same trait and *Rr* hybrid segregates to give equal number of *R* or *r* gametes, they are called *alleles* and they are allelic to each other.

However, it is not true that expression of wild type (++) or normal gene allele is the same always and everywhere. Teemofeef Risowasky and Muller found that wild type *Drosophila* from different natural population had different dominant (red eye) alleles of the gene for white eye. Stern (1973) reported three different wild type alleles in another *Drosophila* mutant, *Cubitus interruptus* showed different degree of dominance over the same mutant. Such gene alleles are called *isoalleles* because these alleles differ in their phenotypic expression in different combinations.

MULTIPLE ALLELES

Changes in the gene by mutation are a rule of nature. However, it is also not a rule that gene alleles *YY* may show a change to *y* only but to other stable forms such as Y^1Y^1, Y^2Y^2, Y^3Y^3, ... Y^nY^n or y^1y^1, y^2y^2, y^3y^3, ... y^ny^n. These changes in the 'gene' alleles effects arise due to mutation on the same locus. This is called *multiple alleles*. For example, in rabbit a range of eye colours are found. When a pink eye rabbit is crossed, with mild albino (white) rabbit, F_1 is coloured. When coloured eye F_1 progeny are intercrossed, they produce colour: albino progeny in 3:1 ratio in F_2 generation. Here coloured (c^+c^+) and albino (*cc*) are alleles to each other.

There is another albino type of rabbit known as Himalayan type (c^hc^h) whose eyes are pink, their coat colour (fur) is white but their feet, tail and tip of nose are blackish or dark brown. When coloured (C^+C^+) and Himalayan (c^hc^h) rabbits are crossed, F_1 hybrid is coloured. When F_1 hybrid is intercrossed, in F_2 generation coloured eye and Himalayan eye rabbits appear in 3:1 ratio. These coloured (c^+c^+) and Himalayan (c^hc^h) genes are alleles to each other. Similarly, when Himalayan (c^hc^h) is crossed with mild albino (*cc*) type of rabbits, F_1 hybrid is always Himalayan type, but when F_1 are intercrossed, in F_2 generation, Himalayan and albino are found in 3:1 ratio. It is evident, therefore, that genes Himalayan (C^hC^h) and albino (cc) are alleles to each other. Coat colour variations in rabbits is given in Figure 10.1

(a) Agouti (wild type)
Genotypes: C^+C^+, $C^+ C^{ch} C^+ C^h C^+ C$

(b) Chinchilla
Genotypes: $C^{ch} C^{ch}$, $C^{ch} C^c C^{ch} C$

(c) Himalayan
Genotypes: $C^h C^h C^h C$

(d) Albino
Genotypes: cc

Figure 10.1 Coat colour variations in rabbit due to multiple allele effect.

There are several other coat colour genes in rabbits, which show different intensity of blackness and elimination of yellow coat colour. They behave as alleles of albino and Himalayan. When chinchilla, which is less intense in colour than grey or agouti, is crossed with mild albino, F_1 hybrid is grey in colour, i.e. intermediate between the parents. In F_2 generation, three types of rabbits in the ratio—1/4 chinchilla, 1/2 light gray, and 1/4 albino were observed. The genotypes and phenotypes of multiple alleles and the albino are given in Table 10.1.

TABLE 10.1 Genotypes and phenotypes of multiple alleles and the albino

Genotype	*Phenotype*
$C^+ C^+$	Wild type
$C^+ C^{ch} C^t C^h C^t C^a$	Wild type
$C^{ch} C^{ch}$	Chinchilla
$C^{ch} C^{ch} C^{ch} C^a$	Light gray
$C^h C^h C^h C^a$	Himalayan
$C^a C^a$	Albino

Similar series of multiple alleles of albino genes are also found in rat, guinea pigs and cat. In *Drosophila* about one dozen multiple alleles have been discovered for white eye (*ww*), called white eye series of eye colours with increasing intensity, such as ivory (*wiwi*), pearl (*wp*), tinged (*wt*), buff (*wbf*), honey (*wh*), apricot (*wa*), cherry (*wch*), eosin (*we*), blood (*wbl*), coral (*wco*), and red (wild-*W*).

In plants, snapdragon (*Antirrhinum*) is a good example. Baur found nine alleles, all affecting flower colour leading from deep red to ivory colour with or without red strips. Each allele appears to be dominant over those lighter than itself.

SELF-INCOMPATIBILITY ALLELES

Self-sterility is also called self-incompatibility. East and Mangeldorf in 1925 discovered self-sterility alleles in *Nicotiana sanderace*. In this system, self-incompatibility is controlled by a gene *s*, which is usually characterized by very large number of allelic forms in which it exist. This is also called *gametophytic incompatibility* (Figure 10.2). The gametophytic system gives rise to three types of pollination systems.

1. Fully incompatible ($S^1S^2 \times S^1S^2$). In this system, both alleles are common.
2. Half pollen compatible ($S^1S^2 \times S^1S^3$). In this case, one allele is different.
3. All the pollen grains are compatible ($S^1S^2 \times S^3S^4$). In this case, both alleles differ.

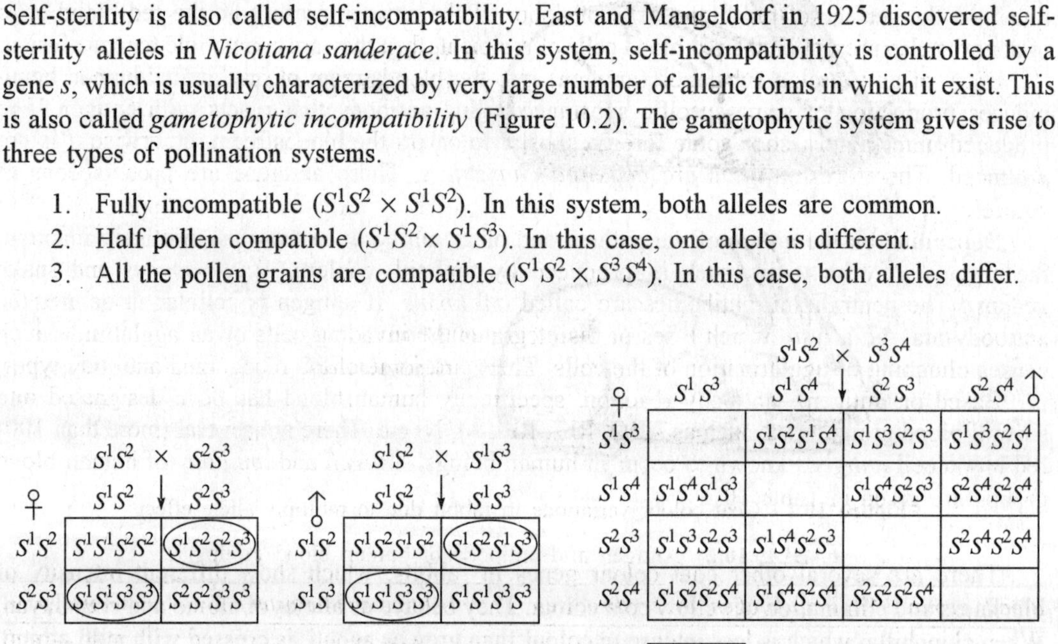

Figure 10.2 Mechanism of self-incompatibility gene alleles.

The pollination could be as under:

A S^1S^2 egg plant discriminates pollen grains S_1S_2; S_1S_3 egg plant discriminates against S_1S_3 pollen and S_2S_3 egg plant discriminates against S_2S_3 pollen by opposing the growth of pollen tube on the similar genetic constitution of style (egg plant).

THE IMMUNE RESPONSE

It is believed that persons once recovered from the infectious diseases like smallpox or influenza are less likely to succumb to the same disease second time. This is because these persons have acquired ability to resist the infection of smallpox or influenza. This process is called *immunity*, and to acquire immunity, vaccines have been developed by man for many diseases. The immunity evoking substances are called *antigens*.

Blood transfusion began in nineteenth century. In 1901 Dr. Karl Landsteiner working in Vienna noticed that red blood cells (erythrocytes) of some persons clump when mixed with the serum of some other persons. He reasoned for this clumping of blood on account of antigen—antibody reaction. It was, therefore, recognized that blood of donor and recipient must match in some way in order for the transfusion to be successful. When the blood of donor and recipient matches with each other and transfusion is successful without any complication, they are called *compatible*. On the other hand, when on blood transfusion, if the recipient develops shocks or dies, in such cases the blood donors and recipients are incompatible. However, blood from some donors might show compatible with some other recipients and at the same time incompatible with some other. The basis for this incompatibility was the presence of some genetically determined antigen found on the surface of red blood cells of donor. If the antigen present on the red blood cells of donor is also present on the red blood cells of recipient, then they are compatible for transfusion.

Entry of any foreign substance (antigen) into the bloodstream of any animal/human being induces production of some specific substance called antibody that reacts with antigen. The produced immediately, soon some foreign substance enters the bloodstream, otherwise it is not produced. Therefore, antigens are *acquired characters*. These antigens are protenaceous in nature.

Depending on the nature of the antigen and of its antibody, in antigen—antibody reaction, the antigen may be toxin (such as produced by typhoid, cholera. *Streptococcus* and snake venom). The neutralizing antibodies are called *antitoxins*. If antigen is cellular in nature, the antibody may be a *lysin* which lyses or disintegrates the invading cells or an agglutinin which causes clumping or agglutination of the cells. There are some of the recognized antibody types.

Based on antigen—antibody reaction, specifically human blood has been designated into several blood group types, such as ABO, Rh^+, Rh^-, M, N, etc. There are several (more than 100) red blood cell *antigens* known to occur in human beings. *Antigen* and *antibody* of human blood groups are given in Table 10.2.

TABLE 10.2 Antigen and antibody of human blood groups

Blood group	*Antigen in erythrocytes*	*Antibody in serum*	*Genotype*
A	A	Anti-B	I^AI^A or I^AI^O
B	B	Anti-A	I^BI^B or I^BI^O
AB	A and B	Neither	I^AI^B
O	Neither	Anti-A, anti-B	I^OI^O

Inheritance studies of a large number of human pedigree indicate that children produce A antigen where at least one parent possesses it. Similar is the case with B antigen. Blood group O individuals may occur in the progeny of A and or B parents. But O parents have only O children suggesting recessiveness of the alleles for group O. In some cases, group A and B parents produce children having both A and B antigens suggesting thereby that alleles for A and B genes are codominant.

Pedigree analysis indicates that individual possesses in homozygous or heterozygous state any two of a series of *multiple alleles*. These alleles are designated as:

1. Genotypes I^AI^A and I^AI^O produce A antigen on red cells. This blood group is called blood group A.
2. Genotypes I^BI^B and I^BI^O produce B antigen on red cells. This blood group is called blood group B.
3. Genotype I^AI^B produces both A and B antigens on red cells. These cells are of AB blood group.
4. Genotype I^OI^O has neither A nor B antigen on the red cells. The cells are of O blood group.

The ABO blood group *antigens* are unique among the several blood group antigens. Because non-A individuals contain naturally occurring antibodies to A and non-B individuals contain naturally occurring antibodies to B.

Naturally occurring antibodies are defined as antibodies that occur normally to the serum without a common source of stimulus.

The ABO blood group antigen poses problems in the blood transfusion because of the presence of the naturally occurring anti-A and anti-B antibodies in the serum. Therefore, before blood transfusion, testing of the blood group of both donor and recipient must be performed and classified. According to antigen—antibody reactions, human beings could be grouped into four blood group.

1. People of blood groups A produce anti-B antibody.
2. People of blood group B produce anti-A antibody.
3. People of blood group AB produce neither anti-A, nor anti-B antibody.
4. People of blood group O produce both anti-A and anti-B antibody.

Since serum and cells can be separated, red blood cells from individuals of blood group O, serves as universal donor. However, the serum of individuals of blood group O, can be given to the individuals of blood group O, AB, A and B. AB individuals are listed as universal recipients because they do not have anti-A or anti-B antibody and, therefore, can receive blood from A, B AB or O individuals. A perfect match is A to A and B to B between donors and recipients in blood transfusion.

Genetically, the AB *antigens* are coded by a locus with three alleles (I^A, I^B and I^O). The I^A and I^B alleles code for *A* and *B antigens* respectively and I^O alleles codes for none. Because the antigens involved one of the type known as *isoagglutinogen* or *isohemagglutinogens*, these alleles are often designated as I^A, I^B and *i* leaving asides the subgroups. According to the dominance relationship, they could be represented as $(I^AI^B) > i$. Recent findings of studies indicate that I^A may occur in at least four allelic forms, viz. A^1 A^2 A^3 A^4, where I^A is dominant to all other I^A

alleles. A^2 is recessive to I^A but dominant to other two and so on. According to their dominance relationship, they can be represented like:

$$[(I^{A1} > I^{A2} > I^{A3} > I^{A4}) = I^B] > i$$

Since A^4 is found very rare, the remaining series of multiple alleles produces 15 genotypes and 8 phenotypes, like:

Genotype	*Phenotype*
$I^{A1} I^{A1}$	
I^{A1}, I^{A2}	A_1
$I^{A1} I^{A3}$	
$I^A i$	
$I^{A2} I^{A2}$	
$I^{A2} I^{A3}$	
$I^{A2} i$	A_2
$I^{A3} I^{A3}$	A_3
$I^{A3} i$	
$I^{A1} I^B$	$I^{A1}B$
$I^{A2} I^B$	$A_1 B$
$I^{A3} I^B$	$A_3 B$
$I^B I^B$	B
$I^B i$	
ii	O

MEDICOLEGAL ASPECTS OF A–B–O BLOOD GROUPS

A number of cases have been reported, when children after delivery have been got mixed up knowingly or unknowingly and parents have made complaints that they were given some one else's child (changed) at the time of discharge of the mother and child from the hospital.

Some time back a case was filed in an Indian court to decide the paternity of two children who were suspected to have got changed by one of the parents at the time of taking the delivery of children from the hospital. The story is like this.

On reaching home, mother of one child detected that the bracelets she had put in the hand of her child was not there. She felt that her child had been changed at the time of taking delivery from the hospital.

The family members contacted the parents of other child who was also discharged almost at the same time suspecting that their real child is with them. But the parents of other child did not agree to return the child in exchange. This led a court case.

But surprisingly, the paternity test, conducted on the basis of blood group did not confirm that either of two children belonged to the two disputing families because their blood groups were quite different from their parents.

The blood groups of suspected parents are given below:

	Parental blood group	*Probable blood group of respective children*
First parents	A × AB	A, B, AB
Second parents	O × O	O

This was a typical case of paternity trials. Therefore, the parents who prefer for the delivery of the children in hospitals should be careful that their genetically well guided and cared children should not get changed in the hospital due to carelessness of maternity hospital staff.

The blood group tests are of vital importance in the cases of illegitimacy. On the basis of A–B–O system, possibility of acquittal of an accused man in a paternity case is only about 0.18. Now with the new DNA fingerprint technique, the accuracy could be raised to 100 per cent.

No doubt A–B–O blood groups have been found most reliable in genetic tests in determining the paternity dispute in human beings, but still in some cases, legal institutions have given their own views in the paternity disputes. One such incidence occurred in 1944 in California. A movie star was accused by a starlet-protége, father of her daughter. The plaintiff's case rested on the oral evidences of date, time and place of their meeting. The results of the blood tests of alleged father, mother and daughter conducted by three physicians are given below:

	Blood group
Alleged father	O
Mother	A
Daughter	B

The above resulting medical report proved that the defender was not the father. But some one else was the real father whose blood group must have been either B or AB. In view of such a clear-cut scientific evidence, the jury in a second trial (the first ended in a hung jury) found defendant's guilty and the defendant was directed to meet the maintenance of the baby for 21 years that was not his own.

A–B–O blood serum is also useful in cases of claim to estates, or in certain criminal proceedings, because blood group similarity could be established from corps, and even from the mummies. Because of these reasons, blood has an important anthropological tool in some investigations.

Rh BLOOD GROUP ANTIGEN SYSTEM

The Rh blood group antigen system was discovered in 1940 by Landsteiner and Wiener, who when infected the rabbit with blood of *Macaca rhesus* monkey, antibodies were formed by the rabbit. The resulting antiserum agglutinated the red blood cells of all the rhesus monkey. The surface of the monkey red blood cells carry a specific antigen called Rh. Test conducted on human blood indicated that most persons also produce this antigen (Rh). In fact, 85 per cent white Americans and more than 91 per cent black Americans produce Rh antigen. The persons, who produce the Rh antigen are called Rh-positive (Rh^+) and those who do not produce the rhesus antigen are designated as Rh-negative (Rh^-).

There is a genetic basis for Rh antigen—antibody reaction, wherein Rh^+ antigen is inherited as dominant trait. Originally, genetics of Rh antigen was thought to be simple that a single pair of genes *R* or *d* controls the system. But this is not the case.

The individuals possessing Rh antigen on their red blood cells are classified as Rh^+ (RR or Rh^+ Rr) and those lack this antigen are called Rh^- (*rr*). However, later studies indicated that Rh blood group system is very complex. There are two hypotheses about the genetic control of Rh blood system:

1. Wicner system, which theorizes that a series of at least 10 multiple alleles, some common and other very rare, are located at a single locus on each chromosome, are responsible to code for the Rh antigen. Each, except allele *rr*, produces one or more Rh antigens.
2. Fisher system, elaborated by Race and Sanger (1968), hypothesizes that a group of at least three very closely linked *pseudoalleles* (*Dd*, *Cc* and *Ee*) code for Rh antigen production. These three genes may give 27 genotypes. Each allele with exception of *dd* determines the presence of a distinct red blood cell antigen that can be detected with an appropriate antibody (Table 10.3).

TABLE 10.3 Rh phenotype and genotype

Phenotype	*Genotype*	
	Wicner	*Fisher*
Rh^+	R^0/r	*DCc/dcc*
Rh^+	R^+/R^1	*DCc/DCc*
Rh^+	R^1/r	*DCc/dee*
Rh^+	R^2/R^2	*DeE/DeE*
Rh^+	R^1/R^2	*DCc/DeE*
Rh^+	R^1/R^2	*DCc/DCE*
Rh^-	r/r	*dce/dce*

These antigens are designated by same letter as the allele, viz. *C*, *c*, *D*, *E* and *e*. For example, genes *CC/cc* produce three genotypes—*CC*, *Cc* and *cc*. The genotype *CC* produces *C* antigen and *Cc* produces both *C* and *cc* antigens, and *cc* produces antigen *c* only. Same is true for gene *E* and *e*. However, gene *DD/dd* produces only two types of distinguishable antigens:

(i) Presence of antigen *D* (these people have genotype *DD* or *Dd*), called Rh-positive (Rh^+).
(ii) Absence of antigen *D* (these people have genotype *dd*), called Rh-negative (Rh^- or rh).

Unlike A–B–O system for which the antigens are common in environmental substances, Rh-like substances are rare. Most people do not produce antibodies against the Rh antigen as they do not themselves possess. However, Rh^- individuals can develop them, if exposed to the appropriate 'external' corresponding Rh^+ antigen which stimulates the persons' immune system and antibodies are produced. Such exposure can occur by blood transfusion. For example, when a Rh^- person has *dd*, genotype exposed to *D*-antigen, anti-*D* antibodies are produced.

Rh^- INCOMPATIBILITY

Anti-Rh antibody development can occur in certain pregnancies which often result in a foetal condition called as *erythroblastosis*. Erythroblastosis foetalis is a haemolytic anaemia of the new born which is often accompanied by jaundice, as liver capillaries become clogged with red blood cells remains and bile is absorbed by the blood. This situation arises due to maternal—foetal Rh incompatibility. The damaged erythrocytes (RBC) are imperfect oxygen carrier. Before birth or soon after birth, death of this baby may occur, if corrective measures are not taken in time.

Although antibodies to Rh$^-$ antigen do not occur naturally in human serum but intensional or unintentional transfusion of blood between mother and child can induce Rh$^-$ persons to produce antibody to Rh antigen. This disorder occurs only when marriages between Rh$^-$ women and Rh$^+$ men and vice versa are held. In addition, there must be also a placental defect where foetal blood, the red blood cell surfaces of which carry the Rh antigen passes from the embryo into the maternal circulation. This occurs largely just or during birth.

The erythroblastosis foetalis occur whenever an Rh$^-$ female (*dd*) carries Rh$^+$ foetus (Dd^+) on account of her marriage with Rh$^+$ male (*DD*/*Dd*). The first Rh$^+$(*Dd*) foetus is usually unaffected because mother does not yet produce anti-Rh(*D*) antibody. However, a few cells from foetus carrying Rh$^+$ antigen (*D*) enter the mother's circulatory system during birth trauma as a result Rh antibody concentration gradually build up in mother and she become sensitized and stimulates production of anti-Rh antibody in body. Once the mother become sensitized (stimulated) by Rh positive cells from the first foetus, in the second and subsequent pregnancy involving Rh$^+$ child, these antibodies may return to foetus where they damage the antigen carrying red cells resulting *haemolytic disease* (anaemia) of new born. However, it is interesting to note, those carrying *dd* foetuses are never at risk. It is *D* antigen (Rh$^+$) ie almost always responsible for maternal foetal incompatibility that result haemolytic disease of new borns.

Anti-A and anti-B antibodies normally do not cross the placental barriers, but antibodies against Rh antigen do so because they are smaller molecules than the ABO antibodies.

Rh incompatibility between mother and foetus is shown in Figure 10.3. A small number of red blood cells (+) from an Rh$^+$ foetus may enter the blood stream of the mother. If the mother is Rh$^-$, then cells stimulate the production of antibodies against the Rh$^+$ antigen. In subsequent pregnancies, these antibodies can cross the placenta and if the foetus is again Rh$^+$, attack the foetal red blood cells and agglutinate them.

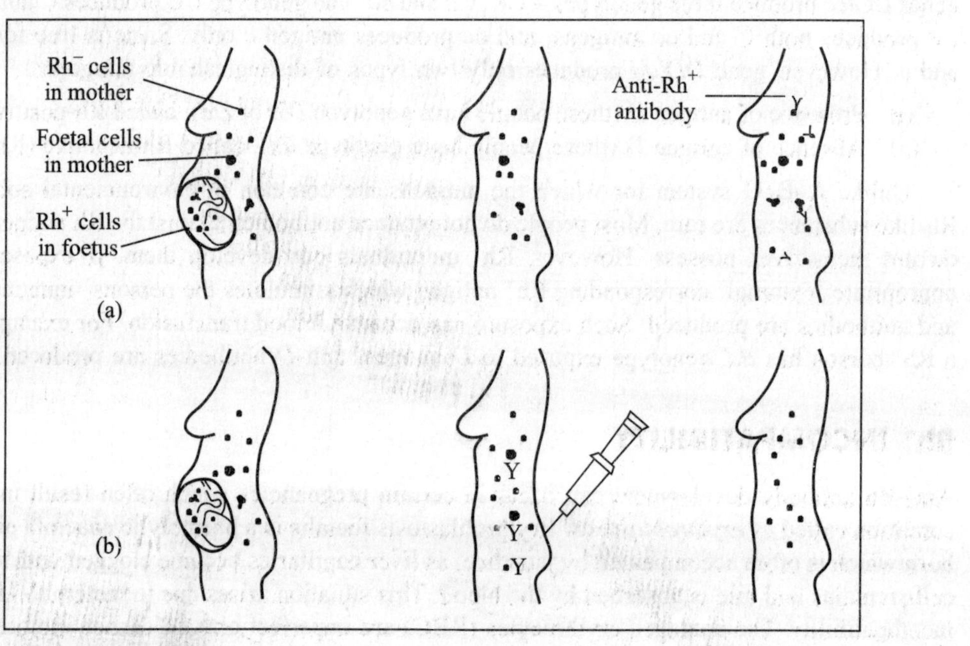

Figure 10.3 Immunization against Rh$^+$ effect on the foetus in the womb.

After birth of an Rh^+ child, stimulation of the immune system in the mother can be prevented by injecting anti-Rh^+ antibody, which rapidly removes the Rh^+ red blood cells from the circulation.

The discovery of two types of anti-Rh (anti-*D*) antibodies simultaneously only in 1994 by Wiener and by Race provided a simple preventive measure against Rh erythroblastosis (anaemia), causing infant death. These two types of anti-Rh antibody are: (a) complete which agglutinates red cells carrying anti-Rh (anti-*D*), and (b) incomplete which does not. The incomplete antibody, however, attaches to the receptor site on Rh^+ erythrocytes 'sensitizes' them and prevents them from acting antigenically. Rh^- incompatibility is a rather rare medical complication and can be completely prevented in mother (Rh^-), who do not have antibodies by intramuscularly injection of human *D* (Rho) immune globulin containing incomplete anti-Rh^+ gamma globulin preparations into Rh^- women within 72 hours after giving Rh^+ child which effectively blocks the child's red blood cells already in the mother's circulatory system for inducing antibody production in her body. The incomplete antibody is often obtained from Rh^- males who have been injected with Rh^+ blood.

The relative infrequency of erythroblastosis is partly due to defective placenta and partly due to ABO incompatibility. ABO, compatible marriage is that in which the husband is of suitable ABO group to donate blood to his wife whereas incompatible marriage is its reverse, where husband is not of ABO group, suitable for the transfusion of blood to his wife (Table 10.4).

TABLE 10.4 Indicates the compatible and incompatible matches (male and female) for marriages based on blood groups

	Female	*Male*
Compatible marriage	A	A, O
	A B	A, B, AB, O
	B	B, O
	O	O
Incompatible marriage	A	B, AB
	B	A, AB
	O	A, B, AB

There are indications that fewer Rh erythroblastic children are born to ABO incompatible marriages than to compatible ones.

M, N BLOOD GROUPS

While research on human blood antigen was going on, Landsteiner and Levine in 1927, discovered antigen M and N which when injected in rabbit or guinea pigs, stimulated antibodies production in their serum. Since human beings do not produce antibodies for M and N, therefore, they are not important for blood transfusion. But they do have genetical interest because inheritance of the trait depends upon a pair of codominant alleles referred as M and N phenotypes, where phenotype M produces antigen M only; N phenotype produces antigen N only and MN phenotype produces antigen M and N both.

GENE MODIFIERS

Gene modifiers are the genes that change the phenotypic effects of other genes in a quantitative fashion: The mammalian coat colour is modified through the dilution or enhancement of the effect of the colour gene. Similarly, the inhibitors gene that prevent the appearance of feather colour in chickens, can be considered as modifiers of normal fowl colour genes.

Pigs have a variety of colour such as red, white and light red (sandy). Wentworth and Lush crossed red and white pigs. The F_1 hybrids were red in colour. When F_1 hybrids were intercrossed, because of gene modifiers, black coat colour pigs were also obtained. When black pigs were removed, 20 red, 14 light red and 3 white pigs were obtained which were close to the expected ratio of 9:6:1. Wentworth and Lush symbolized black with genetic constitution *AABB* and white with genetic constitution *aabb*. Thus, whose all the dominant alleles of both gene pairs were present in homozygous (*AABB*) or heterozygous (*AaBb* or *AABb* or *AbBB*) condition, the pigs were blackish. However, when only two dominant gene alleles of either gene pair (*AaBb*) were present, the pigs were red; when one is homozygous recessive and the other is dominant (*Aabb*, *aabb*, *aaBB*, *aaBb*) the pigs were light (sandy) and where all the two pairs of genes alleles were homogenous recessive (*aabb*) the pigs were white.

REVIEW QUESTIONS

1. Is it possible to cross two agouti rabbits to produce both chinchilla and himalayan progeny?
2. How many of the 3459 persons in the sample of the preceding problems should be A^2B?
3. A hypothetical series of 20 multiple alleles is known for a certain locus. How many phenotypic classes are possible?
4. Considering human blood group A to include three subtypes and groups B and O to include one each, how many phenotypes are included in the A–B–O series.
5. A paternity case involves these facts. The woman is A_1, her child is O and alleged father is B: could he be the father? Explain.
6. In a paternity case, it is determined that the woman is A_1, MSNS and a secretor. The man, she charges is O MS/NS and a secretor. The child is A_2 NS/NS and a non-sector. Using only this information, can you eliminate the man as father of this child? Explain.
7. What are the phenotypes of each of the following genotypes in *Drosophila melanogaster*?

   (i) I_z^{Bs} +++/+$I_z k$ ++ (ii) I_z^{Bs} I_z^{Bk} ++/++++
8. In which of the following marriages is the risk of erythroblastosis foetalis greater:

   (i) A Rh^- O × O Rh^+♀, or (ii) O Rh^- × OA Rh^+♀
9. A woman of group A_2 charges that a man of group A_2 is the father of her group A_3 Rh^+ child. Could he be?
10. Both the men and the women of preceding problem 9 are Rh^-. Does this change your judgement?
11. A couple believes, they have brought the wrong baby from the hospital. Wife is O^+, her husband B^+ and child is O^-. Could the child be theirs?

REFERENCES

Bach, F.H. and J.J. Van Road (1976): The major histocompatibility complex: Genetics and biology, *New England J. Medicine*, 295–806.

Hartl, D.L. (1991): *Basic Genetics*, Jones and Bartlett, Boston.

Hughes, M.B. and E.B. Babcock (1950): Self-incompatibility in *Crepis foetidas*, *Genetics*, **35,** 570–588.

Leder, P. (1982): Genetics of antibody diversity, *Scientific American*.

Landsteiner, K. and P. Levine (1927): Further observations on individual differences of human blood, *Proc. Soc. Expl. Biol. Med.*, **24,** 241–22.

Landsteiner, K. and A.S. Wiener (1940): An agglutinable factor in human blood recognized by immune sera for rhesus blood, *Pro. Soc. Expl. Biol. Med.*, **43,** 223.

Ram, M. (2005): *Unfolding Reality of Life*: *The DNA World*, Dominant Publications, New Delhi.

Stern, C. (1973): *Principles of Human Genetics* (Chapters 12, 13), 3rd ed., W.H. Freeman, San Francisco.

Watkins, M.W. (1966): Blood group substances, *Science*, 152–172.

CHAPTER

11 Sex Determination and Sex Linkage

In the history of evolution, sex appeared first because even single-celled bacteria too show sexual fusion and exchange of genetic materials. The man has been keenly in search of such a miracle. In the Atherva Veda, it is mentioned that sex appeared first in the life of living beings, "Kamo, Janmo, Prathmo", Veda says, Kam(sex), Artha(wealth), Tapa(penance), and Moksha(salvation) are the four objectives of human life.

Before the rediscovery of Mendel's laws of inheritance (1900), sex determination was considered due to materials unrelated to the hereditary factors, such as phases of moon, time of day during fertilization occurs, wind direction and whether 'semen' of right or left testis was involved in the fertilization. Though after rediscovery of Mendel's law of inheritance, the genetical researches led to the material and cytological events taking place in the processes of sex determination. However, in some cases like sex in some Geckos is determined by the temperature. For example, hatching of eggs at 25°C determines the female and hatching of eggs at 30°C determines the male and hatching of eggs at 28°C or 29°C produces half males and half females reptiles. In some marine worms and gastropods, the sex of individuals depends upon the substrate on which they land and the surroundings.

For example, in *Crepudula*, the individuals are found stacked upon one another. The reproductive system can develop as either male or female depending upon the sex of other organism in the cluster. If an organism is attached to a female, the reproductive system will develop as male. Similarly, isolation or presence of a large number of males in a cluster will induce some males to become females. However, a transformed female does not change reverse back to male.

SEX CHROMOSOMES

At the end of 19th century, Henking while examining the spermatogenesis made distinction between the sex chromosomes and the autosomes, by deeply stained body, equally distributed, half-and-half among sperms of the grasshoppers.

After McClung (1900) made cytological observation that in some diploid male insects, half of the spermatozoa possess an accessory chromosome which stained darkly (heterochromatin) but was absent in the other half. McClung believed that only male contained the sex chromosome, but he could not ascertain that female too possesses sex chromosomes. However, Wilson, Stevens and Montgomery (1906) demonstrated a wide variety of chromosomes determining the sex are present in both the male and female organisms in the insects.

Our immediate objective is to examine the mechanisms that determine sex of an organism. There are two basic systems—chromosomal and genetic mechanisms, which determine sex in an individual.

CHROMOSOMAL SEX DETERMINATION

So far, four types of chromosomal sex-determining mechanisms have been reported in diploid organisms. These are XY, XO, ZW and compound chromosomal mechanism. The other chromosomes present in the cell are called *autosomes*.

XX–XY System

In the year 1905, Wilson and Stevens observed different types of arrangement in determining sex in some insects. The female had XX chromosome but the male had, in addition to one X chromosome, an odd chromosome different in size, which was designated as Y chromosome. The females were designated as homogametic because they produced one type of eggs, all with X chromosome and male were designated heterogametic sex because male had XY chromosomes and on meiosis, half the sperm carry X chromosome and other half Y chromosome. However, female produces always only one type of egg carrying X chromosome. This is called XY type of sex determination. This type of sex determination occurs in a wide variety of animals, including *Drosophila melanogaster* and in mammals including human beings, and also in some plants such as *Lychnis* belonging to the genus *angiosperm*. Here again, since female produces only one type of gametes, bearing X chromosome and male produces two types of gametes bearing X and Y chromosomes, the female are called *homogametic* and the male are called *heterogametic*. In *Drosophila*, system is the same, as found in human beings, but in *Drosophila* 'genic balance' between autosomes and X chromosomes determines the sex. In human beings, Y chromosome is smaller than X chromosome. Besides, sex-determining X and Y chromosomes, there are other types of chromosomes, which are alike in both sexes called autosomes.

XX–XO System

The chromosomal differences between the sexes of the several dioecious species were observed much ahead of the discovery of Mendelian laws of inheritance. In 1891, H. Henking, a German biologist observed that half of the sperms of certain insects contained an extra nuclear body in X structure, however, importance of this structure was not ascertained by that time. But in the year 1902 McClung, an American biologist reported that somatic cells of female grasshopper possesses 24 chromosomes whereas the male possesses only 23 chromosomes. In 1905, Edward. B. Wilson and N.M. Stevens while analyzing both the processes of oogenesis and spermatogenesis in several insects, noted that there are chromosomal differences between the two sexes; the female possesses two XX chromosomes and the male possesses only one X chromosome (XO). As a result of meiosis, every egg of the species contains an

X chromosome but only half of sperms possess one X, and other half do not. This system of sex determination is called XX–XO type of chromosomal sex determination.

The bug 'protener' is the simplest example of XX–XO type of sex determination, where presence or absence of a single chromosome determines the sex of an individual. Since male 'protener' forms two types of gametes, i.e. with and without the X chromosomes, therefore, it was called *heterogametic sex*. The female producing only one type of gametes, was called *monogametic*, sex.

ZZ–ZW System

In this type of system, male is homogametic while female is heterogametic. The sex chromosomes in this type of sex determining system are designated as Z and W to avoid confusion with instances in which female is homogametic. Birds including domestic fowls, butterflies, moths and some fishes belong to this category, where females have ZW and males have ZZ chromosomes.

Compound Sex Chromosomes

There are some species where more than one X chromosome or a pair of XX chromosome are present to determine the sex. In such species, there are large differences in the number of chromosomes in males and females. In nematodes, *Ascaris incurva*, the number of autosome pairs is 13, but the number of compound X chromosome is 8. A male, therefore, forms two types of gametes:

(i) 8X + A (21 chromosomes)
(ii) Y + A (14 chromosomes)

Female forms only 8X + A (21 chromosomes). On fertilization, the diploid number of chromosomes in the female is 42 and in the male it is 35, which can be explained as under:

- Sperm (8X + A) + egg (8X + A) = 16X + 26 = 42 chromosomes—female.
- Sperm (Y + A) + egg (8X + A) = 8X + Y + 2A = 35 chromosomes—male.

where A stands for autosomes (13).

Y chromosomes may also exist in compound form and in some species both X and Y chromosomes may exist in the compound form. The compound sex chromosome occurs in beetle, *Blaps polychresta*. In the beetle, male possesses 12X and 6Y chromosomes, in addition to 18 autosomes.

Table 11.1 represents the summary of the mechanism of chromosomal sex determination found in a variety of organisms—human beings, fruit fly, moths, birds, reptiles and plants etc.

Meiotic Behaviour of Chromosmes

A very small segments of X and Y chromosomes are homologous where pairing takes place, as a result X and Y chromosomes forms bivalent during meiosis and move to opposite poles at anaphase I. In case of compound sex chromosome, say XY_1Y_2 system, both the arms of X chromosome possess pairing segment. At the time of pairing, Y_1 and Y_2 chromosomes pair in

TABLE 11.1 Chromosomal mechanism of sex determination in a variety of organisms

Sex chromosome	*Female*	*Gamete*	*Male*	*Gamete*	*Organisms mechanism*
♀–XX–XY–♂	XX	X	XY	X← Y	Human *Drosophila* Plant *Melandrium* *Rumex acetosa* *Cannabis* *Bryonia.*
XX–XO	XX	X	XO	X O	Grasshopper spider plants— *Dioscorea sinulata*
XO–XX	XO	X O	XX	X	Moth
XY XX (ZW) (ZZ)	XY	X Y	XX	X	Birds, reptiles Silk-worm
Compound sex chromosome					
XX XY_1Y_2	XX	X	XY_1Y_2	X⁝ Y_1Y_2	*Rumex acetosa* *Humulus japonica*
$X_1X_1X_2X_2$ $X_2X_2Y_1Y_2$	$X_1X_1X_2X_2$	X_1X_2	$X_1X_2Y_1Y_2$	X_1X_2 ⁝ Y_1Y_2	*Humulus lupulus* var *cardifolius*

such a way that the X chromosomes move to one pole and the Y_1Y_2 chromosomes move to opposite pole (Figure 11.1).

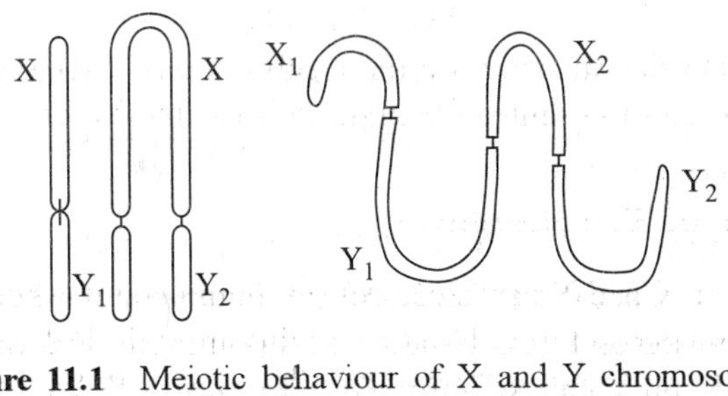

Figure 11.1 Meiotic behaviour of X and Y chromosomes.

Liverworts

Liverworts belong to division *Bryophyta*. In this organism, a haploid is asexually-reproducing phase or generation called *gametophyte*; the alternates in the life history, the diploid is an asexually-reproducing individual called *saprophyte*. The chromosome complements of the saprophyte of the liverwort, *Sphaerocarpus* consists of seven matching pairs plus an eighth pair in which one of the two chromosomes was much larger than the other. The larger chromosome of the eighth pair is designated as X and the smaller one is designated as Y chromosome. At meiosis, which terminates the diploid saprophyte generation; X and Y chromosomes are segregated, so that of the four meiospores produced from each meiocyte, two receive an X chromosome and two receive Y chromosome. The meiospore containing an X chromosome develops female gametophyte and those with Y chromosome develop male gametophyte. Thus, males have Y and females have X chromosomes. Asexually reproducing liverworts are saprophytes.

Primary Nondisjunction of X Chromosome

Normally, during oogenesis, the two sex chromosomes (XX) in female and XY chromosomes in male during spermatogenesis, separate and form females and males gametes with n chromosomes (haploids). But in some cases, two XX chromosomes in female fail to disjoin during oogenesis and give rise to unusual progeny, of course frequency is very rare, i.e. one in 2000–3000 offsprings. Such abnormality is called *primary nondisjunction*. Such flies produce three kinds of eggs; the majority of such eggs will contain normally a single X chromosome; a small number with two XX chromosomes or no X chromosome at all.

Parent	AAX^vX^v (♀) (Vermilion)		×	AAX^+Y (♂) (Red-eyed)	Parent
Gametes	AX^v AX^vX^v AO	(Numerous) (Rare) (Rare)	↓	AX^+ AY	Gametes

F_1	AAX^+X^v	Red female (numerous, normal)
	$AAX^+X^vX^v$	Red meta-female (rare, die)
	AAX^+O	Sterile red (rare, male)
	AAX^vY	Vermilion (male)
	AAX^vX^vY	Fertile vermilion (rare, female)
	AAOY	Die (rare)

AA = autosomes

The individual fly carrying AAXXX chromosomes were called *metafemales*. Such flies are weak and rarely survive beyond pupal stage; AAOY individual die in the egg stage. The flies carrying XXY sex chromosomes are female. The presence of Y chromosome determines maleness while male without Y chromosome (XO) are sterile (Figure 11.2).

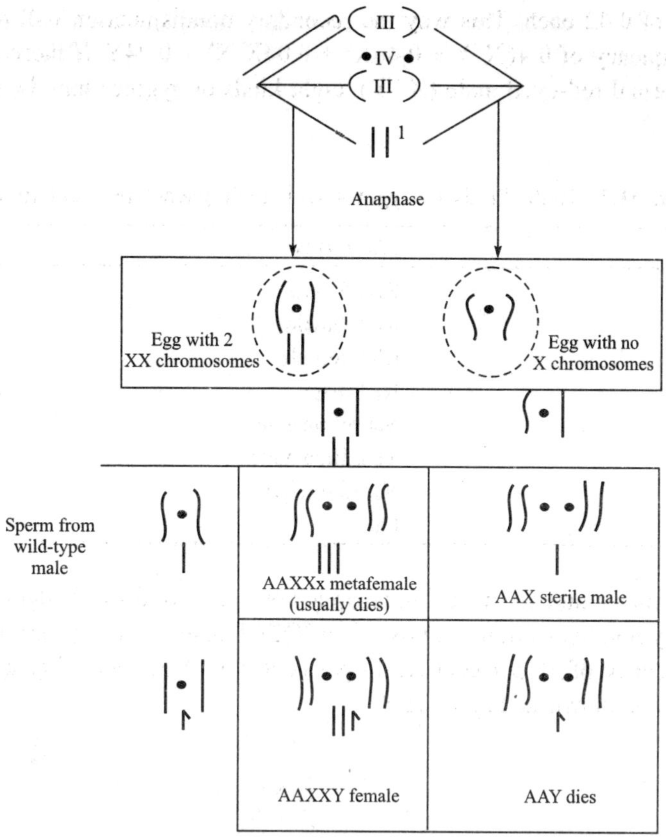

Figure 11.2 Nondisjunction in *Drosophila* and the zygotes from fertilization by the normal male (AAXY). The AAXXY females and the AAX males were the exceptional flies obtained by Bridges. The flies without X chromosome (AAY) did not survive.

Secondary Nondisjunction

Bridges crossed the exceptional vermilion-eyed female *Drosophila* flies (AAX^vX^vY) obtained from the primary nondisjunction to normal red-eyed males (AAX^+Y) and the progeny obtained were in the proportion of 0.46 red females, 0.02 vermilion female, 0.02 red male, 0.46 vermilion female, 0.02 metafemale (die) and 0.02 '0YY' (die).

The occurrence of vermilion-eyed females and red-eyed males is due to secondary nondisjunction. The meiosis may be abnormal in XXY females. In oogenesis, either two XX chromosomes may synapse and Y chromosome may remain unsynapsed or one X and Y chromosome (XY) may synapse and other X may remain unsynapsed. In both the situations, unsynapsed X and Y chromosomes are free to go to either pole. If unsynapsed X may go to either pole, the XY synapsis produces four kinds of eggs—XX, Y, X, and XY with the frequency of 0.04 each. However, if two XX chromosomes synapse, on disjunction, each may move to the opposite poles; the free Y may go to either pole, but results only two kinds of eggs X and XY, with a frequency of 0.42 each. This way the secondary nondisjunction will produce four kinds of eggs in the frequency of $0.46X^vY + 0.46X^v + 0.04X^vX^v + 0.04Y$. If these eggs are fertilized with sperms of normal red-eyed male (X^+ Y), eight kinds of zygotes may be produced given as under in Table 11.2.

TABLE 11.2 Eight kinds of zygotes with their phenotypes and frequencies

Zygote	*Phenotype*	*Frequency*
X^+X^VY	Red female	0.23
X^+X^V	Red female	0.23
$X^+X^VX^V$	Metafemale	0.02
X^+Y	Red male	0.02
X^VY	Vermilion male	0.23
X^VYY	Vermilion male	0.23
X^VX^VY	Vermilion female	0.02
YY	Die	0.02

The terms, primary and secondary nondisjunction were used by Bridges in the year 1916.

The secondary nondisjunction in *Drosophila* XXY female showing formation of egg during oogenesis in frequency of 46 per cent X, 46 per cent XX, 4 per cent Y (based on the work of Bridges (1916(a)) is shown in Figure 11.3.

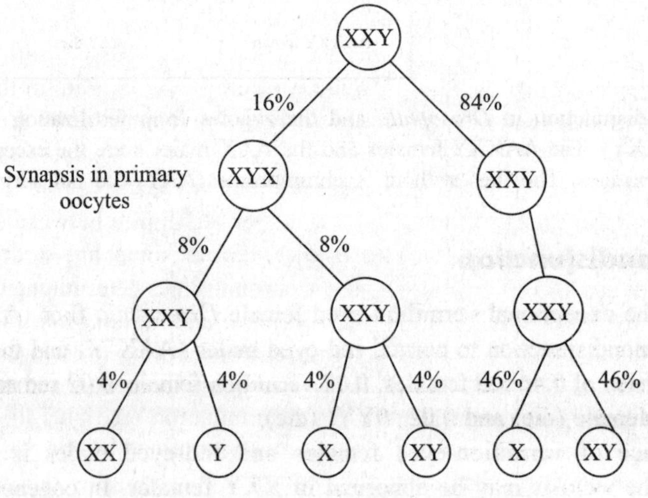

Figure 11.3 Secondary nondisjunction.

Genic Balance Theory of Sex Determination

Bridges in 1921 observed that the sex in *Drosophila* is determined by the balances of genes present on the X chromosomes and autosomes. He gave this theory by raising the progeny of crosses made between triploid females and diploid males of *Drosophila*. According to Genic Balance Theory, the female-determining genes are located on X chromosomes and the male-determining genes are located on autosomes in *Drosophila melanogaster*. A balance between number of X chromosomes and number of sets of autosomes determines the sex of the individuals. If the ratio of X/Autosomes (A) = 1, then fly will be female; if the ratio of X/Autosomes is 0.5, then fly will be male, however, if X/A ratio falls between 1.0 and 0.5, the flies will be intersexes. The experimental results which forms the basis for Bridges to propose this theory are given in Table 11.3.

TABLE 11.3 The ratio of X/autosomes in *Drosophila* in determining sex of the fly

X chromosomes	*Autosomes*	*Ratio X/A*	*Sex of the fly*
1X	2A	0.5	Male
2X	2A	1.0	Female
3X	2A	1.5	Metafemale
4X	3A	1.53	Metafemale
4X	4A	1.0	Triploid female
3X	3A	1.00	Triploid female
3X	4A	0.75	Intersex
2X	3A	0.67	Intersex
2X	4A	0.5	Tetraploid male
1X	3A	0.3	Metafemale

Bridges justified that irregular arrangement of chromosomes results from the *nondisjunctions*, i.e. the failure of the paired chromosomes to separate in anaphase I of meiotic division during oogenesis. The X chromosomes which pair during the meiotic prophase I, should travel to opposite pole during anaphase I, remain together and migrate to the same poles, as a result some female gametes receive two XX and some receive none. Thus, fertilization of female gamete (AXX) with normal male gamete AX or AY gives rise to AAXXX metafemale or AAXXY normal female. In this case, Y chromosome does not play any decisive role. This situation occurs only in *Drosophila* where sex determination is a matter of balance between the X chromosomes and autosomes (A). However in *Melandrium*, Y chromosome has a decisive role in sex determination and it is a balance between X and Y chromosomes determining the sex. Let me give a comparative view of sex determination in *Drosophila* and *Melandrium* (Table 11.4).

The 'Genic Balance Theory' of sex determination according to Bridges is based on genes located on X chromosomes determine femaleness and genes located on autosomes (A) determine maleness. Hence, it is a separate system of sex determination restricted to *Drosophila* only.

The main conclusion is that sex in *Drosophila* is determined by the ratio of autosomes and X chromosome. Females are produced in the ratio of 1:1, where this ratio is altered in favour of X chromosome (3X 2A), flies will show exaggerated form and described as superfemales. If the ratio is altered in favour of the autosomes (1X 3A), the flies will be supermales. Normal males are formed when ratio is X:2A. The intersexes combining male and female organs in one individual are formed when the ratio of X chromosomes and autosomes is in intermediate state, i.e. 3X 4A and 3X 2A composition. Sex determination in *Drosophila melanogastor* could be summarized as:

TABLE 11.4 Chromosomal constitution on *Drosophila* and *Melandrium* showing different kinds of sex determination based on Bridges genic balance theory

Drosophila			*Melandrium*		
Chromosome constitution	*Sex*	*X/A ratio*	*Chromosome constitution*	*Sex*	*X/A ratio*
2AXXX	Superfemales	1.5	2AXX	Female	0.00
2AXX			2AXY		
3AXXY	Females	1.0	3AXY		
3AXXX			4AXY	Male	1.0
4AXXXY			4AXXYY		
3AXX			4AXXY	Male	1.5
3AXXY	Intersex	0.67			
4AXXX	Intersex	0.75			
2AX			2AXXY		
2AXY			3AXXY		
2AXYY	Male	0.50	4AXXY	Male	2.0
4AXX			4AXXXXYY	(occasional blossom)	
3AX	Supermale	0.33	3AXXXY	Male	3.1
			4AXXXY	(occasional blossom)	
			4AXXXXY	Hermaphrodite	4.0
				(occasional blossom)	

1. The sex is governed by the ratio of number of the X chromosomes to the sets of autosomes. Thus, female has ratio of 1.0, and male 0.5. This ratio is applied in polyploid flies also.
2. The genes for maleness *per se* are supposed to be carried on autosomes and those femaleness on the X chromosomes.
3. Y chromosomes govern male fertility rather than sex itself because AAXY and AAXO flies are males with regard to the secondary characters, but sperm is produced only by AAXY males. But it has no effect on AAXXY flies which is a female with X/A ratio 1:0.
4. An X/A ratio greater than 1.0 is a metafemale and less than 0.5 is a metamale.
5. An X/A ratio less than 1.0 but greater than 0.5 produces intersexes. The degree of femaleness is greater when X/A ratio is closer to 1.0 and degree of maleness is greater when ratio is closer to 0.5.

Sex Determination in Human Beings

Sex in human beings is determined by XX–XY system. The mammalian embryo has the potential to become either sex, depending upon the kind of sex chromosome composition. If the embryo contains XX chromosome composition, the gonads differentiate as ovaries, the Mullerian ducts differentiate into fallopian tubes, uterus and upper vagina but the Wolffian duct system regresses. However, if the embryo contains XY chromosome composition, the gonads develop into testes; the Wolffian duct system differentiates into epididymides, Vas differentiates into seminal vesicles but the Mullerian duct system regresses. The individuals with Y chromosomes may develop external genitalia but may be sterile. Similarly, those possessing single X chromosome may develop phenotypic expression as female, but sterile.

1. The autosomes play no part in sex determination.
2. The gene governing maleness is located on Y chromosome in man and other mammals and the femaleness on X chromosome. The individuals with several X chromosomes, e.g. XXXY are male. This indicates that chromosome Y carries male-determining genes.

Hermaphrodite

Hermaphroditism (bisexuality) means development of ovarian and testical tissues in the same individual. According to the Vedic literature, Manu, progenitor of Aryans (Hindus) was hermaphrodite. This is basically wrong. Due to chromosomal abnormalities, hermaphrodite individual may be born, but will be sterile because of rudimentary ova—testes development; while for normal birth of a child normal male and female organs development needed in two individuals.

Therefore, history of origin of Aryan Hindus so called the upper caste in India is biologically wrong. However, hermaphrodite flowers (bisexual) in plants are rule in the self-pollinated crops.

GENETIC SEX DETERMINATION

C.M. Rick and G.C. Hanna collected evidence to show that in *Asparagus*, sex is determined by a single gene pair where maleness (staminate flower) is a dominant character. *Asparagus* is a dioecious plant where male (staminate) and female (pistillate) flowers are produced on separate plants. Rudimentary non-functional pistils occur in staminate flower and abortive stamens occur on the pistillate blossoms. However, some time in pistil, the staminate flower produces viable seeds. It is actually due to self-pollination but from the genetic point of view, it is a male × male cross. If maleness is governed by gene *Aa* and femaleness by gene *aa*, the male × male cross would be:

$$Aa\ (♀) \quad \times \quad Aa\ (♂) \qquad \text{Parents}$$

$$\downarrow$$

$$F_1 \quad A{:}aa$$

$$(3{:}1)$$

where 1*AA* and 2*Aa* would be male and 1*aa* would be female like monohybrid ratio.

Honeybee

In honeybees, workers and the queens are diploid females with 32 chromosomes. Drones, on the other hand are males and possess 16 chromosomes. The femaleness in hymenopterans is determined by heterozygosity on a number of loci on several chromosomes, however, the haploid individuals that hatch from the unfertilized eggs are males.

Corn

Corn is a monoecious plant, where tassel consists of staminate flowers and the cob consists of female flowers. Here sex is controlled by the two pairs of alleles.

1. Genotype b_sb_s (barren stalk) gives no cob, but a normal tassel. This is called staminate flower plant.
2. Presence of gene t_s, in homozygous condition (t_st_s) converts the tassel into pistillate flower, as a result cob develops in the flower, at the top of the plant. Let me show the results of different types of genotypes and phenotypes in corn:

B_sT_s — Normal monoecious corn plant →

$b_sb_sT_s$ — Staminate corn plant →

$B_st_st_s$ — Pistillate, cobs terminal and lateral →

$b_sb_st_st_s$ — Pistillate, cobs terminal only →

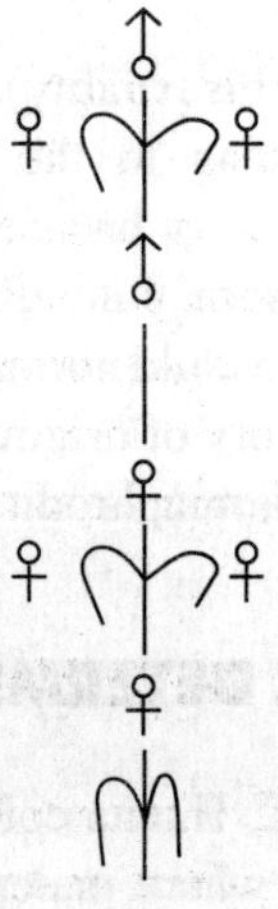

Plants with genotype $b_sb_sT_st_s$, either $B_st_st_s$ or $b_st_st_s$, compromise a dioecious corn race. They can be obtained by crossing $b_sb_st_st_s \times t_st_sT_st_s$ plants, which can produce pistillate ($b_sb_st_st_s$) and staminate ($b_sb_sT_st_s$) corn plants.

SEX LINKAGE

Thomas Hunt Morgan (1910) had identified a white-eyed male in a wild type (red-eyed) population of *Drosophila melanogaster* maintained in his laboratory. When he crossed this white-eyed male with wild type red-eyed female all the F_1 offspring, of both the sexes (male and female) had red eyes, which showed that alleles for white eye (*ww*) are recessive to red eye (W^+W^+). In F_2, offsprings from matings of above eye colour males and females, he obtained 2459 red-eyed females; 1011 red-eyed males and 762 white-eyed males. The white-eyed character appears to be linked with the sex chromosome because only male offspring had white eyes (Figure 11.4).

However, when F_1 progeny of a cross, wild-type female × white-eyed male were back-crossed with white-eyed male parent, the progeny obtained were red-eyed and white-eyed females as well as red-eyed and white-eyed males, in equal proportions.

In a reciprocal cross between white-eyed females and red-eyed males, all the female offsprings were red-eyed and all the male offsprings were white-eyed. The reciprocal cross gave different types of offsprings because the alleles for white eyes is located on X chromosome (X-linked). Sex in *Drosophila* is determined by XX (female) and XY (male) chromosomes (Figure 11.5).

As against above results, an opposite pattern of non-reciprocal inheritance could be found in organisms where male is homogametic and female is heterogametic, such as birds, reptiles, fish, moth and butterflies. For example, some breeds of poultry possesses barred feathers (alternating transverse bands of light and dark colour). The reciprocal crosses between the true breeding barred and non-barred (uniform colour of feather) gave the following result:

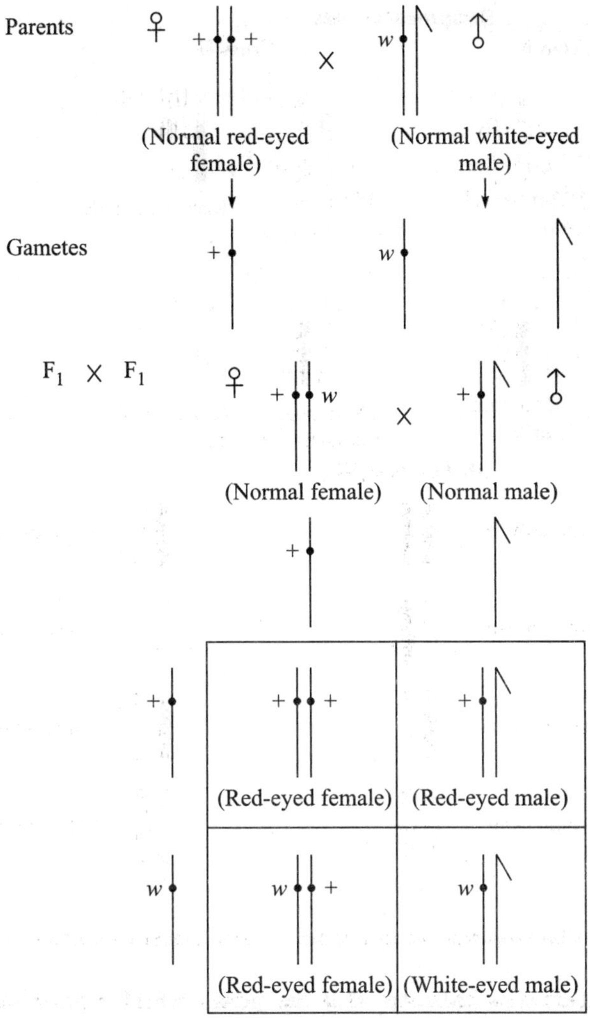

Figure 11.4 First experimental evidence of sex linkage in *Drosophila melanogaster*.

Non-barred female × Barred male

↓

Barred female and barred male

Barred female × Non-barred male

↓

Non-barred female and barred male

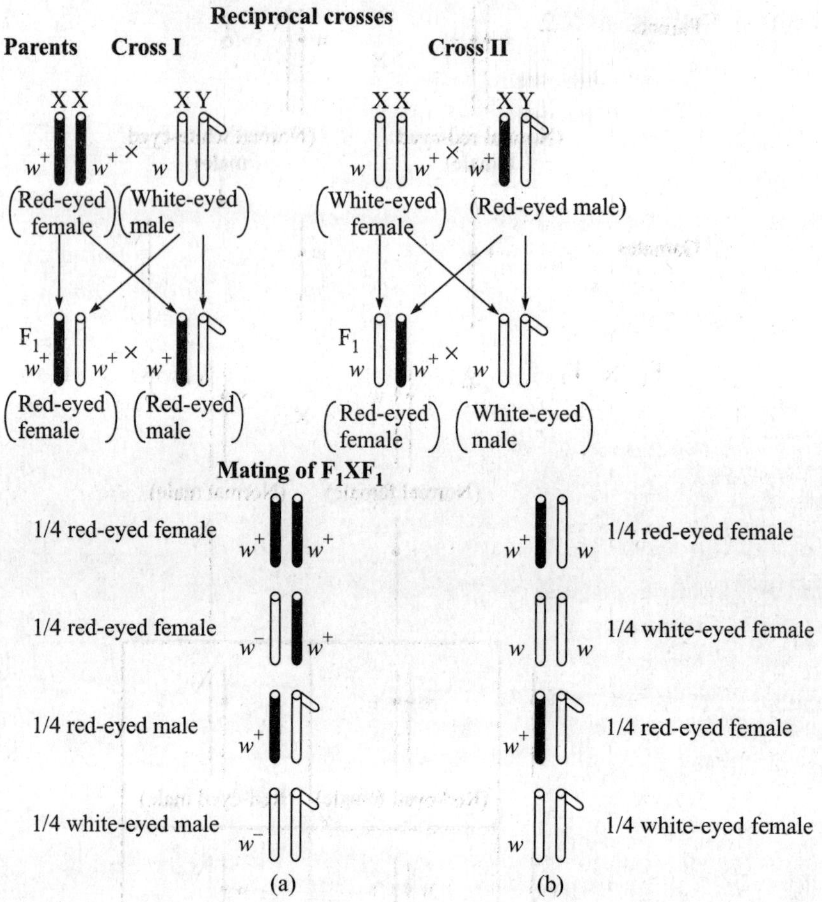

Figure 11.5 Chromosomal interpretation of sex-linked characters in *Drosophila*.

This breeding experiment indicates that the genes which determine the barred feather character is linked with X chromosomes, and is dominant whereas Y chromosome does not carry such type of genes.

Sex-linked Genes in Man

1. In human beings, haemophilia (causes severe loss of blood after injury, blood fails to clot normally) is a sex-linked character, which is governed by a recessive allele. The affected person lack a blood protein needed for normal clotting after incision.
2. Queen Victoria of England was a carrier of this gene allele. One of her sons was a haemophilic (*hh*) and two of her daughters were heterozygous carrier (*Hh*) of the gene (*HH*), who produced three haemophilic sons (*hh*) and four heterozygous (*Hh*) daughters. The two carrier grand daughters were married in the royal families of Russia and Spain. The present royal family of England is a progeny of a normal son of Queen Victoria and all the descendants are free from the disease.

Sex-influenced Dominance

The dominance of alleles may differ in heterozygous of two sexes. This phenomenon is called *sex-influenced dominance*. For example, autosomal genes are responsible for horn in some breeds of sheep; may behave differently in the presence of male and female sex hormones. Generally, female sheep does not bear horn. More than a single pairs of genes are responsible for horn. Among doreset breed of sheep, both the sexes are horned and gene for horn condition is homozygous (b^+b^+) in all animals of the breed.

Sex-limited Gene Expression

One sex may be uniform in expression of a particular trait, but if a gene is transferred in a different genetic background produces a different phenotype in the offspring of the other sex. This is called *sex-limited gene expression*. *Sex-hormones* are responsible for the expression of some genes.

Colour blindness in man

Morgan's work with white-eyed mutant in *Drosophila* could be equally applied to human beings. The colour blindness in man is a genetically-controlled character. Inherited defect in the ability of individuals to distinguish green or red-coloured objects are known to be sex-linked, whereas the defects in blue colour discrimination are autosomal.

Colour perception is controlled by three types of cone-shaped cells in retina of the eye. Each cell carries light-absorbing proteins: blue-absorbing, green-absorbing and red-absorbing, which are coded by genes. The genes that encode red- and green-receptor proteins are located on X chromosomes and the genes coding the blue-receptor proteins are located on the autosome.

A man defective in green colour vision has a single recessive gene (g) located on his X chromosome since Y chromosome carry no colour vision, the single allele is expressed causing colour blindness. If wife of this man is homozygous for dominant allele (g^+) for normal colour vision; all of their daughters will receive an X chromosome from mother carrying g^+ and an X chromosome carrying a g_1, from their father and will be heterozygous. Sons with only one X chromosome receiving from mother (homogametic) will have only one allele (g^+) from mother will have normal colour vision they chromosome carry no allele for this character, will be contributed to his sons only. In the next generations, half of the sons of the carrier females will be normal and half will have defective colour vision because X chromosome carrying gene g^+ will segregate to about half of heterozygous mother. Figure 11.6 shows a criss-cross inheritance, from father to daughter to grandson.

Deleterious Sex-linked Genes

1. *Duchene-type muscular dystrophy:* It is a sex-linked recessive gene-controlled disease. If mother is known to be carrier of this gene, about half of her male children are expected to be affected. Duchene-type muscular disorder afflicts males, usually before they reach to their teens; with muscular deterioration the muscles of legs and shoulder become stiff, and children become paralyzed and crippled, and die before reaching 21 years of age. This has been recently reported in India also.

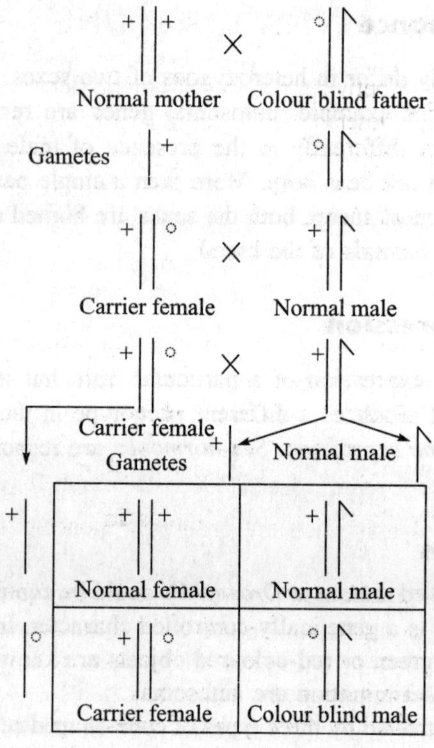

Figure 11.6 Criss-cross inheritance in determining the colour blindness in man.

2. *Hunter syndrome:* It is another type of sex-linked recessive inheritance which causes mental retardation; abnormal hairiness; a broad bridge of nose and a large protruding tongue. It can be detected from the amniotic fluids present in the foetus. Hunter cells (immature cells) are found in the fluid. But process is risky.

Human Y chromosome as dominant male determinant

Human Y chromosome carries genes controlling *holandric* traits that are transmitted exclusively through male-line. For examples, *holandric* genes show the pattern of hairy pinna of the ear. This is autosomal pattern of inheritance. Human chromosome contains at least two genes: (a) *H–Y*, and (b) *TD*/*f*, encoding testis determining factors. Based on these observations, it is concluded that:

1. Sex-linked recessive traits occur more frequently in males than in females.
2. Traits are transmitted from an affected man, through his daughter to half of his grandsons.
3. An X-linked allele is never transmitted directly from his father to his son.
4. All affected females have an affected father and a carrier or affected mother.

REVIEW QUESTIONS

1. What is the difference between male- and female-determining sperms in animals with heterogametic males? Describe sex determination system in man.
2. In line with Bridges genetic balance theory for sex determination, what is the expected sex of individuals with each of the following chromosome arrangements?
   (a) $4 \times 4A$ (b) $3 \times AA$
   (c) $2 \times 3A$ (d) $1 \times 3A$
   (e) $2 \times 2A$, and (f) $1 \times 2A$
3. In the plants with genus *Melandrium*, which sex will be determined by the following chromosome arrangement?
   (a) XY
   (b) XX
   (c) XY with region I removed and XX with region II removed.
4. How could maize plants which are ordinarily monecious give rise to plants that are dioecious?
5. In sheep, the gene b^+ for horned condition is dominant in male and recessive in female. If a hornless ram mated to a horned eye what is the chance that (a) F_2 male sheep will be horned or (b) F_2 female sheep will be horned.
6. In chicken, the gene *b* which distinguishes hen-feathering from cock-feathering, is sex-limited. Male may be feathered (cock-feathered) but female will be always hen-feathered. If cock-feathered males (*bb*) were mated to hen-feathered females (b^+b^+), what pattern of segregation might be expected among the (a) male F_2 and female F_2 progeny.
7. In *Drosophila*, the genes for bobbed bristle (recessive allele *bb*, bobbed bristles, wild-type allele b^+b^+, normal bristle) is located on the X chromosome and on the homologous segment of the chromosome. Give the genotypes and phenotypes of the offspring from the following crosses:
   (a) $X^{bb} \times X^{bb} Y^{bb+}$ (b) $X^{bb} X^{bb} \times X^{bb+} \times X^{bb+} \times X^{bb+} Y^{bb}$
   (c) $X^{bb+} \times X^{bb+} Y^{bb}$ (d) $X^{bb+} X^{bb} \times X^{bb} Y^{bb+}$
8. Consider two different sex-linked traits—recessive and dominant with equal effect on viability and equal frequency in the same population. If males are heterogametic, would the recessive trait or the dominant trait be expressed more frequently in (a) males (b) females?
9. If a father and son both are defective in green colour vision; is it likely that the son inherited the trait from his father?
10. A normal woman, whose father has haemophilia, married to a normal man. What is the chance of haemophilia in their children?

REFERENCES

Bridges, C.B. (1925): Sex in relation to chromosome and genes, *American Naturalist*, **59**, 127–175.

Burns, G.W. and P.J. Brunitto (1989): *Science of Genetics*, Macmillan, New York.

Hartl, D.L. (1991): *Basic Genetics*, Jones and Bartlett, Boston.

Montgomery, T.H. (1906): Chromosomes in the Spermatogenesis of the Hemiptera, Heteroptera, *Trans. Amer. Phil. Soc.*, **27,** 97–173.

Morgan, T.H. (1910): Sex-limited inheritance in *Drosophila*, *Science*, **32,** 120–122.

Morreale, S.J., G.J. Ruiz and E.A. Standora (1982): Temperature-dependent sex determinations, *Science*, **216,** 1245–1247.

Ram, M. (2005): *Unfolding Reality of Life: The DNA World*, Dominant Publishers, New Delhi.

Simpson, E. (1982): Abnormal sex differentiation in human beings, *Ann. Review of genetics*, **16,** 193–224.

Stevens, N.M. (1905): *Studies in Spermatogenesis with Special Reference to the Accessory Chromosome*, Carney Institute, Washington, Publication no. 36.

Warmke, H.W. (1946): Sex determination and sex balance in *Milandrium*, *Am. J. Botany*, **33,** 648–660.

Wilson, E.B. (1906): Studies on chromosomes. The sexual difference of the chromo groups in Homiptera with some considerations on the determination and inheritance of sex, *Jor. Expl. Zool.*, **31,** 1–40.

CHAPTER

12 Inheritance of Cytoplasmic–Genetic System

The mode of inheritance of a particular character (trait) controlled by nuclear gene(s) as a segment of nucleic acid molecules (DNA) packed in the chromosomes has been discussed in the previous chapters. This indicates that the transmission of a character from parents to progeny can be predicted from the knowledge of chromosome segregation and the assortment of genes in meiosis.

No doubt DNA (Deoxyribonucleic acid) is considered as the basic genetic material, packed in the chromosomes, but it does not direct any biological process in isolation, i.e. gene (DNA) alone cannot determine any trait (character), but its function depends upon the internal environment of the cell as well as external environment of the earth atmosphere, as explained in Chapter 13. In other words, the cytoplasm surrounding the nucleus forms the internal environment. The behaviour of the genes will not be the same in the different cytoplasmic backgrounds if placed, because cellular components vary from individual to individual. This is greatest exception to the Mendelian theory of nuclear 'Genes' controlling transmission of characters 'unmixed' (purity of gametes) from parent to progeny. For example, if two strains with genetic constitution *AA* and *BB* are crossed together reciprocally, may produce genotype *AB* with X_1 egg cytoplasm in one case and *AB* with X_2 egg cytoplasm in other case. Thus, these two cross-fertilized diploid zygotes may behave differently depending upon the genotype–cytoplasmic combinations—$AB–X_1$ and $AB–X_2$. If both cytoplasm X_1 and X_2 produce their unique phenotypic effect, the genotypic identity may be disguised by the differences in the cytoplasm i.e. $AB–X_1$ would be different from $AB–X_2$. In the other words, the genotypic differences may be disguised by the effect of different cytoplasm, i.e. $AB–X_2$ would have the same X_2 phenotype as $BB–X_2$ individual. This indicates the abnormal (non-Mendelian) segregation patterns. It is also called extra-chromosomal inheritance or cytoplasmic inheritance or non-chromosomal inheritance or maternal effect, showing inheritance from a single parent, i.e. 'uniparental inheritance' or more specifically maternal inheritance. When genotype of a female parent is only inherited, it has to be considered that cytoplasmic system has an additional genetic system.

Maternal effects are the influence of a mother's genotype on the phenotype of its offspring such as snail coiling and pigmentation in the moths. Though both male and female parents contribute equally as regards chromosomal (nuclear) genes, but zygotic development begins with a maternal milieu so that maternal cytoplasm directly affects zygotic development.

Cytoplasmic inheritance is controlled by non-nuclear genomes found in cytoplasm, such as mitochondria and chloroplast in the plant cells and plasmids in bacteria and viruses.

HOW TO DETERMINE THAT A TRAIT IS CYTOPLASMICALLY INHERITED

1. The environmentally- and socially-induced traits are unstable but can mimic the inherited phenotypic effects such as of vitamin D resistant rickets is mimicked by lack of vitamin D in the diet. Inherited rickets disease does not respond to vitamin D until about 150 times the normally adequate amount is administered.

   Some environmentally-induced traits persist for several generations. This phenomenon is termed *Dauer's modification.*
2. Extrachromosomal inheritance is determined by the odd results of reciprocal crosses. The techniques have been developed to identify chromosomal inheritance by the nuclear transplantation; wherein the nucleus of a cell is removed by microsurgery or destroyed by radiation and another nucleus could be substituted.
3. Chromosomal genes in a particular cytoplasm could also be isolated by repeated back-crossing of offspring with the male parent. In each cross, chromosome content of female parent is halved but cytoplasm remains similar to female parent. Thus, after several generation (5–6) of back crossing male genes could be placed in the female cytoplasm. The phenotypic behaviour of final cross (5th or 6th) will indicate whether inheritance was chromosomal or extrachromosomal (cytoplasmic).

EVIDENCES FOR EXISTENCE OF CYTOPLASMIC–GENETIC SYSTEMS

According to the principles of hereditary patterns, there are four points, which indicate the existence of cytoplasmic–genetic systems.

1. The female gamete contributes all the cytoplasm to the zygote and the male gamete contributes only a nuclear genes, therefore, differences in the inheritance pattern between *reciprocal crosses* indicate the contribution of female cytoplasm, i.e. maternal effect.
2. Where character(s) fail to segregate in a classical genetic pattern and deviate from the standard Mendelian ratio (3:1). In this case, cytoplasmic markers do not segregate along with the known nuclear genetic markers.
3. When the character fails to show linkage to any known nuclear linkage groups and assorts independentally from nuclear genes.
4. Many types of mutants that fit the above said criteria show segregation during mitotic division. This is very common in variegated plants that carry more than one type of plastids (chloroplasts) per cell. This leads variegation suggesting somatic or vegetative segregation.

DISTINCTION BETWEEN MATERNAL EFFECT AND MATERNAL INHERITANCE

Maternal effect occurs when the nuclear genes of the mother determine the phenotype of the progeny. The hereditary determinants are the nuclear genes transmitted by both the sexes in a suitable crosses, the trait undergoes Mendelian segregation (Hartl, D., 1991).

Maternal inheritance occurs when the hereditary determinants of a trait are extra-nuclear and transmission is only through the maternal cytoplasm.

MATERNAL EFFECT

Maternal effect has been thoroughly studied in snail (*Limnaea peregra*), in flour moth (*Ephestia kuhnilla*) and in water flea (*Gammarus*).

The direction of coiling of the shell in snail may be either to the right (dextral coiling) or to the left (sinistral) direction. The direction of shelling is determined by a pair of nuclear gene alleles *DD* (dextral), where *DD* (dextral) is dominant to *dd* (sinister). But the expression of trait depends upon the maternal genotype. The reciprocal crosses between homozygous strains give the results as follows:

(i) Dextral female (rightward) × Sinistral male (leftward)

(*DD*) ↓ (*dd*)

(*Dd*)

All F_1 hybrids coiled rightward (dextral)

(ii) Sinistral female (leftward) × Dextral male (rightward)

(*dd*) ↓ (*DD*)

(*Dd*)

All F_1 hybrids coiled leftward (sinistral)

Thus, a cross between a dextral female (♀) and a sinistral male (♂) gives rise to all F_1 dextral (rightward coiling) and a cross between a sinistral female (♀) and a dextral male (♂) gives rise to all F_1 sinistral (leftward coiling). This is a typical type of maternal inheritance. However, when F_1 are selfed (because snails are hermaphrodite), all of the F_2 progeny of both reciprocal crosses were alike, i.e. 3/4 dextral (rightward coiling) and 1/4 of the progeny were genetically sinistral. This result was consistent in the cytoplasmic inheritance (Figure 12.1, after Hartl, D., 1991).

When each of the F_2 genotypes were again self-fertilized, the F_3 progeny with genetic constitution (*DD*) were dextral and those with *dd* genetic constitution were sinistral. This indicates that inheritance of coiling direction is governed by nuclear genes rather than extra-nuclear factors, because 3/4 of the F_3 progeny had rightward coiled shells (dextral) and 1/4 had leftward coiled shell (sinistral). The reason that reciprocal crosses give rise different results and that the occurrence of 3:1 Mendelian ratio is delayed by one generation indicate that mother determines the initial direction of the orientation of the spindle in the first mitosis of the zygote. The spindle orientation, in turn, is controlled by the genotype of the oocyte from which the egg is developed and appeared to build into the egg before meiosis or syngamy occurs (Figure 12.2). The exact basis of this rather unusual control is unknown. This type of inheritance is due to the maternal effect and not due to the maternal inheritance:

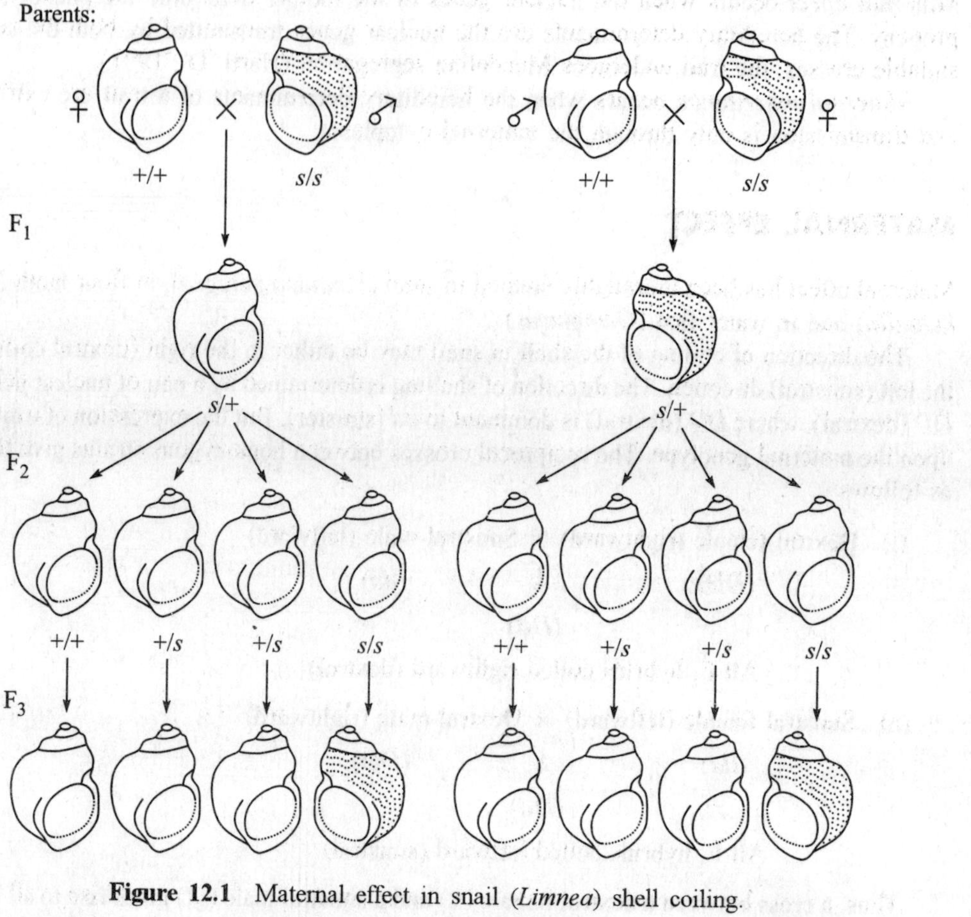

Figure 12.1 Maternal effect in snail (*Limnea*) shell coiling.

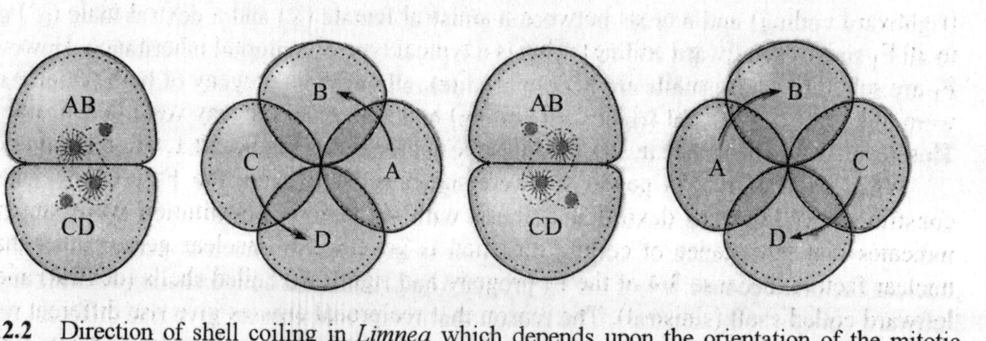

Figure 12.2 Direction of shell coiling in *Limnea* which depends upon the orientation of the mitotic spindles during the second cleavage division in the zygote.

Parent	♀	DD^{++} (Dextral)	×	*s/s* (*dd*) (Sinistral)	♂	
F_1		++	(Dextal)	(*D/d*)		
			↓			
F_2	*DD*(++) (Dextral)	*Dd*(+/*s*) (Dextral)	↓	*Dd*(+/*s*) (Dextral)	*dd* (*ss*) (Sinistral)	3:1
	↓	↓		↓	↓	
F_3	Dextral	Dextral		Dextral	Sinistral	3:1

Parent	♀	*DD* (Sinistral)	×	*dd* (Dextral)	♂	
F_1			+/*s* (*Dd*) (Sinistral)			
			↓			
F_2	*DD*(++) (Dextral)	*Dd*(+/*s*) (Dextral)	↓	*Dd*(+/*s*) (Dextral)	*dd*(*ss*) (Sinistral)	3:1
	↓	↓		↓	↓	
F_3	Dextral	Dextral		Dextral	Sinistral	3:1

Water Fleas and Flour Moth

The pigment production in the eyes of young individual depends upon a pair of nuclear gene alleles *A* and *a*. The dominant gene alleles control the production of *kynurenine*, a diffusible substance which is essential for the pigment synthesis. All the progeny of a cross '*Aa* (Female) × *aa* (male)' will have dark eyes while young, but on reaching the adult stage, those offspring with *aa* genotypes become light-eyed. The reason is that kynurenine diffuses from *A* mother into all young enable in young to produce pigment irrespective of their genetic constitutions. At the same time, progeny with *aa* genotypic constitutions become light-eyed because kynurenine is not produced in them. This is due to the cytoplasmic maternal effect and not because of the cyplasmic genetic system (cell organelles).

In Chapter 4 of the book "Unfolding reality of life—The DNA world" (Ram, 2005) dealing with evolution of cell and cell organelles, it has been discussed that cell organelles such as mitochondria and chloroplast (in plants) and leucoplast (in animals) were evolved from prokaryotic cells that lived inside the primitive eukaryotic cell as symbiont. This process is called *endosymbiosis*.

CYTOPLASMIC INHERITANCE

Cytoplasmic inheritance is due to independent self-replicating (perpetuating) nucleic acid (DNA) present in the cytoplasmic organelles, such as mitochondria (respiratory organelles) or chloroplasts (photosynthetic organelles) and leucoplasts in humans. Because these organelles are located outside the nuclear chromosomes, therefore, these are called *extranuclear* or *extrachromosomal* genetic factors like plasma genes, plasmids and cytogene. On hybridization, these traits are not transmitted in the Mendelian ratio. In contrast to the maternal effect, the traits controlled by extrachromosomal genes do not disappear after one generation but may continue from generation to generation as long as the extrachromosomal factor may continue to self-perpetuate (self-replicate).

Chloroplast DNA

Correns (1908) who was one of the rediscoverers of the Mendelian laws of inheritance (in 1900) was first to report the inheritance of cytoplasmic genes (plasma genes) in *Mirabilis jalapa* (4 O'clock). This plant exists in three forms—normal green, white (no chlorophyll) and variegated. Sometimes, branches with green, white and variegated (green–white) leaves are found on the same plant (Figure 12.3, after Correus, 1908). From these three classes of branches, nine distinct crosses could be made, using pollen or egg source. The results of the crosses could be summarized as:

Phenotype of branch bearing egg parent	*Phenotype of branch bearing pollen parent*	*Phenotype of progeny after hybridization*
White	White	White
White	Green	White
Green	White	Green
White	Variegated	White
Green	Green	Green
Green	Variegated	Green
Variegated	White	Variegated
Variegated	Green	Green or white
Variegated	Variegated	Green or white

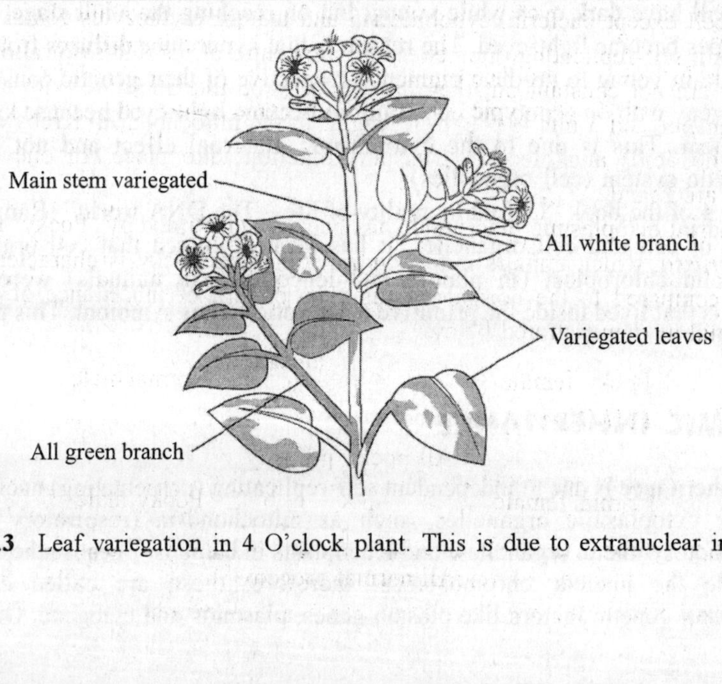

Figure 12.3 Leaf variegation in 4 O'clock plant. This is due to extranuclear inheritance.

From these observations, following results are drawn:

1. Reciprocal crosses give different results. For example, crosses between green (♀) and white (♂) produce green plants and crosses between white (♀) and green (♂) produce white plants, which usually die after formation because of lack of chlorophyll.
2. Phenotype of the female determines the phenotype of progeny; male makes no contribution. The reason is that flowers on the green branches produce eggs with normal chloroplast, therefore, all progeny are green. Similarly, flowers on the white branches produce eggs without plastids hence white and variegated branches produce all the three kinds of eggs, viz. those with lacking green chloroplast (white) and those containing both normal and chlorophyll-free chloroplast. The third type of egg produces variegated plants because in the development of plants, cytoplasmic segregation sorts out normal green and mutant (white) chloroplast. The pollen grain has no effect on the phenotype of the progeny because it carries no chloroplast. This indicates that there is a likelihood of genetic continuity between chloroplast as there is genetic continuity between nuclear genes. The fact that chloroplasts have their own DNA, strongly supports the view that at least some changes in the chloroplast structure can be perpetuated similar to the changes governed by the nuclear DNA. However, since chloroplasts are found in the cytoplasm, their transmission to new generation occur mainly through the maternal gametes. But this does not mean that nuclear genes have no effect on the chlorophyll developments. No part of an organism develops in isolation from other, especially from the nuclear gene effect.

Mitochondrial DNA

Every living cell except bacteria, cynobacteria and mature erythocytes contains mitochondria, which are small, self-replicating organelles and are centre of aerobic respiration. The shapes of mitrochondria are not constant but they are slender, rod-like structure with an average about 0.5 mm in diameter and 3 mm to 5 mm in length. In the mitochondria, Krebs cycle (citric acid or tricarboxylic acid) and electron transport reaction take place. In one cell, 10–10,000 mitochondria are found.

Mitochondrial cytoplasmic inheritance has been demonstrated by 'Pocky' mutant of fungus *Neurospora crassa*, which impairs mitochondrial function. Pocky is characterized by its very slow growth compared to the normal fungus. The pattern of inheritance of pocky mutant is strictly maternal as demonstrated below:

Pocky female × Normal male
↓
All pocky progeny

Normal female × Pocky male
↓
All normal progeny

Drug Resistance in Chloroplast DNA

When Ruth Sanger (1965) placed *Chlamydomonas* algal cells on a culture medium containing the antibiotic *streptomycin*, most of the cells were killed but about one per million survived and multiplied to form a streptomycin-resistant colony. Mutants with resistance to *streptomycin* were being selected from predominantly *streptomycin* susceptible algae. About 90 per cent mutants were found due to nuclear genes (*Sr*-1) and about 10 per cent were unapparent and non-chromosomal. Non-chromosomal mutants were recovered from almost every colony. There was no difference between the phenotypic expression of chromosomal DNA mutants and non-chromosomal DNA mutants. The non-chromosomal genes are presumed located in the chloroplast.

Reciprocal crosses demonstrated that antibiotic resistance was controlled by non-chromosomal genes and was uniparental in inheritance.

Respiration of defective mitochondrial mutation

Occasionally, very small colonies of yeast (*Saccharomyces cerevisiae*) are observed when cells are grown on solid medium. Ephrussi (1953) and collaborators (Ephrussi and Sorieul, 1962) have reported varieties that are defective in their ability to utilize oxygen in carbohydrate metabolism. In presence of glucose, they grow into a small size colonies and, therefore, called *petites* mutants. On the search, it was noted that they lack components (cytochromes b,c and cytochrome oxidase a, a3) necessary for respiratory activity which are found associated with the inner membranes of mitochondria. The *petite* phenotype is the result of large deletions in mitochondrial DNA. For unknown reason, yeast mitochondria frequently fuse and fragments that is cause of deletion of DNA. Such deficiencies prevent *petites* to produce spore, a process that is possible in presence of oxygen respiration. Thus, *petite* diploid cell neither can sporulate nor two haploid *petite* mutants can mate and produce zygote that may sporulate. However, *petite* mutant can be maintained indefinitely in vegetative state as haploid or diploid and can be mated with normal yeast cells. According to mating types, there are three classes of *petites*:

(i) *Segregational (nuclear) petites:* Such *petites*, when crossed with wild type produce ascospores, which segregate in the ratio of 2 petite:2 normal (1:1).

(ii) *Neutral petite:* In a cross with wild type, all spores produce wild-type colonies (no segregation). In further generation, the petite characters never reappear and seen to have been lost physically. Such petites are caused by an extra-chromosomal particle. Extra-chromosomal particles are carried by egg-cytoplasm.

(iii) *Suppressive petite:* Such petites have been seen to suppress normal respiratory behaviour in crosses with normal strains so that in some cases 99 per cent of the diploid cells are derived from a zygote consisting of petites. The frequency of petite mutation is 10–5 per generation. The petite is the result of a large deletion in mitochondrial DNA. For unknown reasons yeast mitochondria frequently fuse and fragments which may cause occasional deletions of DNA.

Behaviour of petite mutation in genetic crosses is shown in Figure 12.4.

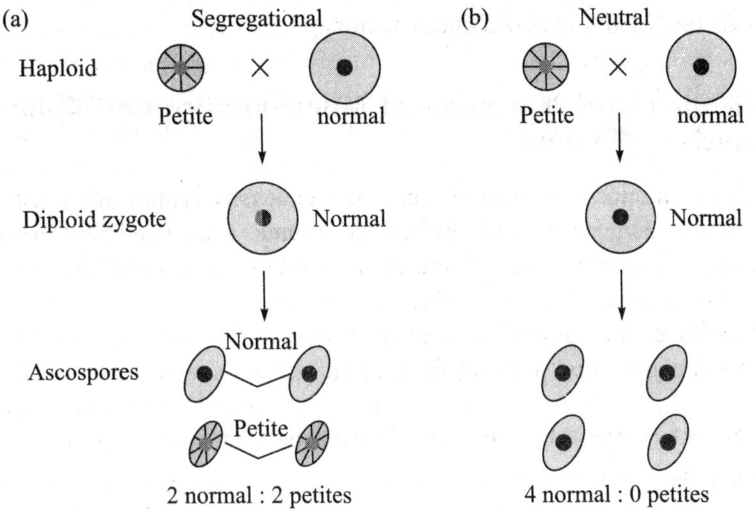

Figure 12.4 Behaviour of petite mutation in genetic crosses.

Molecular Biology of Chloroplast and Mitochondria

Presence of deoxyribonucleic acid (DNA) in chloroplast has been established by several investigations. DNA in *Acetabularia mediterranean* is a closed circular double-stranded molecule. In maize, each molecule of DNA is about 1,39,000 base pairs long and is repeated about 50 times per chloroplast. Each molecule is long enough to encode about 140 proteins. Though very few have been identified yet, plastids contain their own DNA polymerase RNA polymerase found in plastids resembles RNA polymerase II found in nucleus.

DNA molecules of different sizes have also been obtained in the mitochondria of yeast, plant and animal cells. The human mitochondrial DNA has been thoroughly studied and the sequences of 16,569 nucleotides have been known. Yeast mt DNA is reported to be five times larger than the human DNA and maize DNA is about five times larger than the yeast mitochondrial DNA. Mitochondrial DNA encodes several proteins as well as messenger RNA and transfer RNAs.

Mitochondrial DNA consists of 1 per cent of somatic cell DNA of higher animals. This mitochondrial DNA usually exists as relatively small circular molecule. Considerable information is available about the structure of mitochondrial genome. The sequenced mt. DNA of humans, mice and cattle were 16,569, 16,275 and 16,338 nucleotide pairs in length respectively.

It is interesting to know that chloroplast and mitochondrial DNA represent separate genetic system in multicellular organism. The DNA present in them encodes proteins that determine the phenotype whose pattern of inheritance does not follow conventional Mendelian patterns. However, they are not independent of nucleus as many of the proteins in plastids and mitochondria are encoded by nuclear genes.

Extrachromosomal genetic materials are also found in all bacteria, which are collectively designated as *plasmids*. There are three important classes of plasmids: F(fertility) factors; R (resistance) factors, and Col (colcigenic) factors.

Gene Duplication and Migration of Mitochondrial and Chloroplast Genes to Nuclear Genome

Recent progress of structural analysis of gene has demonstrated that there are common gene clusters in mitochondrial genomes and chloroplast genomes such as ATPase subunits genes and NADH dehydrogenase subunit genes. There are nine subunit genes of ATP synthasis in *E. coli*. Plant mitochondrial genomes have only three subunit genes (*atpA*, *atp*6, *atp*0) of ATPase and chloroplast genomes possess six subunits genes (*atpA*, *atpB*, *atPE*, *atpH*, *atpL* and *atpF*); of which three subunit genes (*atpA*, *atpI*, *atpH*) are interestingly common to those of mitochondrial genomes of *E. coli* out of 9 counterparts (Figure 12.5, after Osawa and Honjo, 1991). This indicates that the rest of the genes have been transferred from organellar genomes into nuclear genomes during evolution.

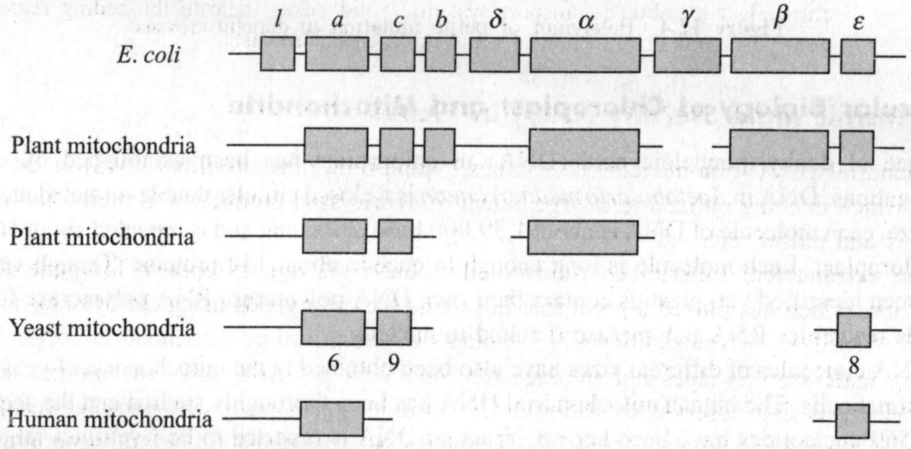

Figure 12.5 Arrangements of ATPase subunit genes. Subunits *a*, *c*, *b*, α, β, ε, correspond to gene products of *atpI*, *atpH*, *atpF*, *atpA*, *atpB* and *atpE* in liverwort chloroplast genomes. Plant mitochondrial genes, *atp*6, *atp*9 and *atpA*, correspond to *atpI*, *atpH* and *atpA* genes in the chloroplast genomes respectively.

In case of ATPase subunit genes, mitochondrial genomes transferred six subunit genes to nuclear genomes during evolution.

The duplication of mitochondrial subunit genes are encoded by the nuclear genomes. If this is not, chloroplast genomes have to transfer their own subunit genes to nucleus to replace the promoters functioning in nuclear genomes and add the signal peptide specific for the chloroplast transportation (Figure 12.6, after Osawa and Honjo, 1991) (Kanji, et al. 1991).

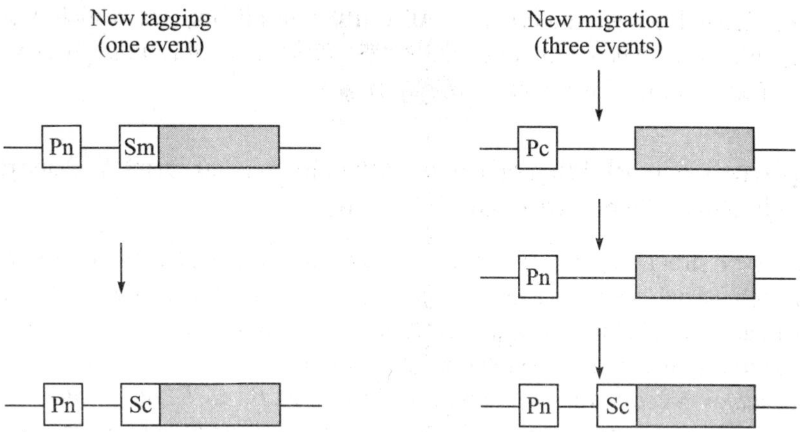

Figure 12.6 Migration of organella genes to nuclear genomes. New tagging genes indicate one event; new migration of a chloroplast gene to a nuclear genome requires three steps in evolution. Pn, promoters functioning in nuclear genomes; Pc, promoters functioning in chloroplast genomes; Sm and Sc are signal peptides required for mitochondrial and chloroplast membrane transport systems. Solid boxes indicate the coding region of a gene, the product of which is transported to chloroplasts.

Cytoplasmic Male Sterility (CMS) in Plants

Mitochondrial DNA (non-nuclear) determines cytoplasmic male sterility (CMS) in the higher plants, which is being utilized to a large extent in hybrid seed production in maize and millets, oil seeds and pulses, etc.

The extranuclear inheritance transmitted through the egg cytoplasm for generation to generation, which has played a great role in raising the crop yields using the cytoplasmic male sterility (CMS). It is presumed that CMS in plants is determined by mitochondrial DNA. A plant showing male sterility does not produce the functional pollen grains. The pattern of CMS inheritance was first observed by Rhoades (1933) in maize. A cross was made between a male sterile variety and normal male-fertile; the F_1 was male-fertile plants for several generations. The resultant plants after several generations of backcrosses no doubt possess all the nulcear genes from the donor parent (male-fertile parent) and cytoplasm of recipient parent (CMS) but still male sterile (Figure 12.7). Thus, male sterility in this case was maternally inherited. Such pollen parents which maintain the male sterility in CMS lines (A) are called Maintainers (B).

However, there are fertility restorer nuclear genes (*R*), which can suppress the male sterility effect of the cytoplasm. There are three types of male sterility systems operative in maize; these are CMS-S; CMS-C and CMS-T. The first two types, viz. CMS-S and CMS-C can be reversed by nuclear strorer genes but CMS-T cannot. Because it causes susceptibility to 'T' race of *Helminthosporium maydis*. Molecular studies indicated that CMS trait in maize is located in the mitochondrial DNA. It is presumed that DNA rearrangement that fuses two mitochondrial genes, creating a novel protein that causes male sterility.

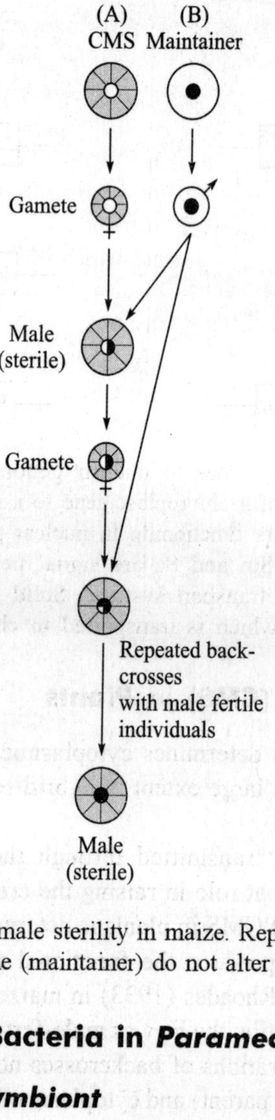

Figure 12.7 Maternal inheritance of male sterility in maize. Repeated back-crosses of CMS: (A) line with male fertile, (B) line (maintainer) do not alter the male sterility character of A line.

Transmission of Symbiont Bacteria in *Paramecium* Cytoplasm

Cytoplasmic transmission of symbiont

It is believed by the biologists that the cell organelles have been evolved originally through the symbiogenesis (endogenesis) of the members of different species/taxa of the micro-organisms which remain associated together for a long time and got integrated in the cells. Now these organisms are living as symbiont in the prokaryotic and eukaryotic cells.

In the eukaryotes, a variety of cytoplasmically-transmitted traits result from the presence of bacteria and viruses residing in the cytoplasm of certain cells. One of the best known is the killer phenomenon found in the certain strains of protozoa (*Paramecium aurelia*). Some strains of *Paramecium* produce substances called *Paramicin* which is lethal to other strains (sensitive) of *Paramecium*. *Paramicin* is water-soluble and diffusible and its production depends upon the presence of 'cytoplasmic bacteria' called *kappa particles* found in the cytoplasm of killer

Paramecium whose maintenance depends on the dominant nuclear gene *KK* present in the cell. *Paramecium* is a diploid protozoa. It undergoes sexual exchange through conjugation. The conjugation cycle is shown in Figure 12.8(a) which proceeds in different stages.

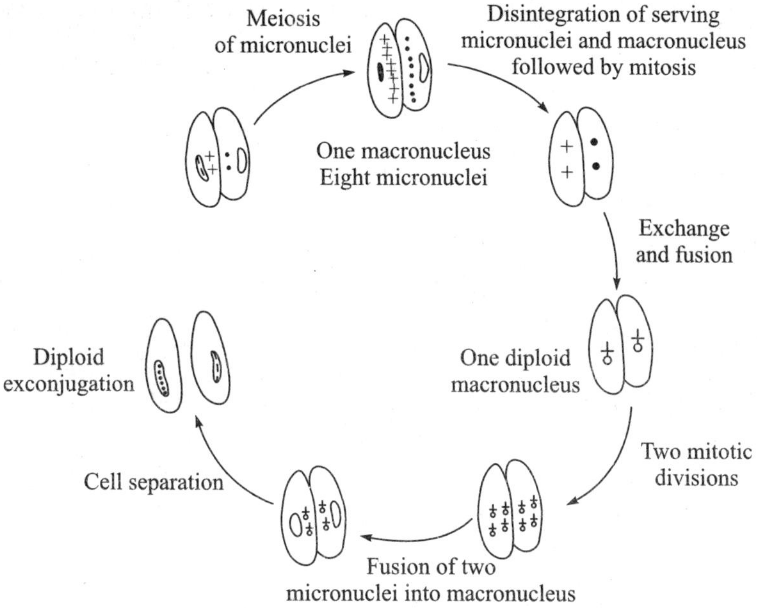

Figure 12.8(a) Conjugation in *Paramecium*.

Initially, the cell possesses two diploid micronuclei. But many *Paramecium* cell possess 8 micronuclei and one macronucleus.

In the first stage meiosis begins; one macronucleus and eight micronuclei (n) are formed. In the second stage, seven micronuclei and one macronucleus disintegrate. This follows mitosis. In the third stage, the cell membrane between the two contacting cells breakdown slightly and the two cell exchange a single micronuclei. After exchange of micronuclei, the micronuclei fuse and form a nucleus; the two cells become genetically identical. In sequence of events only micronuclei are exchanged, no mixing of the cytoplasm takes place.

Autogamy

Autogamy is a second step in the life cycle of *Paramecium* shown in Figure 12.8(b). The *Paramecium* cell occasionally undergoes a typical nuclear phenomenon. The single diploid micronucleus undergoes meiosis and seven micronuclei disintegrate. During meiosis, the *Aa* alleles in heterozygous cell segregate as a result the progeny cell become homozygous.

Second step is mitosis as a result cell divides and form two nuclei. This follows with nuclear fusion to form diploid cell (2n) containing alleles *AA* and *aa* (Figure 12.8(b)). In the third step, two mitotic division produces four micronuclei in each cell. In the fourth step, two micronuclei fuse and form one macronucleus in each cell.

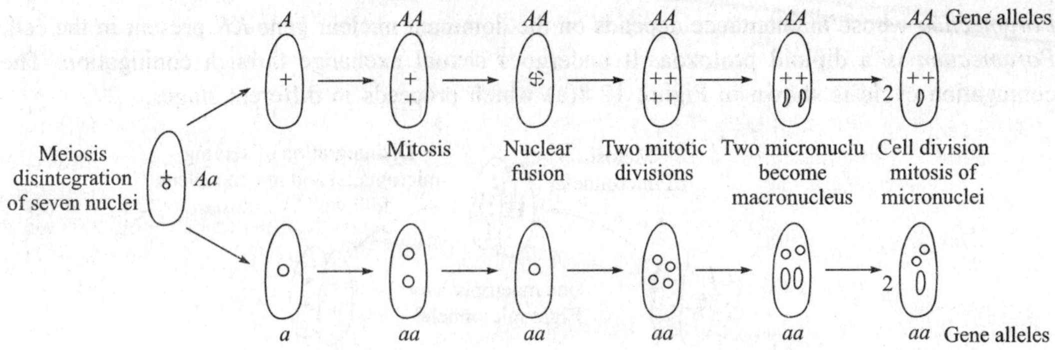

Figure 12.8(b) Autogamy in *Paramecium*. The allele of heterozygous *Aa* cell undergoes segregation in meiosis as a result the daughter cells become homozygous for *A* and *a*.

Conjugation is actually exchange of micronuclei when two cells come in contact, but occasionally cytoplasmic mixing also takes place followed by autogamy. Figure 12.9 shows two situations (a) and (b). In situation (a), no mixing of cytoplasm takes place. The killer and sensitive

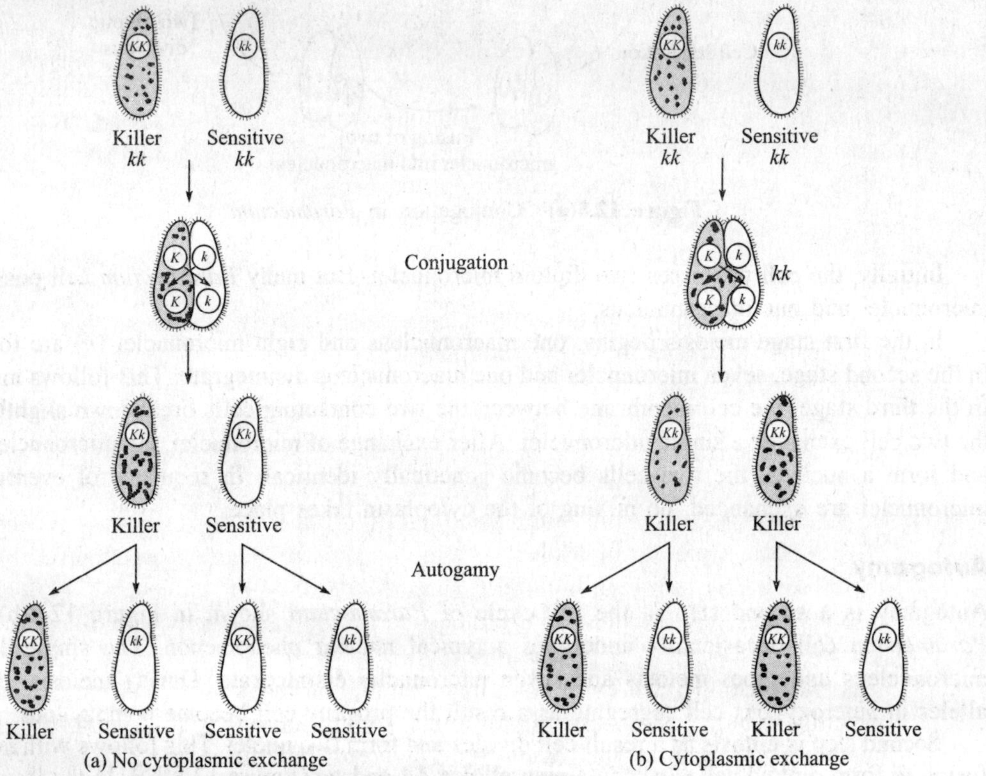

Figure 12.9 Conjugation between killer (*KK*) and sensitive (*kk*) paramecium: (a) Without exchange and (b) with cytoplasmic exchange. Kappa particles are bacteria present in the cytoplasm of killer (*KK*) strain of paramecium. The killer bacteria present in the paramecium is known as symbiont which exudes paramecium.

cell divides in 1:1 proportion. When autogamy occurs, the killer (*Kk*) segregates into homozygous *KK* (killers) and *kk* (sensitive) cells. Sensitive cells possess no kappa particles.

However, another situation is different when mixing of cytoplasm of killer cell (*KK*) and sensitive (*kk*) cell takes place (Figure 12.9(b)). In this case, conjugation produces two killer cell (*Kk*) and (*Kk*) because each cell has kappa particles derived from killer cell. However, as a result of autogamy, the killer cell *Kk* segregates into one *KK* (killer) and one sensitive cell (*kk*). Thus, it makes clear that of killer cell possesses kappa particles (bacterial cell) in its cytoplasm which exude *Paramecium* which kills the sensitive *Paramecium* (*kk*) cells.

Importance of Cytoplasmic Inheritance

1. The economic utility of cytoplasmic male sterility (CMS) system has been extensively explored in plants for the development of high-yielding hybrids in maize, pearl millet, rice, sorghum, sunflower and in many vegetable crops. The increases in the yields have been recorded varying from 30–50 per cent. In some crops, flower size is very small and pollination system (pearl millet and sorghum) being cross-pollinated, it was not possible to evolve high-yielding varieties. Therefore, cytoplasmic male sterility (CMS) provided opportunity to raise their productivity by evolving hybrids.
2. However, beyond utilization of cytoplasmic male sterility (CMS) in the crop plants, cytoplasmic genetic system has not been properly studied in the other life forms. For example, cytoplasmic genetic system may have unique effect on the expression of the alien genes of transgenic plants and animals and also in the organism cloning as a whole.
3. When a bacterial gene or gene(s) of the other species is transferred through the bacterial 'plasmids' or through other means in an unrelated species. The effect (expression) of transferred gene in the new and unrelated genetic background may not find 'Happy Home' and thus get eliminated or its effect may be altered or mimicked by the maternal effect of the recipient cytoplasm/plasma genes. Hence, transgenic plants may or may not give desired results or the transferred gene may show adverse effect such as turning the recipient plant/animal sterile or wild type.
4. Similarly, the clone of animal or human beings may not be 100 per cent phenocopy of stem-cell donor (parent) due to maternal effect of surrogate mother. The surrogate mother can transfer some good/bad traits in the child through her plasma-genes located on the 'leucoplast DNA' and mitochondrial DNA. Because mitochondria are considered 'power house' and seat of oxidation of food-items and energy generation for the body, therefore, gene-controlling breathing must be located on the mitochondrial DNA. This also indicates that, 'asthama' is caused due to alteration or inactivation of mitochondrial genes.
5. It is widely known that there are a large number of people who show susceptibility to pencillin, tuberculostic and malarial drugs. Author himself is susceptible to pencillin, sulphur-based drugs and neomycin-based eye-drop. In India so far no attention has been paid to screen persons for drug susceptibility and analyses their causes–genetic, cytoplasmic (protoplasmic) or physiological, etc. and recommend the medicines accordingly.

Similarly, many children, young and old show allergy to pollen grains of a number of crop plants and trees; hair dye, nylon-based clothes and plastic shoes, etc. Some show allergy even to wheat chapati (gluten), fruit (papaya) and banana, etc. Smoke and house dusts are very common allergents. All these allergenic materials cause frequent cough and cold and give breathing trouble by excessive mucus formation in the windpipe. Some show skin allergy also. In the drug allergy, and also for pollen grains, dust and smoke in humans, blood plasma may be playing important role; of course, we need systematic studies and data analysis.

REVIEW QUESTIONS

1. In a cross between dextral (++) and sinistral (*ss*) snail, what will be the phenotypes from each of the following genotypes in snail, +*s*, *s*+, *ss*, ++ if there is maternal influence?
2. If reciprocal cross gives different results, how would you determine whether this trait is sex-linked, purely maternal effect or the results of extranuclear genetic system?
3. What kinds of progeny with respect to eye colour will result from these crosses in the flour moth (Ephestia): (a) light ♀ × homozygous dark ♂, (b) homozygous dark × light ♂?
4. A 4-O'clock plant with three kinds of branches (green, variegated and white) is used in breeding experiment. What kinds of progeny are to be expected from each of these crosses: (a) green ♀ × white ♂, (b) white ♀ × green ♂, (c) variegated ♀ × green ♂?
5. For the following crosses in the 4-O'clock plant give the phenotypes: of the progeny

Pistillate parent (♀)	*Staminate parent* (♂)
(a) White	Green
(b) Green	Variegated
(c) Variegated	Green

REFERENCES

Ephrussi, B. (1953): *Nucleo-cytoplasmic Relations in Microorganisms*, Oxford University Press, London.

Ephrussi, B. and Sorieul (1962): Mating of somatic cells *in vitro*, *Univ. Mich. Med. Bull.*, **28,** 347.

Hartl, D. (1991): *Basic Genetics*, Washington University School of Medicine, Jones and Bartlett, Boston.

Jacob, F. and J. Monod (1961): On the regulation of gene activity, *Cold Spr. Harb. Symp. Quant. Biol.*, **26,** 193–211.

Ram, M. (2005): *Unfolding Reality* of Life: *The DNA World*, Dominant Publications, New Delhi.

Rhoades, M.M. (1933): The cytoplasmic inheritance of male sterility in *Zea mays*, *Science*, **73,** 340–341.

Sagar, R. (1965): Genes outside the chromosome, *Scientific American*, January.

CHAPTER

13 Gene–Environment and Co-evolution

If the organic world is the play of 'gene', the earth served as the playground, which provided space, raw-materials and the conducive growing conditions for the development of life. Adaptation means, ability of a genotype to fit in an environment. Earth consists of four physical environments—the plain, the ocean, river and lakes, the hills and ice-covers of Tundra and Arctic regions (North and South Pole). Barring Arctic (South Pole) region, all other regions are inhabited by one or other forms of life.

In the past, some studies were conducted to know the genotype × environment interactions. It was observed that all organisms, irrespective of their genetic make-up, interact with their environments. This is highly meaningful for the plant breeders to know that a particular gene allele may dictate phenotypic responses to an environment that is strikingly different from the responses directed by other members of an allelic series.

However, broadly the environment which affect the different life forms are of following types:

1. *External environment:*
   - (a) Physical environment
   - (b) Biotic environment
2. *Internal environment:*
   - (a) Genetic environment
   - (b) Cytoplasmic environment

The physical environmental factors consist of topography, water, temperature and sunlight. Food/feed/plant nutrients are also important factors, which determine growth and development of life forms. This could be further divided into microenvironment within a region or macro-environment within a country or across the country.

The biotic environment consist of pests and diseases; their races and biotypes.

Taking both these environmental factors into account; Darwin (1959) gave the theory of "Natural Selection and Survival of Fittest" which could also be termed 'adaptation'. The basic principle of Darwin's (1859) theory of natural selection and survival of fittest has been discussed briefly in many books. *In nature, a large number of living forms arise continuously through sexual and asexual reproduction and they face competition for food and nutrients, water, sunlight and diseases and pests infestation and out of them only a few survive; rest are eliminated by the natural forces. This process is known as adaptation competition and survival of fittest.*

The conscious man is able to face both physical and biotic (diseases) environments to a great extent by producing building materials, goods (food grains, milk, meat, cloth, etc.), shelter to overcome physical hazards and using drugs and vaccines to control the various diseases and pests causing health hazard.

However, animals and plants have developed special morphological, anatomical and physiological features to survive under the adverse environmental conditions. For example, camel, which is called ship of 'desert', has developed internal body organ (hump) to store water. On the other hand, animals found in the cold regions possess thick furs to protect them from the freezing due to cold. The animals of hills possess small and light body so that they can walk easily on the hilly terrains and graze grasses.

The animals of the economic importance of man (cow, buffalo, goat, sheep, horse, etc.) are provided with all the need-based facilities like feed and fodder, shelter, water and also medicare to protect them from the harmful diseases and pests.

The fishes are adapted to deep sea (brackish water) and sweet water conditions because they can swim in the water and through gills, they can breath. However, frogs and crocodiles can live on the land and in the water both.

Plants are stable living forms of life. They are dependent on soil nutrients water, temperature, and sunlight for their growth and development. Therefore, plants according to their 'adaptation' to different physical topography have been classed into three types:

(a) Xerophytes (desert/dry land)
(b) Mesophytes (normal moisture)
(c) Hydrophytes (water submersible plants)

The gene/gene complexes play important roles to create variability in the morphological, anatomical and physiological features of the plants and animals to adopt to the varying environmental niches.

XEROPHYTES

Those plants which can grow under water-stress conditions such as on dry lands or desert conditions are called xerophytes. They possess specially deep roots, thick cuticle, thick, short spiny leaves and dense hairs as well as special anatomical and physiological features which can conserve water in the cells to protect from intense heat. Cacti, agave, acacia, ber (*Zyzyphus jujuba*), etc. are adapted to water-stress xerophytic conditions.

MESOPHYTES

Such plants grow where sufficient moisture is available. Annual crop plants such as wheat, rice, pulses, oil seeds, vegetables and ornamental plants belong to the categories of mesophytes. The morphological and anatomical structures of leaves and other parts of the plant of mesophytes are quite distinct from xerophytes and hydrophytes.

HYDROPHYTES

Hydrophytes are water-loving plants such as lotus, water lily, hyacinths and rice, etc. which are adapted to water logged conditions.

TEMPERATURE

Temperature has great influence on the growth and development of plants. The crop plants in their distribution are temperature- and season-bound. Accordingly, the crop plants are divided into:

(a) Thermosensitive plants
(b) Thermoinsensitive plants

Winter wheat and winter barley are temperate crops and can grow where prolong winter (low temperature) conditions are available for growth and development. They cannot flower in the spring season. Similarly, spring type cannot survive in too cool climate. However, rice is a crop which can grow under wide array of climatic and geographic conditions from tropical hot humid region of equator to the temperate climate of Japan, Italy and Spain. It can also grow under wide range of edaphic conditions from upland (rainfed) mid-land (irrigated); low-land (water-logged) and deep water and flood-prone areas.

Sunflower is a thermoinsensitive crop, which can be grown throughout the year in India.

The development of white or red flower in *Primula* is largely affected by temperature. *Opunia delleni* when raised in water culture may not develop thorns.

However, environment does not change the genetic make up. Bonnier planted some plants from the plains to the alpine region. These plants changed their morphological appearance in the new environment. However, when these plants were brought back to the plain area after about 18–20 years, i.e. in the original habitat, they reverted back to their original features which indicated that environment does, affect growth and development of plants temporarily but does not alter genetic composition of the organism.

Among the rabbits, there are three-coat colour types—black coat colour, white coat colour and Himalayan, which are also white but with coloured extremities—nose, ears, tail and tips of feet. Gene *AA* determines black, its allele $a^h a^h$ determines Himalayan and its allele $a^h a^h$ determines albino. Albino individuals do not produce any enzyme that occurs in *AA* individuals, which catalyses pigment production. Himalayans $a^h a^h$ produce a corresponding enzyme that is active at temperature under 34°F, but it is inactive at temperature above 34°F. Normal pigmentation take place only in those parts of body that are relatively cool, because of lesser circulation of blood.

Temperature-sensitive mutations have also been reported in microorganism. One mutant in *Neurospora crassa* produces spreading type of mycelium at 25°C, but assume highly restricted colony growth habit at 30°C. Some nutritional mutants were also found temperature-sensitive. Pyrimidine requiring mutant that grows normally on minimal medium at 20°C but at 35°C required the supplement (Srb, 1967).

SUNLIGHT

Plants can be considered as a factory. They manufacture carbohydrate through the chlorophyll in presence of light. In the absence of light, cholorophyll does not develop and plant becomes albino and die. Some plants such as tea, cardamum and betle need protection from light through shades for their proper growth. According to the requirement of sunlight, crop plants are classified as:

1. *Photosensitive:*
   (a) Short-day plants
   (b) Long-day plants
2. *Photoinsensitive or day-neutral plants*

Spring wheat, spring barley and short duration (90–110 days) to mid-duration (120–130 days) rice crops are short-day crop plants, which need limited hours of sunlight per day to mature. On the other hand, winter rice, winter wheat, winter barley are long-day crops, which require 200–240 days to mature. Almost all the high-yielding varieties of wheat, barley, rice and maize are 'photoinsensitive', because of this reason, new high-yielding varieties of wheat and barley can be planted up to middle of January though normal time of planting is the second fortnight of November. Similarly, rice and maize were grown earlier in the monsoon season but, now new varieties and hybrids are being grown in winter season also. The yields of winter maize in Bihar and spring rice in West Bengal and coastal regions of Andhra Pradesh are just double of the monsoon season yields.

STUDIES ON THE ROLE OF ENVIRONMENT ON GENE ACTION

Johannsen was probably the first person who critically studied the role of environment in determining the gene action in the development of phenotype. He conducted his experiment in 1909 soon after discovery of Mendel's laws of inheritance:

Johannsen chose 19 genotypes of princess bean derived from different pure-lines (true breeding) differing in seed size. The smallest seeded-line had 35.1 centigrams, while the largest seeded-line had mean seed weight of 54.2 centigrams. The different lines were found to maintain their seed weight when grown in uniform environment. Johannsen concluded that differences in seed size is governed by genes. He also concluded that observed differences in seed weight within a pure-line were non-heritable and were attributed to the environmental factors and selection on environmentally created variability is not effective due to non-heritable in nature (Table 13.1).

The environmental effects on phenotypes are additive and each environment adds or detracts the same amount from the phenotype independent of the genotype. When this is not the case, the environmental effects on phenotype differ according to the genotype and genotype–environment (G × E) interaction is said to occur. The G × E interaction change the relative rank of the genotypes in the trials; the genotype which is superior in one environment, may become inferior in others. In Figure 13.1, the G × E interaction effect has been given based on the data

TABLE 13.1 Variation in the seed size of pure-lines of two princess beans

Generation	*Mean seed weight of two pure sub-lines*		*Mean seed weight of the progeny of two sub-lines*		*Differences in the seed weight of two sub-lines*
	(a)	(b)	(a)	(b)	(b – a)
1	60	70	63.15	64.85	+1.70
2	55	80	75.19	70.88	–4.31
3	50	87	54.59	56.68	+2.09
4	43	73	63.55	63.64	+0.09
5	46	84	74.38	73.00	–1.38
6	56	81	69.07	67.66	–1.41

from W.A. Russel (1974) on two maize hybrids, developed by crossing different pairs of inbred lines and their overall means are the same, but strain designated 'A', out yielded 'B', in the stress environment (limited moisture and soil and fertility) while strain 'B', out yielded 'A', in better quality environment (where moisture and soil fertility were not the limiting factors).

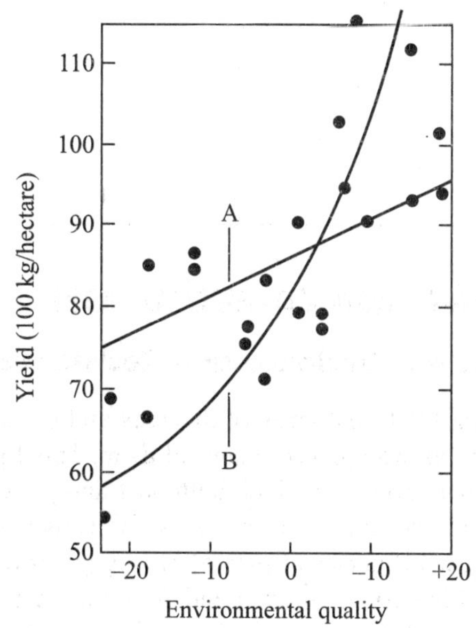

Figure 13.1 Genotype–environment interaction in maize. Strain A is superior when environmental quality is low (negative numbers), but strain B is superior when environmental quality is high.

When different genotypes are not distributed at random in all the possible environment, there is 'genotype–environment association'. In dairy husbandry, farmers feed their cattle according to the milk yield per day. This way cows genetically superior with respect to the milk production also receive better environmental management in the form of more feed. In the plant and animal breeding, genotype–environment association is minimized by appropriate randomization of genotype in the experiments.

RELATIVE CONTRIBUTION OF HEREDITY AND ENVIRONMENT

Phenotype is the interaction product of genotype and environment because both genes and environment are involved in determination of a character. If the environment produces observed variability because of the differences in soil fertility, moisture, then the greater and more uniform yields may be obtained by optimizing the environmental factors such as agricultural inputs. If the genes are responsible for variability under uniform environment, then it should be possible to select a high-yielding variety depending upon the seed yield per plant or per plot. The relative contribution of heredity and environment could be evaluated by growing different varieties of a crop in the same environment (heredity) and a single (pure-line/inbred) variety under different environments. The relative contribution of genotype and environment are given in Table 13.2.

TABLE 13.2 Relative contribution of heredity and environment

Character	*Percentage heredity*	*Determined by environment*
Conception rate in cattle	5	95
Ear length in corn	17	83
Egg production in poultry	20	80
Yield in corn	25	75
Oil content in corn	65	35
Egg weight in poultry	60	40
Root length in radish	65	35
Slaughter weight in cattle	85	15

BIOTIC ENVIRONMENT AND CO-EVOLUTION

Co-evolution of Genes for Virulence and Resistance

Crop plants are highly susceptible to the array of diseases and pests and man has been fighting since he started domestication of crop plants to control them. This fight is still on. However, the co-epicenters are the geographic areas, in which both host and parasites have co-evolved, most accurately depict the story of the evolution of genes for resistance and virulence. *Solanum demissum* (non-tuber-bearing wild species of potato) having genes resistant to late blight fungus (*Phytophthora infestants*) in Mexico; *Tripsacum* and *Teosinte* are the (close relatives of maize) and northern leaf blight fungus (*Helminthus maydis*) were evolved in Mexico and old land races of rice and rice blast fungus (*Pyricularia oryzae*) co-existed since time immemorial vividly depict the classic saga of the co-evolution of the host and parasite. This way both the plants and parasites are involved in the thrill of victory and the agonies of defeat since their origins. This is not merely a philosophical or teleological speculation only but there is a scientific explanation embodied in the co-evolution and co-existence of host and parasite in natural ecological niches. We can examine the beginning of their unique affair to learn how they arrived at their current status of equilibrium of co-existence. The initial resistance of a host to a new parasite probably was accomplished by a genetic change at a single locus. Plants with that resistance gene probably reacted to the parasite in a hypersensitive manner. Subsequently, new populations of parasites were evolved and strains pathogenic or virulent to the resistant-host were selected. The host generated a new source of effective resistance and so on. Such events must have been the case, since both the host and parasite still exist.

The process of co-evolution proceeded perhaps step-by-step in a gene for gene manner and all the genotypes that existed throughout the process of co-evolution remained a part of the current populations of the host and parasite.

Gene and Biotic Environment

The plant breeders and geneticists have so for identified two types of genes governing resistance to the diseases and pests, i.e. major genes governing the race-specific resistance (vertical resistance) and minor or polygenes governing the non-race specific resistance (horizontal resistance). But the terms 'major' and 'minor' appear to be obvious. Nelson, et. al. (1970) proposed the concept that genes for race-specific and non-race-specific resistance are the same genes. The two kinds of resistance are not the result of different genes but rather are the expression of different actions of the same genes in the different genetic background (Nelson, 1975, 1978). There are infact no major and minor genes; there are only genes for resistance. Certainly, different genes may confer varying levels of resistance and expression of resistance genes may be influenced or modified by other resistance genes in different genetic background. Background is the key to gene action.

Effect of genetic background

In *Drosophila melanogaster* position effect of genes due to unequal crossing over has been explained in many textbooks. Ford (1940) demonstrated that a dominant gene might behave like recessive gene when placed in an altered genetic background. Ford had observed that in moth, the wild type allele at *lutea locus* produces yellow colour, which is dominant over white brown colour of its mutant allele. This is because one set of modifiers enhanced the expression of *lutea*, while wild type gene allele, in another set in the second line suppresses it. Accordingly on reviewing the literature, some intriguing facts have been observed as regards nature of gene action, which is summed up as below:

1. Major gene effects are modified by minor genes and minor gene modify minor gene effects.
2. Major genes together exhibit additive effects and so do the minor genes.
3. Major genes and minor genes together express modifying and additive effects.
4. Major genes mask the effect of minor genes and minor genes mask the effect of major genes.
5. Major gene resistance is enhanced by minor genes and major genes enhance minor gene resistance.
6. One dose of major gene confers susceptibility, whereas two, three or four doses confer increasing levels of resistance.
7. Several genes collectively conditioning a rate-reducing resistance; individually condition race-specific resistance and single gene controlling race-specific resistance collectively condition rate-reducing resistance.
8. Genes may be major in one background and minor in another.

Cytoplasmic background

In Chapter 12, it has been described in detail that cytoplasm too possess genes which interact with nuclear genes. Plasmagenes are called plasmones. These plasmones sometime mask or modify the function of nuclear genes. Best citation may be T-cytoplasm of maize and kappa particle in killer *Paramecium strain.*

Utilization of Genotypic Adaptation to Varying Environments in Varietal Improvement

Genotype–environment (G × E) interaction effects have been studied deeply and extensively in the plant breeding programmes to find out varieties suited to different locations and agroclimatic conditions. Adaptability is the ability of a genotype, variety or population (homogenous) to fit to a given environment.

Simmonds (1962) classified adaptation into four categories:

1. Specific genotype adaptation to a limited environment, e.g. rice varieties suited to water-logged conditions.
2. General genotypic adaptation is the ability of a genotype to grow in a range of environments like semidwarf wheat and rice varieties can grow in a wide range of environmental conditions.
3. Specific population adaptation is the ability of a heterogenous population (composite or varietal mixture) that is attributable to interaction between components rather than to the adaptation of components themselves to environment.
4. General population adaptation is the ability (capacity) of a heterogenous population to adapt in a variety of environments. Synthetic varieties of maize or forage crops are the best example.

In India, genotype × environment interaction effects have been extensively studied and exploited under All India Coordinated Multidisciplined, Multilocation Research Projects, formulated on all the major crops such as rice, wheat, barley, maize and millets (pearl-millet and sorghum), pulses, oilseeds, cotton, jute, sugar cane and vegetables, etc. to identify location-specific; region-specific, season and agronomic management-specific crop varieties and hybrids. For example, India is a very large country consisting of five agroclimatic zones, viz. north-western plain; north-eastern plain; central India; peninsular India and northern hill-valleys (Jammu and Kashmir). In each region, rice is grown under five topographic conditions (Figure 13.2).

(a) Upland rainfed condition
(b) Mid-land irrigated condition
(c) Low-land, water-logged condition
(d) Deep water, flood-affected condition
(e) Spring season

However, yields are highest in mid-land irrigated condition and lowest in upland rainfed and deep water, flood-affected conditions.

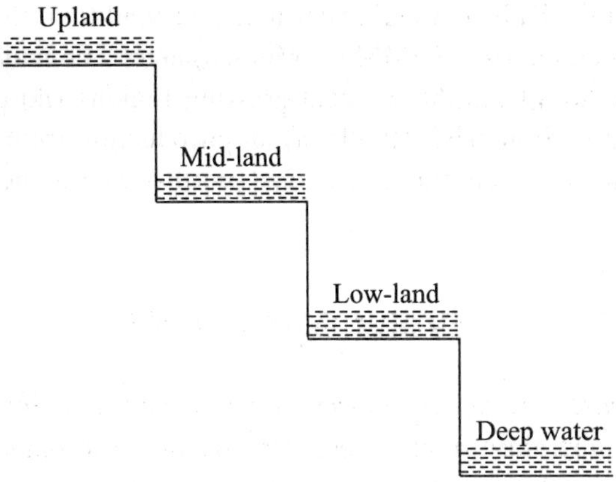

Figure 13.2 Rice ecology in eastern India.

Similarly, wheat and barley varieties are grown under three cropping conditions:

(a) Rainfed, low-input condition (last week of October)
(b) Irrigated, timely-sown condition (mid-November)
(c) Irrigated late-sown condition (December)

Like rice, yield of wheat is highest under timely-sown (mid-November) irrigated condition because of good and controlled management of inputs.

However, even on the same inputs level, yield of crop varieties, for example yield of wheat varies a lot for region-to-region as may be seen in Figure 13.3 prepared on the basis of multilocation regional trials (Ram, 1974).

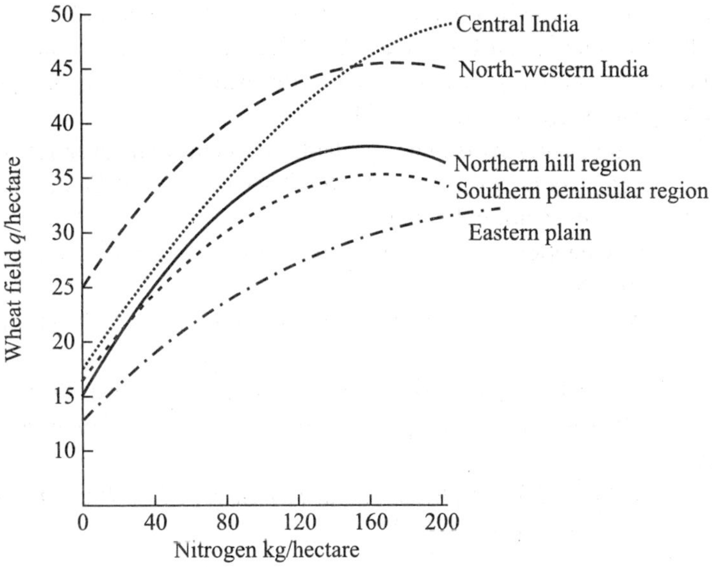

Figure 13.3 Response of wheat varieties to the nitrogen doses in different plain zones of India.

Among wheat varieties, 'Kalyan Sona' under irrigated condition, 'Sonalika' under late-sown condition (both are introduction from CIMMYT Mexico) and C306 under rainfed condition have shown wider adaptations to all the three wheat-growing regions (north-western, north-eastern and central zones) of India. Similarly, IR8 (Jaya) an introduction from I.R.R.I. Philippines was found adapted to all the four geographic regions (north-western, north-eastern, central and peninsular zone) of India.

REVIEW QUESTIONS

1. Define xerophytes, mesophytes and hydrophytes and their importance is agriculture.
2. Based on the temperature requirement, classify the crop plants.
3. Define short-day and long-day plants and their economic importance.
4. What is adaptation? Give Simmonds (1962) classification of adaptation.
5. What do you mean by co-evolution of genes? Describe in brief giving some examples on the genetic mechanism of co-evolution.
6. Explain the concept of genotype × biotic environmental interaction.
7. How genetic background affects gene expression?
8. Describe giving examples how G × E interaction could be utilized by man in crop plants and animal husbandry.

REFERENCES

Darwin, C. (1859): *The Origin of Species*, John Murray, London.

Ford, B. (1940): Genetic research in lepidotera, *Ann. Eugenics*, **10,** 227.

Nelson, R.R., D.R. Mackenzic, and G.L. Scheifle (1970): Interaction of genes for pathogenicity and virulence in *Trichometasphaeria twicica* with different number of genes for vertical resistance, *Phytopathology*, **60,** 1250–1254.

Nelson, R.R. (1975): Horizontal resistance in plants: concept, controversies and applications, *Proc. Seminar on Horizontal Resistance to the Blast Disease of Rice*, Galvez, G.E. (Ed.), 1–20, Series CE 9, CIAT Cali, colombia:

Nelson, R.R. (1978): Genetics of horizontal resistance to plant diseases, *Ann. Review Phytopathology*, **16,** 359–378.

Ram, M. (1974): *Cultivation of High-Yielding Varieties of Wheat*, ICAR, Krishi Bhawan, New Delhi.

Russel, W.A. (1974): Annual Corn and Sorghum Research Conference 29, 81.

Simmonds, N.W. (1962): Variability in crop plants: its use and conservation, *Biological Review*, **37,** 422–465.

Srb, A.M. (1967): Genetic control of physiological processes, *Plant Breeding Symposium* held at Iowa State University, Iowa, 355–390.

CHAPTER

14 Genetics of Fungi, Bacteria and Viruses

Ram (2005) while dealing with the origin of life has made mention that earth arose about 4.50 billion years ago and life in form of 'obcell' arose around 3.50 billion years ago, and all life forms from 'bacteria to man' are interrelated. The causes of such a vast diversity are primarily mutation, recombination and natural selection wherein sexual reproduction has played a great role in creating such a vast genetic diversity among the living forms. In microorganisms such as fungi, bacteria and viruses many steps associated with sexual reproduction, e.g. gamete formation and fertilization, etc. differ drastically, but evidences are available that genetic materials of two different cells when come together show the process of gene recombination. These organisms show a sort of parasexual form of reproduction and most of these show predominance of a haploid phase during their life cycle.

FUNGI

Neurospora crassa

Neurospora is a fungal species, which has been extensively utilized in genetical studies because the products of meiosis, the 'ascospores' are arranged in orderly array. The nuclei possesses a haploid (n) complement of 7 chromosomes which have the same kind of structure as found in the higher organisms. A zygote nucleus contained in a sac-like structure (an ascus) undergoes meiosis almost immediately after it is formed. The four chromatids from a paired homologous chromosomes are recovered without mixing with similar products of other meiotic cell. The four chromatids in *Neurospora* are recovered in an ordered tetrad, i.e. the two upper pairs of ascospores receive chromatids of one homologous chromosome and the lower pairs those of the other homologous chromosome. This is because zygote divides to produce two cells, one lying on top of the other. The second and third divisions are mitotic, the eight cells produced this way continue to lie in a linear order. Thus, each pair formed out of eight ascospores (four pairs) corresponds to one of the four chromatids of the two homologous chromosomes. The ascospores could be removed in the order of their sequence from an ascus and could be germinated in individual culture tubes to determine their genotypic constitution (Figure 14.1).

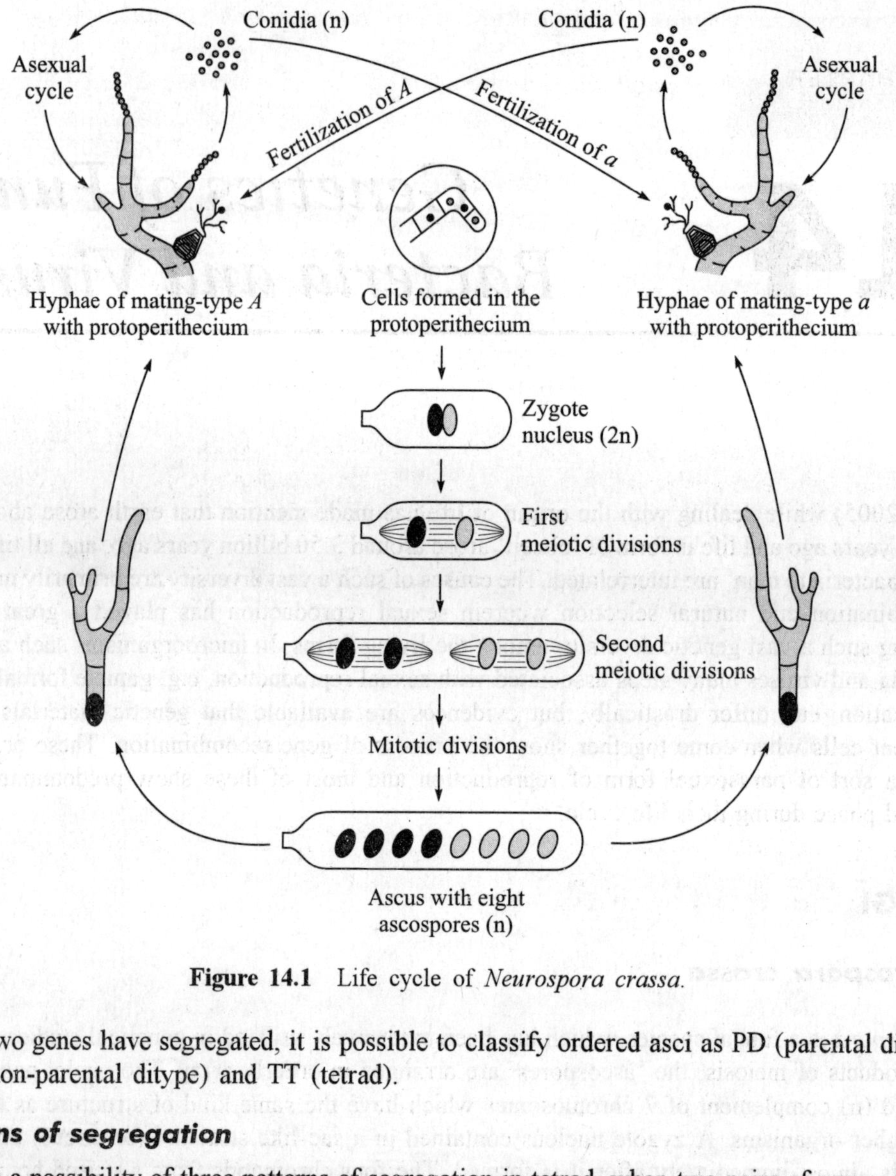

Figure 14.1 Life cycle of *Neurospora crassa.*

If two genes have segregated, it is possible to classify ordered asci as PD (parental ditype), NPD (non-parental ditype) and TT (tetrad).

Patterns of segregation

There is a possibility of three patterns of segregation in tetrad, when two pairs of genes *AB*, *ab* are segregating. Let us say a cross *AB* × *ab* has taken place in *Neurospora*, where three types of tetrads are possible:

(a) *AB AB ab ab* are parental ditype (PD). Only two genotypes are represented and their alleles will have same combinations found in parents.

(b) *Ab Ab aB aB* are non-parental ditype (NPD). Only two genotypes are represented but their alleles have non-parental combinations.

(c) AB Ab aB ab are tetratype (TT). All four possible genotypes are possible. Tetrad analysis indicates whether two genes are linked.

In absence of crossing over, random arrangement of chromosome pairs at metaphase I gives PD and NPD tetrads.

When crossing over occurs between one gene and its centromere, the two chromosome arrangement yields TT tetrads. When both genes are closely linked to their centromeres (crossing over is rare) few TT tetrads are produced.

When genes are unlinked, the parental ditype, tetrads and non-parental ditype tetrads occur with equal frequencies.

In Figure 14.2, two pairs of alleles *Aa* and *Bb* are located on different chromosomes, i.e. they are unlinked. In absence of crossing over, between either genes and its centromeres, the two chromosomal configurations are equally likely at metaphase I, so PD = NPD. If crossing over occurs, a tetra-type tetrad results. But this does not change the fact that PD = NPD.

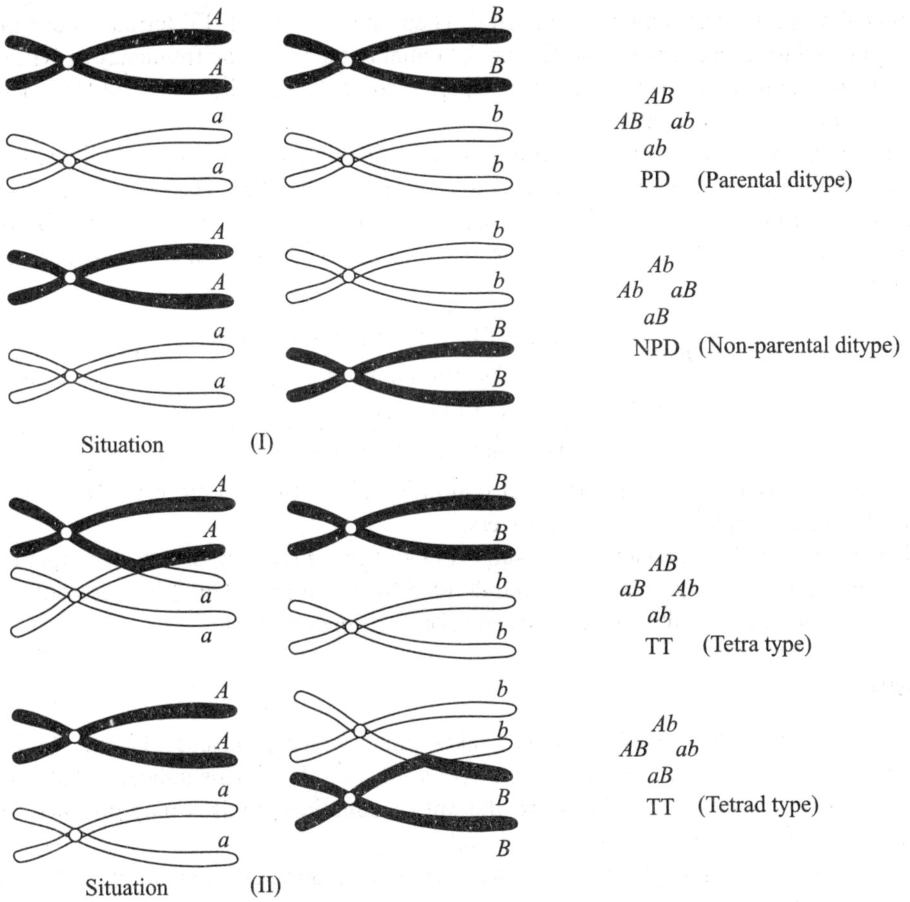

Figure 14.2 Types of unordered asci produced with two genes in different chromosomes.

(I) No crossing over. Two different combinations of chromatids—one yields PD tetrad and other NPD

(II) When crossing over occurs, between one of genes and centromere, two chromosome arrangement yield TT tetrad. If both are closely linked to the centromeres, crossing over is rare and few tetrad are formed.

In contrast, when genes are linked, parental ditypes are more frequent than non-parental types. Assume that genes are linked, then three types of tetrads may be formed. When crossing over occurs between the genes, a PD tetrad may be formed. If a single crossover takes place between the genes, a TT type tetrad may be formed. However, the occurrence of a two-strand, three-strand or four-strand double crossover may result into a PD, TT or NPD respectively (Figure 14.2). When the genes are linked together, the meiotic cells with no crossover may in principle be higher in number than those with four-strand double crossovers. Thus, *when the genes are linked, the recombinant types (non-parental ditype tetrads) appear lesser in number than the parental ditype tetrads. This indicates linkage between the genes under consideration.* The relative frequencies of the different types of the tetrads can be used to determine in the map-distance between the two linked genes. The number of NPD tetrads is used to estimate the frequency of double crossovers. The four types of double crossovers as given in Figure 14.2 are expected to occur with equal frequencies. Thus, the number of TT tetrads occurring from three-strand double crossovers (D and E) may be equal to the twice the frequency of NPD tetrads (F) and so the number of TT on account of single crossover may be TT-2 (NPD) wherein the brackets denote observed number.

Similarly, the number of tetrads resulting from the double crossing over (C D E and F) should be equal to four times of the number of NPD tetrads (F) or 4 [NPD]. The map-distance between two genes (*A–B*) equals to one-half of the frequency of single crossover tetrads (B) plus the frequency of double crossovers. Following is the equation to calculate the map-distance between two genes:

$$\text{Map-distance} = \frac{(1/2)\,(\text{TT}) - 2(2\,\text{NPD}) + 4(\text{NPD})}{\text{Total number of tetrads}} \times 100$$

$$= \frac{(1/2)\,(\text{TT}) + 6(\text{NPD})}{\text{Total number of tetrads}} \times 100$$

This equation takes into notice the fact that only half of the meiotic products in TT tetrad produced by single crossover are cross-overs.

When genes are close enough as a result NPD = 0 (no double crossover) the map-distance may be equal to the percentage of TT tetrad, divided by 2 (we divide by 2 because only half the sperms are TT tetrads, which are recombinant for the gene-alleles).

Example

Suppose, a two gene cross (*AABB* × *aabb*) takes place which produces 112 PD, 4 NPD and 24 TT. It is very clear that NPD is lower than PD tetrads (4 < 112). This indicates that two genes are linked. Substitution of the values into the above equation yields a map-distance of (1/2) [24 + 6(4)] (100)/112 + 4 + 24 = 17.1 map units.

This mapping procedure differs from the earlier procedure given in Chapter 8 (linkage), where recombination frequencies are not calculated directly from the number of recombinant and non-recombinant chromatids.

Gene mapping on the basis of ordered tetrad

It has been indicated above that if two genes are segregating, it is possible to classify ordered asci as PD, NPD and TT and linkage is carried out as given in the preceding pages. However, it is possible to map each gene with respect to its distance from centromere through ordered arrangement of the meiotic products in the ascospores which also indicate the recombination frequency between gene and centromere.

Principle: *The homologous centromeres of parental chromosomes separate at the first meiotic division, the centromeres of sister chromotids separate at the second meiotic division.*

In absence of crossing over, between gene and its centromere, the alleles of a gene (*A* and *a*) separate in the first meiotic division. This is also called *first-division segregation*, viz. *AAaa* or *aaAA* (Figure 14.3(a)). If crossover takes place between a gene and its centromere, the alleles *A*:*a* do not separate until second meiotic division takes place. This is called *second-division segregation* (Figure 14.3(b)). The distinction between first- and second-division segregation is shown in Figure 14.3. The four arrangements in part (b) are—*AaAa*, *aAaA AaaA*, *aAAa*, which are products of random arrangement of homologous chromosomes at metaphase I and of the chromatids at metaphase II.

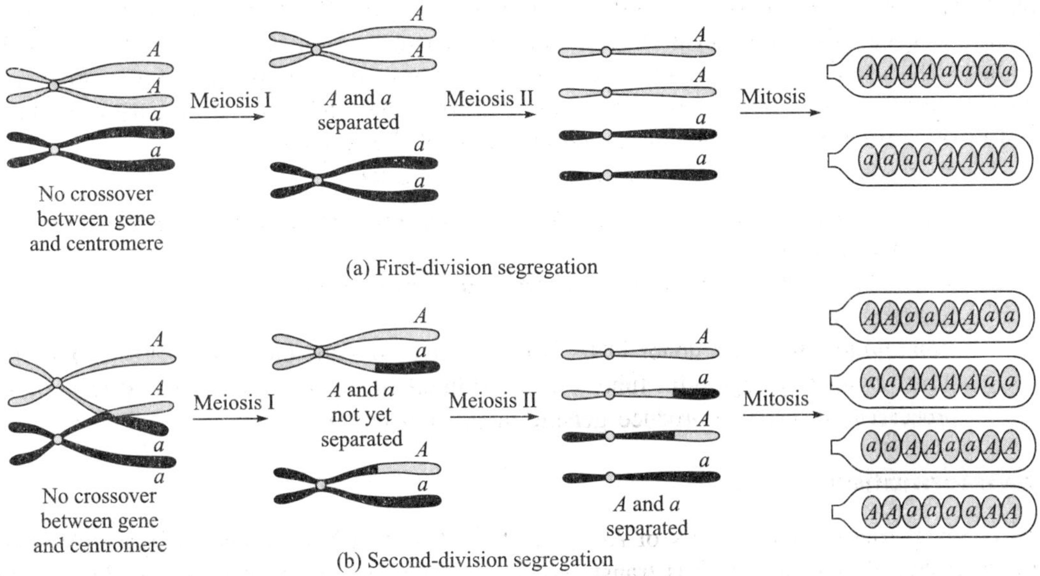

Figure 14.3 First and second division in *Neurospora* and segregation of genes.

From the percentage of asci with second-division segregation, pattern of a gene can be used to map the gene with respect to centromere. Let us say 30 per cent asci of a sample of asci from a cross have a second-division pattern for *A* and *a* alleles. This means 30 per cent of the cells undergoing meiosis had a crossover between gene *A* and its centromere. Moreover, in each cell in which a single crossing over takes place, two of the four chromatids are recombinant and two are non-recombinant, i.e. half are recombinant and half non-recombinant. Thus, a frequency of 30 per cent crossover corresponds to a recombination frequency of 15 per cent. As a principle map-distance refers to the frequency of recombinant of meiotic products, not the frequency of cells in which crossing over takes place. Therefore, map-distance between a gene and its centromere is calculated by the equation:

$$\frac{1/2\,(\text{Asci with second-division segregation pattern})}{\text{Total number of asci}} \times 100$$

This equation is valid if the gene in question is close enough to centromere when few multiple crossover takes place. Thus, reliable linkage values are best determined for genes that are fairly near the centromere.

BACTERIA

A bacterial cell contains a major DNA molecule that does not encounter another complete DNA molecule. The genetic exchange takes place between a chromosomal fragment and an intact chromosome. In fact, a 'donor-recipient' mechanism exists wherein 'donor' cell is the source of a DNA fragment which is transferred to the recipient cell by one of the several mechanisms and exchange of the genetic material takes place in the recipient cell. Incorporation of a part of transferred donor DNA into chromosome of recipient cell need at least two exchange events because recipient DNA molecule is circular, only an even number of exchanges give rise to viable product. The usual outcome of these events is the recovery of only one of the crossover products. But in some cases, transferred DNA is also circular and a single exchange results in total incorporation of this DNA into the chromosome of the recipient.

There are three major types of genetic transfers in bacteria:

1. *Transformation*, in which DNA is taken from the surroundings and incorporated into the bacterial genome.
2. *Transduction*, in which DNA is transferred from one bacterial cell to another by a bacterial virus.
3. *Conjugation*, in which donor DNA is transferred from one bacterial cell to another by direct contact through pilus tube. The recombination frequencies that result from these processes are used to produce genetic maps of bacteria.

Transformation

The classical transformation work of Avery, et al. (1944) in which 'rough-surface' colony of *Streptococcus*, *Pneumococcus* was transferred to 'smooth surface' by exposure of the 'rough surface' strain cells to DNA from 'smooth surface' strain. This experiment also proved that DNA is genetic material.

Bacterial transformation is a process in which the recipient cells acquire genes from the free DNA molecules available in the surrounding medium. In laboratory, donor DNA is isolated, broken and added in the suspension of recipient cells, but in nature DNA become available on the lysis of the donor cells (Bacteria). Transformation starts with uptake of a DNA fragment from surrounding by the recipient cell and terminates with one-strand of donor DNA replacing the homologous segment in the recipient DNA. But DNA is able to penetrate only in some of the cells of growing population. The intake efficiency can be enhanced by treating the recipient cells with some chemicals, which break the cell wall.

Gene mapping

Transformation is a convenient method for gene mapping in some species. When DNA is isolated from donor bacterium, it is broken into small fragments. In most species, transformation occurs at a frequency of about one transformed cell per 10^3 cell. If two genes *a* and *b* are so widely separated in donor DNA that they are always located in two different DNA fragments, then probability of simultaneous transformation (co-transformation) of an *ab* recipient into wild type is the product of the probabilities of transformation of each marker or about $10^{-3} \times 10^{-3} = a^+ b^+$ transformants per 10^6 recipient cells. But, if the two genes are located on the same DNA fragment, the probability of transformation of recipient cell is the same as the probability of a single gene transformation = 10^{-3}.

Co-transformation of linked markers

Markers *a* and *b* are near enough to each other that they are often present on the same donor fragment as markers *b* and *c*. Markers *a* and *c* are not near to each other to undergo co-transformation; the gene order may be *abc* (Figure 14.4).

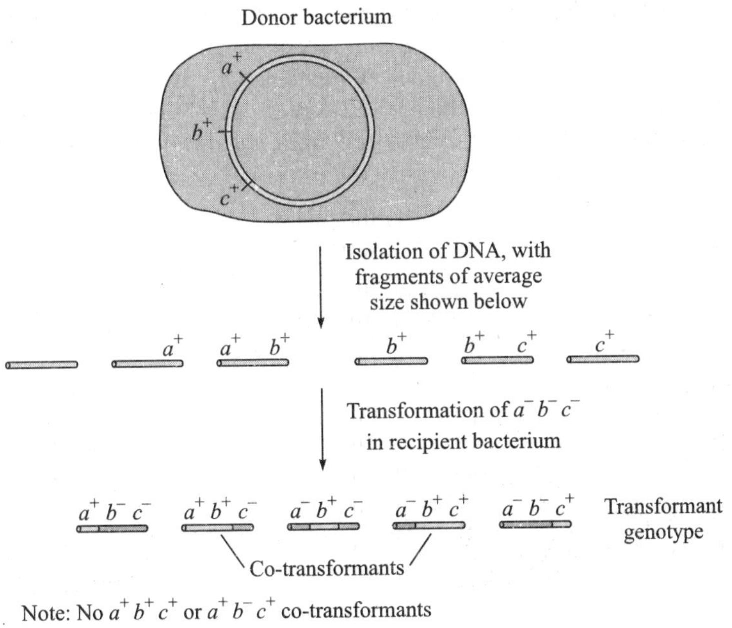

Figure 14.4 Co-transformation of linked markers *a* and *b*, are quite near to each other as a result they are frequently present on the same donor fragment as are markers *b* and *c*. Markers *a* and *c* are on quite long distance hence they do not undergo co-transduction. Hence gene order may be *a*, *b*, *c*.

The general principle is that *co-transformation of two genes at a frequency substantially greater than the single gene transformation indicates that the two genes are close together in the material chromosome.*

Therefore, if genes *a* and *b* could be co-transformed and genes *b* and *c* could be co-transformed, but gene *a* and *c* cannot, then gene order must be *a*, *b*, *c*.

Note: Co-transformation frequencies are not equivalent to recombination frequencies used in mapping gene order in eukaryotes. Here, gene order is determined by the size, distribution of the donor DNA fragments and the likelihood of recombination between bacterial DNA molecules rather than the occurrence of chiasmata in synaped homologous chromosomes.

Transduction

In transduction, bacterial DNA is transferred from one bacterial cell to another by phage particles containing the DNA. Such a particle is called *transducing phage*. There are two types of phages—generalized phage and specialized phage.

Generalized phage produces some particles that contain only DNA obtained from host bacterium rather than phage DNA. The bacterial DNA can be derived from any part of bacterial chromosome.

Specialized phage produces particles containing both phage and bacterial genes linked in a single DNA molecule and bacterial genes are obtained from a specific region of the bacterial chromosome.

If we examine the events that follow infection of a bacterium by a generalized transducing particle say by growth of phage P_1 on wild type *E. coli* containing a *leu*$^+$ gene (Figure 14.5). If such a particle adsorbs to a bacterial cell of *leu*$^-$ genotype and injects the DNA that it contains into a cell, the cell survives because phage head contains only bacterial genes, no phage genes. A recombination event takes place, if the *leu*$^+$ allele carried by phage for the *leu*$^-$ allele carried in the host converts the genotype of the host cell from *leu*$^-$ into *leu*$^+$, which occurs in one out

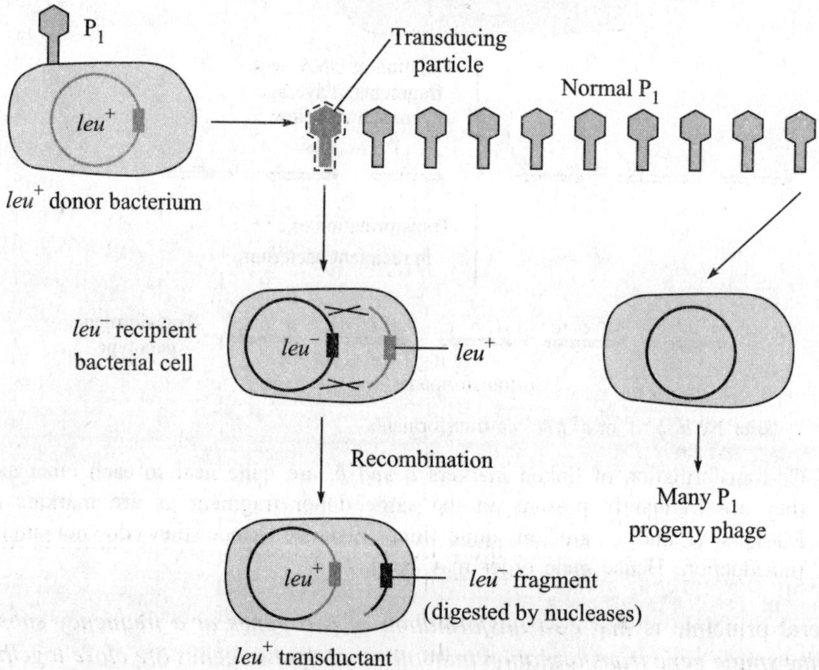

Figure 14.5 Transduction. Phage P_1 infects a *leu*$^+$ donor yielding normal P_1 phage with occassionally one carrying bacterial DNA instead of phage DNA.

of 10^6 cells. Such cells could be detected on selective growth medium. If the infected cell on a solid medium lacking leucine it is able to multiply and a *leu*$^+$ colony can be formed. A small fragment of bacterium contained in a transducing particle includes about 50 genes.

Co-transduction

If the genes are closely linked like *gal* and *bio* genes, as a result both the genes are sometimes present in a single DNA fragment, then they are carried in a transducing phage particle together named as *gal–bio* particle. But all *gal*$^+$ transducing particles do not include *bio*$^+$ and all *bio*$^+$ particles do not contain *gal*$^+$ genes. Therefore, the possibility of both marker genes remains together in a phage particle depends how close they (genes) are. The simultaneous transduction of both marker genes is called *co-transduction*; closer the genes are, greater the frequency of *co-transduction*. Co-transduction of *gal–bio* pair can be detected by planting infected cells on the appropriate growth medium (Figure 14.6).

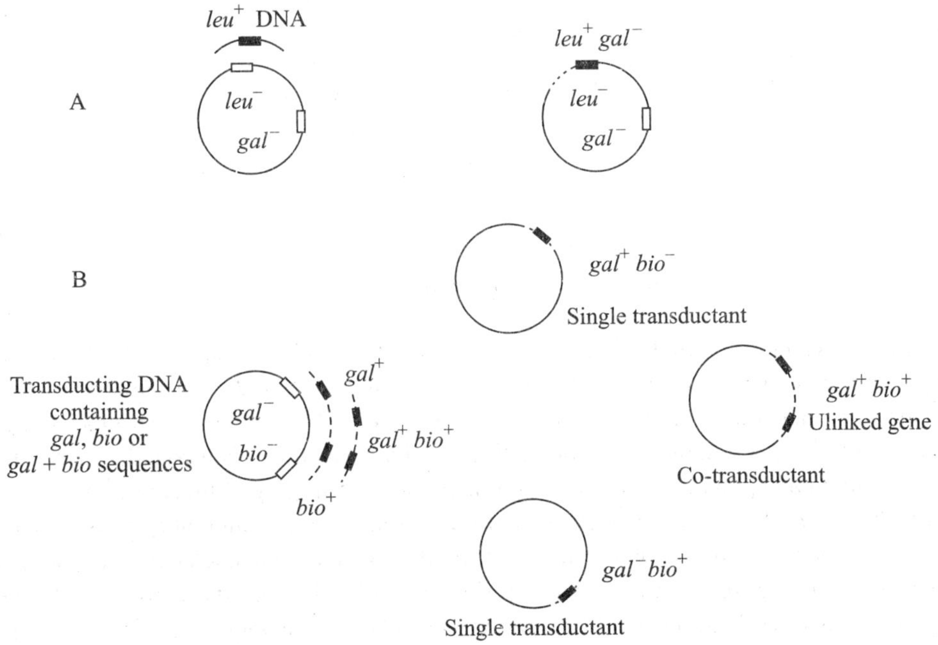

Figure 14.6 Co-transduction of linked gene *gal*$^+$ *bio*$^+$.

Conjugation

Conjugation is a process through which DNA is transferred from donor bacterial cell to the recipient cell via cell-to-cell contact. In Figure 14.7, conjugation process is shown in two strains of bacteria. This concept was given by Joshuva Lederberg in the year 1951. He made a cross between two auxotrophic mutant bacteria by growing them in a joint culture on complete medium and then plating samples of the culture on the minimal medium to isolate any prototrophic recombinant that might arise. But he felt that reverse mutation may also be the reason of the appearance of the prototrophic type instead of recombinant. He, therefore, conceived the idea

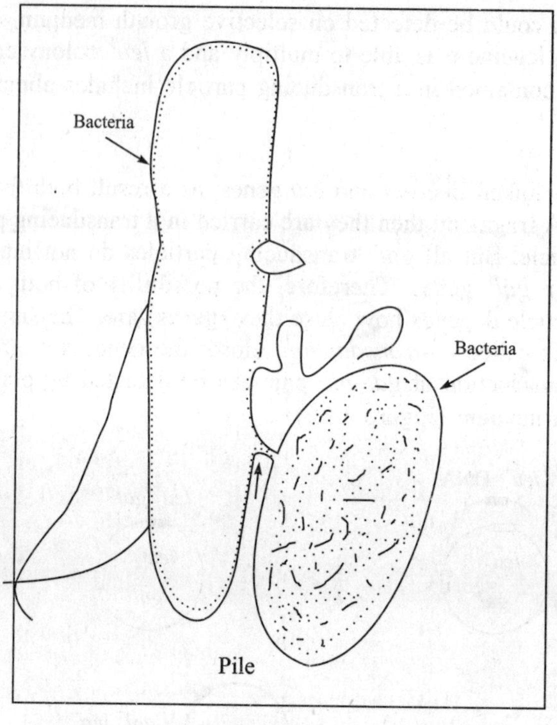

Figure 14.7 Conjugation in bacteria.

of making crosses between multiple auxotrophs. Lederberg took the support of Tatum who did famous experiment, which consists of two strains of auxotrophic *E. coli*-mutant A, (lacking $Meth^-$ and $Biot^-$) and other strain B, mutant for Thr^-, Leu^-, Thi^-. Both these strains were grown together overnight in a complete medium. Mixed culture was then centrifuged, washed free of complete medium and planted on minimal agar medium. The finding of this experiment was that one Met^+, Bio^+, Thr^+ Leu^+ Thi^+ prototrophic colony appeared on the minimal agar for about every 10^7 parent cell plated while no such colony appeared in the parental stocks grown separately on the control plating on the minimal medium (Figure 14.8). It was, therefore, concluded that prototrophs recovered from the mixture culture were genetic recombinants of some kind. This means these prototrophs must have incorporated into their genomes the Met^+, Bio^+ genes from strain B and Thr^+ Leu^+ Thi^+ genes of strain A.

When bacteria conjugate, DNA is transferred to a recipient cell from a donor cell under the control of a set of genes that provide capability to the donor cell to transfer its genes. These genes are present either on the plasmids or a block of genes in the chromosome. In the latter case, plasmid is said to have been integrated into the chromosome.

Conjugation begins with the physical contact between a donor cell and a recipient cell. Then, a passage is formed called *pilus* between the cells and a copy of donor DNA is passed from donor to recipient. Finally, the segment of DNA of the donor cell becomes integrated with genetic complement (DNA) of the recipient cell. If the donor cell contains a free plasmid, only plasmid DNA is transferred to the recipient cell (Figure 14.9).

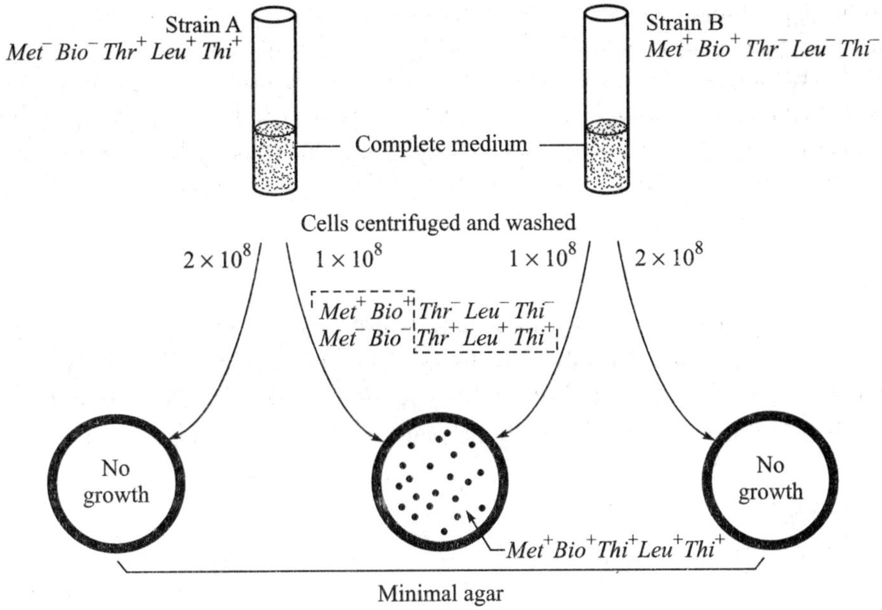

Figure 14.8 Schematic summary of Lederberg and Tatum's classical genetic cross of two auxotrophic mutant strains of *E. coli* K12.

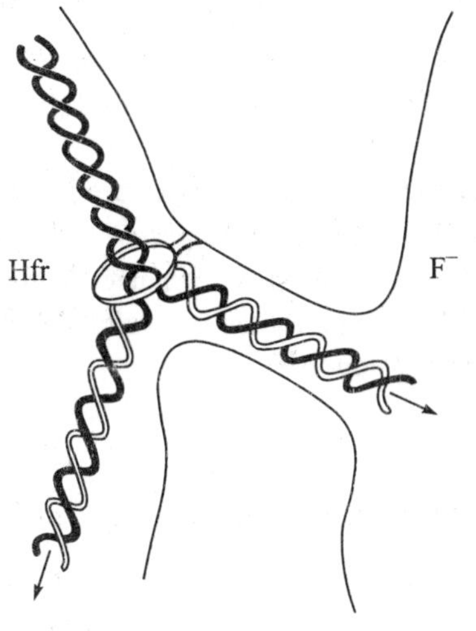

Figure 14.9 Chromosome transfer by replication during conjugation.

Plasmids

Plasmids are circular DNA molecules, capable of replicating independent of chromosome and ranging in size from a few kilobase to a few hundred kilobase. They have been found in many bacterial species.

Plasmids depend upon the DNA replication enzymes of the host cell for their reproduction; but initiation of replication is controlled by plasmid genes. Number of copies of plasmids per cell may be 2–50. From genetic point of view, F-plasmid (F for fertility) is of greatest significance. The cells that contain F^+ plasmids are donors and those lacking F^- plasmids are recipients. F-plasmid contains a set of genes which assist conjugation for transferring genes from donor to recipient cell. Transfer is always accompanied by replication of plasmid. After transfer of F^+ genes, both cells can function as donors.

HFr cells

HFr stands for *high frequency of recombination*, which refers to the relatively high frequency with which donor genes are transferred to the recipient. Integrated F mediates the transfer of DNA from the bacterial chromosome in an HFr cell. Thus, a replica of the part of the beneficial chromosome, the 'F-plasmid' occasionally becomes integrated into the *E. coli* chromosome.

Genetic analysis

Genetic analysis needs that recombinant recipient cell could be identified. For this purpose, use an F recipient cell that contains an allele that can be selected, but it should be located at such a place in the chromosome that most mating cells will have broken apart before transfer of the selected allele occurs and selected allele should not be present in the HFr cell (donor cell). The selective agents can be used to select F^- cells and eliminate HFr donors. Should not be present in the HFr cells and the selected alleles (selected marker). The gene conferring resistance to antibiotics are quite useful for this purpose. Because after mating between HFr leu^+ *str-s* and F^- leu^- *str-r* cells, the HFr *str-s* cells could be selectively killed by planting the mating mixture on a medium containing *streptomycin*.

A selective medium that lacks leucine can then be used to distinguish between recombinant and non-recombinant recipient cells. The F^- leu^- parent cells cannot grow in medium lacking leucine, but recombinant F^- leu^+ cells can grow because of the presence of F leu^+ gene. Here, only recombinant recipient cells having genotype leu^+, *str-r* form colonies on a selective medium containing streptomycin and lacking leucine.

Gene mapping

By HFr × F^- mating, genes can be mapped. For example, make cross like

$$\text{HFr } a^+\ b^+\ c^+\ d^+\ e^+ \times str\text{-}s \times F^-\ a^-\ b^-\ c^-\ d^-\ e^-\ str\text{-}r$$

in which a^- cells require nutrient A; b^- requires nutrient B and so on. Mix the cells, agitate violently in a kitchen blender; take out samples and plate on a series of medium containing *streptomycin* and different combinations of five substances A to E (in each medium one of five substances is lacking). The colonies that form on a medium lacking A are 'a^+ *str-r*' and those growing without B are 'b + *str-r*' and so on so forth. All these can be plotted on a graph as shown in Figure 14.10. Number of recombinants in each curve increases with length of time of mating and for each marker, there is a time of entry before which no recombinants are detected by

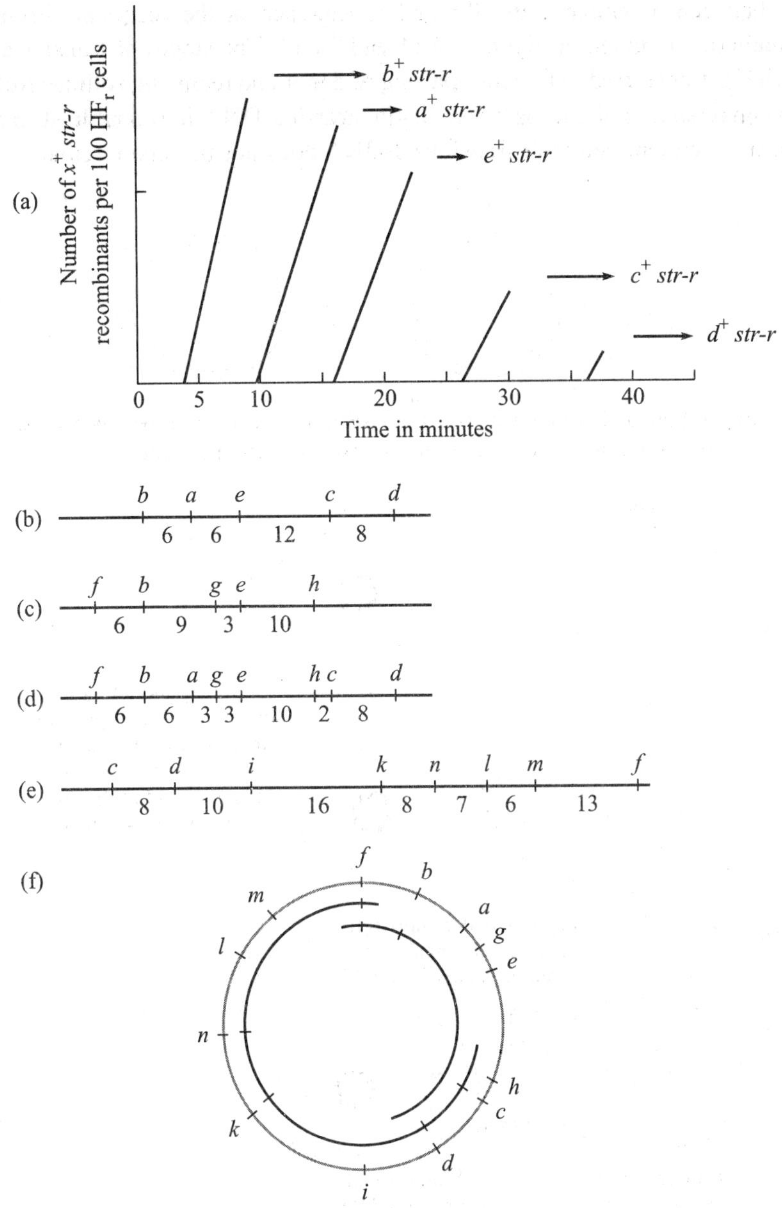

Figure 14.10 Time of entry maping.

(a) Time of entry curve for one HFr strain of bacteria.
(b) Linear map derived from A.
(c) Linear map derived from HFr strain but different F^- recipient strain containing alleles b^-, e^- f^- g^- h^-.
(d) A composite map formed from B and C.
(e) A linear map from another HFr.
(f) Circular map (outer ring) obtained by combining the two map points D and F.

an exchange between a sequence in HFr and a sequence in the recipient chromosome. (F^-) HFr × F^- conjugation is given in Figures 14.11 and 14.12. The stages of transfer are much like those by which F is transferred to F^- cell—pairing of donor and recipient, rolling circle replication in donor and conversion of the transferred single stranded DNA into double stranded DNA by replication in the recipient cell. But transferred DNA does not become circular.

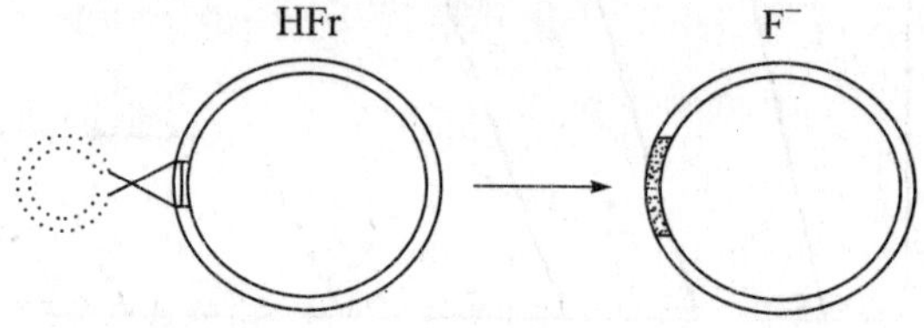

Figure 14.11 Integration of F (dotted circle) by a reciprocal exchange between a base-sequence in F^- and homologous sequence in the bacterial chromosome.

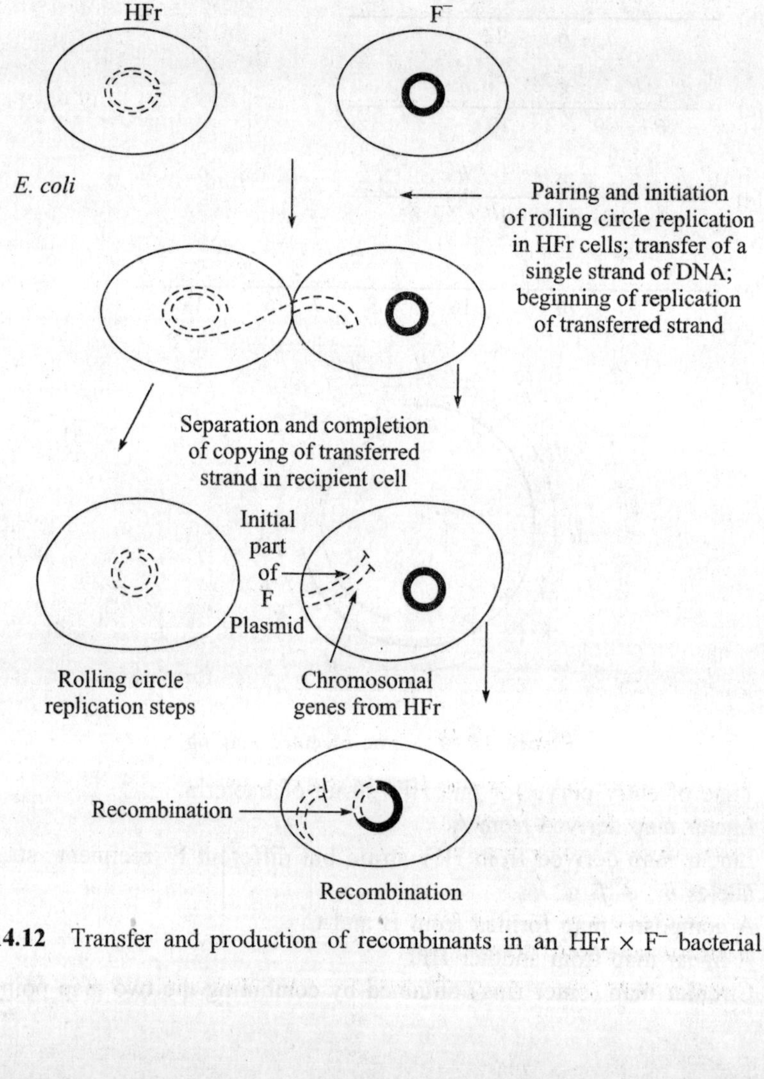

Figure 14.12 Transfer and production of recombinants in an HFr × F^- bacterial mating.

VIRUSES

Bacteriophage Genetics

Phages have two types of life cycles—lytic and lysogenic (Figure 14.13).

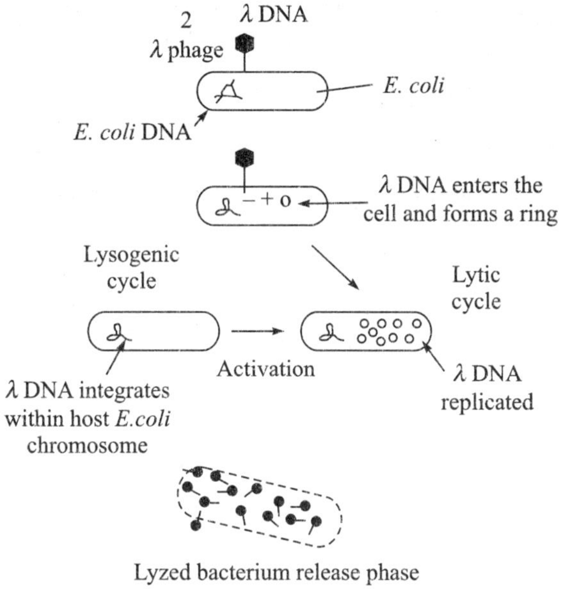

Figure 14.13 Infection of *E. coli* by λ bacteriophage showing lytic and lysogenic cycles.

Lytic cycle

In the lytic cycle, phage nucleic acid enters the bacterial cell and replicates therein repeatedly; the bacterium is killed and thousands of phage particles are released. The lytic phages are also called *virulent* form of the phages. All phages species can undergo lytic cycle.

Lysogenic cycle

In lysogenic cycle, no replication of phage particle takes place in the bacterium; bacterium survives and a phage DNA molecule is transmitted to each bacterial daughter cell on division. In most cases, phage DNA gets integrated with bacterial chromosome. A phage capable of such a life cycle is called *temperate*.

Plaque formation

If a phage is present, it infects a bacterial cell; after some time, bacterial cell lyses and releases thousand of phages which again infect the near by bacteria in the vicinity and lyses the cells. This process of phage multiplication continues and after some time, all the bacteria are killed which give rise a clear transparent region in the otherwise turbid layer confluent bacterial growth called a plaque (Figure 14.14). Phages can multiply in growing bacterial cells. Life cycle of a bacteriophage is shown in Figure 14.15.

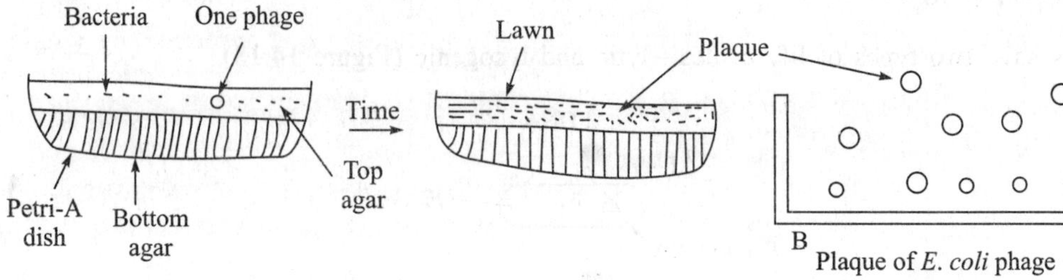

Figure 14.14 Plaque formation in bacterial culture *E. coli* phage on agar media.

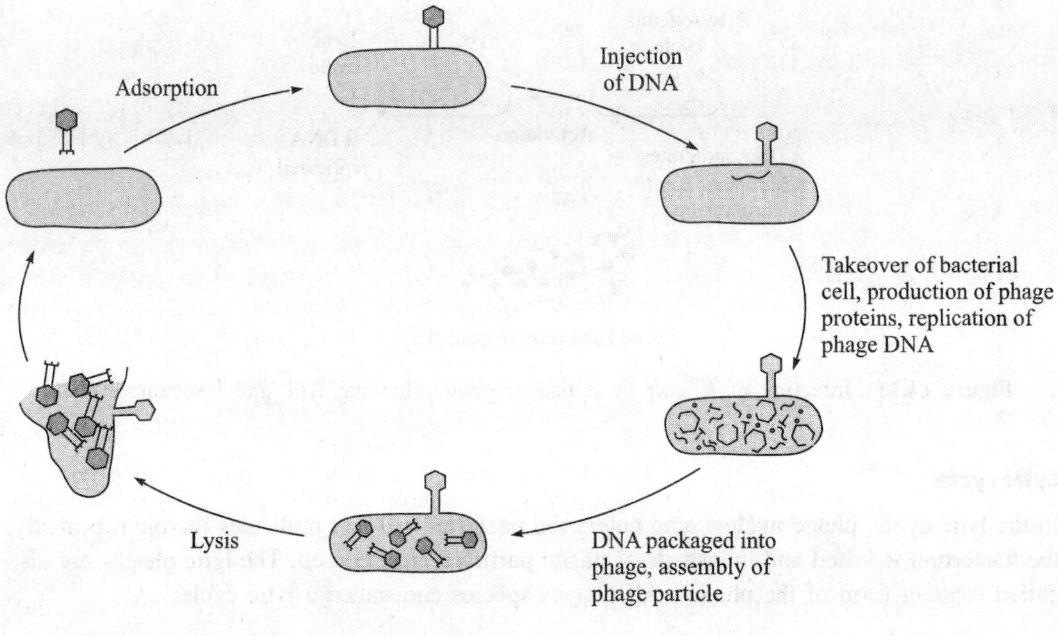

Figure 14.15 Life cycle of a typical bacteriophage.

Genetic recombination can be studied in the lytic cells, if two phage particles with different genetic constitution infect a bacterium which may yield some recombinant types.

$$tu^- \; r^- \text{ (turbid, large)} \times tu^+ \; r^+ \text{ (clear, small)}$$

From this cross, two parental phages and the tu^- r^+ (turbid, small) and tu^+ r^- (clear, large) recombinants can be obtained. The recombination frequency in percentage can be expressed as:

$$\frac{\text{Number of recombinant genotypes}}{\text{Total number of genotypes}} \times 100$$

Gene conversion

We have discussed in detail the recombination of genes that arise due to reciprocal exchange of the parts of the homologous chromatids. When crosses are done in Ascomycetes (fungi) using genetic markers that are very closely linked (such as two mutations at different sites within the same genes) and tetrad data are collected and analyzed, recombination is frequently found to be non-reciprocal. Experimental evidence suggests that if crosses are made between two very closely linked mutant genes in *Neurospora*; *asci* containing wild type recombinant do not show the reciprocal double mutant recombinant. Let us see a cross involving two very closely linked mutations m_1 and m_2. In a cross m, $m_2^+ \times m_1^+ m_2$, asci of the following types of genotypes are frequently observed:

- Spore pair 1: $M_1^+ m_2$
- Spore pair 2: $M_1^+ m_2^+$
- Spore pair 3: $M_1^2 m_2^+$
- Spore pair 4: $M_1^2 m_2^+$

Here wild type $m_1^+ m_2^+$ spores are present, but reciprocal type $m_1 m_2$ double mutant spores are not present in the asci. The $m_2^+ m_2$ ratio is 3:1 instead of 2:2 as expected. This is because one of the allele m_2 is converted to m_2^+ allelic form as a result reciprocal types did not appear. The most striking feature of gene conversion is that the 1:1 allele ratio expected from Mendel's laws of segregation (2:2 in yeast and 4:4 in *Neurospora*) is not maintained. This can be briefly explained that during recombination process, short segments of strands of parental DNA are degraded and then resynthesized (Figure 14.16), using a strand of DNA containing the other parental allele as template (so called excision repair process). Several DNA repair mechanisms are known in different organisms. Several gene conversion are the results of processes involved in recombination. This is detectable, when closely linked genetic markers are used.

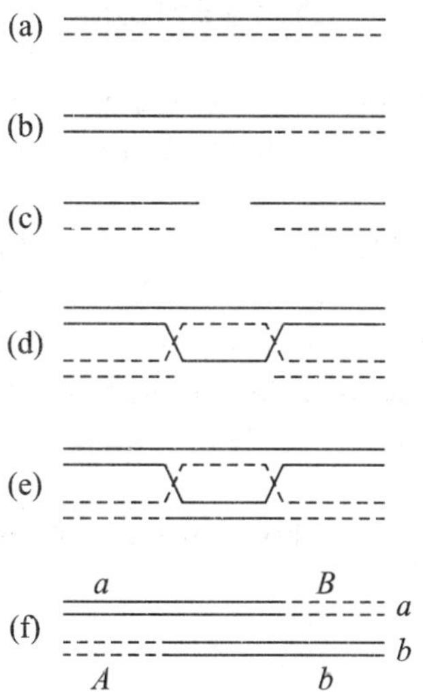

Figure 14.16 Mechanism of gene conversion in lower organisms (Fungi).

REVIEW QUESTIONS

1. A nutritionally-defective *E. coli* strain grows only on a medium containing thymine, whereas another nutritionally-defective strain grows only on medium containing leucine. When these two strains were grown together, a few progeny were able to grow on a minimal medium with neither thymine nor leucine. How can this happen, explain?
2. (a) What are genotypic differences between F^+ cells, F^- cells and HFr, cells?
   (b) What are phenotypic differences?
   (c) By what mechanism F^- cells are converted to F^+ cells and to HFr cells.
3. What is the importance of 'F' factors in genetic analysis? How are they formed? By what mechanism does sexduction occur?
4. What are the basic differences between generalized transduction and specialized transduction, explain?
5. How gene can be mapped by interrupted mating conjugation experiment?
6. What does mean co-transduction? How can co-transduction frequencies be used to map genetic markers?
7. In carrying out specialized transduction of a gal^- bio^- strain in *E. coli*, using bacteriophage λ from a gal^+ bio^+ lysogen, what medium would you like to select for *gal* transduction without selecting for bio^+?
8. How many plaques can be formed by a single bacteriophage particle? A bacteriophage adsorbs a bacterium in a liquid growth medium. Before lysis, the infected cell is added to a suspension of cells that are plated on a solid medium to form a lawn. How many plaques will result?

REFERENCES

Birge, E.A. (1981): *Bacterial* and *Bacteriophage Genetics*, Springer-Verlag.

Campbell, A. (1976): How viruses insert their DNA into the DNA of the host cell, *Scientific American*, December.

Edgar, R.S. and R.H. Epstein (1965). The genetics of bacterial virus, *Scientific American*, July.

Hartl, D.L. (1991): *Basic Genetics*, Jones and Bartlett, Boston.

Hayes, W. (1968): The *Genetics* of *Bacteria and Their Viruses*, Wiley.

Novick, R.P. (1980): Plasmids, Scientific American, December.

Ram, M. (2005): *Unfolding Reality of Life—The DNA World*, Dominant Publications, New Delhi.

Stent, G.S. (1971): *Molecular Genetics*, Freeman and Company, San Francisco.

CHAPTER

15 Human (*Homo sapiens*) Genetics

Hippocrates (460 B.C.) known as Father of Medicine made some observations on human characters and diseases such as epilepsy and baldness. Francis Galton applied statistical tools to study human problems. Mendel discovered Laws of Heredity in pea, but major genetical principles were derived from *Drosophila melanogaster* (fruit fly), and molecular genetic principles were derived using bacteria (*Escherichia coli*) and viruses as investigating materials. For many reasons, study of inheritance of characters in human beings is very difficult because due to social reasons, strict pedigree methods are not applicable. But still, we can study genetic principles in human beings to a limited extent.

Human beings are capable of controlling their own environment to a large extent. For example, diseases like diabetes and haemophilia (bleeder disease) are due to genetic causes, but they can be suppressed by use of drugs. Same is the case with the susceptibility to diphtheria. Artificial immunity can be involved in susceptible individuals using toxin–antitoxin treatments.

Each somatic cell of man possesses 23 pairs of chromosomes, out of which 22 pairs are autosomes and one pair is called sex chromosomes. A female possesses one pair of sex chromosome (XX) and it produces only one type of female gamete (egg), hence female is called homogametic sex. On the other hand, male possesses one X chromosome and one Y chromosome, thus male in human being produces two types of gametes, one with X chromosome and other with Y chromosome, hence male is called heterogametic sex.

The genes which determine the character of a person are linearly located on the 23 pairs of chromosomes. On each homologous pair of chromosomes, a person may have identical alleles AA (dominant) or aa (recessive) at both loci, or a person can carry different forms of gene alleles Aa called heterozygous at that locus. Let us examine some examples of human characters governed by genes.

GENETICALLY CONTROLLED CHARACTERS

Colour of Skin

The recessive genes for albino check the development of pigments in eyes, skin and hairs. Skin of black and yellow races seems to be controlled by multiple genes. Negroides seem to differ from white by two pairs of genes. The mulattoes of AO, BB, AABb combinations are dark, those with AaBb, AAbb, aaBB are intermediate, and those with Aabb, aaBb are between intermediate and white, yellow and white also differ by several genes. There are modifying genes which modify the skin colour in human beings.

Skin Perspiration

Some persons can not perspire in hot weather to prevent rise of body temperature. These individuals do not posses sweat glands. This is an inherited character, Such persons can not shed tears. This is conditioned by a pair of recessive genes.

Eye Colour

Exact number of genes determining eye colour is yet to be discovered, but so far information derived indicates that one pair of gene governs brown eye or blue. Congenital displacement of lens (*Ectopic lentis*) is caused by dominance gene. Congenital contract is caused by dominant gene with incomplete penetrance.

Taste

Taste is physiological character. PTC (Phenylthiocarbomine) tastes bitter for seven out of ten persons and tasteless those out of the ten persons. Ability to taste is governed by a pair of dominant gene, while tastelessness is conditioned by a double recessive genes.

Blood Group

Genetic control of blood group has been already given in chapter dealing multiple alleles. The blood group of human beings is governed by multiple alleles. There are two agglutinogen (A, B) present in every person; any person may possess both AB or neither. Persons with agglutinogen A belong to A blood group and persons with agglutinogen 'B' belong to B blood group, those with neither belong to O blood group. When a person lacks an antigen, he/she always possesses the corresponding antibody; no body lacks both, the antigen and antibody. Some persons possess antibody for 'A', some for 'B', and have neither. If blood of group A person is transferred into A and AB, no harmful effect may be observed, because the agglutinin is absent. Lansteiner (1900) who discovered different types of blood groups in Vienna observed that if red blood cells of one person are mixed with blood serum of other agglutinations may occur, but it may not happen in every case. Later researches revealed that two antigens in human red blood cells and two corresponding antibodies are found in the serum. The genetic constitution of the four groups is given in Table 15.1.

TABLE 15.1 Genetic constitution of four blood groups

Blood group	*Genotype*	*Strandskov's gene (symbols)*
A	AO or AA	I^Ai or I^AI^A
B	BO or BB	I^Bi or I^BI^B
AB	AB	I^AI^B
O	OO	ii

The blood reaction is shown in Table 15.2.

TABLE 15.2 Blood reaction

Blood group	*Genotype*		*Agglutinogen present*	*Agglutinin present*	*Group whose serum will agglutinate cells of the group*	*Group whose cell will be agglutinated by serum of group*
	Bernsteins	*Strandstein*				
O	OO	ii	None	$\alpha\beta$	None	AA AB
A	AA AO	I^AI^A or I^Ai	A	β	OB	B AB
B	BB BO	I^BI^B or I^Bi	B	α	OA	A AB
AB	AB	I^AI^B	AB	None	OAB	None

MN Series

In addition to the AB blood group mentioned earlier, two other forms of agglutinogens are found in human beings. According to MN blood group types, the population could be classed as M: MN: N: types. Human beings rarely carry agglutinins for these antigens. By injecting the blood into rabbits, antibodies are produced and serum from such rabbits are again used to test human blood. MN group is genetically controlled inherited character and three alleles are involved—M type is homozygous for M; N type for N, and MN type is for MN. An individual of M group cannot produce a child of N group. Hence, it is used to test the parentage of children in dispute case.

Rhesus blood

If blood of rhesus monkey is transferred into rabbit, immune serum is produced in the blood of rabbits. If the serum is mixed with human blood, agglunation results in some cases. This agglutinogen is designated as Rh. It is governed by dominant gene 'R'. If a child is born to a Rh positive (RR) and Rh-negative (rr) homozygous couple, the child will be 'Rr' genetically and Rh positive. In this case agglutinogen of foetus may pass through placenta to mother and antibody may again diffuse back to the foetus.

NOSOLOGY OF GENETIC DISORDERS

Because study of genetics of human characters is very complex and cumbersome process due to social reasons, the limited genetic information on the genetic disorders are given as under.

1. Single-gene disorders
2. Polygenic disorders
3. Chromosomal disorders
   - (a) Autosomal disorders
     - (i) Trisomic (2n + 1), tetrasomic (2n + 2)
     - (ii) Monosomic (2n – 1), nullisomic (2n – 2)
     - (iii) Deletion, duplication, translocation and inversion
   - (b) Sex chromosome anomalies (XYY, XXYY, XXX, XXXY)

Single Gene Disorders (Monogenic)

The single gene disorder may be autosomal or sex-linked, according to the locus of the gene allele causing a particular disorder/disease. In addition, single-gene controlled disorder/disease may be due to dominant gene allele (AA) or due to recessive gene allelles (aa). If a particular genetic disorder is caused by dominant gene allele, the person carrying 'AA' (homozygous) and 'Aa' (herterzygous) gene-allele will be equally affected, but those carrying recessive gene alleles (aa) will not be affected.

Dominant gene disorders

1. The best known autosomal dominant gene disorder is achondroplastic dwarfism. This is found one in 5000 births. This is characterized by short limbs, a normal or enlarge head, with a prominent forehead and depressed nasal bridge. The average height of such persons may be 4 feet to 4 feet 4 inches. The affected person lives normal life. Therefore, this is actually a genetic disorder, not a disease.
2. Another comparatively (one in 20,000 persons) dominant gene disorder is 'Huntington chorea'. This is a late-onset neurological degenerative disease, often cannot be detected before a person reaches 35–40 years age. In this disease, involuntary twisting and writhing movement become progressively difficult. Patients often lose the ability to speak and frequently suffer from dementia. The primary metabolic defect responsible for it is still not known.

Recessive gene disorders

The recessive gene caused disorder/disease appears when both the deleterious gene alleles are found in a homozygous (aa) condition. The persons with heterozygous Aa conditions do not suffer from the disease because dominant gene allele A mask the action of recessive gene allele. If two heterozygous parents marry, there are 25 percent chances that their one child will be affected with this disease (aa); 25 percent chances that one child will be completely normal (AA), and 50 percent chances that two children like both of its parents will be heterozygous (Aa). For example:

Father × Mother
Aa Aa

AA (25%) Aa (25%) Aa (25%) aa (25%)

(Complete normal) (Normal but carrier of disease) (Diseased)

There are several other examples of recessive gene disorders viz., sickle-cell anaemia, betathalassemia, cystic fibrosis, phenylketonuria (PKU), galactosemia and tay-sachs are well known autosomal recessive gene control diseases. In some population like black in U.S.A., autosomal recessive gene disorders (sickle-cell) are excessively higher (one in 500) than white population. While in white, one in 2500 children born suffers from cystic fibrosis. The heterozygous persons are normal. Similarly, one in 3600 children born to Ashkenazi Jew suffers from tay-sachs disease. In parts of Africa, four out of 10 children suffer from sickle-cell anaemia. Though many die due to sickle-cell anaemia, the heterozygous persons are usually resistant to malaria in the regions of the world where sickle-cell anaemia is endemic.

Sickle-cell anaemia and betathalassemia both are most common haemoglobin disorders and have no cure though a lot of research have been done. In person with sickle-cell disease, the haemoglobin molecules do not properly bind oxygen because of small structural defect. The person with betathalassemia suffers from severely reduced production of haemoglobin, although what is produced is normal. The persons carrying betathalassemia gene, like those carrying the sickle-cell trait, are normal. They do run a small risk of being misdiagnosed as having minor iron deficiency anaemia and they are being given iron by doctors. But iron can injure such persons. It is probably more important for them to know that their patients carry the gene for betathalassemia than it is sickle-cell anaemia (Stern, 1973).

In case of cystic fibrosis, the gene affects pancreatic endocrine and respiratory function. A patient with phenylketonuria (PKU) suffers from an ability to metabolize the amino acid "phenylalanine", which leads to a toxic accumulation in the tissues and inexorably to mental retardation. If intake of dietary phenylalanine is sharply reduced, the affected person would not accumulate toxic material and would not become retarded. But success depends upon early identification of babies carrying genes for PKU.

Cancer cell

Cancer is a most dreaded disease. It exhibits uncontrolled cell growth and division. In non-circulatory tissues, such uncontrolled cell growth produces cell masses called tumours. Cancerous or malignant tumours are those from which cells detach and migrate to other parts of the body giving rise to secondary tumours—a process called metastasis. Non-cancerous or benign tumours do not metastasize.

Human cancers are responsible for an enormous amount of sufferings. There is now extensive evidence for involvement of over 40 different oncogenes in the occurrence of various types of cancer in the animals.

There are tumours inducing viruses 'viral oncogenes'. Most of the information have come from studies of RNA tumour or retrovirus. The name retrovirus is derived from the fact that these viruses store their genetic information in a single-stranded RNA genome and then convert it into a homologous double-stranded DNA form after infecting a host cell. For checking the growth of cell chemotherapy or radiotherapy is done to stop mitotic division of cells. If growth is not stopped, removal of tumour by surgery is the treatment of malignant cells.

Leukaemia

Blood cancer claims more than 5000 victims in a year. Hans Stauss of Imperial School of Medicine London, has identified a gene called 'WT-I' that becomes overactive in cells that cause leukaemia. There is a ground breaking discovery of an immune 'cell' that can seek and destroy leukaemia cells and could lead to a new treatment for this disease in near future. Lord Robert Winston, Director of Research at Hummersmith Hospital N.H.S. Trust (London) has recently said, that scientists have identified a target which allows "T-cells" to selectively destroy cells that cause leukaemia.

Asthma

In a recent study, the medical scientists have found that asthma is caused by mutation in a gene named "ADAM-33", of course environmental pollution enhances the severity of this disease. Prof. Stephen Holgate's (of Southampton University Britain) identification of asthma gene ADAM-33 could lead to new treatment and also genetic screening of new born for likelihood of developing asthma. Parents of children with a high level of "ADAM-33" mutation could be given advice to minimize the risk of their child being exposed to the environmental triggers.

Every one has this gene. The normal function of one of the proteins it carries is to help to operate the airways in lungs. It is these airways which work badly in people with asthma; the airways become thickened, contract too strongly and too easily causing coughing, wheezing and shortness in breath.

Sex Chromosome-linked Gene Disorders

The sex-linked disorders are caused by genes located on X chromosomes. For the reasons not well understood, Y chromosome seems to carry a very few genes determining the sexual differentiation of males. In the X-linked disorders, the sex-ratio of the affected persons depends on which parent transmits the 'gene' and whether the 'related' gene functions as dominant or recessive gene in the female. All X-linked disorders act as dominants in males because they have only one X chromosome. There is no normal gene to counter the deleterious one. There are a few X-linked disorders which act as dominants in females, but because these are rare, we need to be concerned only with diseases which are recessive in females and dominant in males. For example:

Mother (XX°) (Carrier) × Father (XY) (Normal)

↓

XX (Normal) | X°X (Normal, but carrier female) | XY (Normal, male) | X°Y (Carrier male)

Inheritance of X-linked Diseases

A woman who is XX° (a carrier), if marries a man who is XY normal, their children will be either XX, XX°, XY or X°Y. No daughters will have the genetic disorder, but half of them (on the average) will be carriers of the deleterious gene, at risk for bearing affected sons. About half of all the sons will have the genetic disorder; others will be normal. If the affected sons are able to reproduce, all their daughters will be carriers and all the sons will be normal assuming they marry a normal female.

The X-linked disorders are colour blindness, haemophilia (bleeding disease). Because the affected persons produce an abnormal clotting protein even on minor injury often results in serious haemorrhages. Fortunately, disease can be controlled by infusion of clotting factor extracted from plasma of normal persons.

The other X-linked disease is "Glucose 6 Phosphate Dehydrogenase (G6PD)" deficiency. The affected persons lead normal life except when they inhale pollen of certain bean or ingest certain drugs, which can cause destruction of their red blood cells.

The "Muscular Dystrophy" is an other serious X-linked disease in which affected males apparently normal at birth but begin to show disease symptoms soon he grows 10–12 years of age.

Autosomal Chromosome Disorders

Down syndrome

The patients suffering from Down syndrome possess short stature (about 4 feet tall) and possess an epicanthal fold (thus earlier named mongolism), broad short skull, wide nostrils, large tongue with distinctive furrowing, stubby hands, (particularly fifth digit) with a simian crease on the palm, and a single crease on the fifth digit and general loose jointedness observed particularly in the ankle.

They are rated as low in mental power. But they can be trained in routine mechanical work and skills.

Down syndrome is caused due to addition of a small autosome, chromosome 21, in the normal chromosomal complement (47 + No. 21). Trisomy + No. 21 (2n + 1) is the result of primary non-disjunction which can occur at either of the two meiotic cell divisions or in either of the two parents. The paired chromosome do not separate to poles at anaphase stage as a result one gamete will receive two No. 21 chromosomes and other none.

Down syndrome occurs one in about 100 live births. About one-six (1/6) children die within a year; the expectation of life is about 16 years.

Edwards syndrome

Trisomy-13 syndrome and trisomy-18 syndrome: Trisomy-13 syndrome (47 + No.13) has been reported by Patau in 1960, which occurs in one of the 20,000 new born. It is rare in children and non-existent in the adults because due to serious anomaly patients die early within 3–4 months after birth or in some cases they have survived up to five years. The symptoms include small brain, apparent mental deficiency, deafness and numerous other external and internal abnormalities.

Trisomy-18 syndrome (47 + No. 18) shown in Figure 15.1 described by H.Edwards and his colleagues in 1960 includes mental deficiency and multiple congenital malformations. All the affected children die within 6–12 months. The incidence of Trisomy-No. 18 is one in 8000 births. Possibility or probability of such deformity is greater in the older women because prime cause of this trisomy is non-disjunction of chromosomes in meiosis.

Sex Chromosome Disorders

Turner syndrome (45 + X)

It generally happens in females. Such monosomic abnormality arises when all the 44 autosomes and one X chromosome are present, while other X chromosome is absent. It occurs one in 2500

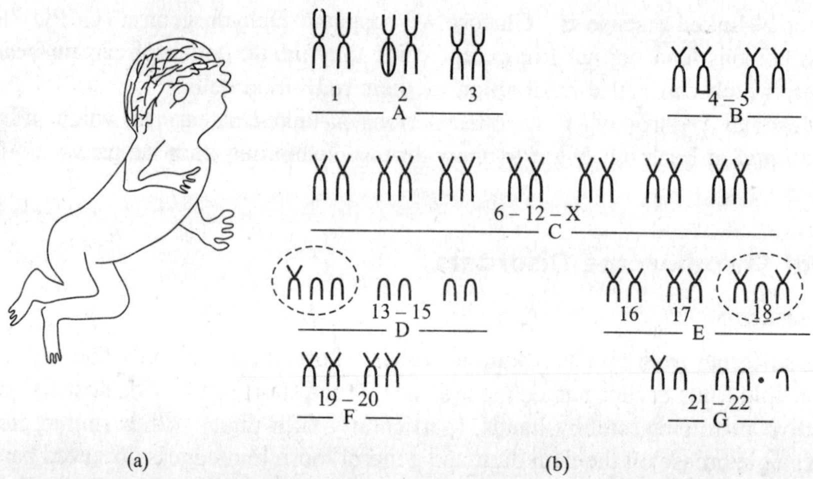

Figure 15.1 Edwards syndrome (trisomy–18). (a) Young triplo–18 child, showing low-set ears, small receding lower jaw, and flexed clenched fingers. (b) Karyotype of same child. Note the seven E-group autosomes.

live female births. The adults have no ovaries and have limited expression of secondary sexual characteristics and are sterile. The ovaries show fibrous streaks of tissues. The affected females possess short stature, low set ears, webbed neck and shield like chest. Mental deficiency is not associated with this disorder. (Figure 15.2).

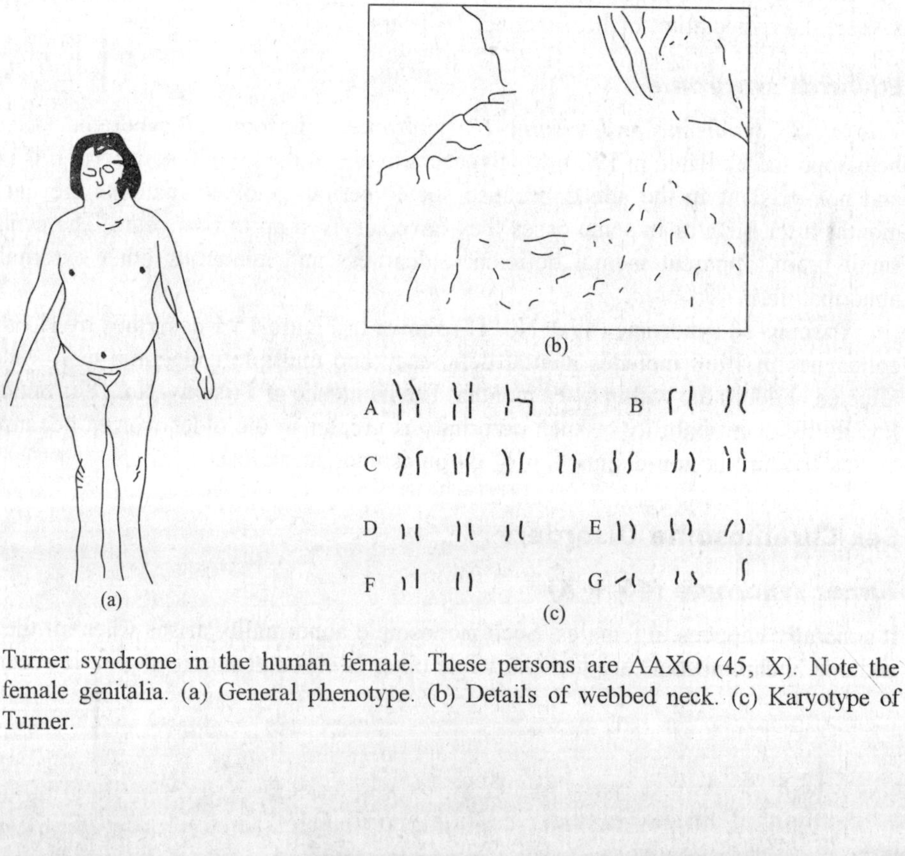

Figure 15.2 Turner syndrome in the human female. These persons are AAXO (45, X). Note the female genitalia. (a) General phenotype. (b) Details of webbed neck. (c) Karyotype of Turner.

Turner syndrome also occurs in male too. In this case, the patients are usually mosaic for 45 x /46 + XY. Persons with one X and a Y fragment, not including the Y short arm, have only streak ovaries but are normal. This suggest that male determining genes are located in the short arm of the Y chromosome and those that prevent the turner syndrome are in the Y long arm as well as X short arm. The male turner syndrome is characterized by defective development of testis, sterility and limited male secondary sexual characteristics along with somatic features of turner phenotype. These persons have normal male karyotype (46–XY).

Klinefelter syndrome (47, XXY)

An extra X-chromosome in addition to the usual male (XY), chromosome complement (47, XXY) has been associated with the abnormal male syndrome described by H.F. Klinefelter (47, XXY) is known as Klinefelter syndrome (Figure 15.3). Its probability of occurrence is one in 500 male births. Individuals with this syndrome are phenotypically males with some tendencies towards femaleness, particularly in secondary sex characteristics. Such features as enlarge breast under-developed body hairs, small testes and small prostate glands are a part of syndrome. The XXY condition arise either by fertilization of an exceptionally XX egg by a Y sperm or of an X egg by an exception 'XY' sperm. The studies on Klinefelter syndrome and Turner syndrome indicate that Y chromosomes determine male sex in human beings. Mental retardation occurs when there are more than two XX chromosomes. XXY male possess less physical and reproductive abnormalities. The females with XXX (47, XXX) show highly irregular menstruation and deficiency in the ovarian follicle formation. The individuals with tetrasomic X chromosomes (48, XXXX) are mentally defective and degree of mental deficiency increases with the increase in the number of X chromosomes present. (Table 15.3)

Figure 15.3 Klinefelter syndrome in man. Such persons are AAXXY (47,XXY). External genitalia are male type, but there is usually some female-like breast development. (a) General phenotype. (b) Karyotype of another Klinefelter individual.

TABLE 15.3 Frequency of sex chromosomes disorder

Karyotype	*Frequency*	*Sex affected*
XYY	1.5	Males
XXX	1.1	Females
XXY (Klinefelter syndrome)	1.2	Abnormal male
XO (Turner syndrome)	0.2	Abnormal female

Most children with chromosome disorders are born to the normal parents. There is a general feeling among the people that such defects are either due to God/Goddess or according to the deeds of past birth. But these defects are caused most frequently by errors in the formation of parental germs cells (egg or sperm) during meiotic division. Normal gamete (germ cell) formation requires the orderly separation of chromosomes to the opposite poles during meiotic division. If both members of a chromosome pair go to one pole, one of the two gametes (germ cell) will be having one extra chromosome (2n + 1) and other will be lacking that chromosome (2n – 1). If either of these two gametes (germ cell) is fertilized with a normal germ cell (sperm) a trisomic (2n + 1) or monosomic (2n – 1) zygote will be formed. The female lacking an X chromosome, (2n – X0) is compatible with life.

A second cause of chromosomal disorders may be "postzygotic non-disjunction." The fertilization of normal sperm and egg may be occurring but shortly after fertilization, the chromosome halves do not separate properly during cell division and two cell population are formed. Non-disjunction at the first cell division would result in a two-cell embryo, of which one cell would be trisomic and the other would be monosomic. In most cases, the monosomic cell would be inviable and a trisomic foetus would develop from the viable cell (Redding and Hirschhorn, 1970). A very high percentage of abnormal embryos are aborted during first trimester of pregnancy as natural selection (Carr, 1972).

The study conducted on Down syndrome (2n + 1) indicated that a maternal age factor in chromosomal disorder was discerned. Age of father also contribute to the disorder. On an average, one out of every 2500 birth to 18 years old mother but one out of 50 babies born to mothers aged over 45 years may have this kind of disorder. The older women produce one-third of all the children with Down syndrome.

Therapy of genetic disorders

1. Some genetic disorders could be treated with traditional methods. For example, cleft lip, cleft palate, pyloric stenosis, oesophagal atresia and some congenital heart defects by surgical techniques. Retinoblastoma, a hereditary cancer can be controlled by surgically removing the eye; the legion of spina bifida can be closed.
2. Some genetic disorders which are result of deficient production of an essential enzyme by some organ can be treated by transplanting a healthy organ into the patient. Persons with Gaucher's disease lack an enzyme that is concentrated in the spleen of a healthy person. The spleen is a very tender and difficult organ to transplant. Lack of donor organs and intricate immunological problems are another difficulties in the organ transplant.
3. In some cases, body produces or accumulates a compound in toxic quantities. An attempt should be made to flush the system by supplying a drug which will either destroy

the excess or carry it away as a waste. Cystinuria is an autosomal recessive genetic disorder characterized by excessive urinary exertion of four amino acids and formation of cystine stones in the urinary passage. Patients are often protein-depleted and at a very high risk for renal damage. Restricting intake of the amino acids which body convents to cystine; maintaining a high urine volume, and using penicillin to bind and flush cystine may be helpful to the patient. (Friedman and Roblin, 1972.)

4. The most dramatic development in enzyme therapy is a single injection of the purified spleen enzyme "glucocerebrosidase" which causes rapid and long-term reduction of the deleterious molecules found in person with Gaucher's diseases. A biannual injection may prevent toxic accumulation and permit them to lead normal life. Because many genetic disorders involve a specific enzyme defect causing toxic accumulation of molecules in the body, the success with Guacher's disease offer hope for developing other treatments. Recent developments in the molecular genetics offer great hope.
5. The most effective developments in the centre of genetic disorders have been found to manipulate the diet, one of such approach is treatment of PKU (Phenylketonuria). Patient fail to metabolize the amino acid phenylalanine which leads to a toxic accumulations in the tissues and lead to mental retardation. If intake of dietary phenylalanine is reduced, the affected person would not accumulate it and would not be mentally retarded. Special diet has shown good effect in controlling mental retardation.
6. Some of the congenital defects associated with Down's syndrome (2n+1) are amenable to surgical repair, but the mental retardation due to presence of extra X chromosome cannot be prevented. Babies born with other autosomal disorders usually live for only a short time and are profoundly retarded.
7. Similarly, significant improvement has been found in the control of haemophilia but affected persons require expensive therapy and lead restricted life.
8. Two most common haemoglobin disorder—sickle-cell anaemia and betathalassemia have no proper cure. The patients with betathalassemia do not manufacture enough haemoglobin molecules; they are very anaemic and die in childhood. The patients with sickle-cell anaemia manufacture haemoglobin molecules incorrectly, but this disorder can be managed somewhat more easily than betathalassemia. The affected persons sometime live up to 20 years or even more. The only proper control of genetically caused disease/disorders is to inform the affected people about the risk of genetic disease before they produce children. Unfortunately, people often learn about these risks after they have had an affected baby. But after diagnosis, future pregnancies could be prevented.

There are two ways to prevent occurrence of birth of even the first affected child.

(i) Premarital identification of carriers of genetic disease. Although quite healthy themselves, these people may bear affected children. The two persons (male and female) who carry a recessive gene allele (aa) for the same disease, if marry they have a 'one-in-four risk' of bearing a child who has inherited both recessive alleles and is consequently affected with the disease.

Any female, who carry the allele for an X-linked disorder has a one-in-two risk that a son will have the disease. Any person who is affected with a late-onset dominant disorder (such as Huntington's chorea) has a one-in-two chances of transmitting that disease to the offspring. A few persons having unusual chromosome rearrangements known as "balanced translocations" look normal, but their children could be affected with serious chromosomal disorders.

(ii) Prenatal diagnosis of the foetuses early enough to permit women to get aborted the defective pregnancies. This can be done by examining the foetal cell or of the amniotic fluid. Let me first explain what is amniocentesis.

Amniocentesis

This involves penetrating the uterine wall with a large bore hypodermic needle (Figure 15.4). The syringe is inserted cautiously to avoid the placenta and the foetus with help of ultra sonographic technique a few millilitres of aminiotic fluid are withdrawn. The fluid, which usually contains sloughed-off foetal cells is then used to begin tissue cultures. After the foetal cells have multiplied enough to provide sufficient material, certain diagnostic tests can be performed (Miunsky, 1973).

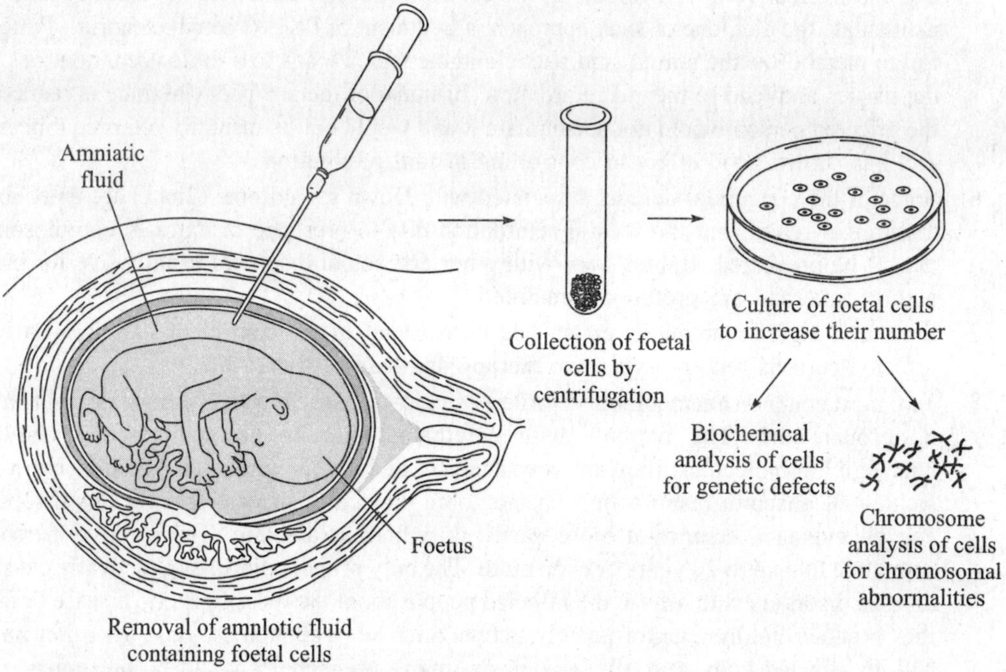

Figure 15.4 Procedure for amniocentesis and screening of foetal cells for genetic defects.

The timing of amniocentesis is also very crucial for a successful result. The tap cannot be performed much earlier than the fifteenth week pregnancy and diagnosis may have to wait 21 days while foetal cells grows. However, diagnosis must be completed before the end of second trimester of pregnancy; it is quite difficult to perform abortion after that time. There is also risk of maternal and foetal morbidity from amniocentesis. According to recent studies, maternal morbidity such as vaginal bleeding, amniotic leakage and amnionitis are of infrequent occurrence. The risk of foetal loss may be 3–5 percent.

(a) ***Chromosomal disorder:***

1. All mothers above 40 years should have amniocentesis unless there are medical or obstetric contradictions.
2. Genetic amniocentesis is recommended when either parent is carrying a chromosomal abnormality whether it is a balanced chromosome rearrangement or mosaicism.
3. Genetic amniocentesis is recommended in all patients where previous pregnancy has terminated in a conceptus with any autosomal trisomy.
4. In absence of general screening program, in individual cases when because of a chromosomal abnormally in another member of the family, amniocentesis may be done.

(b) ***Biochemical disease:*** Amniocentesis is recommended when there is evidence to possible transmission of an autosomal or X-linked recessive gene conditions where the disease is detectable *in vitro* and when affected offspring will suffer from severe diseases like Congenital malformation or mental retardation etc.

Prenatal diagnosis of sickle-cell anaemia, betathalassemia and cystic fibrosis would play an important role in reducing these diseases. Prenatal diagnosis of haemoglobin is available. A tiny sample of foetal blood is withdrawn from the placenta and foetal cells are separated from maternal cells. Appropriate biochemical studies of globin molecules synthesis then permit a diagnosis.

EUGENICS AND EUTHENICS

We know that almost all human traits—mental, physical and physiological are controlled by the genes. Out of these some may be desirable and some undesirable. Francis Galton (1822–1911) was first to conclude that men and animals are the product of evolutionary processes. He coined the word eugenics in 1833 which may be defined as "application of the laws of inheritance/genetics for the improvement of human race". Eugenics is a Greek word which means "well born".

The concept of eugenics is very old and can be traced is the work of Plato and the sages of antiquity. Their primary aim was to produce a race of physically perfect and strong human beings. Only best and the healthiest were permitted to bear children. Plato preferred that liberty to live and bread is the right of those who could distinguish themselves either in war or in scholarship. Until 19th century there was little interest in the eugenics. Francis Galton and Karl Pearson started study of inheritance of human traits and application of laws of inheritance of human population. As a result of there studies, a large number of diseases, defects, errors in the metabolisms abnormalities, intelligence, stature and special abilities or talents were found to be hereditary.

Only such persons should be allowed to produce children. The children unfit for life should not be brought in this world. Persons sufferings with hereditary defects and diseases should be prevented from reproduction.

Objectives of Eugenics

There are four objectives of eugenics:

1. ***Robust health:*** Physically and mentally robust health is a desirable character and it can be presented by following eugenic principles. Genetically robust health means freedom from all hereditary defects, weakness and diseases which impairs efficiency and shortens the life.
2. ***Intelligence:*** In a competitive scientific world intelligence is a most sought character. A reasonably high level of intelligence plus some superior ability is numerous individuals essential for building high state of civilization.
3. ***Moral character:*** Morality is always linked with honesty, dignity, and fidelity. Beside these good conduct or fellow-feelings is directly needed to build up a good human society. But it is my observation that in a free and democratic society, morality is always lacking. Morality is also genetically controlled character.
4. ***Variability:*** Specialization as by different people in different fields of life is needed for human progress and prosperity. If all human beings were casted in the same mold, human civilization could have not reached to a platform, where it has reached today. Special talents are conditioned by heriditary/genetic factors.

There are two fundamental methods to bring betterment in human race.

1. Improving the race by following the eugenic laws, i.e. selective marriage to improve the genetic constitution of the progeny by acquiring desirable genes.
2. Improving the individuals after birth, i.e. by following the principles of euthenics. Euthenics mean by providing good and congenial social environment. By providing good food, good living conditions and good education. Thus euthenics mean "Art of learning to live well in a scientifically awakened society". Both these methods are very important for improvement of human race because an individual is the result of genotype and environmental interaction. Eugenics are of two general kinds:
   (a) Restrictive measures of negative eugenics.
   (b) Constructive measures of positive eugenics.

"Dry up the stream that feed the torrent of defective and degenerative protoplasm." (Davenport). This means the right of living of defective does not mean the right of reproduction. The betterment of human race can be brought by restrictive measures such as

1. Control on immigration of defective and diseased persons.
2. Discriminatory marriage laws. Scientifically marriage is a legal breeding system. In many countries including USA, eugenic laws have been invoked as a result epileptics, insane, habitual drunkards, paupers, idiots, feeble minded persons are not allowed to marry.

In seventies, United States therefore proposed to introduce "Eugenic law". Eugenic means selective marriages aimed to produce good generation free from the genetically-controlled diseases. In this system, the laws are framed for the sterilization of the mentally retarded persons, criminals, epileptic persons, insane and drunkard, etc. However, there were strong political objections to the eugenic ideals on the ground that "marriage and procreations are the fundamentals to the very existence and survival of our race. The power of sterilization if exercised, may have subtle far-reaching and devastating effects on the society. There is no redemption for the individual whom the law touches. He is forever deprived of a basic liberty (Willian O Doughs wrote to the court). The chief justice of U.S. Supreme Court, Mr. Harlan Stone seems even to reaffirms the notion that it may force institutionalized retardates to be sterilized. Stone noted: *"Science has found and law has recognized that there are certain types of mental deficiency associated with delinquency which are inheritable and that may protect itself from the demonstrately inheritable tendencies of the individual which are injurious to the society."*

In Washington State a law was passed that:

> *"No marriage shall take place between two persons in which one or both are common drunkard, habitual criminals, imbecile, feeble minded, idiot or insane person who has therefore been afflicted with pulmonary tuberculosis in its advance stage or any contagious venereal disease shall hereafter inter-marry or marry any other person within this state unless it is established that procreation is not possible by the couple intending to marry."*

ARRANGED MARRIAGES AND THEIR PROBLEMS

A young girl asked her parents how she could marry a boy to whom she did not know? This is because in a close society like India, Pakistan and other Islamic countries, girls were not allowed to choose their life partners freely and cannot marry beyond their caste and religion; except in the rare cases inter-caste marriages take place. While in Islam, it is purely against the religion (Shariat); in Hindus, caste, religion, family status, etc. are the major obstacles. Even within the cast, family status, horoscope and dowry are the main consideration. In addition, even though horoscope is matching, family status is good, but crucial factor is dowry, how much, dowry will be obtained in the marriage is the main demand of the bridegroom and his family; girl's education, and her family status is of least importance.

SEXUAL SEPARATION OF DEFECTIVES

The defective or undesirable persons who possess inherited diseases (haemophilia, colourblindness, diabetes, epilepsis, feeble-mindedness, etc.) should not be allowed to reproduce so that undesirable genes are gradually eliminated from human society. Followings are the methods

(a) ***An eugenic sentiments:*** The public should be educated about the consequences of defective genetic effects on their progeny. If the people are awakened about the ill effect of genetic disorders, they will not oppose the invoking of eugenic laws.

(b) ***Segregation:*** Genetically-defective persons have right and freedom to enjoy life as good as other persons, but society has right to protect itself against repetition of hereditary blunders but segregation is difficult.

(c) ***Sterilization:*** Sterilization is a drastic step to prevent reproduction of defective genes. But after sterilization, they can marry and live normal life. Sterilization is called vasectomy.

(d) ***Birth control and termination of pregnancy:*** Scientific methods are available for birth control and termination of pregnancy.

(e) ***Selective fusion of gametes:*** Sperm and egg banks can be set up to keep them alive for considerable period of time by freezing. This way sperms and ova from most desirable persons can be stored in such banks and can serve as donor for a considerable period even after their death. Successful implantation of artificially fertilized eggs in the womb is must.

(f) ***Education:*** Knowledge of fundamental principles of genetics with special reference to human genetics and appreciation of desirable attributes.

(g) ***Consanguineous marriages:*** Consanguineous (union of gametes within same ancestory) marriages between intellectuals could help in bringing good generation (assembly of desirable genes).

GENETIC COUNSELLING

Genetic counseling is a communication process which deals with the human problems associated with the occurrence of risk or risk of occurrence of a genetic disorder in a family. This process involves an attempt by one or more approximately trained persons to help the individual or family to:

(a) Comprehend the medical factors including diagnosis, probable course of disorders and the available management.

(b) Appreciate the way heredity contribute to disorders and the risk of occurrence in specified relation.

(c) Understand the alternatives for dealing with the risk of occurrence.

(d) Choose the course of action which seems to them appropriate in view of their risk, their family goals and their ethical and religious standards and to act in accordance with that decision.

(e) To make the best possible adjustment to the disorder in an affected family members and/or to the risk of reoccurrence of that disorders.

Francis Galton was first to give the idea of genetic counselling. The concept behind it has been to promote biological race betterment—betterment in human brain structure, and mental endowment and therefore in behaviour. The basic philosophy is that a qualified person/doctor can explain to the newly married couple or yet to be married what the chances would be to get an abnormal child or to the parents the chances of getting another abnormal child so that they could adjust according to the advice given to them. The parents never want to repeat their misfortune.

The development in genetic diagnosis techniques especially in the prenatal detection of diseases combined with the family planning and availability of safe, legal abortions have increased the market of genetic counseling services. A mass screening should become wide spread and fetal diagnosis should become routine to provide genetic hygiene.

Appreciating the idea and philosophy behind genetic counselling, Reed (1955) wrote:

> "*Even the clumsiest counsellor has an excellent chance of relieving the deep anxiety of his client. The counsellor can also help to washing away the guilt feelings of the parents/or boys and girls planning to marry the affected child to be born by pointing out that genetic calamities can occur in any family and that parents are not alone in their problem*"

The philosophy of genetic counselling will be one which is person-centred and not psychologically oriented. The emphasis should be placed on the person or patient within the frame of reference of manifold psychological (feeling, emotion and attitudes) and medical problems (natural history of diseases, diagnosis, prognosis and treatment).

Technique

The main task of the counsellor is to communicate accurate genetic information to a person or concerned couple about risk and burden of specific diseases. The primary allegiance of the medical geneticist may be to reduce the number of children born with the serious genetic disorders. If wife is carrier of Duchenne (X-linked) muscular dystrophy, the decision to whether or not to accept the risk is left to the woman. A counsellor may be disturbed if the client decide to risk bearing child with cystic fibrosis than if risk involve is tuberculosis. However, with mass screening and rise in the awakening of the genetic disorders, the genetic counsellor can play important role in making a person aware of risk involved in the child bearing in the genetically controlled diseases.

Genetic Counselling before Tie of Nuptial Knot

So far, I have dealt above the importance of genetic counselling of the married couples who want to make genetic inquiry about prospect of getting a healthy/unhealthy child, if first child was born abnormal. I am more and more concerned that two persons (boy and girl) who intend to marry or their parents are intended to tie them in the nuptial knot, must be aware of the details of genetic as well as behavioural information of each others. Marriage should not be considered a sacrosanct or decided by some asteric power in heaven. It is purely a bridge of friendship which is standing on two pillars, love and faith; love and faith are non-genetic (cultural) factors connected with the social environments in which they live, but intelligence, health, beauty, skin colour, eye size and colour, nose, height and muscular built-up, etc. are genetically-controlled characters. Here, health means—free from genetically-controlled (heritable) disorders etc., such as mentally retarded, sickle-cell anaemia, betathalassemia, cystic fibrosis, haemophilia, etc.

However, habits like drinking, gambling, criminals, imbecile prostitution etc., are non-genetically controlled habits and primarily controlled by social environment, but for marriage purposes display of information about these bad habits are also equally important because any of these bad habits say drinking, gambling or prostitution can make the married life of the couple hell specially of women. Therefore, genetic counsellors should prepare a "Questionnairc" wherein, information from both the parties intending to settle the marriage of their son and daughter from at least three generation back i.e., starting from grandfather, grandmother; father–mother, uncles and aunts (paternal and maternal both) as well as boys and girls (to be married) and then brothers and sisters should be displayed so that after examination of the biodata, the genetic counsellor could advice properly whether the intending marriage would be happy or unhappy; successful or unsuccessful.

Nothing should be suppressed from either side for sake of love or relative's pressure, otherwise for this very small mistake entire coming generation will suffer. This information must include Rh positive or negative blood group; HIV and AIDS, etc. or any other genetic diseases, etc.

Why these information are desired to be examined by the counsellor from three generation back? Yes, son is "rebirth" of his grandfather; his father is only a conduit and a daughter is "rebirth" of her two "grandmothers"; her mother is only a medium for transfer of her mother's genes. Therefore, any genetic disorder linked with Y chromosomes in grandfather and X chromosome in grandmother will appear in their grandsons and granddaughters respectively.

Matching Blood Groups

I am of the view that before setting marriages, blood group of both the boy and girl must be examined, because blood group has wider repercussion on the longevity, peace, happiness and child bearing, etc. For example:

(a) A marriage between a boy of blood group Rh^+ should not be solemnized with a girl of Rh^- blood group and vice versa. Such marriage will lead premature abortion of the child in the womb.

(b) If a boy of blood group O solemnize the marriage with a girl of blood group B, the marriage life would be very happy because girls of blood group B are foresighted, cooperative in family life.

(c) However, a boy of blood group O if solemnize the marriage with a girl of O blood group, the life would be under tension because the girls of blood group O are of hard temperament. She will not be cooperative with her husband in all matters.

(d) If a boy of blood group B solemnize the marriage with a girl of same blood group B, then she will be best life partner of the husband because both will be having almost similar idea and ideology, thinking and activity. They will be more cooperative for each other in their behaviour.

(e) However, if a boy of blood group B solemnize the marriage with a girl of blood group A, then marriage life would be very painful, quarrelsome and may break.

(f) If a boy of blood group A solemnize the marriage with a girl of blood group A, then marriage life would be very happy because their idea about value of life would be almost same and both will try to keep each other happy.

(g) However, if a girl of blood group A is married with a boy of blood group O, their marriage life would be full of tension because such boys are of obstinate in nature. He will always try to dominate his wife but such boys/persons are wise and responsible.

(h) If a girl of blood group O solemnize her marriage with a boy of blood group A, then marriage life would be full of struggle. To continue their married life happy, they have to compromise a lot.

(i) For a girl of AB blood group, a boy of blood group B would be ideal. Because the girls of blood group AB are generally intelligent and intellectually of high order. The boys are always attracted towards them. They take every step with wisdom and precaution.

(j) The marriage between boy and girl of blood group AB would be also good. Both will give regards to each other and walk cautiously so that their married life should always happy and peaceful.

Risk at Workplace

In addition, there is a growing awareness that cancer may be caused by chronic exposure to the chemicals in the workplace, has provoked much interest among industrial hygienists. For example, liver cancer has been caused by exposure to "Venyl chloride." The goal of current research is to develop a system that would alert employees to the special risk that the potential employees would face at the workplace. Though efforts are being made to provide every worker safe healthy workplace but this information must be known to the girl and her parents about the health hazard.

Similarly, susceptibility to lung cancer seem to be associated with the high level of enzyme that is involved in metabolizing the constituents of cigarette smoke. Screening for hyper lipidemia has made it clear that persons are at usually high risk for heart attacks.

Recently, it has been disclosed that persons working at T.B. (Tuberculosis) hospitals in India are no high risk to get infected. In Delhi, many employees have already died and there are a number of employees who arc undergoing treatments. There is also some stir in the Geological survey of India that several persons have died due to radiation caused skin diseases. These information are equally important for the girls to take decision to marry a person working on these risky places.

Who Should be the Genetic Counsellor?

The ideal genetic counsellor should be a professional who should be fully aware of principles of heredity and should be trained in organizing the candidate's sources of information for a genetic problem by exercising an art for getting that information properly and helping to develop a schedule of action for the families/candidates to follow. In fact, there should be special programs in the medical institutions/universities to educate and train the Genetic Counsellors. The trainees should be taught, about medicines, human genetics, psychology and sociology etc. A medical degree with genetics may be a good for a counsellor to serve the purpose.

Legal Problems

When a counselor educate the client and the conditions under which the counsellor may communicate confidential information to third parties, may invite some legal problems. Because a client may take serious and irreversible decisions, after genetic counselling. There are risk that misinformation may spawn a law suit against the counsellor. Though chances are rare. Yet an overly defensive posture which may protect the counsellor should be always taken. The conflict if any must be clarified by examination of the relevant issues. Failure to properly identify or to explain a genetic risk poses a few new question in legal medicines. However, birth of a child with some serious genetic disorder inspite of the future forecast of a physician or non-physician to alert the mother has greatly troubled medical geneticists and has began to perplex the judges. The possibility of damage suit for wrongful information is always there because abnormal baby may be a great problem to the married couple. An astrologist does not face such legal issue because he leaves everything on God.

What are Essential Things for a Genetic Counsellor to Know?

A genetic counsellor must have a profile of genetically controlled diseases and their symptoms. Let me present the information available on some of the genetically and physio-logically controlled diseases identified in India.

GENETICAL DISORDERS DETECTED IN INDIA

Vast population multiracial origin, diverse cultural, dietary, and living habits, marriage laws and traditions and extreme variation in the environmental conditions make the genetical studies very difficult in India. But some information has been derived which are presented here briefly:

Genetically-inherited Diseases

1. Cystic fibrosis	1 in 2000
2. Huntington's chorea	1 in 2000
3. Muscular dystrophy	1 in 5000
4. Phenylketonuria	1 in 10,000
5. Sickle-cell anaemia	1 in 10,000
6. Betathalassemia	1 in 20,000
7. Retinoblastoma	1 in 20,000
8. Cystinuria	1 in 20,000
9. Tay-Saxs-disease	1 in 200,000

Disorders Caused by Chromosomal Aberrations

Chromosomal aberrations lead to alternations in the phenotype at birth and to mental and physical retardation later in the life. In assessing the caused of congenital defects in the new born, Chandra and Harilal (1978) found Down syndrome in 1.04/1000 birth. Puri et al. (1978) reported 1.33/1000 birth, Verma (1978) placed 1/800 births. The incidence of Down's syndrome detected are given in Table 15.3.

Most of the cases of Down's syndrome were of trisomy 2n+1 (Choudhari et al., 1970; Phadke et al., 1975).

Translocation defects occurred in 6.9 percent cases and only 1.2 percent of birth occurred in mothers aged above 30 years.

Causation of mental retardation remained undiagnosed in 35–65 percent cases in most surveys (Galjuard, 1982).

TABLE 15.3 Incidence of down syndrome

Type of survey	*Number screened*	*Incidence per 1000 births*	*References*
New born	66,684	109	Verma, 1978
Congenital defects	30	50	Talukder, 1974
Mental retardation	45	200	Walter et al., 1971
Mental retardation	104	170	Kulkarni et al., 1980
Mental retardation	100	100	Narayanan and
	700	620	Rama Rao, 1978
	215	37	Thomas, 1979

Sex Chromosomal Anomalies

Such anomalies account for cases related to mental and physical retardation, as well as subfertility hypogonadism and sexual infantilism. The sex chromosome anomalies reported in India are given in Table 15.4.

The classic Turner syndrome reveals varying chromosome patterns from 45 X0 to mosaic involving 46 XX, 45X0 and 46 X / X0; XXP- and 45 X /46XXQ Karyotypes. (Nigli et al., 1980). In 468 cases of mental retardation, Datta et al., 1982, observed 3.94 percent of sex-chromosomal defects with 1.97 percent; 45 XO, 0.56 percent; 47 XXY and 47 XYY. Differentiation of various types of intersexuality and determination of the hormones levels were carried out in some cases (Patel et al., 1983): 79 cases (37 males and 42 females) studies revealed congenital adrenal hyperplasia as the possible cause; in 21 cases, there were metabolic defects in steroidogenesis.

TABLE 15.4 Sex chromosome anomalies in India

Type	*Place*	*Number*	*Abnormalities detected*	*Reference*
New born	Bombay	389	0	Naik and Sah, 1962
	Delhi	3007	1.9	Verma, 1978
	Delhi	1000	82	Majumdar et al., 1975
Mental retardates	Vellore	45	44	Walter et al., 1971
	Delhi	100	0	Dhandial and Verma, 1978
	Calcutta	468	39	Datta et al., 1982
	Bombay	74	94	Mohe et al., 1982
	Bangalore	100	100	Joseph and Thomas, 1980
Primary amenorrhea	Maharastra	115	103	Hiremath et al., 1980
	Calcutta	127	237	Chakravarthy, 1978
Subfertility	Nagpur	831	65	Sen and Thombre, 1971
Psychiatric cases psychosis		67	44	Dutta et al., 1982
Schizophrenia (male)		500	30	Dasgupta et al., 1973
Criminals (males)		400	0	Verma et al., 1978

Single-Gene-Inherited Diseases

Diseases due to autosomal dominant inheritance occur sporadically and common one are rarely reported (e.g., achondroplasia or polydactyly). Over all, data on the autosomal dominant diseases are very limited, but hospital records points to a fairly high incidence of achondroplasia, aperts, polydactyly, osteopetrosis imperfect, etc. (Master Notani et al., 1968). In certain families, marriage of those with minor defects may lead to a pedigree with dominant mode of inheritance of defects.

The detection of autosomal recessive gene defects are very difficult and which may be cause of many death in childhood. The incidence of aminoacidurias and mucopolysaccharidosis are given in Table 15.5.

TABLE 15.5 Incidence of aminoacidurias and mucopolysaccharidosis by screening

Area	*No. of cases*	*Type of error*	*References*	*Number of incidence*
Bangalore	1600	Phenylketonuria	Rao et al., 1978	16
		Hartnup disease		2
		Homocystine		3
		Histidine		1
		Mucopolysuccharidosis		9
Calcutta	355	Phenylketonurea	Datta et al., 1982	1
		Tyrosine		1
		Histidine		2
		Mucopolysaccharidosis		2

Mucopolysaccharidoses (Puri et al., 1968), muscular dystrophies (Goswamy and Chaurasia, 1971); Glucose-6-Phosphate Dehydrogenase Deficiency (G-6-PDD) have been reported to occur frequently in India, though regional variations are there.

PHARMACOGENETICS

Besides the normal genetical studies related to human traits traits and diseases, etc. pharmacogenetics, i.e. genetic variation for drug metabolism as well as genetic responses to the ecological (environmental) factors though very important, have been paid almost no attention in India. For example, a few reports are available which deals with either the identification of chromosomal aberrations in blood from the patients given a particular drug or clastogenic action of a drug on human leukocytes. The action of tuberculostatic drugs on chromosomes in vivo has been recorded for isoniazid and thiacetazone and the gaps and breaks induced in 'leucocytes of patients have been attributed to a possible synergistic action (Ahuja et al., 1981 and Menon et al.,). There is an awareness among the general practitioners of the differential susceptibility of individuals with "Glucose-6-Phosphate Dehydrogenase Deficiency" to antimalarial drugs, but no attempts have been made to survey and find out differential susceptibility of various population groups to these drugs. Screening of population for susceptibility for reaction to anti-malarial and antituberclostatic drugs should form essential component of Primary Health Centres.

ECOGENETICS

Similarly, there is an increasing concern regarding the disturbance of ecosystem following rapid industrialization and the uses of chemical fertilizers, herbicides, and pesticides, etc. in agriculture. Pest control consists of uses of a large number of pesticides including organochloride, organophosphorus and carbamates, etc. Significant amount of residues of the chlorinated hydrocarbons have been recorded in human placenta, cord blood and breast milk by Industrial Toxicology Research Centre Lucknow. The agents better known to cause both cancer and birth defects in human beings are tobacco, ionizing radiation; stilbestrol, alcohol, hydantoin, androgens, antimetabolites and herpes virus (Mulvihill, 1982). Therefore, thorough investigation of mutagenic effect of various chemicals on human system is must using suitable parameters, like mitotic index, chromosomal alternations, micronuclei formation and in some cases sister chromatid exchange (Sharma, 1983). The development of screening methods for haemoglobinopathies and Glucose-6-Phosphate-Dehydrogenase deficiencies (G-6-PD) has yielded good results. Establishment of a few laboratories to provide necessary technology for detection at Family Welfare Centres and Childcare Centres with genetic counselling units are must. It is also essential to transmit these information to the grass root level for the information of the people specially rural people who are not aware of the consequences of the injudicious uses of chemical fertilizers, herbicides and pesticides on their crops.

NUTRIGENOMICS

Nutrigenomics is a new field of science which is investigating whether a change in diet can prevent cancerous growth (tumour formation) in the body. Government of U.S.A. is conducting a trial with about 3200 middle aged men to see if selenium or vitamin E can prevent prostrate cancer. In fact, preliminary but intriguing genetic research supports that certain nutrients may prove more cancer protective for one person than other.

Cancer does not occur overnight. A few tiny cells gone genetically wrong (mutation) grow over years and form tumour which depends upon the gene and the environment including food and living style. Dr. John Miller, Head of National Cancer Research Institute, Washington (U.S.A.) says that upto 35 percent cancer cases are related to the dietary habits. His research program on nutrition and cancer prevention indicate that diets high in plant foods lowers the cancer rates. The people who eats a lot of fruits and vegetables are generally skinnier. Obesity increases the risk of cancer of uterus, gall blader, colon and prostrate glands with increase in body weight at the adulthood is linked to breast cancer. Now scientists are working out, why at genetic level, nutrients do the job of cancer protection? In China, people genetically deficient in an enzyme that metabolize certain nutrients in cruciferous vegetable were found with lowest cancer risk.

It has been also observed that the women who drink green tea daily can reduce the risk of ovarian cancer by almost 60 percent as compared to non-tea drinkers.

Similarly, nitrite a common in the body previously believed to have no major function has been found to greatly increase blood flow indicating it has potential as a treatment for illnesses like heart and blood vessels diseases and sickle-cell anaemia. Nitrite can open blood vessels and improve flow in the parts of body. Nitrite is converted into nitric oxide, a compound which is known to expand the blood vessels.

HOW TO PRODUCE A NOBLE GENERATION

The basic aim of marriage should be to produce a good generation specially gifted children and also to maintain gifted character if a family is lucky to have. Francic Galton, Darwin's cousin published his first book on the study of human intelligence, wherein he wrote that *man's natural abilities are derived by inheritance under exactly the same limitations as are the form and physical features of the whole organic world. It would be quite practical to produce a highly gifted race of men by judicious marriages during several consecutive generations. I shall show that social agencies of an ordinary character, whose influences are little suspected are at this moment working towards the degradation of human nature, and that each generation has enormous power over the natural gifts of those that follow and maintain that it is a duty. We owe to humanity to investigate the range of that power and to exercise it in a way that without being unwise toward ourselves, shall be most advantageous to future inhabitation of the earth.*

NOBLIZATION OF HUMAN RACE NEED NOBLE GENES AS 'DOWRY' NOT MATERIAL WEALTH

It has been amply emphasized that primary aim of the marriage should always be to have a good and noble generation. The men ofcourse are more and more fascinated to have a beautiful wife for sensual satisfaction, but thinkings of the women since very early years are quite different. Their aims are always oriented to have gifted children—son or daughter such as intelligent (high IQ), excellent health, handsome, good look and attractive complexion of boys and fair colour, beautiful eyes, nose and attractive face cut, smart and slim body of girls.

There are women who are fond of to have son(s) with good physique to become good players or wrestlers; some desire to have children who should become scientists, writers and poets; some want their sons and daughters to become doctors and engineers; some want their sons/daughters should become officers; some want their daughters should be singer and dancers, etc. No woman wants her son/daughter should become criminal, corrupt, mentally retarded, drunkard and robber, etc. Thus, main characters of choice for the noblization of human race should be—to have children intelligent, with good health, good height, beautiful face cut and good behaviour, etc. In fact, choice of the family vary from place to place and from community to community. In Punjab and Haryana boys with tall, good health, smart and good personality are the main requirement for military services, because in both these states every family has at least one man in the army. Among Hindus, Brahmins and Kayastha in general prefer intelligence and Kshatria muscle built up, etc., Jats and Yadavas prefer robust physique.

All these characters are genetically controlled. Behaviour is no doubt environmentally controlled, but this is acquired by the boys and girls from the family environment. For example, caste hatred in India is transferred from generation-to-generation culturally. The sons and daughters of the actors and actresses are becoming actors and actresses in the bollywood; sons of players get home grooming for playing. A time is not for when cricket would be dominated by sons of present generation of cricketers. Take field of politics; it has become a family dynasty otherwise "keeps" are dominating the show because they are groomed by their lovelorns. Therefore, bridegroom and his family and parents have to choose one between the two: to enhance "genetic wealth" and produce and good generation or demand the ephemeral material goods like car, gold, T.V., fridge, microoven, sofa-set, etc. as dowry on the cost of good generation. Similarly, bride and her parents before settling the marriage should also choose, whether they want their daughter should be happy with the material wealth of an old, obese body, drunkard, mentally dull bridegrooms or their daughter should bear a good generation—intelligent, talented, healthy boys and girls with robust personality. The doctors, engineers and lawyers are professionals; they may or may not be intellectual/scholarly persons because their main objective is to earn money. Intellectuals are scientists, poets, novelists and philosophers who are thinkers and who invent new ideas, laws and principles which give new twist to the society. Look at scientists and technologists who have almost reshaped the human destiny by bringing revolution in agriculture, industry, medicine and electronics. Therefore, marriage should always be aimed to produce intelligent and healthy children and for this purpose, families of both boys and girls intending to marry their children should be selective to choose good pair where intelligence and health should be the primary considerations. Dowry should be rejected atleast by the bridegrooms family. I am of the view that parents of the boys should pay dowry to the girl's parents and buy the girls with good health, intelligence, etc. As regards beauty, author is of the view that beauty is negatively correlated with good conduct and behaviour which is also an important consideration with the girls author does not advise to choose a very beautiful, intelligent and healthy girl if her conduct is not good. Beauty and scorpians are alike particularly at a time when girls have developed a wrong notion why not to encash the beauty. Do not choose a girl who is working in the filmy world or in T.V. serials because their area of thinking is career advancement to earn money and enjoy life. They are in general "spoiled class".

DISTANCE MARRIAGES INTERCASTE SHOULD BE ENCOURAGED FOR ACQUIRING NOBLE GENES

Marriages within caste (endogamous) is a rule in India. However, close marriages should always be avoided because this system leads to homozygosity and too much homozygosity is most dangerous for the survival of a race. Close marriages also lead perpetuation of certain hereditary diseases from generation to generation. Due to lack of genetic flexibility, during the outbreak of epidemic forms of major diseases and their new races, the entire homozygous human race can get eliminated. Kesvan et al., (1978) studied 23,000 children in Madras and found a definite increase in the rate of congenital malformations in the offsprings of consanguineous marriages: 1.54 percent among non-consanguineous and 2.36 percent among cousin marriages: 2.18 percent among niece-maternal uncle marriages. This was corroborated by Sangunabai et al,. (1982) who found 3.59 percent malformation in the offsprings of consanguineous marriages.

Noble genes have no caste and religious boundary. Noble genes could be found in any caste, race, religious groups, on this basis author is of the view that intercaste, interracial, interreligious marriages should be encouraged, because *prolong isolation of caste/races have accumulated some useful genes in different castes/races* and by intercaste marriage, some good and desirable genes could be comb out and a new recombinant generation having desired intelligence, health and conduct could be produced within 2–3 generations by selective marriages. Distant marriages lead to heterozygosity and heterozygosity enhances robust brain and body stature. This is also known as "Positive eugenics".

Vidyarthi in 1983 reported increasing stature of the people in some of the European countries; in some cases the increase has been 9 c.m. among Sweedes, may have been due to higher standard of living and similar factors, but it may be due to genetic reason instead of breaking the social isolation on account of the open marriages. Stature is a polygenic controlled character and in the heterogenous people, the chances of genes coming together with cumulative or complementary genes effect are greater than homozygous people born out of close marriages.

Though, there is wide variation for height cranium capacity, etc. within each endogamous caste, but as we proceed from Punjab to West Bengal in east and Punjab to Peninsular India in south, there is a gradual decline in the stature and physical build up. For example, average height in Punjab is 168.4 cm, it gradually lower down in UP (165.5 cm); in Bihar (164.0 cm); in West Bengal (163.0 cm); and in Darjeeling (161.05 cm). This lowering down of sature is partly due to nutritional factors and partly due to genetic factors. Inter-caste-marriages also break social and economic isolation and lead mutual respect which is need of the hour in India, where communal hatred is corroding the social harmony.

However, intercaste, interracial marriages must be selective and aimed to:

(a) Maintain the gifted character of the family such as 'intelligence', health, body complexion, height and beauty, etc.

(b) Gradual improvement in the intelligence, health, height, body built up and beauty etc., in which family is lacking can be achived through 'gene-pyramiding' i.e., one-after other through selective marriage for several generations.

(c) Care should be taken that in no case, deleterious genes should enter in the family. Drinking, prostitution, crime, corruption etc. are the acquired characters but marriages in such families should be avoided.

(d) Playing, acting and dancing etc., are also acquired characteristics, (non-genetic) hence cannot be inherited by their progeny. Players and actors in general possess blunt brain. In no case "Intelligence" should be compromised by marrying a player or film actor until he is also an intelligent man. If a married lady cohabitate with an 'intelligent' man with aim to have an intelligent son/daughter, should not be taken as a sin or illegal. It is her fundamental right to produce an intelligent progeny being a progenitor if her husband has some anomalies such as 'retarded' brain.

WHY INTELLIGENCE (IQ) IS PRIORITY?

Though human intelligence has created more problems for survival of this race itself as well as maintaining the beauty of earth, yet, it is most sought for by man. The reason, today the world has turned a most materialistic and competitive in all fields of activities concerning to life and livelihood. Intelligence is infinite resource with man which makes the man competitive by generating aptitude, skill and personality to translate his ability into performance better at workplace. The measurement of intelligence is Intelligence Quotient (IQ), which was coined by Stanford Binet more than 70 years ago. Since then this term has been used to separate the intelligent from the not so intelligent; successful from unsuccessful and has been a stigma to have a low IQ. Most people feel that low IQ is the indicator of poverty or relatively less successful career. But IQ is really a good indicator of competence which does indicate that a person with high IQ can solve a mathematical problem or solve a puzzle quickly than a person with relatively lower IQ. But according to the renowned psychologists, IQ contributes only about 20 percent to the factors that determines the life's successes; rest 80 percent are determined by other factors. Gardener in one of his articles observes "*One's ultimate niche in society is determined largely by non-IQ factors, ranging from social class to luck.*"

In the mid-1990s Daniel Goleman gave the concept of Emotional Quotient (EQ)—a degree of emotional intelligence, awareness of one's own and other people's feelings, such as empathy, compassion, motivation and ability to respond to pain or pleasure appropriately. His basic hypothesis was that for an effective use of IQ, EQ is a necessity. He writes, "*in a sense, we have two brains, two minds and two different kinds of intelligence, rational and emotional.*" How do we do in life, is determined by both—it is not just IQ but EQ that matters?

Emotional Quotient (EQ)

It is ability to get along with the people and situations; a positive and pro-active attitude towards all aspect of life; the ability to command respect by building relationship.

EQ consist of motivating ourself, motivating others and empathizing with others which indicates the functional areas which determine the EQ. Henry Ford puts it "*If there is any secret to success, it lies in the ability to get other person's point of view and see things from his angle*". While IQ emphasizes the ability of mind. EQ focuses on the softer skills of building and maintaining human relationship. This aspect of life assumes a lot of importance since nowhere a person is detached from the human element; be it workplace, the home front, or the social circle, human interactions are inevitable and our success depends to a large extent on what we make of these interactions and relationships. Business intelligence has so far been dominated by IQ, treating everything and everybody in an inanimate, sequential style, almost mathematic in its precision. However, emerging trend focuses attention treating people, markets, ideas and organizations as interactive, alive and capable of learning creativity and growth. The relationship management can get you to be anywhere any times but success with EQ lies in handling the human environment around you and making a place for yourself. Thus, power of entering into others personality and emaginatively experiencing his experience is the key point of EQ.

At workplace, top people often have IQs which are not as high as the IQs of those working under them, but their emotional intelligence is higher and better developed. Market forces today are putting an unprecedented premium on emotional intelligence (EQ) for on the job success. A Swedish study published in 1993 indicated that if the people with high blood pressure, heart ailment, and high serum cholesterol had a dependable web of intimacy to support them (family friends, wife or husband), they could live with their afflictions confortably and for an indefinite number of years.

Spiritual Quotient (SQ)

Danah Zohar and Ian Marshall have coined the term SQ (Spiritual Quotient) for spiritual intelligence. The Indian philosophical treatize—the Upanishads—attributes perception to the mind, conception to the mind, conception to the intellect and "illumination" to the self. They speak of the ascent from intellect (IQ) through meditation.

However, in author's opinion, there are people who believe more and more in God and Goddess for success in their career; business and trade, so they believe in SQ. They believe, worship and meditation are the routes to the highest knowledge (illumination) received from God/Godess.

Genetic Control of IQ and EQ

Emotional Intelligence (EQ) is not controlled by genes. It is purely a socially and culturally determined (acquired) character. However, intelligence is a genetically-controlled character, because a correlation between IQ scores of the same person tested as a child and again as an adult is approximately 0.8 i.e., the score is stable although score on an IQ test is a statistical abstraction rather than a definite measure of intelligence. However, IQ scores of children are shaped in mother's womb and the environment of mother's womb plays important role in determining child's intelligence (Hartl, 1991). The pregnant women therefore can enhance the child's potential intelligence by avoiding toxin (smoking and drinking) and living healthy by taking nutritively rich diets during pregnancy. Prof. Bernard Devlin of University of Pittsburgh Pennsylvania has published result of his studies on IQ on 50470 twins in *Nature*. His studies indicates that IQ is determined both by genes as well as by environment. On the other hand, Charles Murray and Richard Herrinstein in their book "Bell Curve" stated that IQ was so strongly determined by the genes that increase of low IQ means people would lead to a growing "Under-Class". It means by third and fourth generation, descendents of gifted individuals may be reduced to the descendents of ordinary people. Yes, he is very right and author support his views.

Modern approaches on genetic and cultural heritability of IQ in American white estimate are:

$$h^2 = 0.297 \pm 0.023 \text{ (Genetic)}$$
$$b^2 = 0.289 \pm 0.016 \text{ (Cultural)}$$

where h^2 indicates the narrow sense heritability (additive genes) and b^2 is the corresponding used for familial transmission of strictly cultural effects (cultural inheritance). Genetic and cultural effect are the same, i.e., 30 percent accounting for variation in IQ among whites. However, largest single contributor to variation in IQ among whites is that environmental influences.

This speaks that IQ (Intelligence Quotient) is governed by quantitative genes (polygenes). Secondly, genes govern the potential and environment provides the congenial atmosphere for the expression of the genetic potential of an organism or a person. For example, high-yielding varieties of crops and cross bred animals do possess high-yielding genetic potential but for harvest of potential, proper nutrition and good management conditions are must. Therefore, for expression of genes governing intelligence proper nutrition and good living conditions for the pregnant woman (while the child is in the womb) are must even though the husband and wife both are intelligent.

Race and IQ

The results of IQ studies are frequently misinterpreted. The most common mistake is to apply the genetic and cultural heritability differences between the population. For essentially, the some reasons, the two identical twins reared in the different environment may differ in their phenotypic expression. In addition, two genetically identical populations say siblings may differ phenotypically if rearing environments are different.

Heritability parameters have been mistakenly applied to racial differences in averaging IQ. These tests have been standardized so that average score of American white is 100; American blacks average about 85 on the same tests, and in some Japanese studies. the average is about 110.

Now one important question arise whether these averages reveal real genetic differences between races? Of course, not because of following reasons:

1. *Differences in socio-economic status*. IQ scores are correlated with socio-economic status and so comparisons of population differing in socio-economic status will inevitably lead to differences in IQ. Marx said, "*social condition of a man decides his consciousness*."
2. *Differences in language skills*. IQ tests are largely verbal, so differences in language skills certainly reflect in test performance.
3. Test performance is affected by the attitudes and skill of the examiner, mood and motivation of the subject etc.
4. *Methodology adopted for such studies*. So far no proper methods have been developed for conducting such studies. Though, because of social and racial factors, it is very difficult to design methods of identifying racial differences for IQ, however, it is not impossible provided society is prepared to agree to take up such studies.

IQ Score is not Correlated with Glittering Career

However, it has been observed that gifted children with high IQ are not destined for the glittering career. In a research conducted by the office of standard education indicated that those who scored highest marks in IQ tests and in the examinations often failed to fulfil their potential in their career. A study conducted by Howard Gardener of 11 persons, who changed the world, at the Haward Graduate School of Education in the USA found that of those studied, even by the age of 20, only Pablo Picasso's world status was apparent. Others in the review included composer Igor Stravinsky, Martha Graham, the American dancer and choreographer Sigmund Freud and Mahatma Gandhi.

Famous thinkers such as Thomas Edison, Winston Churchill, naturalist Darwin and physicist Albert Einstein did badly at school. World's richest person—Bill Gates is a high school dropout. However, the key to their success has been delligence, dedication, motivation (EQ) and support of their family and friends than in IQ scores or school achievements. Professor Joan Freeman, a visiting Professor at Middle Sex University and author of the report quotes a study in San Francisco which had traced 856 boys and 672 girls "Geniuses" from 1925 to the present day. 2000AD. All had IQ scores of at least 135 putting them top 2 percent of the population, but their latest review shows that they have not been noticeably more successful as adults than if they had been randomly selected from others of the same social background. A study of more than 200 former pupils of the Hunter School for the gifted in New York (Mean IQ 157) found that by middle age, all had failed to reach the top of their chosen professions. This is because a person with a very high IQ but with low emotional intelligence can be very badly adjusted taking failure personally and viewing set backs as life threatening. A constant struggle and battle against these feelings can cause depression and paranoia. Many criminals do possess high IQ, but little or no emotional intelligence. IQ is not an immutable factor.

Hence personality, motivation and opportunity to use the talents are very important, provided proper working environments are available.

In India, situation is quite different. It is commonly said that *some are born great*; *some are made great* and *some are thrust upon the greatness*. This phrase is very well applicable to the Indian science and politics both. Those who are deserving, are not given congenial working atmosphere and those who are corrupt, criminal and sychophants are occupying top positions both in science and politics.

Health

There are two well known phrases about health: "Health is wealth" and "A sound mind in a sound body." Both phrases are very important. A labour with good health can earn his livelihood and can bring up his family, but a person with poor health may die of food and medicine. Health also include "workplace" too.

However, here health means free of genetic disorders—In the pre-pages, details of genetically controlled human diseases have been given. Some of them are controlled by major genes and some of them are controlled by minor or recessive genes given in Table 15.6.

TABLE 15.6 Dominant and recessive genes

Dominant gene	*Recessive gene*
Diabetes	Haemophilia (sex-linked)
Congenital cataract	Albinism
Supernumerary teeth	Blue or gray eye
Wooly hair	Deaf mutism
Pie balding	Red-green, colour blindness
Dwarfism	Hare-lip

A genetic counselor has to examine and tell the pair intending to marry, the consequences of these genetic diseases and chromosomal abnormalities, in their future generations. Example, let me cite here impact of 'Rh' factor on the child bearing by a Rh^+ mother.

Behaviour

Behaviour of an individual boy or girl is important wealth for a successful life not only in married life, but also at workplace both in services, trade and business. The most cool, cardial temperament and pleasing behaviour reflect a magical on the other persons in the family, surrounding and also in office. Behaviour count a lot for the success of trade and business. However, so far no concrete opinion has been formed that socially unacceptable behaviour—feeble mindedness, habitual drunkard, criminality, prostitution and so forth are heritable characters or environmentally (culturally) governed. Familial association of certain characters such as English grammar and pronunciation are considered genetically rather than socially transmitted. Children tend to speak with the same quirks and vocalisms as their parents and other close relatives such as maternal uncle, etc. Therefore, some people believe that English speaking is genetically transmitted, but some environmentalists have opposite view whose thinking is typified as follow:

> *So let us to admit—yes, there are heritable differences in form, in structure. These differences are in germplasm and are handed down from parent to children. But do not let these undoubted facts of inheritance lead as stray as they have some of the*

biologists. The mere presence of these structures tell us not one thing about function...... Our heredity structure lies ready to be shaped in a thousand different ways—the same structure depending on the way in which child is brought up..... We have no real traits (J.B. Watson, 1925).

Prostitution, smoking and drinking, etc. are environmentally (culturally)-influenced characters. But as regards habitual criminals, it was first reported from Britain in 1961 that persons with XYY karyotypes are presumed to possess criminal tendencies. This finding prompted a group of geneticists to wonder whether an extra Y chromosome predisposes its career to unusual agreesive behaviour. To explore this hypothesis they studied chromosomes of mentally abnormal patients who had records of violent and criminal propensity. However, no definite correlation have been obtained. This is still an area which deserve attention.

HARASSMENT OF WOMEN FOR NOT BEARING SON IS UNETHICAL

In general, women are mentally tortured and harassed by the family members particularly mother-in-laws and husband also if they deliver only daughters, not sons. Also sometimes, if the pregnancy is delayed after marriage, women are subjected to taunts and tortured mentally because mother-in-law wants a child should roam in her house within a year of marriage of her son. Let me justify scientifically that in none of these two circumstances, the woman is on fault, if she is physically and biologically fit to bear the child. How?

First, as I have stated earlier that life of a woman is like an earth, whatever a man sows, he reaps, How? Woman is sexually homogametic (XX), she produces only one kind of gametes (egg) all containing X chromosome, which determines femaleness. On the other hand, man is heterogametic, i.e. man's semen contains two types of sperms—sperm containing X chromosome which determines femaleness and sperm containing Y chromosome determines maleness. Therefore, if sperm containing X chromosome fertilizes the egg (X) of female, the resultant child would be female (XX) and if sperm containing Y chromosome fertilizes the egg (X) of female, the child would be male (XY). Thus, it depends upon the mobility and vitality of two kinds of sperm the male produces and which one of the two per chance fertilizes the egg of the female (wife) first which results into male (son) or female (girl) child. In this reproductive process, nowhere a woman is at all on fault. Therefore, her torture on this ground is unethical and illegal.

Second, for delayed pregnancy, there may be several reasons such as physical and physiological reasons within husband or wife or both which need medical checkup of both. However, if lady is physically and biologically fit to bear a child but still being tortured by her in-laws or husband for non-bearing of the child, she is not at all on fault. She can get child by the process given in the guideline such as "in vitro fertilization" with her husband's semen or from other sources. But she can also get a child if she takes the risk the way Pandvas of Mahabharata were born.

In this respect, I was told a story by an old man. A young married lady could not get child for about 10 years after her marriage. She was biologically fit but defect was in her husband. However, she was subjected to taunts and mental tortures regularly from her mother-in-law that she is sterile. One day intensionally or just to lower down her mental tension, she told her story to a young man. The young man said, it is too easy to get a child if you are mentally prepared.

The lady asked how? The man replied that through natural course of cohabitation. The lady agreed and they met once or twice. The lady became pregnant and got a son. Again after 5–6 years, her mother-in-law desired to get one more child, son or daughter. She again talk to the same man. They again met and she became pregnant and got second time too a son. Mother-in-law and other family members changed their outlook against her and she got respect from all corners of her family and relations. She gave thanks to the man who had sired her. This is too risky for a married woman. In such cases, the woman should not convert into a extramarital relation. She should consider that she has begged semen of a person as an alm. But while begging this alm, she should choose intelligent and healthy person; no drunkard, no smoker, no criminal or a diseased person, nor a politician. However, this is socially not acceptable and hence it may lead problem to the lady if her husband comes to know that a son or a daughter born by his wife does not carry his blood (genes).

GUIDELINES FOR THOSE HAVING REPRODUCTIVE DIFFICULTIES TO BEAR CHILDREN

Those couples who desire to have children but who realize that they are on high risk for transmitting a serious disease or genetic disorder to their progeny it is not advisable to them to bear a child. For such couples, it is advised to take following guidelines to have child:

1. For most couple, adoption remain a most preferable solution. Adopting a baby often fulfills a deep personal need of the couple and provide satisfaction too, but the adopted child must be safe from HIV, Rh^+ factor and genetic-controlled diseases, otherwise happiness of adoption of an infant may create more problems.
2. Artificial insemination by husband's semen (AIH) or Artificial insemination by donor's (AID) sperm. Artificial insemination with the semen of the preferable donor might be used in a program of the positive eugenics for those couple who desire experience pregnancy, a mean to circumvent the risk of bearing a child with either an autosomal recessive gene or dominant gene disorder when husband is affected or at known risk. One drawback to the use of AID is the psychological stresses, it places to the husband.
3. Prenatal detection of genetic disorder offer a couple third alternative. Selective abortion of affected foetus is now possible for all chromosomal abnormalities as well as for more than 60% single-gene disorders. It can be used to abort all males if the mother is a known career of an X-linked recessive gene disorder. Prenatal screening is already used frequently. More severe strain is placed on a couple who abort a child preferably male because of genetic disorder than upon those who terminate pregnancy from other reasons.

   However, scientists at Imperial College London have made an attempt to cure genetic disorder in the foetuses as early as 10 weeks old by genetic therapy after isolating and treating stem cells taken from the foetus blood. The abnormalities in the nerves, bones, cartilages, and muscles which at present are irreversible might be prevented long before birth by gene therapy in the womb.

Screeing methods for the Down Syndrome (2n + 1) and other chromosomal disorders can be spoted in the foetuses within three months of conception. But big improvement in Ultrasound equipment and needles to collect blood from pregnant women would be needed as well as further advances in the knowledge about sort of genetic transfers to control different diseases. The report published in the British Journal of Obstetrics and Gynaecology suggests that cells could eventually be removed from foetus at 8 weeks; worked on in the laboratory and retransfused back into the foetus within two weeks. The cells start circulating in the blood of foetuses 8–14 weeks after conception. But application of this technique in human beings will still take long time.

4. Women over the age of 35 years known career of X-linked gene disorders would be at risk for or affected with certain dominant gene disorder and even couple at risk for having children with autosomal recessive gene disorder may use in vitro fertilization (IVF). This is the female counter part of artificial insemination by donor. For example, a woman known to be carrier of haemophilia could ask the physician to secure an egg donation from a healthy non-career of any genetic disorder, the egg could be fertilized in a test tube with the sperm of her husband. After a few days, the embryo could be implanted in the uterus of the career female. She would experience a normal pregnancy and give birth to a child with only a very remote risk of haemophilia.
5. Cloning. After successful cloning of a Scottish sheep—Dolly, researches have been started in many part of the world, for the cloning of the human being from the single somatic cell. This method could be applied where due to some reason husband is not able to sire a baby or wife is not able to conceive due to problem in her uterus (Figure 15.5). The steps are as under:

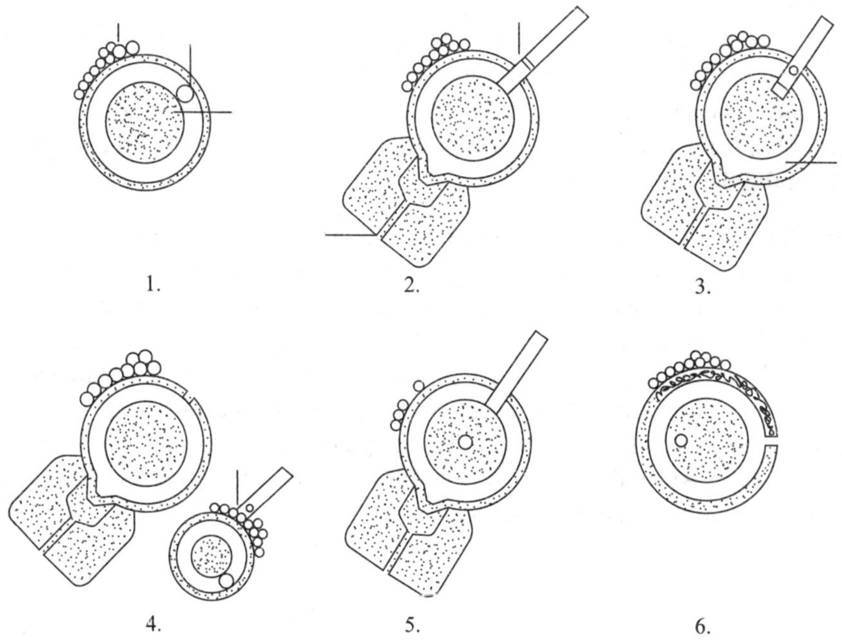

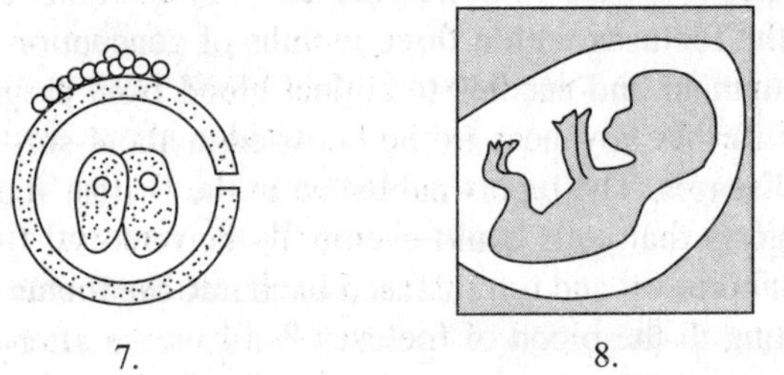

7. 8.

Figure 15.5 Shows how to clone a human female baby in eight easy steps.

Step 1: Eggs are coaxed to mature in a culture dish. Each has a remnant egg cell called the polar body and cumulus cells from the ovary clinging to it.

Step 2: While an egg is help still with a pipette, a needle is used to drill through the zona pellucida, removing a plug.

Step 3: After ejecting the zona plug, the needle is inserted back in the egg through the hole to withdraw and discard the polar body and the egg's genetic material.

Step 4: A cumulus cell from another egg is taken up into the needle. Cells called fibroblasts (or their nuclei) can also be used in this step.

Step 5: The cumulus cell is injected deep into the egg that has been stripped of its genetic material.

Step 6: The injected egg is exposed to a mixture of chemicals and growth factors designed to activate it to divide.

Step 7: After roughly 24 hours, the activated egg begins dividing. The cells contain genetic-material only from the injected cumulus cell.

Step 8: The single-genetic material-infused egg, after fertilization and the usual delivery time of about 9 months is supposed to bring forth the first baby with an exact genetic blueprint of another human: a clone.

In a Fertility and Mother Care Hospital, Jaipur (India), a child was delivered by her maternal grandmother at the age of 48 years who served as surrogate mother for her daughter's child. In this case, the biological mother (real mother) of the child had lost her uterus at the time of first delivery. After a cesarean operation in 1992, she developed severe postpartum haemorrhage therefore, to save her life, a hysterectomy was done. But next tragedy to her life came when she lost her child soon after birth. With no uterus, she was unable to conceive and bear child. However, her eggs were insiminated with her husband's sperm as "*in vitro*" fertilization and embryo was transferred in the uterus of her own mother who served as surrogate mother. Surrogate mother delivered the child in October, 2000.

SURROGATE MOTHER

'In vitro fertilization' (IVF) promises a new kind of motherhood. Human ova fertilized in the test tube can be transferred to any human womb that has been properly treated by the hormone. The recipient need not be same woman who donated the egg.

For a number of working women, pregnancy may pose unacceptable physical burden. If such women are eager to have a child, she could find out a healthy woman to accept the transfer of an embryo into her womb and to carry the foetus to term for the genetic parent. But surrogate motherhood raises some issues about the legal relationship of the involved party to the resulting child. It also raises question pertaining to the control of such activities such as what would be a reasonable price for undergoing a pregnancy for someone else. The estimate should be attractive to women of limited means. Fresh ova are crucial for a successful test-tube fertilization and later on the embryo will be only implanted in the uterus that has been primed by. Since surrogate motherhood raises immediate issue about the legal relationship of the involved parties to the resulting child, therefore mostly relations are being consented to serve as surrogate mother such as sister-in-law, etc. Sometime surrogate mother refuses to give the child.

NEW REPRODUCTIVE TECHNOLOGY

New reproductive technology have been developed which are useful in child bearing provided semen of genuine donor-free from genetic diseases are obtained and stored properly.

Storage of Semen

Storage of sperm for a long period of time can be done in liquid nitrogen at low temperature at –196°C in the sperm bank. Sperm remain viable over longer period of time when stored at low temperature. Deep-freeze storage of human sperm offers major benefit.

Availability of Ova

A successful pregnancy accomplished with extra uterine fertilization (IVF) requires egg from the ovary of a woman (which necessitates surgical intervention), cultivation of ova in a medium that mimics the natural environment, fertilization by sperm, transfer of a normally cleaving human embryo into the uterus implantation and a normal gestation.

The exact timing of ovulation is critical to the collection of eggs that are fit for IVF (*in vitro fertilization*). Pre-mature harvest of ova jeopardizes normal embryonic development; delayed harvest allows egg to be lost in the oviduct. Hormones are used to cause superovulation in the female. Fortunately, laparoscopy (a technique utilizing a thin tube with a light source and thinner tubes for ova separation) can be used to obtain the egg, thus avoiding a major abdominal surgery.

LEGAL PROBLEMS IN ARTIFICIAL INSEMINATION

Some legal problems do arise with Artificial Insemination Donor (AID). In 1954, Illinois Court dealt a sever blow to the use of AID. The Court stated that AI (Artificial Insemination) by donor with or without permission of husband is contrary to the public policy and good morals and constitute adultery on the part of mother. A child so conceived is not child born in the wed-lock and is therefore, illegitimate. In 1963, a New York Court also concluded that a child conceived by A.I. was illegitimate, nevertheless, it decided that woman's husband owed the child, a duty of support of his/her bringing up. The California law was a direct response to 1968 Court decision. It simply declared that if a woman gives birth to a child conceived by A.I. (Artificial Insemination) that child is legitimate, if it is born during or within 300 days of dissolution old marriage. In 1974, a Law was passed that AID children are legitimate if the husband has consented. New York State enacted a similar Law in 1975.

However, ABO blood type and Rh type of proposed donors must be recorded. The most important provision should be that persons with tuberculosis, brucellosis or veneral diseases may not make sperm donations. The AI doctors are also required to record the name's and addresses of the donors and recipients as well as the date of insemination and the results of donor screening. A semen donor should not transfer any infectious diseases to the foetus.

A written consent signed and witnessed should be the binding upon the husband. Selection of donor requires much more consideration and care. The doctor should also conduct thorough family history of donor search for genetic disorders. If donor is married, it would be wise to take consent of his wife also. The anonymous donor of the sperm should not be considered the natural father as he is no more responsible for the use made of his sperm.

NEW GENETIC RESEARCHES

Mutant Gene for Super Body Growth

In the new England Journal of Medicine, a news has been reported that in Berlin (Germany) a baby "superman" is born with bulging arm and leg muscle, who is not yet 5 years old, can hold 3–15 kg weight with arm extended. He has muscles twice the size of other children of his age and half their body fat. The DNA testing indicated that the boy has genetic mutation that boosts muscle growth. The boy's mutant DNA segment was found to block production of a protein called "myostatin" that limits the muscle growth. The scientists think that this finding could help to formulate drugs for treating humans suffering with muscular dystrophy and other muscle-destroying disorders in addition, this type of drug can help in producing athletes to bulk upto their muscles.

Just decreasing the "myostatis" protein by 20, 30 and 50%, the muscle growth could be limited as predicted by Dr. Lou Kunkel, Director of Genomics Programme at Boston Children's Hospital and Professor of Pediatrics and Genetics, Harward Medical School. Looking the huge potential market for such drugs, researches at Neurology Department Charite/University Medical Centre, Berlin and many pharmaceutical companies are already trying to find a way to limit the amount and activity of "myostatin" in the body (Treton June 24, 2004, HT).

The Genes for Infidelity in Men

There are many married men who want to keep sex relations with many ladies besides their own wife. Such kind of infidelity caused in men is due to genetic disorders. It has been reported that "vasopresin" hormone determines the cohabitation tendency in men. In the brain there are "receptors" which control the production of "vasopresin" hormone. This hormone is produced after cohabitation which gives satisfaction in the men for sexual thrust. In those men who have less number of receptor in the brain for "vasopresin" hormone, feel less satisfied and hence they try keep sexual relations with many ladies.

Those men who have many receptors in their brain for "vasopresin" can feel satisfied with one lady (wife). However, if "vasopresin receptor" is injected from outside, then may be change in the men to be fidel to their spouse. The scientists feel that by treating with "vasopresin receptors" the infidelity in men to their spouse could be reduced (London, Hindustan June 19, 2004).

Sexual Desire in Women

A research team in Australia has reported that diminished libidos in women can be due to low-level of a chemical called "DHEA." Earlier it was believed that "testosterone" levels are related to women's libidos. Now it has been reported that low-level of "DHEA" a chemical that can be turned into "testosterone" are associated with decreasing sexual desire in women. Dr. Susan and R. Davis who led the research for 15 months, involved 1432 women between 18 and 75 years; who gave blood sample for hormone testing and filled a sexual function questionnaire on the same day. The analysis focused on the women below 45 years of age who had greatest range in the sex hormone levels. The sexual function which included desire, arousal, responsiveness, orgasm and pleasure declined from mid 20's onward. This study found a low "DHEA" level was associated into low-sexual desire and arousal, in contrast 'testosterone' levels were not associated into these factors. The women who feel low libido but to have normal 'testosterone' levels need to get a treatment their DHEA level check up. DHEA is produced natural way in the body and must be under control of some "gene" (New York/Sydney June 18, 2004, HT).

FURTHER READING

Baxi, A.J. (1974): Glucose-6-phosphate dehydrogenase deficiency, *in: Human Population Genetics Vol. I*, p.60.

Baxi, A.J., V. Balakrishnan, and L.D. Sanghvi (1961): Deficiency of glucose-6-phosphate dehydrogenase—observations from a sample from Bombay, *Curr. Sci.*, **30,** 16.

Baxi, A.J., N.P. Parikh, and H.I. Jhala (1969): Incidence of glucose-6-phosphate dehydrogenase deficiency in three Gujarati populations, *Humangenetik,* **8,** 62.s.

Charlton, R.W. and T.H. Bothwell (1961): Primaquine sensitive red cells in races in South Africa, *Br. Med. J.*, **1,** 941.

Da Costa, H., I.S. Pattani, V. Dandekar, K. Kotnies, K. Mehendale, G.P. Sakrekar and S. Merchant, G-6-PD deficiency, *Indian J. Med. Sci.* **21,** 809.

Das, B.N., O.N. Bhakoo, and J.G. Jolly (1974): Neonatal hyperbilirubinaemia with G-6-PD deficiency, *Indian Pediatr.*, **11,** 645.

Dasgupta, J. and M. Balasubramanyam (1973): XXY syndrome XY/XO mosaiasm and acentric chromosome fragments in male schizophrenics, *Indian J. Med. Res.*, **61,** 62.

Datta, S., A. Banerjee, G. Talukder and A. Sharma (1982): Haptoglobin phenotypes in a group of mental retardates, *Indian J. Pediatr.*, **49,** 659.

Datta, S., A. Sharma and G. Talukder (1982): Sex-chromosomal abnormalities amongst mental retardates, *in: Perspectives in Cytology and Genetics*, B.K. Manna (Ed.), Vol. IV.

Deshmukh, V.V. and K.D. Sharma (1968): Deficiency of erythrocyte G-6-PD and sickle-cell trait: a survey at Aurangabad., *Indian J. Med. Sci.*, **22,** 302.

Dhadial, R.K. and I.C. Verma: (1978) *in: Medical Genetics in India*, I.C. Verma (Ed.), Auroma Enterprises, Pondicherry.

CHAPTER

16

Mutation and Mutagenesis

Mutation is defined as, *a sudden heritable change in the genetic material of an organism, as a result progeny may exhibit an altered form, size and composition.* Undoubtedly, stability in the genetic material (DNA) is one of the basic requirements, but sudden heritable changes in the genetic material are not only the possibility, but it is one of the properties on account of errors in the DNA replication which give rise to occasional aberrant types. In fact, mutation is one of the potent sources of creating new gene/allele not existing earlier in the natural population. Whereas, the variability created by 'recombination' is on account of rearrangement of the existing genes through the crossing over during meiosis. Mutation provides raw material for the evolution. Without mutation, all genes would have existed in only one form. Alleles would have not existed and under such a circumstance, genetic analysis would have not been possible. The reality is that mutation and natural selection have played unique role in the evolution of a large number of life forms. This DNA world is the result of mutation balanced by natural selection.

On the basis of occurrence and nature of genetic changes, mutations can be classified into following major types:

1. Spontaneous mutations (natural)
2. Induced mutations
3. Cytoplasmic mutations
4. Conditional mutations

SPONTANEOUS MUTATIONS (NATURAL)

Spontaneous mutations are constantly occurring in nature and also in the experimental materials randomly without any apparent care; of course, there are some sites which are more susceptible to the mutational changes than others in the same organism. These mutations may be due to errors in the replication of genetic materials (DNA) or due to cosmic radiation or due to chemical agents such as herbicides like '2,4-D'. However, rate of mutation is always very low as compared to induced mutations.

In addition, spontaneous mutations are usually caused by insertion of 'transposon'. These insertions, either disrupt the coding region (amino acid placement in the polypeptide chain) of the

DNA or the promoter. The ability of a transposon to 'jump' into a new position is dependent on the presence of an appropriate 'transposage'. The frequency of transposition depends upon the organism's genetic make up. The insertion of a transposon (segment of DNA) in a new location/site can inactivate the existing gene or can enhance the activity of a gene or put it under the control of new promoter gene the 'transposition'.

McClintock (1951) concluded that *the type of mutation occurring at a locus is a function of the locus or is transposed to it, and does not involve changes in the components of the genes themselves. Rather, it is the chromatin that functions to control how genic material may operate in the nuclear system.* This and other conclusions have been ingeniously derived from a detailed investigation of the nature of the 'Dissociation—Activator' (Ds—Ac) system. This system composed of two extragenic chromatin elements. 'Ds and Ac' called 'jumping genes' modify the genic expression of a number of different loci in maize.

Direction of Mutations

Mutations may be either recessive or dominant. The dominant mutations are generally deleterious and hence screened awav soon after they appear. In the haploid organisms like viruses and bacteria, both recessive and dominant mutations can be recognized on account of their visible effects on the phenotypic appearance of the organism, in which they originated. In the diploid organisms, the recessive mutation could be detected only in the homologous (*aa*) condition. If they are present in the heterozygous (*Aa*) condition, they could not be detectable with exception of sex-linked recessive mutations where they are present in the hemizygous state in the heterogametic sex; (males in humans as well as females in birds) screened always where they appear (Figure 16.1).

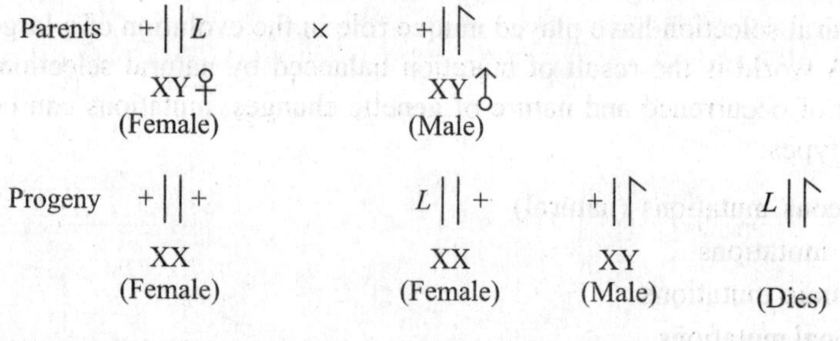

Figure 16.1 An altered sex ratio as a result of sex-linked recessive lethal gene (*L*) in the heterogametic sex (human, fruit fly). The sex-linked recessive lethal mutation will alter sex-ratio because hemizygous individual (male) carrying lethal recessive gene will not survive.

Most of the mutations have been found to be deleterious or recessive. From this, it could be understood that genes control some metabolic activities. The metabolic activities take place by the sequence of biochemical reactions, each step of which is catalyzed by a specific enzyme encoded by one or more genes. Mutation in these genes may produce block in the metabolic pathways (Gardener, 1968) (Figure 16.2).

Precursor → Intermediate → Intermediate → Final products

Enzyme *A*	Enzyme *B*	Enzyme *C*
↑	↑	↑
Gene *A*	Gene *B*	Gene *C*
↓	↓	↓
Mutation Mutant allele *a*	Mutation Mutant allele *b*	Mutation Mutant allele *c*
↓	↓	↓
Inactive enzyme *a*	Inactive enzyme *b*	Inactive enzyme *c*
↓	↓	↓
aa → Intermediate	*bb* → Intermediate	*cc* → Products

Figure 16.2 Blocks in the metabolic pathways due to recessive mutant alleles. One metabolic pathway may be only a few step long or many step long. The wild type form of each genes (allele) codes for an active enzyme that efficiently catalyzes the appropriate reaction.

Explanation

A wild type alleles (+ +) of a gene encoding wild type 'enzyme' or 'structural proteins' carry optional metabolic activity for many generations. Mutation resulting in the 'amino acid' changes in the polypeptide chain that increases the efficiency of 'enzyme' in carrying out a particular function is preserved by the natural selection as most fit. Mutation which alters the amino acid sequences in the polypeptide chain will either make the enzyme less-effective or non-active, will be deleted through the natural selection.

Over a long period of time, spontaneous mutants have provided basis for the improvement of all our agricultural crops. Consciously or unconsciously, plant breeders have been working with the spontaneous mutation

Spontaneous mutations can be further classified into:

1. Somatic mutations
2. Germinal mutation

Somatic Mutations

The somatic mutations occur in all the somatic cells other than those which form the gametes. A somatic mutation yields an individual that is genotypically unaltered but phenotypically carries a mixture of normal and mutant tissues called *chimera.* The chimeras are also of two types—*periclinal* (epidermal cells) and *mericlinal* (below epidermal tissues) (Figure 16.3(a)). Because reproductive cells are not affected, such a mutation (mutant alleles) will not be transmitted to the progeny and may not be identified for genetical studies. However, buds coming out of the such altered tissues can be propagated vegetatively. In the fruit industry, some of the most useful somatic mutations are colour sports of apple, early and late sports of peaches and thornless sports of black berries, which are being commercially utilized. Most famous 'delicious apple' and 'novel orange' are the results of somatic mutations, which have been extensively utilized for production of numerous progeny through the grafting and budding have perpetuated original mutation. 'Delicious apple' is most favoured and widely spread variety worldwide.

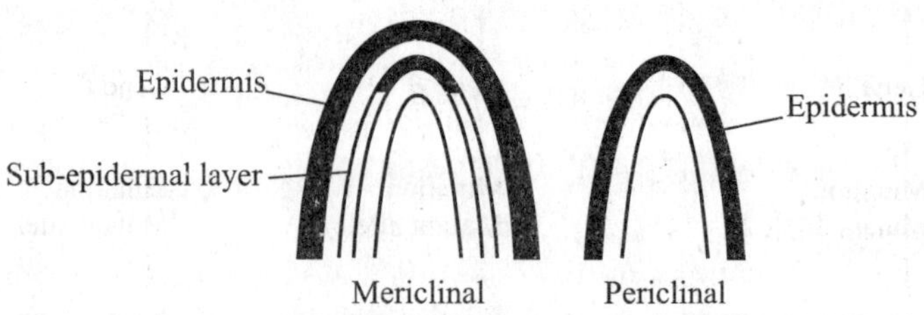

Figure 16.3(a) Apical meristem of plant showing tissue layers with mericlinal and periclinal chimeras. Such chimeric mutations give rise to bud mutation.

Somatic mutation in red gram (*Cajanus Cajan*) is shown in Figure 16.3(b).

Figure 16.3(b) Somatic mutation in red gram (*Cajanus Cajan*). On a same plant two branches coming out from same site showing early maturity of pods (Right) while entire plant has not yet come in flowering (left). Author himself got it.

In the vegetatively propagated plants, somatic mutations may originate as *chimera* in one or more of the ontogenic tissue layers of the shoots apices or from the adventitious buds, which develop from one or a small number of initial cells. The mutations only in epidermal layer (outermost layer) will be expressed phenotypically; mutations in the sub-epidermal cell layer may be expressed only when they pass through a sexual stage. *Chimeral* mutants affecting part of one or more tissue layers are common when multicellular meristmatic buds are irradiated.

Germinal Mutations

Germinal mutation occurs in the cells which form the gametes. It may be chromosomal or genic. If the mutation is dominant in nature, it is immediately expressed in the progeny; however, if the mutation is recessive, its effect may be masked in the diploid organisms. Germinal mutation may occurs at any stage in the reproductive cycle. If it occurs at gametic stage, then only individual gamete/progeny (after fertilization) may be carrying the mutant gene, however, if the mutation occurs in germinal cells, several gametes may be carrying the mutant gene. Thus, the nature of mutation (dominant or recessive) and stage at which mutation occurs in the reproductive cycle is important as regards the size of population it affects.

Germinal mutations can be further classified into:

1. Chromosomal mutations
2. Gene mutations

Chromosomal mutations

Chromosomal mutations are the result of chromosomal aberrations, which may be structural or numerical. For numerical, chapter dealing with polyploidy and aneuploidy can be consulted.

Structural chromosomal aberrations or structural rearrangements: Changes in the linear order of genes detectable cytologically or after special genetic procedures are commonly linked with chromosome breakage. They are of four types:

Deficiencies: A deficiency is the deletion of one or more genes (segments) in a chromosome. The loss of a chromosome segment can be intercalary or terminal. For the survival of deficiency, telomere is essential (Figure 16.4).

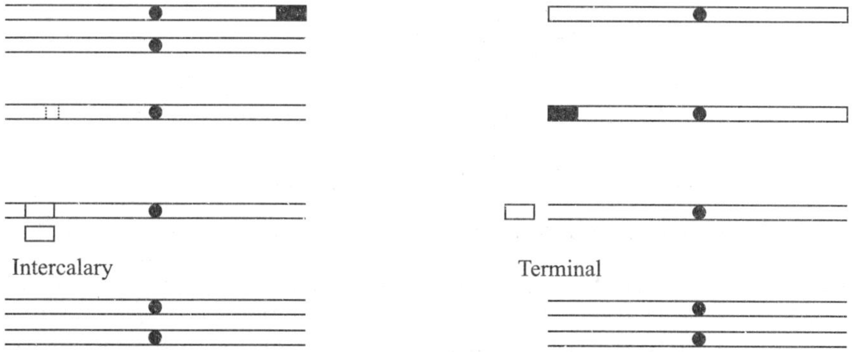

Figure 16.4 Different types of deficiency.

A deficiency can include different size of segments. An entire chromosome arm, if there is misdivision of the centromere; long segment of the chromosome arm (hexaploid wheat, Mackey, 1954(c)) down to a single chromomere of a chromosome (maize, McClintock, 1942) or it may be an individual gene or a nucleotide base with no detectable change or a cistron of a gene may be lost on account of unequal crossing over. Depending upon the gene involved as well as on the ploidy level of a species, deficiency may result in lethality, semi-lethality or loss in viability or no effect in the homozygous condition. Consequently, not all deficiencies are lethal or semi-lethal. Similarly, not all lethals/semi-lethals are due to deficiencies. All deficiencies may be effective to reduce viability in the homozygous or in haploid state but ineffective in the heterozygous state.

Duplication: Duplication is a means of increasing the gene numbers with a diploid organism. Duplication within 'locus' has been described by Rasmuson in 1962. Duplication arise in the different manner; by chromosome breaks followed by a transposition of a segment: by unequal crossing over at meosis; by recombination of chromosomes in the case of translocation or as a consequence of the bridge breakage fusion cycle in the inversion heterozygote or as a consequence of induced mutation. Some duplications decrease fertility and viability, but some increase both (Gustaffson and Von Wettstein, 1958). There are two types of duplication:

1. Chromosome segments are transferred between homologous chromosomes. The duplicated chromosome segment may be lying in the different parts of the chromosome or close to one another; In both cases, they may be having same linear gene orders (tandem repeat) or inverted (inverse repeat).
2. Chromosome segments are transferred between non-homologous chromosomes.

Besides *Drosophila*, duplication has been studied in crop plants such as in maize (gene locus *A* and *R*).

Figure 16.5 shows duplication.

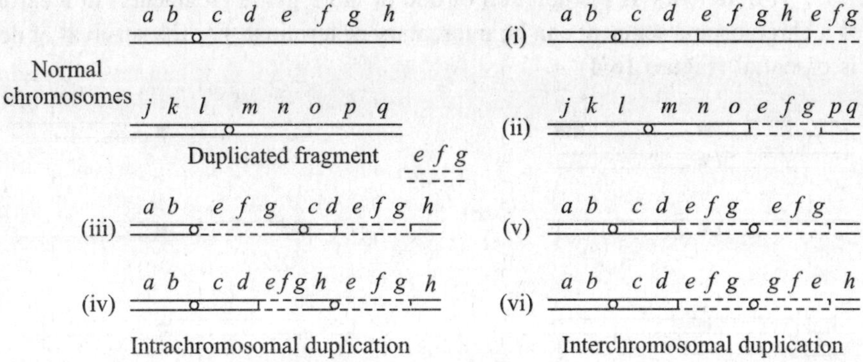

Figure 16.5 Duplication changes gene position which brings mutational effects.

Inversion: Inversion arises as a result of two simultaneous breaks in one chromosome, a 180° rotation of the segment in between and a subsequent reunion at the break point (Figure 16.6). Inversion may be *paracentric*, if centromere lies outside the segment or *pericentric* if it lies in the inverted segment. The *paracentric* heterozygotes usually form reverse loop at the time of pairing in the loop; dicentric bridges and acentric fragments might appear at anaphase I leading to the formation of non-viable gametes. Because of deletion of crossover chromatids and improper pairing of homologous chromosomes, the recombination frequency is reduced to a great extent. In the pericentric inversion heterozygote, crossing over within the inversion give rise to non-centric duplication—deficiency chromatids, which lead to gamete abortion as a result, recombination is reduced. Pericentric inversion causes death of pollen as well as seed also.

Inversion can occur spontaneously or can be induced also by application of chemical and physical agents.

Translocation: A translocation occurs when simultaneous breaks take place in the two chromosomes and again broken chromosomes join in a new manner (Figure 16.7). The chromosome breakage and exchange take place both by spontaneous (cosmic) reasons as well as by radiations.

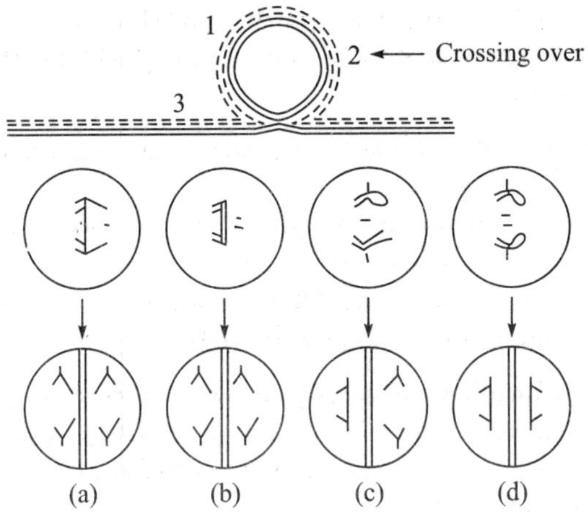

Figure 16.6 Consequences of crossing over in the inverted heterozygote.

(a) Anaphase I and II following single crossover at 1 and 2 within inverted segment.
(b) Anaphase I and II following a four-strand tetrad double crossover.
(c) Anaphase I and II following crossovers at 2 and 3 in top diagram.
(d) Anaphase I and II following crossovers at 1, 2, 3.

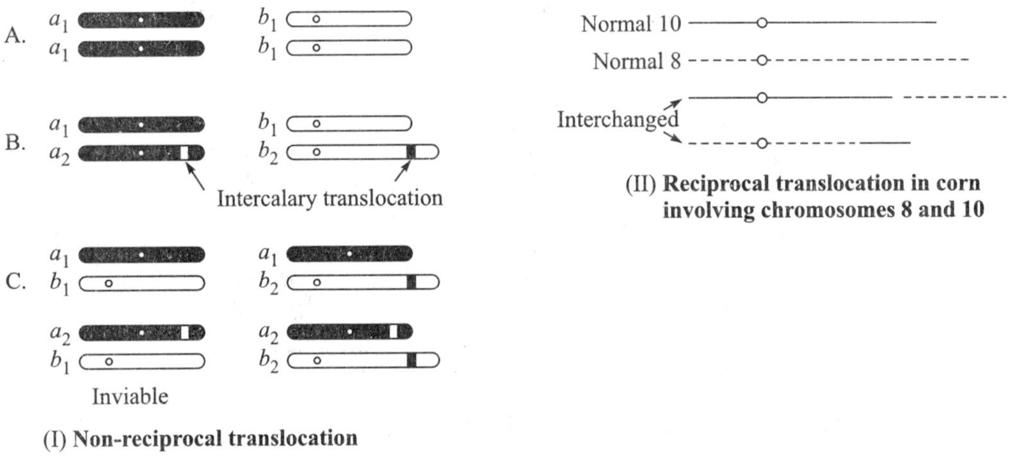

Figure 16.7 Translocations.

A. Normal chromosome pair.
B. Non-reciprocal translocation heterozygote a_1 a_2 b_1 b_2.
C. a_2 deletion causes gamete lethality.

Translocation can be classified into:

(i) Intrachromosomal
(ii) Interchromosomal

In intrachromosomal translocation, a transposition or shift of an *interstitial segment* within an arm or from one arm to opposite arm of chromosome takes place. The interchromosomal translocation are of three types:

(a) *Simple*, in which only one exchange is realized and one fragment of the chromosome is lost.
(b) *Reciprocal*, where a mutual exchange of segment between two non-homologous chromosomes takes place (Figure 16.7(I)).
(c) *Transposition* (*shifts*), in which an interstitial segment is removed and reinserted into a non-homologous chromosome (Figure 16.7(II)).

During meiosis, translocation heterozygote often forms multivalent; the reciprocal translocation forms quadrivalent. In case of several translocations, more than one multivalent are formed. Translocation heterozygotes often give rise to unbalanced gametes leading to male and female sterility due to formation of chromatids with duplication and deficiencies. Irregular pairing and separation may result into aneuploids plants (Figure 16.8).

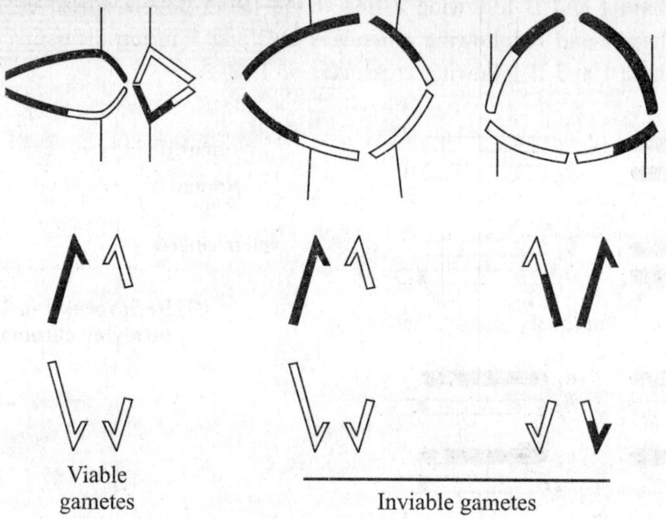

Figure 16.8 Metaphase ring forming reciprocal translocations.

Spontaneous translocations have been reported in several plant (*Datura*, *Oenothera*, *Pisum*, etc.) and animal species. Radiation induced translocations have been used to determine the gene linkage and gene order in the individual chromosomes.

Gene mutations

These are of three types:

1. Macromutations (major gene mutation)
2. Micromutations (minor gene or quantitative gene)
3. Molecular or point mutation.

Macromutations

Dominant gene mutations: Good example of dominant germinal mutation is short-legged male lamb cited in many books of Genetics which was detected by Wright in 1791 on his sheep farm to which he found beneficial because short-legged sheeps were not able to jump over the stone wall fences of his farm. He used short-legged male lamb to breed a new line of short-legged sheep (Figure 16.9).

(a) Dwarf mutant of ram (b) Norkal sheep

Figure 16.9 Dominant gene mutation in sheep (dwarf male).

Recessive gene mutations: Mendel crossed a short stem pea with a long stem pea and in F_1, he found long stem was dominant over short stem. However in F_2 generation, long and short stem pea were segregated in 3:1 ratio. On that basis, some papers were published in rice and wheat where tall vs. dwarf plants were reported monogenially (single gene) controlled character. But height is a quantitative character and quantitative characters are generally polygenic in nature. In rice and barley Ram (1972, 1985) found that plant height is controlled by 3–4 pairs of genes where dwarfness is governed by recessive genes (d_1d_1, d_2d_2, d_3d_3, d_4d_4). Wheat also possesses tall, semidwarf, dwarf, triple and quadruple dwarf plants. Here, dwarfness has been found controlled by the recessive genes (Rht_1, Rht_2). In fact, in wheat—Darum; in rice–Dee–Geo–Woo–Gen (DGWG) and in barley—Azam-4 dwarfing genes have been utilized to evolve dwarf and semidwarf high-yielding varieties which do not lodge even under high-level of nitrogen and irrigation management. Although dwarfing genes (recessive) provide lodging resistance to a considerable extent, however, lodging resistance is not correlated with dwarfness in plant height until and unless dwarf plant types have deep root system and thick stem to avoid root/stem lodging during high-wind velocity.

Norin 10 and Norin 20 carrying Darum, dwarfing genes were the commercially cultivated dwarf wheat varieties in Japan, which Dr. N.E. Borlaug got from A.O. Vogel wheat breeder, at North Pacific Wheat Breeding Station, California (USA) in 1952. Dr. Vogel had already utilized dwarf plant types of Norin 10 and Norin 20 in breeding dwarf wheat and had developed 'Gaines' winter wheat varieties in California, USA. Dr. N.E. Borlaug utilized Norin 10, Norin 20 and Gaines to evolve dwarf spring wheat varieties responsive to the nitrogen management at CIMMYT, Mexico in sixties, which brought wheat revolution in the many spring wheat-growing countries of the world. Dwarfing gene "Dee–Geo–Woo–Gen" (DGWG) in rice was found as natural dwarf mutant in Taiwan in a cultivated variety Dee–Geo–Woo–Gen by a farmer. In barley 'Azam 4' is also a natural triple dwarf mutant collected by author himself from a local barley variety in district Azamgarh, U.P. (India) which was utilized to evolve a number of dwarf high-yielding varieties of hull-less and hulled types. Thus, economic utility of dwarfing genes in restructuring the 'plant types' of wheat, rice and barley is a good example of economic utility of 'recessive' genes, which brought miraculous changes in the productivity of these crops (Figure 16.10).

Rice

(1) Tall (2) Semi-dwarf (3) Dwarf (4) Triple dwarf (5) Quadruple dwarf

(Dr. Mahabal Ram)

Wheat

(1) Tall (2) Semi-dwarf (3) Dwarf (4) Quadruple dwarf

(Dr. Mahabal Ram)

Barley

(1) Tall (2) Semi-dwarf (3) Dwarf (4) T.D. (5) Quadruple dwarf

(Dr. Mahabal Ram)

Figure 16.10 Dwarfness in rice, wheat and barley is due to recessive gene mutations. (Authors own work)

Micromutations: Micromutation has been wrongly classed as point mutation. Micromutation means mutation occurring in the quantitative characters which are polygenically controlled and highly influenced by environmental factors. Since individual effect of quantitative genes are very small and cumulative, therefore mutation taking place in one or two genes may not show visible effect unless entire gene complex is altered in one direction positive (+) or negative (–), or there is deletion, duplication and translocation of entire group (segment of DNA) of genes controlling a character. Micromutation are not generally deleterious. For example, if height is considered a polygenically controlled character, a gradual effect on plant height could be observed if treated with a mutagen.

Molecular basis of gene mutations: A gene is a specific sequence of nucleotide pairs coding for a particular polypeptide chain (protein). A molecular mutation (change) occurring in a triplet 'AUG' of a fragment of DNA (nucleotides chain) of a gene may produce a new mutant form of a new gene 'allele' of that gene. Molecular mutation is called *point mutation*. Because of degeneracy (presence of several synonymous codons) of the specific genetic code, some base pairs changes do not bring corresponding changes in the mRNA code and hence no change in the protein product is noticed coded by that gene. If the gene containing mutation with small effects, that can be detected by some specific technique called 'isoallele'; it may be not visible if it occurs in intron region (non-coding). Other mutations result in the total loss of gene product activity, if occurs in the 'exon' (coding region). If mutations occur in the essential genes required for the viability, they will die. Why it happen so?

Because gene synthesize 'enzymes' designed to catalyze a particular biochemical path. A random change in the amino acid sequence of an 'enzyme' caused by mutation in the DNA structure can make it more capable for carrying out its catalytic activity efficiently, if favourable. If deleterious, it may block the synthesis of that particular enzyme involved in a particular metabolic path.

MOLECULAR MUTATIONS

Molecular mutation means point mutation.

Mechanism of Molecular Mutations

Tautomerism

The stability of the activity of a 'gene' depends upon the fidelity of a specific *complementary base pairing rule* of DNA molecule as theorized by Watson and Crick (1953) through which it transmits the genetic information from generation to generation. However, the structure of base in the DNA is not always static. The hydrogen atom can move from one position in a *purine* or *pyrimidine* base to other position, i.e. from 'amino' group to the 'nitrogen'. Such chemical fluctuations are called *tautomeric-shift*. Although such tautomeric-shift is very rare, but in the DNA replication, it is of significant importance, because it alters the pairing direction of the bases. Normally, in the DNA strand, adenine pairs with thymine (A–T) and guanine with cytosine (Figure 16.11). This base pairing system determines the stability of the gene function. However, the stability of the bases in the specific base pairing rule gets affected when 'tautomerization' results the more stable 'keto' form of thymine and guanine to less stable 'enol' form and 'amino' form of 'adenine' and 'cytosine' into less-stable 'imino' form.

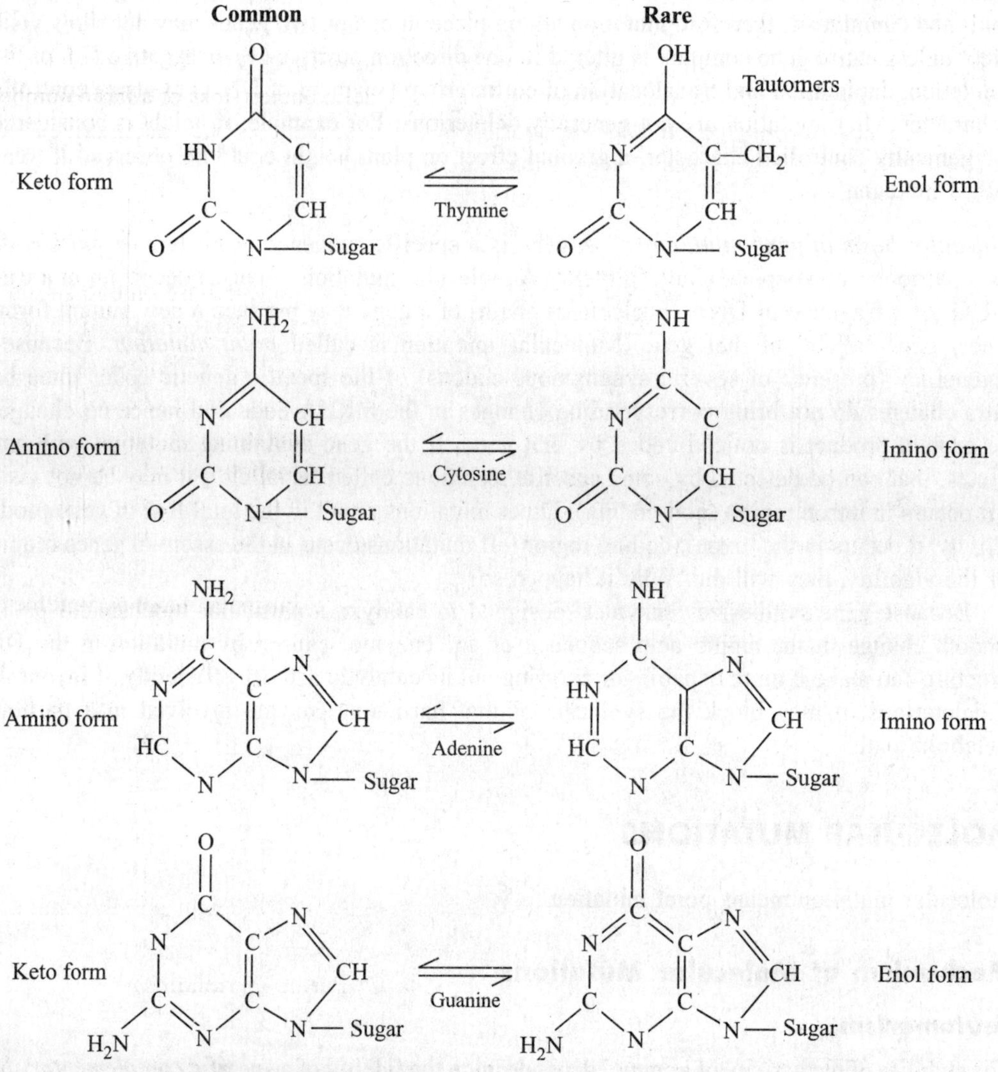

Figure 16.11 Tautomeric form of four DNA bases. The movement of hydrogen atoms between No. 3 and No. 4 position of pyrimidine and No. 1 and No. 6 of the purines, change the base pairing rules.

The bases in the tautomeric forms exist for a moment due to their instability. However, if the bases exist in the tautomeric form for some more time, it may be replicated or incorporated into a nascent DNA chain, which may result into mutation. Because bases present in their rare 'imino' and 'enol' forms can undergo wrong pairing where adenine can pair with cytocine and guanine can pair with thymine bases: C–A and T–G

The net result of such occurrence would be mismatched base pairing in the subsequent DNA replications: AT to GC or GC to AT

Base substitution

Replacement of one base pair by another pair in the DNA is known as base substitution. The base substitution includes a simple base addition or deletion or deficiencies (loss of a large number of pairs), insertion of transposon and duplication of base sequence. The tautomerized bases (unstable) may bring two types of base substitutions:

(i) Transition
(ii) Transversion

Transition: In this type of base substitution, purine replaces a purine base on one strand and a pyrimidine base replaces a pyrimidine base on the other. There will be four possible transition mutations:

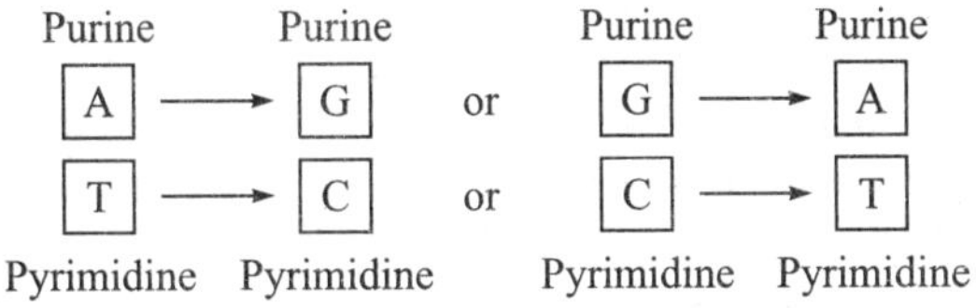

Transversion: In each DNA strand, the tautomerized (unstable) purine replaces pyrimidine base and pyrimidine replaces a purine base:

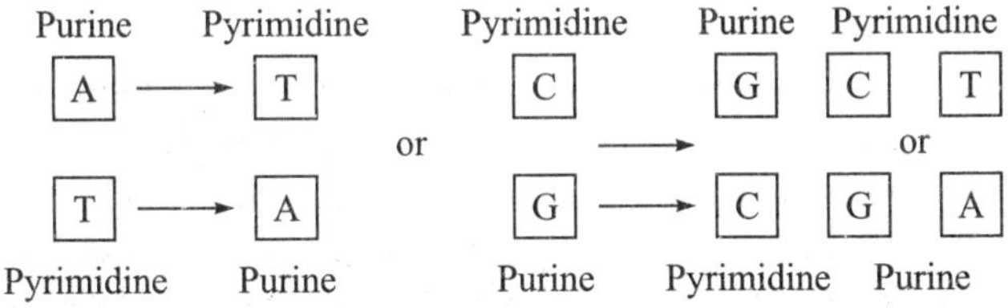

In the transversion type of rearrangement of bases, there may be eight possible mutations:

T→ A→ or T→ G or C→ A or C→ G (Pyrimidine–Purine)
A→ T or A→ C or G→ T or G→ C (Purine–Pyrimidine)

FRAME-SHIFT MUTATIONS

The addition or deletion of a single base in DNA polynucleotide chain brings a corresponding change in the mRNA codon (genetic code) and can introduce wrong and undesirable amino acid substitution in the polypeptide chain.

The correct reading of the long nucleotide sequence of a gene is assured by beginning the translation into amino acid sequence from a fixed strarting point and preceding triplet by triplet to the end. However, if the 'triplet' reading frame happens to be shifted by addition or deletion of one nucleotide base at middle site, the *Reading Frame* gets disturbed, that distort the genetic message by distorting mRNA codon. This is called *frameshift* mutation (Figure 16.12).

Leu Leu Leu Leu
CUG CUG CUG CUG
↓
CUG CAG GCU GCU G
Leu His Ala Ala

For example:

Correct genetic message DNA.	:	CAT	CAT	CAT	CAT	CAT	CAT	CAT
Base added G^+	:	CAT	CAT	G^+CA	TCA	TCA	TCA	TCA
Base A^-	:	CAT	CAT	CAT	CAT	A^-TC	ATC	ATC

Figure 16.12 Effect of insertion or deletion of a single nucleotide on the coding of genetic message. Hypothetical correct message is CAT, CAT.... Insertion of nucleotide G^+ distart the remaining message TCA, TCA TCA... Deletion of a nucleotide A^- again distort the genetic message ATC, ATC and ATC. The frame shift mutation may be effective or may not be, depends on four situations.

MISSENSE MUTATIONS

The accuracy of the incorporation of amino acids into the polypeptide chain to form a primary structure of a protein depends upon the accuracy of the genetic message *blue-print* written in the DNA, which are *transcribed* in form of mRNA codon. It is mRNA codons, which *translate* the genetic message correctly or incorrectly depending upon the direction by the gene (DNA). Therefore, a base substitution in the DNA polynucleotide chain can bring a corresponding change in the mRNA codon and may insert a wrong amino acid in the polypeptide chain. This kind of base substitution called 'missense' or 'missence codon' because it changes the 'sense' of mRNA codon which encode undesirable amino acid in the polypeptide chain.

A good example of 'missense' mutation could be cited from the base substitution in the haemoglobin gene. The amino acid sequence for B chain of haemoglobin A begins with the eight amino acids starting from the NH_2 end.

There are many other types of haemoglobins, each differs from Hb-A at least for one amino acid in its composition.

As shown in Table 16.1, haemoglobin Hb-B differs from Hb-A by substitution of 'valine' for glutamic acid in β-chain (transversion). The codons for glutamic acids are GAA and GAG whereas (transversion) as 'valine' is coded by GUA, GUG, GUC and GUU. If codon GAA (glutamic acid) undergoes a replacement to become GUA 'valine' will be coded. Such replacement in mRNA codon will occur, if a DNA triple CTT changes to CAT, i.e. adenine replaces thymine (A–T). This is called *transition*. Similarly, a *transversion* in DNA, due to substitution of an A–T base pair to T–A, bring a corresponding change in the mRNA codon from 5′ GAA 3′ to

TABLE 16.1 Comparison of first eight amino acid components of β-chain of four haemoglobins

Hb-A	*Hb-B*	*Hb-C*	*Hb-G*
Valine	Valine	Valine	Valine
Histidine	Histidine	Histidine	Histidine
Leucine	Leucine	Leucine	Leucine
Threonine	Threonine	Threonine	Threonine
Proline	Proline	Proline	Proline
Glutamic acid	Valine	Lysine	Glutamic acid
Glutamic acid	Glutamic acid	Glutamic acid	Glycine
Lysine	Lysine	Lysine	Lysine

5′ GUA 3′ due to change (substitution) of a single base. Hb-C containing lysine in place of glutamic acid (Table 16.1) might have arisen from a transition in the DNA from a C–G base pair to a T–A base pair which might have changed the glutamic acid codon 5′ GAA 3′ to 5′ AAA 3′ codon for lysine. The substitution of hydrophilic amino acid lysine for the hydrophilic amino acid glutamic acid at position 6 in the β-chain has shown mild-physiologic effects, as compared to the substitution of hydrophobic amino acid valine, which produces sickle cell anaemia. The mutations with little effects are called 'Quiet' mutation (Figure 16.13).

		← Gene →							
DNA		GTA CAT	CAT GTA	CTT GAA	ACT TGA	CCT GGA	GAA CTT	GAA CTT	AAA TTT
mRNA codons	Haemoglobin-A	GUA	CAU	CUU	ACU	CCU	GAA	GAA	AAA
Amino acid		Val	his	leu	thr	pro	glu	glu	lys
DNA							GTA CAT		
mRNA codon	Haemoglobin-B						CUA		
Amino acid							Val		
DNA							AAA TTT		
mRNA codon	Haemoglobin-C						AAA		
Amino acid							lysine		
DNA								CGA CCT	
mRNA codon	Haemoglobin-G							GGA	
Amino acid								gly	

Figure 16.13 Three mutants of β-chain, haemoglobin derived from haemoglobin-A by a single nucleotide change.

A base substitution of a single base may cause far-reaching consequences on the phenotype. The difference between haemoglobin-A (normal person) and haemoglobin-s (sickle cell) is on account of replacement of one amino acid, i.e. replacement of glutamic acid by a valine in the β-chain, which is on account of base transition which changes one nucleotide pair from A–T to T–A in DNA. Thus, single nucleotide change in DNA may cause a difference of life and death. This is because a change in the amino acid sequence in a protein, which may alter the biological property of a protein. Production of haemoglobin 's' is caused by a single allele. That is co-dominant with the allele for Hb-A and alleles for haemoglobins A, S, C and G of β-chain appear to be multiple alleles. The loci α and β are not a linked, hence mutation in one loci does not affect other.

Silent Base Substitution

It is not always that a *point mutation* (base substitution) may bring visible phenotypic changes of useful or harmful significance. There are two limitations for the missense mutation:

1. Location (site) where missense mutation takes place.
2. Degeneracy (presence of synonymous codons).

Location

While dealing with Fine Structure of Gene (Chapter 20), it has been felt that entire DNA does not possess the coding property. There are coding regions called *exon* and non-coding regions called *intron*. Any addition or deletion or insertion of transposable element in the non-coding regions 'intron' or in the spacer sequences between genes (non-essential) will not show any detectable effects on the phenotype. However, any 'micro' change in the coding regions 'exon' will show visible effect on phenotype.

Degeneracy of codons

Because of degeneracy (presence of synonymous codons) of the code, many a time base substitutions have no effect on the coding of the amino acid sequences by mRNA. For example, arganine is coded by six codons (CGU, CGC, CGA, CGG, AGA and ACG). A single base substitution that produces interconversions among any of these six codons does not show any effect on the amino acid sequence of a polypeptide.

Both kinds of above-mentioned mutations are called 'silent' mutation.

CHAIN-TERMINATING MUTATIONS

There are three codons in the genetic code—UAA, UAG and UGA, which cause termination of a polypeptide synthesis. A base substitution, that can unfortunately convert a codon for encoding any amino acid into a terminating codon, viz. the tyrosine codon UAC into terminating codon UAG causes pre-mature termination of protein synthesis. Occurrence of any such terminating codon near the N-terminating end of a polypeptide chain, the site which initiates the synthesis first, can stop the polypeptide synthesis responsible for synthesis of another protein (enzyme).

However, a single base substitution in the codon UAU which encodes 'Tyrosine' can produce nine possible codons; out of that, one is silent mutation UAC; six codons—GAU (aspartic acid), AAU (aspertine), CAU (histidine), UGU (cystein), UUU (phelanine) and UCU (serine) are missense mutations, which change the amino acid sequence in the polypeptide chain by encoding wrong amino acid not required for protein synthesis. The last two codon UAA and UAG are stop codons which terminate pre-maturely, the translation of genetic message for the synthesis of a truncated fragment of polypeptide chain. A base substitution creates a new stop codon which stops the synthesis of a protein is called 'nonsense' mutation.

Chain-elongating Substitution

There is a possibility that a polypeptide-terminating codon responsible for the normal termination of a polypeptide chain can also undergo base substitutions, as a result normal termination of polypeptide chain is further prolonged. If the mRNA possesses a tailer sequence, an elongated chain is obtained. Human haemoglobin Constant Spring (Hb-CS) mutation is a good example, in which polypeptide chain contains 172 amino acid instead of 141 in normal case, the first additional amino acid is a glutamine residue which is the result of a single base alternation on account of changed terminating codon from UAA to CAA, a codon encoding glutamine. The continued elongation of polypeptide chain takes place on account of presence of a tailer sequence in the mRNA. About 31 amino acids were added before another terminating codon appeared and stop the functioning in the tailer sequence.

Normal α-chain:	137	138	139	140	141	142	143	144... 170	171	172
	Thr	ser	lys	tys	arg	term				
Alpha chain: mRNA:	ACX	UCU	AAA	UAC	CGU	CAA	GGA	GUC	UGA	UAA.
Hb-constant spring:	Thr	ser	lys	tys	arg	glu	ala	gly...	val	phe glu

Protection of 'Blue-Print' of Life from Deleterious Base Substitution

Genetic code is the product of a biological evolution, therefore, it is logical that it must have ability (assigned duty) to protect the genetic message 'Blue-Print of Life' coded in DNA of an organism from distortion from the addition or deletion of the bases, leading to the unwanted amino acid replacement. There are two types of amino acids:

(i) *Hydrophilic amino acids:* They are arginine, lysine, histidine, glutamic acid and aspertic acid which have similar codons but oftenly differ from each other by a single base.

(ii) *Hydrophobic amino acids:* They are phenylalanine, valine, leucine and isoleucine which have similar codons (Figure 16.14). The base substitution either in *first* or *third* position of these codons will either cause no amino acid substitution or will substitute one hydrophobic amino acid for another hydrophobic amino acid.

A transition from A–C in the second position will result in the substitution of an amino acid by another hydrophobic amino acid (proline) or by one of the three ambivalent amino acids (serine, threonine or alanine). However, very rare transversion from U to A or from U to G in position two will substitute a hydrophilic amino acid for a hydrophobic one. This way, the structure of codon within mRNA protects itself against the harmful (deleterious) impact from illogical base substitution mutation. The similarity of codons for amino acids like physical properties speaks of common evolutionary origin for the tRNA's that become charged with these amino acids.

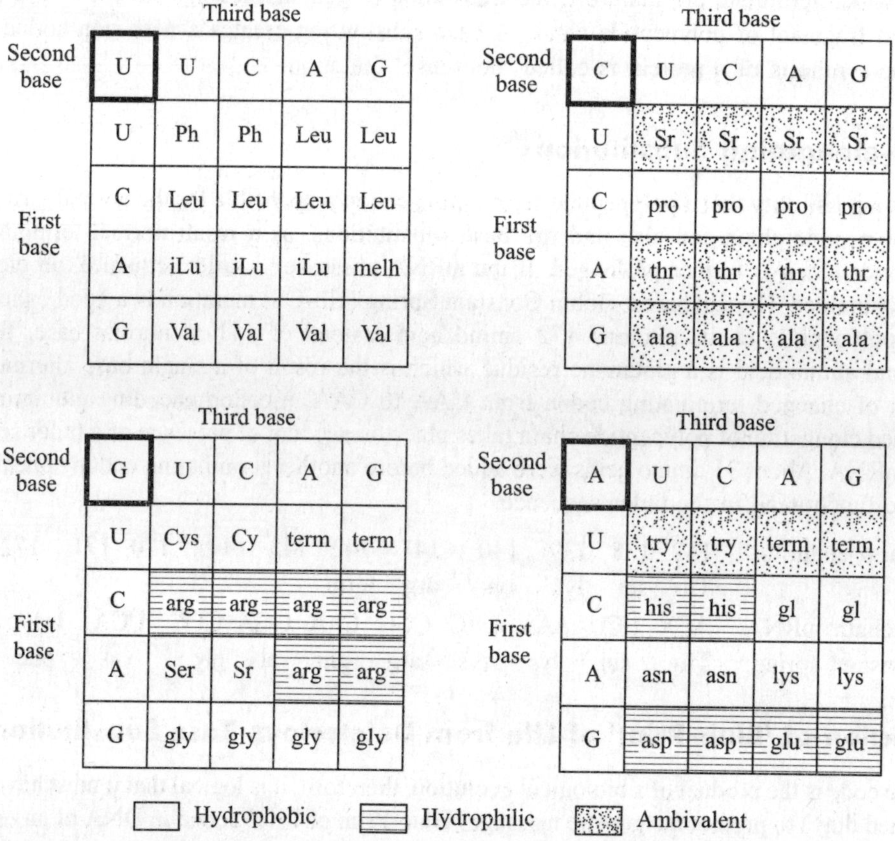

Figure 16.14 Protection of 'blue-print' of life from deleterious mutation.

MEASUREMENT OF MUTATION RATE BY MULLER'S CLB METHOD

Herman. J Muller in 1920 devised ClB method for estimating mutation rates in diploid organisms. This technique measures the rate at which new lethal mutation arise on X-chromosomes of *Drosophila melanogaster*. The presence of recessive lethal in the X-chromosome means that males with that chromosomes and females homozygous for it can not survive. This technique has been designed to detect mutation in normal sex-linked X chromosome (Figure 16.15) of *Drosophila melanogaster*.

In this technique, 'C' represents a *paracentric inversion* in the large part of chromosomes which prevents crossing over in the inverted segment, 'l' represents a recessive lethal. Females with lethal genes can survive only in the heterozygous conditions, 'B' stands for bar-eye which is dominant marker and helps in the identification of the mutant flies.

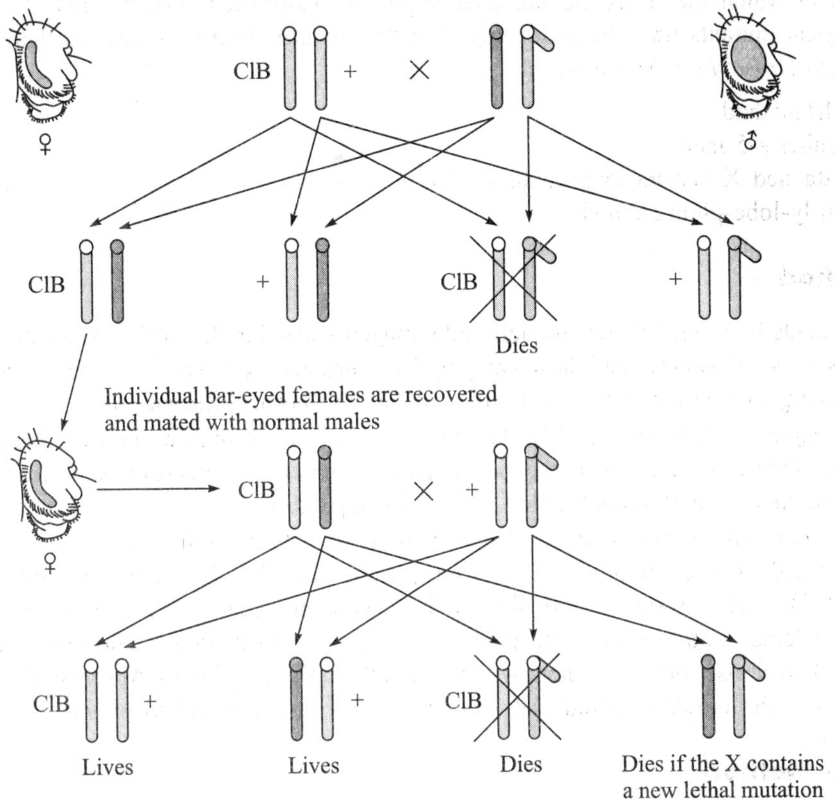

Figure 16.15 ClB technique of measurement of mutation rate in *Drosophila melanogaster*.

Muller crossed males carrying normal X chromosomes with heterozygous females carrying ClB chromosomes. From the progeny produced, females with bar phenotype (narrow eyes) were selected and then individually mated with normal males (presence of bar eyes indicated that females are heterozygous for the ClB chromosome). A ratio of two females to one male (2:1) is expected among the progeny of such cross because ClB males do not survive. This method provides a quantitative estimate of rate at which mutation to a recessive lethal allele may occur at any of the large number of genes in the X chromosome. However, the ClB method does not indicate the rate of mutability of a particular gene because mutation rate vary from one gene to other. For instance, mutation in the yellow body of *Drosophila* occurs at the frequency of 10^{-4} per gamete per generation whereas mutations to *streptomycin* resistance in *E. coli* occur at the frequency of 10^{-9} per cell per generation. The reasons are the presence of "Hot-Spot" regions of the DNA.

DETECTION OF MUTATION IN DROSOPHILA

Mutation brings changes in the morphological, physiological and biochemical reactions in all the living forms. But not all mutations are phenotypically detectable. By and large, the induced mutations are more deleterious and cause the organisms carrying them to be less fit than the organisms in which the traits are not phenotypically manifested. The methods to detect the morphological mutants have been developed in *Drosophila*. There are four methods to detect mutations in *Drosophila*. These are:

1. ClB method
2. Muller's-5 method
3. Attached X chromosome method, and
4. Curly-lobe-plum method

ClB Method

A cross is made between ClB female half and mutagen-treated male. In F_1, half of males having X chromosomes will survive and those carrying ClB chromosomes will die. Among females, half may be having ClB chromosomes and half will be with normal chromosomes.

In F_2, some females having ClB chromosomes and males with normal chromosomes are selected for further testing about the occurrence of mutation. Accordingly, a cross is made between ClB females and normal males to raise F_2 population.

In F_2, there will be two types of females—half with ClB X-chromosomes and half with mutagen-treated (mt) X chromosomes (normal phenotypes). On the other hand, on male side, half with ClB X chromosomes will die and half with mutagen-treated X chromosomes will survive. If a lethal mutation has been induced in the mutagen-treated X chromosome of male, the other half will also die. Thus, there will be no male in the population. Absence of male in the F_2 population is the indicative of induced sex-linked lethal mutations in *Drosophila* (Figure 16.16).

Muller-5 Method

This method is improvement over ClB method to detect sex-linked mutation in *Drosophila*. This method differs from ClB method in two ways.

(a) In this method 'apricot recessive gene' is used in place of recessive lethal in ClB.
(b) The female is homozygous for bar-apricot genes, whereas in ClB method, female is heterozygous for lB.

In this method, mutation is detected by absence of wild male in F_2 progeny. The procedure is given below.

1. A homozygous bar-apricot female is crossed with mutagen-treated male. In F_1, two types of progeny are produced, heterozygous bar females and apricot male.
2. In F_2, on intermating the F_1, four types of individuals are produced.
    - Half of females are homozygous bar-apricot and other half are bar heterozygous.
    - Similarly, among males, half are bar apricot (Muller-5) and half are normal. If a lethal mutation has been induced, there will be no normal males in F_2 progeny (Figure 16.17).

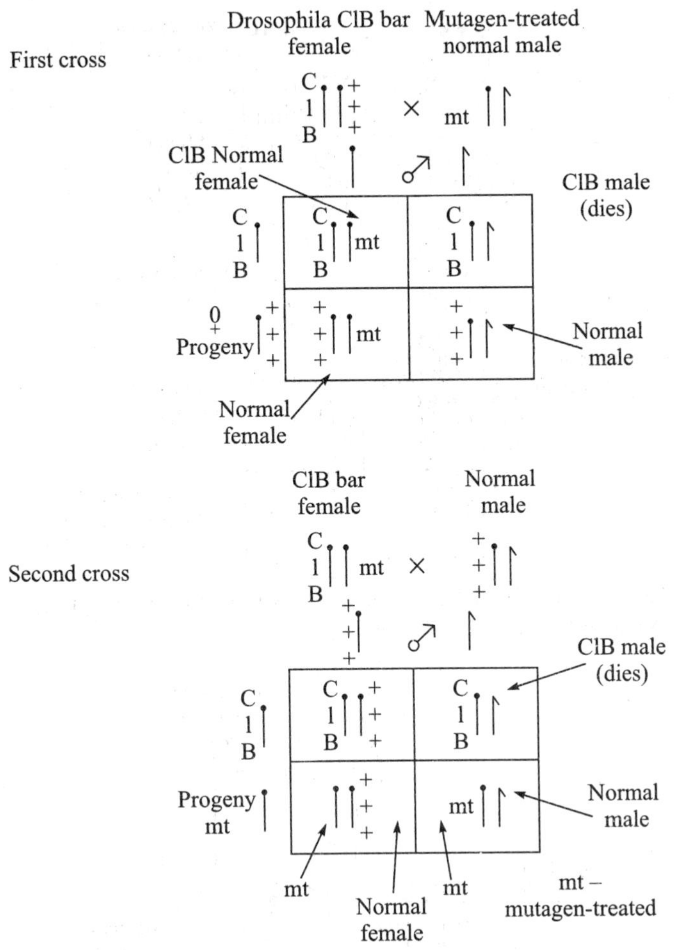

Figure 16.16 Muller's ClB method for detection of mutation in X chromosomes.

Attached X Chromosome Method

This method is used to detect sex-linked visible mutation in *Drosophila*. In this method, a female having two attached (XXY) chromosomes is used to study the mutations.

Procedure

A female fly with attached X chromosome (XXY) is crossed with a mutagen-treated male. This cross gives rise to superfemales (XX-X); females with attached chromosomes (XXY); mutant male (XY) and (YY). The males with YY chromosome die and superfemales (XX-X) also die. The surviving males possess X chromosome from mutagen-treated males and Y chromosome from female with attached female (XX-Y). Because Y chromosome does not possess the corresponding alleles of X chromosome, therefore, even the recessive alleles in X chromosome can be deleted in the surviving males (Figure 16.18).

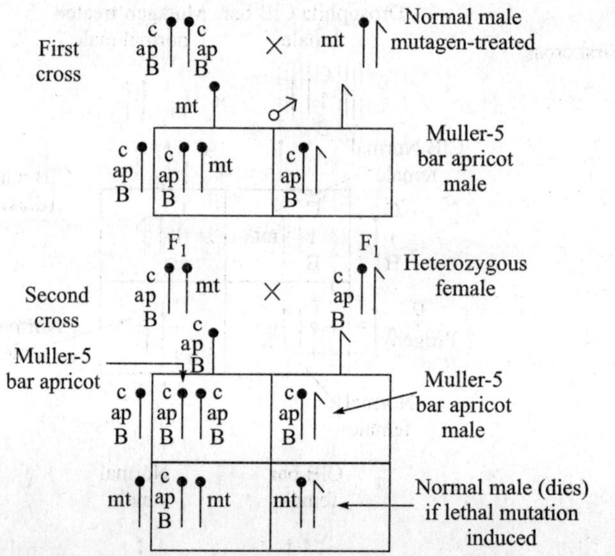

Figure 16.17 Muller's method for detection of induced mutation in X chromosome of *Drosophila*. mt–mutagen-treated, ap–apricot.

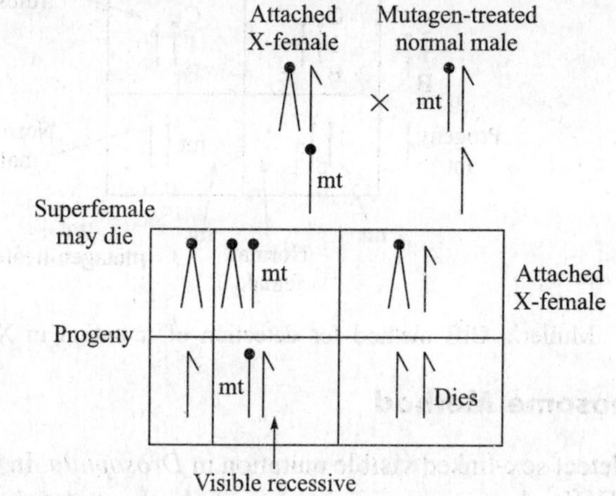

Figure 16.18 Attached x-female for detection of, visible mutation in *Drosophila*. mt–mutagen-treated.

Curly-Lobe-Plum Method

This method is used to detect mutation in the autosomes. Here curly means 'curly wing' (CY) 'lobe' means lobed eye (L), and 'plum' means brownish eyes (Pm). All the three genes are recessive lethal. Curly (CY) and lobed (L) genes are located on one chromosome and plum (pm) on the other chromosome, but on homologous chromosomes. Crossing over is prevented because of presence of inversion. The homozygous individuals with CYL or Pm genes donot survive because of lethal effect. Therefore, only heterozygous individuals survive. This system may also be called balanced-lethal-system (Figure 16.19).

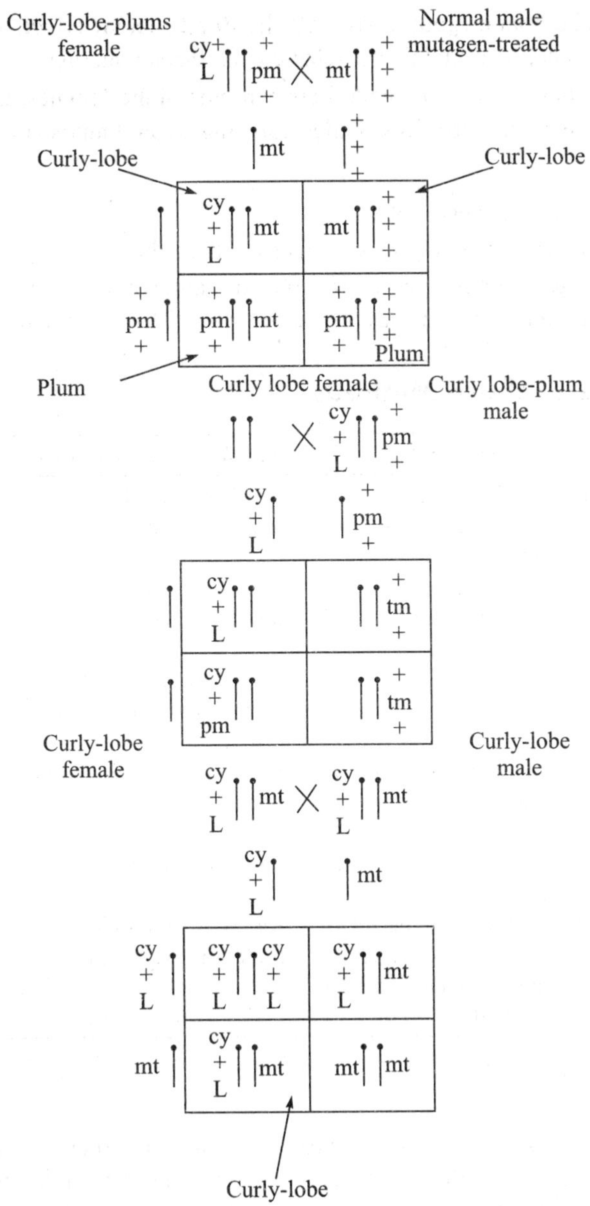

Figure 16.19 Detection of induced autosomal recessive mutation in *Drosophila*.

Procedure

1. A cross is made between curly-lobe-plum (cyl-pm) female and mutagen-treated male. This produces 50% progeny as curly-lobe and 50% as plum in F_1 generation.
2. In the second generation, cross is made between curly-lobe-plum female and curly-lobe-plum male. This gives rise to curly-lobe-plum, curly lobe and plum individuals in 1:1:1

proportion. The homozygous curly will die due to lethal effect. From this population, curly-lobe females and males are selected for further mating.

3. In this generation, a cross is made between curly-lobe females carrying one mutagen-treated autosome and curly-lobe males carrying treated autosome. This will give result as follows:

   (a) 50% progeny as curly lobe
   (b) 25% homozygous curly lobe progeny which dies
   (c) 25% progeny homozygous for treated autosomes. This will give autosomal recessive mutation and may be one-third (1/3) of surviving progeny.

Basic Differences in these Methods

ClB	*Muller-5*	*Attached X chromosome*	*Curly-lobe-plum*
1. To detect sex-linked lethal mutation	To detect sex-linked recessive lethal mutation	To detect sex-linked visible mutation	To detect visible autosomal mutation
2. Heterozygous ClB female is used	Homozygous bar-apricot female is used	Attached X chromosome female is used	Heterozygous curly-lobe-plum female is used
3. ClB male die due to lethal effect	—	Male having YY chromosomes dies due to lethal effect	Homozygous fly for cyl-pm gene die
4. Crossing over does not occur due to inversion	Crossing over does not occur	Crossing over does not occur	Crossing over does not occur
5. Two step crosses are made	Two step crosses are made	One cross is made	Three step crosses are made
6. Mutation is detected due to absence of males in the F_2 population	Mutation is detected due to absence of normal male in the F_2 population	Males with visible mutation are found	Individuals with visible mutations are found

Hot-Spot Region

Certain DNA sequence 'sites' are more liable for mutation than others. These mutation-susceptible regions (sites) are called Hot-Spots regions. These include *monotonous* runs of a single nucleotide or tandem repeats of short sequences. Hot-Spots regions may be located at many sites throughout the genome and within the 'gene' at the base-level also. Hot-Spot mutational sites were located in r-II gene of bacteriophage T_4 which accounted 20 per cent of all the mutations obtained.

The methylated sites of cytosine are highly mutable and mutation occurs from GC–AT transition. In many organisms such as bacteria, mammals, maize (not *Drosophila*) about one per cent of cytosine bases are methylated at 5′-carbon position which are 5-methyl cytosine instead of cytosine (Figure 16.20). Some special enzymes are responsible for addition of methyl groups to cytosine in certain target DNA sequences. When DNA replicate, 5-methyl cytosine pairs with guanine. The real function of cytosine methylation is not known but methylated cytosine regions in DNA slowdown the genetic activity.

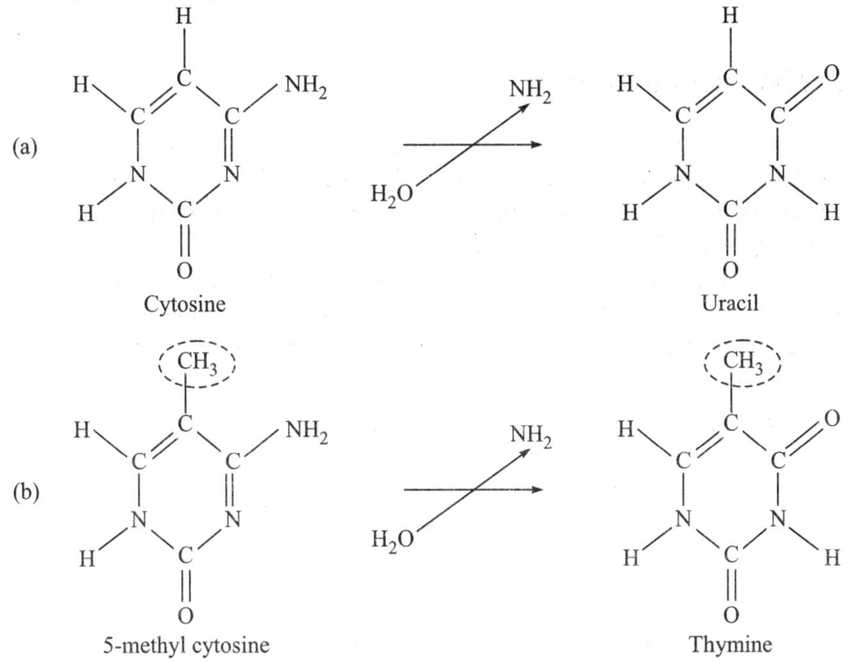

Figure 16.20 Spontaneous loss of amino group of (a) cytosine to yield uracil, and (b) 5-methyl cytosine to yield thymine.

Methylation of cytosine increases the chances of mutation. Both cytosine and 5-methylcytosine occasionally loses an amino (NH_2) group. If cytosine loses amino group (NH_2), it turns to uracil, which pairs with adenine instead of guanine. On the replication of a DNA molecule, a GU base pair may lead to substitution of an AT pair for original GC pair$\rightarrow$ GU $\rightarrow$ AT in the successive round of replication. However special enzymes present in the cell removes uracil from DNA because it is not DNA base. This stops conversion of cytosine to uracil and thereby reduces the chances of mutation. Similarly, loss of amino group from the 5-methylcytosine produces thymine which is not removed because it is a normal DNA base. This way, original GC pair become GT pair which in the next generation on replication give rise GC (wild type) and AT (mutant) DNA molecule.

REVERSION OF MOLECULAR MUTATION

Mutation is a two-way traffic forward $\rightleftharpoons$ revert that is normal $\rightleftharpoons$ mutant phenotype. Normal is also called wild. This kind of mutation (reversion) is on account of second mutational changes that may occur after several generations. But not all kinds of mutations can revert to normal or half-way normal. On this basis, reversion could be classed as complete or full revertants and second the suppressor type. In true reversion, revertant back to full normalcy. But in case of suppression, phenotypic expressions are reduced or expressed as subnormal type, such as an *auxotroph* which reverts back to *prototroph* but still grows abnormally. In reversion, a base substitution can go true reversion by changing GC base pair to AT and back originally to GC.

INTRAGENIC SUPPRESSION

Reversion of *frame-shift* mutation within a gene occurs from the intragenic suppression. The addition or deletion of a nucleotide base pair in DNA changes the *reading frame* of mRNA which if rectified by compensating a deletion (or addition) of a second nucleotide base pair at near by the site in the gene is called *intragenic suppression.*

Another type of intragenic suppression occurs when the loss of activity of a protein (enzyme) caused by a change of one amino acid is restored by addition or deletion of one amino acid. This type of suppression can be cited in the protein product of 'Trp' A gene in *E. coli* as shown in Figure 16.21.

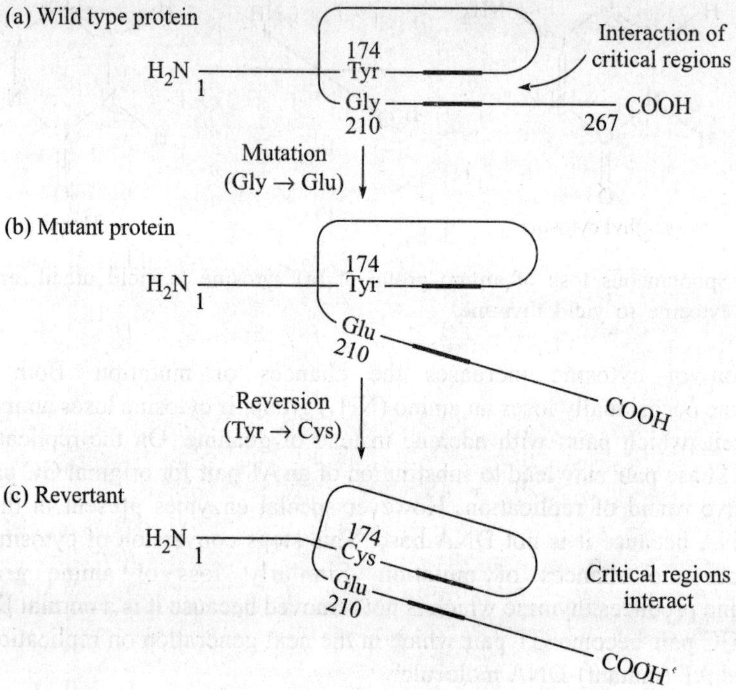

Figure 16.21 Intragenic suppression of mutation, (suppression on the folding and activity of the 'A, protein of tryptophan synthetase in *E. coli*.
(a) Wild type protein in which two critical regions (heavy line) of polypeptide chain interact.
(b) Disruption of proper folding of the polypeptide chain by the substitution of amino acid Glu (210) prevents the critical regions to interact.
(c) Suppression of the effect of the original forward mutation due to subsequent change in the amino acid 174 (Cys).

A polypeptide chain containing 268 amino acids is one of the two polypeptide chains that make-up the enzyme 'tryptophan synthatase' in *E. coli*. Mutation that changes glycine to glutamic acid at number 210 in wild type protein, inactivate the enzyme 'tryptophan synthatase' because glycine is not in the active site of the enzyme. The activity of this enzyme is partly restored by second mutation in which amino acid at number 174 *tyrosine* changes to *cystocine*. A protein in which only the *tyrosine* has been replaced by cystein is also inactive because it changes the shape of protein. This change in the second site, found to occur repeatedly in the reversion of the original mutation; restores activity because two sites containing amino acids 174 and 210 interact fortuitously to produce a protein with correct folding.

INTERGENIC SUBSTITUTION

A change in a second gene can alleviate a mutant condition in several ways. For instance, molecular aggregates such as haemoglobin which is a *tetramer*, an amino acid substitution in the α-chain might suppress the deleterious effects of a substitution in the β-chain by permitting normal interaction of the subunits in the same way as it happens in the case of intragenic suppression.

HOMEOTIC MUTATIONS

The activation and inactivation of promoters, and the regulatory genes provide the basis for the orderly development of an organism. In 1940s, Lewis had studied the mutations in the bithorax-bithoraxoid locus and gave the idea of *pseudoalleles*. The concept of *pseudoalleles* did not stay any more in the genetic literature, when it was discovered that genes are segments of DNA polymer (molecules) and crossing over takes place within both the genes as well as between the genes. Mutations which Lewis described have proved to be of great interest because some of these genes result in the transformation of one body part into another, which is normally determined by a gene located at a different locus. In this way, a homeotic mutations simulate the phenotypic expression of other genes. For example, a mutant gene whose wild allele produces legs may result in the development of an extra antene or an extra wing where eyes are normally formed beside bithorax gene complex. Thus, homeotic gene mutations have been found to impair the morphogenesis of the anterior body part.

In fact, the homeotic genes are not the structural genes, but they are regulatory genes which can 'switch-on' or 'switch-off' the activities of a large number of structural genes. Garber, et al. (1983) were first to isolate homeotic genes from 'antenna podia' of *Drosophila*. They used a cDNA probe complementary to mRNA of the gene to identify the DNA sequences of the coding region. This probe not only hybridized the antenna podia coding sequence, but coding sequences of a number of homofic genes including those of bithortax gene complex.

Homeobox

Walter Gehring and his colleagues designated the common sequence of nucleotides found in more than dozen homeotic genes as the 'homeotic box' (homeobox). The homeotic sequence present in the homeotic genes codes for a related group of proteins. Therefore, they are also called homedomains. These proteins bind a DNA in the regulatory elements of structural genes and right combinations of these proteins bound a regulatory element which signals the activation or repression of a gene. Thus, homeotic genes are the regulatory genes responsible for the regulation of the developmental activities of the organisms.

Evolutionary significance of homeobox

The homeo story is very interesting because evolution of man may be much more similar to his distant relatives, worm and insects than the chimpanzee. Because, the base sequences found in homeobox, are almost similar in *Drosophila*, frogs, mice and man. They form part of *home* genes in invertebrates and in vertebrates; they are designated as hox genes. These are infact sequences of DNA, which have been conserved in the evolution of different species for a long time (Figure 16.22).

	1									30			
Antp	CGC	AAA	CGC	GGA	AGG	CAG	AGA	TAC	ACC	CGG	TAC	CAG	ACT
Hul	GGG	AAA	AGG	GCC	CGG	ACC	GCG	TAC	ACC	CGC	TAC	CAG	ACC
							60						
Antp	CTA	GAG	CTA	GAG	AAG	GAG	TTT	CAC	TTC	AAT	CGC	TAC	TTG
Hul	CTG	GAG	CTC	GAA	AAG	GAG	TTC	CAC	TTC	AAC	CGC	TAC	CTG
				90									
Antp	ACC	CGT	CGG	CGA	AGG	ATC	GAG	ATC	GCC	CAC	GCC	CTG	TGC
Hul	ACC	CGG	CGA	CGG	CGC	ATC	GAG	ATC	GCC	CAC	GCA	CTC	TGC
	120												
Antp	CTC	ACG	GAG	CGC	CAG	ATA	AAG	ATT	TGG	TTC	CAG	AAT	CGG
Hul	CTG	TCC	GAG	CGC	CAG	ATC	AAG	ATC	TTG	TTC	CAG	AAC	CGG
								180					
Antp	CGC	ATG	AAG	TGG	AAG	AAG	GAG	AAC	AAG				
Hul	CGC	ATG	AAG	TGG	AAG	AAG	GAC	AAC	AAA				

Figure 16.22 The DNA sequence of the homeobox region of Antp (from *Drosophila*) and Hul (from humans), demarcated from nucleotides 1 to 183. The degree of conservation within the 183 bp for Hul gene homeobox region relative to that of Antp is 81 per cent.

CYTOPLASMIC MUTATIONS

Cytoplasmic heredity can be classed as *plasmon* and *plastidom* types. The plastidom inheritance is related to the function and properties of plastids. The mutated plastids are in general transferred from generation to generation via the egg cells. Plastids (and mitochondria) contain DNA (and RNA), the possible mutation in the plastids (and mitochondria) DNA are different in those that occurs in nuclear genes.

The biochemical reasons of 'plasmon' mutation are not properly undersrood, but cytoplasmic male sterility has been found of fundamental importance in the crop improvements to utilize it in the heterosis breeding. The sources of cytoplasmic male steriles (sterility) may be grouped as intergeneric, interspecific, intraspecific, spontaneous and induced. The use of cytoplasmic male sterility and restorer gene interaction is important for hybrid seed production. It is presumed that some factors controlling male fertility must be residing on the organelle DNA and some alteration of organelle DNA may be responsible for cytoplasmic male sterility. This kind of change must be due to the *point mutation*. Ethidium bromide is known to convert respiratory sufficient yeast (grandos) to cytoplasmically-inherited respiratory deficient mutant (petites) close to 100 per cent (Goldring, et al., 1970). Mitochondria have been reported responsible for susceptibility of Texas male sterile maize (CMS line) to *Helminthosporium maydis* race 'T'. Therefore, mitochondrial mutation for incorporation of resistance to *H. maydis* race 'T' could be explored by use of chemicals like acridine, dye, ethidium bromide and physical mutagens.

TRANSPOSABLE MUTAGENESIS

In 1940, while studying the mottling of kernel in maize, Barbara McClintock struck to an element, which regulates the mottling and causes the chromosome carrying genes for colour and consistency to break. The element she called *dissociation* (Ds). Mapping data showed that chromosome breakage is always at or very near to the location of Ds. McClintock critically observed that Ds does not have any constant location but occasionally moves to another position (transposition) and causes chromosome breakage at new sites. Further more, Ds moves only when another element called 'Activator' (Ac) is present. But Ac itself has ability to move within the genome and can cause alternation in the expression of genes at or near the insertion site. Similar to the modification caused by Ds, these transposable elements are short DNA sequences of nucleotides, and these are present in multiple copies of several kinds in almost all organisms. These transposable elements are also important cause of mutations in two ways. First, most transposable elements are present in the non-essential region 'intron' of the genome (DNA) and do little or no harm. But, when these elements transpose, it can insert into an essential region (exon) and disrupt its function.

Figure 16.23 indicates the result of transposition into a coding region of DNA (an exon). The insert interrupts the coding region. Since most transposable elements have their own coding regions, the transcription of the transposable element interferes with the transcription of the gene into which it is inserted or transcription of the gene might get terminated within the transposable element. However, if transcription proceeds by the element, the phenotype will be mutant because the coding region then have incorrect sequences.

Another way for the transposable element to cause mutation is recombination. As stated earlier, transposable elements are present in the multiple copies, sometime two or more in the same chromosome. If the copies are closed to each other, they can pair during synapsis and undergo crossing over, as a result, the DNA region between the elements is deleted.

On the other hand, if downstream 'element' in one chromosome can pair with the upstream 'element' in the homologous chromosome, an exchange within the paired 'element' will result gene duplication in one chromatid (top) and deletion in other chromatid (bottom).

Similarly, if downstream 'element' in one chromosome can pair with the upstream element in the homologous chromosomes, an exchanged within the paired 'element' will result in one chromatid with missing region between the element (bottom) and one chromatid in that the region is duplicate (top).

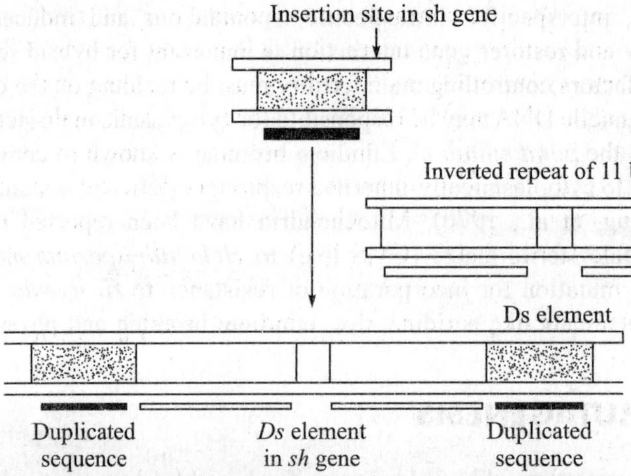

Figure 16.23 Transposable element in maize. Ds is inserted into the maize 'sh' gene at position indicated. In the insertion process, a sequence of 8 base pairs next to the site of insertion get duplicated and flank to Ds.

CONDITIONAL MUTATIONS

There are mutations whose effects can be turned 'on' or 'off' at will. These are called *conditional mutations*. Such mutations show phenotypic expression in one set of environmental conditions called 'restrictive conditions', but do not show expression in another condition called 'permissive condition'. There are some lethal mutations, which express their effects in the restrictive conditions, but are viable in the permission environment. Therefore, mutations carrying lethal genes could be propagated in the permissive condition. There are three classes of conditional mutations:

1. Auxotrophic
2. Temperature-sensitive
3. Suppressor-sensitive

Auxotrophic

Auxotrophic mutants as opposed to phototropic 'wild-type' are the mutations that are unable to synthesize an essential metabolic (amino acid, purine, pyrimidine, vitamins, etc.) materials. That is essential for the expression of 'wild-type' individual. Such type of organism can grow and reproduce when metabolite is made available in medium (permissive). However, these cannot grow when required metabolic is not available (restrictive condition).

Temperature-sensitive

The temperature-sensitive mutations are also of two types. Those which can express phenotypic characters under high temperature (thermosensitive) and those which can grow under low temperature (cold tolerance) under permissive condition.

Suppressor-sensitive

Such mutations are viable when another genetic factor called 'suppressor' is present; but nonviable in the absence of 'suppressor'. The suppressor gene may correct or compensate for the defect in the phenotype that is caused by the suppressor-sensitive or it may render the gene product.

INDUCED MUTATIONS

Spontaneous mutations arise randomly in nature without any human intervention. On the other hand, mutation that occurs by human intervention applying some external agents is called *induced mutation*. Herman J. Muller (1927) was the first man who induced mutation in *Drosophila* by application of x-rays. In 1928, L.J. Stadler increased the frequency of mutation in maize by x-ray treatment. In 1943, Charlotte Auerbach reported induction of mutation by using chemical compounds. She reported that nitrogen and mustard gases used in World War I, induced mutations in *Drosophila melanogaster*.

However, there is some objectives behind inducing mutations in crop plants:

(i) To knock out certain demerits undesirable traits.
(ii) To induce a new gene allele (loci) related to yield component characters, quality and resistance to diseases and pests, etc. It is alternative of back-cross to induce a new character, not available in the population or to knock out a demerit of a variety in a short-time.
(iii) To improve the quality character etc.
Induced mutation causes same type of chromosomal, genic and molecular changes as spontaneous mutation. Hence there is no need to repeat the same.

MUTAGENIC AGENTS

The mutagenic agents are classified into two major groups:

1. Physical mutagens
2. Chemical mutagens

Physical Mutagens

The physical mutagens which are used for inducing mutations especially in crop plants are:

(i) High-energy ionizing radiations—x-rays, gamma rays, neutrons, beta and alpha particles.
(ii) Low-energy, non-ionizing radiations—ultraviolet light.

Ionizing radiations

The spectrum of electromagnetic radiation is a broad one; it extends from the long radio waves which may have wavelength as great as several kilometres, down through the cosmic rays, which may be as short as 10^{-14} cm. As wavelength becomes progressively shorter, the energy that electromagnetic photons emit becomes progressively greater, as a result they penetrate the cells and tissues.

The ionized atoms and molecules in which they occur are chemically more reactive than neutral ones. The mutagenic effects are caused by the chemical reactions undergone by ions as their charges are neutralized. The ionizing radiation causes breakage of the sugar phosphate strands of DNA, which leads chromosomal aberrations such as deletion, duplication, inversion and translocations. If radiation causes loss of one or two nucleotide pair, then only shift in the *reading frame* occurs leading micromutation (point mutation). This ultimately causes change in protein or an enzyme, which loses capacity to function. The homozygous mutation at genic level may be lethal but if it occurs in the germinal cells, and if mutation is dominant, a new genotype (phenotype) of economic importance could be obtained by mutation.

The dosage of an ionizing radiation is based on the amount of energy deposited in the tissue; one unit of measurement is roentgen (r) which is equal to roughly 2×10^{-9} ion per cubic cm of air. The dose of radiation based on the energy absorbed is expressed in rad; one rad is the amount of radiation that deposits 100 ergs of energy in a gram of matter. The rad is slightly larger than the roentgen in as much as one gram of tissue exposed to kr of gamma rays absorbs about 93 ergs. The mutation frequency is not proportional to the dosage but optimum dose for inducing useful mutations differ from species to species among the crop plants and from organ to organ in a plant species. For example, according to Ram (1972, 74), the optimum dose of gamma rays for inducing useful mutations in rice ranges between 25 to 30 kr (unsoaked seed).

The ionizing and non-ionizing radiations are discussed below:

X-rays: X-rays, like gamma rays and ultraviolet light are electromagnetic radiations, emitted as quanta but differ in wavelength (10–0.001 nm for x-rays and gamma rays compared with 2000–3000 nm for ultraviolet light) result in little similarity in practical usage: It is desirable to use x-ray machine producing radiation of short wavelengths (hard x-rays) because their penetration is greater than x-rays with large wavelengths. Soft x-rays of longer wavelength are less-penetrating but more densely ionizing than the shorter wavelength, x-rays. Through the use of high voltages and appropriate filters, to absorb the soft radiation, high-energy radiation, similar to but less homogenous than gamma rays can be obtained. When x-rays irradiation is performed, the kvp, milliampere (m) thickeners and type of filter distance of tube to the target dose and dose rate should be noted.

Gamma rays: In general, gamma rays have shorter wavelengths and, therefore, possess more energy per photon than x-rays. Monoenergetic gamma radiation is usually obtained from radioisotopes in contrast to x-rays which are generally produced in a wide-range of energies. A gamma-radiation facility can be used essentially in the same manner as an x-ray machines, for acute or semiacute exposure. However, gamma-radiation source has distinct advantage for prolonged treatments in that it may be placed in a green house or field so that plants will be exposed as they develop over long period of time. Cobalt-60 and Caesium-137 are the main sources of gamma rays used in radiobiological researches at present.

Ultraviolet radiation: The ultraviolet light has limited tissue-penetrating ability so that in genetic experiments on plants, its use is restricted for treating spores and pollen grains. The increasing use of cell and tissue culture for mutation breeding of plants could mean an increase in the use of ultraviolet light as a mutagenic agent especially when a single gene mutation is sought. The absorption of ultraviolet light depends to a considerable extent on the molecular structure. To make a quantitative assessment of its effect, it is necessary to use monochromatic (or near monochromatic) ultraviolet light, because the biological effectiveness of the light varies with wavelength. Wavelength in the range of 2500–2900 nm are biologically more effective because this is the range of maximum light absorption by nucleic acids.

Beta radiation: Beta particles (electrons) such as those from p-32 and S-35 produce effects in tissues similar to those of x-rays and gamma rays, although penetrative ability of beta particles is lower than that of x-rays and gamma rays. The lower penetrating ability of beta particles can be overcome by putting the radioisotopes in a solution and administering them to the plant materials. In this way the p-32 and S-35 may be incorporated directly into the cell nucleus giving a somewhat greater localization of the site of action (Thomson, et al., 1950). The mode of action of interval emitter is different from the external radiation source.

Neutrons: Neutrons of a wide range of energies could be obtained from a reactor where the U-235 fuel is undergoing nuclear fission. In this process a thermal neutron enters the nucleolus of an atom of U-235 and causes it to split apart or fission. When this occurs, two or three neutron on the average are emitted and a large amount of energy is released. The neutrons released are then capable of reacting with other atoms of uranium to repeat the process. This series of events is called *chain reaction.*

The neutrons have been found to be highly effective for the induction of mutations in the plants.

Chemical Mutagens

The chemical mutagens can be classified into four groups:

1. Base analogues and related compounds
2. Antibiotics
3. Alkylating agents
4. Azide

Base analogues and related compounds

The base analogues are the chemical compounds which are almost similar to one of the four DNA bases that can be incorporated into a DNA molecule during replication. They can pair with a base in the DNA template strand. Some of the base analogues are mutagenic because they induce mismatch pairing. For example, 5-bromouracil (Bu) is a common base analogue, which can be incorporated in the DNA of bacteria and viruses. The 5-bromouracil is an analogue of thymine and bromine atoms which are similar in size as methyl group of thymine. In keto form, it substitutes for thymine and structurally closely resembles. If bacteria or virus cells are grown in the media containing 5-bromouracil during DNA replication, thymine (Figure 16.24(a)) is replaced by a 5-bromouracil, as a result A–Bu pairs at a normal AT site. After incorporation, 5-bromouracil generates mutagenic activity in the changed base configuration (Figure 16.24(b)). This shift is influenced by the bromine atom. In the less-common configuration, 5-bromouracil pairs preferentially with 'guanine' (Figure 16.24(c)), so in the next round of replication, one daughter molecule has a GC pair at the altered site; yielding an AT–GC transition.

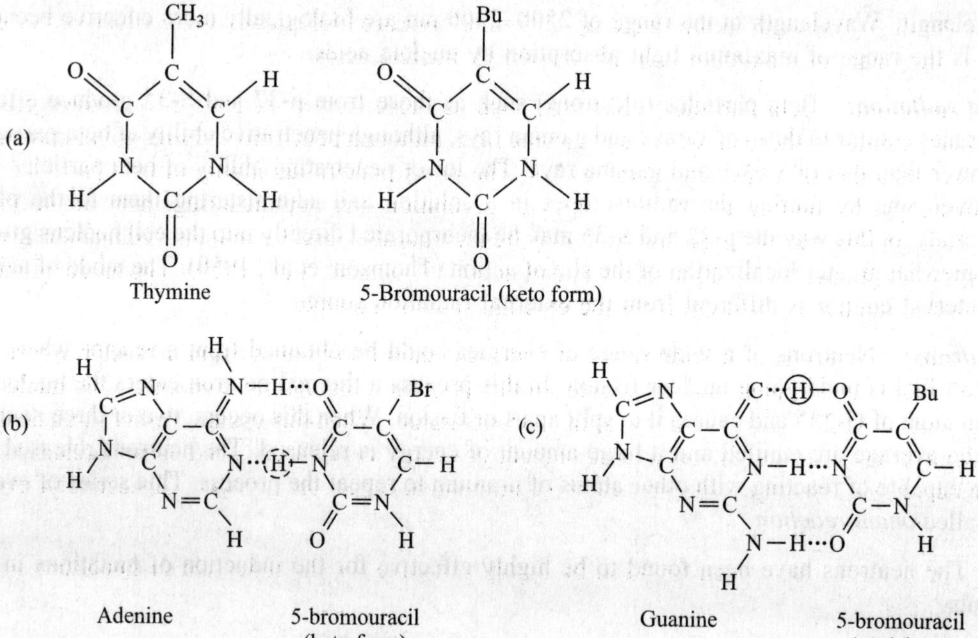

Figure 16.24 Mispairing mutagenesis by 5-bromouracil (H) H-changes the position when 5-biomouracil is in keto-form.

In Figure 16.25 AT–GC transition produced by incorporation of 5-bromouracil (Bu) during DNA replication forming an A–Bu pair is shown. In the mutagenetic condition of replication; Bu (enol form) pairs with G and in the next round of replication G pairs with C completing the transition mutation.

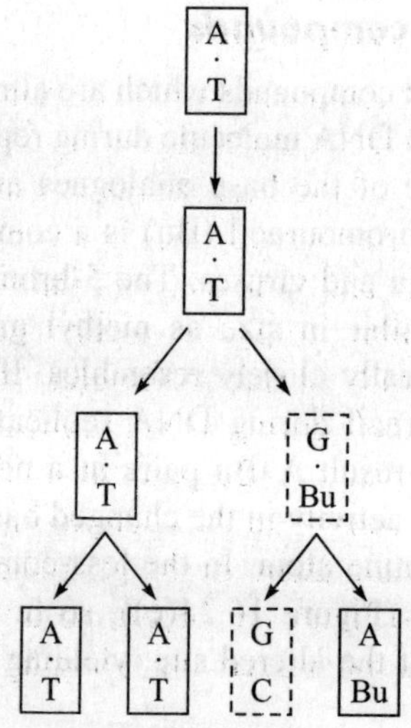

Figure 16.25 Mutagenesis by 5-bromouracil (Bu). An AT–GC transition is produced by the incorporation of 5-bromouracil during DNA replication forming an ABu pair.

Malic hydrazide (MH): A structural isomer of uracil, induces chromosome breaks in the cell of *Vicia faba* root tips. It causes chromosomal aberrations in a heterochromatic segment near the centromere of the chromosome.

Apart from true analogues, N-methylated oxypurines have a chromosomal breaking effect.

Deamination: Many chemical mutagens react with DNA and disturb the hydrogen bonding ability of the bases. For example, nitrous acid (HNO_2) possesses mutagenic property, which converts the amino group 'NH_2' of the nitrogen bases-adenine, cytosine, guanine and thymine into 'keto' (=0) group. This process is called *deamination*, which alters the hydrogen bonding specificity of each base. The application of nitrous acid deaminise the adenine which produces a base (hypoxathine) which pairs with cytosine, rather than thymine, and results AT–GC transition. Similarly, as mentioned in earlier pages deamination of cytosine gives rise 'uracil' which pairs with adenine instead of guanine and this results a transitional mutation frame GC–AT (Figure 16.26).

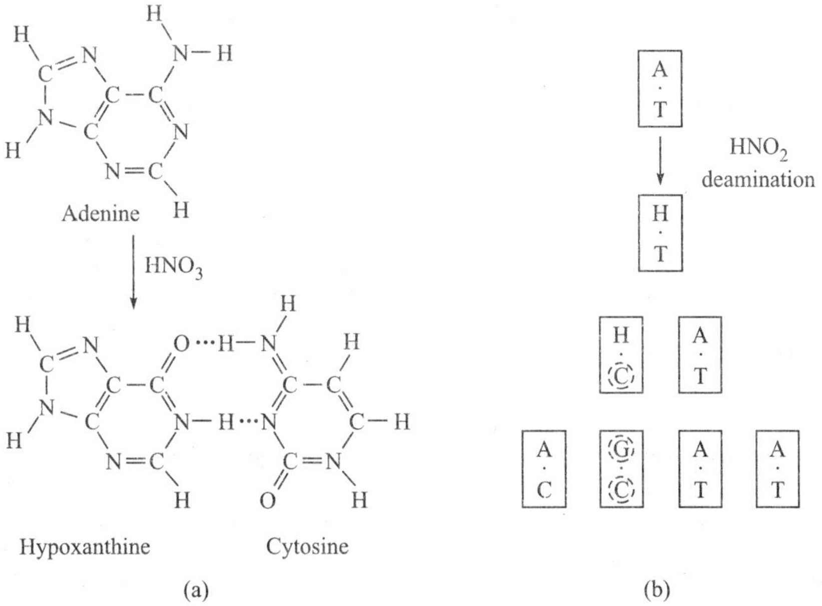

Figure 16.26 Nitrous acid mutagenesis. Conversion of adenine into hypoxanthine, which pairs with cytosine.

(a) Nitrous acid (HNO_2) through deamination convert adenine to hypoxanthine which pairs with thymine. Similarly, spontaneous loss of amino group converts cytosine to uracil.

(b) Deamination: Product of AT–GC transition. In the mutagenic form of replication, the hypoxanthine (H) pairs with cytosine; in the next replication. C pairs with G completing the transition.

Antibiotics

A number of antibiotics like azaserine, mytocin-c, streptonigrin and actinomycin D have been found carrying chromosome breaking properties.

Alkylating agents

Most of the alkylating agents are ethylmethane-sulphonate (EMS), diethylsulphate (DES), ethylnitrosourethane (ENU), ethylnitrosourea (MNH). These alkylating agents (Figure 16.27) possess mutational properties.

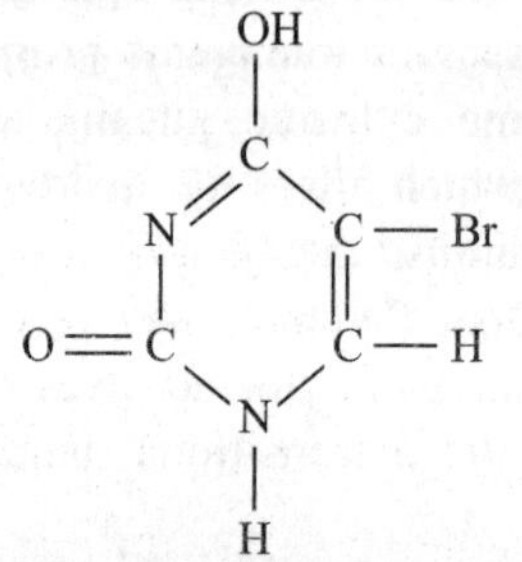

5-bromouracil

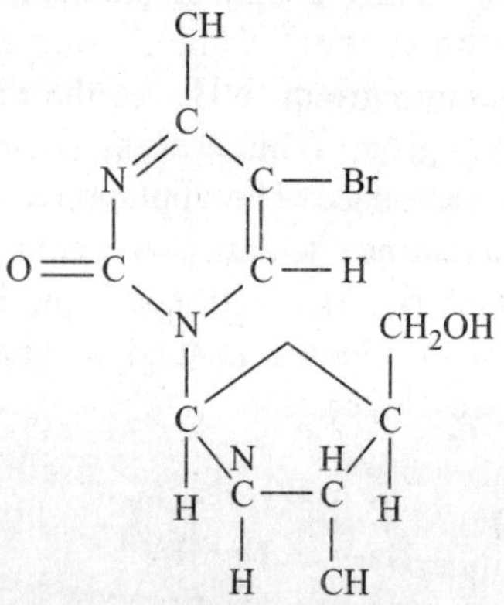

5-bromodeoxyuridine

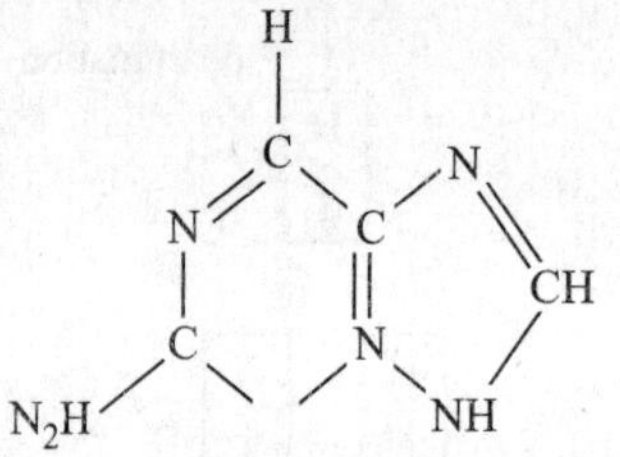

2-amino-purine

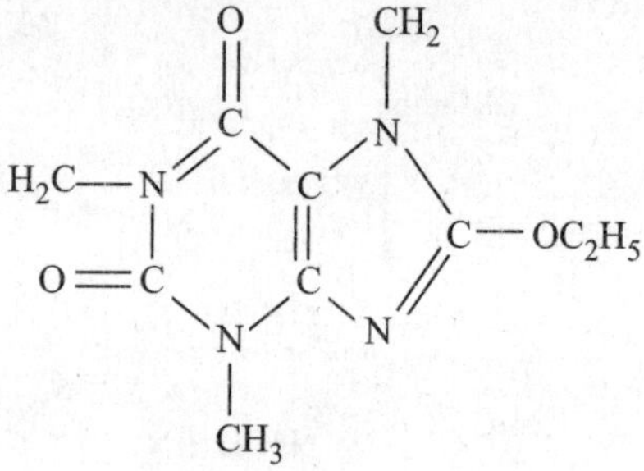

8-ethoxycoffeine

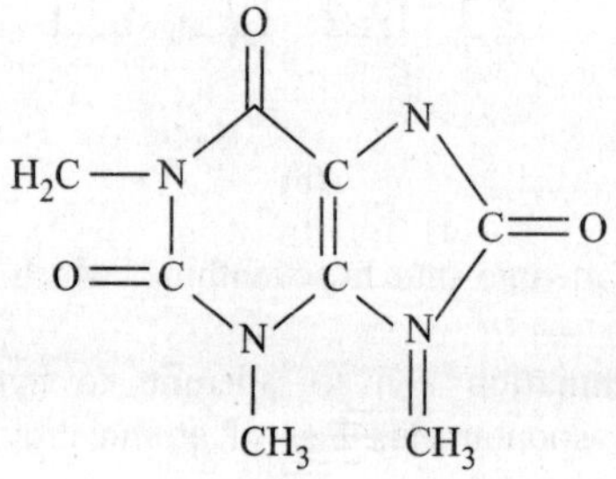

1, 3, 7, 9-tetramethyl-uric acid

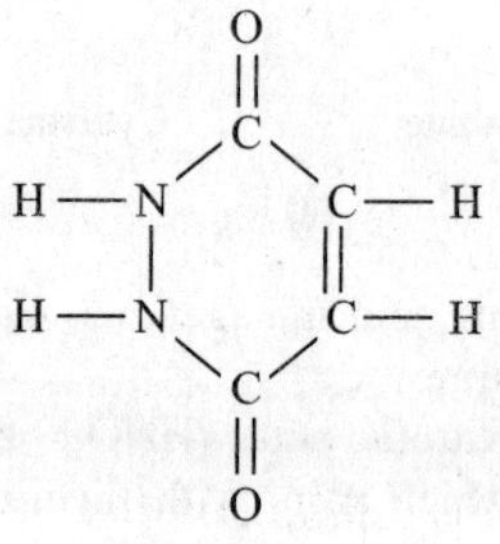

Malic hydrozide

Figure 16.27 Principal categories of alkylating mutagenic agents.

Azide

Azides are effective mutagenic agents. However, these chemicals are carcinogenic, therefore, must be handled carefully.

Cytological effects: When ionizing radiations (α- and β-particles and γ-rays) interact with water, highly reactive ions called 'free-radicals' are formed, which react with other molecules including DNA and show mutagenic effects. Ionizing radiations cause three types of damage in DNA. *First*, single-strand breakage (in sugar-phosphate backbone), *second*, breaking of both DNA strands and *third*, change in the position of nucleotide bases. The breaks in the single strand is get repaired, but breaks in the double strands as well as change in the positions of nucleotide bases induce mutation. Ionizing radiation causes chromosomal breakage such as deletion, duplication, inversion or translocation.

Ionizing radiation such as x-rays, is used in tumour therapy. The basis for this kind of treatment is to increase the frequency of chromosomal breakage (to induce lethallity) in the cells undergoing mitosis compared to the other cells in the interphase. Tumour cells contain a large number of mitotic cells than most normal tissues and this way a more mitotic cells are destroyed than normal cells. Because, all tumour cells are not in the mitotic division at the same time, therefore, x-rays therapy is conducted at the interval of a few days to permit all the tumour cells to enter into mitosis with the idea that almost all the tumour cells are killed. An abnormality in the mitotic cell division is shown in Figure 16.28.

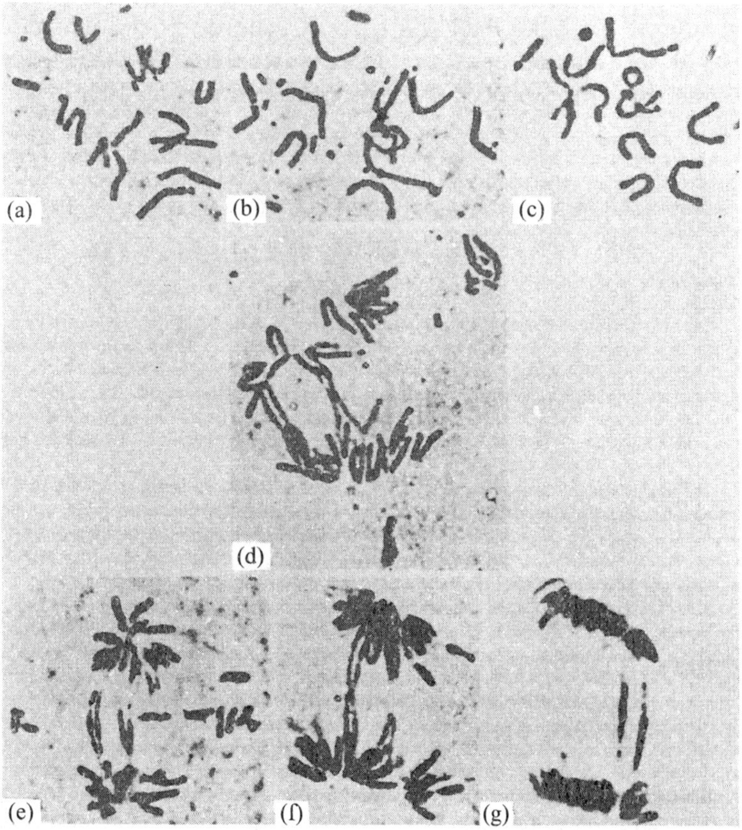

Figure 16.28 Chromosomal irregularity in the meiotic cell after irradiation.

In case of mitosis, metaphase and anaphase are most suited for recording chromosomal aberrations, in the plants. In anaphase, two kinds of cytological aberrations can be determined: bridges and acentric fragment. (1) These bridges may occur as double bridges; paired bridges separate and more or less parallel (2) apparently touching or (3) occasionally interlocked. The acentric fragment may be very minute dots or rod-like.

The reason of the bridge formation is the fusion of two centromere-bearing chromosome fragments. In the anaphase, the diacentric chromatids tend to go to opposite poles as a result bridge is formed which breaks sooner or later.

Meiotic aberration: For the detection of chromosal abnormality, diakinesis and metaphase I are the best stages. Reciprocal translocation can be easily identified by ring or chain formation at metaphase I.

CHEMICAL AND PHYSICAL MUTAGENS

X-rays treatment causes random breaks over all the chromosomes; a regional specificity has been observed in the chemicals. Non-random 'localized' distribution of breaks caused by chemical mutagens is due to the fact that some definite chromosome of the genome and some definite 'sites' of a chromosome are more liable for a break than rest of the chromosome. Normally, heterochromatin of the chromosomes are more susceptible to break than euchromatin. Figure 16.29 shows the x-rays induced chromosomal aberration in *Vicia faba*.

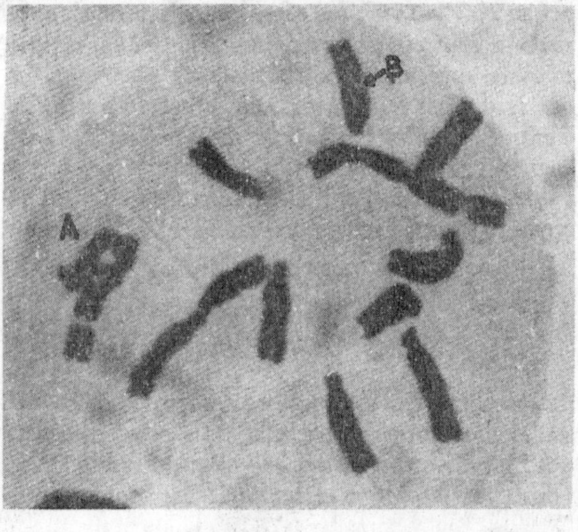

Figure 16.29 X-ray induced chromosomal aberrations in root tips of *Vicia faba*.
(A) An interarm interchange in one of the long arms.
(B) Achromatic legion.

Chemical mutagens induce more chromosome fragments and fewer chromosome recombinations than ionizing radiations. EMS produces more fragments per bridge than x-rays. Chemical mutagens produce relatively more intrachromosomal changes than interchromosomal.

In Figure 16.30, base analogues, chemical mutagens inducing chromosomal aberrations have been shown.

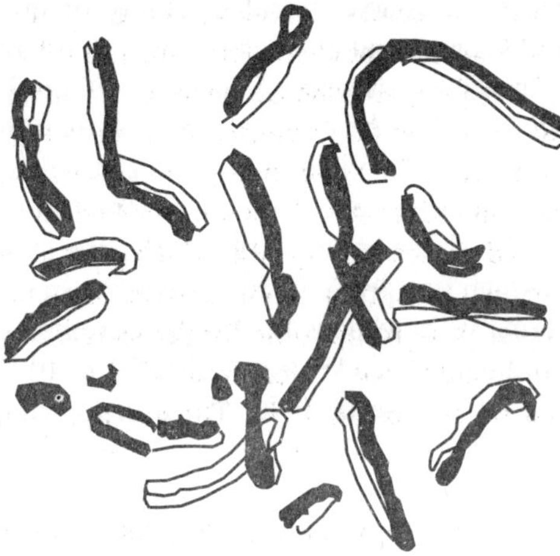

Figure 16.30 Base analogues were found to induce chromosomal aberrational such as segmental exchange between chromatids.

Rate of Induced Mutation

It is not correct to say that mutation rate is correlated with mutagenic dose rates. The mutagenic effects of physical and chemical agents depend on:

(i) Kinds of mutagens
(ii) Biological characteristics of the organism

Kinds of mutagens

Take for example, physical mutagen, the penetrating ability of x-rays and gamma rays in the tissues are deeper than ultraviolet light and beta particles. In addition, these physical mutagens have a narrow range of optimum dose to induce mutation, below that they show indetectable effect and beyond that they show lethal effect. For example, author himself worked on gamma rays and observed that 25–30 kR dose is most effective dose for inducing mutation in rice (dry seed). Gamma rays at 35 kR and above induces seedling lethality (Ram, 1972). In wheat 20–25 kR doses may be effective for inducing mutations of economic interest; 30 kR and above may induce seedling mortality (Ram, 2009 unpublished data). Above dose rates are for dry seed of rice (paddy) and wheat; for water-soaked seeds, the dose rate must be reduced. This is because, in the irradiation process, positive radical ions and free electrons are produced.

ECONOMIC UTILITY OF INDUCED MUTATION

By induced mutations, plant characters can be improved which are discussed here.

Yield Component Characters

Yield itself has no genes, therefore, in literature mention of improvement of 'yielding ability' through radiation is not correct. Of course "yielding ability of the crop varieties have been improved by improving the yield component characters" such as tillering ability, grain per spike/pod and 1000 grain weight. Therefore, statements made by *Gregory* (1957), *Gustafson* (1965) *and Swaminathan* (1968) *that yield can be increased by mutation breeding technique are not correct statements.* If there is no gene for yield, how yield can be increased. Yield is the result of constillation of yield component characters. Author developed an induced mutant (1M-16) by gamma-rays treatment of the seed of rice variety IR8, which gave about 10 per cent higher yield than the parent variety with medium fine grain. In this case increase in grain per spike and tillers/plant with synchrony in flowering were responsible for the increased yield. This induced mutant was tested under all India Coordinated Rice Project from 1971 to 1973 in the name of IET 2570 and this variety was identified for release in 1974. The average yields are given (Table 16.2) (Ram, 1972).

TABLE 16.2 Grain yield (kg/ha) of rice obtained in different crop seasons at different locations

Variety/strain	*Rainy season 1969, Pusa*	*Summer 1969, Pusa, Bihar*	*Rainy season, 1970*		
			Pusa, Bihar	*Faizabad, UP*	*Patna, Bihar*
IR8-1M-16	6700	5332	7225	5024	4612
IR8-1M-1	6025	4666	6475	4364	4513
IR8	5552	4750	–	–	–
Jaya	–	–	6783	4764	4685

Flowering Time

The mutant M-16 (IET, 2570) was about 7–8 days earlier in maturity than the parent variety. Early maturing mutants have been obtained by Teden (1957) too. Early maturing mutants have been developed in wheat also by author. (unpublished yet.)

Adaptability

Wider geographic adaptability has become an important topic of plant improvement. Some practical examples of this trait exist and have been explored by Mari in 1974 and Sigurbjornsson in 1975.

Plant Type and Growth

Dwarf mutants are very frequently observed in different plant species, when treated by physical or chemical mutagenic agents (Zschege and Haaring, 1962; Hansel, et al., 1963). Author induced dwarfness in the height grouped as semidwarf, dwarf and triple dwarf in rice variety–NP 9 (a low-land paddy), NP 49 and NP114 (both scented mid-late group) grown in eastern India (Ram, 1972). The reduction in height was due to reduction in the internode length.

Figure 16.31 shows four categories of dwarfness in the plant height induced by gamma-rays treatment of seed of NP 9.

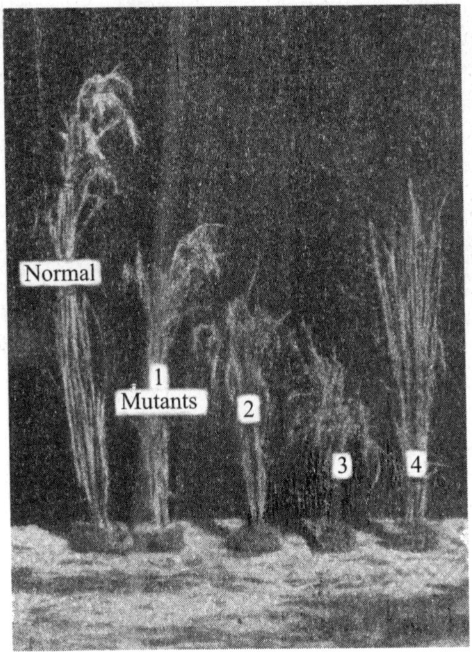

Figure 16.31 Short-culm mutations with gradual reduction in height in rice. Extreme right (4) is a 'grass-clump' (Ram, 1974).

Increasing the Tillering Ability

Through induced mutation, we can increase the tillering ability of the cereal crops. Author increased the tillering ability in wheat using gamma rays 20 kR dose. Figure 16.32 shows increase in tillering of wheat at 20 kR gamma rays (Ram 2009, unpublished)

Resistance to Lodging due to Stem Breakage

Numerous lodging-resistant mutant lines have been evolved through the use of physical and chemical mutagens. For instance, lodging-resistant mutants were obtained by Frey (1955) in oats (*Avena sativa*), Gustafson and coworkers (1963) in barley (*Hordeum vulgare*), Vettel (1957) in rice (*Oryza sativa*), Kawai, Sato and Masima (1961) in bread wheat (*Triticum aestivum*), Natrajan, Sikka and Swaminathan (1950) in bread wheat (*Triticum aestivum*) and durum wheat (*Triticum durum*) in India.

Ram in 1972 has evolved lodging-resistant induced mutant in NP 9; a rice variety popularly grown in water-logging condition in eastern UP and Bihar (India) in low-land area as winter rice. Figure 16.33 shows a comparative view of lodging-resistant culture (right) and lodging-susceptible parent variety NP 9 (left) evolved through the gamma rays treatment of seed. The mutant had thick stiffs straw resistant to stem breakage.

Effect of Gamma rays on tillers/plant in wheat (*Triticum aestivum* L.)

Figure 16.32 Increase in tillering ability in wheat at 20 kR gamma rays.

Figure 16.33 Tall gamma rays induced lodging-resistant mutant of 'NP 9' rice (Ram 1974).

Disease Resistance

Disease resistance is key factor in increasing the yield by keeping the crop safe from epiphytotics of diseases attack. The potential of induced mutation to improve disease resistance in crop plants was reviewed in detail by International Atomic Energy Agency (IAEA, 1971) in Viena. Freisleben and Lein (1942) were the first to isolate disease-resistant induced mutants in Haisa barley against three races of powdery mildew in Germany through x-ray treatment. Thereafter, a number of report were published since 1950 with positive results to evolve mutants resistant to *Erysiphae graminis* in barley (Hagberg, Persson and Wiberg, 1963 and Gaul, 1964); in wheat against *Puccinia graminis* (Sunderman, 1955, Favret, 1959, Bhatia, Swaminathan and Gupta, 1961), *Puccinia recondita* (Konzak, 1956 a, b Favret and Cenoz, 1963), *Puccinia graminis* and in durum wheat against *Puccinia graminis* (Konzak 1954, 1959(a), (b)), oat (against *Puccinia graminis*) in rice against bacterial leaf blight (Ram, 1974), and in corn against *Sclerospora maydis* (Respondek, 1958).

FR-13B is flood-resistant rice grown in the flood-prone areas of eastern India (Bihar). This variety was found susceptible to bacterial leaf blight and tungro virus. Ram (1974) induced resistance to bacterial leaf blight and tungro virus diseases in rice by gamma rays treatment (Figure 16.34).

Figure 16.34 Left: Variety FR-13A (deep-water paddy) severely attacked by tungro virus, bacterial leaf blight and blast. Right: a semi-dwarf mutant culture 11 of NP 9, showing resistance to bacterial leaf blight, and tungro virus 9.

Induced Translocation

Sears (1956) and Knot (1968) induced leaf rust resistance in bread wheat by inducing translocation in an intergeneric cross *"Triticum aestivum* × *Agropyron elongatum*:" and *T. aestivum* × *Aegilops umbullata*.

Triticum aestivum	×	*Agropyron elongatum*
(Chinese spring)		
(AA genome)		(DD genome)

Full scheme is given in Figures 16.35 (Knot, 1968) and 16.36 (Sears, 1956).

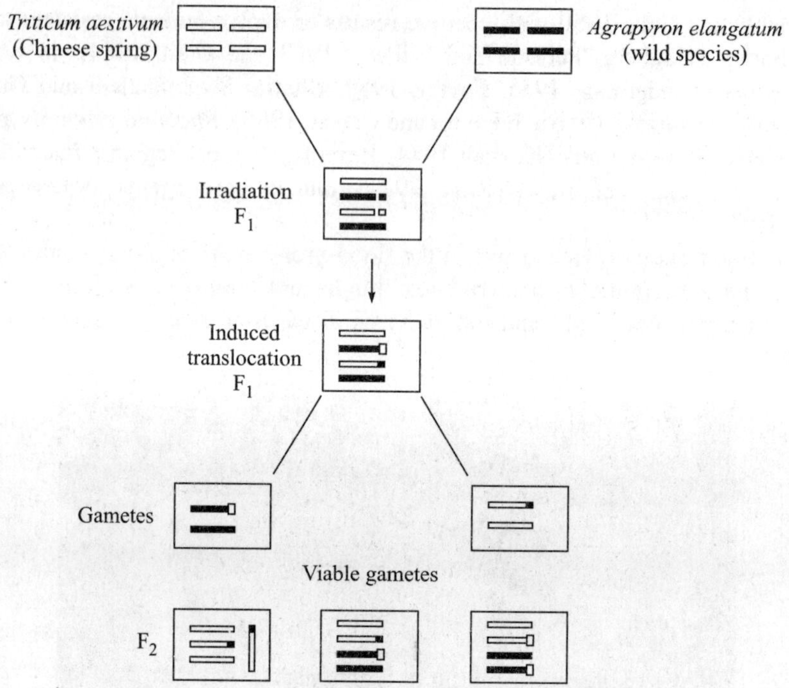

Figure 16.35 Procedure for transfer of stem rust resistance inducing translocation by irradiation. (Knot 1968)

Similarly, by induced mutation, grain quality such as fineness in grain can be improved in rice. Ram (1972) induced fineness in grain quality in the rice variety IR8 through gamma rays; of course, grain was not too fine but its glutinous (sticky) character was removed. Similarly, in NP_9, also fine grain mutants were obtained.

Removal of Toxic Characters

Melilotus alba (white clover) contains a glucoside of the o-oxycinnamic acid, which is converted into coumarins and toxic dicoumarol, which gives bitter taste and harmful to the animals. By treating the seed with mutagenic chemicals and physical mutagens (by ioning radiation) Scheibe and Holsmann (1958) and Mike (1962) were able to select non-bitter mutants.

Brassica napus, *B. compestris*, *B. juncea* and *Eruca sativa* are widely cultivated throughout world, because besides oil, their meal cake and green vegetative parts are also sources of valuable protein. However, presence of glucosinolates which contains degradating products toxic to animals and their milk to the man; Josefisson and Appelqvist, 1968, Konda and Dawney, 1969 and Robbelen, 1973, lowered down the glucosinolate by conventional breeding but this need use of mutation breeding technique to remove the toxic substance from these important crops.

NEUROTOXIN-FREE MUTATION

Lathyrus sativa (Khesari) is consumed in some parts of India (East U.P., Bihar and Madhya Pradesh) as pulse. But the seed contains neurotoxic substance, which causes lathyrism (paralytic

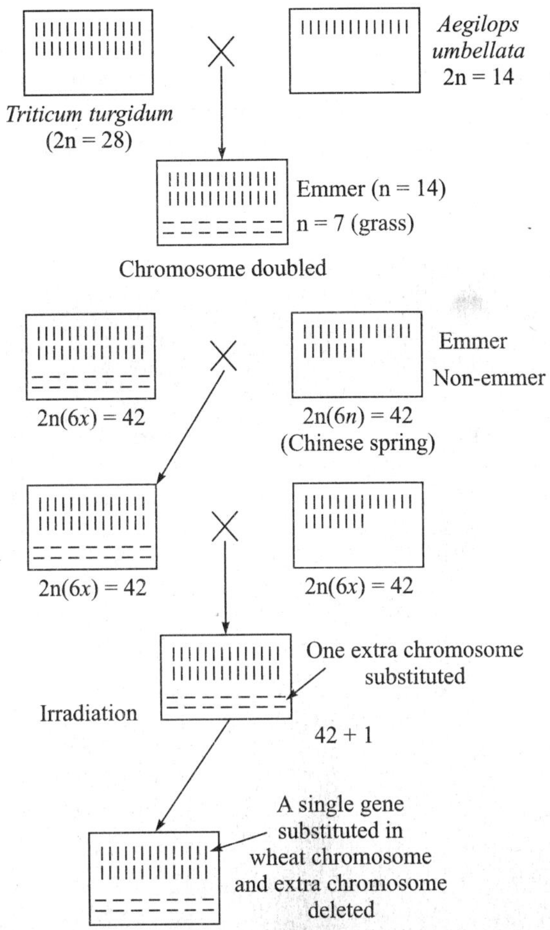

Figure 16.36 Procedure for alien genes transfer in Chinese spring wheat from *Aegilops umbellata* governing resistance to leaf rust (Sears 1956).

effect) on the consumers, if taken for longer time. Natrajan and Gopalan (1968), were able to isolate neurotoxic-free mutants of *Lathyrus sativa* (Swaminathan, 1969).

ADVERSE EFFECTS OF MUTATION BREEDING

Both physical and chemical mutagens have shown adverse effects on the reproductive capacity of the plants, such as severe stunting or growth prevention which inhibit flowering; flowers with aborted reproductive system; reproductive system present but pollen grains are aborted; fertilization occurs but embryos are aborted or seed looks normal but does not germinate or if seed germinates, seedling dies. Albino mutants die within 15–20 days after emergence.

In many cases, root growth is adversely affected which is also a cause of poor seedling survival.

Sterility

The mutagenic induced sterility is caused due to chromosomal abberation, gene mutation, cytoplasmic mutation and physiological effects. Spikelet sterility is observed in M_1 generation also. M_1 sterility could be recorded in terms of aborted pollen grain and poor seed setting. Figure 16.37 shows spikelet sterility in barley (*Hordeum vulgare*). Sterility is caused due to translocation heterozygotes (Burnhim, White and Liver, 1954, Tsuchia, 1960).

Figure 16.37 Sterility in barley spikelet after irradiation.

Grassy Mutants

1. As shown in Figure 16.31(3) author himself recorded grassy mutant which did not flower at all. Similar grassy mutant have been obtained by author in wheat in M_2 generation treated with gamma rays (30 kR).
2. In NP 9, one mutant line was found with spready plant type and shattering type of grains, i.e. reverting to wild-type character such as spready plant habit and shattering grain habit. This was a wrong signal in the sense that in case of transgenic rice, wheat, maize, pulses and vegetable crops, mutation in the alien regulator gene or structural gene may revert back cultivated plants to wild type, i.e. grassy type and shattering seed habit. It is well known that almost all the cultivated crops have originated from wild progenitors.

UTILIZATION OF MUTATIONS

Induced Mutations are not always of direct use as a variety, but it could be utilized as good source of germ plasma for the chromosome reconstruction and transfer of gene etc. Sears (1956, b) transferred leaf rust resistance (*Puccinia recondita*) from *Aegilops umbullata* to bread wheat. *Aegilops umbullata* (2n = 28) was first crossed with *Triticum dicoccoides* (2n = 28) from which an amphidiploids (2n = 42) was obtained which was crossed to common wheat, (2n = 42). This hybrid was back crossed twice to wheat and during this process only plant possessing leaf rust resistant were selected. This resulted isolation for resistant plants having all 42 chromosome plus one extra chromosome from *Aegiolops* carrying resistant (R) gene. However, this extra-chromosome had deleterious effect on vigour and also on fertility. Fortunately, in the progeny of this plant, an individual with an iso-chromosome with duplication of the chromosome arm carrying 'R, gene was found. This plant was irradiated with x-ray before meiosis and the pollen grains were used to pollinate untreated plants. In one case, 40 plants having resistance to leaf rust were obtained wherein a DNA segment transferred carrying R gene was found which was not having deleterious effects on growth and fertility.

By similar irradiation technique, stem rest resistance was transferred from *Agropyran elongatum* to the bread wheat (Elliot, 1957, Knot, 1968) and *Aegilops intermeduim* (Wienhuce, 1966); leaf rust resistance from *Agropyron elongatum* to wheat (*Triticum aestivum*) (Sharma and Knot, 1966) and *Secale cereale* (Driscoll and Jensen, 1964).

To Break the Linkage

Recombination is a key process for increasing the genetic variability but due to close linkages, the recombination of genes are limited to the chiasma formation frequency, However, irradiation of F_1 plants, especially during premeiotic stages has been found to change the recombination frequency by enhancing the genetic variability in F_2 population.

Male Sterility

Male sterility is characterized by non-functional male gametes. Male sterility has been found an important tool in genetic advancement of yield by hybrid seed production. Male sterility are of three types: genetic, cytoplasmic and cytoplasmic genetic. Mutation for male sterility can be induced by irradiation technique.

To Overcome Cross-breeding Barriers

Asynchronous flowering, prefertilization barriers (failure of pollen germination or inadequate pollen tube grown) and postfertilization barriers (zygotic lethality and malfunction of the endosperm) can be overcome by radiation treatment (Tanake, 1957, Swaminathan and Murty, 1959).

Utilization of Induced Mutations as Germplasm

It has been earlier explained that induced mutations are not always of direct use as a variety. But the economic mutants could be utilized in the conventional breeding programmes to evolve high-yielding/lodging-resistant/disease- and pest-resistant crop varieties. Plant breeders at Indian Agriculture Research Institute, New Delhi in 1960s had developed a dwarf mutant in NP_{12} rice by gamma-rays treatment. This dwarf rice mutant was not high yielding. Ram (1974) crossed NP_{12} (induced dwarf) with IR8 and in F_2 segregation, a number of semidwarf ($d_1 d_1$), dwarf ($d_2 d_2$), triple dwarf ($d_3 d_3$) and double dwarf ($d_4 d_4$) plant types appeared. The dwarf types ($d_2 d_2$) were suited to the mid-land (irrigated); semi-tall ($d_1 d_1$) were suited to low land (rainfed, water-logged) conditions. However, triple dwarf ($d_3 d_3$) and quadruple dwarf ($d_4 d_4$) were not economically useful (Figure 16.38).

IR8 NP_{12} dwarf mutant

F_1

Segregation in F_2 generation

Figure 16.38 By crossing dwarf mutant of rice NP_{12} with IR8 dwarf variety gave a range of dwarf types classed as semidwarf ($D_1D_1D_2D_2D_3D_3d_4d_4$) dwarf ($D_1D_1D_2D_2d_3d_3d_4d_4$) triple dwarf ($D_1D_1d_2d_2d_3d_3d_4d_4$) and double dwarf ($d_1d_1d_2d_2d_3d_3d_4d_4$) (Ram 1974).

PARAMUTATION

One of the basic assumptions of the Mendelian genetics has been that alleles in the heterozygotes remain unblended state. This contention of Mendelian law of segregation was seriously questioned by Brink for a particular locus in maize. The locus 'R, bears a number of alleles connected to anthocyanin pigment production of which R^r is the self-colour, R^{st} is for the stippled effect and R^{mb} is for morbled effect.

Normally, one dose of R^r produces dark mottling and two and three doses produce full colour. Brink noticed that allele R^r when associated in heterozygous condition with alleles R^{st} and R^{mb}, alleles produce considerable lighter colour pigment than before. He named this effect as paramutation and alleles R^{st} and R^{mb} were said as mutagenic because this change is passed on to the future generations.

Paramutation were also reported in *Drosophila*, *Oenothera*, tomatoes and fern. Perhaps transposable elements are involved in paramutation, though no clear evidence is available.

POSITION EFFECT

In *Drosophila*, mutation effects associated with genes located within or near heterochromatic regions can be followed closely. For example, a white mottled effect is caused either by placing a heterochromatic region near white$^+$ locus in X chromosome or by placing the 'white' locus near a heterochromatic regions. The females that bear heterochromatic regions next to this locus when heterozygous for 'white' allele *w* (*w*/*w*+) will show variegated eye colour of red and white patches like Ac dosage in corn. An increase of heterochromatin in *Drosophila* through the addition of Y chromosome seems to inhibit variegation in these cases. On the other hand, instances are known in which genes normally located in the heterochromatic areas will when transferred to the euchromatic areas produces variegated effects.

In Mice, Russell found that heterochromatinization of one of the two female X chromosomes normally occurring in mammal will cause variegation in autosomal genes, when X chromosome is translocated to one of the autosomes. This is also called 'variegated position effects'. A variegated position effect may not be mutation in the sense that there is physical change in gene; but rather regulatory change that affect the ability of a gene to make or distribute its product. Duplication of 16A segment of chromosome in *Drosophila* producing 'bar' eye is best example of position effect.

In *Drosophila*, it was observed that removal of heterochromatin from genes proximity or transfer of mutated gene to other position results in re-establishment of the original non-mutated function of gene. It is, therefore, difficult to say whether an observed gene mutation is due to physical or bio-chemical change in the gene itself or due to position effect caused by change in a neighbouring gene.

MUTATION AND CANCER

One of the most exciting discoveries made in the recent past is that of *oncogene* or gene that causes cancer in the human beings and in the animals. These genes were found in the viruses called *retroviruses*. It was found that when virus carrying *oncogene* invades a cell, the growth pattern of the cell get changed and unchecked proliferation of cell starts piling up undifferential cells. It is called cancerous (neoplastic) type of growth. The *oncogene* acquires cancerous property due to mutation occurring in the process of becoming from *protooncogenes* to *oncogenes*.

REVIEW QUESTIONS

1. Define mutation. Describe the different types of mutations occurring in nature.
2. Distinguish between macro- and microgene mutations. What are their causes.
3. Identify the point mutations (i) transition, (ii) transversion, and (iii) reading frame-shift, occurring in DNA and RNA in the following:

   (a) A to G (b) C to T
   (c) C to G (d) T to A
   (c) UAU ACC UAU to UAU AAC CUA (f) UUG CUA AUA to UUG CUG LUA.
4. Both lethal and viable mutations occur when *Drosophila* is subjected to irradiation. Outline a method to detect (a) sex-linked lethals, (b) viable mutation in irradiated fly.
5. If a single, short-legged sheep occurs in a flock. Suggest experiments to determine whether it is the result of mutation or environmental modification. If it is a mutation, how it is possible to determine whether it is dominant or recessive.
6. A single-gene change blocks the normal conversion of phenylalanine to tyrocine. Is mutant gene expected to be pleiotrophic, explain.
7. Why is a sickle cell anaemia called a molecular disease?
8. Why does frequency of chromosome breaks induced by x-rays vary with the total dosage and not with the rate at which it is delivered?
9. One person was in an accident and received 50 roentgen (r) of x-ray at one time. Another person received 5r in each of 20 treatments. Assuming no intensity effect, what proportionate number of mutations would be expected in each person?
10. How does ultraviolet light produce reversible mutation?
11. How does nitrous acid induce mutation? What kinds of changes it can bring in DNA and in RNA of viruses, if treated.
12. How induced mutation is used to improve crops of economic importance.

REFERENCES

Astvet, K. (1965): Induced mutation in barley, *Meld Nory Landbr Hoisk*, **44**, 1–31.

Bekedams (1961): X-rays induced mutation in rice: effect of ionizing radiation on seed, *Proceed. Symp. Karisrube*, 1960, IAEA Viena, 47–58.

Bhatia, C.R. and M.S. Swaminathan (1962): Induced polygenic variability in bread wheat and its bearing on selection procedure, *Z. Pf*1, **48,** 317–36.

Bozzini, A., G.T. Scarascia-Mugnzza (1967): A dominant short strawmutation induced by thermal neutrons in *Durum* wheat, *Wheat Information Service*, No. 23–24.

Brenner, S., L. Barnett, F.C.H. Crick, and A. Orgel (1961): The theory of mutagenesis, *J. Mol. Bio.*, **3,** 121.

Brink, A. (1960): Paramutation and chromosome organization, *Q. Per. Biol.*, **35,** 120–137.

Bruns, G.W. and P.J. Bruttino (1989): *Science of Genetics*, Macmillan, New York.

Chandrasekharan, S.N. and Parthasarthy, G.V. (1953): *Cytogenetics and Plant Breeding*, P. Vardhachary and Co, Madras.

Darlington, I.D. and Upcott, M.B. (1941): Spontaneous chromosome changes, *J. Gene*, **41,** 297–338.

Elliott, F.C. (1957): X-ray induced mutation in *Agropyron-stem* rust resistance to common wheat, *J. Hered.*, **48,** 77–81.

Favret, E.A. and Ryan, G.S. (1966): New useful mutations in plant breeding, *Mutation in Plant Breeding* I (*Proc. Panel Viena*), IAEA Viena, **49,** 61.

Gardener, C.O. (1968): Mutation studies involving quantitative traits, *Gamma Field Symp.*, **7,** 57–77.

Gaul, H. (1964): Mutation in plant breeding, *Radi. Bot.*, **4** (3), 155–232.

Gaul, J.V. (1967): Induced polygenic mutation in hexaploid wheat, *Radi. Bot.*, **7,** 321–331.

Gaul, L. (1958): The concept of macro- and micromutation and results of induced micromutation in barley. The use of induced mutation in plant breeding, Ber Gamon Press, Oxford, 408–426.

Gregory, W.C. (1968): A radiation breeding experiment with peanut, *Radi. Bot.*, **8** (3), 81–87.

Gustaffson, A. (1941): Mutation experiment in barley, *Hereditas*, **27,** 225–242.

Hagberg, A. and G. Hagber (1968): Induced translocation in barley, *Mutation in Plant Breeding* II (*Proc. Panel Viena*), AGA Viena, 129–137.

Hartl, D.L. (1991): *Basic Genetics*, Jones and Bartlett, Boston.

Jain, H.K. (2000): *Genetics: Principles, Concept and Implications*, Oxford and IBH, New Delhi.

Muller, H.J. (1927): Artificial transmutation in gene, *Science*, **66,** 34–87.

Ram, M. (2005): *Unfolding Reality of Life—The DNA World*, Dominant Publications, New Delhi.

Ram, M. and F. Zaman (1972): Induction of induced mutation in NP9 rice, *Indian J. Agri. Sciences.*

Ram, M. and F. Zaman (1974): Artificial transmutation of gene in rice, *Indian J. Agri. Sciences.*

Rhoades, M.M. (1945): On genetic control of mutability in maize, *Cold Spring Habor Symp. Quant. Biol.*, **9,** 138–144.

Sears, E.R. (1956): Transfer of leaf rust resistance from *Aegilops umbullata* to wheat, *Brookhaven Symp. Biol. Genet. and Pl. Breed*, **9,** 370–391.

Stadler, L.J. (1928): Mutation in barley induced by x-rays and radium, *Science*, **68,** 186.

Part III

Molecular Genetics

CHAPTER 17

Nucleic Acids (DNA and RNA): Their Structures and Functions

Science of genetics took a new turn by 1940s with the development of molecular genetics, which made it possible to understand the fine structure of gene; how the genetic information was created and how they control the complex physio-biochemical systems of living organisms including human race.

The principle of continuity is the most important biological phenomenon in the evolutionary history of life, i.e. self-duplication of life. The essential part of self-duplication is the replication of genetic information. Since nucleic acid is the carrier of genetic information, the creation of nucleic acid is the first step towards the creation of life as imagined by the molecular geneticists.

DEOXYRIBONUCLEIC ACID (DNA)

It is stated that by the time earth cooled down to permit the formation of permanent water bodies, the entire earth was like crucible filled with a large amount of organic compounds which was continuously increasing through reactions. Many of these organic compounds would have tend to form in colloidal solution, i.e. soups of organic compounds (ammonia, methane, nucleic acids, etc.).

Nucleic acids are present ubiquitously in all living organisms, plants and animals. By the turn of 19th country, A. Kossel (1868), a biochemist discovered nucleic acid which consists of four nitrogenous bases—adenine, guanine (purines), cytosine and thymine (uracil in RNA) a pyrimidines base, a five carbon sugar and phosphoric acid. The chemical formulas of these nitrogen bases are given in Figure 17.1.

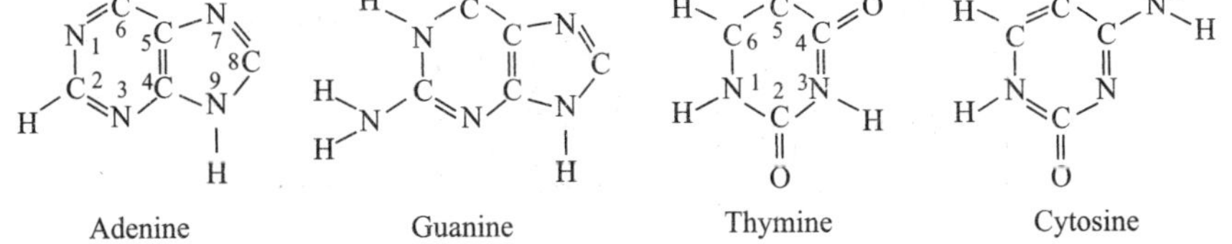

Figure 17.1 The four nitrogen-containing bases in DNA and RNA.

P.H. Levene and W. John in 1920s as mentioned by Stent (1971) reported that there are two kinds of nucleic acids—RNA (ribonucleic acid) and DNA (deoxyribonucleic acid). Both have slight differences in their chemical structures, which are:

1. RNA possesses two hydroxyl groups (–OH), while DNA possesses only one hydroxyl group (–OH) (Figure 17.2). In mRNA uracil replaces thymine. Hence structure of thymine and uracil is also shown below.

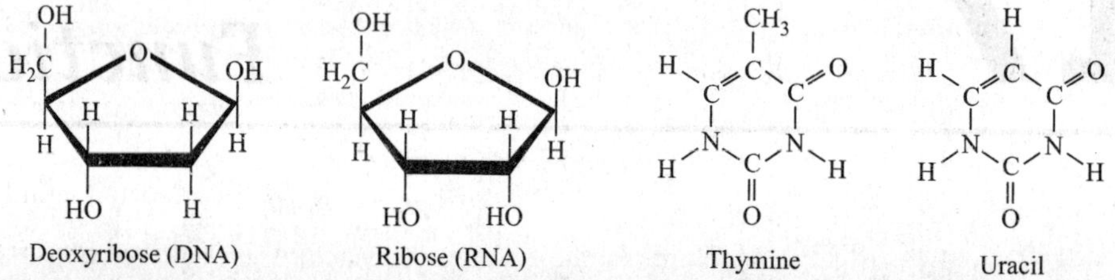

Figure 17.2 Differences between chemical structures of DNA and RNA.

Figure 17.3 shows the chemical building blocks of DNA and RNA which makes the differences very clear.

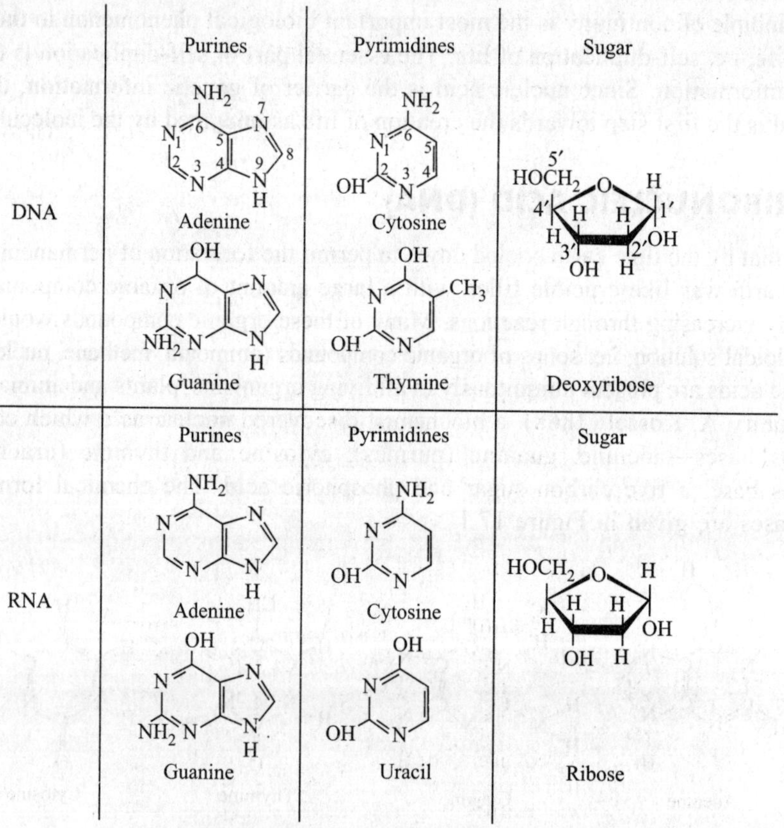

Figure 17.3 The chemical components of the nucleic acids DNA and RNA.

2. The successive ribonucleotides in RNA are joined by 5-3 phosphodiester bonds as in DNA. Both absorb strongly ultraviolet light of wavelength 260 mμ.
3. RNA contains pyrimidine 'uracil' (methyl group), while DNA contains thymine. These two have slight chemical differences (an additional hydroxyl group) on one of the sugar of RNA and an additional methyl group 'uracil' on one of the pyrimidine of DNA. Both have played momentous consequences in the biological evolution by two types of nucleotides.
4. RNA molecules are generally single-stranded polymers of ribonucleotides while DNA is double stranded polymer of deoxyribonucleotides.
5. DNA reacts with the Feulgen reagent to produce an intense coloured stain at its site of location, while RNA is Feulgen negative nucleic acid.
6. RNA is unstable and found in the cytoplasm whereas DNA is stable and found in nucleus.

Small quantity of RNA is found in the nucleus also. DNA is a major constituent of the chromosome in which it is combined with about 3–4 times its own protein. Figure 17.4 shows the manner in which nitrogenous bases, sugar and phosphoric acids are linked together to form a nucleotide. Figure 16.5 shows differences between DNA and RNA nucleotides composition.

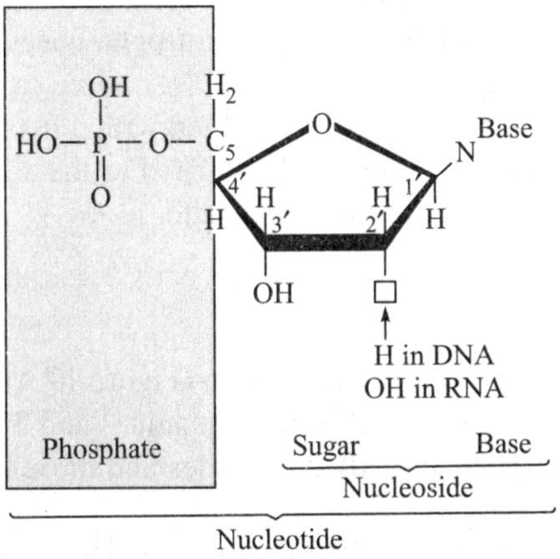

Figure 17.4 A typical nucleotide consisting of three major components (phosphate, sugar and base).

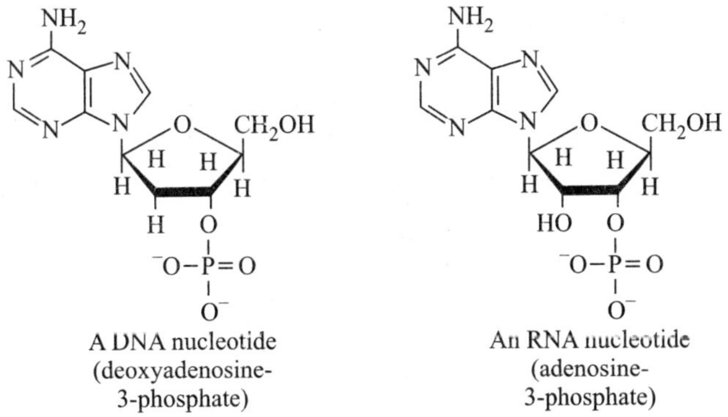

Figure 17.5 Differences between DNA and RNA nucleotides.

The nucleic acid molecules consist of polynucleotides made up of thousands and some times millions of nucleotides in continuous chemical linkage.

Nitrogen Bases

The nitrogen bases are heterocyclic compounds containing nitrogen, carbon and hydrogen in their rings. These bases belong to two groups, viz. purines and pyrimidines.

Purines

The purines possess two carbon–nitrogen rings. They are known as adenine (A) and guanine (G). One ring is six-membered and other five. Adenine contains an amino group (NH_2) at position 6, while in guanine, this position is occupied by keto oxygen (O). Guanine possesses amino group at position 2.

Pyrimidine

Pyrimidine ring is similar to benzene ring except it contains nitrogen in place of carbon at position 1. The two nitrogen bases are cytocine and thymine, which are found in DNA. There is a third pyrimidine nitrogen base 'uracil'. Uracil is found in RNA only.

Both RNA and DNA contain same two types of nitrogen bases adenine and guanine (purine) and pyrimidine (cytosine), but thymine is found in DNA while in RNA, on place of 'thymine', uracil is found. In DNA, each base is chemically linked with a deoxyribose forming a compound called a nucleoside. When a phosphate group is attached to the sugar of each nucleoside called nucleotide. Thus, nucleotide is a nucleoside phosphate.

Nucleosides

A combination of base with sugar is called nucleoside (Figure 17.4). One carbon of pentose sugar is attached to the 4th position of purine (adenine or guanine) and 3rd position of pyrimidine. The nucleosides derived from the ribose are called ribosides and those having deoxyribose are called deoxyriboside.

A nucleotide without phosphate is called nucleoside. The ribonucleoside (those containing ribose) are called adenosine, guanosine, cytidine and uridine; correspondingly, the deoxyribonucleoside (those containing deoxyribose) are called deoxyadenosine, deoxyguanine, deoxcytidine and deoxyuridine. Four different nucleosides are given in Table 17.1.

TABLE 17.1 Four different nucleosides

Nitrogenous base	*Ribonucleoside*	*Deoxyribonucleoside*
Adenine	Adenosine	Deoxyadenosine
Guanine	Guanosine	Deoxyguanosine
Cytosine	Cytidine	Deoxycitidine
Thymine	—	Thymidine (only in deoxyriboside)
Uracil	Uridine	—

Nucleotides

A nucleotide is a compound of nitrogenous base, sugar and phosphoric acid. When nucleosides are attached with phosphoric acid, it is called nucleotide. Phosphoric acid may be attached at either 5′ or 3′ end of pentose sugar and in such condition, nucleotides may termed as either 5′-P-3′-OH nucleotide or 3′-P-5′-OH nucleotide. The chemical formulas of four nucleotides are shown in Figure 17.6.

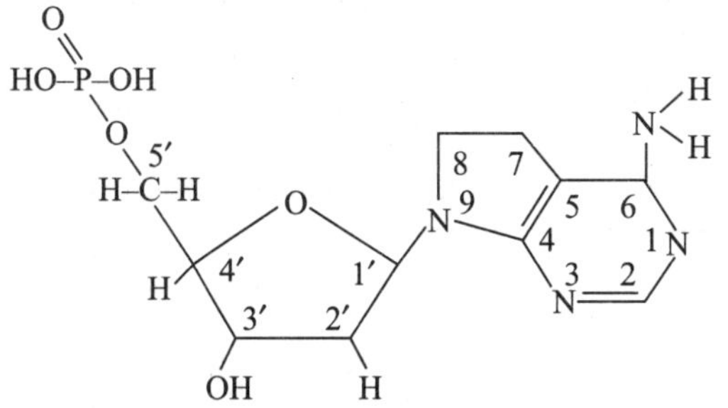

Deoxyadenylic acid (dAMP)

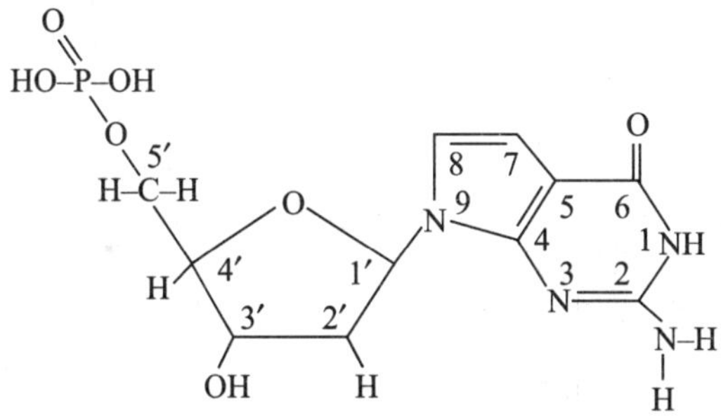

Deoxyguanylic acid (dGMP)

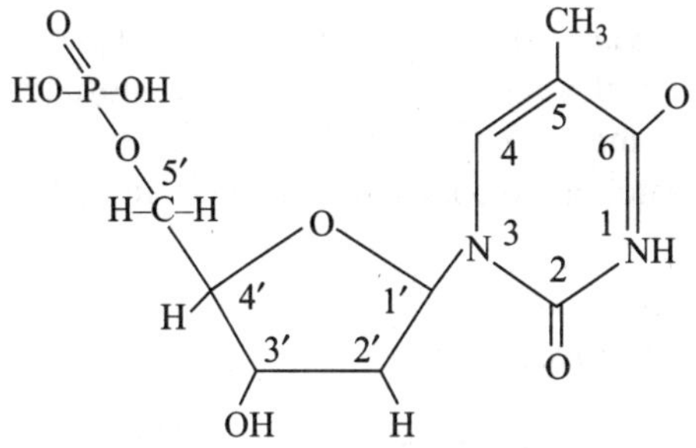

Deoxythymidylic acid (dTMP)

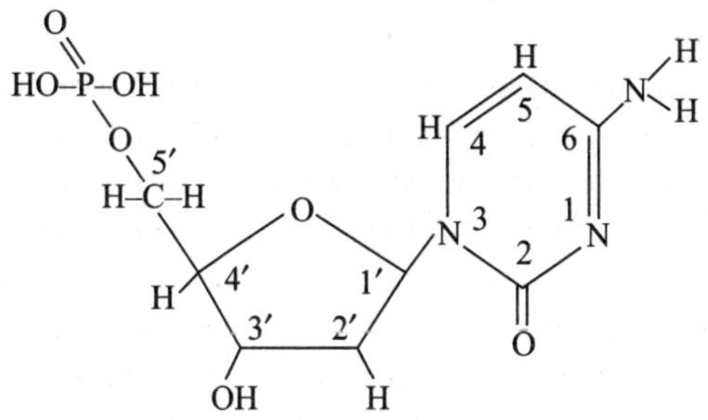

Deoxycytidylic acid (dCMP)

Figure 17.6 Structures of four different nucleotides.

The free nucleotides present in the cell are found in four forms—ATP, GTP, CTP, UTP (ribonucleotide triphosphate) in RNA and dATP, dGTP, dCTP and dTTP (deoxyribonucleotide triphosphate) in DNA. According to DNA nomenclature, DNA may be present in the cell in the form of adenosine monophosphate (AMP), adenosine diphosphate (ADP) and adenosine triphosphate (ATP); guanosine monophosphate (GMP), guanosine diphosphate (GDP), guanosine triphosphate (GTP); cytidine monophosphate (CMP), cytidine diphosphate (CDP) cytidine triphosphate (CTP); thymidine monophosphate (TMP), thymidine diphosphate (TDP) and thymidine triphosphate (TTP) and uridine monophosphate (UMP), uridine diphosphate (UDP) and uridine triphosphate (UTT) in RNA.

Nitrogenous bases and nucleotides found in RNA and DNA are given in Table 17.2.

TABLE 17.2 Nitrogenous bases and nucleotides found in RNA and DNA

Nitrogenous base	*RNA nucleotide*	*DNA nucleotide*
Adenine (A)	Adenylic acid or adenosine monophosphate(AMP)	Deoxyadenylic acid or deoxyadenosine monophosphate(dAMP)
Guanine (G)	Guanylic acid or guanylic monophosphate	Deoxyguanylic acid or deoxyguanosine monophosphate(dGMP)
Cytosine (C)	Cytidilic acid or cytidine monophosphate(CMP)	Deoxycytidilic acid or deoxycytidine monophosphate(dCMP)
Thymine (T)		Thymidylic acid monophosphate(TMP) or deoxythymidine
Uracil (U)	Uridylic acid or uridine monophosphate(UMP)	

Polynucleotides

The nucleotides join together to form a polynucleotide chain through phosphodiester bond, the phosphodiester bond is formed when 3′ OH of a nucleotide reacts with phosphoric acid residue attached to the 5′ C of another nucleotide to make (5′ C–O–P–O–C 3′) bond (Figure 17.7). Since nucleotides occur as triphosphate, it liberates a pyrophosphate (O–P). In this manner, several nucleotide chains are formed.

Molar Proportion in DNA Composition

The tetranucleotide theory of DNA proposed by Leven and others in 1924 had shown that DNA contains approximately equal molar proportion of bases, i.e. adenine (A) = guanine (G) = cytocine (C) = thymine (T). However, Chargaff (1950) based on his compositional studies of DNA disproved the tetranucleotide hypothesis. He observed that despite wide compositional variations, exhibited by different types of DNA, the molar proportion of (A) is always nearly equal to that of (T) and (G) is always equal to that of (C) that is, (A) + (G) = total pyrimidine (T) + (C), which were not far from '1'. This is known as base-equivalence theory of Chargaff. This base equivalence rule was observed in viruses, bacteria and also in plants and animals. The knowledge of (G) + (C) content is of potent value for evolutionary and taxonomical deliberations. The near equality of the DNA base composition informs the close taxonomical relationship of the organisms.

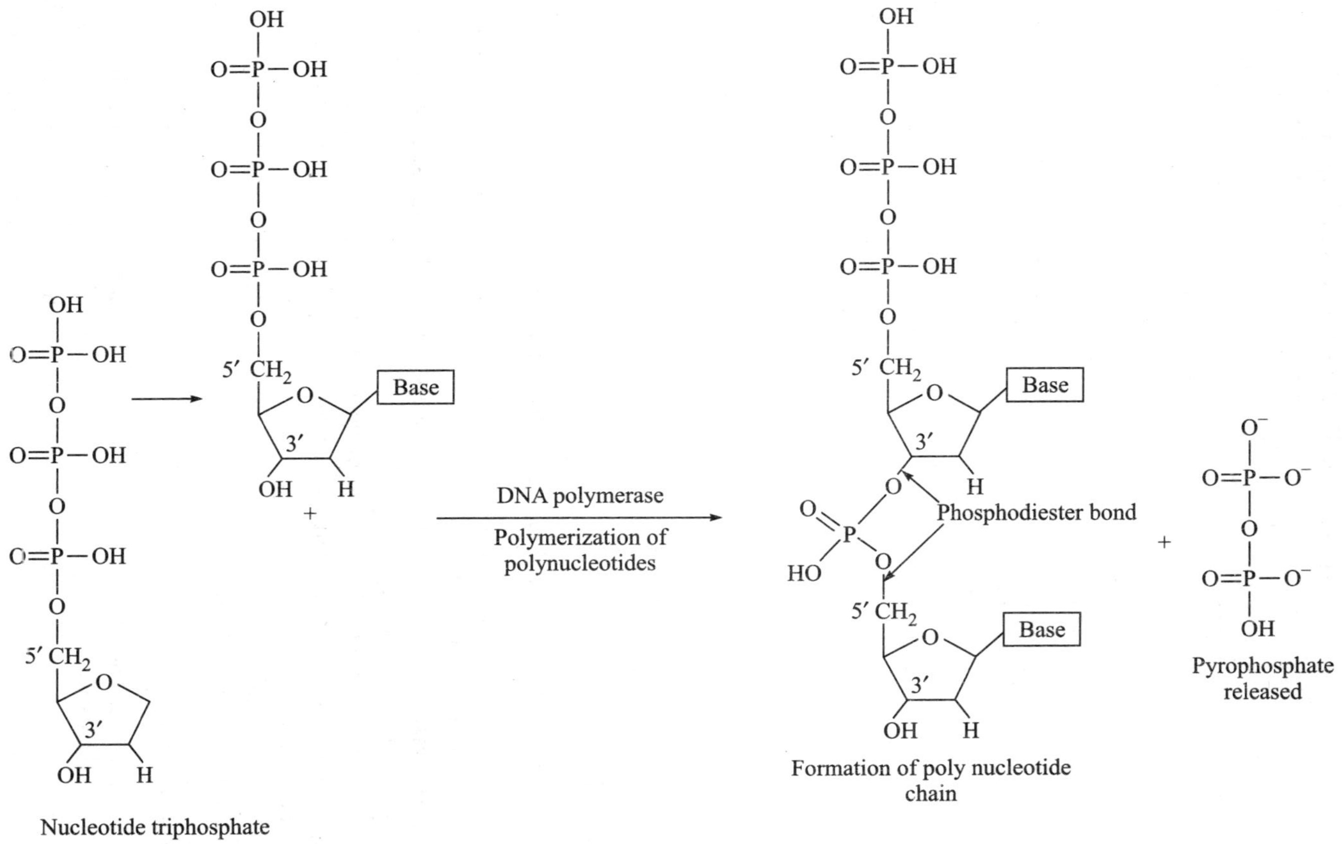

Figure 17.7 Polymerization of nucleotide catalyzed by DNA polymerase enzyme. Nucleotides are added to 3′ end of DNA chain.

The base composition of DNA of different organisms are given in Table 17.3.

TABLE 17.3 Base composition of DNA of different organisms

Organism	*Base (per cent of total)*				*Base composition*
	Adenine	*Thymine*	*Guanine*	*Cystosine*	*GC per cent*
Virus					
Bacteriophage T-7	26.0	26.0	24.0	24.0	48.0
Bacteria					
Clostridium prefenges	36.9	36.3	14.0	12.8	26.8
Streptococcus pneumoniae	30.3	29.5	21.6	18.7	40.3
Escherichia coli	24.7	23.6	26.0	25.7	51.7
Sarcina lutea	13.4	12.4	37.1	37.1	74.5
Fungi					
Saccharomyces cerevisiae	31.7	32.6	18.3	17.4	35.7
Neurospora crassa	23.0	23.3	27.1	27.6	54.7
Higher plant					
Wheat	27.3	27.1	22.7	22.8	45.5
Maize	26.8	27.2	22.8	23.2	46.0
Animal					
Drosophila melanogaster	30.7	29.4	19.6	20.2	39.8
Pig	29.4	29.7	20.5	20.5	39.8
Human being	29.8	31.8	20.2	18.2	39.0

Is DNA Genetic Material

The genetic materials need to possess following characteristics:

1. It must carry all of the information needed to direct the specific organization and metabolic activities of the cell.
2. The genetic material must replicate accurately, so that the information it contains is precisely inherited by the daughter cells.
3. The genetic material must be capable of undergoing occasional mutations.
4. The genetic material must be capable to recombine with the other genetic material to create new variation.
5. The genetic material must have constancy and is inherited from generation to generation without any change or blending.

Experimental evidences

Evidence **I:** Bacterial pneumonia in the mammals including human being is caused by the strains of *Streptococcus pneumoniae*. *Pneumococcus* owes its pathogenecity by forming a polysaccharide capsule around its cell, which protects the bacterium from destruction by the ordinary defence system of the infected organisms/mammals. The capsule polysaccharide is composed of glucose and glucoronic acid. The presence of the capsule gives smooth (S), glittering surfaces to the bacterial colony. However, there are mutants with rough surface (R) of the *Pneumococcus* which have lost capacity to synthesize protective capsule around the bacterial cell wall, therefore, they are not able to cause pneumonia to man or mouse due to presence of inherent defence mechanism. Both S-type and R-type strains of bacteria breed true. Griffith (1928) was the first who discovered that when mice was infected with the mixture of small number of R strains with large number of heat-killed S strain, the mice died. However, when mice were injected with either R strain or with heat-killed S strain, it did not die. He isolated the bacteria present in the heart blood of dead mice and concluded that presence of heat-killed S strain of bacterial DNA, must have *transformed* the living R strain to form capsule by restoring its capacity to infect which it had lost due to mutation. This was termed as *transforming principle* of Griffith.

Avery, Mcleod and McCarty (1944) added pure DNA extracted from heat-killed S strain to the living culture of R strain and obtained a few S smooth strains. On this basis, they concluded that substance responsible for this kind of *transformation* was DNA of the cell and hence DNA is the carrier of hereditary (genetic) information (Figure 17.8). A.E. Mirsky in 1950 found that in the cells of different tissues of same organism, the amount of DNA per haploid chromosome set is *contant*. He concluded that DNA is genetic material on the basis that DNA amount should be constant in per somatic cell and half in the germ cells.

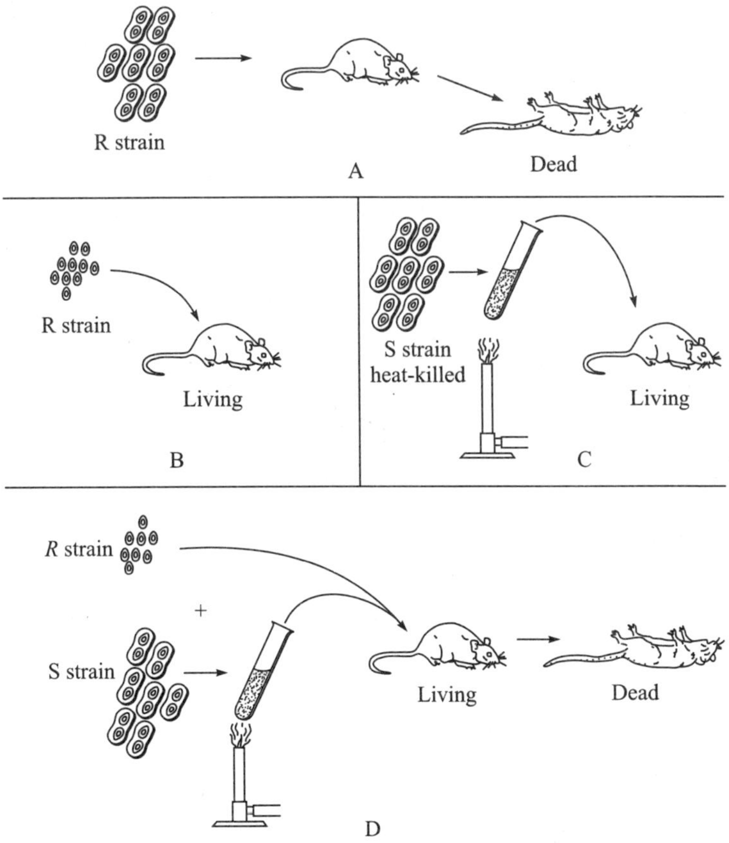

Figure 17.8 Evidence I shows that DNA is genetic material.

By 1950, this theory was suddenly in air and many molecular geneticists embraced it by 1952. The key portion of the theory is, *if DNA is the molecule containing genetic information, then this information can not be carried in any way other than as the specific sequence of four nucleotide bases along the polynucleotide chain. That is, four nucleotide bases of DNA represent the hereditary code, which determines the nature and function performed by every living organisms.* Hershey and Chase (1952) through their studies on *Escherichia coli*, T_2 bacteriophage conferred that DNA is the genetic material of the T_2 bacteriophage by labelling proteins and DNA with radioisotopes of sulphur (S^{35}) and (P^{32}) respectively. Of course in certain viruses, RNA serves as genetic materials such as tobacco mosaic virus (TMV) as reported by Frandel-Corat and Singer in the year 1957.

***Evidence* II:** We know that the entire genome complex (DNA) is enclosed in a nuclear membrane called nucleus. Keck (1963) studied the nuclear control of the metabolic activities in *Acetabularia* algae. It is an excellent algal material to study the aspects of cell differentiation. Apical end of this alga forms a cap, which is determined by the nucleus.

There are two species of *Acetabularia—A. mediterranea* with a round cap and *A. crenulata* with lobbed cap. If a nucleus from *A. mediterranea* is implanted in an unnucleated *A. crenulata*, during second cycle, after nuclear transplant, an intermediary type of cap appears and in the third cycle, the cap would be of *A. mediterranea* type. This experiment and its reciprocal indicate that nucleus produces some diffusible materials which after some time exert a morphogenetic effect at the apical end. If two nucleated cells of different species are grafted together, a hybrid having a cap with intermediate shape is formed. It is presumed that, there may be a kind of mRNA that control the production of specific proteins on the direction of DNA. In presence of ribonucleic acid the unnucleated portion of algae become unable to regenerate the cap. If nucleus is taken out, the algae may die. The findings of both these experiments proved that DNA is the genetic material, i.e. carrier of hereditary information (Figure 17.9).

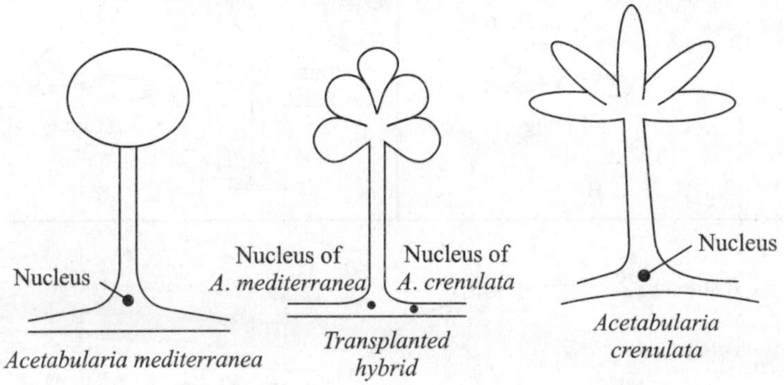

Figure 17.9 Evidence II shows that DNA is genetic material. Role of nucleus in determining the head type in *Acetabularia*.

RIBONUCLEIC ACID (RNA)

DNA is the genetic material of choice. However, some viruses like tobacco mosaic virus (TMV) carry their genes in RNA. RNA like DNA is a long chain of polymers made up of four nucleotides—

adenine, guanine, cytosine and uracil. Uracil replaces thymine as one of the pyramidine base. In the polynucleotide chain with 5′-3′ sugar phosphate bond; 'sugar' molecule in this case is pentose rather than deoxypentose. The essential difference is that RNA is a single-stranded nucleic acid lacking duplex structure of DNA, as a result DNA is more stable because of hydrogen bonding, the sequence of bases. These two rather slight chemical differences (an additional hydroxyl group on the sugar of RNA and an additional methyl group on one of the pyrimidines of DNA) turnout to have momentous consequences in the biological roles played by the two types of polynucleotides. If a sequence of bases happens to find a complementary sequence in the close proximity, a secondary structure may develop between two, through complementary base pairing rule. The transfer RNA (tRNA) seems to have arisen to show complementary base pairing rule. Thus having so much commonality, it is expected that DNA and RNA have their related origin. Alexander Rich in 1962 proposed that in the evolution of life, the parent nucleotide molecule was initially as RNA polymer.

Single-stranded DNA

The evolutionary relationship of DNA and RNA appeared to get the support from the discovery of different kinds of DNA. A single-stranded DNA in phage 'ϕX 174' virus, a minute bacterial virus having 5000 nucleotides have been obtained. Sinsheimer (1959) made the interesting observation that not only the DNA molecule of phage ϕX 174 was much smaller in size, but it was qualitatively different. A striking difference was that it did not obey the Chargaff's law of equivalence of the purine–pyrimidine bases. By infecting *E. coli* cells with N^{15} P^{32} labelled isotopes ϕX 174 virus particles, Sinsheimer collected the proof for his hypothesis and concluded that in the course of its replication, the single-stranded DNA had entered in a double-stranded replicative form. Thus, while ϕX 174 phage DNA is single-stranded (Figure 17.10). However, it cannot escape a duplex structure, if it is to multiply. It also shows that the evolution of a double-stranded structure for most DNA has a sound proposition.

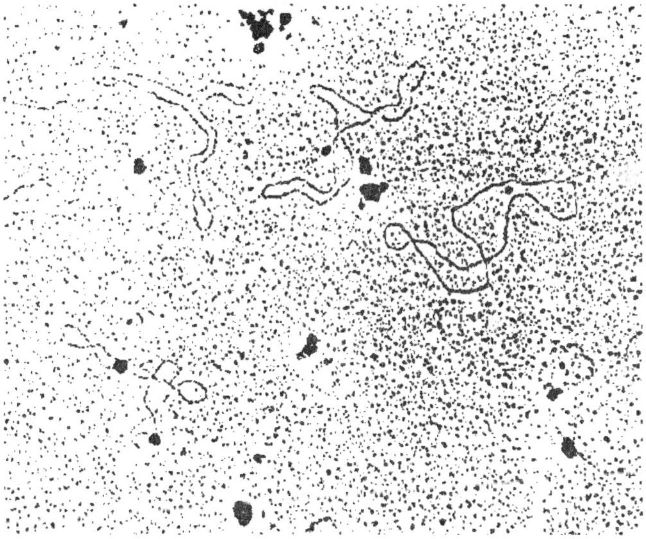

Figure 17.10 A single DNA strand of a typical phage ϕ X 174 (virus) in the replicative form.

STRUCTURE OF DNA

After identification of DNA as hereditary material on account of *Transforming principle*, the determination of molecular structure by the application of x-rays crystallography became matter of interest of the molecular geneticists and biochemists. A team working under the leadership of M.H. Wilkins made an important technical advancement by managing to prepare highly oriented DNA fibres that allowed them to obtain an x-ray diffraction photograph as shown in Figure 17.11. Based on this photograph, Watson and Crick (1953) concluded that (a) DNA polynucleotide chain has helix which makes one complete turn at every 34 Å (Angstrom) along its length, and hence, since the inter-nucleotide distance is 3.4 Å, it contains a stack of ten nucleotide per turn. Based on the density of DNA molecule, James Watson and Francis Crick in 1953 (on the basis of x-ray diffraction photograph of DNA) prepared a hypothetical model of double helical DAN molecule, as shown in Figure 17.12 according to which the helix consists of two intertwined polynucleotide chains, twisted around one another forming a double-stranded helix, in which adenine pairs with the thymine (A–T) and guanine pairs with the cytocine (G–C). Each chain makes one complete turn at every 34 Å distance. The helix is right-handed meaning that each chain follows a clockwise path as it progresses. As stated above, bases are spaced at 3.4 Å, so there are ten bases per helical turn, i.e. each strand has ten bases pairs per turn of the double helix.

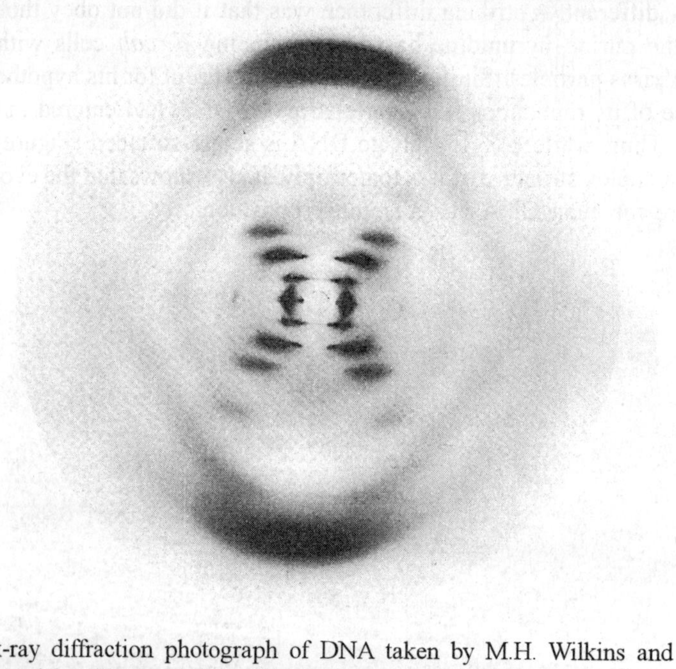

Figure 17.11 An x-ray diffraction photograph of DNA taken by M.H. Wilkins and his team which was used by Watson and Crick to prepare hypothetical structure of DNA (1953).

The DNA double helix is able to assume a number of forms. The most biologically important form is B form. This was the form M.H. Wilkinson observed in x-ray diffraction photograph. It is more hydrated than A form. The A form is more compact with 11 base pair per turn of the

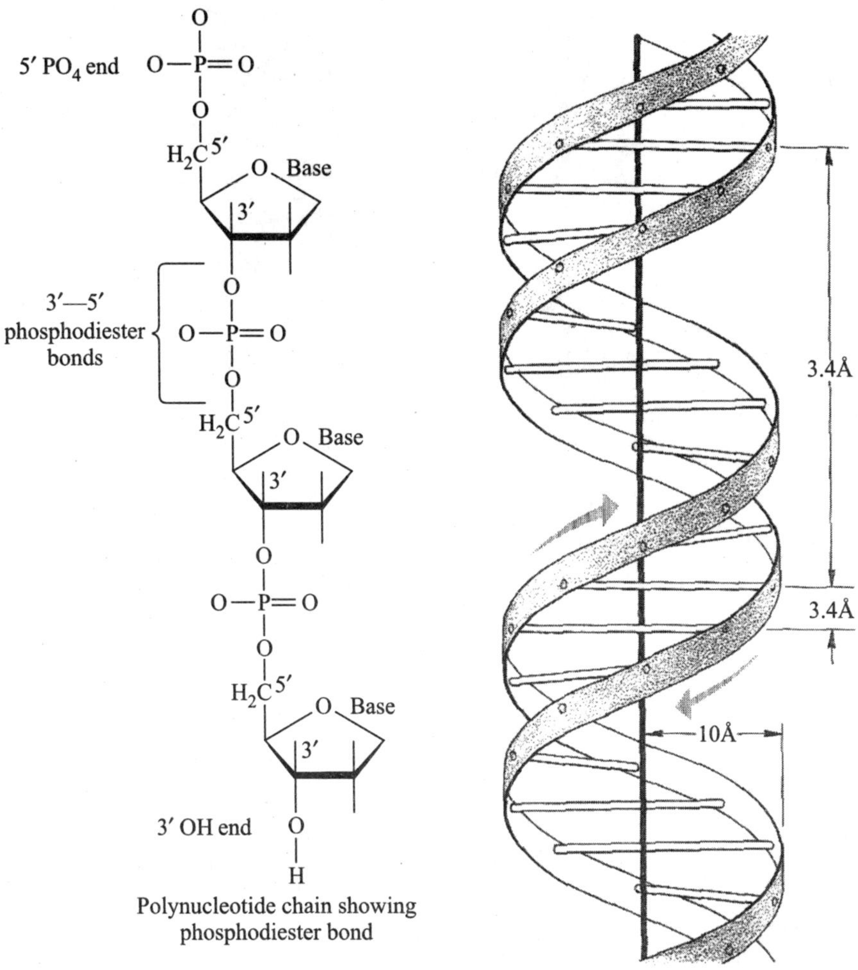

Figure 17.12 Structure of DNA.

helix and 23 Å in diameter. The bases are tilted more in relation to the axis of the helix than in the B form in the cell. In solution of sodium chloride (NaCl), DNA is in a left-handed helical form. This left-handed helical structure is known as Z-DNA because of its zigzag structure. The helix is 18 Å in diameter and there are 12 base pairs per turn (Figure 17.13).

The two chains DNA helix are antiparallel, i.e. have opposite direction so that polynucleotide chain has a direction, which can be seen through the chemical formula in Figure 17.12 or the abbreviated symbolic form as shown in Figure 17.14. Moving along the chain from top to bottom is evidently different from moving along it from bottom to top. The plane of the deoxyribose sugar forming with its esteried phosphate, the backbone of the polynucleotide chain is parallel to the main axis and hence perpendicular to the plane of the base rings. The base point inward, and at each nucleotide residue, the two polynucleotide chains held together by hydrogen bonds formed between a purine on one chain and pyrimidine on the other chain (Figure 17.15).

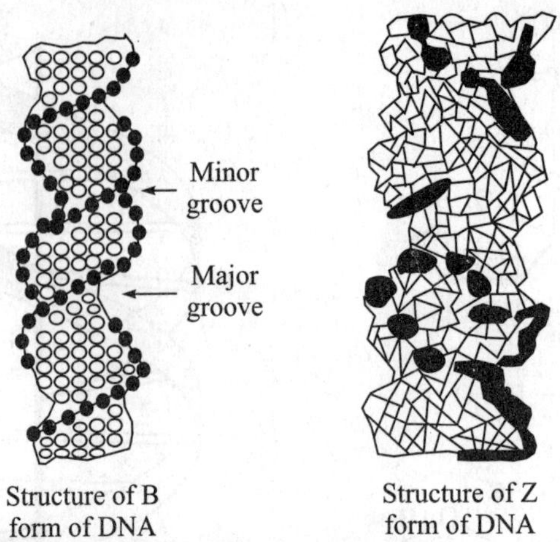

Figure 17.13 A comparative structure of B-form DNA and Z-form DNA.

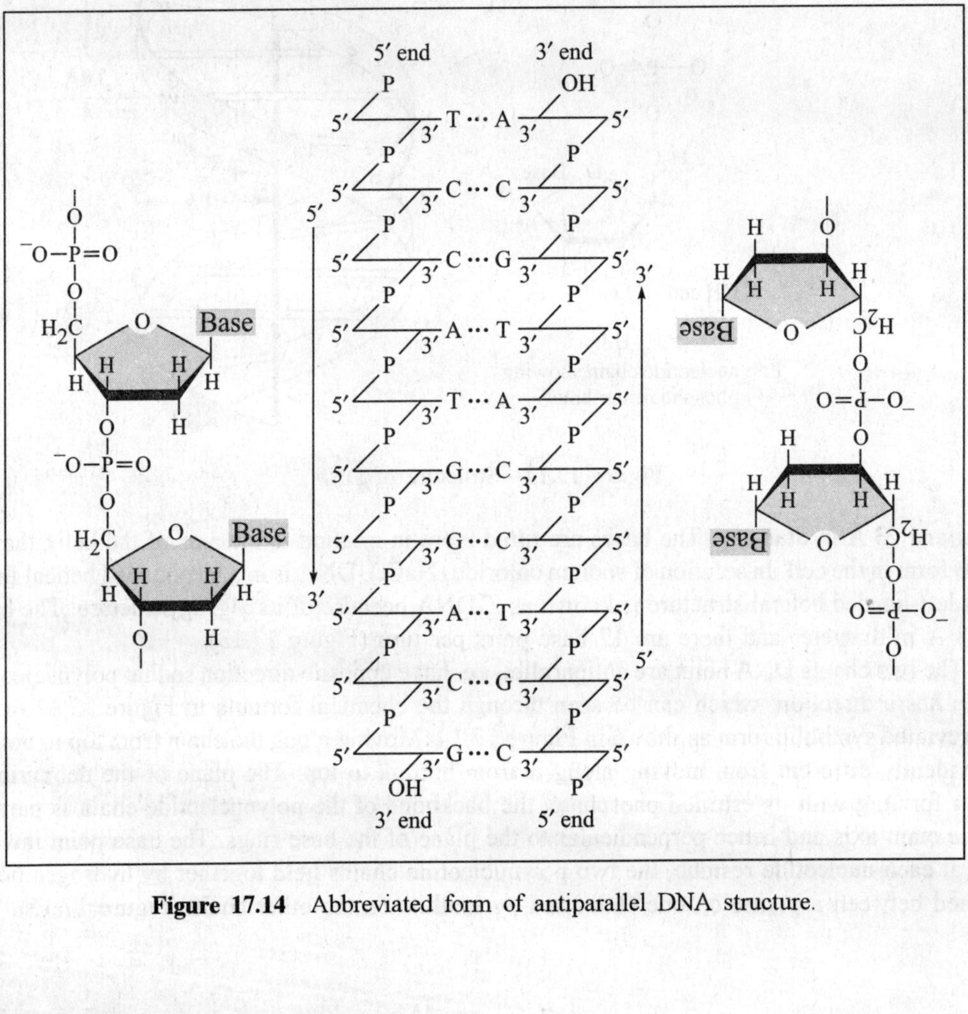

Figure 17.14 Abbreviated form of antiparallel DNA structure.

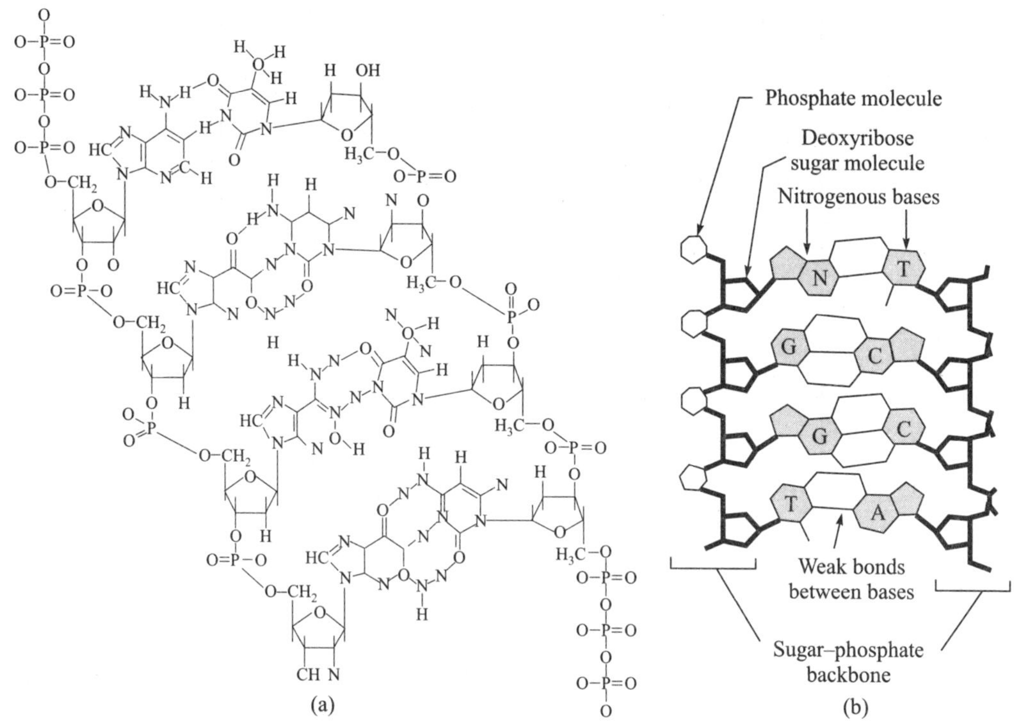

Figure 17.15 (a) Molecular structure of DNA, (b) A slightly more simplified diagram is shown in B. Other forms of DNA in expanded form.

Complementary Pairing Rule of Nitrogen Bases in DNA

The most important feature of this model is the manner in which the hydrogen bonds are formed (Figure 17.16), i.e. opposite to every adenine base on one helix (chain) there is a thymine base at the corresponding position on the other chain and a similar complementary relation holds between the guanine and cytocine bases of the two chains. In this way, it is possible to form the hydrogen bonds and at the same time maintain the constant diameter of the double helix. On the other hand, the opposite arrangement of adenine and cytocine or guanine and thymine would satisfy the spatial requirement but neither would allow the formation of hydrogen bond at proper distance. (Figure 17.17).

Just as guanine pairs with cytocine in the two strands of the DNA double helix: so can a guanine pairs with a cytocine in a ribonucleic acid (RNA) strand (and vice versa). Similarly, adenine of DNA can pair with uracil in RNA. This forms a DNA–RNA hybrid. The basic hydrogen bonding rule that applies to these DNA–RNA hybrid molecules are those that apply to the two strands of DNA itself. The two RNA strands can also form a duplex structure by similar rule. The discovery of structure of DNA by Watson and Crick (1953) opened the gate to understand the secret of life. According to them, genes are located like beads of pearl on the DNA, which serves as biochemical page on which genetic information of life is written. The structure of DNA is like a long helical spiraling strand of complex molecules resembling ladder that is twisted like twist of liquorice.

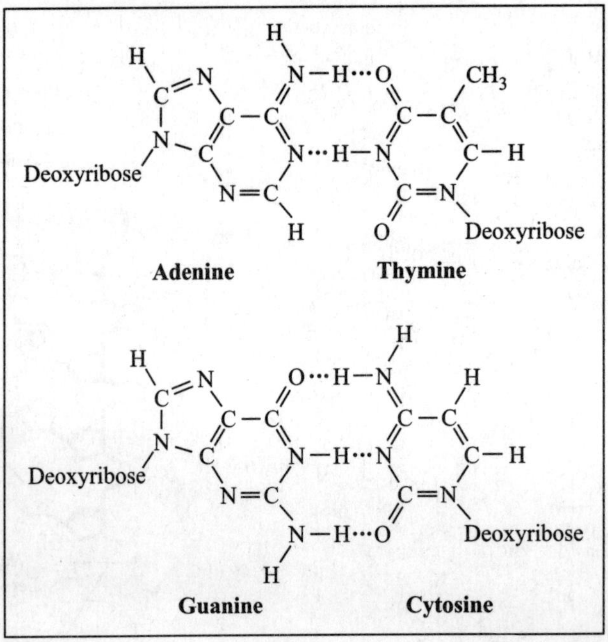

Figure 17.16 Hydrogen bonds formed between purine and pyrimidine chains of two DNA strands have maintained a constant diameter between two double helix of DNA.

Limitation of DNA Structure

1. The DNA structure (double helix) prepared by Watson and Crick (1953) is purely a hypothetical model. They prepared DNA model based on Chargaff's theory of equivalence of the nitrogen bases (wherein concentration of thymine is always equal to adenine and the concentration of cytosine is equal to guanine) and Wilkins and Franklins x-ray diffraction data and inference drawn from the model building. Har Govind Khurana after deciphering the genetic codes, said that his findings are correct, if DNA structure prepared by Watson and Crick (1953) is correct. In science hypothesis has no place until and unless it is confirmed with visible evidences (data). This need further investigation otherwise entire molecular genetic theory is baseless.
2. Later studies indicated that there are different kinds of DNA, viz. A, B, C and Z types. Watson and Crick's DNA model is B type, which can be observed through x-ray diffraction when humidity is 92 per cent and salt concentration is high. This model is right-handed double helix with 10 pairs of bases per turn and each pair is spaced at 3.4 Å apart on DNA backbone. On the other hand, A is found when humidity of the sample is 75 per cent. Here there are 11 pairs of bases per turn, each pair is spaced at 2.3 Å apart on DNA backbone. C type DNA is observed when humidity of DNA is below 66 per cent. It possesses 9.3 pairs of bases per turn and each pair is spaced on 3.1 Å apart. Z type DNA is left-handed and possesses 12 of base pairs per turn, each pair is spaced 4.0 Å apart on the DNA frame. So much variation in DNA is against the genetic stability of a trait it governs.

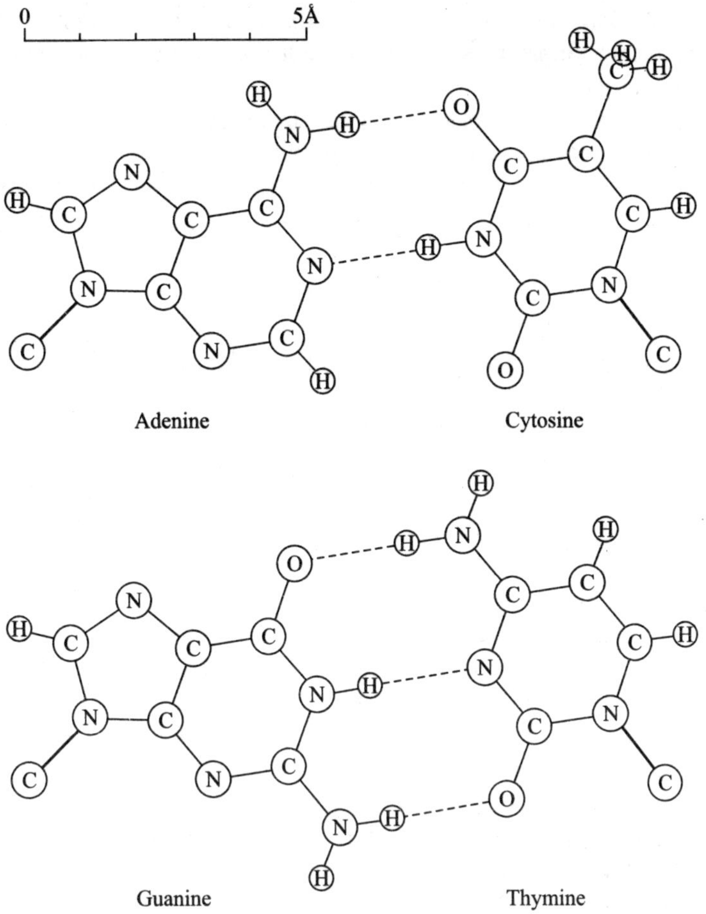

Figure 17.17 Illegitimate pairing of bases as H. bonds are not located at proper distances. In complementary base pairing rule, where adenine pairs with thymine and cytocine pair with guanine in DNA, hydrogen bonds must maintain a constant diameter between two double helix.

Among these, B type DNA structure is stable but variability is very high. This is against the principle of genetic materials because model lack fixity. This needs further investigations so that future generation should not be misled.

DNA performs three main functions in the cells:

(i) Replication to distribute equally the DNA contents among the dividing cells.
(ii) Transcription of genetic message.
(iii) Translation of genetic message.

DNA Replication

DNA replication takes place in the cell in different phases:

1. Initiation of replication occurs at a specific point on the chromosome. This point is known as origin. Sometimes endonuclease make the incision at the origin from where replication starts.
2. Unwinding of two complementary DNA stands takes place with the help of specific protein. This protein also unites both the strands. Hence, it is called DNA binding protein.
3. Formation of RNA primer is essential for initiation of DNA synthesis. RNA primer is synthesized by DNA template near the origin with the help of specific type of RNA polymerase.
4. Synthesis of DNA starts on RNA primer. Deoxyribose nucleotide are added to the 3-OH end of the RNA primer one by one.

RNA primer is degraded by DNA polymerase I. This enzyme also catalyzes the synthesis of short DNA segment to replace the primer.

The primary function of DNA replication is to duplicate the base sequence of the parental molecules. The specificity of the base pairing, i.e. adenine pairs with thymine and guanine pairs with cytosine (in RNA, thymine is replaced with uracil) is maintained throughout the replication mechanism. In this mechanism nucleotides are added one by one to the end of growing strand catalyzed by DNA polymerase. The sequence of bases in each new strand (daughter strand) is complementary to the base sequences in the parental strand. This is also called the 'complementary base pairing rule', as suggested by Waston and Crick (1953), where the DNA molecule could replicate itself directly by having each chain to serve as a *template* for the formation of its own complementary copy. In this act of replication they thought, the two strands of the double helix separate and each purine (A–G) pyrimidine and (C–T) base attracts a complementary free nucleotide available for polymerization in the cell and holds it in place by means of specific hydrogen bond as given in Figure 17.18.

The most important points in the history of the evolution of life are two, *first*, self-reproduction, i.e. autocatalytic property of DNA molecule to regenerate one exact copy of every gene (DNA fragment) with which every daughter cell is to be endowed. In other words, how parental DNA material is replicated to give rise to two identical DNA molecules of purine—pyrimidine base sequences. *Second*, heterocatalytic property of DNA material, i.e. how DNA presides over the biosynthesis of polypeptides whose primary structure is inscribed in DNA or more specifically, how the purine—pyrimidine base sequence of hereditary (polynucleotide chain) is translated into the amino acid sequence that it specifies via genetic code.

Once held in the place on the parental *template* chain, the free nucleotides are sewed together by formation of the phosphate diester bond that link adjacent deoxyribose residue, forming a new polynucleotide molecule of predetermined base sequence polydiester bond (Figure 17.19). Thus, after growth of complementary replica chain has taken place along the full length of both parental polynucleotide chains; the two DNA chains generated so have identical sequences of four bases and hence have the same information content as the parental DNA double helix. This has been achieved through autocatalytic property of DNA molecules.

When the time comes for the next replication cycle in the two daughter cells, that were endowed with the two-daughter DNA molecules generated by the first cycle. The four polynucleotide chains of both daughter DNA molecules again separate. Each chain then acts as a template for the growth of another polynucleotide chain, then by generating four grand daughter DNA double helix each identical in base sequences to grand parental double helix and provide the

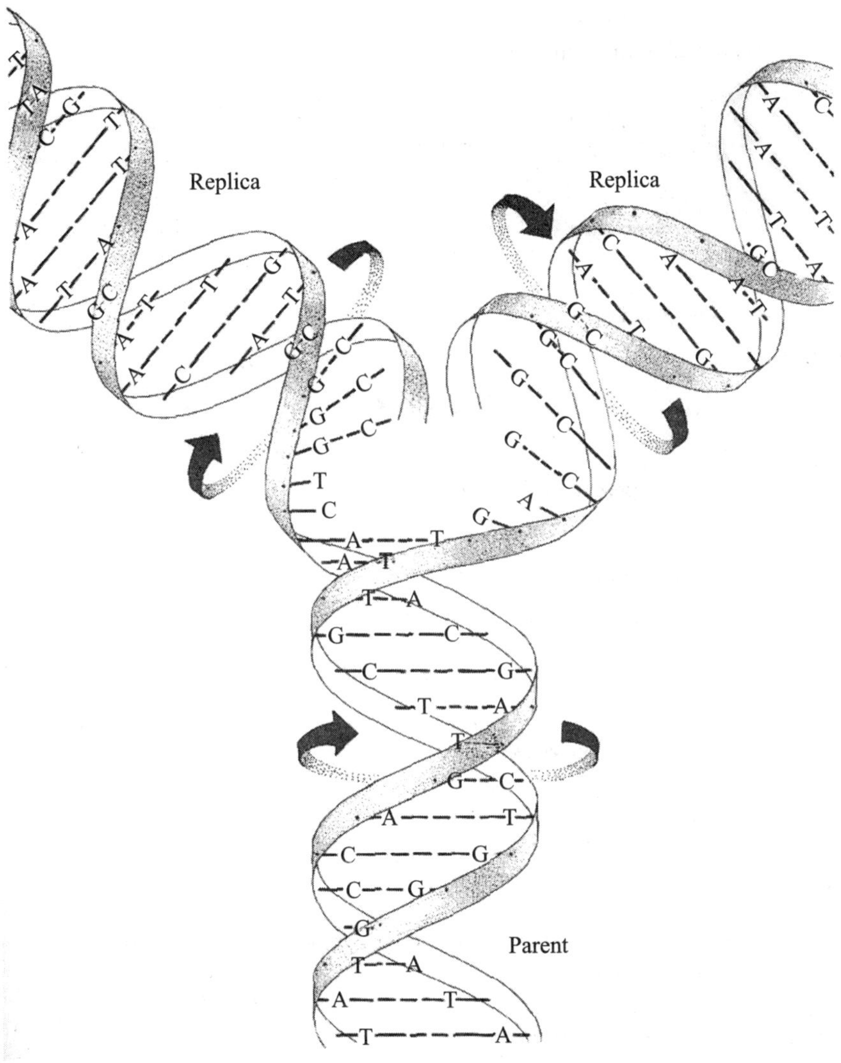

Figure 17.18 DNA replication according to the hypothesis of Watson and Crick (1953).

genetic information (endowment) for four grand daughter cells. Thus, it was realized that there was equal distribution of parental material of DNA molecule over the daughter cells. This equal distribution of the substance of the parental DNA molecule to daughter molecules could have not been possible, had the discovery of the complementary double helix was not proposed by Watson and Crick (1953); because, in the act of serving as template for the growth of new replica chain, the two parental polynucleotide chains separate, the genes of two parent double helix become equally distributed over the two daughter molecules of the first replication cycle.

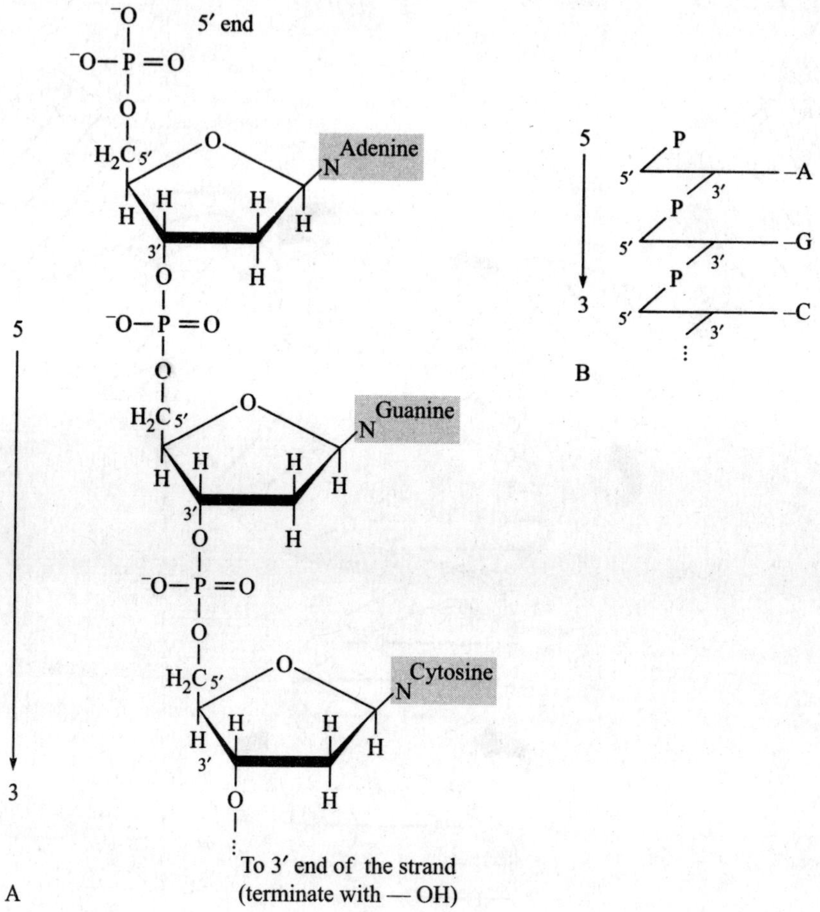

Figure 17.19 Manner in which n-bases, sugar and phosphates bonds linked together to form the polynucleotide chain.

Semiconservative DNA replication

As indicated above, each parental DNA serves as a *template* for new one to be formed and the newly synthesized DNA strand is hydrogen bonded to its parental DNA. As the replication proceeds, the parental DNA double helix continue to unwind and then rewind again with new double helices, each double helix consist of one newly formed DNA strand and one original parental strand. This is called semi-conservation system of DNA replication.

Figure 17.20(a) shows semiconservation, (b) dispersive and (c) conservative view of DNA replication. Among three systems of DNA replication, only semi-conservative is considered valid.

Semiconservative replication of DNA was demonstrated by Mathew, Meselson and Franklin Stahl (1958) using N^{15} isotope of nitrogen in the experiment for physical separation of parental and daughter DNA molecules. The DNA isolated from the bacterium *E. coli* grown in medium containing N^{15} is denser than DNA obtained from bacterium grown in medium containing normal N^{14}. Isotopes N^{15} and N^{14} containing DNA can be separated by means of *Equilibrium centrifugation in a density gradient*.

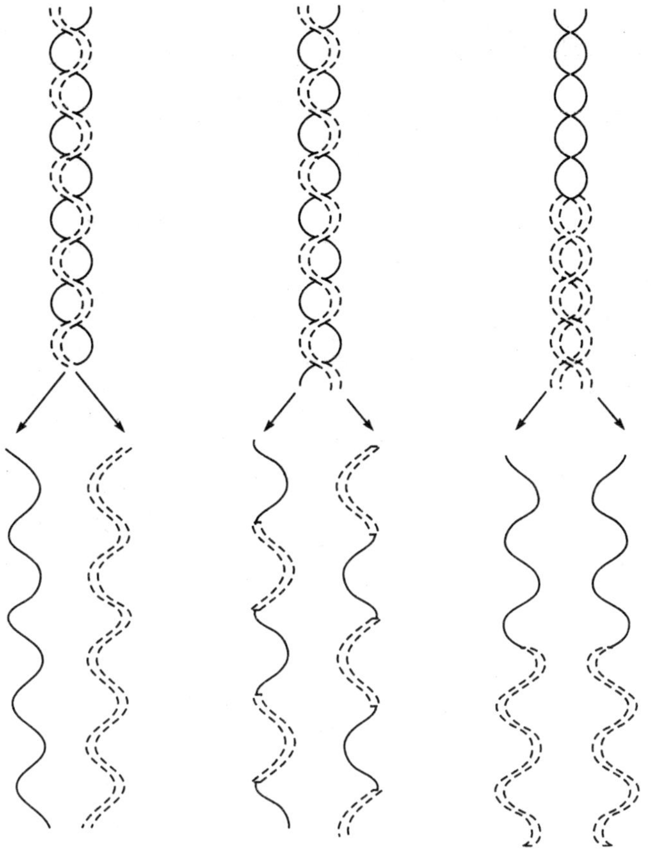

(a) Semiconservative (b) Dispersive (c) End to end conservative

Figure 17.20 Different types of DNA replication.

In dispersive replication, two strands of parental DNA break at several points which make DNA in several pieces. This is not valid concept.

In conservative mode of replications, one molecule has both parental strands and the other contains both newly formed strands. This is also not valid concept.

Replication of circular DNA molecule

Some scientists write geometry of DNA replication. There is no geometry involved in the replication of circular DNA. Through autoradiographic experiments, it was observed that DNA of *E. coli* is circular like θ shape, a Greek word called θ replication. *E. coli* has double helix and has 400,000 turns, must make a full rotation to unwind these gyres.

Replication of circular DNA (Figure 17.21) molecule demonstrates the semiconservative feature of DNA. Current evidence indicates that axis of rotation for unwinding is provided by cut made in the backbone of one of the double helix during replication. These cuts are rapidly repaired during unwinding of the DNA by the enzymes called topoisomerases, which cut and then rapidly repair the cut ends.

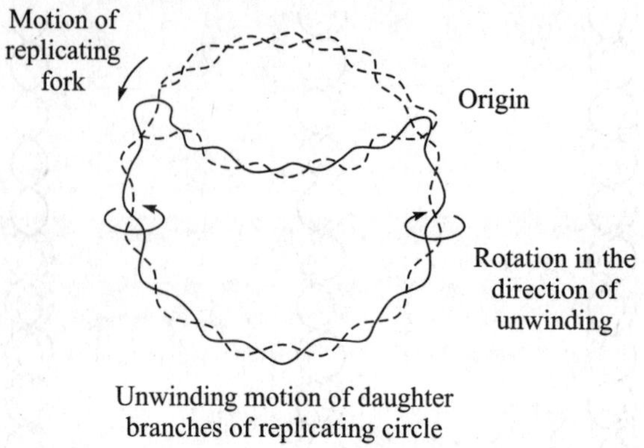

Figure 17.21 Replication of circular DNA molecule.

The position at which DNA replication starts is called *replication origin* and the region in which parental strands are separated and the new strand starts synthesis is called *replication fork*. The replication begins at a unique position and DNA always replicates bi-directionally in the bacteria as well as in higher organisms (Figure 17.22).

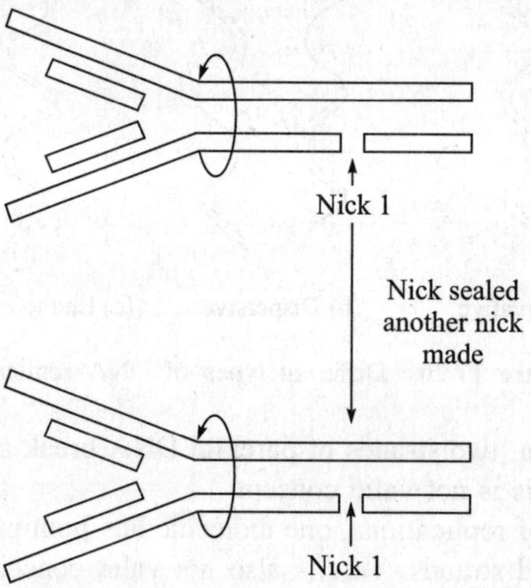

Figure 17.22 Mechanism by which a single-strand breaks (nick).

Multiple replication

The single DNA molecule in the chromosome of higher organism is linear and replicates bi-directionally. Replication starts at a number of sites in the DNA and form multiple *loops* in the molecule. The multiple initiation of DNA replication takes relatively less time than normal unidirectional replication. *E. coli* DNA molecule contains 4×10^6 nucleotides and takes 40 minutes to replicate. Figure 17.23 shows multiple replications in DNA of in *E. coli*.

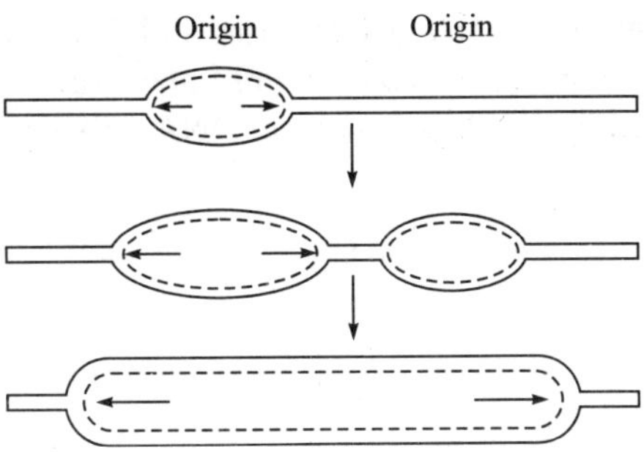

Figure 17.23 Multiple replication of DNA take place in both directions in *E. coli*.

Rolling circle replication

Some circular DNA molecules of bacterial and eukaryotic viruses do not replicate forming θ shaped intermediates. Replication starts with a cut in DNA double-stranded circle at a specific sugar phosphate bond making chemically two distinct ends—a 3′ end (at which nucleotide has a free 3′-OH) and a 5′ end (at which the nucleotide has a free 5′-P group). New DNA is synthesized by addition of deoxyribonucleotides to the 3′ end pushing away in 5′ end from the circle. As replication proceeds along the circle, the 5′ end rolls out as tail of the growing length of DNA (Figure 17.24).

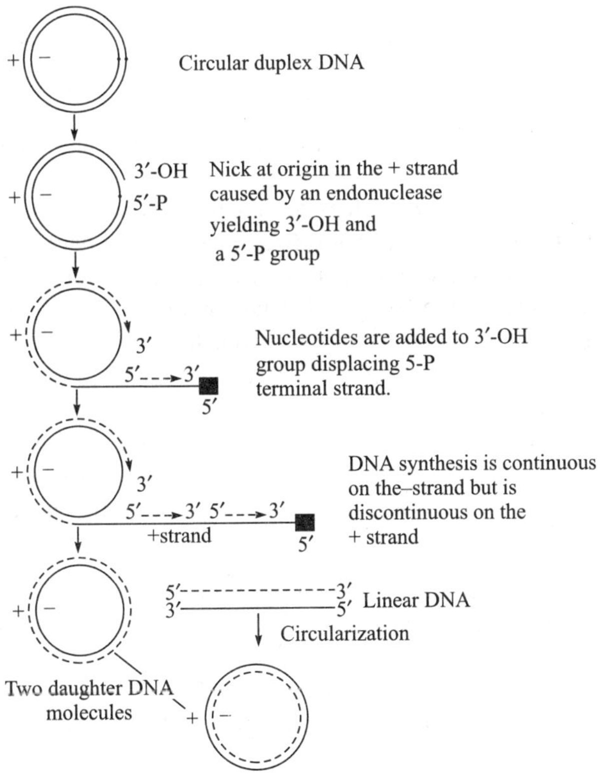

Figure 17.24 Rolling circular DNA replication in *E. coli*.

For replication of DNA, two DNA polymerases Pol I and Pol III in *E. coli* have been determined. Pol III is a major replication enzyme. Pol I plays secondary role in DNA replication. Eukaryotic DNA is replicated by DNA polymerase. DNA polymerase opens the two strands as shown in (Figure 17.25).

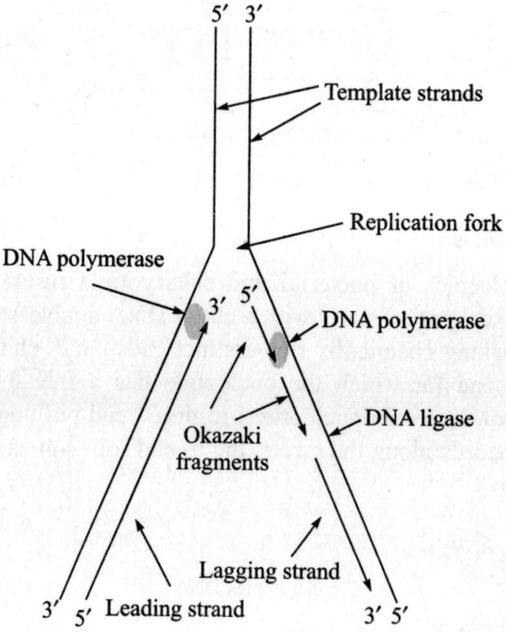

Figure 17.25 DNA polymerase opens the two strands of DNA and prepares for replication.

Besides their ability to polymerise nucleotides, many DNA polymerases function as nucleases that break phosphodiester bonds in sugar phosphate backbones of nucleic acid chains. These are of two types (i) *exonucleases*, which can remove a nucleotide from end of the chain, (ii) *endonuclease*, which breaks the bond within the chain.

Two DNA polymerases are needed for DNA replication in *E. coli*—the DNA polymerase I and III of *E. coli* functioning as exonuclease at 3′ terminus (a 3′ → 5′ exonuclease activity) which serves as built in mechanism for correcting rare errors in polymerization. Occasionally, polymerase adds an incorrect nucleotide, which does not 'base-pair' to the end of growing DNA chain. The presence of an unpaired nucleotide activates the 3′–5′ exonuclease which performs *proofreading* or *editing function*. This activity excises the by unpaired nucleotide from 3′-OH end of growing chain.

Note: *Genetic significance of proofreading function is that it is an error-correcting mechanism that serves to reduce the frequency of mutation resulting from the incorporation of incorrect nucleotides in DNA replication.*

The DNA polymerases perform two functions; first, synthesis along the template strand only in the 3′–5′ direction; second; it check to initiate new chains, which may create problems in replication.

Discontinuous replication

Replication of daughter strands does not take place continuously in a linear order but new daughter strands are made in short fragments which are then joined together (Figure 16.26).

Fragments in the replication fork

All known DNA polymerase can add nucleotides only to a 3′-OH group. Thus, if both daughter strands grow in the same direction, each growing strand would need a 3′-OH terminus. But, two DNA strands are antiparallel, $5' \rightleftharpoons 3'$ and so only one of the growing strands can terminate in a free 3′-OH group; the other must terminate in a free 5′ end. Within a single-replication fork, both strands grow in the 3′–5′ orientation, which requires that they grow in an opposite direction along the parental strands. One strand of the newly synthesized DNA consists of small precursor fragments (Figure 17.26).

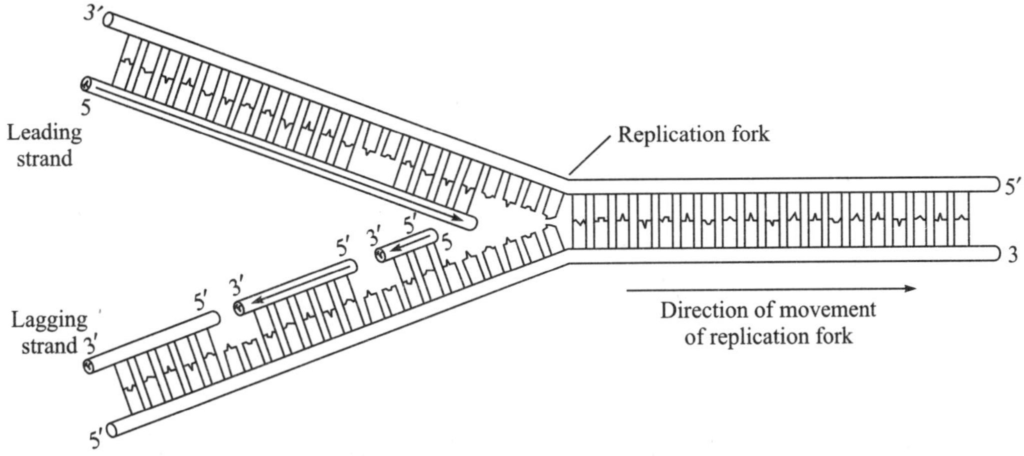

Figure 17.26 Discontinuous replication of DNA and synthesis of short fragments of DNA.

Synthesis of fragmented DNA strand is initiated only at intervals which causes a single-stranded region of the parental strand, present on one side of the replication fork. The 3′-OH terminus of the continuously replicating strand is always ahead of the discontinuous strand, therefore they are called leading strand and lagging strand.

Initiation of DNA replication by RNA primer

DNA polymerase can not initiate synthesis of a new strand; until a free 3′-OH is available. In most organisms, initiation is accomplished by a special type of RNA polymerase. An RNA strand is synthesized by DNA as a template forming a complementary strand in which the bases of DNA strand are paired with those in the RNA. Synthesis is catalyzed by enzymes *RNA polymerases*. These polymerases differ from DNA polymerase in that they can initiate synthesis of RNA without need for a *primer*. A RNA strand is synthesized by using a DNA strand as template and forming complementary strand in which the bases in DNA are paired with those in RNA.

For synthesis of DNA, a short segment of RNA, normally ranging 50–75, nucleotides in length remain bound to the DNA template after the RNA polymerase dissociates from the

DNA. This segment of RNA serves as primer to which a DNA polymerase can add deoxynucleotides. The RNA polymerase that produces the primer for DAN synthesis is called *primase* (Figure 17.27).

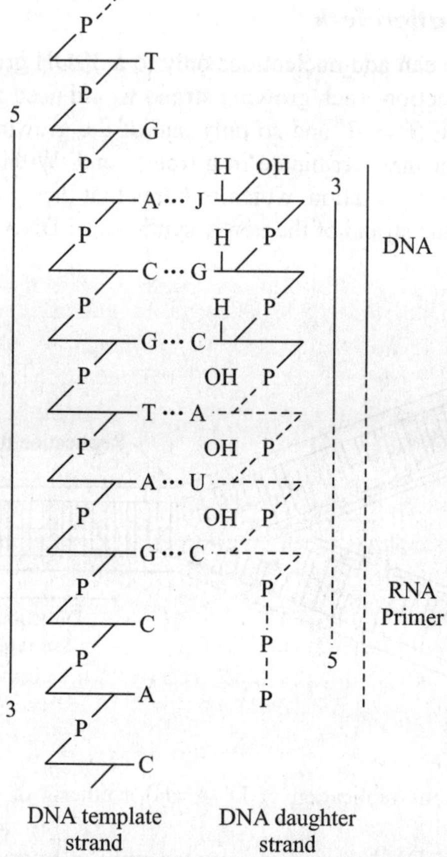

Figure 17.27 Priming of DNA synthesis with RNA primer.

Rules for Replication of Nucleic Acids

The primary function of replication is to duplicate the base sequences of the parental molecule, specifically of base-pairing, i.e. adenine with thymine (uracil in RNA) and guanine with cytosine is the rule of DNA replication. In addition:

(i) Nucleotides monomers are added one by one to the end of growing chain by enzyme DNA polymerase.

(ii) The sequence of bases in each new strand or daughter strand is complementary to the parental base strand being copied, in which, if there is an adenine in the parental strand a thymine nucleotide is added to the end of the growing daughter strand, when adenine is being copied, 'thymine' is present on the old parental strand.

TRANSCRIPTION

In protein synthesis, amino acids do not bind directly to DNA, it means an intermediate step is needed for arranging the amino acids in a polypeptide chain in order determined by the DNA base sequences, the intermediate product is mRNA (Watson and Crick, 1953). This process begins with the transcription of messenger RNA molecules by the base sequence of the coding strand of DNA. In prokaryotes, this RNA molecule (mRNA) is used directly in polypeptide synthesis. In eukaryotes, the mRNA is processed before it becomes mature mRNA. The amino acid sequence is then determined from the base sequences in mRNA by the protein synthesizing machinery of the cell.

Not all the base sequences of mRNA molecules are translated into the amino acid sequences of polypeptides because translation of mRNA molecule rarely starts exactly at one end of the mRNA and proceeds to other end. Instead, initiation of polypeptide synthesis may begins, many nucleotides distant the 5′ end of the RNA. The part of untranslated RNA before the region encoding the first polypeptide chain is called 'Leader', which in some cases contains regulatory sequences that affect the rate of protein synthesis. The 'leader' sequence is followed by a *coding sequence* which specifies the polypeptide chain. A typical coding unit in mRNA consists of about 500–3000 bases (depending on the number of amino acids in a protein). The 3′ end of the mRNA followed by coding sequences is also not translated. Out of two strands, only one DNA strand serves as *template* in each gene, but which DNA strand serves as *template* may differ from gene to gene along a DNA molecule. Thus, except in some small viruses, not all mRNA molecules are transcribed from same DNA strand.

In prokaryotes, most mRNA molecules are degraded within a few minutes after synthesis; in eukaryotes, degradation occurs after several hours; some persists for a day and some stay for a minute. Degradation of mRNA enables cells to dispense with unwanted molecules. It helps in regulating gene activity in prokaryotes.

The heterocatalytic function of DNA is a two-stage process. First is *transcription* of mRNA molecules where one DNA strand of a gene serves as *template* and second is, translation of genetic information in form of polypeptide chains of predetermined primary structure of protein biosynthesis. The basic features of mRNA transcription are:

1. The basic precursors in the transcription of RNA are four ribonucleoside-5′ triphosphates, namely, adenosine triphosphate (ATP), guanosine triphosphate (GTP), cytidine triphosphate (CTP) and uridine triphosphate (UTP). They differ in DNA constitution with respect to sugar, i.e. ribose in RNA and deoxyribose in DNA and base uracil (U) in RNA replaces thymine (T) in DNA.
2. In the formation of RNA, a sugar phosphate bond is formed between the 3-hydroxyl group of one nucleotide and the 5′-triphosphate of a second nucleotide. This is the same chemical formula for the formation of DNA, but enzyme is different.
3. The sequences of bases in RNA is determined by the base sequence of DNA template, i.e. the bases C,T,G and A in DNA strand pairs with G,A,C and U in RNA (mRNA).
4. Nucleotides are added only to the 3′-OH end of growing RNA molecule bears a triphosphate group (3′–5′ direction of chain growth as in DNA synthesis).

In detail, *transcription* and *translation* processes are given in Chapter 21, dealing gene regulation and expression. Here in very brief both mechanism are described.

Mechanism of Transcription

Transcription of mRNA takes place in four stages:

(i) Binding of RNA polymerase to DNA
(ii) Initiation of transcription
(iii) Elongation of RNA chain
(iv) Termination of transcription

Binding of RNA polymerase to DNA (Promoter recognition)

RNA polymerase binds to the 'Promoter site' on the DNA molecule. The promoter site consists of a few specific base sequences which is generally located 'upstream' (on the 3′ side of DNA being transcribed) of the start site but in rare cases it may be located at downstream too. 'Start site' is the point from where transcription begins by RNA polymerase. Promoter consist of 6–8 base pairs specifically called TATA box or Pribnow box; the consensus sequence is TATAAT, which are located about 20–30 base pairs upstream from start site of transcription both in prokaryotes and eukaryotes. An other sequence is TTGACA, in *E coli* which occurs about 35 bases upstream and is called recognition region.

In case of transcription of gene (non-histone protein) the promoter sequence for transcription of 5 sr RNA lies in the middle of gene and promoter sequence is located downstream of the transcription site.

	Recognition site	*Recognition site*	*Transcription site*
Gene	(a) *–35 sequence* TTGACA	(b) *–10 sequence* TATAAT	(c) *Transcription start* TGGAATTGTAC
Lac	TAGGCACCCCGC TTTACA GTGT	TGCTTCCGTCG	TATGTTGTG
trp	(a) TCTGAAATGAGCTGTTGACAATTAA	(b) TCATGCAACTAGTTAACAGTA	(c) CGCAAGTTCACG

Initiation of transcription

Transcription starts from 20–30 base pairs, downstream from the TATA box. A detection of the region between TATA box and the transcription start *site* shifts the transcription about 30 bp downstream from TATA box.

Transcription of mRNA always starts with a purine base—ATP or GTP (adenosine triphosphate or guanosine triphosphate). The nucleoside triphosphate is placed at this site and synthesis begins in a 5′–3′ direction.

Elongation of RNA chain

Messenger RNA synthesis as a rule is complementary to the DNA strand and transcription continue from 5′–3′ direction. During mRNA synthesis DNA double helix get unwind. RNA polymerase moves along the DNA adding nucleotides to the growing DNA chain.

Termination of transcription

The transcription of mRNA is terminated either when DNA receive chain termination signal or a protein called rho (ρ) helps to stop the transcription. The terminator gene sends the signals to stop transcription and release the transcript. When mRNA transcript is detached, the DNA template strand returns to its original position to form a double helix with its complementary strand with support of H-bond.

Transcription of mRNA takes place in the nucleus after transcription of mRNA is over, it migrates to the cytoplasm where it takes part in protein synthesis.

TRANSLATION

Proteins are larger polymers of amino acids connected by peptide bond. Peptide bonds are formed when carboxyl radicals (–COOH) of one amino acid react with the amino acid ($-NH_2$) group of the other amino acid. It consists of one alpha (α) carbon atom that is associated with an amino group ($-NH_2$) with a potential (+) charge, a carboxyl group (–COOH) with a (–) charge, a hydrogen atom and a side-chain 'R' that indicates variation in different amino acids. However, all 20 amino acids need not be present in all proteins. Basic structure of a polypeptide chain is as follows:

$$\begin{array}{ccccccc} H & & H & & O & & \\ | & & | & & \| & & \\ N & — & C & — & C & — & OH \\ | & & | & & & & \\ H & & R & & & & \end{array}$$

PRIMARY STRUCTURE OF PROTEINS

Proteins are long polypeptide chain. The protein molecules containing a single polypeptide chain (monomeric protein) can take primary, secondary or tertiary structures. The primary polypeptide chain consists of at one end, free amino acid ($-NH_2$) while other end consists of free carboxy group (–COOH) and the amino acids of the chains are linked with peptide bond (–C–N–). The number of amino acids and their sequences vary in different types of proteins.

MECHANISM OF PROTEIN SYNTHESIS

The mechanism of protein synthesis (Figure 17.28) is also known as translation of genetic information into amino acid sequence. Protein synthesis involves three steps:

(a) Initiation of protein synthesis
(b) Elongation of polypeptide Chain
(c) Termination of polypeptide chain

Initiation

This involves following steps:

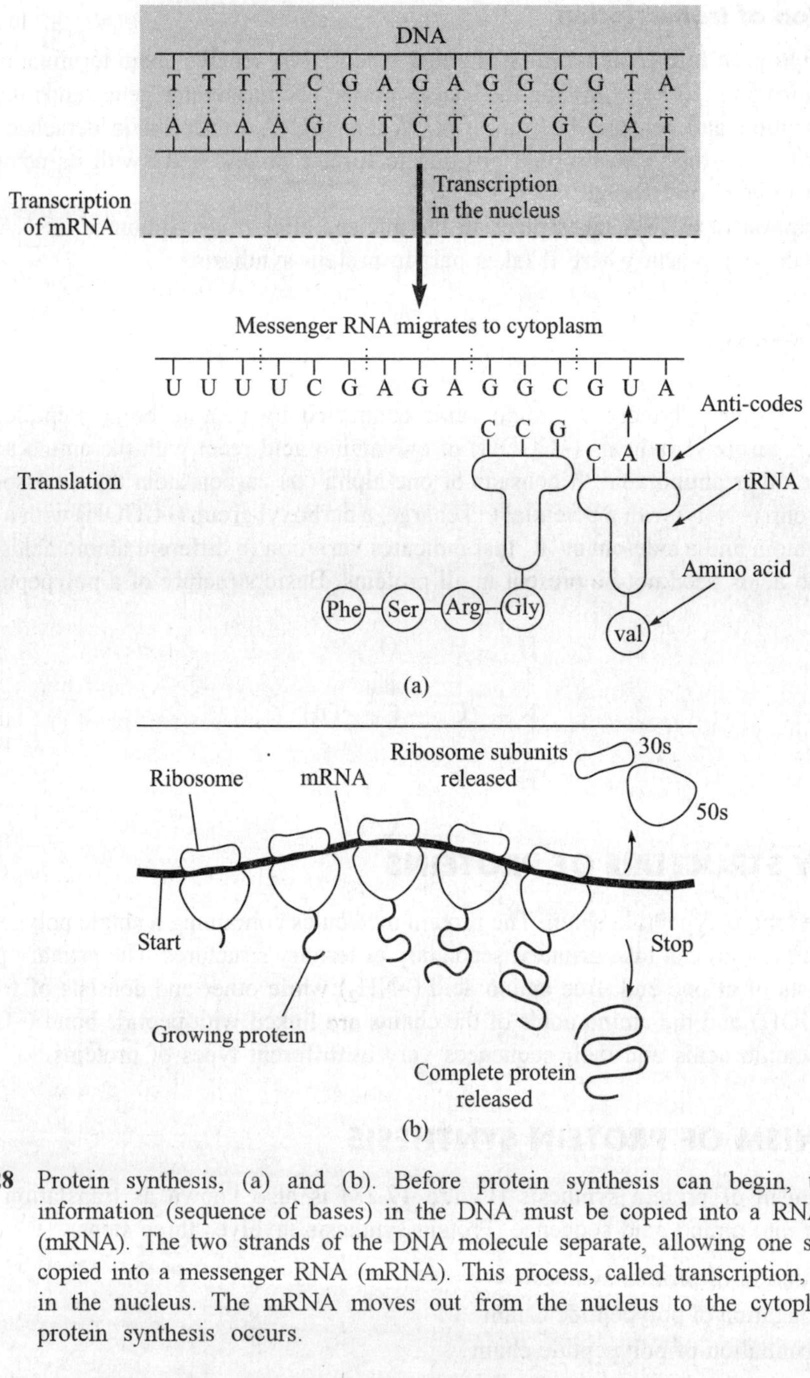

Figure 17.28 Protein synthesis, (a) and (b). Before protein synthesis can begin, the genetic information (sequence of bases) in the DNA must be copied into a RNA molecule (mRNA). The two strands of the DNA molecule separate, allowing one strand to be copied into a messenger RNA (mRNA). This process, called transcription, takes place in the nucleus. The mRNA moves out from the nucleus to the cytoplasm where protein synthesis occurs.

(i) *Amino acid activation:* The first step in the protein synthesis is activation of amino acid by the enzyme "amino acid-acyl synthetase" and energy is obtained by the breakage of ATP (adenosin triphosphate). The –OH group (carboxyl end) of amino acid and –OH group of AMP (adenosin monophosphate) react to form aminoacyl adenylate (AA–AMP) by releasing a molecule of pyrophosphate (P–p).

$$\text{AA} + \text{ATP} \xrightarrow{\text{Aminoacyl synthetase}} \underset{\text{(Aminoacyl adenylate)}}{\text{AA–AMP}} + \underset{\text{(Pyrophosphate)}}{\text{P–p}}$$

(ii) *Transfer of amino acid to tRNA:* Aminoacyl adenylate (AA–AMP) is bound tightly with enzyme "aminoacyl synthetase". Same enzyme catalyses the reaction to transfer amino acid to tRNA, as a result aminoacyl tRNA (AA-tRNA) is formed and AMP is released.

$$\text{AA} \longrightarrow \text{AMP} + \text{tRNA} \xrightarrow{\text{Aminoacyl synthetase}} \text{AA ---- tRNA} + \text{AMP}$$

The enzyme aminoacyl-synthetase possesses two different active *sites*; the site for recognition of side group of amino acid and the site for recognizing the tRNA specific for an amino acid.

(iii) *Binding of mRNA to ribosome:* In unicellular organisms (prokaryotes), the initiation factor 3 (IF3) binds to 30s subunit of ribosome to form IF3-30s complex, which again binds to mRNA to form IF3-30-mRNA complex. The initiation factor 3(IF3) get attached to N–f. methyl-tRNA (N-formylmethionyl-tRNA) and helps to bind the IF3-30s-mRNA complex. A molecule of GTP (Guanine triphosphate) get attached with initiation complex. (IF3-30s-mRNA). It provides energy for association of 50s ribosome with above complex (IF3-30s-mRNA) to make 70s ribosome.

$$\text{IF3} + 30\text{s} \rightarrow \text{IF3} - 30\text{s} \overset{\text{mRNA}}{\text{-----}} \text{IF3} - 30\text{s mRNA} \xrightarrow[\text{GTP}]{\text{IF2} - \text{fmeth} - \text{tRNA meth}}$$

$$\text{IF3} - 30\text{s} - \text{mRNA} - \text{IF2} - \text{fmeth} - \text{tRNA meth} + \text{GTP} \xrightarrow[\text{GDP} + \text{P1} + \text{IF2} + \text{IF3}]{50\text{s}} 70\text{s mRNA}$$

fmeth – tRNA meth

In eukaryotes, first amino acid is methionine and not the formyl-methionine. Initiation factors are greater in number than prokaryotes. GTP get attached with IF2 and makes a complex with aminoacyl-tRNA. This complex get associated with small ribosomal submit '40s'. The 5′ end of mRNA binds with 40s initiation complex. Then initiation complex moves in 3′ mRNA direction to locate AUG initiation codon. At the same time 60s subunit of ribosome unite with 40s subunit to form 80s ribosome. Guanosine triphosphate (GTP) is hydrolyzed to GDP (Guanosine diphosphate), which provide energy to this reaction.

Elongation of Polypeptide

N-formyl-methionyle-tRNA (in prokaryotes) and methionyl-tRNA occupies P site on the ribosome, soon after an initiation complex is formed, the newly initiated polypeptide chain is elongated by addition of amino acid one by one as directed by codon in mRNA.

Termination of Polypeptide Chain

There are three codons—UAA, UAG and UGA which do not specify any amino acid. They are utilized by the organism as 'termination codons' or 'stop codon' for growing polypeptide chain. The mRNA moves in $3' \rightarrow 5'$ direction and amino acids are added one by one. When the stop codon comes at the ribosome, no AA-tRNA comes on 'A' site and, therefore, incorporation of amino acid in the polypeptide chain is stopped.

RNA Translation

The genetical studies of T4 phage carrying rII mutants by Benzer (1955) indicated that primary structure of protein is encoded in DNA in form of sequences of nucleotide triplets or codon, each of which represents one of the 20 amino acids. It has been already explained that sequence of nucleotide triplet codon is transcribed from its permanent repository in DNA onto ephemeral mRNA molecules as the first stage of the heterocatalytic function. In the second stage of the heterocatalytic function, amino acids are actually assembled into polypeptide chains of predetermined order on the ribosomal RNA bodies.

The structure of polypeptide chain is so that at one end of chain, a free carboxyl group (–COOH) and at the other end, a free amino acid group ($-NH_2$) are present. The simple way by which amino acids are assembled into a polypeptide chain can be imagined to occur in that nascent chain grows from one of its ends by gradual addition of one amino acid after other. Finally, when the requisite number of amino acids has been assembled, the growth of new chain terminates and the completed polypeptide chain is released from the ribose to allow it, to assume whatever functional role it has to play in the life of cell/organism.

REVIEW QUESTIONS

1. What are the main characteristics that prove that DNA is the genetic material?
2. Name the four bases in DNA which form base pairs. What is the difference between nucleoside and nucleotide?
3. How many phosphate groups are there per base in DNA and how many phosphates are there in each precursor for DNA synthesis.
4. Which chemical groups are at the ends of single polynucleotide strand?
5. What are some of the genetic phenomena that can be investigated in corn, fruit fly, peas and human beings that cannot be investigated in *Neurospora*, bacteria and viruses.
6. If one strand of DNA is found to have the sequence 5′ AA CG TA CT GC3′ what is the sequence of nucleotides on the 3′–5′ strand.
7. What is the relation between the amount of DNA in a somatic cell and in a gamete?
8. Assume that you have just determined the adenine content of the DNA of *Bacillus hypotheticus* to be 20 per cent. What is the percentage of each of the other bases?
9. If adenine—guanine content of another species to total 25 per cent; should you believe?
10. Differentiate between transformation and transduction.
11. Develop a formula for determining the length in micrometres, of a DNA molecule whose numbers of deoxyribonucleotide pairs are known.

(*Note:* $1 Å = 1 \times 10^{-4}$ μm).

12. For a molecule of n-deoxyribonucleotide pairs, give a mathematical expression of the possible sequences of those nucleotide pairs where only the 'usual' bases are present.

13. Phage T_2 DNA is estimated to consist of about 200,000 deoxyribonucleotide pairs. What is the length in micrometre (μm) of its DNA complement?

(*Note:* Distance from one nucleotide to other is 3.4 Å, 1 Å = 1×10^{-4} μm).

14. Chemical analysis of DNA extract from a formerly unknown virus indicates the following composition. (a) Adenine 40 per cent, thymine 20 per cent, guanine 15 per cent and cytosine 25 per cent. What does this result... about the structure of viral DNA?

15. When the base composition of DNA from bacterium *Mycobacterium tuberculosis* was determined, 18 per cent of the bases were found to be adenine.

(a) What is the percentage of thymine?

(b) What is the entire base composition of DNA and the [G] + [C] content?

REFERENCES

Avery, O.T., C.M. Macleod, and M. Carthy (1944): Inducing transformation of pneumococcal types. Induction of transformation by a deoxyribonucleic acid. Fraction isolated from *Pneumococcus* type, *J. Exp. Med.*, **79,** 37.

Benzer, S. (1955): Fine structure of genetic region in *Bacteriophage*, *Proc. Nat. Acad. Sci.*, **41,** 344.

Brenner, S., J. Jacob, and M. Meselson (1961): An unstable intermediate carrying information from genes to ribosome for protein synthesis, *Nature*, **190,** 5786.

Chargaff, E. (1950): Chemical specificity of the nucleic acid and mechanism of their enzymatic degradation, *Experimentica*, **6,** 201.

Griffith, F. (1928): The significance of pneumococcus types, *J. Hygiene*, **27,** 113.

Hartl Daniel, L. (1991): *Basic Genetics*, Jones and Bartlelt, Boston.

Hartman, P.E. and S.R. Suskind (1968): *Gene Action*, Prentice Hall, Englewood Cliffs, N.J.

Hersey, A.O. and M. Chase (1952): Independent functions of viral protein and nucleic acid in growth of bacteriophage, *J. Gen. Physiol.*, **36,** 39.

Jocab, F. and J. Monod (1961): Genetic regulatory mechanisms in synthesis of proteins, *J. Mol. Biol.*, **3,** 318.

Kech (1963): Nuclear control of synthetic activities in *Acetabularia*, *Proc.* of *XIV International Congress*, Washington, **3,** 203.

Meselson, M. and W. Stahl (1958): Replication of DNA in *E. coli*, *Nat. Acad. Sci.*, Washington, **44,** 671.

Ram, M. (2005): *Unfolding Reality of Life*: *The DNA World*, Dominant Publications, New Delhi.

Sinsheimer, R.L. (1959): A single-stranded deoxyribonucleic acid from bacteriophage Φ X 174, *J. Mol. Biol.*, **1,** 43.

Stent, G.S. (1964): Genetic transcription, *Proc. Roy, Soc.*, London, Ser. B, 164–181.

Stent, G.S. (1971): *Molecular Genetics*, W.H. Freeman and Co., San Francisco.

Watson, J.D. and F.H.C. Crick (1953): A structure of deoxyribonucleic acid, *Nature*, **171,** 737.

18 DNA Synthesis

The nucleic acids are synthesized by enzymes in the living cells through chemical reactions. The enzymes that form the sugar—phosphate bond (the phosphodiester bond) in between two adjacent nucleotides in a nucleic acid chain are called *DNA polymerases*. A number of DNA polymerases have been purified and DNA synthesis has been taken up *in vitro* in a cell-free system prepared by disrupting cells and combining purified components in a test tube under a most precised environment. The principal requirements are:

1. The 5′-triphosphates of the four deoxynucleosides must be present. The compounds are—dATP, dGTP, dTTp and dCTP. They contain all the four bases—adenine, guanine, cytosine and thymine. The chemical structures are given in Figure 18.1.

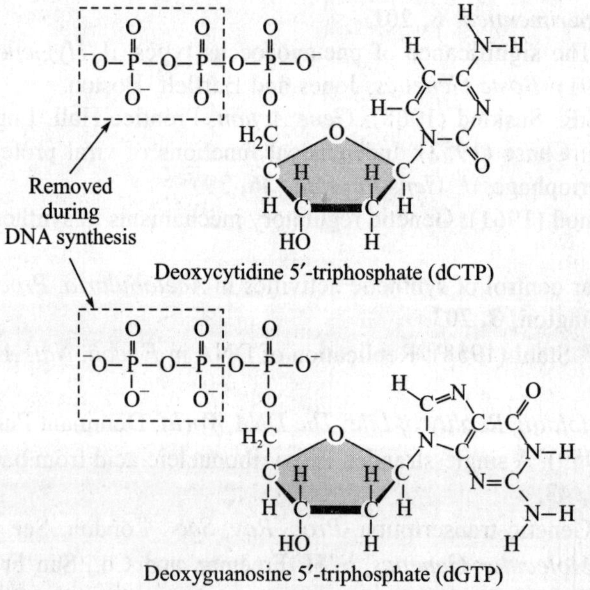

Figure 18.1 Two deoxynucleoside triphosphates used in DNA synthesis. The dotted part of each molecule is removed during synthesis.

2. A pre-existing single DNA strand to be copied must exist. This strand serves as *template* strand. (Figure 18.2).
3. ˙A nucleic acid segment may be very short of either DNA or RNA must be present along with hydrogen bond to the template strand. This segment is called a *primer*. Without primer no known DNA polymerase is able to initiate chain. The presence of a primer chain with a free 3′-OH group is absolutely necessary for initiation of replication. Figure 18.2 shows an addition of nucleotides to the 3′-OH terminus of a *primer*. DNA synthesis occurs by the elongation of primer chains, always in 5′ → 3′ direction.

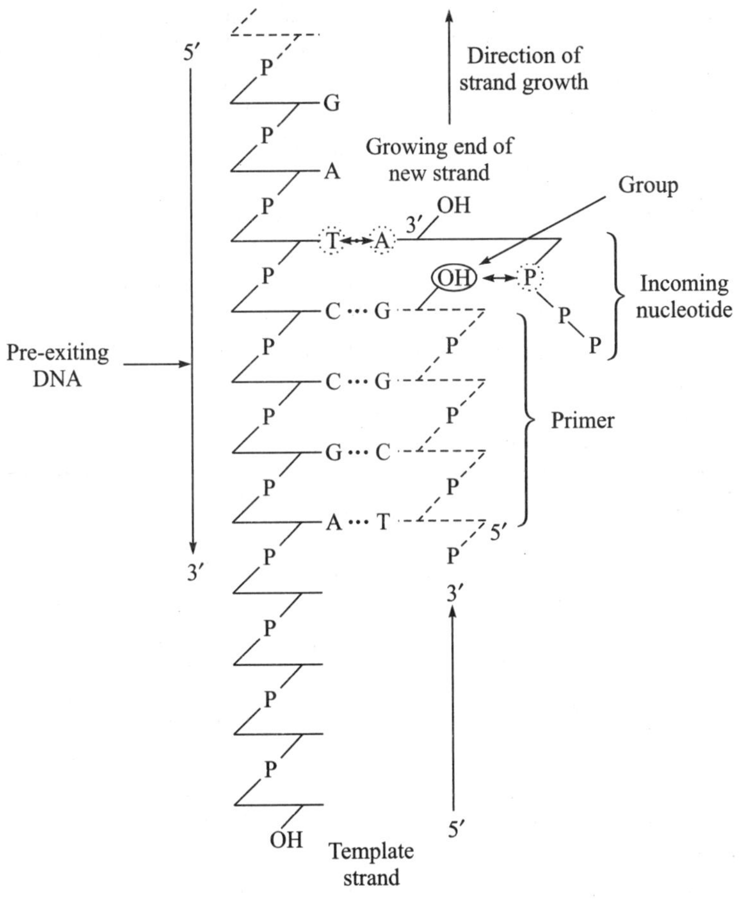

Figure 18.2 Nucleotides are being added to the 3′-OH terminus of the primer.

DNA POLYMERASE

In mid 1950 Arthur Kornberg and his colleagues studied the mechanism of DNA replication through transfer experiments. Based on his findings, Kornberg decided that growth of the replica polynucleotide chain must be catalysed by an enzyme. For this purpose, he extracted protein from *E. coli* and added these extracts to a reaction mixture containing C^{14} or P^{32} labelled deoxynucleoside triphosphate of the type carrying the four DNA bases—adenine, guanine, cytosine and thymine (dATP, dGTP, dCTP and dTTP) mg^{++} and template DNA molecules. He expected that this mixture would produce enzymic conversion of low-molecular weight radioactivity residing in deoxyribo-nucleoside triphosphate into high-molecular weight polynucleotide materials. It so happened that an enzyme catalyzed polymerization of the labelled deoxyribonucleoside triphosphate into high-molecular weight polynucleotide chains in the mixture. This polymerization had resemblance to DNA synthesis because the reaction proceeded only if in addition to protein extract, all four deoxyribonucleside triphosphates as well as DNA template were present. Kornberg purified this enzyme and named it 'DNA polymerase' and after a few years, enormous amount of DNA polymerases were purified from *E. coli*. The purified enzymes consist of a single polypeptide chain containing 1000 amino acids in length in each and about 300 DNA polymerase were found in *E. coli*. cell. The enzyme catalyzed polynucleotide synthesis, which liberates one mole of pyrophosphate (P–P) per mole of nucleoside triphosphate polymerization (Figure 18.3).

The most important evidence in favour of this theory is that DNA polymerase not only catalyzed the polymerization of nucleotides to regenerate exact copy of the original DNA *template* but DNA *template* added to the mixture determines the character of the polynucleotide formed, i.e. the relative frequencies of nucleotide base composition in DNA template were also almost similar to the nucleotide base composition found in the enzymatic reaction product (Table 18.1).

TABLE 18.1 The nucleotide base composition (%) of enzymatically formed DNA product and in original DNA template (%)

Source	*A*	*T*	*G*	*C*	*G+C*
1. *Calf*					
DNA template	29	27	23	21	44
Reproduced DNA product	30	30	20	20	40
2. *Escherichia coli*					
DNA template	25	24	25	26	51
Reproduced DNA product	26	25	24	25	49
3. *Aerobactor aerogene*					
DNA template	22	22	28	28	56
Reproduced DNA product	25	25	25	25	50
4. *Mycobacterium phlei*					
DNA template	16	16	34	34	68
Reproduced DNA product	16	20	29	35	64

The DNA *template* not only governs the overall base composition of the reaction product but also the relative frequencies with which any two bases occur as nearest neighbours in the replica polynucleotide chains. Thus, polynucleotide synthesis catalyzed by DNA polymerase purified from *E. coli.* indicated that as Watson and Cricks in 1953 proposed mechanism for autocalatytic function of DNA, the DNA acts directly as a template for the orderly copolymerization of its

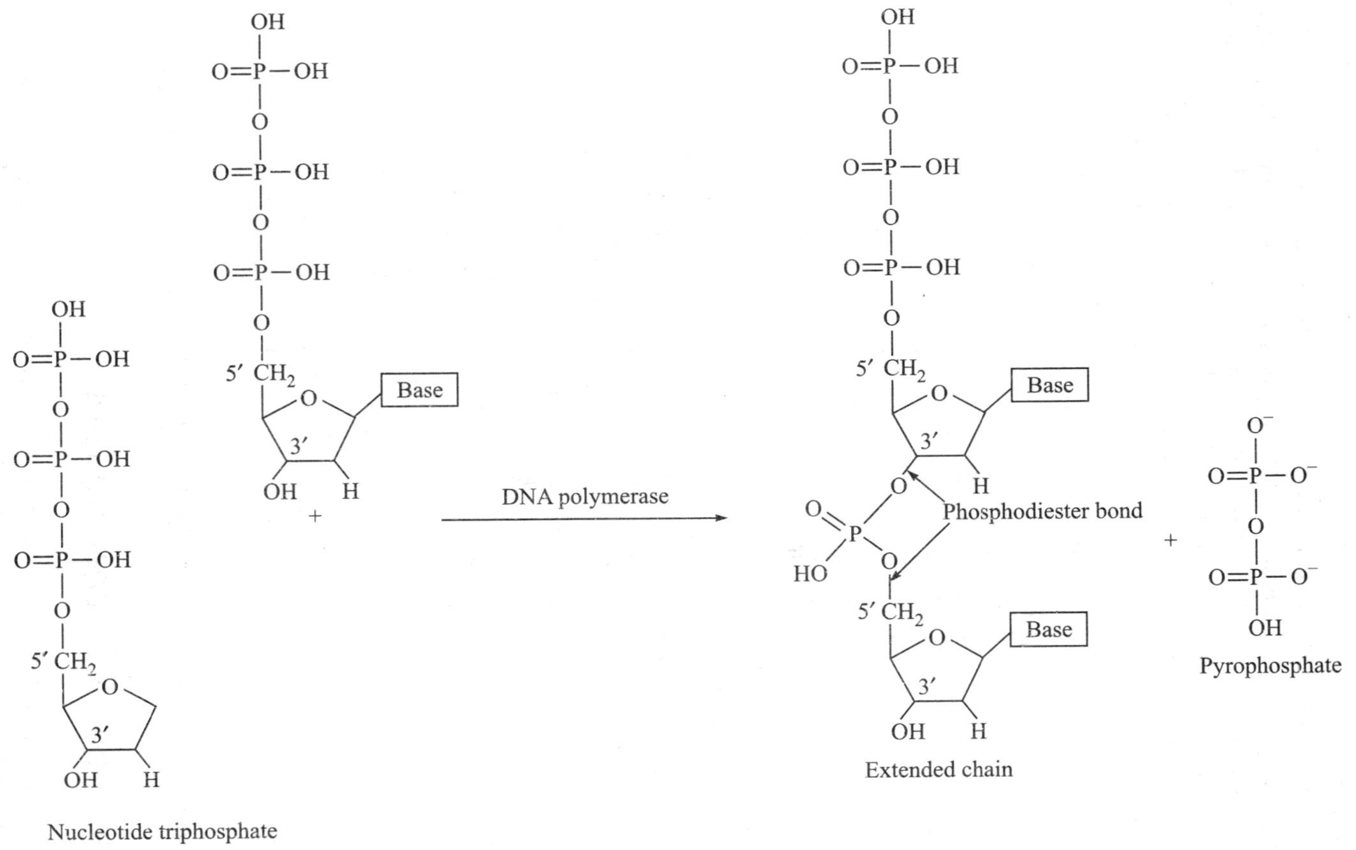

Figure 18.3 Polymerization of polynucleotides in DNA synthesis by DNA polymerase as a result nucleotides are added to the 3′-OH end of DNA chain.

replicas without requiring any non-DNA intermediaries in the synthesis (replication). But it is still unknown whether *in vivo*, replication of *E. coli* DNA is catalyzed by DNA polymerase isolated by Kornberg in 1950.

However, replication of DNA is not only the activity which takes place in *E. coli*. There occurs DNA repair through DNA *ligase* which catalyzes this internucleotide phosphate diester bond between 5′-phosphate and 3′-hydroxyl of two adjacent nucleotides belonging to an interrupted polynucleotide forming part of a double DNA helix.

Since the DNA molecule proceeds at a single replicating Y-fork (Figure 18.4), it is doubtful that all 300 DNA polymerases found in *E. coli* may be involved in autocatalytic functions of DNA.

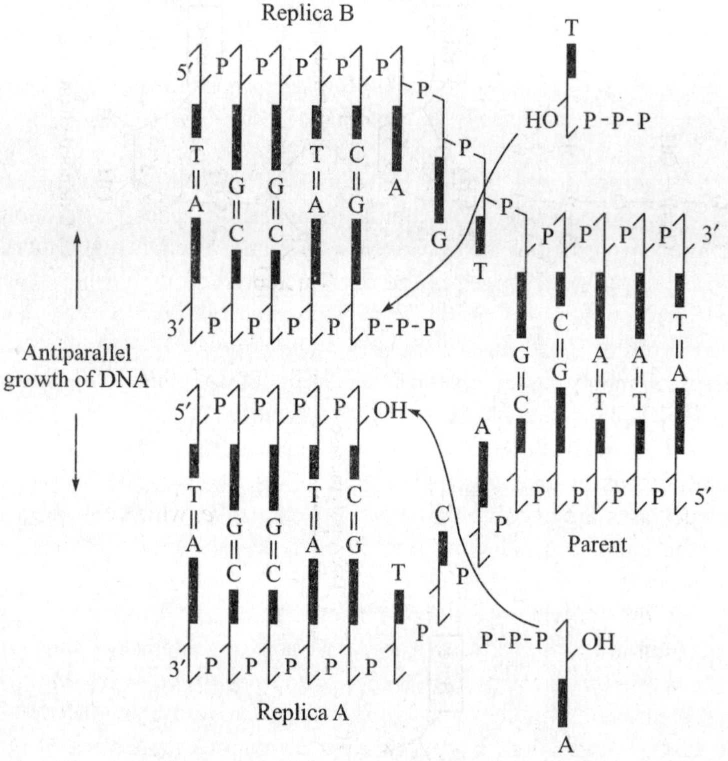

Figure 18.4 Schematic representation of *Y-fork DNA replication*. Here polynucleotide nascent replica A is growing in the direction 5′ → 3′ and that of replica B in the direction 3′ → 5′.

DNA polymerase add nucleotides only to a 3′-OH group.

DNA REPLICATION IS DISCONTINUOUS

But DNA replication does not proceed in a continuous series, but in assembly of short segment of interrupted polynucleotide chains which are later on linked together in a long polynucleotide chain by DNA *ligase*, i.e. one of the daughter strands is made in short segment which are then joined together. This discontinuous Y-fork DNA replication also solves the reason of antiparallel growth of DNA polynucleotide chain. DNA ligase is capable of repairing single strand interruption of the double DNA helix (Figure 18.5).

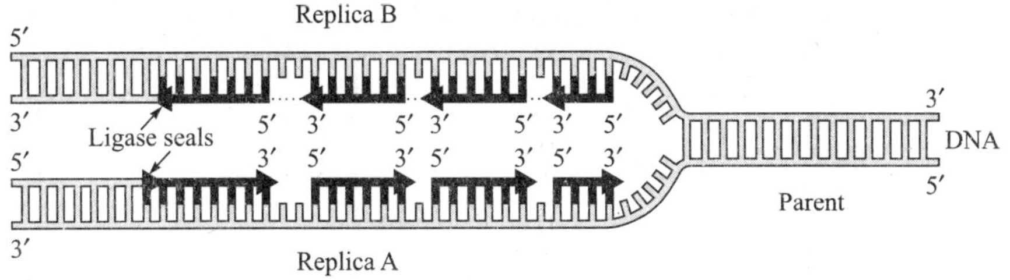

Figure 18.5 Structure of Y-fork discontinuous DNA replication. Replica B is sealed by *ligase* and replica A is discontinuous yet (unsealed).

The DNA polymerase catalyzes the reaction for the formation of phosphodiester bond between free 3′-OH group of the *primer* and the innermost phosphorus atom of nucleoside triphosphate being incorporated at the new *primer* terminus. The DNA polymerase catalyses the polymerization reaction that incorporates the new nucleotide at the *primer* terminus only when correct base pair is present.

There are two types of DNA polymerases which help in the DNA replication in *E. coli*. *First* is DNA polymerase symbolyzed as DNA polymerase I (POL I) and *second* is DNA polymerase III (POL III). POL III plays major part in DNA replication.

Besides, DNA polymerization, some DNA polymerase function as 'nuclease', which breaks the phosphodiester bonds in the 'sugar–phosphate', which serves the backbone of the nucleic acid chain. The nucleases are of two types—first, *exonuclease*, which can remove the nucleotide from the end of the chain and *second*, *endonuclease*, which breaks the bond from within the chain.

Some time, the DNA polymerase adds an incorrect nucleotide that does not base pair to the end of growing chain. The presence of unpaired nucleotide activates the 3′–5′ exonuclease, which terminates the wrong nucleotide. This is called *proofreading* or *editing*. It is an error-correcting mechanism, that reduces the frequency of mutation by incorporation of wrong nucleotide in the DNA replication. But there is no continuous replication of DNA molecule as indicated above, instead, the daughter strand is made in short fragments which are then joined together by enzyme 'DNA ligase' (Figure 18.5).

Use of RNA Primer

The DNA polymerases can not initiate synthesis of a new strand until a free 3′-OH group is available. In some living organisms, the synthesis is initiated by a special kinds of RNA polymerases. RNA is a single-stranded nucleic acid consisting of the four nitrogen bases (nucleotides) joined together by 3′–5′ phosphodiester bonds in the same kind of chemical bonds as those found in DNA deoxyribose, except here thymine pairs with uracil. RNA is synthesized using DNA strand as *template* where DNA bases form a complementary strand with RNA bases. The synthesis is catalyzed by RNA polymerases. The RNA polymerases do not need a *primer* while DNA polymerases need *primer* as a condition (Figure 18.6).

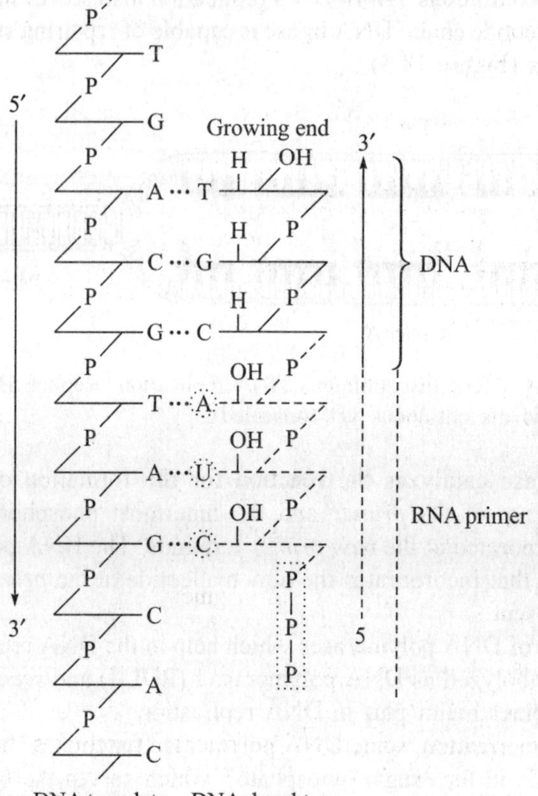

Figure 18.6 DNA synthesis through use of RNA primer by RNA polymerase enzyme, shown in dotted lines. The RNA segment which has been shown by dotted lines has been made by RNA polymerase.

Process

The DNA synthesis is initiated through a short fragment of RNA consisting of 50–70 nucleotides in the length which remain bound to the DNA template after RNA polymerase dissociates from the DNA. This RNA fragment serves as *primer* to which a DNA polymerase can add deoxyribonucleotides. The RNA polymerase that produces the *primer* for DNA synthesis is called *primase*.

PPP-5 ———— 3′-OH

RNA DNA

Structure of a precursor fragment in lagging strand that joined to make a continuous strand of DNA [which contains no RNA sequence] by catalytic action of DNA *ligase*.

Joining of Precursor Fragments in DNA Synthesis

The precursor fragments are finally joined together to form a continuous strand of DNA. This strand contains no RNA sequences and so assembly of lagging strand must require removal of the RNA *primer*, replacement with a DNA sequence and then joining. In *E. coli*, the first two processes are accomplishable by DNA polymerase I and joining of fragments is catalyzed by DNA ligase which can join adjacent 3′-OH and 5′-P group at nick point. The *E. coli*, DNA ligase can not seal the nick because a triphosphate (P′-P′-P′) is present. It can link only 3′-OH and a *5′-monophosphate*. But polymerase I has an *exonuclease* activity that can remove nucleotide from 5′ end of a base-paired segment. It is effective with DNA and RNA. This activity is also known as 5′ → 3′ exonuclease activity. Thus, polymerase I acts at the nick and moves the nick in the 5′ → 3′ direction, removing RNA nucleotides one by one and adding DNA nucleotides to 3′ end of the DNA strand. As soon as, all the RNA nucleotides are removed, DNA ligase joins the 3′-OH group to the terminal 5′-P of the precursor fragment.

Determination of Sequence of Bases in DNA

Determination of sequence of bases in DNA, several techniques are available, one is described here. No technique can determine sequences of bases in an entire chromosome in a single experiment. So chromosomes are first cut into fragments of few hundred base pairs from long DNA. To obtain the sequence of a long stretch of DNA, a set of overlapping fragment is prepared. The sequence of each is determined and all sequences then combined.

Dideoxy of sequencing method (Sanger Method)

This technique employs DNA synthesis in presence of a small amount of abnormal nucleotides that contains the sugar 'dideoxyribose' instead of deoxyribose. Dideoxyribose lacks 3′-OH group which is essential for attachment of the next nucleotide in a growing DNA chain; and so incorporation of a dideoxynucleotide on place of deoxyribonucleotide immediately terminates further synthesis of the strand. To sequence a DNA strand, four DNA synthesis reactions are carried out:

(a) Each reaction contains the single DNA stranded *template* to be sequenced.
(b) A small *primer* fragment complementary to a stretch of the template strand.
(c) All four deoxyribonucleoside triphosphates and
(d) One of the deoxyribonucleoside triphosphates in dideoxy form.

The difference between dideoxyribose and deoxyribose is shown in Figure 18.7.

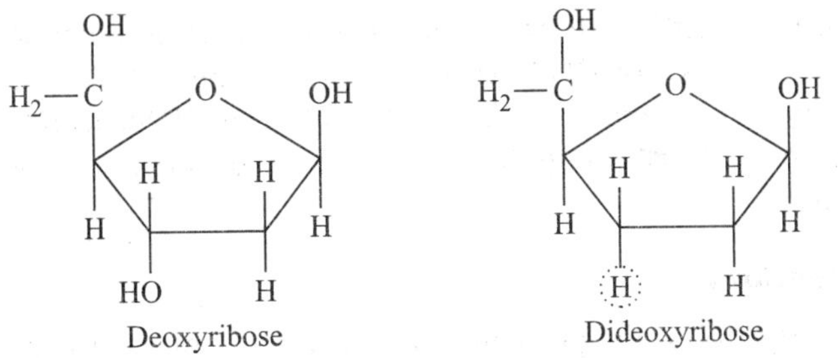

Figure 18.7 Difference between deoxyribose and dideoxyribose.

Each reaction produces a set of fragments that terminate where a dideoxyribonucleotide was randomly incorporated in place of the normal deoxyribonucleotide, and in each of the four reactions, the lengths of fragments are determined by the position in the daughter strand where the particular dideoxynucleotide present in that reaction was incorporated. The sizes of the fragments produced by chain termination are determined by gel electrophoresis and the base sequence is determined by the following rule:

> *"If a fragment containing n nucleotides is generated in the reaction containing a particular dideoxynucleotide, then position n in the daughter strand is occupied by the base present in the dideoxynucleotide. The numbering is from the 5′ nucleotide of the primer."*

For instance, if a fragment consisting of 93 base pairs is present in the reaction containing the dideoxy form of dATP, then 93rd base in the daughter strand produced by DNA synthesis must be an 'A'. Because of complementary nature of double helix, it does not matter whether the sequence of the template strand or the daughter strand is determined. Therefore, sequence of template strand can be deduced from the daughter strand because their nucleotide sequences are complementary. But practically, both the strands of a molecule are sequenced independently and compared to eliminate errors if any observed in misreading the gels (Hartl, 1991).

Gel electrophoresis

Principle: The DNA molecules are negatively charged and thus can move in an electric field. Therefore, if the terminals of a battery are connected to the opposite ends of the horizontal tube containing a DNA solution, the molecules will move toward the positive end of the tube at a rate depending on the electrical field strength and on the shape and size of the molecules.

The most common type of *electrophoresis* used in genetics analysis is *gel electrophonosis*. In this procedure, rate of movement depends on the molecular weight of the molecule as long as molecules are linear and not too large. In Figure 18.8, gel electrophoresis of DNA is shown, where thin slab of a gel is prepared containing small slots (wells) into which samples are placed. An electric field is applied and the negatively charged DNA molecules penetrate and move through the *gel*. A *gel* is a complex network of molecules, containing narrow, tortuous passages, through which smaller molecules pass more easily and rate of movement increases with the decease in the molecular weight. Each discrete region containing DNA is called band.

In Figure 18.9 result of the electrophoresis of a collection of double stranded DNA molecule is shown.

Gel electrophoresis can be carried out with both double- and single-stranded DNA. The short, single-stranded fragments differing by a single base can be separated. For instance, if a reaction mixture contains a set of DNA fragments consisting of 20, 21, 22… 100 nucleotides, electrophoresis will yield 81 bands—one fragment containing 20 nucleotides; the next fragment 21 nucleotides and so on. This extraordinary sensitivity to the size is the basis of the DNA sequencing procedure.

Sequencing procedure

As described by Hartl (1991), the procedure for sequencing of a DNA fragment is shown in Figure 18.10. In this example, *primer* length is 20 nucleotides. Termination at G produces fragments

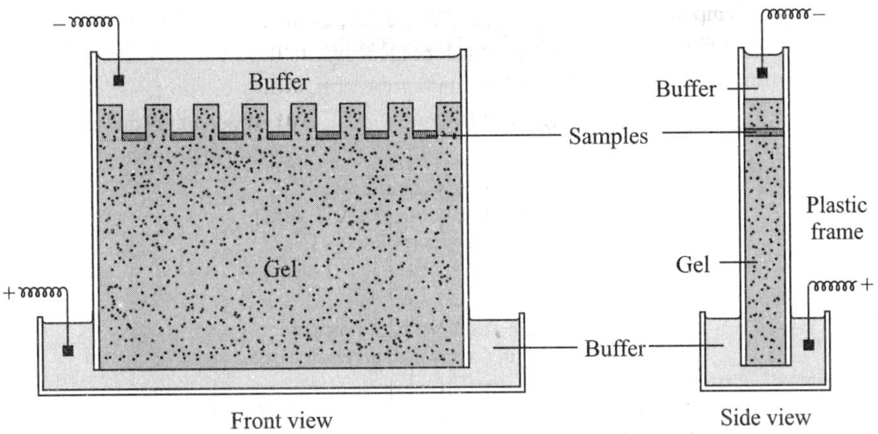

Figure 18.8 Apparatus for gel-electrophoresis, Liquid gel is allowed to harden in place in order to make wells for keeping samples. After electrophoresis, the sample is made visible by removing the plastic frame and immersing the gel in the solution containing a reagent that binds to or reacts with the molecules that were electrophoresed. Which are located at various positions in the gel. The separated samples may be seen as coloured or fluorescent bands when illuminated with fluorescent light, the plastic frame in which sample is kept is also called *lane*.

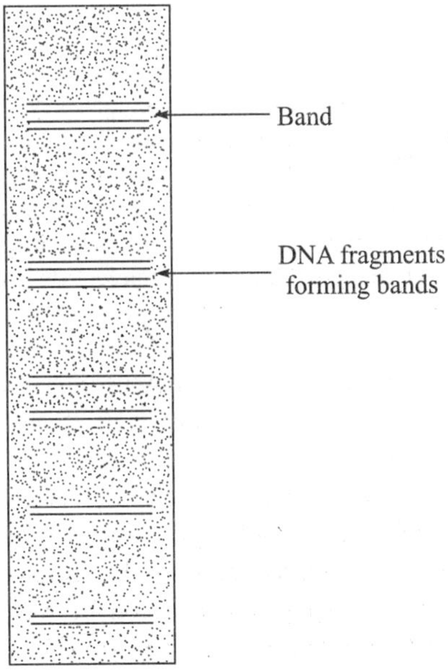

Figure 18.9 Gel electrophoresis of DNA. The molecules of different sizes are mixed and placed in a well. The DNA is made visible by addition of a dye (ethidium bromide) that binds only to DNA and that fluoresces when gel is illuminated with ultraviolet light.

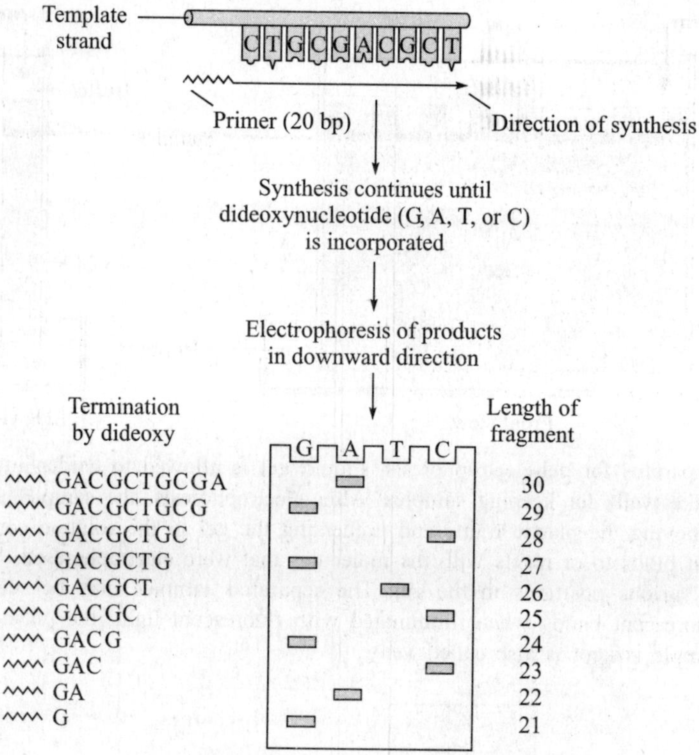

Figure 18.10 Dideoxy method of DNA processing. Four DNA synthesis reactions are carried out in presence of all the four normal nucleotides plus one of the dideoxynucleotides containing G, A, T or C. Synthesis continues along the template strand until a dideoxynucleotide is incorporated. The products resulting from termination at each dideoxynucleotide are indicated at the left. The fragments are separated by size by electrophoresis and the position of the nucleotides are determined directly from the gel.

of size 21, 24, 27, 29 nucleotides; termination at A produces fragments of size 22, and 30 nucleotides; termination at T produces fragments of size 26 nucleotides and termination at C produces fragments of 23, 25, and 28 nucleotides.

Most probably, either the *primer* or one of the deoxyribonucleotides used contains a radioactive atom as a result after electrophoresis, when fragments of different sizes are separated, the fragments can be visualized as bands on the photographic film placed over the gel. After electrophoresis and autoradiography, the position of bases in the daughter strand can be read directly from the *gel* starting with position 21 GAC GCT GCGA. This correspond to the template sequence 'CTG CGA CGCT'. Figure 18.11 shows the autoradiogram of a sequencing gel. The shortest fragment move faster and farthest. Each fragment continues the *primer* fragment at 5′ end of daughter strand. The fragments are read from bottom to top of the gel.

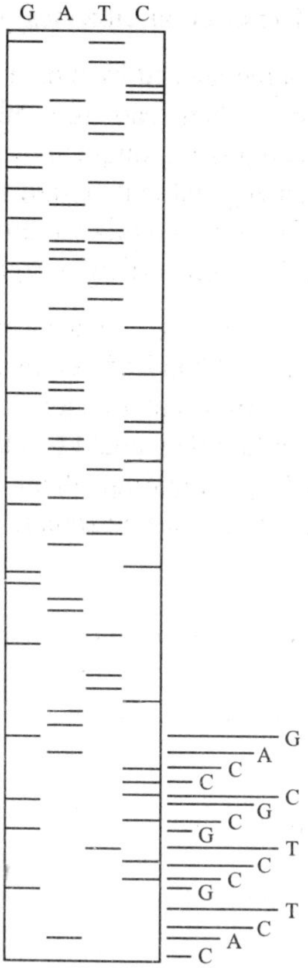

Figure 18.11 Autoradiogram of a dideoxy sequencing gel. Each horizontal row represent a single nucleotide position in the DNA strand synthesized from the template.

DNA Fragmentation and Isolation

DNA sequencing and their utilization needs isolation and identification of a particular DNA fragment. This can be achieved by cutting down DNA fragments at specific sites for isolating particular DNA fragments. The technique given by Hartl (1991) is as follows:

> *"The DNA fragments are obtained by treatment of DNA samples with a specific class of nucleases, the class of nucleases is called restriction endonucleases or more simply the restriction enzyme, which are sequence-specific enzyme. Most restriction enzymes recognize only a short-base sequence, usually 4 or 6 nucleotide base pairs.*
>
> *The enzyme binds with the DNA at these sites and makes a cut in each strand of the DNA molecule resulting in 3′-OH and 5′-P groups at each position. The nucleotide sequence recognized for cleavage by a restriction enzyme is called restriction site of the enzyme. Several hundred restriction enzymes specific for different restriction sites have been isolated from many species of prokaryotic microorganism (Table 18.2)".*

The important characteristics of restriction enzymes are:

1. The restriction site sequences recognized by most restriction enzymes are '*palindromes*'. That is, the sequence is same, when read from other side NUN, MADAM, REFER.
2. Most restriction enzymes recognize a single *site*.
3. The restriction *site* is recognized without regard to the source of DNA.
4. Because, most restriction enzymes recognize a unique restriction *site*; sequence and the number of cuts made in the DNA of a particular organism by a particular enzyme is fixed.

The DNA fragments produced by a pair of adjacent cut is called restriction fragments. A large DNA molecule will be cut into several hundred (a typical bacterial DNA molecule containing about 4×10^6, nucleotide base pairs may cut to more than a million fragments. Normally, restriction fragments are very small so that they could be easily separated by electrophoresis and utilize for different purposes such as legating them into self-replicating molecules eg bacteriophage plasmids or small artificial chromosomes. This procedure is also called DNA cloning.

TABLE 18.2 Some restriction nucleases, sources and their cleavage sites

Enzyme	*Microorganism*	*Target sequence*	*Cleavage site*
EcoRI	*Escherichia coli*	GAA	TTC
		CTT	AAG
BomHI	*Bacillus amylo liquefaciens*	GGA	TCC
		CCT	AGG
Hacll	*Haemophilus aegyptius*	PuGC	GCPU
		PyCG	CGPU
Hind III	*Haemophilus influenzas*	AAG	CTT
		TTC	GAA
Pstl	*Providencia stuartii*	CTG	CAG
		GAC	CTG
Taqp	*Thermus aquaticus*	TC	GA
		AG	CT

Because of sequence specificity, different restriction enzymes cut different set of fragments from same DNA molecule. These procedures constitute "DNA cloning" also.

Figure 18.12 shows the map and unique site of cutting of the phase DNA by enzyme Eco-R1 and BamH1, This is also called restriction map. The fragments cut by a single restriction enzyme can be detected by *gel electrophoresis* of the enzyme treated DNA.

Location of DNA fragments in the gel by Southern blot

One of the most commonly applicable procedures to locate in DNA fragment in the gel is the southern blot. In this procedure, a gel in which a DNA molecule has been separated by electrophoresis is treated with alkali ($CaCl_2$) to render the single-stranded DNA (denaturation of DNA) and then DNA is transferred to a sheet of 'nitrocellulose' in such a way that relative position of DNA bands are maintained. The *nitrocellulose* to which the single-stranded DNA binds tightly is then exposed to radioactive complementary RNA or DNA (*Probe*) in such a way

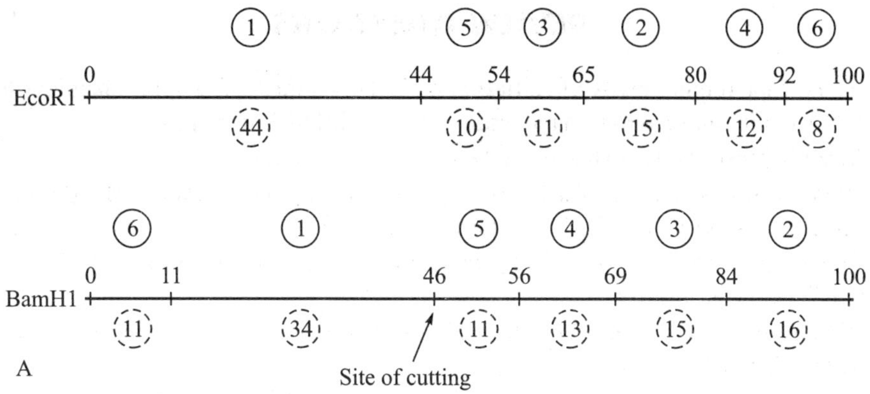

Figure 18.12 A restriction map of λ DNA for the restriction enzymes EcoR1 and BamH1.

The number enclosed in dotted circles indicate the length of each fragment expressed in percentage of the total length.

The number kept in open indicates the approximate percentage of the total length of λ DNA measured from the end of molecule.

that it leads to complementary strands coming together to form a duplex molecule (renaturation). The radioactivity becomes stably bound (not removable by washing) to DNA only at position at which base sequences complementary to the radioactive molecule are present so that duplex molecules can be formed. The radioactivity is located by placing the paper in contact with x-ray film. After development of film, the blackened region indicates the position of radioactivity. Expose to photographic film and develop. A clone DNA fragments from one species may be used as probe DNA in southern blot with DNA from another species. The probe will hybridize only with restriction fragment that contains DNA sequences that are sufficiently homologous to allow stable duplexes (Hartl, 1991) (Figure 18.13).

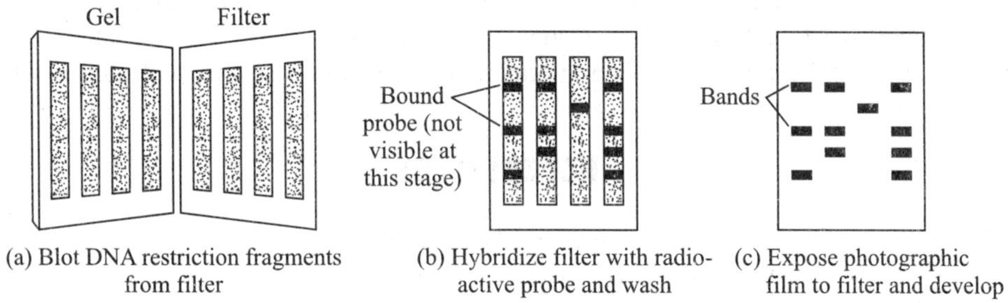

Figure 18.13 Structure of southern blot.

(a) DNA restriction fragments are separated by electrophoresis, blotted from the gel onto a special filter material and chemically attached using ultraviolet light.

(b) The DNA strands are denatured and mix with radioactive probe DNA, which binds the complementary sequences present on the filter. The unbound probes wash out.

(c) The bound probe is revealed by darkening of the photographic film placed over the filter. The positions of binds indicate with restriction fragments which contain DNA sequences homologous with those in probe.

REVIEW QUESTIONS

1. Name four requirements of initiation of DNA synthesis. To what chemical group in a DNA chain, an incoming nucleotide reacts with DNA terminus?
2. In what sense the two strands of DNA are antiparallel?
3. What is base sequence of a DNA strand that is complementary to the deoxyribonucleotide 3′-A-G-G-C-T-C-5′. Label terminus 5′ or 3′.
4. In what direction (5′–3′) or (3′–5′) the template strand stands during DNA synthesis and why? Describe briefly.
5. What is the chemical difference between the groups joined by DNA polymerase and DNA ligase? Describe briefly.
6. What are the three enzymatic activities of DNA polymerase, describe briefly?
7. Why RNA polymerases and primases that initiate DNA replication whereas DNA polymerases cannot?
8. In gel electrophoresis, do smaller double-stranded molecules move slowly or move rapidly than larger molecules, describe briefly?
9. The following sequences of bases are present in one chain of a DNA duplex that has opened up at a replication fork and synthesis of an RNA primer on this template begins by copying the base. T_1 3′...TCTGATA T_1 CAGTACG....5′
   (a) If RNA primer consists of eight nucleotides, what is its base sequence?
   (b) In the intact RNA primer, which nucleotide has a free hydroxyl group (–OH) terminus and what is the chemical group in the nucleotide at the other end of the primer?
10. How continuous and discontinuous modes of DNA replication can be distinguished.
11. What are the differences between DNA and RNA?
12. If a strand or helix of Watson Crick model has a base sequence of 5′-GT CA TG AC 3′, what would be the base sequence of the complementary DNA strand?
13. What is the relation in between the amount of DNA in a somatic cell and is a gamete?

REFERENCES

Hartl, D.L. (1991): *Basic Genetics*, Jones and Bartlett, Boston.
Kornberg, A. (1980): *DNA Replication*, Freeman, San Francisco.
Stent, G.S. (1971): *Molecular Genetics*, W.H. Freeman and San Francisco.
Watson, J.D. (1968): *The Double Helix*, Antheneum, New York.
Watson, J.D. (1970): *Molecular Biology of Gene*, 2nd ed., W.A. Benjamin Inc., New York.

CHAPTER

19

Discovery of Genetic Code

After construction of DNA structure by Watson and Crick in 1953, a new era dawned in the field of genetics. The next phase was the formulation of genetic code. There are two principles related to it. *First*, that it is the exact sequence of nucleotide bases in the DNA polynucleotide chains that represent the genetic information of the gene. That is, the long DNA double helix was to be thought of as a tape on which genetic information is inscribed in a language employing the four letter alphabet, A, G, C and T. This information is inscribed twice in every DNA molecule and hence twice in every gene, since any base on one chain of the two helically intertwined polynueleotides determines its complementary base on the other chain. The *second* principle on account of 'one–gene one–enzyme theory' held that information content of any gene is coded in a triplet code and hence the actual meaning of the four letter alphabetic language recorded in the DNA could not be any thing but a representation of the primary structure of a given polypeptide. Taking together both these principles, the meaning of any purine–pyrimidine base sequence of the DNA polynucleotide chain that constitute a gene, is nothing other than the specification of amino acid sequence of some particular polypeptide chain. This indicated that primary action of a gene consists of the *transcription* of the coded genetic information in DNA into the ribopolynucleotide sequence of mRNA and subsequent *translation* of this information contained in the four letter alphabet AUG and C of mRNA into specific amino acid sequence. That is, gene transcribes its inherited information into messenger RNA and mRNA translates it into the expression of a particular character (phenotype). Thus, the relationship between gene and phenotype is

$$\text{Gene} \rightarrow \text{mRNA} \rightarrow \text{polypeptide} \rightarrow \text{phenotype}$$

This concept of information transfer thus raises some problems, viz:

(a) How many ribonucleotide bases code for a given amino acid, i.e. does each code word a 'codon' consist of one, two, three or more ribonucleotide bases?
(b) Are codes commaless?
(c) If a code composed of two or more nucleotides, is the code overlapping or non-overlapping?
(d) What is coding dictionary?

(e) Are there codes that serve as 'start' and 'stop' signals for translation, i.e chain initiation and chain termination?
(f) Is the code universal for the entire organism?
(g) Does the same codon specify the same amino acid in phages, bacteria, corn, fruit fly and human?

There are 20 amino acids that must be coded by some sequence of four nucleotides bases in the DNA or their complements in the mRNA. The mRNA nucleotides are the nucleotides sequences that code to a particular amino acid called a *codon.* If a codon was to consist of only a single base, the genetic code would have to be very ambiguous, that is, the same codon would have to code for different amino acids under different conditions. This may lead to much confusion in the synthesis of proteins. Therefore, some combination of the bases is needed for each amino acid. Infact more than twenty distinct combinations are needed for polypeptide synthesis because signals are required for starting and stopping the synthesis of each polypeptide chain.

The codons of two bases present some kind of problem by providing only 4 × 4 = 16 codons for the 20 amino acids but three bases group provide 64 codons (4^3), more than enough for the number of amino acids involved. Each codon in mRNA is a sequence of the three adjacent bases that specifies a particular amino acid. The genetic code is set for all the codons (Table 19.1).

TABLE 19.1 Sixty-four codons of the genetic code (hypothetical)

UUU	UCU	UAU	UGU
UUC	UCC	UAC	UGC
UUA	UCA	UAA	UGA
UUG	UCG	UAG	UGG
CUU	CCU	CAU	CGU
CUC	CCC	CAC	CGC
CUA	CCA	CAA	CGA
CUG	CCG	CAG	CGG
AUU	ACU	AAU	AGU
AUC	ACC	AAC	AGC
AUA	ACA	AAA	AGA
AUG	ACG	AAG	AGG
GUU	GCU	GAU	GGU
GUC	GCC	GAC	GGC
GUA	GCA	GAA	GGA
GUG	GCG	GAG	GGG

From theoretical view point, the triplet code includes both 'sense' codon (those that specify a particular amino acid) and 'nonsense codon' that do not specify any amino acid, but have some other functions such as signaling 'start' or 'stop' in the synthesis of polypeptide chain.

CONTEMPLATION OF CODE TABLE (19.1)

The genetic code Table (19.1) reveals the following important features:

1. The code contains many synonyms in that almost all amino acids are represented by more than one code. Three amino acids—arganine, serine and leucine, have six synonymous codons; five amino acids—valine, proline, threonine, alanine and glycine have four synonymous codons; one isoleucine has three synonymous codons and nine amino acids—phenylalanine, tyrosine, histidine, glutamine, asparagine, lysine, aspartic acid, glutamic acid and cysteine have two synonymous codons. Only two amino acids—methionine and tryptophan are represented by a single codon.
2. The code has a definite structure in that synonyms, for the same amino acids are not randomly dispersed in Table 19.1. As is evident from Table 19.1, the synonymous codons are almost always in the same box and hence differ from one another only in the last of their three nucleotides (the only exception to this rule are *arganine*, *serine* and *leucine* which are represented by six synonymous codons and hence are spread necessarily over more than one (codon box).
3. The code contains three nonsense codons—UAG. UUA and UGA leveled as No. 1 amber or signal codons; No.2 and No.3, do not code any amino acid at all. These are also called stop codons.
4. The fourth codon 'AUG' is called *start codon*. It also specifies the amino acid methionine. In a rare case, some other codons in the protein biosynthesis like GUG initiate translation.

The proteins are nitrogenous compounds, which are generally water-soluble by nature but become insoluble or coagulate in the aqueous solution, if denatured by heat or exposed to strong acids. They coagulate approximately half or two-third of the dry weight of the cell. Mulder (1938) was the first who suggested the name protein (Greek *proteos*: Primary) and wrote after six years, "a substance is present in both plants and animals, which performs an important function". It is one of the very complex substances, which under various circumstances may alter their composition and serves as an enzyme for the regulation of chemical metabolism. It is without doubt, the most important of the known components of living matter, and it would appear that, without it, life would not be possible. This substance has been named protein.

By the later half of the nineteenth century, many different kinds of proteins with large molecules were observed which on hydrolysis yielded simpler compounds—amino acids. By 1935, twenty amino acids to be considered essential building blocks of all proteins were discovered. But now, some molecular biologists claim that they have discovered two more amino acids.

NATURE OF GENETIC CODE

Genetic code is still hypothetical in nature. But hypothesis says that code is triplet, commaless, non-overlapping and non-ambiguous (i.e. out of 64 codons, 61 codes for 20 amino acids). Code has polarity, i.e. code has a definite direction for translation of genetic message, i.e. formation of 'reading frame'.

FORMULATION OF CODE

The sequence of four nucleotide bases in DNA molecule store the hereditary potentialities of an organism by synthesing different kinds of proteins. These protein molecules determine different phenotypic characteristics of an organism by their involvement as 'enzymes' and other form of activities like nucleic acids proteins also exist in form of a long chain of amino acid polymers. How the information coded in the linear order in DNA (base sequences) recognizes the different amino acids to be arranged in a protein and the direction of sequence in a polypeptide chain is explained in this chapter?

George Gamov (1954) was the first man who on the theoretical basis suggested that the code was a triplet one, i.e. a sequence of three bases may be coding one amino acid. This was argued that one letter code will recognize not more than four amino acids; a two letter code will recognize not more than 16 amino acids ($4 \times 4 = 16$). However, with a triplet code, the four bases can generate $4 \times 4 \times 4 = 64$ code words. When combined in all possible triplet sequences (4^3) which are sufficient to code for 20 amino acids.

A fundamental question was, whether the different triplet code words in the gene were also separated from one another by spaces or commas? If there were no spaces or commas between the different triplet words, the only way one can make sense out of such a coded message will be to begin reading from a fixed point triplet by triplet. Crick and his colleagues solved this problem.

George Gamov (1954) proposed a overlapping code, when DNA can function as the direct template for the ordered assembly of the specific polypeptides from their amino acid building blocks. But due to internal contradiction in the position, his attention was drawn to the fact that proteins are not synthesized directly on the DNA any way, but some other substance is the immediate template for the protein synthesis. In this system, coding ratio is 1:1.

Crick and his two colleagues then derived a non-overlapping code. In this scheme, each triplet of nucleotide bases (codes) participate in the specification of only one amino acid, where ATT represents one amino acid, GCA another, and so on (coding ratio 3:1), but with such a code new difficulty had to be faced. How is the 'reader' supposed to know which of the triplets to 'read', e.g. in the sequence of ATT, GCA, TCG, TGG,... where ATT suppose to code an amino acid and GCA another, how could this message be read correctly, if the commas were removed? To resolve this difficulty, another restriction was placed on this code. Each triplet code word, such as ATT, TCG, GCA and TGG can be automatically recognized as a word by means of a dictionary of *sense* words. The set of *sense* words would have the property that no *sense* word can be formed by misreading parts of two adjacent sense triplet. Only *sense* words of nucleotide triplets correspond to amino acid in dictionary; all other triplets that might be formed by misreading adjacent sense triplets do not code for any amino acid at all are called nonsense code.

COMMALESS NON-OVERLAPPING CODE

The sequence of amino acid in the polypeptide chain appears to be random. This is possible only when the coding units are 'non-overlapping', no one amino acid always or even occasionally has the same adjacent neighbours. However, if coding units are 'overlapping' in two letters, a given amino acid would always have the same four nearest neighbours (Figure 19.1).

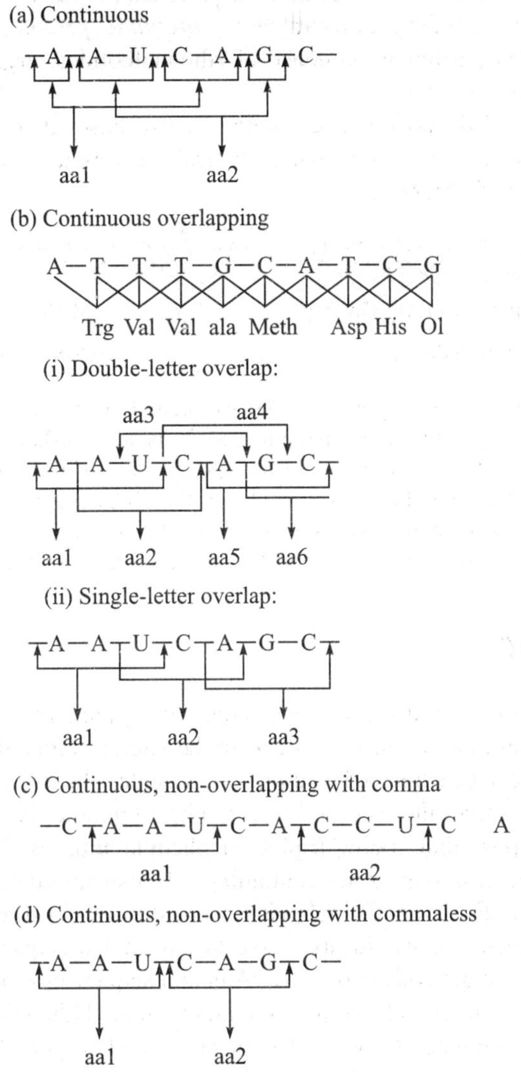

Figure 19.1 Different theoretical coding possibilities.

Even if more than one unit codon may code for a particular amino acid in the mRNA, this constancy in the nearest neighbours would be found with a measurable frequency and randomness in the amino acid arrangement may not be obtained.

One can also presume that a single mutation in the mRNA nucleotide base can affect the position of two or more adjacent, amino acids in the polypeptide chain but only single amino acid change has been found to arise from a single mutation.

A messenger RNA beginning from AACC, GA, GCA, consists of three triplets, AAC, CGA, and GCA, which code for *asparagine*, *arginine* and *alanine* respectively. If the code is overlapping, by two bases, this sequence of nine bases consist of the triplets—AAC, ACC, CCG, CGA, GAG, AGC and GCA, coding for *asparagine*, *threonine*, *proline*, *arginine*, *glutamic* acid, *serine* and *alanine*. In this particular sequence of ribonucleotides, arginine may always occur between 'proline and glutamic' acid.

In the mRNA codons, if the code is overlapping by two bases then, UUC, UGC, CGC code for *phenylalanine*, *serine* and *arginine*, that is entirely different sequence from the original sequence (Figure 19.2) (G.S. Stent, 1963).

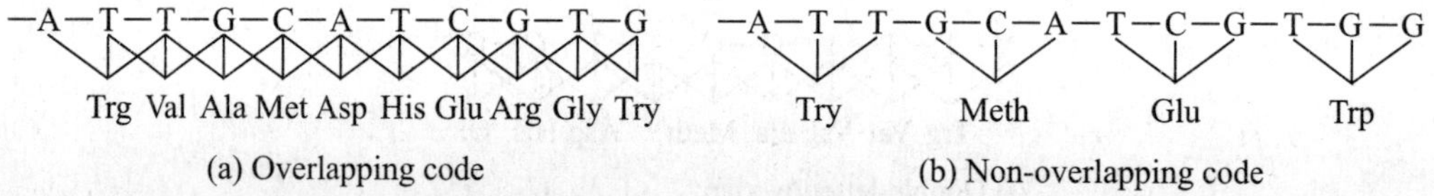

Figure 19.2 Two general type of hypothetical codes related to nucleotide sequence in DNA to amino acid sequence in polypeptide. (a) It is an overlapping code in which every nucleotide of a triplet takes part in the specification of three amino acids. Here, coding ratio is 1:1 of the number of nucleotides of the DNA segment (triplet) to the number of amino acid coded in the polypeptide chain. (b) Non-overlapping code in which three nucleotides take part in the specification of one amino acid and the ratio is 3:1.

COMMALESS CODE

This is a hypothesis that no commas exist between the coding units for an individual polypeptide chain. Introduction of commas may cause problems in the interpretation of data that do not arise if commas are omitted. This can be explained by citing an example. The synthetic polymer polyuridylic acid (poly-4) is known to direct the incorporation of phenylalanine into a synthetic polypeptide 'Polyphenylalanine'. Suppose that coding triplet for phenylalanine is 'UUU' and that 'U' also constitute commas. The other polymers not containing 'U' also stimulate amino acid uptake into artificial polypeptide chain. For example, polyadenylic acid (poly-A) stimulates lysine incorporation. If U are commas, then, we would also have to say that a sequence of A's can serve as commas. Our present view about code in the mRNA is that sequences of nucleotides are arranged in a series of triplets; each triplet code for a single amino acid. Although 'spacer' nucleotides or 'commas' do not occur within each triplet codes for an individual peptide chain, but 'spacer' or 'nucleotide' (commas) must be present between the codes for different polypeptide chains.

CODE DICTIONARY

Nirenberg and Matthei (1961), Nirenberg and Leder (1964) and Khorana, et al. (1967) prepared dictionary of genetic code (Table 19.2), which are responsible for a particular amino acid/acids to be incorporated in a particular polypeptide chain for synthesis of particular protein. From Table 19.2, it is very clear that with exception of *methionine* and *tryptophan*, all other amino acids are coded by more than one codon; as many as six different triplets code for each of three amino acids—arginine, serine and lencine; four different triplets code for each of five amino acids—valine, proline, threonine, alanine and glycine; isoleucine is coded by three codons and nine other amino acids are coded by two triplets each.

TABLE 19.2 Dictionary of genetic code

Amino acids	*Codons*					
Lucine	CCU	CUC	CUA	CUG	UUA	UUG
Serine	UCU	UCC	UCA	UCG	AGU	AGC
Arginine	CGU	CGC	CGA	CGG	AGA	AGG
Proline	CCU	CCC	CCA	CCC		
Threonine	ACU	ACC	ACA	ACG		
Alanine	GCU	GCC	GCA	GCG		
Glycine	GGU	GGC	GGA	GGG		
Valine	GUU	GUC	GUA	GUC		
Isoleucine	AUU	AUC	AUA			
Phenylalanine	UUU	UUC				
Tyrosine	UAU	UAC				
Histidine	CAU	CAC				
Glutamine	CAA	CAG				
Asparagine	AAU	AAC				
Lysine	AAA	AAC				
Aspertic acid	GAU	GAC				
Glutamic acid	GAA	GAG				
Cystine	UGU	UGE				
Tryptophan	UGG					
Methionine		AUG				
Chain terminater			UAA	UAG	UGA(nonsense codon)	

The concept of genetic code is shown in Figure 19.3, which is based on the work of Nirenberg and Matthei (1961) and Khorana, et al. (1967). Based on this concept genetic code can be deciphered.

The final results of code breaking effects by Nirenberg and Matthei (1961) and Khorana, et al. (1967) using artificial messenger RNA are given in Table 19.3.

The features of Table 19.3 are same as that of Table 19.1.

The frequency of proline, arginine, alanine and glycine in primary structure of protein may increase the level of [G] + [C]; whereas more frequent use of phenylalanine, tyrosine, isoleucine, methionine, asparagine and lysine would contribute to a low [G] + [C]; therefore, variations in the [G] + [C] content in the DNA could be utilized as parameters to differentiate the species.

The code contains three nonsense codons—UAG, UAA and UGA, levelled as Non-1, Non-2 and Non-3. Non-1 is called 'amber' and Non-2 'ochre'.

Thus, it fully satisfies the contemplation of theoretical code given in Table 19.1.

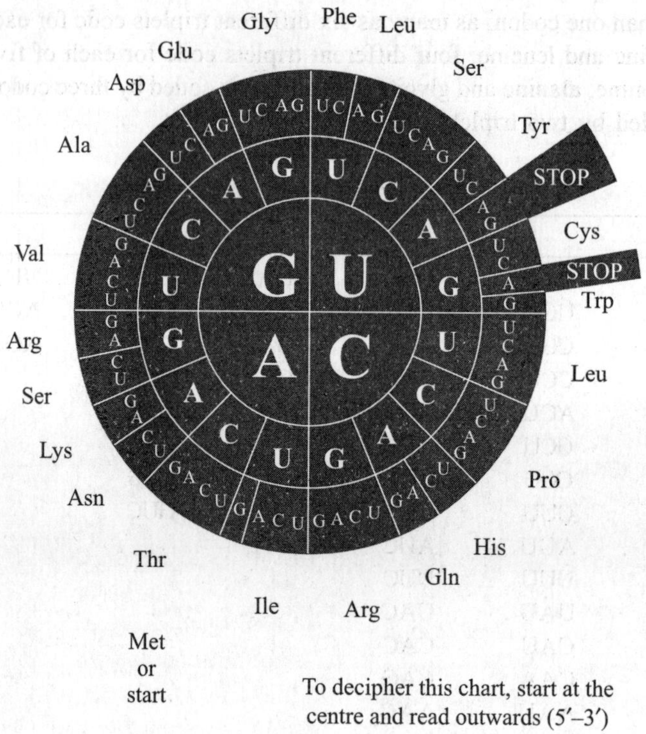

I		II		III	
Alanine	Ala	Glycine	Gly	Proline	Pro
Arginine	Arg	Histidine	His	Serine	Ser
Asparagine	Asn	Isoleucine	Ile	Threonine	Thr
Aspartic acid	Asp	Leucine	Leu	Tryptophan	Trp
Cysteine	Cys	Lysine	Lys	Tyrosine	Tyr
Glutamine	Gln	Methionine	Met	Valine	Val
Glutamic acid	Glu	Phenylalanine	Phe		

Figure 19.3 The concept of genetic code.

TABLE 19.3 Decoding of genetic code

	U	*C*	*Non*-1	*G*	
	UUU	UCU	UAU	UGU	
	Phe	Ser	Tyr	Cys	U
U	Phe	Ser	Tyr	Cys	C
	Leu	Ser	Non 2	Non 3	A
	Leu	Ser		Trp	G
	Leu	Pro	His	Arg	U
	Leu	Pro	His	Arg	C
C	Leu	Pro	Gln	Arg	A
	Leu	Pro	Gln	Arg	G

	Ile	Thr	Asn	Ser	U
	Ile	Thr	Asn	Ser	C
A	Ile	Thr	Lys	Arg	A
	Met	Thr	Lys	Arg	G
	Val	Ala	Asp	Gly	U
	Val	Ala	Asp	Gly	C
G	Val	Ala	Asp	Gly	A
	Val	Ala	Asp	Gly	G

Initiator Codons of Polypeptide Chain (AUG and GUG)

We know that DNA is a polymer of a long chain of polynucleotides and a set of hundreds/ thousands of polynucleotides form a gene and there are many genes on a DNA strand. But there must exist a base sequence separating the adjacent gene; as well as there must exist a type of nucleotide sequence which marks the point from where transcription of genetic information begins in form of messenger RNA. The above two kinds of information assisted in the identification of specific codon involved in initiating the transcription of mRNA molecule at the end of the chain. It was observed that about 45 per cent of proteins synthesized in *E. coli* have methionine as the beginning amino acid of polypeptide. This observation suggested that a large number of proteins synthesized in *E. coli* start their genetic message with AUG, the codon which code for methionine. However, before incorporation into polypeptide chain, methionine is chemically transformed into *formylated* methionine to initiate the chain. Thus, it was proposed that formylated methionine is the initiator of polypeptide chain and the triplet code AUG which codes for methionine should be considered *chain-initiator* triplet in messenger RNA transcription. GUG was also found as an additional codon for initiating polypeptide synthesis, but it codes for valine. But GUG is not as effective as AUG as an initiator codon. Here, Wobble phenomenon was invoked to explain the function of GUG. It was suggested that guanine is the first base of codon GUG, which can also pair with the first base of corresponding 'anticodon' CAC, even if this pairing does not come so naturally as with AUG.

Chain-terminating Codons

Chain-terminating codons are UAG, UAA, UGA, which are also called nonsense codons.

When Crick, et al. in 1961 prepared their view point on the general nature of genetic code, they also argued that there should also be triplets, which may not be coding for any amino acid. They termed such codons as nonsense codons. The evidence in support of such a concept came from the mutants of phage T4 of the rII locus.

The rII locus consists of two adjacent genes A and B whose mutant alleles can be resolved by the gene *complementation test*. One of these mutant alleles got deleted, as a result, their nucleotide sequences came together without leaving any boundary. The result was that rather than directing the synthesis of two separate polypeptide corresponding to gene A and gene B, the welded sequences produced a single polypeptide chain of large sizes. It was realized that deletion had affected the function of gene A, but B was fully functional. This indicated that deletion in A, gene affected incorporation of some of the amino acid and hence stopping the formation of a polypeptide chain.

The deletion of mutant gene A, became subject of interest when Benzer and Champe (1962) observed that if a second mutation was incorporated through recombination in its A segment, B gene should continue to function. But this did not happen so when some of the rII mutant gene alleles were incorporated in gene A locus; this knocked down the function of gene B also. This was only possible when these exceptional rII mutations led to the termination of polypeptide synthesis, as a result, direction of B gene was not translated into action. This is presumed that unlike most of the missense mutations which replace correct amino acid by wrong amino acid the rII mutation ordered a codon which did not recognize any amino acid. Such codons may be called nonsense codons. In this respect UAG, UAA and UGA are considered chain-terminating nonsense codon.

The studies of Brenner, Stretton and Kaplan (1965) and Garen (1968) in *E. coli* provided direct evidence. Brenner and his associates analyzed the polypeptide fragments synthesized under the direction of different amber mutants, the amino acid sequence of these fragments indicated that a nonsense codon in these mutants had replaced a codon which in wild-type strain codes for the tryptophan, serine, tyrosine, leucine and glutamic acids. A reference to the code dictionary revealed that only codon that could arise through a single nucleotide change from the codons of these amino acids is UAG. These studies extended to other amber mutant and their wild type revertant strains revealed that two other nonsense or amber codons are UUA and UGA.

Nonsense Codon

As stated in chapter dealing 'fine structure of gene', Benzer had collected a large number of rII mutants of phage T4. These mutants as he stated were able to grow on the ordinary strains of *E. coli* on which they form plaques of r, mutant morphology; but they can not grow on k-strains, i.e. *E. coli* carrying the lambda phage on which the r^+ wild-type phase is able to grow. In 1960, Benzer, suddenly observed that although no rII mutants were able to grow on any k, bacterial strain, his collection of rII mutants included some ambivalent specimens that can grow on one but not on other variety of k-strain. During course of his studies, he further came across that his collection had three distinct subsets of ambivalent mutants; the members of each subset can grow on one of three k-strains of bacteria but not on the other two. Each subset included about one out of *thirty of the hundreds of mutant sites*, Banzer had identified in both rII A and B genes.

Benzer proposed the mechanism of genetic suppression. He suggested that his ambivalent rII mutants have sustained missense mutations that each of the three permissive k-strains that differ from each other are subsets of ambivalent phage mutants that carry a different missense suppressor mutation. The phage mutant members of that subset would, therefore, be able to grow on only that permissive k-strain whose missense suppressor happens to insert an acceptable amino acid in response to that mutant codon calling for a noxious amino acid. This way activity of the enzyme produced by mutant rII phage gene would be restored.

Among Benzer's collection of rII mutants, there was one mutant in which a long DNA deletion had removed the boundary between the contiguous rIIA and rIIB genes. The consequence of this mutation is to weld the normally separable rIIA and rIIB polypeptide chain into a single polypeptide chain. This long polypeptide chain lack rIIA gene function, but it can supply to the rIIB gene 'substrate' to function in mixed-infection complementation test with ordinary rIIB point mutant.

The ambivalent mutations represent changes of a codon from sense to nonsense rather than to 'missense'. Since the mRNA transcribes from rIIA and rIIB genes, the message is translated from left to right (Figure 19.4). The presence of nonsense codon in the rIIA portion of deletion mutant will arrest polypeptide assembly at that point and rIIB segment can not be synthesized at all. The permissive k-strains, therefore, contain nonsense suppressors, rather than missense suppressors. These nonsense suppressors insert an amino acid in response to a mutant nonsense codon, to which there is no ordinarily corresponding to amino acid at all. The nonsense suppressors of the three permissive k-strains differ with respect to a particular nonsense codon, they mistake for a sense codon or with respect to the amino acid they mistakenly insert in the peptide chain (Stent, 1971).

Wild type:	CAT	CAT	CAT	CAT	CAT	CAT	CAT	CAT	CAT
Base added:	CAT	CAT (+G)	GCA	TCA	TCA	TCA	TCA	TCA	TCA message disturbed
Base removed: A	CAT	CAT (–G)	CAT	CAT	CTC	ATC	ATC	ATC	ATC
Base added:	CAT	CAT	GCA	TCA	TAT	CAT	CAT	CAT	CAT
Base removed:			(+)	(–)					message in phage regained

Figure 19.4 Effect of insertion and deletion of a single nucleotide in the reading frame on the translation of genetic message.

Reading frame and result of shift

Crick, et al. (1961) through their experimental evidences phage mutant rII regions of T4 showed that correct reading of the long nucleotide sequence of a gene is assured by beginning the translation into amino acid sequence from a fixed starting point and proceeding *triplet* by *triplet* to the end. However, if *triplet* reading frame is shifted by one nucleotide at any intermediate point, then the reading of all subsequent triplet is shifted in 'phase' and a totally incorrect translation proceed from that point onward. The results of these experiments suggested that genetic code is rich in synonyms in that most of the 20 standard amino acids must be represented by more than one of the 64 possible nucleotide triplets.

The acridine dyes (proflavin, acridine orange and acridine yellow among others) are the substrates that possess binding property to DNA and at least in phages (rII-T4) act as mutagens by causing additions or deletions in the nucleotide sequence during DNA replication. An acridine molecule may become intercalated between the stacks of base pairs and double the normal distance between adjacent nucleotide pairs. This may allow insertion of a new nucleotide during replication or result in the deletion of a base and, therefore, alter the sequence of bases. Crick and his colleagues studied a number of acridine induced mutations in the rII region of the DNA of phage T4. They found that mutant types arose through the additions or deletions at any of a large number of sites within the rII region; but these may revert to wild type or a very similar one (pseudo wild type), through deletion or addition elsewhere in the rII region (Figure 19.4). That is, in many cases, one mutation may be suppressed by another, particularly if the second change is located relatively close to the first.

Explanation

The *transcription* of genetic message proceeds triplet by triplet from a fixed point to right. The hypothetical 'correct' message of a normal type of polynucleotide triplet code is CAT CAT ... insertion of nucleotide 'G' disturbs the remainder message to TCA TCA ... deletion of nucleotide 'A' disturbs the remainder message ATC ATC. However, insertion of 'G' (guanine) at one site and deletion of A (adenine) restore the correct message CAT CAT (After: F.H.C. Crick, 1962).

The deletion or addition of a single nucleotide alters the reading frame of the entire polynucleotide chain. This alternation is termed a 'frame shift' and a single frame 'frame shift', therefore, may result in a protein so altered in the amino acid sequence that may be non-functional. On the other hand, if the altered reading frame between two opposing events (for example, a deletion followed by an insertion) is of a small magnitude and if the *missense* mutation is of such a nature as to alter slightly or not at all the function of the ultimately produced protein, then second event suppresses the first. The closer the points at which these *opposing* alternation take place, higher is the probability that this suppression will take place. The greatest significance of this work came from the finding that when three additions (+ + +) or three deletion (– – –) mutation occur, the phenotype was not mutant. This is because if 3 (–) or 3 (+) mutation can occur, the phenotype may not be mutant. This is because if 3 (+) or 3 (–) mutation occur, the *reading frame* is altered only between the first and third change. By constructing triple mutant of type (+ + +) or (– – –) it was possible to show that coding is done by nucleotide triplet. Such triple mutants manifest the normal or *pseudo* wild-type phenotype, provided that 'frame-shifted nucleotide sequence' between the two outer most mutant sites does not encompass any 'unacceptable triplet' which may lead to non-functional protein. For example, the triple + + + mutant carrying FCO and two suppressors of FCO form plaque on strain k of the bacteria, even though either singly or in pairs each of these mutant sites produces an rII mutant phase of 'T4' that is quite unable to grow on strain k. It is thus combination of all three mutations of the same sign within same DNA polynucleotide chain that restores the function or partial function to the mutant rII locus (Figure 19.5, Crick, 1962).

This type of result is expected only if the code words are in triplets. This clearly proves that number of nucleotides per code word is a multiple of three (3n), where n is the number of nucleotides insertion or deletion at a time by the mutagenic action of acridines.

It is also unlikely that only 20 of the 64 possible nucleotide triplet represent the standard set of 20 amino acids and that remaining 44 triplets are 'nonsense' and correspond to no amino acid. If that would have been the case, then *the shift in the reading frame* that obtained between (+)

and (–) mutant sites separated by more than a very few nucleotide would invariably lead to atleast one "nonsense" codon, that would interrupt the continuity of the assembly of polypeptide chain and hence would prevent the assembly process from ever reaching those regions of the gene to which the correct reading frame had been restored. It becomes, therefore, quite likely that the code is rich in synonyms in that many of the 20 standard amino acids represented by more than one kind of nucleotide triplet.

Wild type:	CAT	CAT	CAT	CAT	CAT	CAT	CAT	CAT
Three bases added:	CAT	CAT	CGA ↓ (+)	TCA	TGC ↓ (+)	ATG ↓ (+)	CAT	CAT message restore again

Figure 19.5 Insertion of three nucleotides (GGG) at closely situated sites distorts the genetic message over a short DNA segment but leaves remainder of the message unaffected.

NATURE OF GENETIC CODE

The coding problem was solved with the use of bacterial- and mammalian cell-free systems. *In vitro*, it was possible to establish the sequence of different nucleotides triplets directing the incorporation of different amino acids into polypeptide chains. After discovery of RNA polymerase (enzyme) and before it, discovery of polynucleotide phosphorylase, it became possible to synthesize the polynucleotide chains of a defined sequence. The use of synthetic RNA polymers as messengers proved very helpful in identifying the specific nature of genetic code.

Cell-free System

A cell-free protein synthesizing system is worked out in a manner that stimulates the conditions existing in the living cells. For such synthesis, the system is designed to study the functioning of genetic code by extensive use of a single kind of messenger RNA (mRNA) of known nucleotide sequence. The cell's own mRNA of molecules were destroyed by incubating the extract with enzyme ribonuclease and DNA present in the extract was destroyed by deoxyribose nucleage enzyme. It has been explained that messenger RNA carries the information from DNA of gene to transfer RNA to assemble the required amino acid to ribosome the 'work shop' for a particular protein synthesis as desired by the organism.

Deciphering the Codons using Synthetic Polynucleotides

Students must know the work of Nirenberg and Matthei (1961) and Khorana, et al. (1967) who solved the coding problem by using bacterial and mammalian cell-free extracts. *In vitro*, the cell-free extract system assisted in determining the sequence of different nucleotide triplets (codons) directing the assembly of different amino acids into polypeptide chain formation. With the discovery of enzyme RNA polymerase and earlier of polynucleotide phosphorylase, it was possible to synthesize polynucleotide chains of defined sequence with the help of these synthetic RNA polymers as messengers; it was possible to understand the specific nature of genetic code.

The cell-free protein synthesizing system is so prepared that it creates the environment in the living cells for such synthesis. The system set-up for understanding the genetic code was provided with the mRNA of known nucleotide sequence and cells own mRNA were destroyed by incubating the cell-free extract with the enzyme ribonuclease and DNA present in the extract was destroyed by deoxyribonuclease enzyme. Thus, the mRNA which carries the direction from gene's DNA to the ribosomes in cytoplasm for protein synthesis were destroyed.

Nirenberg and Matthei (1961) were the first to decipher the genetic code using cell-free extract. They found that addition of 10 micrograms of polyuridylic acid (UUUUUU... UUUU), a synthetic polynucleotide, containing only the base 'uracil' per millilitre of the cell-free extract brought rapid incorporation of L-phenylalanine. No other synthetic polynucleotide was found to replace polyuridylic acid for the incorporation of chloramphenicol of phenylalanine. Another polymers containing both adenylic and cytidylic acids were completely inactive in such incorporation because incorporation was inhibited by antimetabolites like puromycine and chloramphenicol. The specific binding of phenylalanine by polyuridylic acid was confirmed in protein synthesis by C^{14} labelled 20 amino acids. On the basis of these observations, Nirenberg and Matthei concluded that a sequence of uracil bases (UUU...) constitute the code for phenylalanine incorporation in protein synthesis. Nirenberg and his colleagues as well as in a parallel study, Ochova and his associates by using cell-free extract system containing synthetic mRNA polymers of different kinds identified different code words. For example, use of synthetic messenger in form of polycytidilic acid and polyadenylic acid brought to notice that CCC may be code word for proline and AAA for lysine (Figure 19.6).

UUU	UUU	UUU	UUU	UUU		CCC	CCC	CCC	CCC
Phe	Phe	Phe	Phe	Phe		Pro	Pro	Pro	Pro

AAA	AAA	AAA	AAA	AAA
Lys	Lys	Lys	Lys	Lys

Figure 19.6 In a cell-free extract poly 'U' binds phenylalanine; poly 'C' binds proline and poly 'A' binds lysine for incorporation in fresh protein synthesis. These triplet code words were first derived from the studies of Nirenberg and Matthei (1961).

Role of Copolymers

The studies with polynucleotides were extended with use of copolymers, as synthetic messengers are the cell-free extract system containing two or more bases in a definite ratio. The randomly ordered polyribonucleotides were found to direct the incorporation of appropriate amino acids in protein synthesis. For example, a copolymer containing uridylic and cytidylic acids with its two bases can generate eight different types of triplet sequences, if randomly distributed. Thus, if proportion of two bases in the copolymer is known, e.g. 75 per cent for uridylic acid and 25 per cent for cytidylic acid, it is hoped that UUU will be the most frequently occurred triplet with a probability of $3/4 \times 3/4 \times 3/4 = 27/64$ and UUC will be in frequency of $3/4 \times 3/4 \times 1/4 = 9/64$.

A chain of polyribonucleotides containing uracil and adenine (U–A), uracil and cytosine (U–C) and uracil and guanine (U–G) in different proportions were synthesized and were used as synthetic messengers. The incorporation of a given amino acid by such synthetic polynucleotides was to vary according to the base ratio of the polymer, i.e. amount of different amino acids incorporated in a polypeptide was found to be correlated with the frequency of different triplet sequences in the synthetic copolymers. The use of synthetic copolymers of different kinds described as above could give information on the composition of base sequences of the codons needed for incorporation of different amino acids for the synthesis of different kinds of proteins.

Synthetic Messengers with Predetermined Base Sequences

Nirenberg and Leder (1964) were the first to begin identifying the precise sequence of the bases in the different triplet codes by substituting synthetic messengers containing randomly distributed bases by those in which the sequence of nucleotides had been predetermined. It was noticed that code word could not be smaller than a triplet. Further, it was noted that binding of different amino acids was being done in a selective manner by the different trinucleotide sequences tested in a reaction mixture, i.e. trinucleotide UUU, CCC and AAA specifically directed the binding of phenylalanine, proline and lysine respectively. Similarly, binding of C^{14} valine was directed by trinucleotide CUU not by UCC, UUG or by dinucleotide, UU, UC or UG. Thus, CUU is the codon of valine.

This way, most of the 64 codons (triplet code) could be assigned to various amino acids using this technique. A group of the codons were found to bind one and same amino acids.

Synthetic Messengers with Repeating Base Sequences (Hargovind Khorana, 1969)

The idea of using messenger RNA of predetermined base sequence adapted by Nirenberg was taken a step ahead by Hargovind Khorana and his associates. They used deoxyribopolynucleotide (DNA polymer) of a known sequence. They wanted to study not only the translation of genetic information in form of protein but also transfer of genetic information from DNA to RNA in form of RNA transcript. Khorana used DNA polymer in which base sequence was so designed to synthesize a predictable kind of protein. He started his experiment to genetic message in form of a short deoxyribopolynucleotide chain containing repeating nucleotide sequences. He used these short polynucleotides chains as *templates* for producing/synthesizing larger DNA polymers of repeating sequences. Thereafter, these larger DNA polymers were used as *templates* for synthesis of messenger RNA molecules. This RNA transcript produced through complementary base pairing rule also showed a repeating nucleotide sequence. One of the first synthetic messenger RNA produced by Khorana with repeating sequence was AAG. This was formed to direct the synthesis of three different homopolypeptide containing a repeating sequence of a single amino acid-lysine, arganine and glutamic acid. But the problem was, which one of the three possible codons AAG, AGA, GAA code for which amino acid. By using Nirenberg and Leder's trinucleotide technique of the binding of the amino acid to ribosome, Khorana and his associates were able to find out that AGA codes for arganine; AAG codes for lysine and GAA codes for glutamic acid.

While continuing this experiment, with repeating sequence of nucleotides of different types, they were able to direct the synthesis of polypeptides with an alternating sequences of amino acids. One such repeating dinucleotide sequence was in the form of UCUC UC UCUC UC UC ... UC. This messenger was found to direct the synthesis of two amino acids—serine and leucine coded by UCU, CUC. This finding made it clear that genetic code is a triplet code and neither of the two codons was coding for more than one amino acid.

Similarly, the repeating sequence UGUGUG ... UG was found to code for amino acid arginine and glutamic acid in an alternating sequence; the sequence UGUGUG ... UG was found to direct the synthesis of copolypeptide containing cystine and valine; messenger AC AC AC ... AC directed the synthesis of copolypeptide containing threonine and histidine. This is the finding of Khorana and his colleagues which established many of the characteristics of genetic code. This way 64 codons of genetic code were identified by him.

DECIPHERING THE GENETIC CODE BY AMINO ACID REPLACEMENT

Analysis of the Primary Structures of Wild Type and Mutant Proteins

The gene is pictured as a sequence of adjacent nucleotides in the DNA polymers. Gene is divisible by mutation because any single nucleotide can be replaced by another. The gene is divisible by recombination, which can take place between the pairs of adjacent nucleotide along the polynucleotide chain. Gene is actually the functional unit. It controls the synthesis of protein (structural proteins).

The sequence of amino acid in the polypeptide chain is dictated by nucleotides in the corresponding gene. The mutations at the different position within gene (polynucleotides polymer) ultimately brings changes in the structure of the polypeptide chain. On this basis, we conclude that gene controls the primary structure of individual and specific protein molecules. Thus, mutation in the structure of a gene results in the primary structure of protein molecule. The amino acid replacement gives rise to the mutant proteins by bringing qualitative changes. The particular location at which the amino acid substitution occurs within the polypeptide chain is important in determining the ultimate effect of mutation on the biological activities and other qualities of protein. In addition, the kind of amino acid substitution is also important because substitution of one kind of amino acid by the other containing dissimilar side chain may show a drastic effect than replacement with similar amino acid. The polypeptide chain has a linear structure; same is true for messenger RNA and DNA also. Therefore, above three classes of molecules are collinear.

Quasi-universality of the Code

Genetic code was discovered in 1960. All codons were identified by means of experiments in test tubes with synthetic polynucleotides to produce amino acid sequences. *So far, it has been determined that the sequence that codes for a particular amino acid in a bacteriophage protein is precisely the same sequence code for the amino acid in human being also.* Of course, this is a *tentative* view not universal. The latest discovery is that codes are not frozen but evolving continuously. Let me briefly explain how?

Burn and Brunitto in 1989 resolved that *actual origin and evolution of the genetic code is still unsolved problem*. They are of the view that with the evolutionary view point, the genetic code actually to be considered as tRNA code and not an mRNA code as it is given in Table 18.3. Because amino acids are picked up from cytoplasmic pool by tRNA molecules having anticodons rather than codons of mRNA. These tRNA anticodons find complementary codons to translate the message of gene. There are 64 mRNA codons out of which 61 are translated; some anticodons pairs with 2–3 codons. Therefore, total number of anticodons may be lower than mRNA codons.

It is believed that mitochondria arose from the symbiogenesis of purple respiring bacteria during the evolutionary stage; this simplification in mammalian mitochondria resulted minimum 22 anticodons translating 60 codons for 20 amino acids. This deviation is given in Table 19.4.

TABLE 19.4 Deviation from the universal code by mitochondrial DNA

Codon	*Universal code*	*Mitochondrial DNA code*
UGA	Termination	Tryptophan (yeast), human, termination in plants
AUA	ileu	Methionine (human)
CUA	leu	Threonine (yeast)
CGG	arg	Tryptophan (plants)

Thomas Juke in 1983 suggested that first step in the evolution of code is duplication of the existing *anticodon* followed by nucleotide substitution of G for U in one of the duplicates. This way, new anticodon, GAC pairs with codon CUU and GUC; other anticodon, UAC pairs with GUA and GUG The original *anticodon* again can undergo duplication and nucleotide substitution is one of the duplication to produce *anticodon* UAC and CAC. Later pairs with GUG codon. A third duplication and substitution is GAC *anticodon* results in the formutation of AAC anticodon. Due to change in the aminoacylation site in one of the tRNA duplicates, a quite different amino acid is specified by the changed *anticodon*. This pattern of duplication and substitution in *anticodon* followed by changes in the aminoacylation specificities has resulted increasing number of *anticodons* for amino acids as well as degeneracy of the codon. It is believed that A and C were not the part of *anticodon* in the *primitive* code. These are later addition as evolution proceeded. This process continues until the living system became too complex to tolerate any change and code became frozen.

EVOLUTION OF GENETIC CODE

The studies conducted on both prokaryotes and eukaryotes (including man) indicated that code (Tables 19.1 and 19.3) have universal validity. This concept has been validated by mutant amino acid replacement as summarized in Table 19.5. All these replacements have been obtained by single base substitution of codon assignments in *E. coli* as well as in tobacco mosaic virus (TMV) coat protein and human haemoglobin. This universality of the code among the contemporary organisms shows that code has remained immutable over long evolutionary period. There are two explanations for it:

TABLE 19.5 Amino acid replacement

Amino acid	*Possible replacements**												
ala	asp	*glu*	*gly*	pro	ser	thr	*val*						
arg	cys	gln	*gly*	his	*ile*	leu	*lys*	met	non	pro	*ser*	*thr*	trp
asn	*asp*	his	ile	*lys*	*ser*	thr	tyr						
asp	*ala*	*asn*	glu	*gly*	his	tyr	val						
cys	arg	*gly*	non	phe	ser	trp	tyr						
gin	*arg*	*glu*	his	leu	lys	*non*	pro						
glu	*ala*	*asp*	*gln*	*gly*	*lys*	non	*val*						
gly	ala	*arg*	*asp*	*cys*	*glu*	non	ser	trp	*val*				
his	*arg*	asn	*asp*	*gln*	leu	pro	*tyr*						
ile	arg	asn	leu	lys	*met*	phe	ser	*thr*	*val*				
leu	*arg*	gln	his	*ile*	met	non	*phe*	pro	ser	trp	val		
lys	arg	*asn*	gln	*glu*	ile	met	non	thr					
met	arg	ile	*leu*	lys	thr	val							
non	arg	cys	*gln*	*glu*	gly	*leu*	*lys*	*ser*	*trp*	*tyr*			
phe	cys	ilu	*leu*	ser	*tyr*	val							
pro	ala	arg	gin	his	*leu*	*ser*	*thr*						
ser	ala	*arg*	asn	cys	*gly*	ile	*leu*	non	*phe*	pro	thr	trp	tyr
thr	*ala*	arg	asn	*ile*	*lys*	*met*	pro	ser					
trp	arg	cys	gly	leu	*non*	ser							
tyr	asn	asp	*cys*	his	non	*phe*	ser						
val	*ala*	*asp*	*glu*	*gly*	ile	leu	*met*	phe					

* Italicized amino acids have been observed as replacements in mutations (Ram, 2005).

(i) *Stereochemical theory:* It is stated that, contrary to the chain of reasoning that had led Crick to postulate the amino acid adaptor in the first place, there does exist a steric relation between a codon (or anticodon), nucleotide triplet and amino acid chain that it designates. Therefore, the code table would derive both its structure and its evolutionary stability from the pre-ordained stereochemistry of its elements.

(ii) *Frozen accident theory:* This states that the structure of the code table evolved by chance but that once the complete codon assignments had been made in a common ancestor cell of all present day living form; further evolutionary diversification of the code became well nigh impossible to change (Stent, 1971).

There is a reason for that, any mutation engendering a change in codon or codon assignment would have been lethal to individual in which it could occur.

EXCEPTION TO UNIVERSAL GENETIC CODE

Genetic Code is Evolving, not Frozen

Genetic code was discovered in 1960. All the 64 codons were identified by means of experiments in a cell-free extract with the acid of synthetic polynucleotides to produce amino acid sequences.

However, there are eight departures from the universal genetic code: five of these are in mitochondria and three are in nuclear codes. These changes are preceeded by disappearance of a codon from coding sequences in mRNA of an organism or organelle. The function of codon that disappears is taken by other synonymous codons so that there is no change in the amino acid sequences of proteins. The deleted codon then reappears with a new function. This may be implemented in one of the three ways—(a) by a mutation in anticodon, (b) by change to a different amino acid in amino acylation of tRNA molecule, while retaining same anticodon, and (c) in mitochondria by change in the codon—anticodon pairing.

The *first* procedure occurred in changing codon UGA from stop to tryptophan, codon UAR from stop to glutamine and codon AAA from lysine to asparagine.

The *second* procedure occurred in changing codon CUN from leucine to threonine; CUG from leucine to serine. The *third* procedure was in AGR from arganine to serine, AGR from serine to stop and in CAU from only pairing with AUG to pairing with AUA and AUG. These changes were non disruptive (Thomas, S. Juke and Syzo Osawa, 1991). All these codons were found to be the same as those in Table 19.1, which were considered universal genetic code. But later on, a number of departures/changes from universal genetic codes have been reported; some in mitochondria and some in nuclear codes. These changes are in codons UGA from stop to tryptophan; AUA from isoleucine to methionine; UAR (R = A or G) from stop to glutamine, AGR from arganine to serine; AAA from lysine to asparagine; CUN (N = C, A,G) from leucine to threonine; CUG from leucine to serine and AGR from serine to stop (Table 19.6, Juke, T.H and S. Osawa, 1991). These changes contradict two earlier theories:

1. That the code is frozen and any change could not be tolerated because it may induce changes in amino acid sequences of a protein, which could be injurious and disruptive (Crick, 1968).
2. That the code is stereochemical, i.e. the result of some kind of an affinity between an amino acid and its anticodon with which the codon pairs during protein synthesis (Junke, 1978).

TABLE 19.6 Changes in universal genetic code

Codon in universal genetic code	*Changes to*	*System*	
		Nuclear	*Mitochondria*
UGA stop	Tryptophan	Mycoplasma	All except plants, yeast
AUA isoleucine	Methionine to isoleucine		Echinoderms
AGR arginine	Serine	Metazoa	Except vertebrate
AAA lysine	Stop asparagine		Flat worms
UAR stop	Glycine	*Acetabularia* (*ciliated* protozoa)	Echinoderms
CUN leucine	Threonine		Yeast
CUG leucine	Serine	*Candida cylindracea*	
UGA stop	seCys	Vertebrates, eubacteria	

The new discoveries of changes in the code did not abolish the idea of the universal genetic code because the mitochondria had very small genomes and only a few short proteins. Therefore, mitochondria in contrast to the whole organism can tolerate changes in code.

It was reported from Osawa's laboratory that a complete organism *Mycoplasma capricolum* used stop codon as well as regular codon UGG for coding tryptophan (Yamao, et al., 1985). Thereafter, several other laboratories found that ciliated protozoa used stop codons UAA and UAG to code for glutamine as well as regular codons CAA and CAG (Helftenbein, et al., 1985; Horowitz, et al., 1985; Preer, et al., 1983; Caron, et al., 1985 and Hanyu, et al., 1985). A yeast, *Candida cylindracea*, uses leucine codon, CUG in its nuclear code to code for serine. (Kawaguchi, et al. 1989).

Juke (1985), Osawa and Juke (1989) proposed *the codon capture hypothesis.* This states that codons may disappear during evolution from coding sequences in the DNA of an organism/organelle and then may later reappear with a different assignment and such changes may take place without altering the amino acid sequences in protein. In short, the changes are neutral. Many codons may have been changed to a different assignment once they have disappeared from reading frame. Such a change in codon occurs due to directional mutation pressure leading toward either a greater GC or a greater AT content of DNA (Sueoko, 1982, 1988). Tryptophan will remain in the same position in the proteins of mycoplasma, but it will have a new codon. Many TGA codons in the protein gene of mycoplasma appear TGG tryptophan position in the corresponding *E. coli* protein gene remains unchanged. Various experimental finding substantiate this idea. For example, genes in *Mycoplasma capricolum* for two tryptophan tRNA's one with anticodon CCA and the other with UCA are arranged in tandem, close together on the chromosome. This suggests gene duplication followed by mutation of anticodon CCA to UCA (Yamao, et al. 1988).

The case of *Mycoplasma capricolum* gave the experimental evidence for codon capture. Mycoplasma DNA is very high in AT (75%). This indicates AT replacing GC by directional mutation pressure. UGA stop codon disappeared and replaced by UAA, so UGA became an assigned codon.

Juke and Osawa (1991) gave the model for the changes in the mitochondrial code. According to them, a codon disappear from coding sequences, typically as a result of directional mutation pressure. The code reappears, after sometimes as a result of a change in directional function. This may result from a change in an anticodon or from change in amino acylation of a tRNA or from a change in codon—intercodon pairing in mitochondrial DNA (Table 19.7, Juke, T.H. and S. Osawa, 1991).

TABLE 19.7 Changes in codons (underlined), while amino acid sequence remains constant

Amino acid	*Asn*	*Lys*	*Lys*	*Asn*	*Lys*	*Asn*	*Lys*	*Asn*
Original sequence	AAC	AAG	AAA	AAC	AAG	AAU	AAA	AAU
	GUU	CUU	UUU	GUU	CUU	AUU	UUU	GUU
GC pressure	AAC	AAG	AAG	AAC	AAG	AAC	AAG	AAC
AAA, UUU	GUU	CUU	CUU	GUU	CUU	GUU	CUU	GUU
Disappear	<u>AAA</u>	AAG	<u>AAG</u>	AAC	AAG	<u>AAA</u>	<u>AAG</u>	AAU
Anticodon changes	IUU	CUU	CUU	IUU	CUU	IUU	CUU	IUU
Amino acid	Asn	lys	Lys	Asn	Lys	Asn	Lys	Asn

ROLE OF DIRECTIONAL MUTATION PRESSURE IN THE EVOLUTION OF CODE

It has been already explained that UGA and UGG both originally coded for tryptophan. Under GC pressure, UGA disappeared and anticodon UCA mutated to CCA pairing only with UGG; UGA reappeared when some UAA stop codon mutated to UGA. In addition, it is also proposed that AUA and AUG were methionine codon. Under GC pressure, AUA disappeared and CAU anticodon paired with AUG for methionine.

Juke (1966) proposed that the present code has evolved from an earlier code consisting entirely of the family boxes.

In his co-evolution model, Wong proposed (1976) that genetic code initially contained only the small number of amino acids that were formed in the primordial soup. These amino acids later served as precursor for the formation of other amino acids along prehistoric pathways, which evolved into the amino acid biosynthesis pathways for present day organisms. Thus, amino acids were evolved in two phases:

(i) *Phase* I: Amino acids are—glycine, alanine, serine, aspartic acid, glutamine, valine, leucine and isoleucine.
(ii) *Phase* II: Amino acids are—arginine, histidine, tryptophan, methionine, asparagine, glutamine acid and lysine.

Codons of phase II amino acids replace codons of phase I amino acids and phase II enter the code as a result of biological synthesis from phase I amino acids.

ANTICODON AND tRNA

Soon the results of decoding experiments indicated that, the genetic code contains synonymous codon. Such as lucine is coded by both UUC and UUG, it became matter of keen interest to acertain how such synonymous codons are decoded in the polypeptide assembly process. Since more than one species of tRNA may be affined to the same amino acid, it appeared most plausible that every codon is in correspondence with its own tRNA species one that carries the complementary anticodon.

It has been established that of the 64 possible codons, 61 represent an amino acid. On this basis, one might expect that every cell contains 61 species of tRNA, each endowed with one of the 61 anticodons. And in apparent conformation of that expectation, more and more tRNA species were being discovered upon further efforts to fractionate tRNA extracts from such biological sources as *E. coli* and *yeast*. The adaptor hypothesis leads to the following image of the amino acid assembly process, its incorporation into the nascent polypeptide chain. Each amino acid molecule is fitted out with a nucleotide adaptor (tRNA) that contains a nucleotide triplet or anticodon complementary in its nucleotide sequence to the nucleotide triplet or codon, that codes for the same amino acid in the mRNA. The amino acid–nucleotide complexes then diffuse into ribosome, where they are held in their proper places on the mRNA templatc by pairs of hydrogen bonds between complcmentary purine and pyrimidine of adaptor (tRNA) and messenger RNA molecules. Once lined up in this way along the mRNA in the correct order, the individual amino acid residues are lined to each other through peptide bond by means of chemical rearrangement. That simultaneously liberates the amino acid from its bond to the nucleotide adaptor and turn it to the nascent polypeptide chain.

DEGENERACY OF CODON

It has been shown in Table 19.3 that the code is extensively degenerative with several codons specifying the same amino acid, with exception of AUG for methionine and UGG for tryptophan; a certain order to this degeneracy can be discerned. In many cases, the only two bases of codon are characteristic and critical part of the codon for given amino acid, whereas third one may be read either as a purine only (viz., glutamine, CAA and CAF) or as a pyrimidine only (viz., histidine, CAU and CAC). In other cases, third position is read as any base, e.g. AC for thrconine, CC for proline, etc.

Crick (1966) proposed Wobble theory to account for lack of specificity in the five base of the *anticodon* which may pair. However, possible combinations are not limitless, but restricted to those shown in Table 19.8.

TABLE 19.8 Pairing between codon and anticodon at third position

Anticodon	*Codon*
A	U
C	G
G	U, C
Ic(inosine)	U, C, A
U	A, G

The reason for this is due to the conformation of the anticodon loop. The base at the first position of anticodon influences the number of possible bases recognized at the third position of the codon. If the first anticodon base is G, the third codon base may be either C or U. This is known as *Wobble hypothesis*. The first anticodon bases A or C allows only recognition of U or G in the codon of mRNA since this loses the fidelity of hydrogen bonding 'inosine' is the anticodon may pair with uracil, cytosine or adenine ribonucleotides of codon. Inosine is derived from ademine by deamination of the six carbon to produce a six keto group. It shows pairing with cytosine.

REVIEW QUESTIONS

1. If DNA contained only the bases—adenine and thymine, how long a code word would be necessary to enable coding for each of 20 different amino acids.

2. (a) Can you devise a non-overlapping commaless triplet code for 20 different amino acids in which codes can be translated correctly without necessity of a reading frame?

(b) What theoretical arguments can you present against the existence of this type of code?

3. A particular DNA base sequence transcribed into messenger RNA is TTATCTTCGGG AAAACA:
   (a) If reading begins at left, what amino acids are coded by the sequence?
   (b) If proflavine treatment caused the deletion of the first adenine nucleotide on the left, what changes will occur in the first six amino acids coded by this sequence?
4. If a polyribonucleotide contained equal amounts of randomly positioned adenine and uracil bases, what proportion of triplets will code for (a) phenylalanine, (b) isoleucine, (c) leucine and (d) tyrosine?
5. In what sense and to what extent, the genetic code is (a) degenerative and (b) universal.
6. Why is colearity between codons and polypeptides significant?
7. How was genetic code first decided? What refinements have been incorporated in the technique.
8. What do you mean by deciphering of the genetic code? How many of the 64 possible codon are functional? Whether different species use the same codons or different?
9. Who was first to crack the genetic code? Give a brief account of his work done?
10. Are the results of the experiments cracking genetic code *in vitro* valid *in vivo*? Is the code universal, i.e. do the codons specify the same amino acids in all the organism?

REFERENCES

Bennet, J.C. and W.J. Dryer (1964): Genetic coding for protein structure, *Ann. Rev. Biochem.*, **33,** 205.

Benzer, S. and S.P. Champe (1961): Ambivalent rII mutants of phase T4, *Proc. Nat. Acad. Sci.*, Washington, **47,** 1025.

Benzer, S. and S.P. Champe (1962): A change from nonsense to sense in the genetic code, *Proc. Nat. Acad. Sci.*, Washington, **48,** 1114.

Benzer, S. and S.P. Champe (1962): A active cistron fragment, *J. Mol. Biol.*, **4,** 288.

Brenner, S.A., O.W. Stretton and S. Kaplan (1965): Genetic code the nonsense triplets for chain termination and thier suppression, *Nature*, **206,** 1994.

Caron, F., E. Meyer (1985): Does *Paramecium primaurdia* use a different genetic code in its macronucleus? *Nature*, **314,** 185–188.

Crick, F.H.C. (1962): The genetic code, *Scientific American,* October.

Crick, F.H.C. (1963): On genetic code, *Science*, **139,** 461–464.

Crick, F.H.C. (1966): Codon—anticodon pairing: the Wobble hypothesis, *J. Mol. Biol.*, **19,** 548–535.

Crick, F.H.C. (1968): Origin of genetic code: the Wobble hypothesis, *J. Mol. Biol.*, **38,** 367.

Crick, F.H.C., C.L Brenners Barnett and R.J. Wattstobin (1961): General nature of the genetic code for proteins, *Nature*, **192,** 1227–1232.

Deer, J.R. Deer, L.B. (Jr), B.M. Radman, and A.J. Barnett (1985): Deviation from universal code shown by the gene for surface protein. 51A in *Paramecium*, *Nature*, **314,** 188–190.

Frisch, l cd (1967): The genetic code, Cold Spring Harbour Symposium, *Quant. Biol*, 1966.
Gamov, G. (1954): Possible relations between deoxyribonucleic acid and protein structure, *Nature*, **173,** 318.

Garen, A. (1968): Sense and nonsense in genetic code, *Science*, **160,** 149.

Hanyu, N. Y. Kuchino, S. Nishimura, and H. Beier (1986): Dramatic events in ciliate evolution. Alteration of UAA and UAG termination codons to glutamic codons due to anticodon mutation in two *Terahymena* tRNAs Glu E MBO J.S: 1307–1311.

Hartman, Philip E. and R.S. Sigmund, (1968): *Gene Action*, 123–137.

Helftenbein, E. (1985): Nucleotide sequence of a micronuclear DNA molecule coding for alpha-tublin from the *ciliate Stylanychia lemnae*. Special codon usage. TAA is not a translation termination codon, *Nucleic Acids Res*., **13,** 415–433.

Juke, T.H. (1933): Evolution of the amino acid code, *in*: *Evolution of Gene and Proteins*., M., Alci and R.K. Koehn (Eds.), Sinauer Assoeintes in. Sundleand Mass.

Juke, T.H. and S. Osawa (1991): Recent evidences for evolution of genetic code *in*: *Evolution of Life*, 79–95.

Junke, J.R. (1978): The genetic code as a periodic table, *J. Mol. Biol.*, **11,** 211–224.

Khurana, H.G., H., Bochi, H., Ghos, N., Gupta, T.M., Jacob, et al. (1967): Polynucleotide synthesis and genetic code, *Cold Spring Harbour Symposium*, **31,** 39–49.

Nirenberg, M.W. and J.H. Matthei, (1961): The dependence of cell-free protein synthesis in *E. coli* naturally occurring or synthetic polyribonucleotides, *Proc. Mat. Acad, Sci.*, USA, **47,** 1602–1888.

Nirenberg, M.W. and P.L. Leder (1964): RNA code words and protein synthesis: the effects of trinucleotides upon the binding of sRNA to ribosomes. *Science*, **145,** 1399–1407.

Ram, M. (2005): *Unfolding Reality of life—A DNA world: Dominant,* Delhi.

Sarabhai, A.S., A.O.W. Strotion and S. Brenner (1964): Colinearity of the gene with polypeptide chain, *Nature*, **201,** 13–17.

Stent, G.S. (1971): *Molecular Genetics*, N.H. Fremon and Co., San Francisco.

Sucoka, N. (1988): Directional mutation pressure and neutral molecular evolution, *Proc. Nat. Acad. Sci.*, USA, **85,** 2653–2657.

Thomas, H. Juke and Syzo Osawa (1991): Recent evidence for evolution of genetic code *in*: *Evolution of Life*, Springer, Tokyo, pp. 79–96.

Woese, C.K. (1967): *The Genetic Code The Molecular Basis for Genetic Expression*, Harper and Row, New York.

Woese, C.R. (1970): The problem of evolving genetic code, *Bioscience*, **20,** 471–485.

Yamao, F., A. Muto and Y., Kawauchi I et al. (1985): UGA is read as tryptophan in *Mycoplasma capricolum*, *Proc. Nat. Acad. Sci.*, USA, **82,** 2306–2309.

Yancfsky, C. (1967): Structural relationship between gene and proteins, *Ann. Rev. Gene*, **1,** 117–138.

Zinder, N.D., Englehardt and R.E., Wibster (1966): Punctuation of UN genetic code, *Cold Spring Harbour Symposium Quant. Biol.*, **31,** 251–256.

CHAPTER

20

Gene Regulation and Gene Expression (Transcription and Translation)

According to the classical genetics theory, gene is the unit of recombination with other similar unit and has the ability to mutate into alternative form and function to the endowment of the organism with some particular phenotype effect. But according to the molecular genetic studies, gene is a fragment of DNA (polynucleotide chain) that determines the synthesis of polypeptide chain (proteins), which frequently functions as enzyme in the biosynthesis or degradative process in the cell. The number of protein molecules produced by a gene varies from gene to gene. For example, certain products such as tRNA molecules, rRNA molecules, ribosomal proteins, RNA polymerase components (polypeptides) and other enzyme catalyzing metabolic processes that are known as 'house keeping functions' are very essential components of all living forms. The genes that produce such type of products which are continuously needed for 'house keeping' are known as *constitutive genes.*

However, genetical studies indicate that gene does not directly function to synthesize a product (protein) but its expression (function) is regulated by a set of genes. These genes are called *regulatory genes* and the genes which perform a particular function are known as *structural genes.* The regulatory genes are found both in prokaryotes and eukaryotes, but they differ in the number of nucleotides present in the fragment.

GENE REGULATION

It is quite interesting to know that the replication of DNA during mitosis results genetically identical copy of daughter cells. Similarly, cells derived from the zygote should also be genetically identical. But all cells in the higher animals are not physically identical. For example, the cells of heart and liver/pancreas have common genetic constitution, but phenotypically and physiologically they are different. It means all the genes may not be functioning together or encoding the product which may be translated into functional protein together in all the cells every time; there may be differences in the gene expression in the different tissues. Let us examine this differential mechanism in gene expression in both, prokaryotes and eukaryotes separately.

A. Gene Regulation in Prokaryotes

The prokaryotes are free-living unicellular organisms, which grow and divide indefinitely till they get proper environment. The adaptability of bacteria and other prokaryotes depend upon their capacity to 'turn on' and 'turn off' the expression of a specific set of genes to meet the demand of environmental changes the bacteria encounter for its survival. The expression of a gene is 'turn on', when the products of a gene/genes is in demand and 'turn off', when their products are not needed. Same situation may apply to the free-living unicellular eukaryotes, such as yeast, algae and protozoa also but not much information is available on these organisms. Bacteriophages are less-responsive to the environmental changes because of very short-life span; of course, they are responsive to the changes in the host.

The ultimate product of a gene is a polypeptide chain, which frequently acts as an enzyme. There are three levels of gene regulation for the synthesis of these enzymes.

First: Gene regulates the synthesis of enzyme by controlling the *transcription* of mRNA for these enzymes. It is called *transcriptional control.*

Second: Control occurs at *translation* level.

Third: There is control over the activity of the enzyme whose synthesis is not regulated. It is called *posttranscriptional control.*

One of these, is through the *operon.* A group of adjacent genes whose transcriptions are regulated by other sites 'on' and another is end product inhibition 'off'.

Transcriptional regulation through operon (lac operon in E. coli)

Among many genes of the colon bacteria (*Escherichia coli*) two have been widely studied to understand the transcriptional regulation of prokaryotic genes. Two enzymes are needed for *lactose* metabolism in *E. coli*. First is *galactoside permease* which is responsible for transport of lactose into the cell and maintains its concentration in the cell. Second is *betagalactosidase*, which catalyses the hydrolysis of *lactose* to *galactose* and *glucose*. The wild-type *E. coli* produces these enzymes. There is a third enzyme, known as *thiogalactoside transacetylase* present in a very small amount (ten molecule per cell), when lactose is not present. In the presence of *lactose*, the rate of synthesis of these enzymes is increased. Thus presence of *lactose* induces the synthesis of enzymes. The enzymes whose synthesis is increased with the supply of the substrate are called *inducible enzymes*.

On the *lac* segment in the *E. coli* chromosome (DNA), three genes are located—*z* for *betagalactosidase*, *y* for *galactoside permease* and *a* for *thiogalactoside transacetylase*. All the three genes *z*, *y*, *a* are closely linked and regulated together. Although François Jacob and Jaques Monod in 1961 developed a novel model of genetic system that regulates the *lactose* metabolism in *E. coli* to which they named *operon* but later on, several other scientists developed similar regulatory system in bacteria and other multicellar organisms but every one reported that bacterial genes are regulated through *operons*.

Each of the three genes *z*, *y* and *a* is responsible for the synthesis of one enzyme and they are linked on the DNA in the same order and all the three *transcribe* genetic information into a single mRNA molecule that is subsequently translated into the polypeptide chains of the three enzymes. The RNA polymerase get attached to the promoter site (p) and then transcription proceeds in $5' \rightarrow 3'$ direction beginning with the operator (o) that lies right of p or downstream of p, and proceeds towards structural genes, like

$$\underline{p, o, z, y, a}$$
Operator

The three genes together with promoter and operator sites form the *lac operon*, thus, an *operon* consists of a system of genes (structural), operator (o) and promoter (p) sites. The basis for the regulation of synthesis of the three enzymes lies still another site of the regulator (i) and interaction of its protein product with lactose and with operator site. The regulator gene is located at the site of promoter in the *lac operon*, i.e. the genetic map of operon could be *pozya* which together consist of 4700 nucleotides (Figure 20.1).

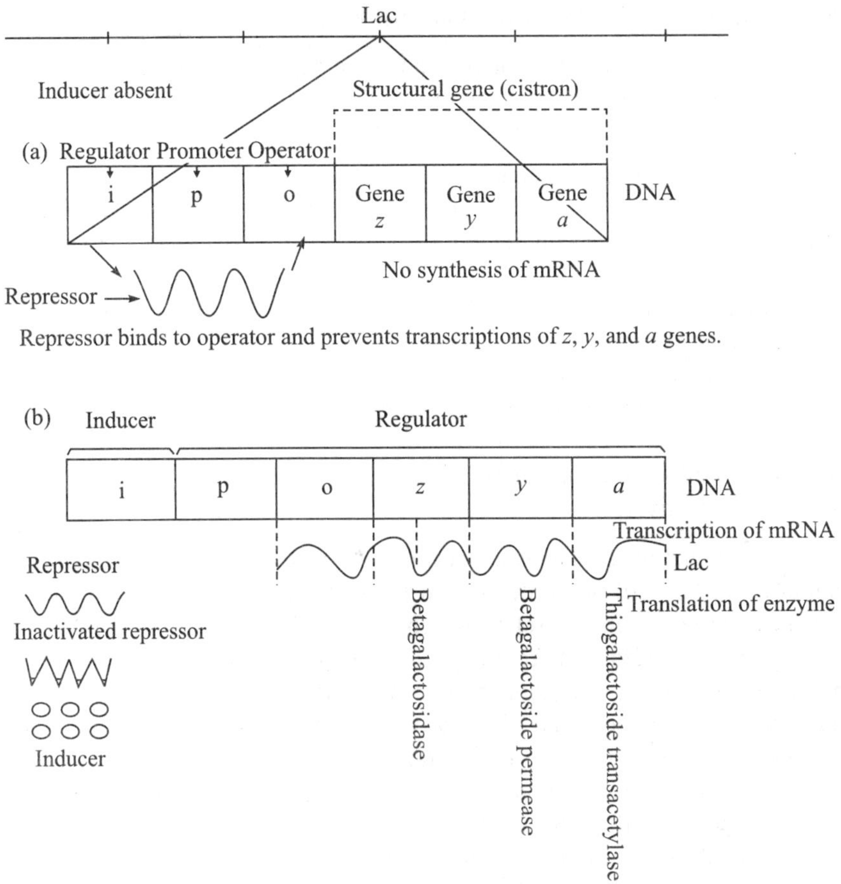

Figure 20.1 Model of *lac operon* of *E. coli*: Transcription is initiated at operator site (o).

The regulator site produces a protein called repressor which in absence of lactose (inducer) binds the operator and inhibits the attachment of *RNA polymerase* with promoter and without polymerase, the genes *z*, *y*, *a* are unable to *transcribe* the required mRNA. But when lactose is present, it binds the 'repressor', changes its shape and prevents it from binding the promoter and the three structural genes *transcribe* the mRNA. The lactose is a substrate for the synthesis of *betagalactosidase* and *galactoside permease*; thus it (lactose) acts as inducer of the operon. *Transcription* is initiated at the *operator* site and *translation* begins in the gene z, at eleventh nucleotide instead of first.

Since *lactose* is the main substrate, which is to be metabolized, for the production of galactose and glucose through hydrolysis, the theory of repressor (protein) and inducer (lactose) appears to be additional burden imposed by its researchers. Because without lactose (substrate), no metabolic activity can take place in the same way as without air in the tubes, how one can drive the bicycle. However, this needs further investigation.

The operon model

1. The lactose metabolic system is controlled by two kinds of genes:
   (i) Structural genes needed for transport and metabolism of lactose.
   (ii) Regulatory genes (lac *i* genes), the lac operator o and the lac promoter (p) together are designated as 'lac operon'.
2. The genes lac *z* and lac *y* encode a single polycistronic mRNA molecule. This mRNA contains a third gene lac *a*, which encodes the enzyme *transacetylase*. This enzyme is used in metabolism of certain *betagalactosides* other than *lactose* and performs no other functions.
3. The promoter for genes lac *z*, and lac *y*, and lac *a*, of mRNA is located adjacent to lac *o*, operator region.
4. Lac *i* gene produces the 'repressor' (protein) which binds to the operator.
5. When 'repressor' binds to the operator, initiation of transcription is prevented by RNA polymerase.
6. The inducer (substrate) stimulates mRNA synthesis by binding to operator and thereby repressor is inactivated.

The regulation of enzyme synthesis is possible only when operator gene is present adjacent to the structural genes (*z*, *y*, *a*) of the operon. At the same time proximity of repressor '*i*' is not necessary because lac *i*, is soluble protein which is diffusible throughout the cell.

Since repressor (*i*) is diffusible from $i^+ p^+ o^+ z^+$ chromosome to the other chromosome $i^- p^+ o^+ z^-$ and represses i^- gene; even though later is linked to i^- gene, the regulator gene is thus functions in either *cis* (on the same chromosome) or *trans* (on the different chromosome) position. But operator always function in *cis* state.

Since the direction of *transcription* is from promoter (p) to operator (o) to gene *z*, *y*, and *a*; mutation in one gene will affect the expression of genes further along the operon. Such types of mutations are called *polar mutations*. For example, mutation in gene *z* will affect functions of all the three genes and mutation in *y* will affect *y* and *a*, and mutation in a will affect a only.

There are some strains of colon bacteria (*E. coli*) derived by mutation from *inductive* wild type which produces enzymes regularly whether *lactose* is present or not. These are called as *constitutive* strains. The *constitutive* strains of *E. coli* are designated as defective regulator (i^-) or a mutant operator that is unable to bind repressor such as $i^+ p^+ o^+ z^+ y^+ a^+$ is an inductive strain where i^+ produces a repressor (protein) that binds to o^+ and prevents transcription of z^+, y^+ and a^+. Presence of lactose inactivates the repressor as a result o^+ allows *transcription*. On the other hand, $i^- p^+ o^+ z^+ y^+ a^+$ genes are constitutive because repressor is defective, which is unable to bind o^+ and stop *transcription*. Therefore, an $i^+ p^+ o^+ z^+ y^+ a^+$ is called *operator constitutive* strain of bacteria which is also constitutive because repressor binding site is defective. Here gene i^+ is dominant over gene i^-. The lac *i* gene is apparently a regulatory gene whose product is an inhibitor that keeps the system turned off, where lac i^- mutant lacks ability to produce inhibitor and stop *transcription*. The product of lac-*i* hence is a protein molecule called the lac repressor.

The regulator gene may function in either *cis* (on the same chromosome) or *trans* (on the different chromosome). However, operator (o) functions only in the *cis* position.

Mode of regulation

The enzyme being synthesized may be degradative or biosynthetic in metabolic pathways. The molecular mechanism for each of these regulatory patterns vary quite widely but usually fall in one of the two major categories—*negative* regulation and *positive* regulation. In a negatively-regulated system, a *repressor* is produced which blocks the expression of *lac* genes present in the cell and prevents *transcription* of mRNA. This *inhibitor* is nothing but *repressor* and its antagonist is *inducer*. In positively-regulated system, an *effector* molecule is present which is also protein, which activates promotor. But negatively- and positively-regulated systems are not mutually exclusive. There are some systems where both are regulated to respond different conditions in the cell. A degradative system may be regulated either *negatively* or *positively*. The presence of product decreases enzyme synthesis and the absence of product increases enzyme synthesis (Figure 20.2).

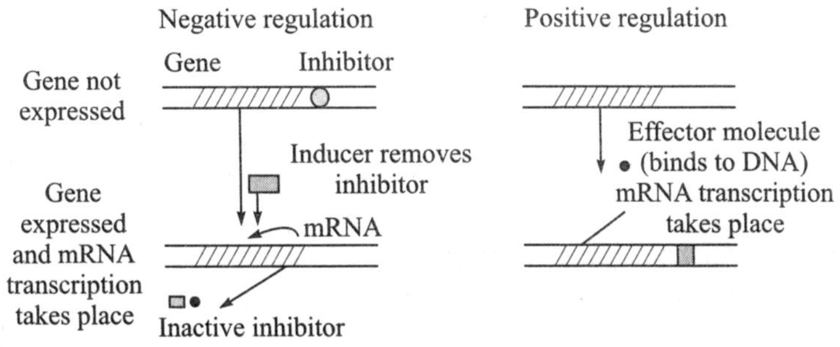

Figure 20.2 Differences between positive and negative regulation.

E. coli lactose system

Escherichia coli and many other bacteria can grow on any of the several carbohydrates, e.g. sucrose, glucose, lactose, galactose and arabinose etc. as carbon source. The *E. coli* cells growing in sugar lactose medium as carbon source synthesize two proteins (enzyme), necessary for lactose metabolism—the enzyme *β-galactosidase* that cleaves lactose (a β-galactoside) to produce galactose and glucose and a carrier *β-galactoside permease*, which is required for passage of lactose through the cell membrane into cell. These enzymes are not needed to *E. coli*, if lactose is not present in the environment.

Positive regulation of lac operon

The function of *β*-galactosidase in lactose metabolism is to form glucose by cleaving lactose. The other cleavage product 'galactose' is also ultimately converted to glucose by the enzymes of the 'galactose' operon. Therefore, if both glucose and galactose are present in the growth medium, activity of lac operon is not needed because no *β-galactosidase* is formed until all the glucose in the medium is consumed. Non-synthesis of *β-galactosidase* is the result of non-synthesis of *lac* mRNA, i.e. no *lac* mRNA is synthesized in presence of glucose because glucose functions as *inducer* as well as inactivate lac i, repressor; the inhibitory effect of glucose on the expression of lac operon is indirect.

Cyclic adenosine monophosphate (cAMP) system

The small molecules of cyclic adenosine monophosphate (cAMP) are found in the cells of animals and many eukaryotic organisms, which regulate the function of many hormones (Figure 20.3). It is also found in *E. coli* and in many other bacterial organisms.

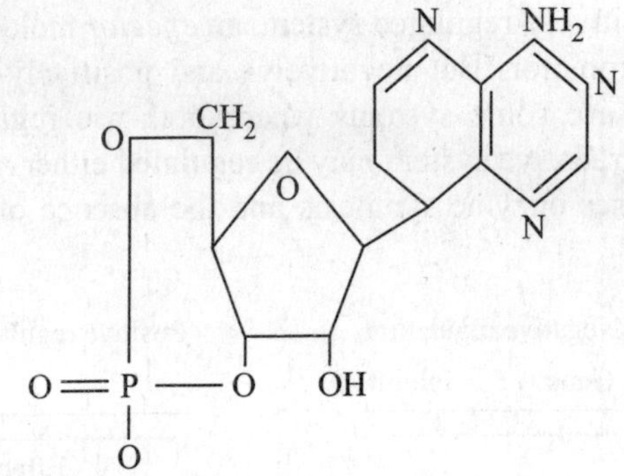

Figure 20.3 Structure of cAMP.

Cyclic adenosine monophosphate (cAMP) is synthesized by an enzyme *adenyl cyclase* and its concentration is regulated by glucose metabolism. When bacteria are growing in medium containing glucose, concentration of cAMP in the cell is low.

'Turning on' of the *lactose operon* (lac operon) also requires molecules of catabolic receptor protein (CRP), which is bound to the molecule of cAMP. CRP is synthesized by a gene *crp* and cAMP is made from ATP (adenosine triphosphate) with the aid of an enzyme adenylate cyclase, synthesized by gene *cya*. Both genes *crp* and *cya* are on the left of promotor (Figure 20.4). Mutation in any gene fails to synthesize *lac* mRNA. This indicates that both CRP function and cAMP are required for *lac* mRNA synthesis. CRP and cAMP bind to one and other and form unit cAMP-CRP and transcribe the synthesis of *lac* mRNA. The presence of glucose interferes the synthesis of cAMP and as a result CRP–cAMP cannot be formed. This halts the transcription of lac operon.

Because *z*, *y*, *a* genes are 'turned on' in presence of cAMP–CRP complex, this complex is a positive regulator of *lac operon* and if regulated negatively called 'turned off' by repressor. This CRP–cAMP complex is a positive regulator in contrast with repressor (Figure 20.5).

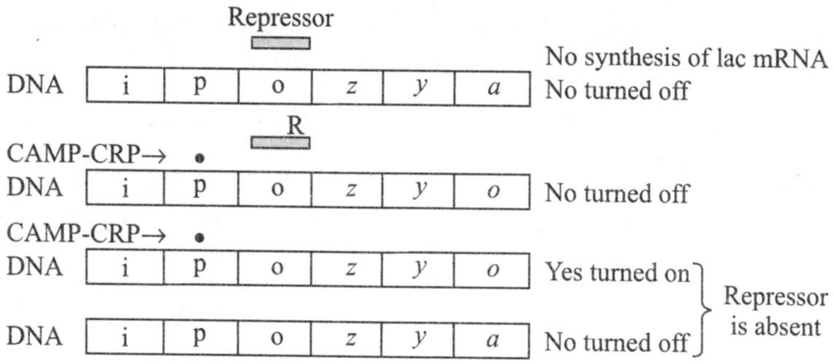

Figure 20.4 Three states of lac operon. It shows that *lac* mRNA is transcribed only when cAMP–CRP is present and repressor is absent.

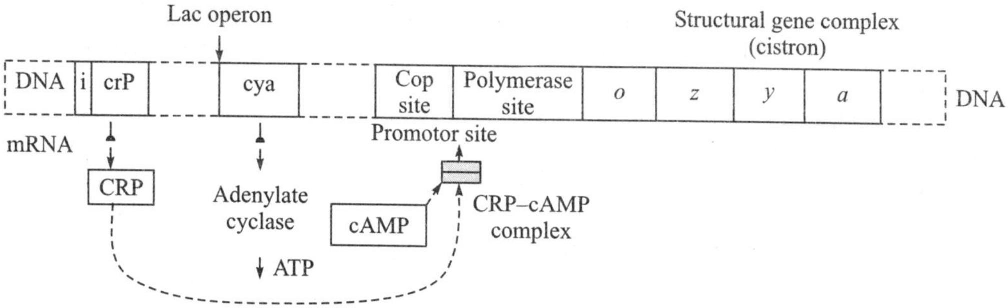

Figure 20.5 The CRP–cAMP positive control of *lac operon* in *E. coli.*

There are differences in the number of copies of *Betagulactosidase permease* and the *lac acetylase* produced in the cell (ratio—1.0: 0.5: 0.2). These differences in the *translational* regulation are due to.

1. The lac *z* gene is translated first and lac mRNA molecules are frequently detached from its translating *ribosome* following chain termination. This probably occurs due to AUG codon.
2. All bacterial mRNAs are degraded to nucleotides after several rounds of translation. Degradations of lac mRNA is initiated more frequently in the lac '*a*' gene than in the *y* gene and more often in lac *y* gene than lac *z* gene. $z < y < a$ and the number of copies per cell are in order of $z > y > a$.

The polypeptide synthesis takes place from 5′ terminus to 3′ terminus of mRNA molecule. This is called 'polarity', which occurs with most polycistronic mRNA molecules. Thus in prokaryotes, the over all expression of the activity of an '*operon*' is regulated by controlling the *transcription* of a polycistronic mRNA and the relative concentration of proteins encoded in the mRNA are determined by controlling the frequency of initiation of translation of each *cistron*. The mechanism by which *transcription* is regulated varies from one system to other. For example, an '*inducer-repressor*' system is common to many 'operon' responsible for degradative metabolism and cAMP–CRP is an element in many carbohydrate-degrading system. Similarly, regulatory system also differs. For example, two promoters are present in the galactose-operon whereas 'arabinose' is regulated by positive mechanism (Figure 20.6).

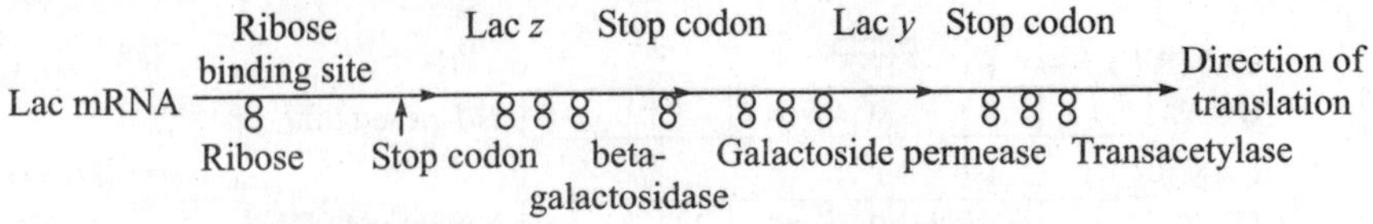

Figure 20.6 Structure of galactose operon.

Tryptophan galactosidase operon: a biosynthetic system

The 'trp operon' of *E. coli* is responsible for the synthesis of amino acid tryptophan. The regulation of 'trp operon' occurs when 'tryptophan' is absent, i.e. when adequate amount of tryptophan is present, transcription of 'trp operon' is inhibited, but when supply is insufficient, transcription occurs. In this way 'trp operon' is distinct from the 'lac operon'. Here, tryptophan acts directly as *repressor* than as *inducer*.

Secondly, since 'trp operon' encodes a set of biosynthetic rather than degradative enzyme; neither glucose nor cAMP–CRP plays any role in operon activity.

A simple 'on–off' system as in the 'lac operon' is not optimal in tryptophan biosynthesis.

Tryptophan is synthesized in five steps; each step is catalyzed by a particular enzyme. In *E. coli* chromosome, the genes encoding these enzymes are translated from a single polycistronic mRNA. The trpE gene is first translated. In between 'promoter' and 'operator', there are two regions designated as leader (trpl) and attenuator (trp A) and its repressor is symbolized as 'trpR, which is located far away from the 5 structural genes (Figure 20.7).

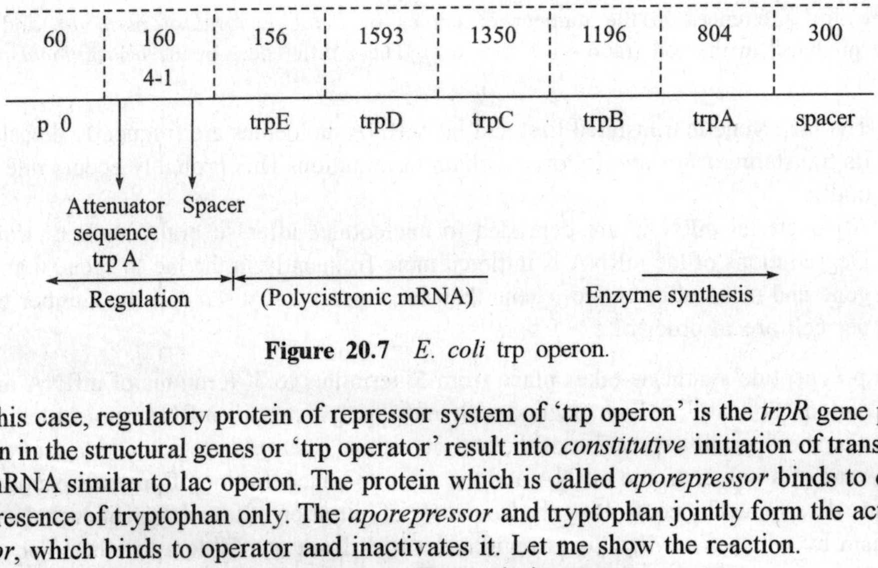

Figure 20.7 *E. coli* trp operon.

In this case, regulatory protein of repressor system of 'trp operon' is the *trpR* gene product. Mutation in the structural genes or 'trp operator' result into *constitutive* initiation of transcription of trp-mRNA similar to lac operon. The protein which is called *aporepressor* binds to operator in the presence of tryptophan only. The *aporepressor* and tryptophan jointly form the active *trp-repressor*, which binds to operator and inactivates it. Let me show the reaction.

Aporepressor alone → No repressor → Transcription occur

Aporepressor + Tryptophan-active repressor → Inactivates the operator → Transcription does not occur.

This indicates that 'repressor' is active only when tryptophan is present which inhibits the transcription of trp-mRNA. When supply of tryptophan from outside is reduced, operator becomes free and transcription begins. This is the basic 'on–off' regulatory mechanism.

Histidine-operon system

The *lac operon* involves an active repressor but not all '*operons*' function in this manner. '*Histidine-operon*' in *Salmonella typhimurium* closely related to *E. coli* is responsible for the biosynthesis of amino acid *histidine.* In *S. typhimurium,* synthesis of *histidine* is catalyzed by ten steps controlled by nine enzymes, synthesized by a group of nine genes. Regulator produces an inactive repressor protein that is not able to suppress the activity of '*operator*' gene in the absence of *histidine*, as result the RNA polymerase moves through the nine genes and transcribes a long chain of mRNA that codes for all the nine enzymes needed for *histide* synthesis. If *histidine* is produced in excess, it combines with the repressor and converts it into active form. In this state histidine called as '*core repressor*', it binds the operator gene and blocks the transcription of '*operon*' but histidine synthesis occurs. It *(his operon)* looks like '*lac operon*' but it differs in the sense that it exerts a negative control because here *repressor* protein is responsible for turning the 'operon-off'.

Arabinose-operon system

Here in *E. coli*, an interesting variation occurs in the operation of operon. In a segment of DNA molecule, three genes, (each responsible for synthesis of one enzyme), an *inhibitor* site, an *operator* and a *regulator* are present. The regulation of three genes which synthesize three enzymes which catalyze the breakdown of pentose sugar '*arabinose*' to *xylose*-5-*phosphate* (another pentose sugar) is a three-step process. For example, the regulator gene produces a protein that serves both as *activator* and also as repressor of three genes. In absence of '*arabinose*', this regulatory protein serves as an active repressor by binding to the operator. However, when *'arbinose'* is present, it binds to the regulatory protein and removes it from the operator. Thus, in absence of the *inducer* (*arabinose*), the *regulatory* protein functions as *repressor* and prevents transcription of Ara-mRNA by binding to the operator. This is called negative control. But same protein in presence of inducer (arabinose) is converted into an activator and on binding to the operator allows the transcription of mRNA. Here it plays a positive control.

Conclusively, the above system indicates three types of 'operon' control over gene activity/ expression:

1. *Lac operon* shows negative control in which the active repressors inhibit the gene expression by binding to the operator gene. In order to allow the gene to express, the repressor (protein) must be removed from the 'operator' by adding inducer (lactose). Glucose (a catabolite) demonstrates a positive control over '*lac operon*'.
2. *His operon* demonstrates negative control because 'his-repressor' binds to the operator only after binding to histidine.
3. *Ara operon* is an example of both negative and positive control because in absence of inducer (arabinose), the regulatory protein represses the 'operon' to transcribe mRNA and in presence of inducer (arabinose), it stimulates the operon to transcribe mRNA.

Post-transcriptional control

Sugar is the basic material which is broken down by different enzymes to produce amino acids and finally protein and the entire process is controlled by genes which transcribe a particular mRNA and the mRNA encodes the translation of metabolites into protein (enzyme), which catalyze the amino acid synthesis and finally protein.

However, in *E. coli*, it was seen that biosynthesis of isoleucine is blocked at threonine → isoleucine pathway (end product of five-step conversion by five enzymes), if isoleucine is added exogenously in the culture medium containing the bacterium. This indicates that the bacterial cells use the added 'isoleucine' and cease to synthesize their own isoleucine. However, synthesis of five enzyme is not inhibited, except inhibition of activity of last (5th enzyme responsible for deamination of threonine to alpha-ketobutyrate) is inhibited by the end-product, i.e. isoleucine (Figure 20.8).

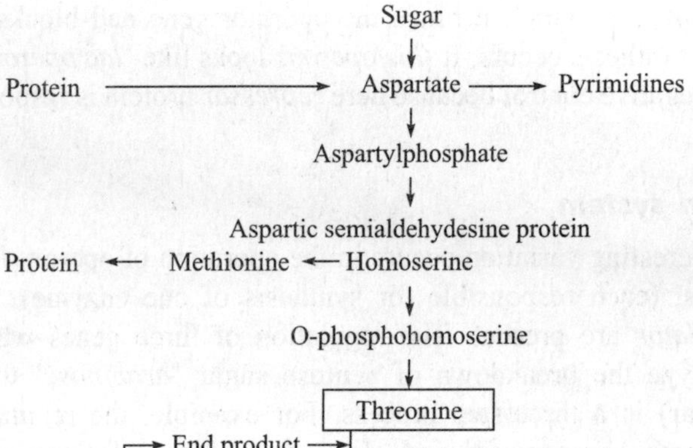

Figure 20.8 Exogenous addition of 'isoleucine' in the bacterial culture of *E. coli*, which inhibits the deamination of 'threonine' to *a*-ketobutyrate. This is called end-product inhibition.

But how this happens. In fact inhibition occurs due to binding of isoleucine (end product) to an enzyme, which changes the shape of the enzyme and thereby change the enzyme activity. Thus, only when tryptophan is present, an active repressor molecule inhibits the transcription and termination of transcription depends upon concentration of the tryptophan in the cell.

It has been studied that a 162 base leader (non-coding DNA) sequence is present at the 5′ end of the trp mRNA. A mutation in which 28 bases from 123–150 are deleted, which synthesize the 'trap' enzyme in both the repressed cells and constitutive mutants with higher rate than normal, stop the synthesis of tryptophan. This indicates that bases 123–150 possess regulatory activity. In normal bacteria, after intiation of transcription, most of the mRNA terminates in this region (123–150), except that when tryptophan is present. The region where termination occurs contains 140 nucleotides which stop the trp gene to code the tryp-enzyme. The 28 base region of mRNA where termination occurs and regulater is called *attenuator*, which forms stem and loop figure in the mRNA (Figures 20.9 and 20.10).

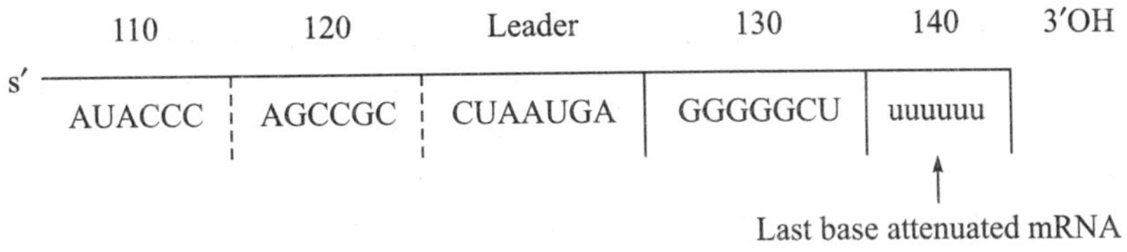

Figure 20.9 Terminal region of trp leader mRNA (L).

	Leader polypeptide													trap, E. protein				
	met	lys (dotted)	Ala	lie	phe	val	leu	lys	Gly	trap	trap	arg	thr	ser	stop	met	gln	thr
CCC AAG (23)	AUG	AAA	GCA	AUU	UUC	GUA	CUG	AAA	GGU	UGG	UGG	CGC	ACU	UCC	UGA	AUG	CAA	ACA

Figure 20.10 Sequence of trp leader mRNA code for leader polypeptide.

Mechanism of premature terminations of transcription

Termination of *transcription* of 'trp' mRNA takes place on account of translation of the leader *peptide* because with the two 'trp' codons, present in this sequence (dotted), the translation of the sequence is sensitive to the concentration of charged tRNA trp. This indicates that when supply of tryptophan is reduced, the translation will be lower down at 'trp' codon (UGG, UGG), i.e. in bacteria, transcription and translation are coupled.

Repression and attenuation

The trp '*repressor–operator*' system may not be operating as single 'on–off' switch but may be producing some intermediate product. Repression is dominant regulatory mechanism in higher concentration of tryptophan, whereas attenuation is not relaxed until starvation for tryptophan become severe, i.e. charging of tRNA of trp gene occurs, when concentration of tryptophan reached quite low in the cell.

Many *operons* responsible for amino acid synthesis are regulated by atteunuator equipped with *base-pairing* rule. In *E. coli*, *salmonella typhimurium* except for *phenylalanine* operon, each amino acid lacks repressor–operator system and is regulated by attenuation.

A common intermediate in the sythesis of the aromatic amino acid is made by three distinct enzymes, each having same enzymatic activity. These three enzymes may be needed when all the three aromatic amino acids are also synthesized. Here, each enzyme is independently regulated by an amino acid specific-repressor.

Autoregulation

A variety of proteins are made from transcripts that are initiated at instant. With some gene products, the requirements of a cell vary greatly and the rate of transcription of the corresponding gene matches the need. One mechanism for the regulation of synthesis of monocistronic mRNA is autoregulation. In the autoregulation systems, the gene product is also a repressor; it binds to an operator site adjacent to promoter. When the concentration of gene product exceeds the product occupies by the operator, transcription of mRNA is stopped and when gene product will have the site; promoter will be free and transcription will begin.

Gene fusions

The '*pur-operon*' is reponsible for synthesis of purines. It is located downstream (in the direction of transcription) from lac-operon in the *E. coli* chromosome (Figure 20.11). Between these two operons is gene *tsx* which governs sensivity to the phage T6. The lac z^- is a mutation of lac z^+, the *tsx-s* culture of phage T6 is due to deletion in the lac *z* gene. This mutation removes the RNA polymerase termination sequence *z* in the 'lac-operon' and brings it near to 'pur operon'. The joining of two operons is called gene fusion.

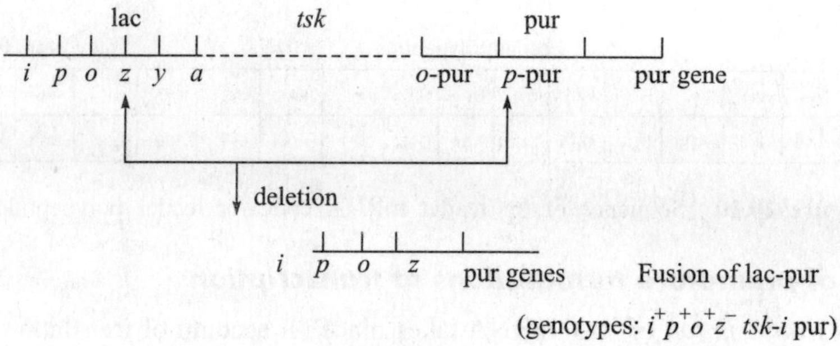

Figure 20.11 Gene fusion in T6 phage.

B. Gene Regulation in Eukaryotes

The eukaryotes particularly higher organisms well known as multicellular ones possess genomes whose size, structure and regulation differ from prokaryotes drastically. There is no polycistronic mRNA in eukaryotes, each mRNA is responsible for the synthesis of one polypeptide chain. Another striking difference between eukaryotes and prokaryotes is that in higher eukaryotes comparatively little of DNA (about 10 per cent) codes for structural information.

Because of equatorial distribution of chromosomes during mitosis, all the cells in eukaryotes presumed to have same genetic constitution, yet all cells in higher animals are not physically and functionally identical. For example, liver, lungs and pancreas have common genetic constitution but phenotypically and functionally different. While discussing gene regulation in prokaryotes, we have seen that all genes may not function in all cells at one time or the encoded products may not be translated into functional proteins in all cells at one time. There may be differential gene expression in different tissues. This indicates that some system exist in the cell to turn gene 'on' and 'off' or in prolonged suspension.

So for the regulation of structure and function of two prokaryotes the *E. coli* and *Salmonella* genes have been discussed. Let us now discuss the structure and regulation of viral genes that parasitizes the eukaryotic cells.

Genome structure of simian virus 40

Simian virus 40 (SV 40) is a small, spherically 45 nm in diameter with a protein composed of 70 subunits and a circular, double-stranded DNA chromosome possessing 15260 base pairs. Nucleosomes are present containing histones H2A, H2B, H3 and H4. All are provided by the host cell.

SV 40 lyses monkey cells but causes oncogenic transformation of other mammalian cells. The isolated culture of SV 40 from monkey infects other monkey cells in "*in-vitro*" culture. In about 1 per cent of infected cells SV 40 chromosome rapidly replicate and about 1 million viruses are produced in the cell nucleus, within 48 hours after infection. The infected cell lyses and viruses are released into medium. *If the tissue culture of human or mice is infected, oncogenic transformations take place.* The viral DNA become integrated into DNA of at least one chromosome of human cell preferably chromosome 7, may be through crossover mechanism in the same manner as *h* virus DNA get integrated with *E. coli* DNA as in RNA of mammory tumour virus. Integration of viral DNA with host DNA is essential for oncogenic transformation. This way at least one cell out of 10 becomes transformed, which induces tumour.

SV 40 genome (DNA) is divided into two transcriptional regions:

1. Early gene transcription for protein t and T (T for tumour) which as in early phage function in viral DNA replication. Protein of SV 40 is needed for the oncogenic transformation of host cell.
2. Late gene transcription (encoding) of protein for virus assembly. Late gene encodes viral coat protein, called viral proteins (VP_1, VP_2, VP_3).

Early gene transcript is about 2650 nucleotides long and is transcribed in a anti-clockwise direction. Its transcription starts first at nucleotide pair 5248 and later at nucleotide pair 5252 and extend to pair 2604. The promoter for the early gene transcription resides near the origin of replication (Figure 20.12).

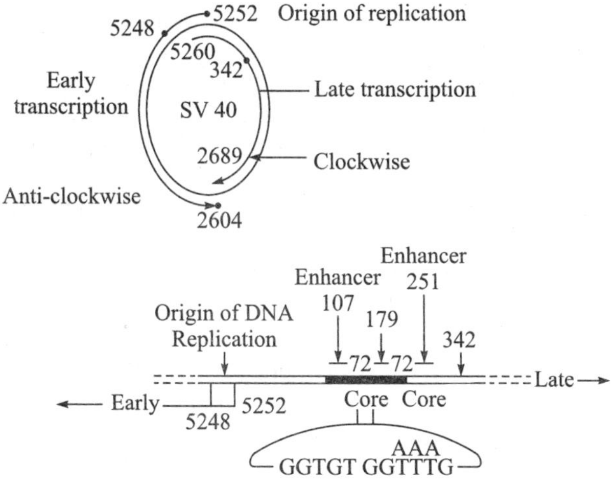

Figure 20.12 Early and late transcription of genetic message in SV 40.

The late gene transcript is about 2350 nucleotides long and is formed in clockwise direction.

Starting approximately at nucleotide pair 342 and extends to nucleotide pair 2689. The two-transcription units overlap at their 3′ ends by 85 base pair.

The adequate transcription of the SV 40 chromosome depends upon a 72 base pair enhancer sequence. An enhancer is a 50–100 base pair sequence that increases the transcriptional efficiency of neighbouring structural gene, producing more mRNA. In SV 40 chromosome there are two 72 base pair copies of transcriptional regulatory sequence called *enhancer*. These viral enhancers contain a core region with consensus sequence $\overset{\text{AAA}}{5'\text{TGGTTTG}3'}$. These enhancers are also present in the terminal repeats of RNA retroviruses such as mammary tumour virus. The SV 40 *enhancers* reside between the origin of transcription of both early and late genes.

We can increase the transcription of many eukaryotic genes by using recombinant DNA manipulation to place an *enhancer* sequence near a eukaryotic gene. When introduced into an animal cell by *transinfection*, such a gene exhibits increased transcription compared to the gene that lacks nearby enhancer. The enhancer sequence can enhance the rate of transcription of the gene when it is located at 5′ or 3′ end to the gene and when it is inserted in either of orientation (inverted or not) with respect to 5′ end of gene (Figure 20.13). Enhancers can act at a distance, as much as 10 kb from a gene and can still increase its transcription.

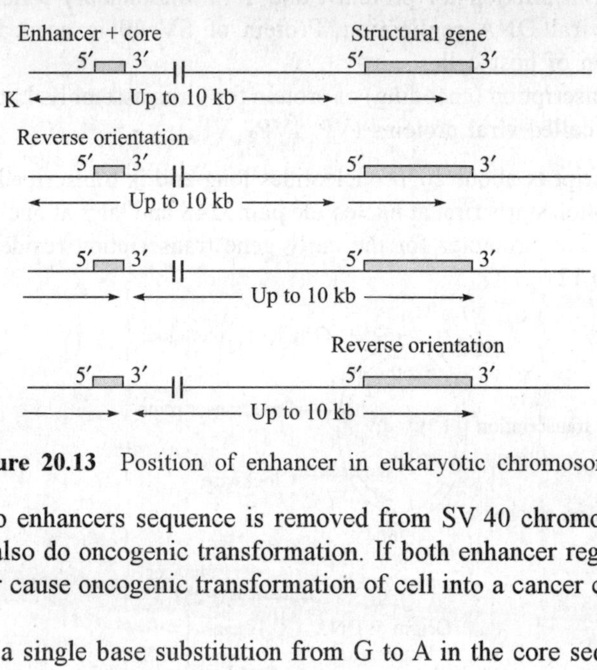

Figure 20.13 Position of enhancer in eukaryotic chromosome.

If one of the two enhancers sequence is removed from SV 40 chromosome, the virus can cause infection and also do oncogenic transformation. If both enhancer regions are removed, it can neither infect nor cause oncogenic transformation of cell into a cancer cell called neoplastic transformation.

Not the least, if a single base substitution from G to A in the core sequence of the SV 40 chromosome that contains a single enhancer take place, it is sufficient to destroy the infectivity of the virus and to greatly reduce early gene transcription (Figure 20.14).

The synthesis of each viral protein depends on the correct processing of primary RNA transcript. How? The cutting and legating of the primary transcript into different correct functioning of mRNA units. There are five known viral proteins and only two primary RNA *transcripts*; *none of the proteins is synthesized by the use of a polycistronic mRNA. For example, Early genes.*

↓

GGTGTGGAAAG   Fully functional wild type

↓ G is substituted by A nucleotide

GGTGT AG AAAG   Non-functional mutant.
Some transcription but no infectivity

Figure 20.14 A single G in the core region of the SV 40 virus enhancer is essential for function.

The coding regions for protein t extend from nucleotide 5180 to 4658 and for protein T, including intron, from 5180 to 2710 nucleotides. Synthesis of both proteins is initiated from codon AUG and both share a common N-terminal amino acids sequence of about 80 amino acids, while t has total of 146 amino acids and T has 708. The coding sequence of mRNA that produces t protein is collinear to the DNA. Non-coding unit, the intervening sequences are removed after transcription. For synthesis of t protein a 66 base pair intron is removed and for T a 346 nucleotide introns corresponding to base pairs 4934 to 4589 are removed from the primary mRNA transcript. This sequence contains terminating codon for t protein. Removal of terminating codon and t *in-frame* splicing allow polypeptide synthesis to continue which results into production of large T polypeptide. This way, two different polypeptides are produced on two mRNA's each derived from same primary transcript.

OVERLAPPING GENES

In Chapter 19 (Figure 19.2) it has been demonstrated that genetic codes are not overlapping, i.e. *reading frame* for one genes is independent from other neighbouring genes. This is because, it would be evolutionary mistake for one region of a genome to encode two different proteins, wherein, the mutational changes if any, that may be advantageous for one protein, would be disadvantageous for other protein.

However, in certain prokaryotes such as in small bacteriophages- ϕX174 and its close relative 'G_4', overlapping genes have been discovered. For example, ϕX174 bacteriophage possesses nine proteins coded by nine genes A→H and I . In 1977, Sanger and his colleagues reported that the entire base sequence consists of 5375 nucleotides on the ϕX174 chromosome. This bacteriophage possesses only one DNA strand, which is used for synthesis of mRNA. Therefore, direction of transcription must be the same for every gene. The greatest surprise in the functional organization of ϕX174 genome is the presence of two pairs of overlapping genes on the ϕX174 chromosome (Figure 20.15).

Gene B is entirely included within gene A, and gene E resides within gene D. The initiating methionine codon 'AUG' for gene J uses the last A of the codon GAA in terminating codon UGA of gene D. All these overlaps are possible because the different sequences are translated by the use of different *reading frames*. For instance, gene A starts coding with nucleotide No. 1 of the code 'AUG' and gene B with nucleotide No. 972 in the *reading frame* (Figure 20.16). The nucleotide 972 is the third nucleotide of the codon GAA, (*glu*) for the residuc number 324 of the A protein and is the first base in the initiating AUG (meth) codon of the gene B, therefore, two polypeptides must be formed.

With different reading frames, *Overlapping* genes have been found in genomes of $\phi \times 174$.

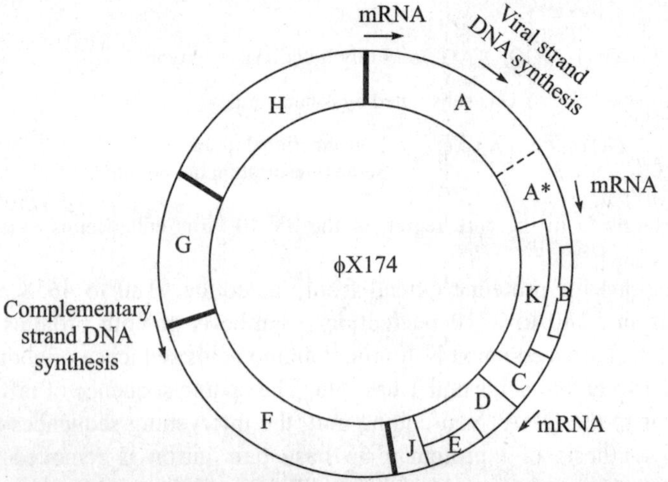

Figure 20.15 A model overlapping genes in ϕX174 bacteriophage.

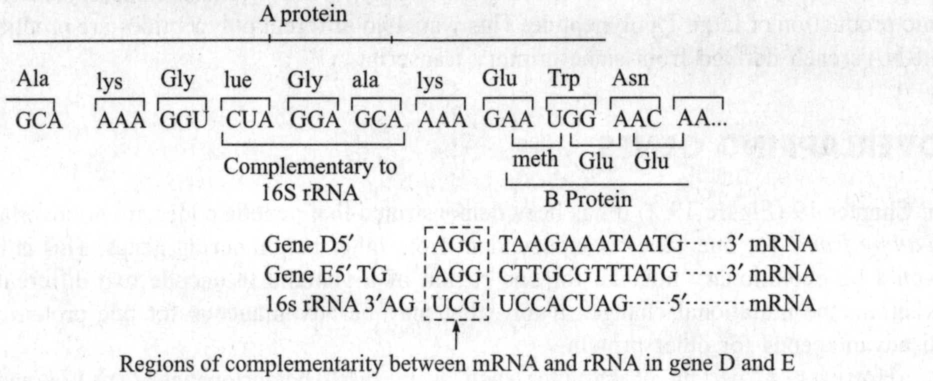

Figure 20.16 Initial nucleotide sequence of gene B is within gene A, is a single ΦX174 mRNA.

ANIMAL GENOMES

The genomes of prokaryotes differ from eukaryotes in following ways:

1. The genomes of eukaryotes are larger in size than prokaryotes.
2. These are composed of highly repeated sequences as well as double copy of sequence, in contrast to the genomes of prokaryotes that are of single copy sequences.
3. The genomes of eukaroytes consist of DNA sequences whereas in many viruses RNA sequences are present.
4. Eukaryotes need much more structural information than bacteria in order to live and adapt to natural forces. The flies, plants, animals and human beings encode more structural information than viruses and bacteria.

However, highly repeated sequences are mostly located in the central heterochromatin region of eukaryotic chromosome but they do not encode structural information. The heterochromatin region of X chromosome in mammalian females lack function. It is, therefore, concluded that highly repeated sequences which constitute 20–40 per cent of the total DNA in higher organisms—plants and animals do not encode structural information. In fact, function of highly repeated sequences are not yet known (deciphered). Thus, only 60–80 per cent of the genomes of the higher organisms are available for coding structural information; moderately repeated sequences consists of only 2–4 per cent of the total animal genomes. Because of the complexity of higher organisms—plants and animals; much more DNA is required for regulation of the function than required by prokaryotes.

Chromomere vs Gene Relationship

Recent studies indicate that there is no simple one-to-one relationship between genes and chromomeres. According to earlier concept, one-to-one relationship between chromomeres and genes existed but now this concept is considered invalid, because there are about three times as many transcripts as there are chromomeres, i.e. chromomeres on an average, encode more than one transcript. It is not possible to discard earlier notion that a chromomere is somehow a unit of function. By comparing the molecular result at the 87D to 87E of the third chromosome of *Drosophila melanogaster*, it was observed by Spierer and colleague that there are many genes whose function is not yet known or they are nonfunctional. Thus, chromomeres on average encode more than one transcript.

GENE EXPRESSION

The gene expression is accomplished through a sequence of events in which the information contained in the base sequences of DNA is first copied into a mRNA which then used to determine the amino acid sequence of a polypeptide chain (protein molecule). For mRNA synthesis, one of the DNA strands serves as template in a polymerization reaction that is catalyses by an enzyme called RNA polymerase. The process by which the segment of a DNA synthesizes the mRNA molecule is called *transcription*. The protein molecules are then synthesized with the help of base sequences of mRNA molecule which direct the sequential joining of amino acids in a particular order which is in fact a direct reflection of the DNA base sequences. Thus, arrangement of amino acid sequence in a polypeptide chain from mRNA base sequence is called *translation*.

Thus, gene expression consists of four main processes:

1. Unwinding of DNA strands of which one strand serves as template.
2. Transcription of mRNA polymerase for mRNA synthesis.
3. mRNA processing.
4. Translation of mRNA, i.e. sequential arrangement of amino acids into polypeptide chain which forms a polypeptide chain (protein).

Unwinding of DNA Strands

The two DNA strands, which are in the helical form get unwinded and one of the two strands serves as template on which the mRNA synthesis takes place, forming a DNA–RNA hybrid. After synthesis of mRNA, the unwinded two DNA strands get restored. This is difference from *replication*. In replication, two unwinded original DNA strands do not again get rewinded. (Figure 20.17(A)).

Transcription

Transcription of mRNA consist of following characteristics:

(a) The precursors in the mRNA synthesis are the four 5′ triphosphates of the ribonucleotides—adenosine, guanosine, cytosine and uridine. A riboside is a base, which is covalently-linked to the sugar ribose.

(b) In the synthesis of a mRNA, a 3′ OH group of ribose at the 3′ end of the growing RNA molecule, reacts with a precursor nucleoside 5′ triphosphate. The two terminal phosphate groups are displaced as inorganic pyrophosphate (pp_1) and a sugar phosphate bond results, extending the RNA molecule by one nucleotide unit (Figure 20.17(B)).

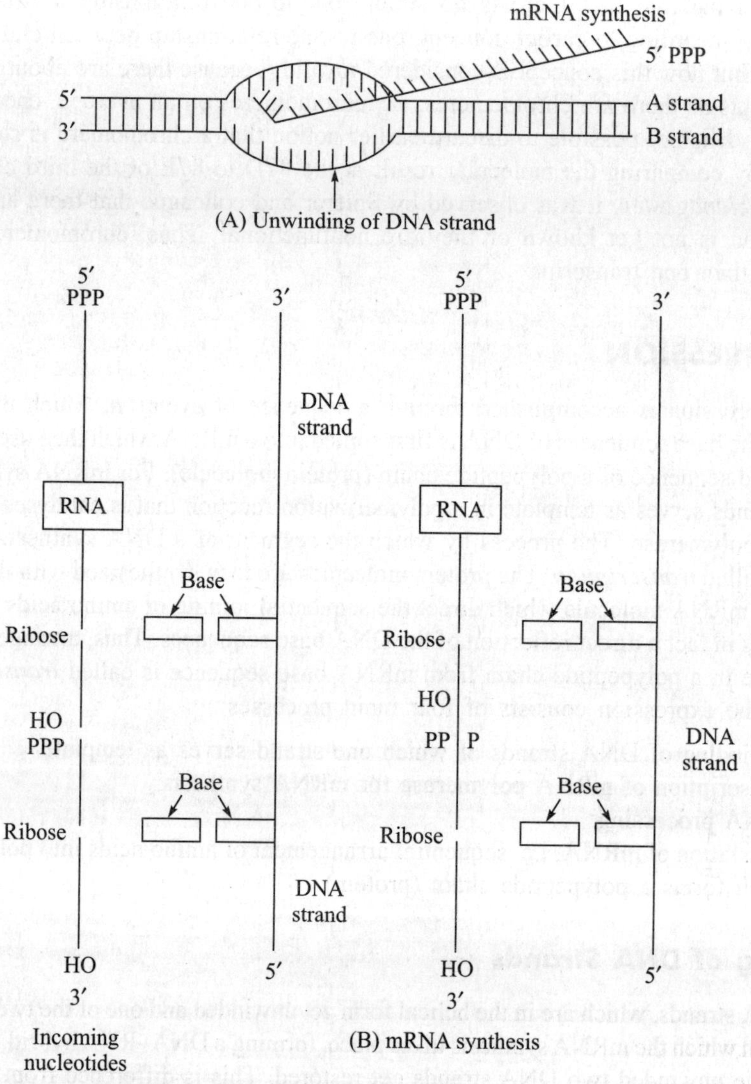

Figure 20.17 Mechanism of transcription.

(c) The sequences of bases in RNA are determined by the sequence of bases in DNA template. Each base to be added to the growing mRNA chain is done according to the base-pairing rule in DNA, e.g. A pairs with T, and C pairs with G in DNA. Similarly C pairs with G and A pairs with U in RNA respectively.

(d) The bases viz. ribonucleotides are added to the 3′ OH of the growing RNA chain; the 5′ end of the growing RNA molecule is a triphosphate and the two chains—RNA and DNA template are antiparallel to one another like two strands of DNA, like

$$5' \rightarrow 3'$$
$$3' \leftarrow 5'$$

The most important thing to be learnt from mRNA synthesis is that *DNA molecule being copied is a double-stranded*, winded molecule but before RNA synthesis starts, it gets unwinded at the region where mRNA synthesis takes place and only one strand of DNA serves as template.

However, synthesis of mRNA is not so simple; it has several stages, such as (a) local unwinding of DNA, (b) promoter recognition, (c) chain initiation, (d) chain elongation, and (e) chain termination. These stages are described below.

(i) The two DNA strands remain in a winded form before initiation of mRNA synthesis, the local unwinding of the two DNA strands occurs, and RNA polymerase forms an 'operon-promoter complex'.

(ii) RNA polymerase binds to DNA template at a specific base sequence (20–200 bases long) called *promoter*.

(iii) The RNA polymerase gets attached to a specific *start site* from where *polymerization* begins which is very close to the initial binding site at DNA. The synthesis starts at this *site* soon after the first nucleotide is placed here.

(iv) The RNA polymerase starts moving along the DNA *template* adding nucleotides to the growing chain of RNA.

(v) The RNA polymerase soon reaches at the completion of its function gets terminated and both the mRNA chains and polymerase is set free. RNA polymerase is an enzyme which assists in RNA synthesis.

The promoter sequences could be identified by mixing RNA polymers with DNA fragments and the digesting DNA with enzyme, which digest the DNA but leave intact region bound to RNA polymerase. After enzymatic digestion, RNA polymerase could be removed and base sequence of DNA could be determined. The laboratory examination of a large number of promoter sequences of a variety of genes found in different species of bacteria have proved that they possess almost some common features, e.g. sequence *TATAAT* box also called *Pribnow* box or –10 region (10 base pair) is found in the *promoters* of almost all unicellular microorganism (prokaryotes). An other sequence a –35 region or –35 base pair long is also common and is located about 35 base pair away from direction of transcription site from where RNA polymerization starts. These *promoter* sequences direct the RNA polymerase from where to initiate the RNA synthesis.

The potential of binding of RNA polymerase to different 'promoter' genes is not the same but varies greatly and this differentiated variation is fundamental of gene regulation mechanism. Because some promoters have binding *sites* for proteins instead of RNA polymerase and these proteins serve the binding sites for polymerases to bind correctly. Cyclic Adenosine monophosphate (cAMP) and cAMP–CRP complex are the most common sites of this type of binding, where the activity of the promoter is regulated by cAMP concentration.

Termination of mRNA transcription

So far two types of termination events have been identified. *First*, self-termination, which depends on DNA base sequence (AAAA tail). *Second*, requires presence of 'Rho' a sort of protein for termination of RNA synthesis.

Self-termination occurs at base sequence in DNA template consisting of a series of adjacent adenines (poly-AAAAA tail) preceded by a nucleotide *palindrome*, which is usually interrupted by a few nucleotide bases. It is believed that RNA molecule fold back to form a stem loop-like structure at termination site (Figure 20.18). Rho-dependent DNA sequences are yet to be recognized. However, initiation of second round transcription does not await completion of the first round because the promoters become available once polymerase has polymerized 50 to 60 nucleotides.

Direction of transcription

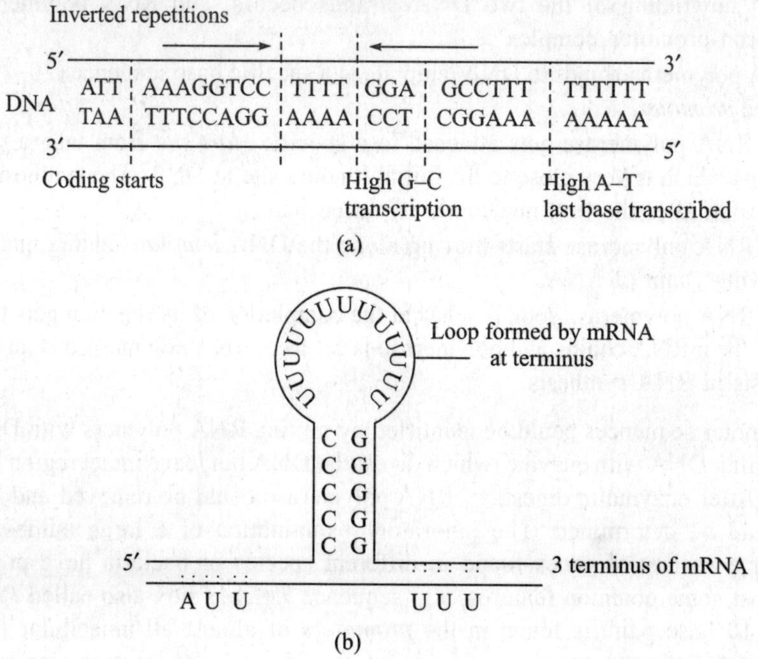

Figure 20.18 Base sequence of (a) DNA of *E. coli* trp-operon at which transcription terminate, (b) mRNA molecules is folded to form a stem and loop structure.

Role of messenger RNA

In the protein synthesis, amino acids do not directly bind to the DNA but intermediate steps are needed for arranging amino acid in a polypeptides chain. Therefore, it is proposed that heterocatalytic function of DNA is a two-step process. In the first step, each DNA (gene) serves as a *template* for the synthesis of mRNA molecule onto which the precise nucleotide sequences making up that gene is *transcribed.* This transcription of mRNA takes place in the nucleus. After their transcription on the DNA template, the mRNA molecules are imagined to migrate to the cytoplasm, where they initiate translation of amino acid, i.e. sequential arrangements of the amino acid into polypeptide chain of the pre-determined primary structure.

Under the early version of informational essence of protein synthesis, the RNA *transcription* was considered to provide the RNA moieties for the newly formed ribosomes. Hence each gene was being thought to give rise to a special kind of *ribosome*, which directs the synthesis of one and only one kind of protein. Brenner, Jacob, and Meselson (1961) gave the hypothesis of *one-gene one-ribosome*. But later on, Cohen in 1948 found that the synthesis of new kinds of ribosome is not a precondition for the synthesis of new kinds of protein. Contrary to 'one-gene one-ribosome'; one-gene one-protein hypothesis was given by Francois Jacob and Jacques Monod, (1961). Instead of ribosome, RNA forming template for orderly assembly of amino acids, they proposed that the nucleotide sequence of each gene (DNA) is *transcribed* onto mRNA molecules. They further envisaged that these mRNA molecules then enter into temporary combination with unspecialized ribosomes already endowed with their own ribosomal RNA and this ribosomal RNA-mRNA complex that is able to synthesize *the protein encoded into* the DNA (gene) represented by its mRNA transcript. This indicates that ribosome is the 'workshop' for protein synthesis and mRNA is the 'blue-print'. The mRNA has a very limited functional life and can serve for the construction of only a few protein molecules. Thus, during active protein synthesis, mRNAs are in the state of rapid turn over. One and the same ribosome may therefore, synthesize one kind of protein at one moment and another kind at next depending upon the kind of mRNA–ribosome complex (Stent, 1971).

In prokaryotes, mRNA molecules contain information for amino acid sequences for several different polypeptide chains. Such a molecule is called *polycistronic mRNA*. The polycistronic mRNA molecule often encodes different proteins needed for a metabolic pathway. For instance, in *E. coli*, ten enzymes are needed to synthesize 'Histidine', which are encoded in one mRNA molecule. Since with a single polycistronic mRNA molecule, synthesis of several related proteins (enzymes) can be regulated by a single signal, as a result appropriate quantity of proteins are synthesized at the same time. This is called a *coordinated regulation*.

However, not all the base sequences in a mRNA molecules take part in the *translation* of amino acids needed for the formation of polypeptide chains. For instance, *translation* of a mRNA molecule rarely starts exactly at one end of RNA molecule and proceeds other end but initiation of polypeptide chain may start from hundreds of nucleotides away from 5′ p terminus of mRNA. The fragment of mRNA not involved in the *translation* of the first polypeptide chains is called *leader* which in some cases contains regulatory gene that influences the protein synthesis (e.g., trp-gene system). The untranslated sequences are generally located at both 5P and 3′ OH terminal. The *polycistronic* mRNA molecules contain *spacer sequences* at least 10 base pairs long which separate the *coding sequences*; each *coding sequence* corresponds to polypeptide chain.

Though coding sequences of each gene is obtained by *transcription* of only one DNA strand, but there are very rare exceptions to this rule also, where two complementary base sequences in a particulate gene are to be transcribed. There are some small *phages* in which all mRNA are not synthesized from same DNA strand, and also not from same region of DNA. The mRNA may be growing in two directions (Figure 20.19) depending upon which DNA strand functions as template.

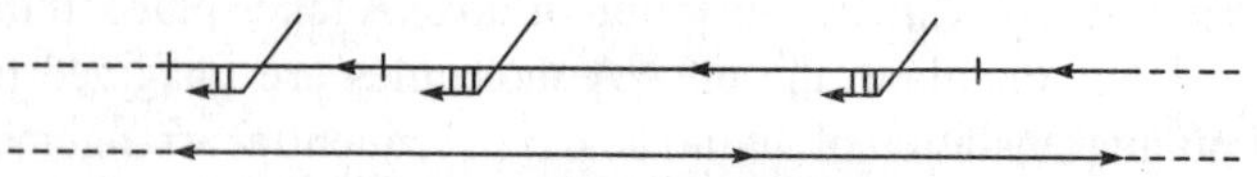

DNA

Figure 20.19 The *transcription* of mRNA by both DNA strands but not from the same region. Arrow head indicates promoters and black bars the termination sites. Promoters are present in both the strands, where complementary DNA helix are involved in mRNA transcription.

There are differences in the *transcription* of mRNA synthesis between *prokaryotic* and *eukaryotic* organisms. In prokaryotes, the immediate product of transcription is mRNA (the primary *transcript*). In contrast, in eukaryotes, the primary transcript is heterogenous RNA (hnRNA), must be converted to mature mRNA. This conversion is called mRNA *processing* that consists of two types of events—modification of the *termini* and *excision* of untranslated sequences embedded within coding sequences. The DNA fragments (genes) are divided between coding (*exon*) and non-coding regions (*introns*). The untranslated sequences within 'genes' are called *intron*. *Introns* are present in almost all the eukaryotic transcripts, but rarely found in the free-living unicellular eukaroyotes, e.g. yeast. The *introns* are excised from the primary transcript and the remaining fragments (*exons*) rejoin together to form mRNA. The excision of *introns* and joining of coding sequences (*exons*) to form mRNA is called *RNA splicing*. This is of course is not important in the prokaryotic organisms (classical microbial genetics). But, it is important in the cloning of eukaryotic genes by recombinant DNA construction.

Processing of messenger RNA (mRNA)

The synthesized RNAs molecules are longer in size than functional RNA. The prokaryotic RNA is polycistronic, which does not undergo processing to produce individual mRNAs.

Processing of heterogenous nuclear RNAs (hn RNAs)

The eukaryotic mRNAs are monocistronic. The primary transcript of mRNA is much larger than mature mRNA or processed mRNAs. This is called *heterogenous nuclear RNA* (hnRNA). The heterogenous mRNA undergoes processing and finally mature mRNA is produced which is 10 per cent of heterogenous mRNA. This is also called pre-mRNA or precursor mRNA. The main characteristics of hn RNAs are: size is not uniform and sedimentation value ranges from 20s to 100s. Base composition is similar to DNA template; hnRNA is located outside the nucleus. Master mRNA is derived from hn RNA through enzymatic processing.

The newly synthesized hn RNAs are packaged with protein called *ribonucleoprotein* complex. After transcription, the poly (A) tail—AAAA is added at the 3′ end, the polymerization reaction is catalysed by the enzyme '*Poly (A) polymerase*'. This enzyme does not require DNA template and adds about 200 adenine nucleotide at the 3′ OH end of hnRNA. The function of Poly (A) tail is to transport mRNA from nucleus to the cytoplasm (Figure 20.20).

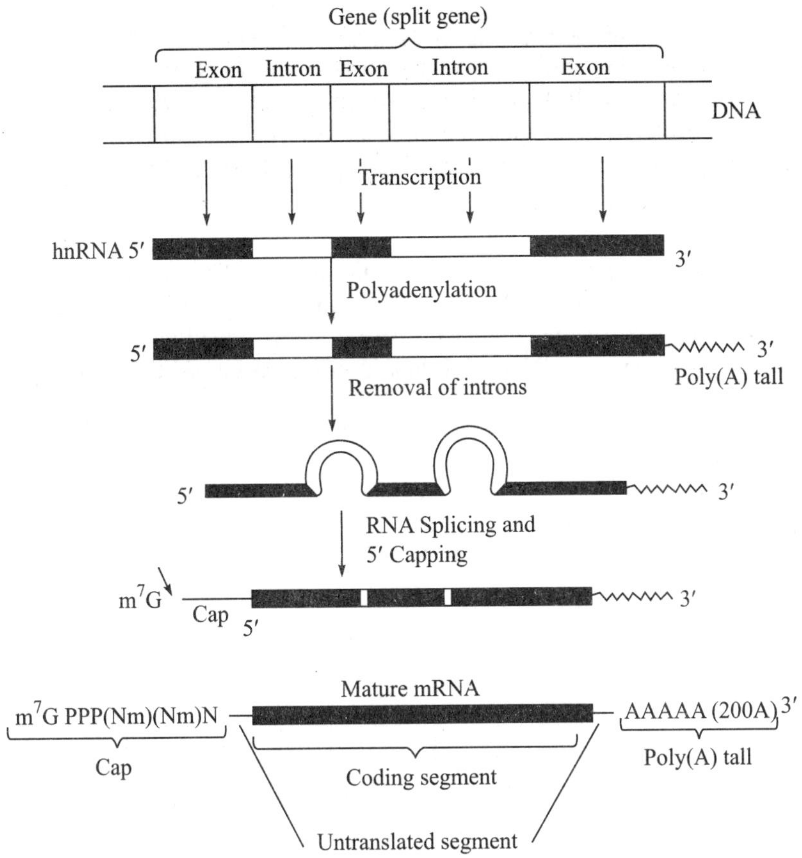

Figure 20.20 Processing of heterogenous mRNA in eukaryote m = methylated nucleotides and nm = non-methylated nucleotides.

RNA splicing

The heterogenous RNA (hnRNA) consists of coding sequences (*exons*) and non-coding sequences called intervening sequences (*introns*). These sequences are found alternate to each other. Such genes are called *split genes*. The number of *exons* and *introns* differ in different genes. For example, in eukaryotes the numbers of 'exons' and 'intron' found in genes vary from species to species (Table 20.1).

TABLE 20.1 Number of *exons* and *introns* in organisms

Organism	*Gene*	*Number of exon*	*Number of intron*
Human	Insulin	3	2
Chicken	Insulin	3	2
Chicken	Actin	4	3
Chicken	Ovalbumin	8	7
Chicken	Collagen	51	50

In extreme cases, actual coding region of mRNA is hardly 10 per cent of the length of heterogenous nuclear hnRNA. During processing of RNA, '*introns*' are cleaved and the *exons* are united in the correct sequence. It is called RNA-splicing. Some nucleoprotein particles are called 'snurps', which are involved in the cleaving of *introns* and splicing of RNA. One base sequence at the end of *intron* is similar in all *introns*, which is called *splice-junction*.

Capping

During processing of RNA, a cap is formed at the 5′ end. The cap consists of one or two nucleotides, whose sugar is methylated at the 2′ position. The methylated nucleiotide connected by a triphosphate bridge to 7-methyl guanosine (m^7G); the 7-methyl guanosine is connected with 5′ triphosphate at the 5′ end of mRNA and thus it terminate into 3′ OH end (Figure 20.21).

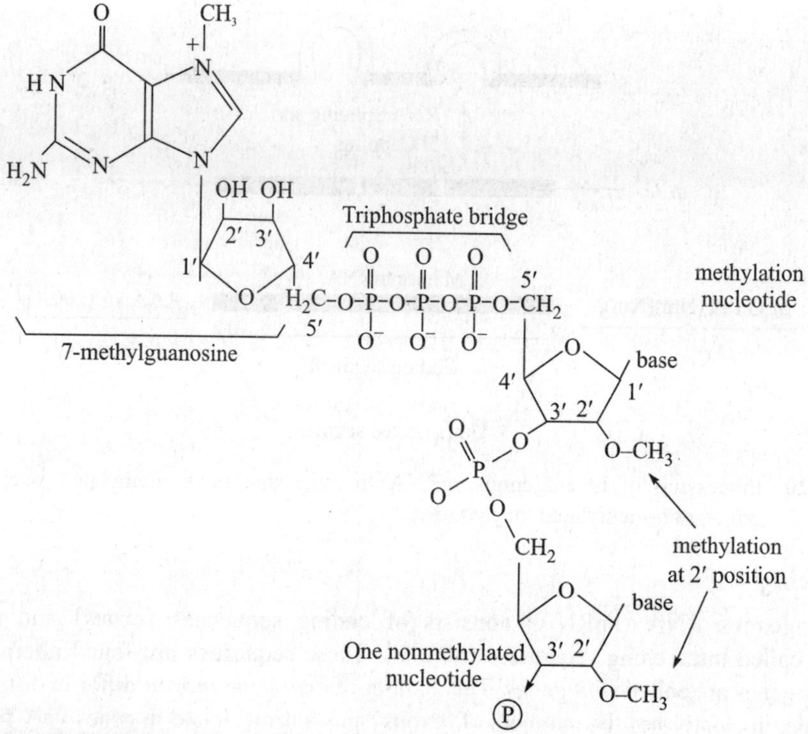

Figure 20.21 Formation of caps at 5 end of RNA consisting of one or two nucleotides.

Transfer RNA (tRNA)

Transfer RNAs (tRNAs) are also called soluble RNAs (sRNAs). They are small molecules varying from 75 to 100 nucleotides in length. They carry amino acids and bring them to the ribosome (workshop of protein synthesis) to form polypeptide chains. There are 20 amino acids but number of tRNA is greater than amino acids because some amino acids are adapted by more than one form of tRNAs. The transfer RNAs are designated as tRNA-gly (tRNA-glycine); $tRNA_1$ leu, $tRNA_2$ leu. Such tRNAs are called *iso-accepting tRNAs*. Transfer RNAs are synthesized by DNA templates and the reaction is catalyzed by DNA-dependent RNA polymerase. The mature

RNA is produced after precursor tRNAs are produced. Transfer RNAs help in the decoding of the mRNA codons by which the base sequence within a mRNA molecule becomes translated to an amino acid sequence of a protein.

Structure of transfer RNA: Universally accepted two-dimensional tRNA model is *clover leaf shape* formed on the pairing of short complementary sequences and the formation of unpaired loops (Figure 20.22). On the down side loop 'anticodons' are attached; on the upper side is the site for amino acid attachment.

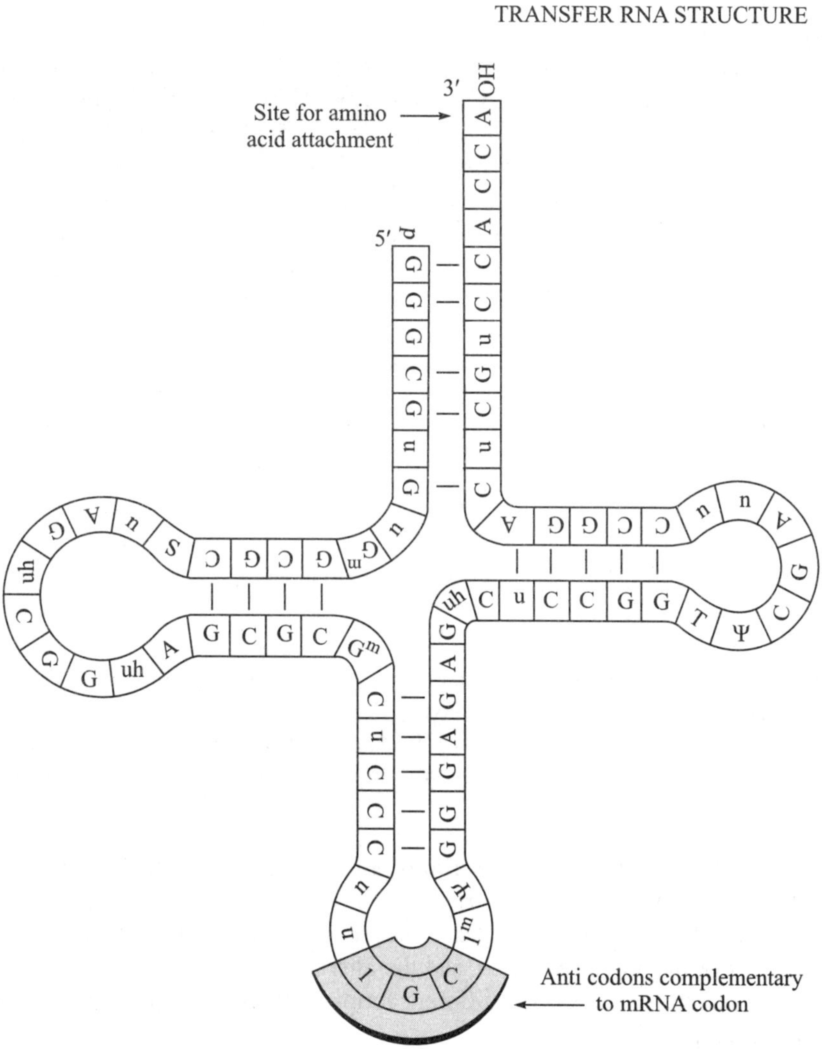

Figure 20.22 Clover leaf shape structure of transfer RNA (tRNA). The kind of protein formation is the function of (tRNA) to which amino acid it carries to the 'workshop' (ribosome).

Alternate splicing

Even though when the same promoter is used to *transcribe* a gene, different cell types can produce different quantities of proteins or even different proteins because of the differences in the mRNA produced in processing. The reason is that the same transcript can be *spliced* differently from one cell type to the next. The different *splicing* patterns may include exactly, the same protein coding *exons* in which case the protein is *identical* but the rate of synthesis differs because the mRNA molecules are not translated with same efficiency. In other cases, the protein coding part of transcript has a different splicing pattern in each cell type and the resulting mRNA molecules code for proteins that are not identical even though they share certain exons.

For the synthesis of amylase in the rat, different mRNA molecules are produced from the same gene because of different patterns of *introns* removal in RNA processing. The rat salivary gland produces more enzymes than does the liver, although the same coding sequence is transcribed. In each cell type, same *primary transcript* is coded but two different splicing mechanisms are used. The initial part of *primary transcript* is shown in Figure 20.23. The coding sequence begins 50 base pairs inside exon 2 and is formed by joining exon 3 and subsequent exons.

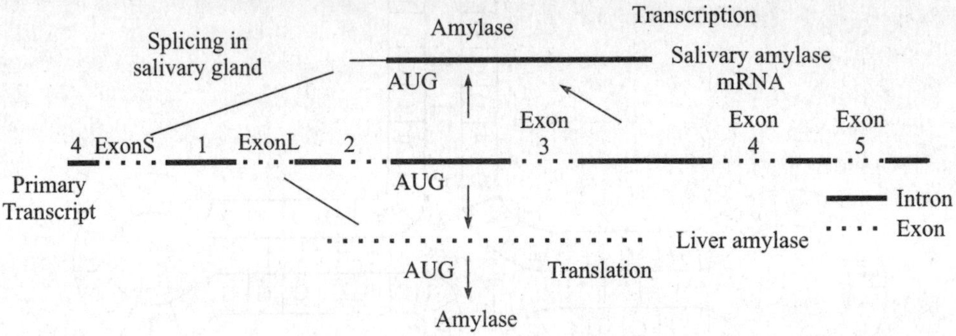

Figure 20.23 Synthesis of distinct amylase mRNA molecules by different splicing events in the cells of the salivary gland and liver of the mouse. Leader L and *introns* in black solid line. The *exons* are in dotted line. Coding sequence begins at the AUG codon in exon. Exon S and L form part of untranslated 5′ end of the mRNA in salivary glands and liver.

In the salivary gland, the primary transcript is processed such that exon 5 is joined with exon 2 (i.e., exon L is removed as part of *introns* 1 and 2). In the liver, *exon L* is joined with exon 2 and exon 5 is removed along with *intron* 1 and the *leader L.* The exon 5 and L become alternate 5′ leader sequences of the amylase mRNA and the alternative mRNA are translated at different rates. In chicken, skeletal, muscle, two different forms of the muscle protein myosin are made from the same gene.

Primary transcript is made from each of the two promoters and these transcripts are processed differently to form mRNA molecules encoding distinct forms of the protein. In *Drosophila* the myosin RNA is processed in four different ways, the precise mode depending on the stage of development of fly. One class of *myosin* is found in *pupas* and another in later *embryos* and *larval* stages.

Alteration of DNA

Some genes in eukaryotes are regulated by alteration of DNA. For instance, certain sequences may be amplified or rearranged in the genomes or the bases may be chemically modified. Some of the alteration are reversible but most of them permanently change the genome of the cell. But permanent changes take place in the somatic cell only, therefore, these changes do not affect the germ cells.

Gene dosage and gene amplification

Some gene products are required in larger quantities than others. This can be maintained by gene dosage. If two genes X1 and X2 are transcribed at the same rate and if the translation efficiencies are also the same, the synthesis of the product A could be increased by increasing the number of copies of the gene A in proportion to the copy of gene B, i.e. if 10 times more product A is desired then 10 copies of gene A per copy of gene B is needed. The histone genes exemplify a gene dosage effect to synthesize the huge amount of histone required to form chromatin; most cells contain 100 times or many copies of histone genes as of genes required for DNA replication.

The gene dose is increased temporarily by a process called gene amplification, in which the number of genes increases in response to some signal. The gene amplification is noted in the development of oocytes of the toad, *Xenopus laevis*. The oocyte is a complex process and requires a huge amount of protein synthesis for that a large number of ribosomes are needed in the development of oocytes. The number of rRNA genes increases by about 4000 times. This increase exists only during and for the purpose of development of the egg. The precursor to oocytes contains about 600 rRNA–gene (rRNA) units; after amplification, about 2×10^6 copies of each unit are present.

RNA–DNA hybridization

Soon after realization of messenger RNA (mRNA) concept that RNA formed in T4 virus-infected bacteria is infact transcript of nucleotide sequence of phage DNA, the methods were designed for producing specific molecular hybrids between complementary RNA–DNA polynucleotide chain. On heating double helical DNA molecules at temperature above their melting point cause the rupture of complementary purine—pyrimine hydrogen bond. On rupturing, the two complementary polynucleotide chains get separated. However, on cooling the two disassociated DNA complementary chains (strands) gradually get associated, by matching homologous base-sequences and finally reconstitute the original double helical structure. This is also called *denaturing* and *renaturing* of DNA.

However, if a single-stranded mRNA molecule with purine and pyrimidine bases complementary to the some of the DNA strands is added to the hot solution; on gradual cooling (through annealing process) RNA–DNA hybrid having double helices will be formed in addition to the original DNA double helix (Figure 20.24) The formation of double helix could be detected by several ways, but one of the most easiest and simplest way is the retention on special filters that permit free passage of single-stranded RNA molecule.

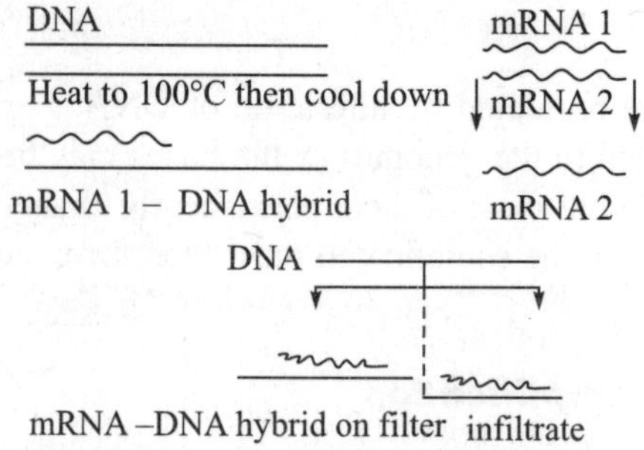

Figure 20.24 Schematic diagram of the formation of specific molecular hybrids between DNA–RNA polynucleotide chain. Two kinds of mRNA molecules were added to the solution of DNA molecules that have served as transcript templates for mRNA 1 but not for mRNA 2. On heating at 100°C the complete DNA strands get unwinded and on slow cooling mRNA 1 matches its nucleotide base sequence with the complementary base sequence of DNA which served as template for its transcription and both strand showed RNA 1—DNA hybrid (double helix). However, since mRNA 2 did not find matching nucleotide base in the solution, therefore, it remained single-stranded.

After successful trial of this method, the p32 labelled post-infection RNA extract was removed from T4-infected *E. coli*, grown in a medium containing $PO_4^{32} - 3$ and tested in a RNA—DNA molecular hybridization reaction. For this purpose, the samples of the radioactive RNA extract were added to a hot solution of either phage DNA extracted from infective T4 phage particles or *E. coli* DNA extracted from uninfected bacteria. The solutions were then annealed in order to allow formation of hybrid double helices between any DNA and RNA strands of complementary base sequence. Thereafter the solution was allowed to cool down; the extract was filtered and the amount of p32 radioactivity retained on the filter was measured. The results of this trial experiment indicated that filters retained most of the p32 labelled post-infection RNA added to the solution containing T4 DNA and very little of the added RNA to the solution containing *E. coli* DNA. In other words, the nucleotide base sequences of the post-infection RNA are complementary to those of the phage DNA and not to those of the host cell DNA. This also confirms that phage DNA serves as template for the synthesis of mRNA transcript of the phage gene (Stent, 1971).

RNA polymerase

The details of the mechanism of DNA replication came to be worked out after discovery of the *DNA polymerase* enzyme, so was the mechanism of DNA transcription into RNA elucidated upon the nearly simultaneous discovery in 1960 by S. Weiss, J. Hurwitz and Audery Stevens of RNA *polymerase* enzyme in the extracts of various eukaryotic and prokaryotic cells including *E. coli*. In the presence of DNA template molecules, RNA polymerase catalyses the conversion of ribonucleoside triphosphates into ribopolynucleotide chain according to the following reaction as given in Figure 20.25.

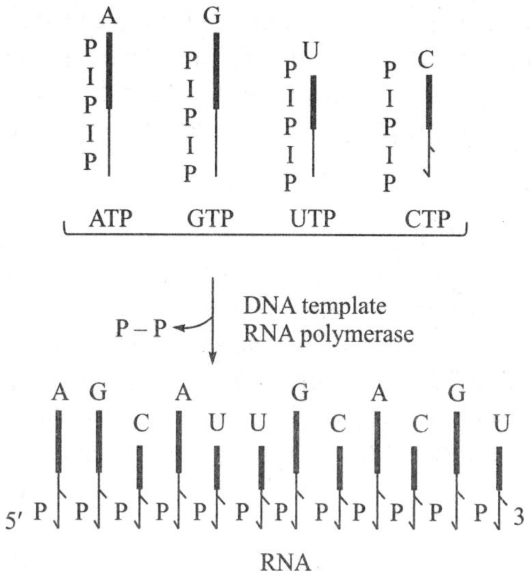

Figure 20.25 How RNA polymerase catalyzes the conversion of ribonucleoside triphosphates into ribonucleotides chain.

That is to say, ATP, GTP, CTP and UTP are polymerized into RNA by formation of an easter-link between the inner most of the three phosphates attached to the 3′ hydroxyl of one ribonucleotide and the 3′ hydroxyl group of another; with the attendant liberation of the outermost two phosphates as inorganic pyrophosphates. Thus, chemistry of RNA synthesis, closely resembles the chemistry of DNA synthesis. As regards direction of the chemical reaction of RNA it may take place in two ways (Figure 20.26).

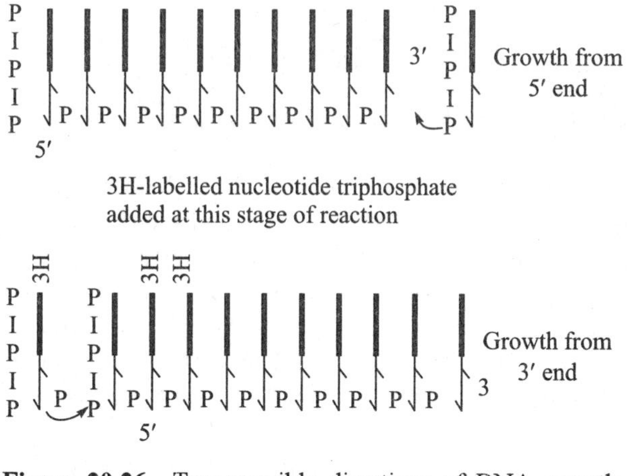

Figure 20.26 Two possible directions of RNA growth.

The growth of new RNA molecule may proceed by incorporating next nucleoside—triphosphate base either linking its 3′ hydroxyl group to the 5 triphosphate of the last nucleotide added to the growing chain which may be called the chain is growing from 3′ end or linking its

5 triphosphate to the 3′ hydroxyl group of the last nucleotide added to the growing chain. In this case, it is said that chain is growing from 5′ end. However, experimental findings so far have indicated that growth of RNA chain begins from their 3 end.

Following is the manner in which DNA transcription is proceeded by RNA polymerase.

Ribosome

Ribosome is actually the 'workshop' of protein biosynthesis and mRNA is the 'blue-print'. During active protein synthesis, the mRNAs are in a state of rapid turn over. One and the same ribosome may therefore, synthesize one kind of protein molecules, at one moment and another kind at the next depending upon the kind of message the ribosome has happened to have engaged. The ribosomes are small particles, not resolved in the light microscope, that contain RNA and proteins. In bacteria, ribosomes are packed throughout the cytoplasm. In certain cells of higher organism, some ribosomes are found free and others are attached to the surface of the tubular lipoprotein membrane, collectively called endoplasmic reticulum. Ribosome is composed of two subunits 30s and 50s together 70s. This categorization is according to the velocity at which each of these bodies sediments in a centrifuse under set of a standard condition.

When the cell is broke opened, the portion of membrane with ribosomes attached are released. The attachment of tRNA to ribosome does not require an enzyme, and also it does not require that the tRNA be charged with an amino acid. It requires the presence of C–C–A end and magnesium ion. There are some 20 to 50 additional sites to which tRNA can bind ribosome with lower affinity. There are ribosomal workbenches from which tRNA molecules can shuttle, carrying amino acid for the synthesis of the polypeptide chain. The specific order in which these charged tRNA molecules are utilized, however, depends upon the mRNA molecules bind to ribosome (Figure 20.27).

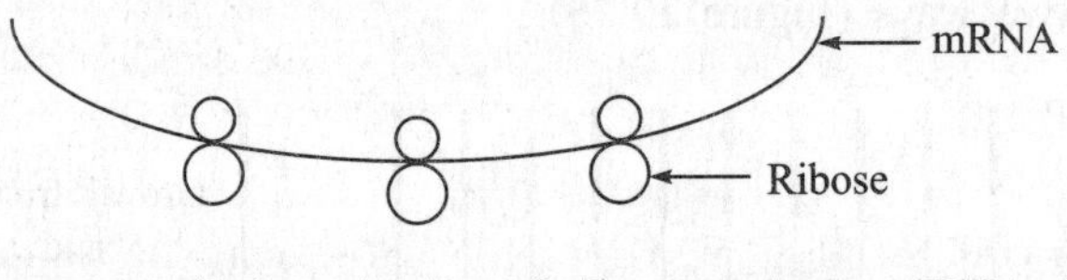

Figure 20.27 Polymerization of ribosome on the mRNA chain.

The prokaryotic ribosomes have a diameter of about 20 n.m. and contain about twice as much RNA by weight as protein. Eukaryotic ribosomes are slightly larger and contain a somewhat lower proportion of RNA than prokaryotic ribosome. The compositional equivalence rule valid for DNA [A] = [T] and [G] = [C] does not hold here, since neither [G] is equal to [C] nor [A] is equal to [U] (U being the structural analogue of T). Consequently, the ribosomal RNA molecules can not (and do not) exist in the complementary base-pairing rule of Watson-Crick double helix; the exact conformation nor the nature of their function in the intact ribosome has yet been elucidated.

Messenger RNA (mRNA) carries the blue-print of life dictated by DNA/genes seated on the nuclear throne. But there are basic differences between the mRNA of prokaryotes (bacteria and viruses) and eukaryotes (multicellular organisms such as plants and animals) (Table 20.2).

TABLE 20.2 Difference between mRNA of prokaryotes and eukaryotes

Prokaryotic mRNA	*Eukaryotic mRNA*
1. It is polycistronic (governed by many cistrons)	1. It is monocistronic (governed by a single cistron)
2. Possesses short-life (2 minutes)	2. Possesses long-life (few hours and days)
3. Translation begins when mRNA is formed on DNA template	3. Translation begins after transcription is over
4. There is no processing of mRNA after synthesis	4. Newly formed mRNA is heterogeneous, very long and undergoes processing

Translation (Protein Biosynthesis)

Translation of genetic message means protein biosynthesis. The primary structure of proteins is encoded in DNA in the form of sequence of nucleotide triplets or codons and each of which represents one of the 20 amino acids. In protein biosynthesis, the sequence of nucleotide triplet codons is transcribed from its permanent repository in DNA on to ephemeral mRNA molecule in the first step of heterocatalytic function. In the *second* step, the amino acids are linearly assembled into polypeptide chain of predetermined order on the ribosomal work benches. *Translation* machine consists of four important steps but actual *translation* of the genetic information is embodied in the tRNA sequences. Messenger RNA (mRNA) serves as intermediary. These steps are:

1. *Ribosome:* This has been already described earlier. These are particles on which the protein biosynthesis is carried out. They contain the enzyme needed to form a polypeptide bond of the amino acids. Ribosomes are the *sites* for binding of mRNA molecule, and sites for bringing in and aligning the amino acids in an assembly to form polypeptide chain. It is, therefore, called '*workshop*' for protein synthesis.
2. *Transfer RNA (tRNA):* The amino acids floating in the cytoplasm do not bind directly to mRNA but for synthesis of a particular polypeptide, these are ordered by the base sequence (codons) in the mRNA molecule. This ordering is accomplished by a set of adopter molecules (tRNAs). A tRNA 'reads' the message (of the base sequences of mRNA) and as per message of base sequences written in mRNA, called *genetic code*, they pickup the desired amino acid and brings to the ribosome (workshop).
3. *Aminoacyl tRNA synthetase:* These are sets of enzymes which catalyze the attachment of an amino acid to its corresponding tRNA molecule by activating them.
4. *Initiation, elongation and release factor:* These are protein molecules needed at a particular stage of polypeptide synthesis.

An outline of protein synthesis is given in Figure 20.28.

A mRNA molecule binds to the surface of a ribosome (protein synthesizing particle). An appropriate tRNA—amino acid complex, which is made elsewhere in the cytoplasm by the aminoacyl-tRNA synthetase, binds sequentially one by one to the mRNA molecule that is attached to the ribosome. The peptide bonds are made between successively aligned amino acids and

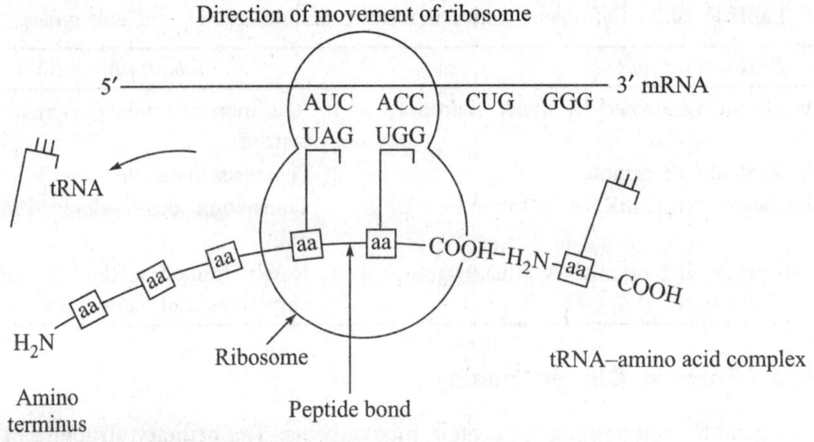

Figure 20.28 Assembly of polyribosome and synthesis of polypeptide.

amino acids join each time with the carboxyl group–COOH of the comming amino acid at the growing end, wherein the second step of the heterocatalytic function of DNA (the nucleotide sequences) are translated into polypeptide chain of predetermined *primary structure*. In the prokaryotes, the RNA molecule which is called mRNA, is directly involved in polypeptide synthesis. The amino acid sequence is obtained from mRNA by protein synthesizing machinery of the cell.

In the prokaryotes, the mRNA contains information for the amino acid sequences of different polypeptide chains. Such mRNA molecule is called polycistronic mRNA (a cistron encodes a single peptide chain).

In the end, the chemical bond between the tRNA and its attached amino acid is broken and the completed protein is released.

The most important features of *translation* to know are:

(i) RNA is translated from the 5′ end of the molecule towards the 3′ end, but neither from the 5′ terminus nor all the way to the 3′ end.
(ii) Polypeptides are synthesized from the amino terminus towards the carboxyl terminus, by adding amino acids one by one to the carboxyl end.

For instance, a protein synthesis could be started having NH2. Met-Pro... Gly-Ser-COOH would have started with from methionine and serine would be the last amino acid added to the chain. This rule can be illustrated by Figure 20.29.

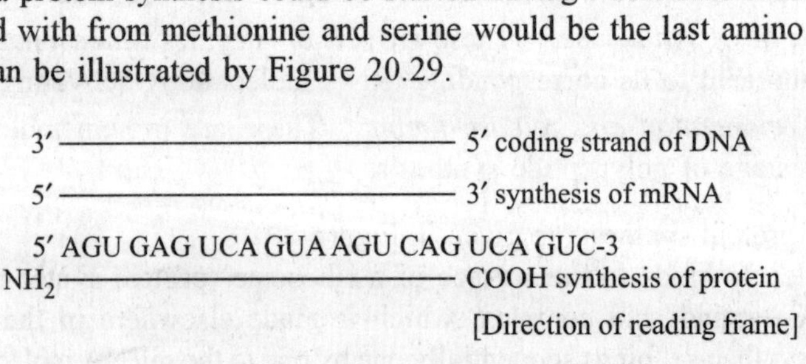

Figure 20.29 Direction of synthesis of mRNA according to the coding strand of DNA and protein according to mRNA (coding) sequence which form a reading frame.

Polyribosome

Of the total bacterial RNA, about 80 per cent is made up of ribosomal RNA (rRNA) and only 5 per cent of mRNA. The mRNA might run through the ribosome in the same way as a magnetic tape runs through the head of a tape recorder. In this way only a short-segment of mRNA (about a dozen nucleotides) will be in contact with the ribosome at any time and among these dozen nucleotides are the nucleotide triplet codons which specify the next amino acid to be added to the nascent polypeptide chain under construction on that ribosome. As soon as the next amino acid is added, the mRNA tape advances through its client ribosome by three nucleotides in order to bring the second-next triplet codon into position. In this way, a single mRNA runs through the several ribosomes simultaneously in the manner of a magnetic tape running through the tandem heads of a tape recorder designed to generate an *echo* chamber effect.

The reality of this concept was ascertained later on, on finding that clusters of ribosomes are in fact, attached to one and the same molecule of mRNA that is, the polyribosomal clusters are actually connected with a continual molecule of mRNA. As regards interrelationship between the formation of bacterial polyribosome and its attachment with mRNA, there are two possibilities:

First, that the mRNA molecules are first transcribed in their entirety before any ribosome is attached to it and translating its nucleotide sequence into the polypeptide chain.

Second, that the ribosomes attach themselves one by one to the head or 5′ end of the newly formed mRNA chain while that chain is still being transcribed from its own DNA template. In this case, formation of polyribosome upon mRNA proceeds concomitantly with its synthesis and the head of newly formed mRNA serves as template for the polypeptide assembly. The general feature of the heterocatalytic function of DNA that evolved from the preceding insights can be summarized as in Figure 20.30. Figure 20.30 illustrates how ribosomes are clustered on the messenger RNA during protein biosynthesis.

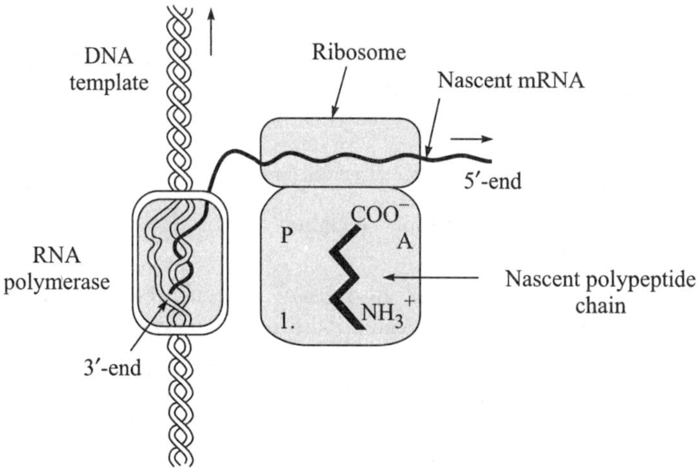

(a) Transcription

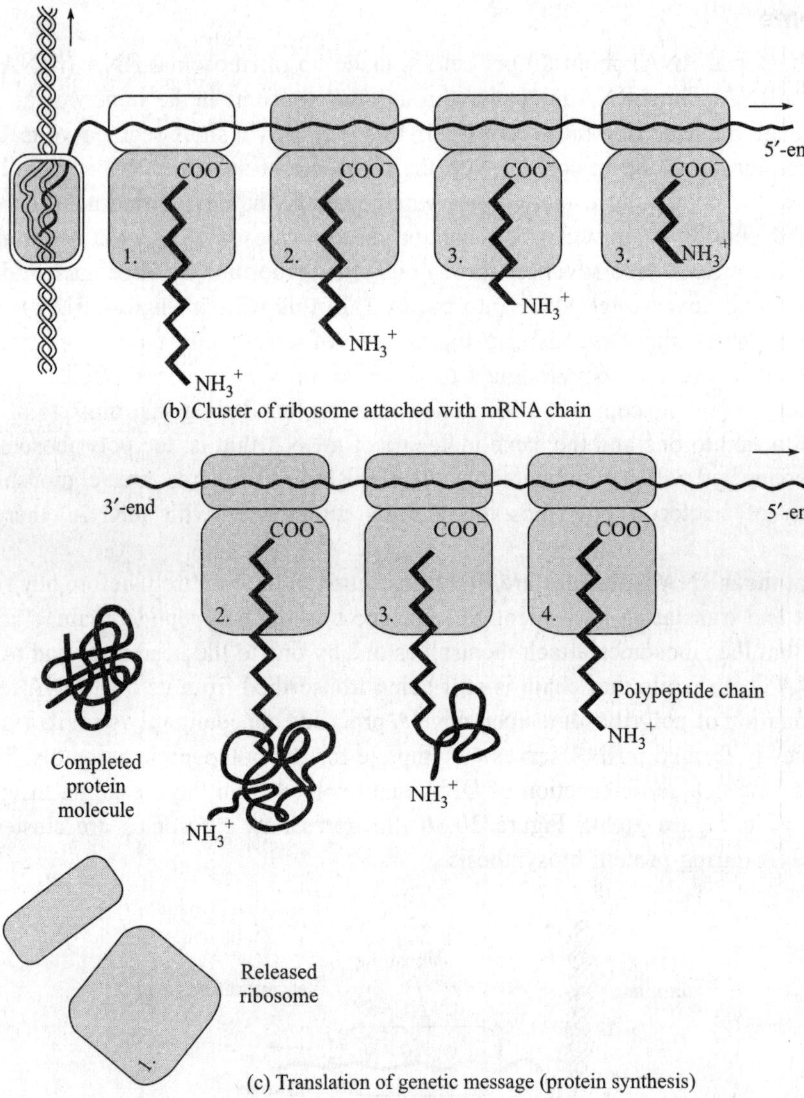

(b) Cluster of ribosome attached with mRNA chain

(c) Translation of genetic message (protein synthesis)

Figure 20.30 Transcription and translation together in bacteriophage DNA during protein synthesis.

(a) The head or 5′ end of an mRNA molecule has just been transcribed and nucleotides are being added to the newly polynucleotide chain at this tail or 3′ end.

(b) Second, third and fourth ribosomes have been engaged at the 5′ end of the newly formed mRNA chain. This way polyribosomal cluster has been formed.

(c) After a few second, *transcription* of mRNA, has been completed and it is no longer in contact with the DNA template, where once it arose. The first ribosome has now completed the assembly of newly formed polypeptide chain; has released that chain to the cell and has itself been released from the mRNA, molecule; other ribosome's are nearing to the completion of their polypeptide chain.

Transfer (tRNA)

It is well understood that the mRNA molecules provide the information for protein synthesis in the ribosomes but how the amino acids are actually assembled by the mRNA into a sequence of predetermined order was first analyzed by Crick (1958) who gave the idea that RNA chain forms the 'cavities' for uptaking the amino acid, i.e. twenty cavities for twenty amino acids. But on physical–chemical grounds this idea did not fit to the hypothesis. Therefore, another hypothesis was proposed by him that amino acids are carried to the template by *adopter* molecule. This hypothesis partly fits to the RNA.

> *"Crick wrote about the nature of a Adopter. He said there is one possibility which seems more likely than any other which might contain nucleotides. This would enable them to join onto the RNA template by the same pairing rule of bases as found in DNA or in polynucleotides. Crick thought further that a separate enzyme will be required to join each adopter to its own amino acid and that the specificity required to distinguish between (say) leucine, isoleucine and valine would be provided by these enzyme molecules, instead of by cavities in RNA. Enzymes being made of proteins can probably make such distinction more easily than can nucleic acids".*

This adopter hypothesis led to the new image of amino acid assembly in the polypeptide chain that is, before its incorporation into the newly formed polypeptide chain, each amino acid is fitted to a nucleotide adopter (tRNA) that contains a nucleotide triplet called 'anticodon', which is complementary in its nucleotide sequence to the nucleotide triplet or codon of mRNA that codes for the amino acid. This nucleotide triplet-adopter (codon-anticodon) complex after picking the amino acid diffuses into the ribosome where they are held in the proper places on the mRNA template by the pairs of hydrogen (H) bonds between the complementary purine and pyrimidine bases of adopter and mRNA molecules. Once lined up along the mRNA template in 'correct' order, the individual amino acid residues join to each other through the peptide bond, by means of a chemical arrangement that simultaneously liberates amino acid from the adopter nucleotide (anticodon) and the free amino acids to join the polypeptide chain (Figure 20.31).

AGA	CAG	UCA	GUC	AGU	CAG	UCA	GUC	3′ mRNA
UCU	GUC	AGU	CAG	UCA	GUC	AGU	CAG	Adopter (tRNA)

Figure 20.31 Complementary base pairing rule.

While these speculations were going on, a short RNA molecule about 80 nucleotide in length known as *transfer RNA* or *tRNA* was obtained by Paul Zamecnik in his laboratory, which acts as transfer agent of amino acid synthesis. The *E. coli* consists of 15 per cent tRNA, 80 per cent rRNA and 5 per cent mRNA.

Activation of amino acids

The involvement of transfer RNA (tRNA) in protein synthesis came to light when M. Hoegland, Zamecnik and their colleagues found that very first step in the utilization of amino acids in protein synthesis is their 'activation' by *amino acid activating* enzyme called *aminoacyl tRNA synthetase* that catalyses the reaction between amino acid and ATP (adenosine triphosphate) (Figure 20.32).

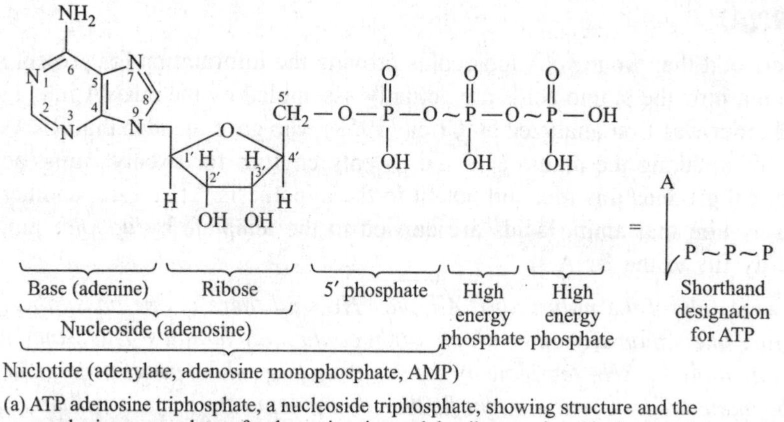

(a) ATP adenosine triphosphate, a nucleoside triphosphate, showing structure and the numbering nomenclature for the purine ring and the ribose moiety (the latter and listed as, 1′, 2′, etc., to differentiate them from the ring carbons of the base).

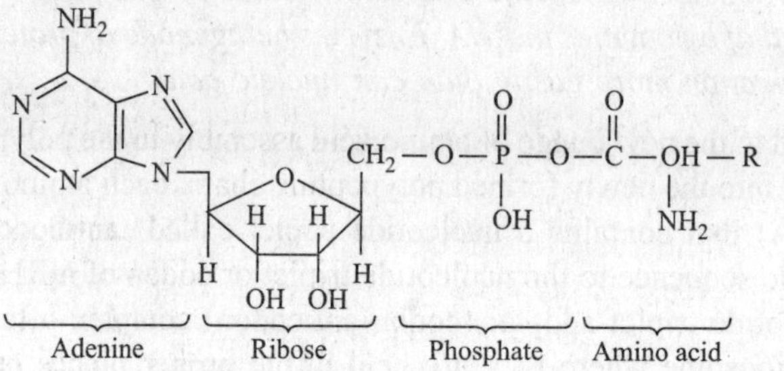

(b) Amino acid adenylate. R designates the side chain for any alpha amino acid

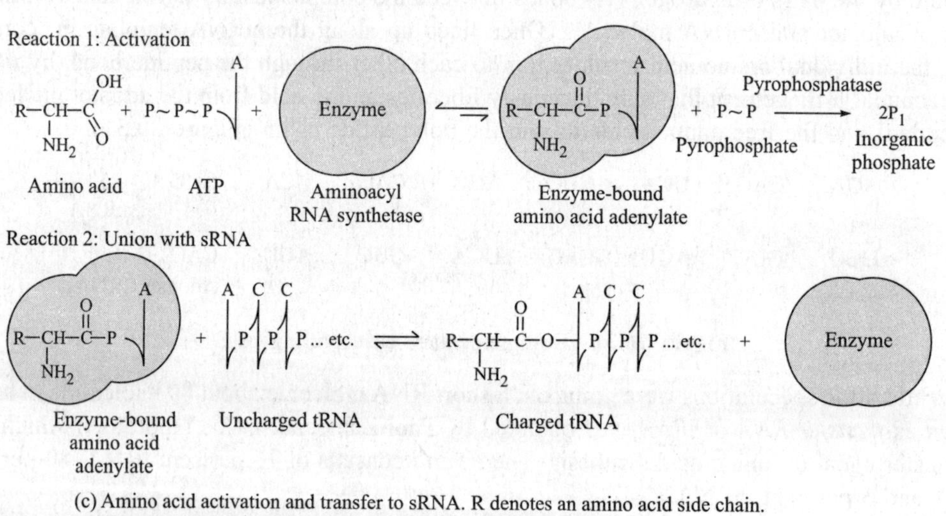

(c) Amino acid activation and transfer to sRNA. R denotes an amino acid side chain. The shorthand designation for ATP, a 5′ nucleoside triphosphate, is that symbolized in the figure configuration of the amino acid-acceptor portion of the sRNA molecule follows the same shorthand convention. The nucleotides adjacent to A in the sRNA are cytidylates, designated as C.

Figure 20.32 Activation of amino acid by *aminoacyl tRNA synthetase* before its involvement in protein synthesis.

The product of this reaction is 'amino acid adenylate' in which amino acid is linked to adenosine triphosphate (ATP), i.e. ester-bond. At next step, very same activating enzyme molecule with which the amino acid adenylate is still attached, was found to catalyze the attachment of *another* amino acid to a tRNA molecule. *More accurately, the amino acid adenylate was found to react with the free '3 OH' group of the ribose of adenylic acid that tRNA molecules carry at their* 3 *end* so as to link the α-carboxyl group of amino acid by an acyl bond to tRNA molecule. It is through the aminoacyl-tRNA complex that the amino acid reaches to its final destination, i.e. to polypeptide chains. Quantitative observations indicated that attachment reaction of amino acids to tRNA is highly specific in nature in that for each and every one of 20 amino acids, there is a corresponding one species of tRNA molecule that will accept only that amino acid. This indicates that transfer RNA will accept only the targeted or assigned amino acid irrespective of the presence of remaining 19 amino acids in the pool.

Amino Acid Assembly

It has been already stated that ribosome is a 'workshop' and mRNA (codon) is the 'blue-print' for protein synthesis. Ribosome consists of two subunits—30s and 50s. Figure 20.31 focuses light on how amino acids are assembled to form a polypeptides chain. First of all, the mRNA molecules come in contact with 30s ribosome subunits, not at 50 ribosome subunits. On the other hand, tRNA molecules come in contact with the 50s ribosomal subunits, which possess two subunits to which a tRNA can be bound. First site is called *aminoacyl* (A) *site* at which a tRNA carrying as amino acid can attach, provided that the *anticodon* borne by the tRNA molecule matches with codon displayed by that section of mRNA tape which faces the aminoacyl site.

The second site is the *peptidyl* (P) *site*, which accepts tRNA molecule only from *aminoacyl site*. Growth of a peptide chain is initiated when aminoacyl site of a ribosome displaying mRNA codon corresponding to the first amino acid of the peptide chain is occupied by corresponding tRNA and its attached amino acid. Transfer RNA and first amino acid now move to the peptidyl site.

The movement is attended by the growth of mRNA through the ribose and by the display of next codon at the *aminoacyl site*. The *second* tRNA carrying the second amino acid now occupies the *aminoacyl* site, which brings the amino group of the second amino acid into juxtaposition with the ester bond in which the caboxyl group of the first amino is linked to the 3′ adenylic acid terminus of its own tRNA.

Polypeptide formation in eukaryotes

Polypeptide formation consists of three stages:

1. Chain initiation
2. Chain elongation, and
3. Chain termination.

1. ***Chain initiation:*** The main steps in the chain initiation are—the binding of messenger RNA to the ribosome, the selection of initiation codons (AUG) and the binding of *acylated* tRNA carrying first amino acid methionine.

2. ***Chain elongation:*** In the elongation stage there are two steps:
   (i) Joining together of two amino acids by peptide bond formation.
   (ii) Moving of the mRNA and ribosome with respect to one and the other so that codon could be translated properly,
3. ***Chain termination:*** In the termination stage, the completed protein is dissociated from the ribosome and the ribosome is released to begin another cycle of synthesis.

The chain elongation continues until a termination codon (UAA, UAG or UGA) reaches the ribosome. The chain is terminated on the reach of termination codon and released from the ribosome. This process needs release factors RF_1, RF_2, and RF_3 in prokaryotes and RF in eukaryotes.

The termination codon gives signal to the ribosome for attachment of the released factors. The release factors interact with peptidyl transferase causing hydrolysis of the bond between tRNA and polypeptide chain.

Hydrolysis of GTP also assist in the dissolution of release factors from the ribosome.

Polypeptide synthesis in prokaryotes (Bacteria)

The first step in the initiation of polypeptide synthesis in both prokaryotes and eukaryotes is the use of a specific initiating tRNA molecule. In prokaryotes, this tRNA molecule is acylated with the modified amino acid *N-formyl methionine* (f meth). The tRNA is also designated as $meth^{fmeth}$. Structure of the N-formyl methionine is given in Figure 20.33.

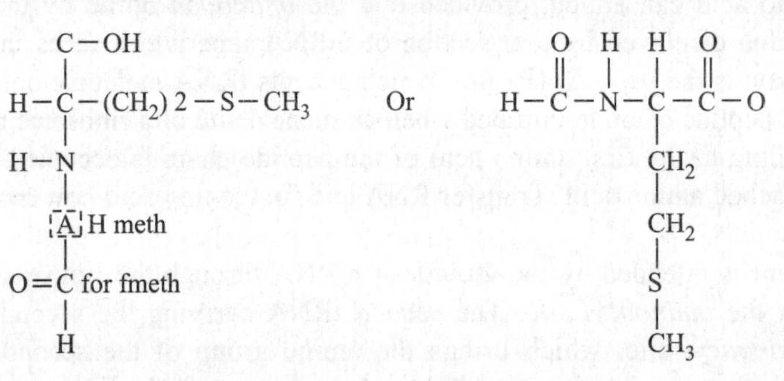

Figure 20.33 Structure of N-formyl methionine.

Both $tRNA^{fmeth}$ and $tRNA^{meth}$ recognize codon AUG but only $tRNA^{fmeth}$ is used for initiation. In this process, tRNA is first acylated with methionine and then an enzyme (found only in prokaryotes) adds formyl group to methionine. It means all the prokaryotic proteins have 'fmeth' at the amino terminus. However, this amino acid is quickly 'deformylated' or removed later.

In eukaryotes, the initiating tRNA molecule is charged with methionine, but there is no formylation.

Polypeptide synthesis in bacteria begins with the association of one 30S subunit (of the 70S ribosome) with mRNA molecule, fmeth tRNA, three proteins known as *initiator* factors (IF_1 IF_2 IF_3) and guanosin 5′ triphosphate (GTP). These molecules constitute the 30S *pre-initiation complexes* (Figure 20.34). Since polypeptide synthesis begins at AUG (starting codon) and AUG codons are found within the coding sequences (i.e. methionine occurs within a polypeptide chain) some signal must be present in the base sequence of the mRNA molecule to identify a particular AUG codon as a starting codon.

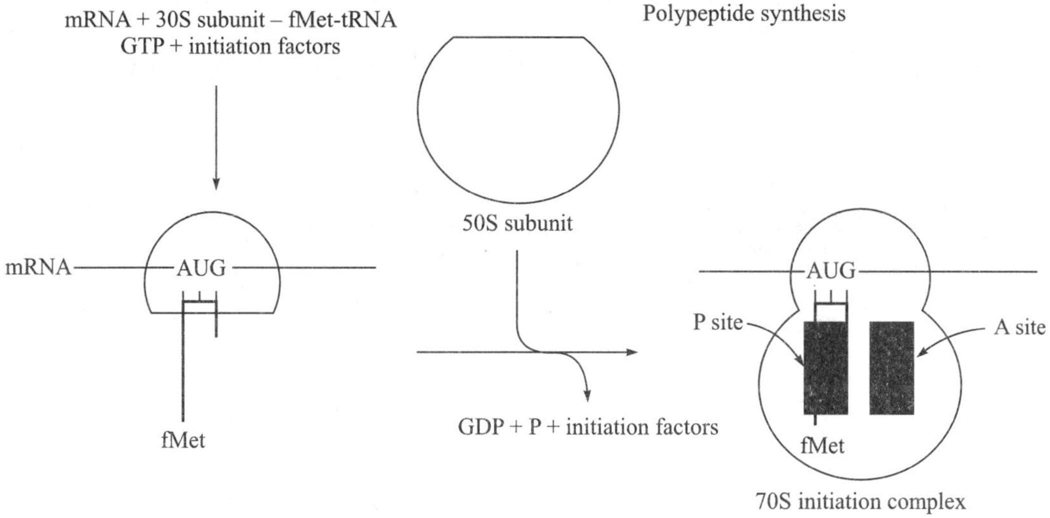

Figure 20.34 Early steps in protein synthesis in prokaryotes: formation of the 30S pre-initiation complex and of the 70S initiation complex.

The means of selecting correct AUG codon differs in prokaryotes and eukaryotes. In prokaryotes the ribose-binding site near AUG codon is used for initiation.

The 50S subunits contain two binding sites. These sites are called amino acyl (A) site and the peptidyl (P) site. The fmeth-tRNA fmeth which was bound to 50s pre-initiation complex (A) moves to the P site of the 50S subunit in the 70S initiation complex. The placement of fmeth-tRNA in the P site fixes the position of the fmeth-tRNA anticodon such that it pairs with the AUG initiator codon in the mRNA. This way, *the reading frame is unambiguously defined upon completion of the 70S initiation complex.*

Once the P site is filled, the A site of the 70s initiation complex becomes available to any tRNA molecule whose anticodon can pair with the codon adjacent to the initiation codon. After occupations of A site, a peptide bond between fmeth and the adjacent amino acid is formed by an enzyme complex, called *peptidyl transferase*. Soon the bond is formed, the fmeth is cleaved from the tRNA in the P site.

After the formation of peptide bond, an uncharged tRNA moves to the P site and a dipeptidyl tRNA occupies A site. At this moment three activities take place:

(i) The tRNA fmeth in P site is not linked to any amino acid leaves this site.
(ii) The peptidyl tRNA moves from A site to P site.
(iii) The mRNA moves a distance of three bases in order to position the next codon at the A site. This is called *translocation*, which need the presence of an elongated protein EF-G and GTP (Figure 20.35).

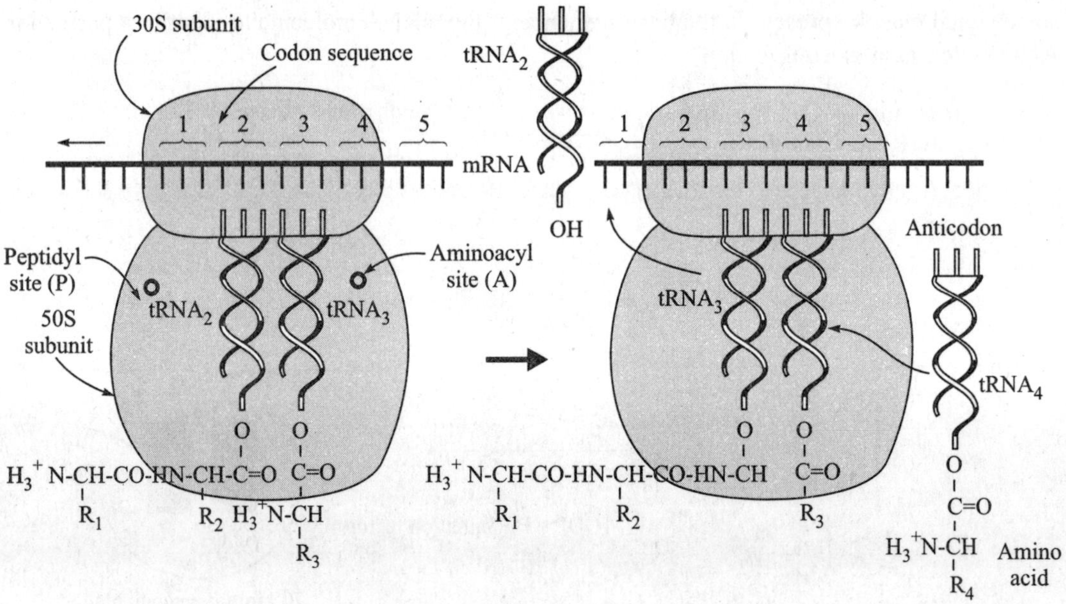

Figure 20.35 Scheme of amino acid assembly in protein synthesis.

After mRNA movement has occurred, the A site is again available to accept a charged tRNA molecule having a correct anticodon. When a chain termination codon UAA, UAG or UGA is reached, no tRNA exists that can fill A site so chain stops. But polypeptide chain is still attached to the tRNA occupying P site. The proteins released by release factors also cause disassociation of 70S into 30S and 50S.

However, if mRNA molecule is polycistronic and the AUG codon initiating the second polypeptide chain is not too far from the stop codon of the first cycle, the 70s ribosome will not dissociate but will reform on initiation complex with the second AUG codon.

Complex translation units

The mechanism of translation is not so simple that one ribosome is traversing an mRNA molecule, but it is a somewhat complex character.

After joining of about 25 amino acids together in a polypeptide chain, an AUG (initiating codon) becomes completely free of the ribosome and it can form second initiation complex. The overall configuration is that two 70S ribosomes may be moving together along a mRNA at the same speed. When second 70S ribosome has moved along a distance similar to that traversed by the first, a third ribosome can attach to the initiation site. The process of movement and reinitiating continues until the mRNA is covered with the ribosomes at a density of about one ribosome per 80 nucleotides. This large translation unit is called *polyribosome* or *polysome*. This is the usual form of translating mechanism.

The direction of translation is $5' \rightarrow 3'$ of the mRNA. The ribosome binding site is transcribed first followed by initiating AUG codon, then region encoding amino acid finally stops codon.

Differences in protein synthesis of prokaryotes and eukaryotes are given in Table 20.3.

TABLE 20.3 Differences in protein biosynthesis of prokaryotes and eukaryotes

Prokaryotes	*Eukaryotes*
1. A single RNA polymer controls the synthesis of three types of RNA (mRNA, ribosomal RNA and tRNA).	1. Several RNA polymerases are involved in the biosynthesis of cellular RNA.
2. 70S ribosome (30S and 50S subunits) is involved in the mRNA synthesis.	2. 80S ribosome (40S and 60S subunits) is involved in the mRNA synthesis
3. Process of transcription and translation can overlap.	3. No such overlapping takes place.
4. N-formyl methionine is initiator amino acid.	4. Methionine is initiator amino acid.
5. There are several initiation and termination sites.	5. There is a single initiation and terminator site
6. There are three initiation factors, IF1, IF2 and IF3.	6. There are several initiation factors, elf2, elf2', elf-2al, elf-a2, elf-a3.
7. There are three chain elongation factors, EF-T4, T5 and EF-T6.	7. Only two chain elongation factors, EF-1, EF-2 are involved'
8. DNA and RNA both can synthesize mRNA.	8. Only DNA can synthesize mRNA

ANTIBIOTICS AND ANTIBIOTIC RESISTANCE

There are about hundred antibiotics, which inhibit gene expression. Some have been used as valuable tools in the study of microbial genetics.

Rifampicin antibiotic inhibits transcription in bacteria. It binds the subunits of core enzyme and prevents initiation of RNA synthesis. But it has no effect on RNA polymerase, if that has initiated transcription. Other classes of antibiotics such as 'streptolydigin' bind RNA polymerase and inhibit all the movements of RNA polymerase.

Most of the antibiotics inhibit polypeptide synthesis, viz. *streptomycin* (str), kanamycin (kan) bind to 30s ribosome and inhibit mRNA synthesis. *Tetracycline* (tet) inhibits binding of *acylated tRNA* to the 30s ribosome. *Chloramphenicol* (cam) inhibits peptidyl transferase and *puromycin* (pur) causes premature chain termination. In microbial genetics, alleles for sensitivity and resistance to these antibiotics are common genetic markers. *Penicillin* does not affect protein synthesis, but interferes production of bacterial cell walls.

Some bacteria are naturally resistant to certain antibiotics but others can acquire resistance by mutation. Natural resistance to *penicillin* is conferred by a gene that synthesizes lactamase—an enzyme that cleaves the lactam ring in penicillin and ampicillin (Anonymous: the source from where this material is taken is not available therefore word Anon. is written).

REVIEW QUESTIONS

1. Are 'operons' more common in prokaryotes (bacteria) than eukaryotes (higher organism)?

2. If the operon theory had been established at the time of the two classical hypotheses for the mechanics of differentiations were being considered that is (i) segregation of nuclear elements, and (ii) intervention of cytoplasm. Which hypotheses would have been most acceptable?
3. Many of the RNA transcripts of prokaryotes are polycistronic. Are polycistronic RNA transcripts common in eukaryotic organisms?
4. How do histones regulate gene activity? When in the cell cycle are histones synthesized? How is the synthesis of histones themselves regulated?
5. How is the synthesis of tryptophan regulated?
6. Why are microorganisms such as bacteria, viruses and molds are favourable materials for biochemical genetic studies? What kind of experimental materials would be most suitable for (a) operons, and (b) repressing effect of histones?
7. How inducible and repressible enzymes of microorganisms can be distinguished?
8. Distinguish between repression and feedback inhibition caused by end product of a biosynthetic path. How do these two regulatory phenomena complement each other to provide efficient regulation of metabolism.
9. In the 'lactose-operon' of *E. coli*, what is the function of each of the following genes or sites:

   (a) Regulator (b) Operator
   (c) Promoter (d) Structural gene *a*
   (e) Structural gene *y*
10. What would be the results of inactivation by mutation of the following genes or sites in the *E. coli* lactose operon:

    (a) Regulator (b) Operator
    (c) Promoters (d) Structural gene *z*, and *a*
    (e) Structural gene *y*
11. Groups of alleles associated with the 'lactose operon' are as follows (in order of dominance for each allelic series): repressor i^s (super repressor), i^+ (inducible) and i^- (constitutive) operator, O^c (constitutive, *cis*-dominants and O^+ (inducible *cis*-dominant), structural Z^+ and Z^-:

    (a) Which of the following genotypes will produce β-galactosidase and permease, if lactose is present:

    (i) $i^+ o^+ z^+ y^+$ (ii) $i^- o^c z^+ y^+$
    (iii) $i^s o^c - z^+ y^+$ (iv) $i^s o^c z^+ y^+$
    (v) $i^- o^+ z^+ y$

    (b) Which of these genotypes will produce β-galactosidase and permease, if lactose is absent.

REFERENCES

Berner, H. and M.W. Konrad, (1964): A complete 30: 293 of enzymatically synthesized RNA and template DNA, *Proc. Nat. Acad. Sci.*, Washington, **51,** 801.

Brenner, S., F. Jacob, and M. Meselson (1961): An unstable intermediate carrying information from genes to ribosomes for protein synthesis, *Nature*, **190,** 576.

Hartl, D.L. (1991): *Basic Genetics*, Jones and Bartlett, Boston.

Ram, M. (2005): *Unfolding Reality of Life*—The DNA *World*, Dominant Publications, New Delhi.

Stent, G.S. (1971): Genetic transcription, *Proc. Roy. Soc.*, London, Ser. B., 164.

Stent, G.S. (1971): *Molecular Genetics*, W.H. Freeman and Co., San Francisco.

Wastson, J.D. (1963): The involvement of RNA in synthesis of protein, *Science*, 140.

Zamecnik, P.C., E.B. Keller, J.W. Littlefield, M.B. Hoagland, and R.B. Loftfield, (1956): Mechanism of incorporation of labelled amino acids into proteins, *J. Cell Comp. Phyciol*, 47 suppl. **1,** 81.

CHAPTER

21 *Fine Structure of Gene*

After rediscovery of Mendel's laws of inheritance, gene was pictured to perform some function, which determines the phenotypic appearance of an organism; creates variation in the population by recombination of similar units on the homologous chromosomes and also creates new genes by mutations. However, after identification of DNA as the genetic material by Avery, et al. (1944) through *transformation* principle, the molecular geneticists became interested to know at molecular level, the gene as unit of function; unit of recombination and unit of mutation. In this respect, before quoting Seymour Benzer's (1955) work who carried out genetic analysis on a gene of the bacteriophage T4 and prepared *fine structure* of gene and gave the terms 'cistron' unit of function, 'recon' unit of recombination and 'mutan' unit of mutation. It is worth to make clear here once again, what is molecular make-up of a gene? Actually, *gene is a sequence of nucleotides arranged in a linear fashion on the DNA polymer* and unit of mutation and unit of recombination are evidently the individual nucleotides of a DNA polymer; whereas unit of function consists of hundreds and thousands of nucleotides that specify sequence of 20 amino acids that make the primary structure of the proteins coded by the gene. The reason is that polypeptide is a linear structure of amino acids. DNA and RNA are also linear structures where polynucleotides are linearly arranged, thus all the three molecules, viz. 'DNA–RNA–Proteins' are *colinear* to each other. Any point mutation, i.e. change in the nucleotide level at any site within the gene will ultimately alter the structure of polypeptide chain because a single change (mutation) in the DNA polynucleotide chain caused by addition or deletion of a nucleotide replace one amino acid by another in the polypeptide chain also. This may result changes in the 'primary' structure of protein molecules and finally affect the biological activities of the organism.

Benzer coined the new terms *cistron*, *recon* and *mutan*, through his genetical studies in T4 bacteriophage and collected about 2000 mutants of T4 bacteriophage, all with rapid lysis ability. Normally, phage T4 lyses *Escherichia coli* and produces a small circular area in the medium where all bacteria have been killed. This area is called *plaque*. In general *plaque* formation takes place in several hours. But Benzer found some phages of T4 with ability to accomplish the task of lysis within about 20 minutes. These were termed 'rapid lysis' mutants. Benzer also discovered some specific lysis mutants, which were conditional lethal, i.e. they could grow in one specific conditions but can not grow under some other conditions. He termed such specific type of mutant as 'rII' (T_4, rII) mutant, which enters, replicates and causes lysis of *E. coli* B-cell but can not cause lysis of k-strains of *E. coli*. On k-strains, the r^+ wild-type phage as well as other r mutants that do not belong to the rII class can grow perfectly but rII can not. These differential growth characteristics of rII mutants were considered a very rare genetic property of the mutants. After collection of 2000 mutants of T4, all with same rapid lysis property, his next question was to know: do all these mutants belong to the one and same functional unit? He utilized the inability of rII mutants to grow on k-strain of *E. coli* to find out answer of his question and the failure of different rII mutants to grow on the *non-permissive* k-strain of *E. coli* also need not necessarily reflect one and the same functional defect in their hereditary material. In order to determine whether the two different rII mutants belong to the same functional unit, to which he termed 'cistron' Benzer conducted *cis-trans gene complementation test* (Figure 21.1). Cistron is nothing but functional unit of gene, which became popular by 1955, but later on gene was considered better word for the unit of heredity than *cistron*.

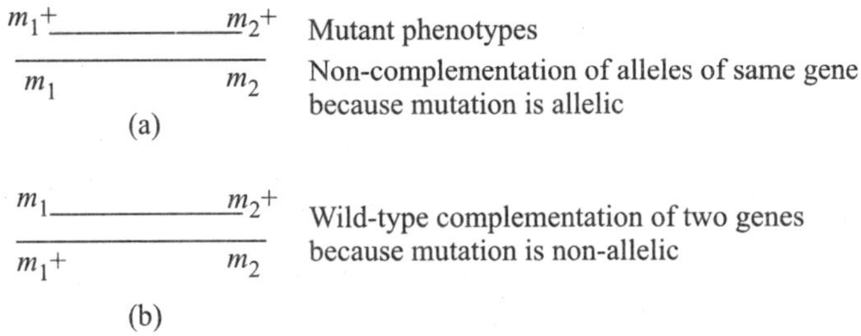

Figure 21.1 Test for complementation with heterozygous mutants in *trans* arrangement. Two mutations that are allelic will not complement each other in *trans* arrangements (a). Two mutations that are not allelic will complement each other in the *trans* arrangement (b).

However, Benzer proceeded with the construction of *fine structure* of genetical map. His concept was based on the findings that a k-strain bacterium infected jointly with rII mutant and rII+ wild-type phage, both types can grow normally. Here normal rII+ gene of wild-type parents is, therefore, able to supply the 'substrate' necessary for growth in the k-strain of bacterium not only for itself but also for the growth of defective rII mutant. Or, in the classical genetic term in the mixed infection, the r^+ gene (RR) may be *dominant* over its rII mutant gene allele(rr).

Phages of T4 carrying rII (locus) mutations can not grow at all on special *E. coli* strains to which he called k-strains. On k-strains, the rII^+ wild-type phage as well as other mutants that do not belong to rII class can grow perfectly. The rII mutants adsorb to infect and even initiate the synthesis of phage DNA and other products of phage metabolism in k-strain host cells, but such rII infected k-strain bacteria do not produce infective progeny phages nor do they ever lyse.

PROCEDURE OF THE COMPLEMENTATION TEST

In the complementation test, an *E. coli* k-strain is infected with a pair of 'rII mutants' in order to examine whether the two mutants, each unable to grow alone in a strain of k-bacterium, are able to cooperate in the mixedly infected cell to produce a litter of infective progeny. *If mutant pair is able to cooperate and grow in this way, then it may be concluded that two different mutations have not occurred on the same site in the rII phage genome, governing the same function.* The failure of one mutant to grow in a strain of k-bacterium (its rII phenotype) must mean that it can not carry out one particular function or elicit the synthesis of one particular protein necessary for growth in the infected strain 'k-cell' at the same time it must not mean that rII mutant can not carry out some other necessary function pertaining to a different functional unit of the genome. If two mutants are able to reproduce in joint growth because each supplies that necessary 'substrate' (perform function) in which other is lacking. Then it is believed that mutation has taken place on two sites and they complement to perform a function. But if two mutants are unable to cooperate to grow, then it can be concluded that both mutations have taken place at the same genetic site affecting the same functional unit of the phage genome. In that case, the rII phenotype of both the mutants is caused by a deficiency in exactly the same function which neither mutant can supply to other in joint growth.

Benzer proceeded with the construction of a fine structure of the genetic map of T4 genome. For this purpose, he crossed members of his rII colllections two × two and selectively scored for the frequency of rII^+ type recombinants produced by plotting the phages progeny produced by each cross on plates seeded with k-strain of *E. coli*. In this way, he collected rare recombinants between adjacent genetic sites.

On this basis, Benzer found that rII mutants fall in two categories—A and B. All mutants belonging to one group (A) complement with any member of the other group (B) in the production of infective progeny in a joint growth in k-strain, but do not complement with the member of their own groups (A – A or B – B). It is, therefore, considered that two groups A and B can be assigned definite positions on the *genetic fine structure map* of the T4 phage. Benzer referred to the genetic unit of function on the basis of *cis-trans test* as *cistron* is genetic unit of function (Figure 21.2).

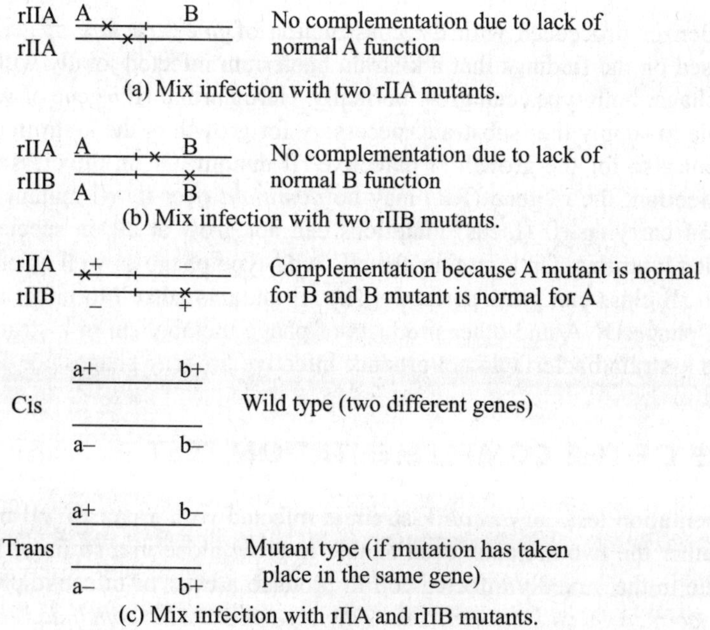

Figure 21.2 Complementation test in rII locus in T4 bacteriophage conducted by Benzer.

CONFLICTING VIEW ON GENE

Pseudoalleles

The gene complementation test by constructing heterozygotes has made it possible to determine the functional and structural complexity of genes. In 1940, Lewis and other *Drosophila* geneticists made it possible that two mutant genes which appear to be allelic in the complementation test can be resolved from each other through crossing over. This also suggests that gene, as a unit of function, cannot be unit of recombination, which is against the earlier belief.

Green and Green (1949) studied the structure of lozenge locus in *Drosophila*, which possesses large number of recessive alleles associated with the varying degrees of rearrangement of eye facets, reduction in the intensity of eye pigmentation and reduced female fertility. The flies heterozygous for different alleles showed a mutant phenotype, but some heterozygous females produced in their progeny some wild-type flies. By using the marker genes on both the sides of the locus, it was possible to confirm that wild-type flies appeared in the test cross progeny due to intralocus recombination.

From these and other studies, it was determined that lozenge locus consists of four subunits each of which on mutation can give rise 'mutant phenotype'. These four subunits can be resolved by crossing over; the amount of recombination between them would be very small. But on the basis of recombination values, this subunit can be arranged in linear order (Figure 21.3).

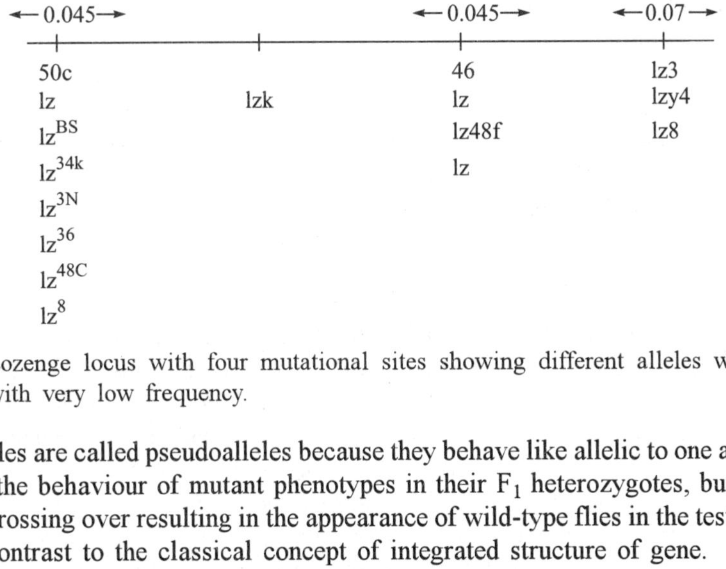

Figure 21.3 Lozenge locus with four mutational sites showing different alleles which recombine with very low frequency.

These alleles are called pseudoalleles because they behave like allelic to one another as could be seen from the behaviour of mutant phenotypes in their F_1 heterozygotes, but they could be separated by crossing over resulting in the appearance of wild-type flies in the test cross progeny. This was in contrast to the classical concept of integrated structure of gene.

The concept 'pseudoalleles' is no more valid now.

The above said conflicting view about gene was resolved with the discovery that gene with its linear DNA structure has many mutational sites. These can mutate independently and give rise to a series of different alleles of a gene. However, these subunits are closely linked, therefore, they cannot be easily separated by crossing over. It is very difficult to recover rare recombinant types because in plants and animals, we cannot raise so high segregating population. But genetical studies conducted in the microorganisms solved this problem because very large population we can raise here. Now gene is the unit of function but not at the same time unit of mutation and unit of recombination. The concept of pseudoalleles was soon replaced with the notion that gene can mutate at several different sites. Then, another question arises, how many mutational sites can be expected to find in a gene. On purely theoretical bases, it could be argued that a gene may consist of as many different mutational sites as there are number of nucleotides base pairs in its DNA segment, i.e. nucleotide is the unit of mutation.

BENZER'S NEW CONCEPT OF GENE

Benzer's (1955) second approach was to revise the classical concept of gene by discovering the intragenic recombination at the rII locus in the T4 bacteriophages. He recognized that no single unit in the linearly arranged genetic material in the chromosome has the status of a superstructure behaving as a unit of function, recombination and mutation. The classical gene had been considered unit of function, recombination and mutation. According to the modern concept 'genetic material in the chromosome can be resolved into three kinds of units, which are described below.

Cistron

First unit of genetic material described by Benzer was *cistron*, the unit of function, which could be characterized by *cis-trans* test or the complementation test. The mutation arising in a *cistron* produces a defective protein; the different alleles of a *cistron* may not show complementation effect in *trans* heterozygote. For better understanding of gene function, *cistron* has been defined as *a segment of DNA in the chromosome which carries information for directing the synthesis of a single polypeptide chain*. The size of *cistron* can be determined by the recombination analysis. In the term of map units, the size of *cistron* is given by the amount of recombination value between the most distantly located mutational sites marking the two ends of cistron.

Recon

Second unit of genetic material was described by Benzer as *recon*, which could be defined as ultimate unit of recombination, which is indivisible by futher crossing over. The size of *recon* must be equivalent to one or more than one very closely linked mutational sites/units, which cannot be resolved by crossing over. The size of *recon* cannot be measured, in term of recombination values. The *recon*, therefore, would be smaller than the smallest recombination value, observed between two of the mutational sites within a *cistron*. This smallest recombination value helps to identify the two mutational sites in a gene which are closer than any other. However, they are not so close that recombination cannot occur between them. The *recon* is the smaller segment of DNA than that occupied by two mutation sites. Benzer observed 0.01 per cent recombination in the rII locus.

Benzer carried out crosses between rIIA and rIIB mutants. This was done to infect *E. coli* B-strain with a large number of rIIA and rIIB mutant of T4 bacteriophage. Both these mutants were able to infect, replicate and lyse the B-strain of *E. coli*. However, these rapid lysis progeny were not able to infect k-strain. But some normal plaques were obtained from B-strain which were able to infect k-strain. These results point out that some normal recombinations were produced due to crossing over between mutant sites A cistron × B cistron. It may be concluded that cistron consists of linear arrangement of polynucleotides that can be separated by recombination. The recombinational analysis reflected smallest map distance in the rII region was 0.02 map unit and entire T4 chromosome is 1500 map unit long consisting of 1.8×10^5 base pairs. Thus, recon was called by Benzer the unit of recombination which may be equivalent to a nucleotide (Figure 21.4).

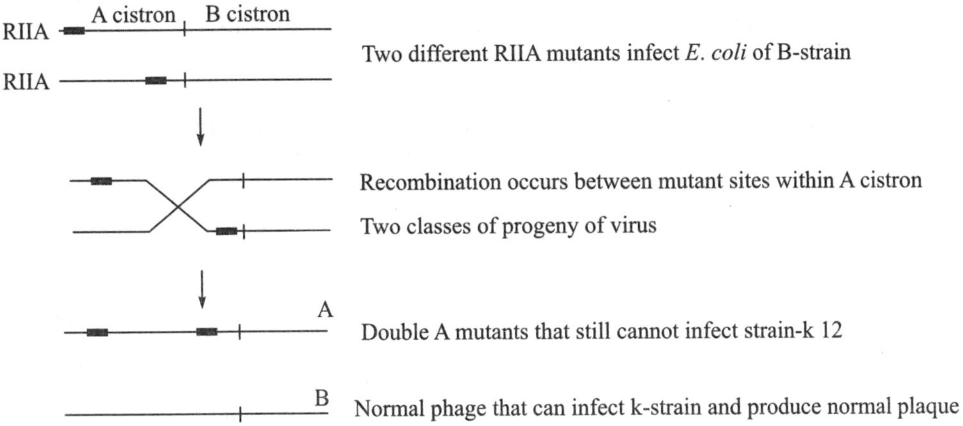

Figure 21.4 Recombination within cistron. The frequency of crossover is determined by the distance between the mutant sites. The cis-trans (complementation test of rII mutant is *E. coli*. k-strain in detail.

Muton

The third unit was described as *muton* by Benzer. Muton is the smallest subunit in the cistron which can produce mutation. The unit of mutation could be as small as a single nucleotide base pair. Based on this study, a new definition of gene was given, *a gene is a distinct chromosomal region responsible for a single cellular function consisting of a linear array of potentially mutable sites between which recombination can occur.* Thus, unit of mutation (point mutation) is a single nucleotide.

However, these new names did not survive for a long time in genetic literature. In my opinion, there is some merit in retaining the older terminology of a gene as the unit of function. However, the new knowledge has provided a better understanding of gene because DNA in a chromosome could be resolved in the form of a large number of discrete genes as different units of function. Now genetic cloning technique has assisted man to clone a desirable gene and transfer it in a desired genetic background bypassing taxonomic barriers and put them in market for sale like any other commodity.

rII CISTRON

The rII mutants belonging to two different functional groups mapped in adjacent but non-overlapping parts of the rII region. On the basis of complementation tests, the rII locus could be resolved in two different segments of DNA—A and B. The mutants which map in the same segment (A or B) do not complement one another, while those located in two different segment do show complementation. For example, rII mutants 47 and 101 of T4 bacteriophage did not multiply on k12-strain (lambada) by mixed infection, while another pair 47 and 51 did multiply and extensive lysis of the bacteria take place. This is inspite of their similar phenotypic expression and close linkages. The different rII mutants belongs to two different functional units. Benzer called A and B segments as two *cistrons*. The term *cistron* was used, as synonym of gene, owes its origin to the *cistron* test, which assisted to resolve the functionally different mutants. The *trans* configuration describes the heterozygote when mutants and their corresponding alleles are located on separate homologous chromosomes; the cis-configuration refers to the heterozygote when mutant alleles are located in one chromosome and the wild-type alleles in its homologous chromosome (Figure 21.5).

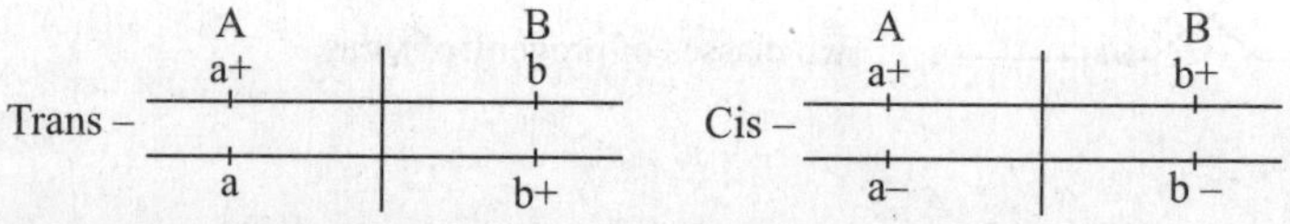

Figure 21.5 A and B cistrons in rII locus of phage T4. Mutants of two cistrons complement one another to produce a wild-type phenotype, but within a *cistron* there is no complementation. Therefore, phenotypic expression remains that of mutant type.

FUNCTIONAL DIFFERENCES OF rII LOCUS

Structurally, the rII locus is similar to the lozenge locus in *Drosophila*. But in the matter of function, different rII mutants show a greater degree of differences. Benzer carried out complementation test to varify whether different mutants were alleles of same gene or of different genes. For this purpose, he got infected the bacterial cells of strain-k12 with a mixture of λ phage particles of two mutants to examine their functional relationship. The mixture of two mutant bacteriophages were equivalent to the heterozygote of a diploid organism. When such heterozygotes were constructed, two types of observations were recorded:

(a) Many of the rII mutant pairs failed to multiply in the infected cell. They were, therefore, concluded mutant alleles of same gene.
(b) However, some other combinations gave extensive lysis in the same manner as these mutants normally do on the B-strain of *E. coli*.

It was interesting finding that some pairs of rII mutants were able to multiply on strain-k12 of bacteria in the same way as the wild-type recombinant did. This showed that two rII mutants belonged to two different functional groups unlike lozenge alleles of *Drosophila*. In other words, all the rII mutants did not arise from mutation at one locus.

LINEAR STRUCTURE OF rII LOCUS

The recombination analysis of 20000 rII mutant led to the conclusion that mutational sites could be placed in linear sequence on the basis of observed crossover values. The use of

'3-point cross' and concept of additivity of map distances between closely linked mutational sites was found to be as valid in determining the relative position of different rII locus mutants as given by Morgan and his colleagues in mapping different classical genes in *Drosophila*. These observations showed that *intragenic* recombination did not invalidate one of the concepts of genetics that genetic elements in chromosome are linearly arranged.

FINE STRUCTURE OF rII LOCUS

The viral particles of T_4 even series phages (bacterial viruses) consist of DNA molecules enclosed in a protein sheath (coat). They multiply infecting *E. coli* as host cell. After infection, the viral DNA replicates to produce a pool of vegetative DNA molecules. They mature into phage particles with the synthesis of protein coat. The growth cycle is completed with the bursting (lysis) of infected bacterial cells. This way, it liberates the newly synthesized viral particles in the form of colonies or plaques as they are called. During multiplication inside the host cell, the newly formed pool of viral DNA molecules 'pair' and undergo crossing over, showing recombination in the same way as homologous chromosomes do in the eukaryotes.

T4 phage offers several 'marker' characters to study intragenic recombination. One of the phenotypic marker is *host specificity* and another is *delayed* lysis of the infected bacterial cells. Wild-type T4 particles show delayed lysis (take more time for multiplication) and produce plaque which are small and fuzzy in their outline.

However, a group of mutants were traced out to the locus called rII locus which gave rapid lysis and produced larger plaques with more sharply defined outline. Like rII^+ (wild-types), the mutant rII infects and multiplies readily on B-strain of *E. coli*. However, they failed to multiply on k12-strain. The k12-strain thus provides an ideal screening media for detecting rare wild-type recombinants from a cross between the two different rII mutants. The lysate progeny of a cross between the mutant can be plated on a petri dish having *E. coli* cell of strains k12 of bacteria. The two parental types mutants particles failed to multiply on the k12 of bacteria cells, however, the wild-type recombinants if any produced from their cross did multiply on these cells and lyses the bacterial cell to produce rII+ plaques. In this way, wild-type recombinants can be distinguished from rII mutants on both strain-B and strain-k12. On B, they form smaller plaques (Figure 21.6).

RECOMBINATION STUDIES OF rII LOCUS

Benzer did pioneering work on phage T4 which led better understanding of the *fine structure* of gene by making crosses between different rII mutants in many different combinations. These crosses led to understand that rII mutants are recessive: This was indicated by the observations that rII mutant cannot multiply on the k12-strain, however, if bacterial cells are infected with the mixture of rII mutant, wild-type phage particles (colonies) appear. For making crosses, the bacterial cell (B) was infected with the mixture of two rII mutants proposed to be mated. After lysis, the newly multiplied viral particles from the cross were plated on k12-strain. As neither of two rII mutant types can multiply on this strain, this gave an opportunity to find out wild-type recombinants from the proposed cross which could be able to multiply on k12 λ-strain (Figure 21.7).

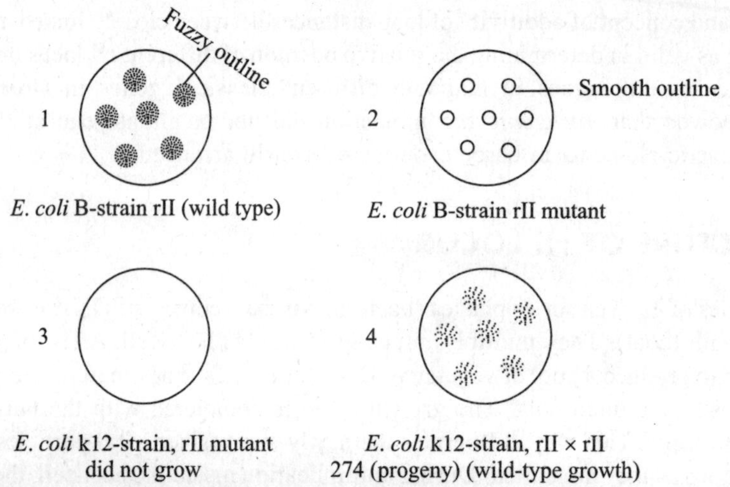

Figure 21.6 (1) rII mutant of phase T4 grew on B-strain of *E. coli*; forming Fuzzy outline (wild) type (2) rII mutants of phage T4 grew an strain B of E. coli eith smooth outline (3) rII mutants of phase T4 failed to grew on k_{12} strain of *E. coil* (4) when two rII mutant were crossed ($rII \times rII_{274}$) they produced some wild type recombinant progenies which formed normal plaque an k_{12} strain of *E.coil*.

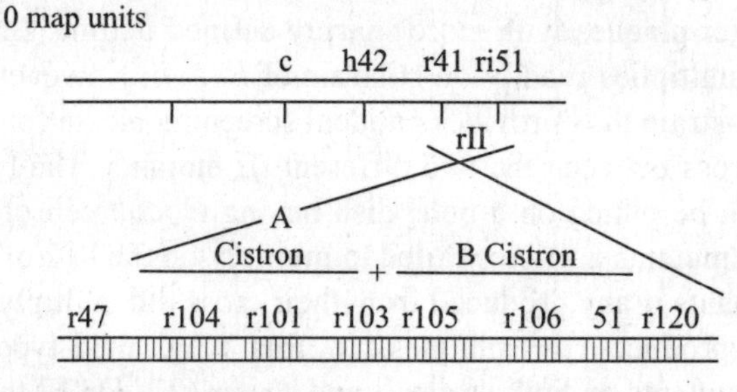

Figure 21.7 Fine structure of genetic map of the rII locus of T4, based on frequencies of r^+ recombinants produced in pair crosses between a set of 60 rII mutants of independent origins.

By making crosses of 2000 mutants of independent origin, Benzer was able to identify about 300 different mutational sites in rII locus. Total map length of rII locus was found to be 8 units and minimum recombination value between any two mutational sites 0.01 per cent. In higher organism, not more than 10 mutational sites were noted in a gene (Figure 21.8).

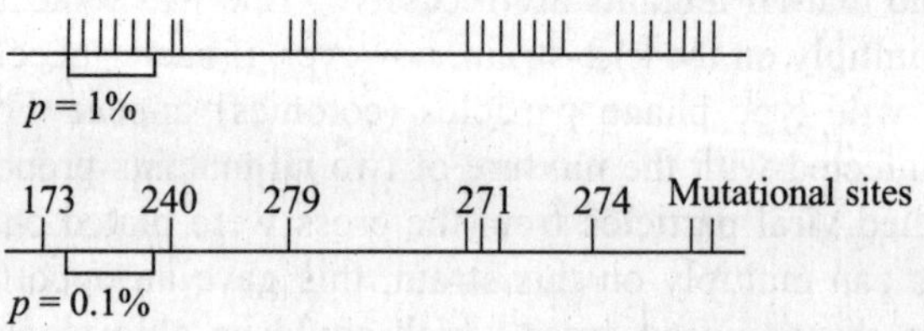

Figure 21.8 Mutational sites in a segment of rII locus based on Benzer's work.

MOLECULAR MECHANISM OF MUTATION

The mechanism by which new changes arise due to molecular mutation in the DNA nucleotide sequences has been dealt in Chapter 19 dealing with the genetic code. For the *fine-structure* studies, the bacteriophages are the better material than bacteria because it is possible to study mutation of the phage DNA both in its resting and extracellular states in the infective phage particle as well as in its replicating or intracellular vegetative state.

Infact, rate of spontaneous mutation of the resting phage DNA is very low and extracellular phage particles do not mutate at all over the period of months and years. It is mainly during the vegetative multiplication of the phage in the host cell, the new mutants appear. Because of quick multiplication of phages in the *E. coli*; until all bacteria have been lysed, the phage liter (population), may rise to a final level of 10^{10} particles per millilitre; this way, the proportion of r, mutants among the whole population rise in each cycle from an initial level of suppose 10^{-4} r mutant per r^+ wild type to a final level of about 10^{-3} r mutant per r^+ wild type. It is believed, therefore, that phage mutation arises as a result of *copy errors* during the intracellular replication of the genome. Although replication of DNA of the parent phage is highly accurate but occasionally an *error* may occur in the replication process, which affects the correct sequencing of the polynucleotides in the DNA. This cause mutations in the vegetatively replicated phage particles. The mutant replica (r mutant), when infects another host bacterial cell in course of next growth cycle, it is the mutant information that is replicated faithfully to produce stack of mutant phage progeny.

How *copy errors* arise in the DNA during replication has been given a molecular explanation by Watson and Crick in their paper published in Nature (1953). As stated in Chapter 17 dealing with schemes of DNA replication given by Watson and Crick, the DNA molecule consists of two complementary polynucleotide strands arranged in a helical manner. These two complementary polynucleotide strands of the DNA molecule separate from each other by the catalytic action of DNA *polymerase*. Each parental strand then serves as a *template* for the ordered synthesis of a new complementary polynucleotide chain through the formation of specific hydrogen (H) bonds between purine–pyrimidine bases of the parental nucleotides and the mononucleotide building blocks are available for polymerization in the intracellular metabolic pools. However, formation of the specific Watson–Crick hydrogen bonds between adenine–thymine (A–T) and between guanine–cytocine (G–C) requires that these nitrogen bases should be in *tautameric* forms, i.e. A–T and G–C. But in the aqueous (cytoplasmic) medium, each base exists in an equilibrium state between various tautomeric forms, but some time, shift of a hydrogen atom causes *error* by formation of specific hydrogen bond to a base different from that of normal (legitimate) complementary base pairing rule. For example, rare *imino* form of adenine and rare *enol* form of thymine allow illegitimate complementary bonding of these bases, adenine–cytosine and guanine–thymine instead of 'legitimate' base pairing of adenine–thymine and cytsine–guanine respectively. In this way, 'illegitimate' base pairing results into the insertion of a wrong nucleotide in the growing polynucleotide chain. This illegitimate base pairing may bring a long-run change in the polynucleotide chains that is called 'molecular' mutation Figure 21.9 is most probable. However, this mutation could be 'reversed' by another molecular accident in which one or two bases involved in the replication happens to be in the rare *tautameric* form which restores the original nucleotide to its rightful position in the polynucleotide chain by an other 'illegitimate' base pairing.

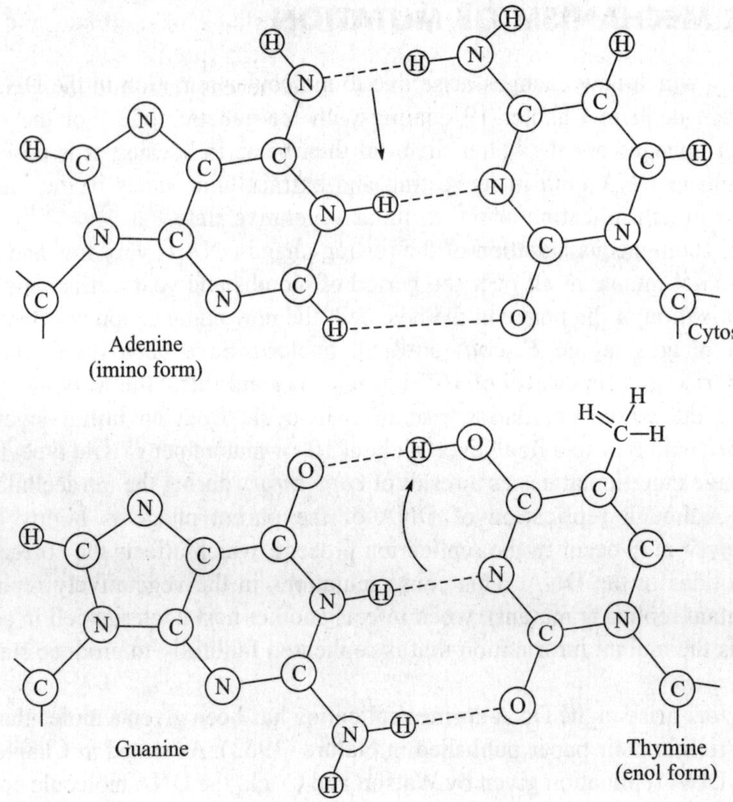

Figure 21.9 Illegitimate base pairing of rare *imino* form of adenine with cytosine and of rare *enol* form of thymine with guanine.

STRUCTURE OF EUKARYOTIC GENE

In the microorganism, it is easy to carry out 'complementation' test and decipher the structure and function of a gene because a large population of the organisms could be raised, as a result detection of rare genetic events become possible. But this is rather difficult in case of eukaryotic life forms. Because of limited population, detection of rare event is not possible. However rosy locus in *Drosophila* and waxy gene in barley have been detected as the two 'complementation cistron'.

SPLIT GENE

Almost all the eukaryotic genes have one structural aspect that is not detected in the prokaryotes so far. By 1970, it was reported that many eukaryotic genes contain nucleotide sequences that are not translated into amino acid sequences in the polypeptide chain. This gave the idea of *split genes*. According to this concept, the eukaryotic genes (polynucleotide chain) consist of two regions of sequences, *coding sequences*; which are translated into protein and *intervening sequences* (non-coding); which are not translated into protein (polypeptide chain). The coding sequences were termed as *exons* and the intervening sequences (non-coding sequences) were termed as *introns* (Figure 21.10).

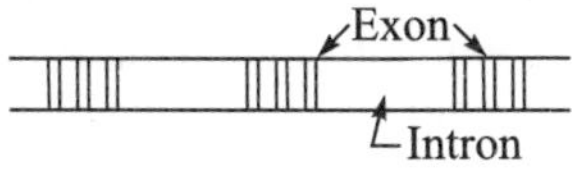

Figure 21.10 Fine structure of gene–split gene.

The immediate product of transcription in eukaryote is a molecule of many more ribonucleotides than that are required as functional mRNA. This primary transcript in eukaryotes is referred to as 'heterogenous nuclear' RNA (hnRNA). The molecules of hnRNA may range from 500–1000 nucleotides in length (Figure 21.11). These are rapidly processed into smaller molecules. It is estimated that only 10 per cent of the total transcript (polynucleotide chain) reaches to cytoplasm as functional mRNA. The processing of already transcripted hn RNA takes place in three steps.

First, soon after initiation of transcription, the 5′ end of hnRNA gets modified by formation of a 'cap', which consists of an additional methylated guanosine in a rare 5′–5′ linkage. This cap is required perhaps for the formation of mRNA–ribose complex in the eukaryotes and its subsequent role in the polypeptide synthesis.

Second is RNA splicing. This step involves controlled cutting (excision) of large intervening sequences or introns from the segments called as coding sequences or exons together to produce coding mRNA. Actual mechanism of splicing and excision is not known.

Third, RNA processing involves enzymatic addition of a segment of up to 200 adenylic acid residues called 'Poly-A tail' to the 3′ end of the molecule. A poly-A tail seems to be present in all eukaryotic mRNA except those responsible for histone synthesis. Poly-A tail is not required for polypeptide synthesis.

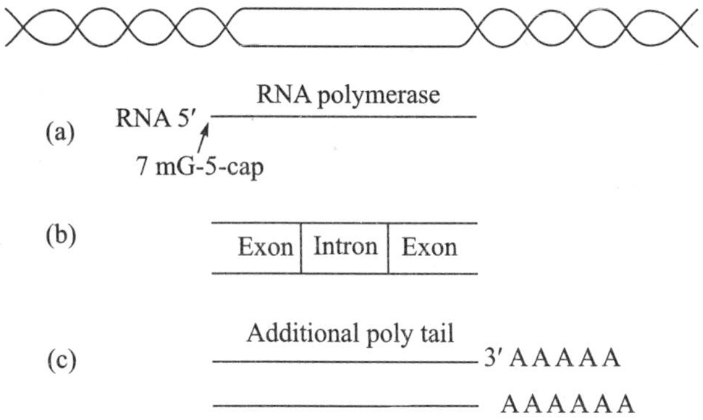

Figure 21.11 *Transcription* of an heterogeneous nuclear RNA (hmRNA).

(a) Addition of 7 methyl guanosine at 5 end.
(b) Removal of 'intron' in the *splicing* step.
(c) Addition of 'poly-A' (adenylic acid) tail on 3 end.

Movement of processed mRNA through pores of nuclear membrane into cytoplasm.

The intervening mRNA (polynucleotide) sequences are found in eukaryotic genes only with two exceptions of a few genes in viruses known as 'adenovirus', which can replicate in eukaryotic nucleus. These with the exception of two genes governing 'histone' and 'interferon', all other eukaryotic genes that code for protein possess 'intron'. The number of 'intron' varies from two introns in the 'globin gene'; seven in the chick albumin gene and about fifty in 'alpha-collagen' gene. In the lower eukaryotes, all ribosomal RNA (rRNA) genes are interrupted, while in higher eukaryotes, most of rRNA genes are interrupted.

But the concept of 'split' gene in eukaryotes is against the colinearity of DNA→mRNA→protein molecular assembly. Because, this is widely accepted view that gene (cistron) is the DNA sequence responsible for the synthesis of a single protein. In reality, the DNA polynucleotide sequence consists of *mutans* (mutable sites), which can be separated by recombination but presence of *exon* and *intron* in the DNA sequences do not change our views about gene. The *split gene* may be preserved by *exon-shuffling* or exchange of *exon* within or between genes through either normal recombination or through chromosome breakage (deletion or duplication of exon), which may create new proteins with new functions. In rats, there are two insulin genes, one gene has two 'introns', while other gene has only one. In other animals, there are gene-governing insulin. The single insulin gene has two introns in the mammals.

Ancestral gene of globin family had three introns. These are oxygen-binding proteins in animals (haemoglobin), fishes (myoglobin) and in the leguminous plants, such as soya bean (leghaemoglobin). The genes governing haemoglobin, myoglobin and leghaemoglobin possess introns. The leghaemoglobin genes possess three introns, two of which occur at identical points in the globin genes of the animals. As shown in Figure 21.12, the middle introns of leghaemoglobin separates two exons that together code for a protein corresponding to that produced by a central myoglobin exon. It is, therefore, highly possible that their ancesters had three introns, which consisted of four exons, encoding four modules, respectively characterized as shown in Figure 21.12.

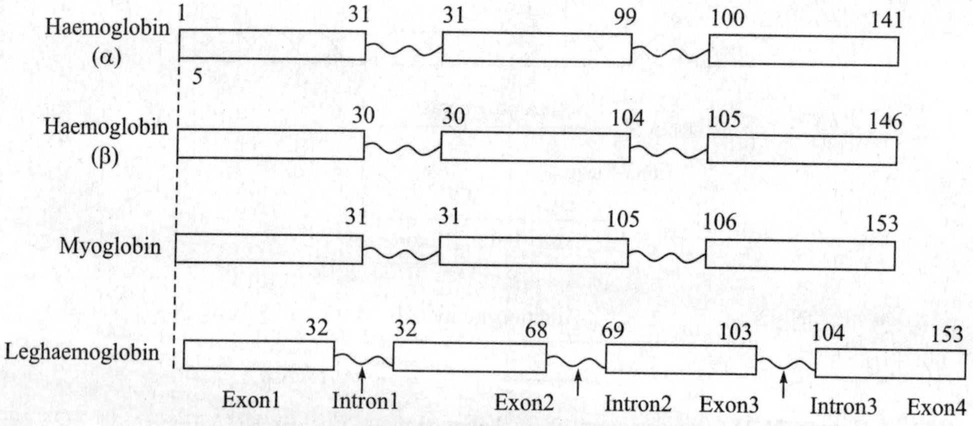

Figure 21.12 Exon–intron organization of the gene of globin family. Numbers show amino acid residues of a number of encoded proteins corresponding to the intron positions in their genes. The length of intron is not proportional to the actual length of the compact structural units in contemporary Lb; Mb and Hbs.

In mouse, (α and β globin genes), three *exons* correspond to polypeptide segment 1–31, 31–99 and 100–141 and 1–30, 30–104 and 105–146 respectively are found (Leder, et al., 1978). The two introns at the homologous positions to the mouse were reported in the genes of human, rabbit, sheep, goat, chicken and *Xenopus laevis* haemoglobin (Hb) subunits. The intron positions are conserved not only among mammals but also between highly diverged organisms such as amphibians.

The invertebrate globin (Gb) gene, from *arthropoda* and *annelids* were determined by the disclosure of the three introns in the Lb gene. The globin gene of the insect *Chironomus thummi* surprisingly contains no *introns* (Antone, 1984). The organization of a gene coding for an extracellular haemoglobin of the earthworm *Lumbricus terrestris* was investigated to see if it is similar to intron organization to those of vertebrates, plants or the insects. *Chironomus* (Jhiang, et al., 1988). This Hb is composed of about 200 polypeptides of six kinds out of which four major chains (a, b, c, and d) are homologous to vertebrate 'Hb'. The gene encoding chain c of *Lumbricus terrestris* Hb has two introns which are characteristics of vertebrate Gb genes and exact positions of exon–intron junctions are conserved. This finding suggests that the two intron and three exon organization of genes encoding vertebrate Gb, was established by the time of divergence of *annelids* and the common ancestor of *chordates* and *echinoderms.* Therefore, loss of the third intron must have occurred early in the evolution of invertebrate Gb genes.

The globin gene exists not only in the leguminous crop plants but also found in non-legume plants known to form nitrogen-fixing root nodules in symbiotic association with *Rhizobium* or *Frankia. Parasponia andersonii* (*Ulmaceae*) shows homology with the Hb genes of legume plants and has three introns at identical positions to the Lb introns.

Comparison of animal and plant globin sequences suggest a possible common ancestor about 1.3–1.5 billion years ago (Figure 21.11) (Dayhoff, et al., 1972). This suggests that one–one correspondence of 'intron' position to module boundary had been already established 1.3 billion years ago. This is not far from the time of divergence of prokaryotes and eukaryotes of both kingdoms. A homologous gene Lb with Gb was surprisingly found in bacteria *Vitreascilla* (Wakabayashi, et al., 1986) suggesting a common origin of Gb genes in eukaryotes and prokaryotes.

GENETIC ORGANIZATION (GENE FAMILY)

It was believed that each gene is present singly in the genome. But this is not always true. A gene may undergo duplication and form a 'Gene Family' to control a function. Such genes may be present in cluster or may be scattered in the entire genomes. The members of a gene family have identical function. A classical example is ribosomal RNA genes. They are present in cluster form at the nucleolar organizing region (MOR) and are expressed together.

The α and β haemoglobin genes are also present in cluster in human beings. The gene family is located on chromosome 11 and extends 60,000 base pairs. The α gene family is located on chromosome 16. The functions of these genes are controlled and regulated by growth at the time of development. All the globin genes are structurally similar in the animals with three exons and two introns, all located at the constant positions. It is, therefore, concluded that all the globin genes, 'globin gene family' have common ancestral origin through duplication and mutation.

The histone possesses five genes in the family, but they do not contain intron. All the five genes exist in chusters around the genome. In *Drosophila* about 100 copies of these five genes exist; in other organism such as in human beings, all the genes are linked on the same chromosome, but histone genes do not exist in tandemly repeated clusters. Gene families are presently classified into three groups:

(i) Simple multigen families in which only one or very few genes are repeated in tandem manner.
(ii) Complex multigen families consist of related genes in a cluster, but transcribe independently such as five histone genes.
(iii) Developmentally controlled complex multigen family codes for a number of different kinds of some proteins, which are expressed at the different stages of developments. The α and β haemoglobins are expressed at different developmental stages such as at embryonic, foetal and adult stages.

Immunoglobulin Gene Family

The genes, which code for the synthesis of immunoglobulins, show a very typical case. The proteins responsible for the immune response in vertebrates are called *antibodies*, which the body synthesizes against an *antigen*. These molecules possess tremendous variability. They may occur in million or in billion forms with slight differences, as a result they can respond to millions/billions of foreign antigens with which organisms come in contact throughout life. Now the question is how does an organism produce such a huge varieties of proteins? Is there a gene for each variety of antibody?

Structure of Antibody

Each *antibody* molecule is made up of four immunoglobulin chain. There are two identical heavy chains consisting of 330–440 amino acids, and identical light chains of 220 amino acids each. The four *immunoglobulin* chains are held together in such a way to form Y-shaped molecule (Figure 21.13). Actually, all antibody molecules are identical except on the arms of the Y, at the amino acid terminal ends of the four polypeptide molecules each antibody differ slightly from other antibody. This is also where the antigenic specificity of each antibody exists. Based on the amino acid sequence, it has been observed that each polypeptide chain of antibody consists of smaller subunit proteins known as *domains*, the constant part of each polypeptide is made up of same repeating 110 amino acid *domain*. The variable part of each polypeptide contains different amino acids *domain* depending on the specificity of the *antibody*; conclusively, each antibody is made up of four polypeptide chains; containing large constant regions of amino acid sequences and small variable regions of amino acid sequences. The sequences of specificity of each antibody lie in the diversity of variable regions that can be present.

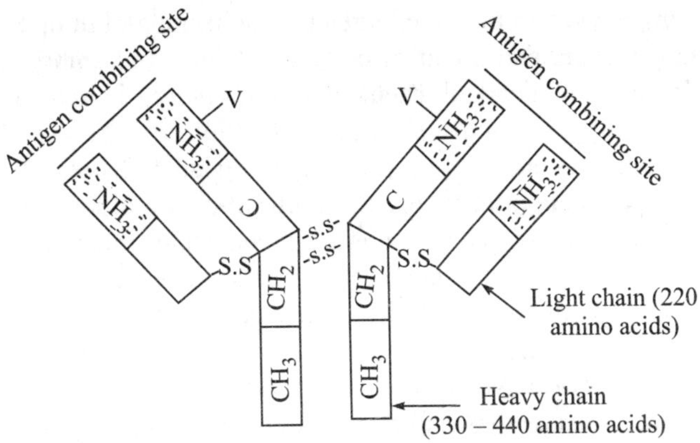

C = constant domain—common element (110)
V = variable domain—variable element antigen specificity.

Figure 21.13 Structure of four immunoglobulin chains into functional antibody.

Gene and Immune Response in Mammals

The genes, which govern the immunity in the mammals provide strong support for the construction of new genes. This is a challenging question how millions of antibodies are produced by the body to fight against the vast array of fungal, viral and bacterial cells or other invading antigen. Each antibody is made up of different proteins and there are not enough genes to code for all these proteins. For example, human genome consists of about 60,000 to 1,00,000 genes, most of them perform different functions required by the body, thus there are not enough genes to produce millions of antibodies required by the human body to generate immunity against an array of disease organisms or invading antigens. There are three components to immune response in the mammals:

First, there are immunoglobulin proteins, which are produced by the B-lymphocytes (they originate in the bone marrow). It is immunoglobulin, which binds with the invading antigen in the process and destroy it.

Second, the T cell receptors. They are produced in the thymus, which recognize the antigen and in this way have a complementary relationship with the B-lymphocytes.

Third, component takes the form of a cellular protein manufactured on the direction of the histocompatibility loci. One function of histocompatibility protein is to bind with a peptide fragment of the invading antigens and present it to the receptor proteins for recognition and after recognition, immunoglobulin proteins start binding with the antigen. This is a hypothesis narrated by Jain in 2000. However, it is still to work out about the reality and original haemoglobin gene, which gives rise to additional copies through duplication followed by independent mutation of ancient phylogeny (Jain, 2000).

Duplication and Haemoglobin Gene

The new genes arising through duplication by unequal crossing-over followed by mutation has been provided by the loci involved in the synthesis of globin proteins of haemoglobin molecules responsible for carrying oxygen in the blood stream. The haemoglobin of a person is a *tetramer* made up of an alpha (α) chain and a pair of beta (β) chains. The embryonic and foetal blood cells carry a similar but not identical haemoglobin *tetramer*. Lewin (1994) has explained haemoglobin genes from a gene family showing tight linkage, where alpha (α) gene family consists of cluster of linked genes. Alpha (α) cluster includes 2 beta, 2 alpha, 1 sigma and 3 pseudogenes. The amino acid sequence and composition of alpha, beta and other globin proteins indicate common origin of genes coding for them. Alpha chain is made up of 141 amino acids and beta chain is made up of 146 amino acids. The sequence of amino acids in alpha, beta, gamma and delta chains demonstrate similarity among them at great extent (Brauneitzer, et. al., 1979), Among the first 40 amino acids of alpha, beta, gamma and delta haemoglobin of the human blood, more than 10 amino acids have been found identical on corresponding position in the four polypeptide chains. This indicates that different haemoglobin-coding genes might have evolved through series of duplications of same gene, viz. common evolution and haemoglobin gene cluster should be treated a gene family rather than individual gene (Figure 21.13).

Chemical Synthesis of Gene

Chemical synthesis of gene was first accomplished by Agrawal and Khurana in 1970. It was infact a small gene that codes for an alanine transfer RNA (tRNA). However, on an average structural genes possess one thousand nucleotides. For the purpose of chemical synthesis, the gene is divided into small 50–100 nucleotides long subunit called *oligonucleotides,* which are synthesized by an automatic, commercially available machine. The chemical synthesis gives a single-stranded DNA fragments of the desired oligonucleotides sequences. Several oligonucleotides are then assembled together employing a PCR (Polymerase Chain Reaction) to obtain double-stranded DNA. Such DNA designed entirely by theoretical consideration can be made to code for specialized *promoter* (regulatory regions) or protein is synthesized for the desired purposes.

UNSTABLE LOCI (TRANSPOSON)

McClintock (1965) made a series of crosses in maize between lines possessing contrasting characters and observed instability of some traits. McClintock's work led discovery of two different *loci,* which were responsible for the observed mutation. She named the first loci as *Activator* (Ac) which acted as master switch. In absence of Ac, plants showed normal behaviour (no mutations). Second locus was described as *Dissociator* (Ds), acting under the control of Ac, creates an unstable situation in the genome (a). A chromosome carrying Ds locus shows a high frequency of breakage at this locus and (b) Ds tend to move out of its normal position and get inserted at a new site in the chromosome. The gene located at new site start to mutate in an erratic manner, causing instability in the expression. Both Ac and Ds could be positioned on the linkage map. Ac occasionally gets lost while changing its position.

Gene C^+ in maize controls the synthesis of purple pigment in the aleurome layers of seed. But frequently variegated kernels are found which indicates forward and reverse mutation. Soon Ds locus gets integrated with C^+ gene, the C^+ gene lost its function and when Ds moves away, C^+ gene starts functioning.

Findings of McClintock (1965) were observed in bacteria also in which DNA sequences were found to move out of their normal position in the chromosome and get integrated into another locus in the genome (Figure 21.14).

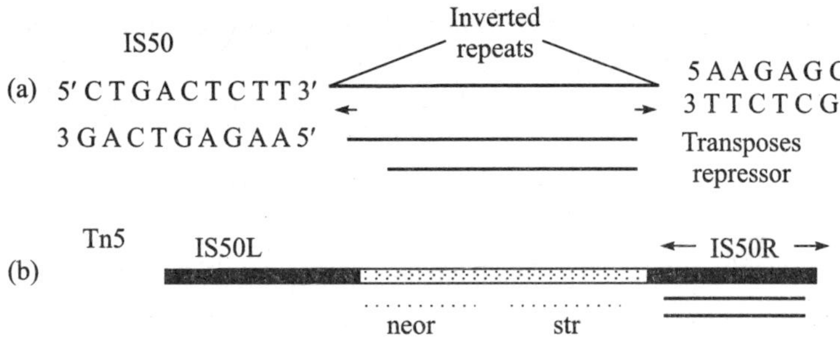

Figure 21.14 Structure of bacterial transposon: (a) insertion sequence IS50, (b) composite transposon Tn5. The central sequences possesses gene for neomycin 'neo-r' and streptomycin 'str-r'.

Importance of Transposon

Since these mobile DNA sequences are widely distributed in the genomes of prokaryotes and eukaryotes, and create new variability; it is essential to know their importance in plants and animals of economic importance.

Spontaneously, arising gene mutations and chromosomal aberrations are well-known sources of creating genetic variability in nature, in addition to the recombination of genes. These mobile DNA sequences are the cause of genes as well as chromosomal rearrangement of genes. Thus, transposons are a built-in some of mutations.

CONSTRUCTION OF NEW GENE

Most interesting question oscillating in the mind of the geneticists have been: how do new gene arise? One of the answers of this question was that new gene arise through duplication on account of unequal crossing over and as a mutation. The duplicated gene acquires quite a new function. There are many gene families, which appear to have arisen through this mechanism. The haemoglobin molecules found in different animal species show little difference with the substitution of amino acids in their globin proteins. It is believed that all these globin proteins are related on account of their origin to a single gene (Figure 21.15).

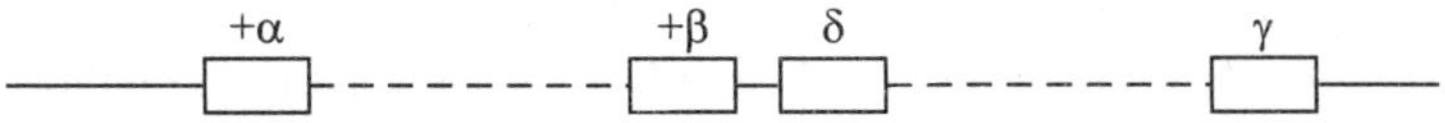

Figure 21.15 Evolution of genes through duplications for the different polypeptides of the haemoglobin molecules.

REVIEW QUESTIONS

1. In what ways does our present concept of gene differ from the pre-1940 or classical concept of gene?
2. What was the first evidence which indicated that unit of function and unit of structure of genetic material were not the same?
3. What is currently accepted operational definition of gene? Give fine structure of gene citing an example.
4. Are the following statements concerning the genetic element referred to as the gene, true or false?
   (a) The classical (pre-1940) concept of gene was that it was
      (i) unit of physiological function or expression
      (ii) the smallest unit that could undergo mutation, and
      (iii) a unit, not subdivisible by recombination
   (b) The cis-trans test provides an operational definition, by which we can often identify a gene as the unit which specifies one mRNA molecules.
   (c) Our present knowledge of structure of gene indicates that unit defined by criteria, (ii) and (iii) in statement (a) above are both equivalent to a single nucleotide pair.
   (d) Studies in the 1940, demonstrated the existence of heteroalleles (pseudoalleles), clearly indicate that many mutations which were allelic by functional criterion could be separated by recombinations, thereby indicating that unit of function, mutation and recombination are not equivalent.
   (e) Homoalleles are functionally and structurally allelic; heteroalleles are functionally allelic but structurally not allelic.
5. What is the difference between a pair of homoalleles and a pair of heteroalleles?
6. Why are mutant (alternative) alleles essential for identifying the wild-type alleles locating the position of genes on the chromosome.
7. Two different inbred varieties of a particular plant species have white flowers. All other varieties of this species have red flowers. What experiment might be done to obtain the evidence to determine whether the difference in flower colour in these varieties is the result of different alleles of single gene or the results of gametic variation in two or more genes?
8. (a) What are some of the genetic implications of overlapping genes or gene within genes?
   (b) What is the maximum number of different amino acid sequences that can be produced (a) from the same segment of one strand of DNA, and (b) from both the strands of a DNA double helix?
9. In *Drosophila*, the white, cherry and vermilion are all sex-linked mutations affecting eye-colour. A white-eyed female crossed with a vermilion-eyed male produces a white-eyed male offspring and a red-eyed female offspring. A white-eyed female crossed with a cherry-eyed male produces white-eyed male offsprings. Do these results indicate whether or not any of these mutations affecting eye-colours are functionally allelic? If so, which mutations are functional alleles?

REFERENCES

Avery, O.T., C.M. McCleod and McCarty (1944): Studies on the chemical nature of the substance including transformation of *Pneunococcus* types: induction of transform by a deoxyribonucleic acid fraction isolated from *Pneumococcus* type III, *J. Exp. Med.*, **79,** 137.

Baunitzer (1979): *Introduction to Evolutionary Genetics*, David Parkein Ealward Arnold Publication.

Benzer, S.L. (1953): Fine structure of a genetic region in bacteriophage, *Proc. Nat. Acad. Sci.*, **47,** 403.

Dayhoff, M.O., L.T. Hunt, P.J. McLaughlin and D.D. Jones (1972): Gene duplication in evolution: the globin, *in:* Dayhoff, M. (Ed.), *Atlas of Protein Sequence and Structure*, Nat. Biomed. Foundation, Washington DC, Vol. 5, 17–30.

Green, M.M. and K.C. Green (1949): Crossing over between alleles at lozenge locus in *Drosophila melanogaster*, *Proc. Nat. Acad. Sci.*, **35,** 586.

Jain, H.K. (2000): *Genetics: Principles, Concept and Implication*: Oxford IBH, New Delhi.

Jhiang, S.M., J.R. Garey and A.F. Riggs (1988): Exon–intron organization in genes of earthworm and vertebrate globins, *Science*, **240,** 334–336.

Khurana, H.G. (1979): Total synthesis of a gene, *Science*, **203,** 614.

Klecker, N. (1981): Transposable elements in prokaryotes, *Annual Review of Genetics*, **15,** 341.

Leder, P. (1982): Genetics of antibody diversity, *Sci. Amer.*, **256**(5), 102.

Lewin (1994): *Genes V*, Oxford University Press, Oxford, Cytoplasm, gene and organelles, Academic Press.

McClintock, B. (1965): Control of gene action in maize, *Brookhaven Symposia on Quantitative Biology*, **18,** 162.

Sanger F., S. Nicklen and A. R. Coulson (1977): DNA sequencing with chain terminating inhibition, *Proc. Nat Acad. Sci. USA*, **74**, 5465–5467.

Stent, G.S. (1963): *Molecular Biology of Bacterial Viruses*, Freeman and Co.

Wakabayasi, S., H. Matsubara and D.A. Webste (1986): Primary sequence of a dimeric bacterial haemoglobin from *Vitroscilla, Nature,* **322,** 481–483.

Watson, J.D. and F.H.C. Crick (1953): Genetic implication of the structure of deoxyribonucleic acid, *Nature*, **171,** 964.

CHAPTER

22 Gene–Enzyme–Protein Relationship

The ensemble of cellular proteins functions to make the cell that it is called an engine built of highly specific structural members and enzymes which carry out a complex network of catalystically facilitated metabolic reactions. The basic problem of cellular self-replication can be put in the terms of the question. How in the course of cell generation period, does the entire apparatus of cellular proteins double, so that each of the two daughter cells generated by mitotic division of the parent cell can be endowed with its own complete outfit? According to the fundamental principle, the primary structure of polypeptide chain is solely responsible for secondary, tertiary and quarterly protein structures. This can be reduced to the following query. How is it that a particular sequence of 20 amino acids assembled which make up the primary structure of any given one of the one to two thousand different enzyme molecules.

SYNTHESIS OF AMINO ACIDS

Amino acids, the building blocks of protein themselves are of course, synthesized by the metabolic pathways; the complete description of metabolic reactions of 20 amino acids and nucleotide synthesis is beyond this book, also not needed but a cursory view of a few representative pathways of amino acids and nucleotide synthesis is must for better understanding of the genetic principles. In any case, it can be slated as a general principle that the point of departures of most of these pathways is one or another of the intermediate compound of the degradative sequence of glycolysis and citric acid cycle.

This principle is exemplified by the pathways of amino acid biosynthesis. As shown in Figure 22.1, the twenty amino acids can be grouped into four families according to the metabolic provenance of their skeletons of carbon atoms.

Synthesis of the aromatic and histidine family depart from glucose at top (Figure 21.1), and that of pyruvic family from phosphoglyceric acid in the middle and from pyruvic acid at the bottom of the glycolytic pathway. Synthesis of glutamic family departs from α-ketoglutaric acid in the middle, and that of the aspartic family from oxaloacetic acid at the end of the citric acid cycle (Figure 22.2).

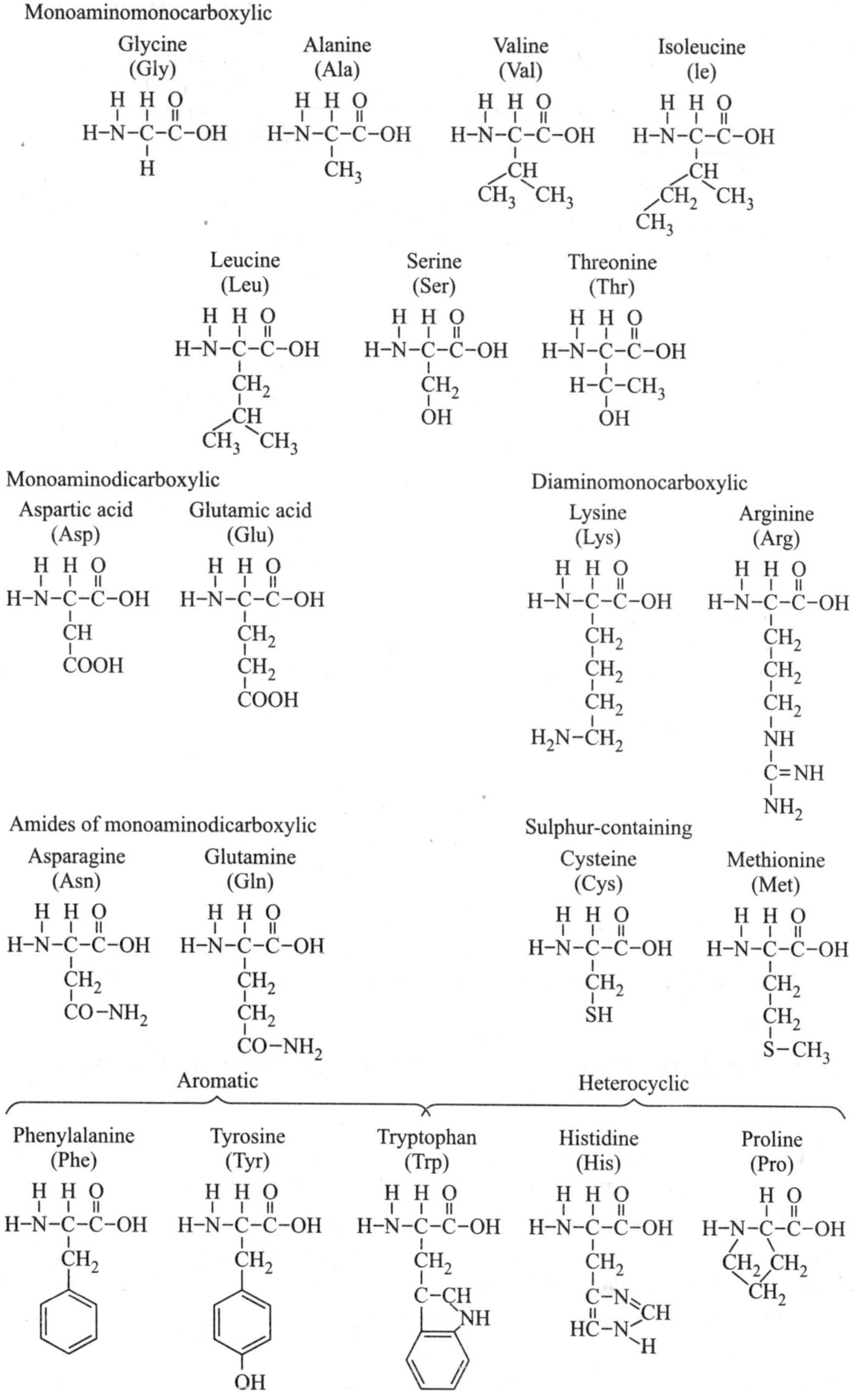

Figure 22.1 List of 20 amino acids with their chemical structures.

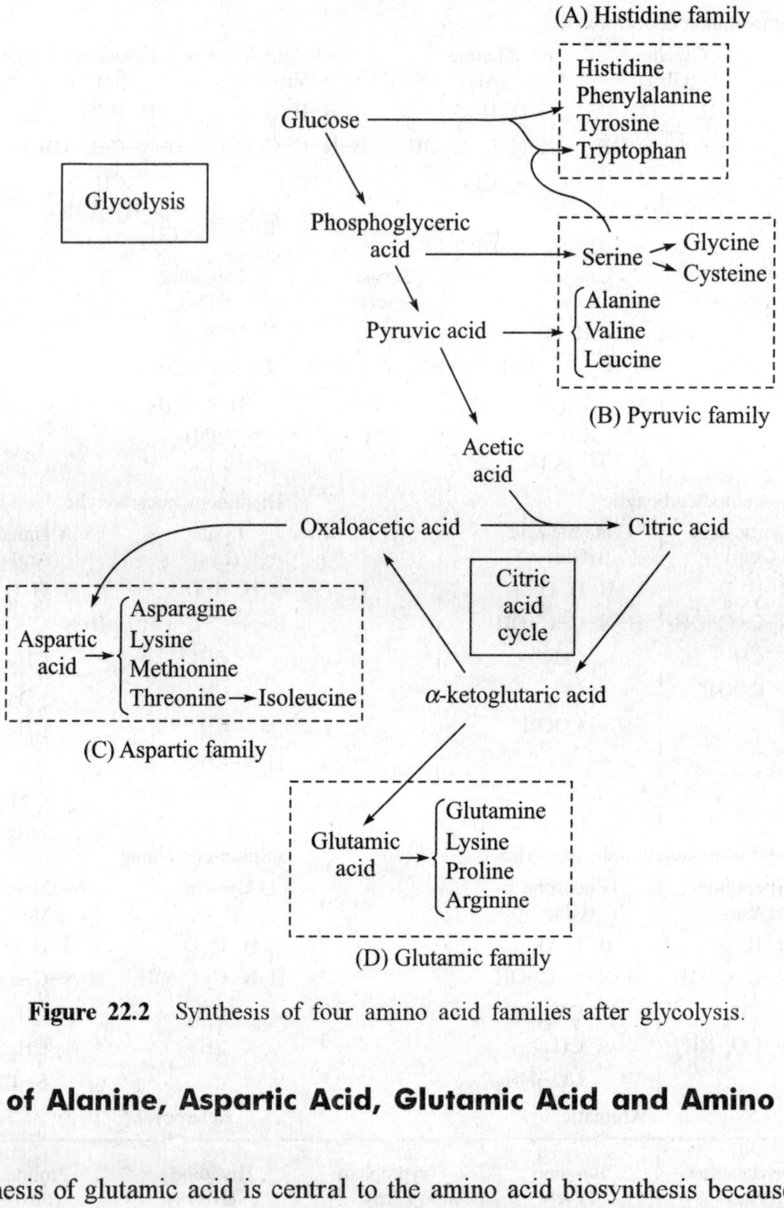

Figure 22.2 Synthesis of four amino acid families after glycolysis.

Synthesis of Alanine, Aspartic Acid, Glutamic Acid and Amino Serine Acids

The biosynthesis of glutamic acid is central to the amino acid biosynthesis because it is in its synthesis that inorganic ammonia is first elevated to the status of an organic amine. In the midway in the citric acid cycle, there arises the five-carbon dicarboxylic acid, α-ketoglutaric acid. As shown in Figure 22.3, the enzyme glutamic dehydrogenase catalyzes the replacement of the α-keto group of α-ketoglutamic acid by an inorganic ammonium ion, thus giving rise to glutamic acid. This amination involves a reduction of the α-keto group, which is accompanied by coupling of the reaction to the oxidation of one molecule of NADH to NAD^+. Nicotinamide adenine dinucleotide (NAD) is reduced to NADH by α-ketoglutarate).

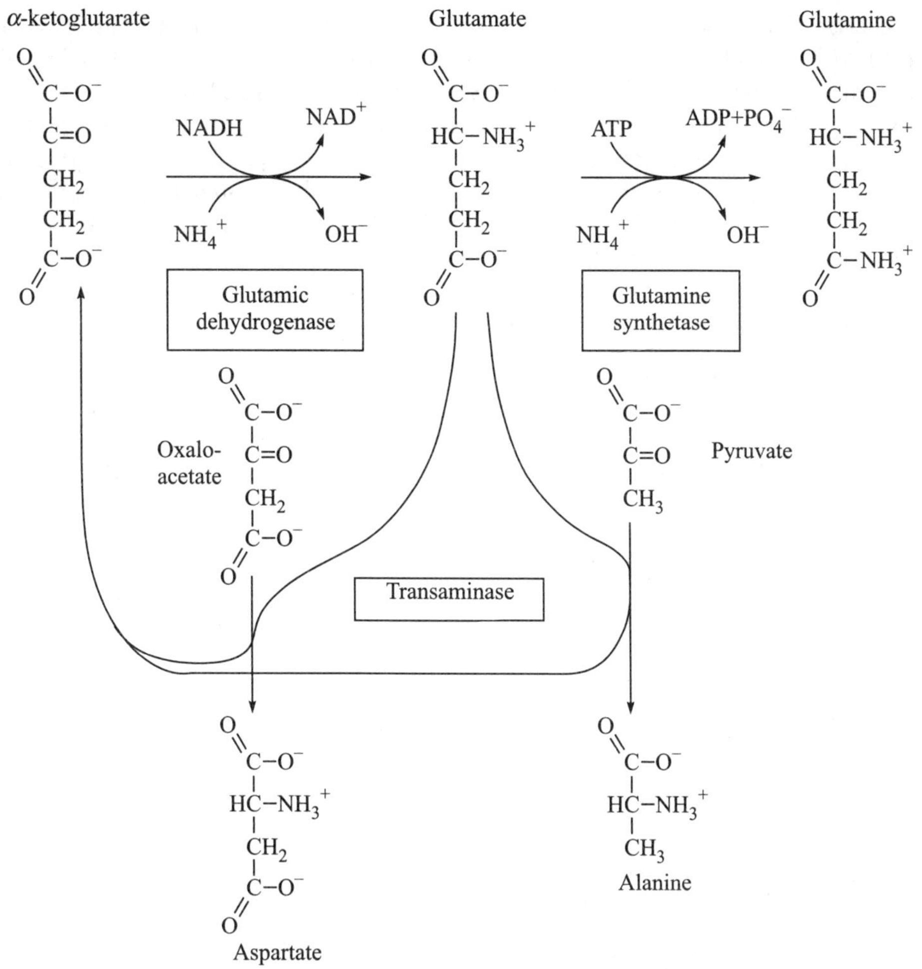

Figure 22.3 Synthesis of four amino acids—glutamic, glutamine, aspartic and alanine.

Whenever, an intermediate such as α-ketoglutaric acid is removed from the citric acid cycle, an equivalent number of oxaloacetic acid molecules are stored is the cycle, if the cycle is to continue operating in the long run. This replenishment of oxaloacetic acid is accomplished by the addition of CO_2 to pyruvic acid according to the reaction given below:

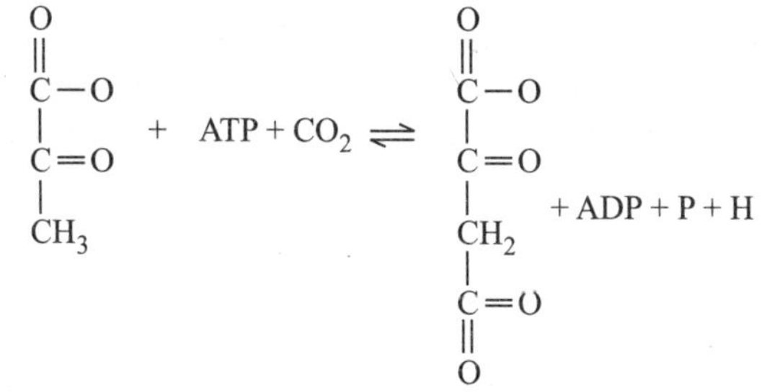

The energy requirement of this reaction is met by the concomitant hydrolysis of one molecule of ATP to yield ADP and inorganic phosphate. The energy cost of synthesizing alanine is 18 molecules of ATP and 19 molecules of ADP for aspartic acid synthesis.

Serine is synthesized via a slightly more complicated synthetic pathway. The starting point midway in glycolysis is 3-phosphoglyceric acid. As shown in Figure 22.4, phosphoglyceric acid is oxidized to the α-keto acid phosphohydroxy pyruvate with the attendant reduction of one molecule of NDA^+ to yield NADH. A transamination exchanges the α-amino group of glutamic acid for the keto group of the phosphohydroxy pyruvic acid and give rise to phosphoserine and α-ketoglutaric acid. Hydrolysis of the phosphate ester of phosphoserine finally yields serine and inorganic phosphate. The energy cost of synthesizing one molecule of serine arises from the loss of one molecule of ATP.

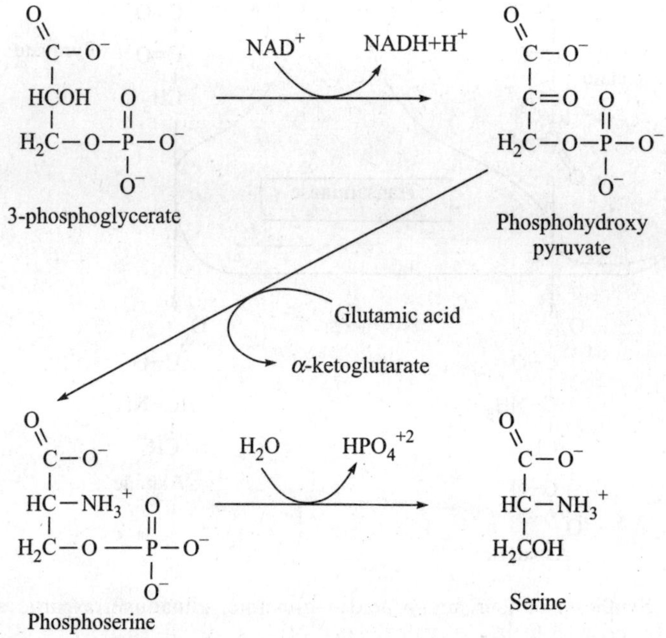

Figure 22.4 Synthesis of amino acid serine.

A second pathway of assimilation of inorganic nitrogen into organic nitrogen also involves glutamic acid, namely the amination of 'y' carboxyl of glutamic acid by ammonium ions to yield glutamine. This reaction is catalyzed by the enzyme glutamine synthetase and requires one molecule of ATP to yield ADP and inorganic phosphate via hydrolysis. The glutamine nitrogen atoms assimilated in this way may then proceed in a series of further metabolic relations and have as one of their principal ultimate destination, two of the four purine ring nitrogen of adenine and guanine.

Summary

1. The 20 different basic cellular building blocks in particular, the amino acids and the nucleotides are synthesized from glucose and ammonia in a limited set of chemical

reactions numbering about 1 to 2000. Each step in this reaction sequence leading from nutrient to cell building is catalyzed by the specific enzyme.

2. The chemical energy required for these syntheses met by the step-wise oxidative degradation of glucose to CO_2 through glycolysis and citric acid cycle. This energy is first invested in the phosphorylation of ADP by inorganic phosphate to produce ATP, whose coupled hydrolysis to regenerate ADP is later used to derive synthetic reaction. The various chemical intermediation of the degradative reaction (oxidation) sequences serve as the starting materials (substrate) for the synthesis of other amino acid.
3. When the carbon compound other than the glucose serves as a nutrient for cell growth, the compound is first converted into glycose or into one of the intermediates of the glycolytic pathway by means of a few reaction steps.
4. Each reaction step involves only a small chemical changes, e.g. addition or deletion of H_2O, H, CO_2 or phosphate. It is the collective reaction sequence of small steps; transforms glucose into cellular building blocks, i.e. amino acids. And the dehydration reaction by means of which these amino acids are joined by peptide bonds into polypeptide chain is as may readily imagine, catalyzed by one or more of specific cellular enzymes.

GENE–ENZYME RELATIONSHIP (ONE-GENE ONE-ENZYME THEORY)

In 1902, Archibald Garrod based on his studies concluded that alkaptonuria, an arthritic condition of man who excretes wine colour urine is a hereditary disease which on account of the alternation in the normal nitrogen metabolism causes excretion of dark substance in place of normal urine. In 1908, he again proposed that alkaptonuria persons are homozygous for 'recessive gene' and those possess this gene fail to enzymatically catalyzed the metabolic reaction, the substrate which should be normally destroyed by the reaction get accumulated in the urine. This was the first case reported the gene controlled enzymatic reaction, i.e. inborn errors of metabolism.

The continuity of study of the biochemical effect of genes began with a series of investigations on two recessive eye colour mutations of *Drosophila melanogaster*—vermilion and cinnabar. The eyes of the normal (wild type) fly is deep red but eyes of the homozygous mutant flies—vermilion and cinnabar were not so deep red as in wild type.

In 1935, G.W. Beadle and B. Ephraussi implanted embryonic eye tissues from larva of vermilian and cinnabar mutants into the eyes of larva of the normal *Drosophila* flies and observed that upon metamorphosis of these larva into mature flies, the implanted eyes tissue developed into supernumerary eyes with normal eye colour (deep red). It could be concluded that body tissues of the normal flies supply some substrate which the vermilion and cinnabar mutants eye tissues were unable to synthesize but which they can convert into normal eye colour pigment.

Beadle and Ephrussi then implanted in a reciprocal manner some of the embryonic mutant tissues into the larva of vermilion and cinnabar mutant flies and observed that vermilion eye tissues implanted into cinnabar larval host developed the normal eye colour; whereas cinnabar eye tissues implanted into vermilion larval host developed the mutant of cinnabar eyes. Beadle and Ephraussi inferred from these observations that synthesis of the deep red pigment colour arises by a metabolic chain.

Precursor → Substrate I → Substrate II → Brown pigment eyes.

The vermilion mutant would thus carry a block in the reaction that converts the precursor to substrate I, whereas cinnabar mutant would carry a block in the reaction that converts substrate I to substrate II. Thus, in the wild-type host larva both mutant eye tissue transplants were provided with substrate II, which they can convert into the brown eye pigment.

On the other hand, vermilion mutant eye tissues transplanted into the cinnabar host larva is provided with substrate I, which it can convert to substrate II and to brown pigment. But the cinnabar mutant eye tissues implanted in the vermilion larva is not provided with substrate II that it lacks, and hence fails to form brown pigment. After some time, the biochemical studies indicated that the 'precursor' is amino acid *tryptophan* and that substrate I is *formylkynurenine* and substrate II is *hydroxykynurenine*. The genetically-controlled eye colour could be explained as in Figure 22.5.

Precursor ⟶ Vermillion ⟶ Substrate I ⟶ Cinnabar ⟶ Substrate II ⟶ Brown pigment

(Gene–Enzyme) ↑ (under Precursor) (Gene–Enzyme) ↑ (under Cinnabar)

Precursor (Tryptophan) ⟶ Vermillion ⟶ (Gene–Enzyme) Substrate I ⟶ Formylkynurenine

Cinnabar ⟶ Gene–Enzyme Substrate II ⟶ Hydroxykynurenine Brown eye pigment

Figure 22.5 Biochemical pathway leading to the synthesis of brown eye pigment in *Drosophila melanogaster*.

This is an excellent example of physiologic role of gene in enzyme synthesis. The normal wild-type allele of the vermilion *Drosophila* gene could be envisaged as presiding over the synthesis of an enzyme which catalyzes the conversion of *tryptophon* to *formylkynurenine*. The mutant alleles have lost the capacity to form the enzyme hence the tissues of a homozygous mutant gene on both of its homologous chromosomes lack enzyme as a result the metabolism of such mutant flies is blocked at the reaction step normally catalyzed by that enzyme. The tissues of heterozygous fly carrying one mutant allele (v) and one wild-type allele (V^+) of the vermilion gene, would contain the enzyme and hence capable of forming formylkynureinine. Similarly, the mutant allele of cinnabar gene has lost the capacity to form enzyme that catalyzes the conversion of formylknurenine to hydroxykynurenine, the enzyme that is normally formed under the direction of wild-type genes allele. From this experiment, it was concluded that the recessive character (one alleles) of both vermilion and cinnabar mutations is responsible for the absence of an enzymatic function, which is performed by dominant alleles of wild-type flies.

NUTRITIONAL DEFICIENT MUTANTS

Finding some difficulties in *Drosophila*, in 1940, Beadle and Tatum, shifted their studies on a fungus, the bread mold, *Neurospora crassa*. They found so many mutants which were unable to synthesize vitamins, amino acids and other essential components of the protoplasm and growth.

The procedure they used in the isolation of these mutant strains was to inoculate with single *Neurospora* spore to a large number of test tubes containing *complete* growth medium. Their choice was shifted from *Drosophila* to bread mold because latter can be grown in a simple synthetic medium containing vitamin, biotin, etc. could be said to be the simple synthetic growth medium 'fortified with yeast extract'.

The samples of fungal growth appearing on the complete medium were then shifted to the test tubes containing minimal medium fortified with only 'biotin' in the synthetic medium. Beadle and Tatum found that though wild-type fungus strain can grow on both 'complete' and 'minimal' medium, but about one in two hundred *Neurospora* spores gave rise to a fungus, which could grow only on complete medium. From this it was concluded that the usual fungi, which fail to grow on the *minimal* medium, are the descendants of the mutant spores in whose hereditary material, there is a gene mutation, which has blocked an essential metabolic pathway. Each such mutant is able to grow on the complete medium because it finds ready-made the required substance in the yeast extract whose capacity it has lost for synthesis from glucose and inorganic salt. The nutritional substances which the mutant can find in the complete medium are mainly the twenty amino acids, two or three vitamins, purine and pyrimidine bases in order to identify the exact nature of the growth requirement of any particular mutant. The samples of fungal spores are inoculated into a series of *minimal* medium in which the various putative growth factors were added. The particular growth requirement is then identified by adding the substance which promoted the growth of the mutant.

The crosses made between the wild-type *Neurospora* and many mutants revealed that most of them owe their growth factor requirement to the mutation of a single gene in the *Neurospora* genome. On the basis of these findings Beadle and Tatum promulgated the *one-gene one-enzyme theory*. This theory asserts that each gene has only one primary function, which usually consists of indirectly the formation of one and only one enzyme. The idea that gene controls a single metabolic function had been adumbrated by Garrod about 30 years earlier. Beadle and Tatum's clear formulation and experimental evidence in favour of *one-gene one-enzyme theory* had a profound impact on subsequent thought about real *core* of genetics. But this theory had some serious limitations:

(i) This theory attributes the power to direct the formation of a single enzyme is still invisible, formal and abstract, Mendelian hereditary unit whose physical nature had no relevance to the interpretation of these experimental result.
(ii) This theory has nothing to say, how the gene actually manages to direct the formation of enzyme under its control.
(iii) It did not include the idea that gene directs the assembly of amino acids into polypeptide chain of a given primary structure of protein.

AUXOTROPHY

After working with Beadle on *Neurospora* nutritional mutants, Tatum began a systematic research for analogous nutritional mutants in *E. coli*. In order to isolate the rare bacterial mutants that are deficient in nutritional requirements for their growth on which the wild type can grow, Tatum followed the same procedure that Beadle had followed for isolating *Neurospora* nutritional mutants. For this purpose, a large number of *E. coli* cells were planted on a nutrient broth agar, that is on complete medium. The colonies that came up after overnight incubation on the complete medium, were then picked up and a part of each bacterial clone was tested to ascertain, whether, it would grow on a minimal medium that is agar containing a synthetic medium. In this way, Tatum observed that where on majority of colonies consisted of bacterial clones capable of growing on both medium; about one per cent of the colonies that grew on the complete medium were made up of bacteria that were unable to grow on the minimal medium. These rare colonies were made up of clones of the nutritional *E. coli* mutants which require for their growth, some element present in the complete medium but absent from minimal medium. Such bacterial mutants were known as *auxotrophs*. (aux: increased; trophe: food). Prototrophs means which can grow on minimal medium.

In the minimal medium for *Neurospora* mutant, growth factors such as amino acids, vitamins, purine and pyrimidims were added. For *E. coli* nutritional mutant, proline, threonine, tryptophan, adenine and thiamin were added; therefore, auxotrophic mutants were designated as thr^-, pro^-, trp^-, thi^- and Ade^- respectively. This supports one gene–one enzyme theory because the mutation: which created auxotroph out of prototrophic/wild type pertained to a bacterial gene, that controls a single enzyme catalysing a step in the biosynthesis of growth element.

ENZYME

An enzyme is a polypeptide chain (polymer molecule) of the 'ordered' amino acid assembly with catalytic properties. This polypeptide chain acts as the catalyst of a specific chemical reaction. Thus, an enzyme is proteinaceous in nature. An enzyme acts on a substrate and converts it into the reaction product per unit of time:

$$\text{Lactose} \xrightarrow[H_2O]{} \beta\text{-galactosidase} \longrightarrow \begin{cases} \text{Galactose} \\ \text{Glucose} \end{cases}$$

There are about two thousand enzymes which catalyse the cellular metabolic pathway, each of which catalyses a single reaction step. These enzymes are proteinaceous in nature and determination of the chemical structure of a protein can be done by the quantitative analysis of the amino acid composition of its polypeptide chain. The results of such studies (by hydrolysis followed by ion-exchange resin chromatography methods developed by W.H. Stein and S. Moore 1950) of the two *E. coli* enzymes—*β*-galactosidase and tryptophan synthetase showed that enzyme *β*-galactosidase and tryptophan synthetase are composed of polypeptide chain containing 1173 and 267 amino acids. Tryptophan synthetase is composed of two different polypeptide chains called 'A-protein' and 'B-protein'.

The two enzymes differ in their relative content of the 20 amino acids. For example, tryptophan synthetase is nearly twice richer in 'alanine' but possesses only half of the methionine content; *β*-galactosidase contains no tryptophan at all. These enzymes differ in their specificity of catalytic power because of the differences in the amino acid composition in their polypeptide chains. However, each polypeptide chain must contain at least one residue of the least abundant amino acid. Such as tryptophan synthetase contains 1.1 mole per cent of the least abundant amino acid 'methionine' and hence the polypeptide chain must contain a total of at least $100/1.1 = 90$ amino acid residues.

In 1949, Frederick Sanger decided to determine the exact amino acid sequence of the polypeptide chain of bovine insulin. Insulin is not an enzyme, but a 'harmone' that regulates the function of some of enzymes active in the glucose metabolism of vertebrate animals. Sanger reported that 51 amino acids of insulin are present in two different polypeptide chains—chain A contains 21 amino acids and chain B contains 30 amino acids as shown in Figure 22.6. The A–B chains are held in covalent linkage by two disulphide (S–S) bonds formed between the sulphur atoms of pairs of cysteine residues in both the chains.

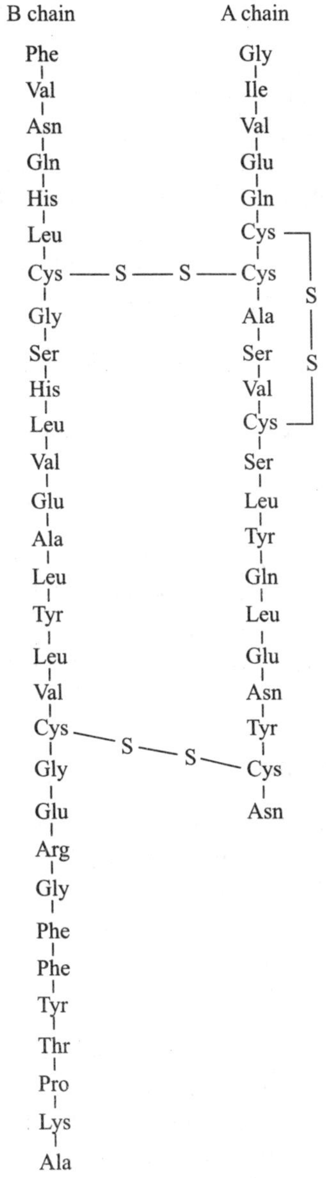

Figure 22.6 The primary structure of bovine insulin.

Enzyme Action

An enzyme temporarily forms a chemical bond or enter into a transient complex with substrate whose reaction it catalyzes. It enhances the rate at which an enzyme catalyzes reaction; the formation of the enzyme—substrate enzyme complex and the actual chemical reaction which transfers the substrate held in the complex into reaction products:

$$\text{Substrate—enzyme} \rightleftharpoons \underset{\text{complex}}{\text{Substrate–enzyme}} \xrightarrow{\text{Turnover}} \text{Product enzyme}$$

The extent to which the first of the above said steps occurs depends on the fraction of enzyme molecules which bind the substrate at any given substrate concentration is called *affinity of enzyme*, for its substrate. Each enzyme has active site. Many enzymes particularly those composed of only a single polypeptide chain possess only a single active site. The specificity of substrate recognition and the formation of the complex and catalytic chemical transformation of the substrate is due to the existence of active site of the enzyme, which is estimated (active site) to in area of about 400 A, i.e. only a few per cent of the total surface area of an enzyme.

The most fundamental feature of the mechanism of catalysis of enzyme is that (a) the active site can hold the substrate by formation of weak chemical bond and (b) the presence of catalyst can affect only rate at which the reaction reaches to its equilibrium, but never the state of that equilibrium.

Thus, an addition of galactosidase to the lactose solution will cause hydrolysis to reach at equilibrium in as many minutes as the spontaneous reaction requires years. But upon attaining this catalytic equilibrium, the residual lactose level in the solution is still exactly the same as in the spontaneous equilibrium. The reason for this is that the enzyme accelerates the hydrolysis of lactose into glucose and galactose just as much as it accelerates the eventual reformation of lactose from high concentration of the reaction products. *Thus, enzymes do not perform chemical miracles*, whose occurrence would be impossible in their *absence*; they merely speed up the rate of spontaneous chemical reaction which would not be otherwise have reached equilibrium in our life time (G.S Stent, 1971).

Enzyme cannot Make Enzyme

The imagination that the 'ordered amino acid' assembly is the work of another 'ordering enzyme', raises the confusion that for each particular protein of given primary structure, there exists a corresponding specific ordering enzyme, which knows how to assemble that particular protein. However, if that 'ordering enzyme' also turns out to be a protein of specific amino acid sequence, then it further complicates the question that instead of providing answer, we generate paradox. In fact, the informational elements responsible for the synthesis of proteins which have 'knowledge' how to assemble the particular amino acids into predetermined polypeptide sequences cannot reside in ordinary enzyme but carried by other cellular elements (chromosomes) which have property to achieve their own self-replication to equally distribute these 'element' in their two daughter cells so that each daughter cell must be endowed not only with the complete parental outfit of the enzyme but also with the complete set of parental informational elements 'genes'. This makes two things clear, (a) 'enzyme' cannot make enzyme, it would lead paradox (b) cells of all the living organisms possess their self-reproducing informational elements 'genes' that govern enzyme synthesis. These informational elements are nothing but 'genes'. Now, the question arises on how gene manage to control over the specific cellular physiological processes from its nuclear throne.

Genes direct the assembly of amino acids into a polypeptide chain of a given primary structure (protein). Thus, enzymes are proteinaceous in nature.

GENE–PROTEIN RELATION

The human disease like alkaptonuria gave the first inkling that genes control the synthesis of enzymes. Another human disease sickle-cell anaemia, provided evidence that genes exert their control by directing the primary structure of polypeptide chains. In case of sickle-cell anaemia, the red blood corpuscles (RBC) of persons affected with sickle-cell anaemia assume abnormal sickle shape (Figure 22.7) at low-oxygen level, which lead difficulty in the blood circulation, Sickle-cell anaemia is caused due to a single recessive mutation in the human genome. In 1949, Linns Pauling and his associates detected that persons suffering from sickle-cell anaemia carry mutant genes in the homozygous recessive condition and these mutant genes are responsible for the presence of abnormal 'haemoglobin' molecule in the blood cells. These sickle-cell haemoglobin molecules carry lower negative charges than normal haemoglobin molecules. However, persons carrying heterozygous condition of recessive mutant genes (Hb hb) do not suffer from sickle-cell anaemia, because they possess mixture of (RBC) and abnormal haemoglobin (RBC). Sickle-cell anaemia is, therefore, called gene-controlled disease or molecular disease. In 1956, Ingram reported that the difference between normal and sickle-cell haemoglobin is that mutant haemoglobin carries a valine residue at a site on its polypeptide chain and the normal haemoglobin carries a glutamic acid residue at that site (Figure 22.8). This explained the observed difference in the 'negative charge', since one ionizable–COOH group is lost upon the replacement of the glutamic acid side chain by the valine side chain.

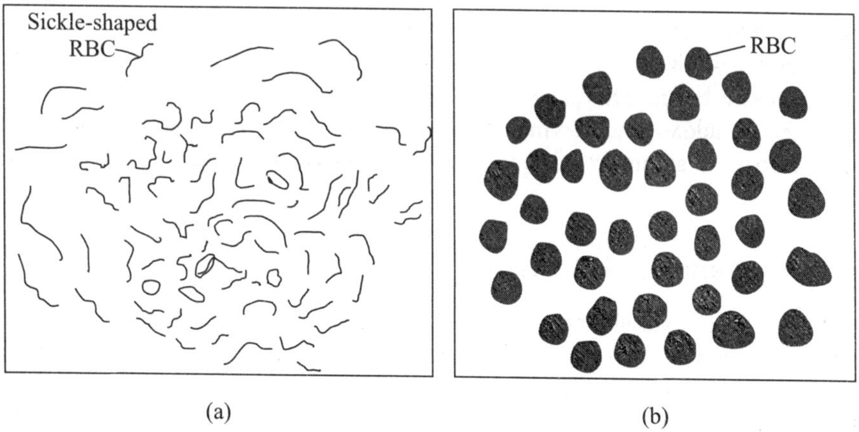

Figure 22.7 Sickle-cell anaemia in man: (a) shows red blood cells with low-oxygen tension of individual carrying the sickle-cell anaemia, (b) shows red blood cells of normal individuals under similar condition.

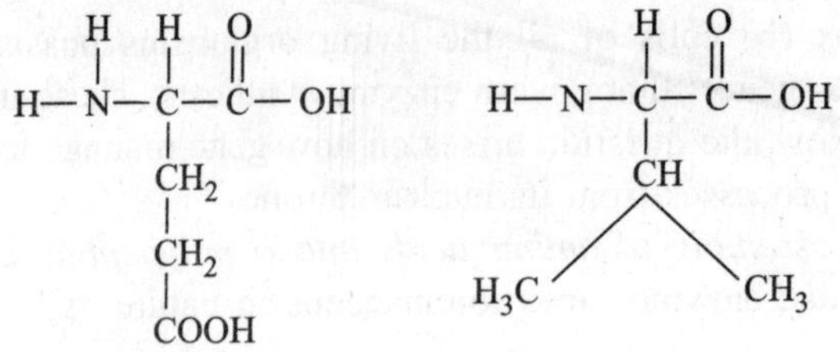

Glutamic acid Valine

Figure 22.8 How molecular mutation causes sickle-cell anaemia.

Sickle-cell abnormality is caused due to defective haemoglobin synthesis as a result of molecular mutations from glutamic acid to valine in the *β*-chain (Figure 22.9).

Sickle-Cell Mutant Protein

α-chain	Amino acids						
	1	2	16	57	58	68	141
Normal	Val^{-}	leu^{-}	lys^{-}	Gly^{-}	His^{-}	Asp^{-}	Arg^{-}
Sickle cells	Val^{-}	leu^{-}	lys^{-}	Gly^{-}	His^{-}	Asp^{-}	Arg^{-}

β-chain	Amino acids									
	1	2	3	6	7	26	63	67	125	150
Normal	Val	His	leu	Glu	Glu	Glu	His	Val	Glu	His
Sickle cells	Val	His	leu	Val	Glu	Glu	His	Val	Glu	His

Figure 22.9 Sickle-cell abnormality.

The mutant haemoglobin carries a valine residue at a site on its polypeptide chain. In alkaline solution, the sickle-cell haemoglobin molecule carries lower negative charge. Sickle-cell mutant is recessive in nature hence heterozygous escape the disease.

By 1956, the paradox, that enzyme cannot make enzyme, was widely appreciated and the idea that gene direct the assembly of amino acid into 'polypeptide chain' became most accepted theory.

Colinearity of Gene and Protein

While discussing the general nature of genetic code in Chapter 19, it has been emphasized that nuclear DNA base sequence is colinear with the protein—amino acid sequence that it specifies. That is, it is presumed that the order of the bases specifying an amino acid sequence is the same as the order of the amino acids being specified in protein synthesis. For a long time, this theory was not well confirmed. However, this concept was verified by Yanofsky in 1967 based on his *transductional fine structure of genetic mapping* of the *E. coli* trp A gene. While carrying out this genetic mapping experiment, Yanofsky himself got involved in the extraction, purification and

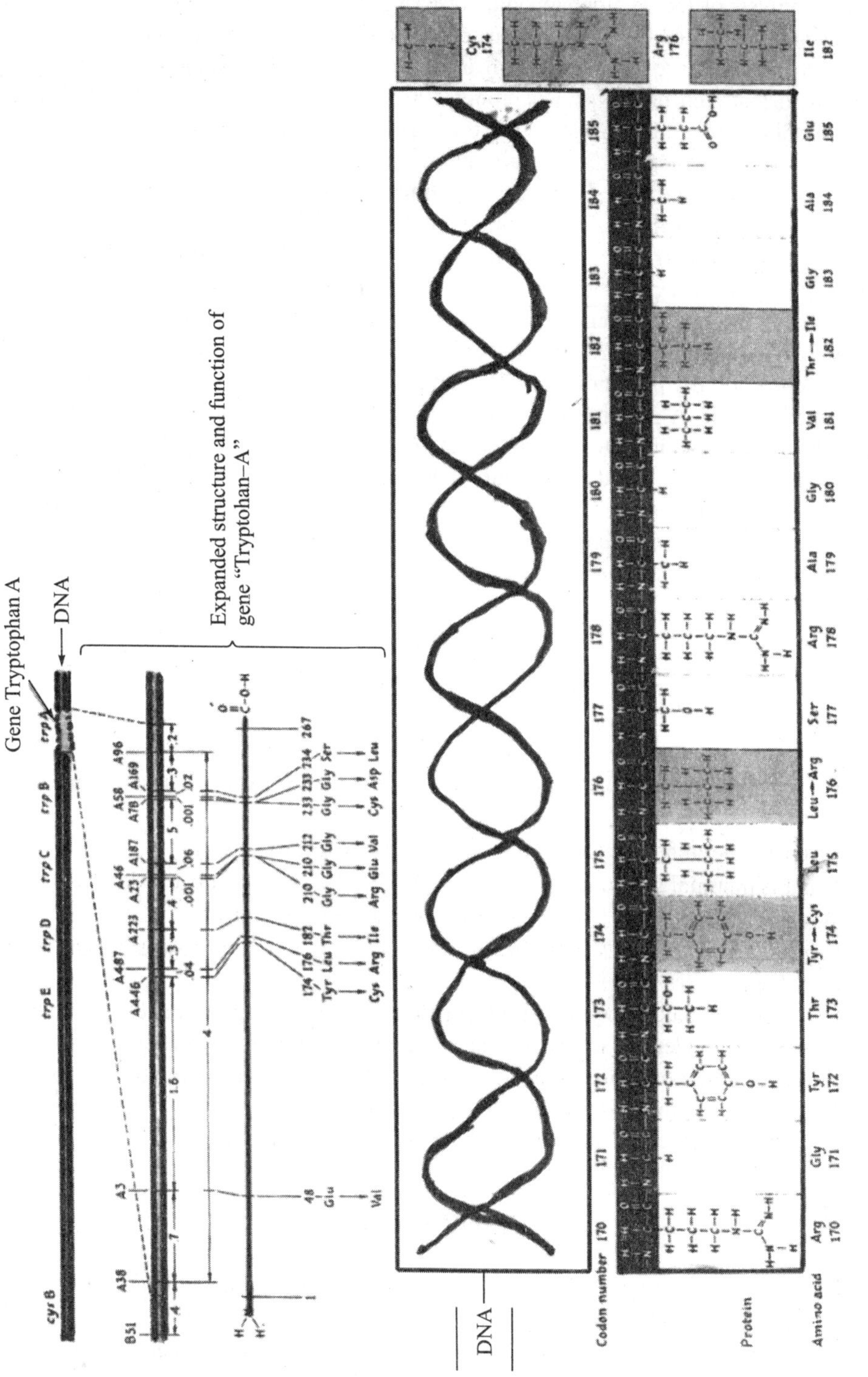

Figure 22.10 Colinear structure of DNA chain and protein synthesis.

structural analysis of the tryptophan synthetase, a protein coded by gene trp A. The findings of this experiment indicated that, the protein A, consists of a polypeptide chain of 267 amino acids arranged in the exact sequence as shown in Figure 22.10. Thus, length of gene trp A should correspond to a stretch of 267 × 3 or about 800 nucleotides. Yanofsky also indicated through transductional mapping of trp A gene that eight different sites of mutation of this gene in which two sites trp A-46 and trp A-23 are quite close but not identical.

Yanofsky also determined primary structure of the functionally-defective tryptophan synthetase A proteins of the eight trp mutants of *E. coli*. He noted that in every mutant, the non-functional A protein carries an abnormal amino acid of some definite polypeptide site, i.e. mutation; try A-446 has resulted in a change of 174th amino acid of the protein A polypeptide chain whereas tyrosine is normally present at that site, mutant trp-446 carries cysteine there. Similarly, trp A-46 and trp A-23, which occurred very closely have resulted the replacement of glycine residue normally present at the 210th amino acid in the protein A whereas trp A-46 carries glutamic acid and trp A-23 carries arganine as the 210th amino acid of the mutant protein A polypeptide chain. This indicates that the relative order of the mutant sites on the fine structure map of the trp A gene (nucleotide representing the .nutant sites) is exactly the same as the relative order of the amino acid residues in the polypeptide chain to which these mutations have affected. Thus, *according to the molecular genetic article of faith, it is beyond any doubt that the nucleotide sequence of the gene is colinear with the polypeptide amino acid sequence that it specifies*.

REVIEW QUESTIONS

1. How gene controls the biosynthetic function in the plant and other organism?
2. Give the biosynthetic pathway of amino acids—alanine, aspartic acid, serine and glutamic acid.
3. What is one-gene one-enzyme theory? Who gave this theory?
4. What is minimal medium? How is it helpful in identifying 'nutrient deficient mutations'?
5. What is auxotrophy? How does it differ from phototrophy?
6. What is enzyme? Does it control a biochemical function or accelerate the biochemical activities?
7. Can enzyme make enzyme? Is it fact that enzymes are proteinaceous in nature?
8. Give some evidences to prove that gene controls protein synthesis.
9. What is sickle-cell anaemia? Is it fact that sickle-cell haemoglobin molecules are responsible for this disease?
10. Explain that gene and protein synthesis is collinear process.

REFERENCES

Beadle, G.W. and B. Ephraussi (1936): The differentiation of eye pigments in *Drosophila* as studied by transplantation, *Genetics*, **21,** 225.

Beadle, G.W. and E.L. Tatum (1991): Genetic control of biochemical reactions in *Neurospora*, *Proc. National Acad. of Sciences*, Washington, **27,** 499.

Pauling, L.R., B., Corey and H.R. Branson (1951): The structure of proteins: two hydrogen bond helical configuration of the polypeptide chain, *Proc. National Acad. of Sciences*, Washington, **37,** 205.

Sanger, F. (1949): The structure of insulin, *in*: *Currents in Biochemical Research*, D.E. Green (Ed.), Inter Science, New York.

Stent, *G.S. (1971): Molecular Genetics*, W.H. Freeman and Co., San Francisco.

Yanofsky, C., B.C. Cartn, J.R. Guest, D.R. Helinski and V. Henning (1964): On the colinearity of the genetic structure and protein structure. *Proc. Nat. Acad. of Sciences*, Washington, **51,** 266.

Yanofsky, C. (1967): Gene structure and protein structure, *Science American.*

CHAPTER

23 Biotechnology and Transgenic Plants

Biotechnology is not a new science/technology as many people believe so. Biotechnology began with fermenting fruit juice and alcoholic beverages and improving the taste-adding microorganisms (fungi). In agriculture, it has found application in cell and tissue culture for rapid propagation of plant species, in diagnosis for detecting plant pest and diseases based on the uses of monoclonal antibodies and nucleic acid probes and in the genetic engineering of the plant species, to introduce new traits of interest and also in aiding conventional plant breeding using molecular makers (Persley 1992). Thus, biotechnology can be defined as a technique that uses living organisms or substances from those organisms, to make or modify a product, to improve plants or animals or to develop micro-organisms for specific uses.

Agriculture biotechnology began with the plant tissue culture and mass production of biological control agents. In 1960, a German plant physiologist discovered the idea of hydroponics, i.e. growing plants without soil. He grew many plants in the aqueous solution containing macronutrients (calcium, magnesium, potassium, sulphate, nitrate and phosphate) and supplemented with eight micronutrients. Hydroponics is practiced in many countries specially in the urban areas to grow higher values crops such as tomato, chillies, egg plants and vines of pepper and bitter gourd. For better root development, three vitamins, e.g. thiamin, necotinic acid and pyrodoxine could be added in the solution. However, micro-propagation brought a turning-point in tissue culture technique or biotechnology.

MICROPROPAGATION

Meristem Culture

The embryonic leaves and the tiny masses of the dividing meristematic cells could be cultured to grow a whole plant. When placed in a culture medium, the embryonic leaves grow into a mature leaf and a tiny apex shoot cell can produce a mature shoot (stem with leaf).

Occasionally, roots appear at the base of the cultured shoot and a whole plant develops. If a small amount of growth hormones (auxin and cytokinin) are added in the proper and tolerable concentration, many shoots can develop from a single apex shoot cell culture (Figure 23.1). After a few weeks, these nascent shoots could be separated and planted separately in the pots, first, before planting to the field. This method of *clone* multiplication is called *clonal propagation,* because plants are genetically alike. This method of clone multiplication through tissue culture is being widely utilized worldwide by micropropagation industries of economically important plants of commercial value, who need millions of plants every year to replace the old and aging plants/trees.

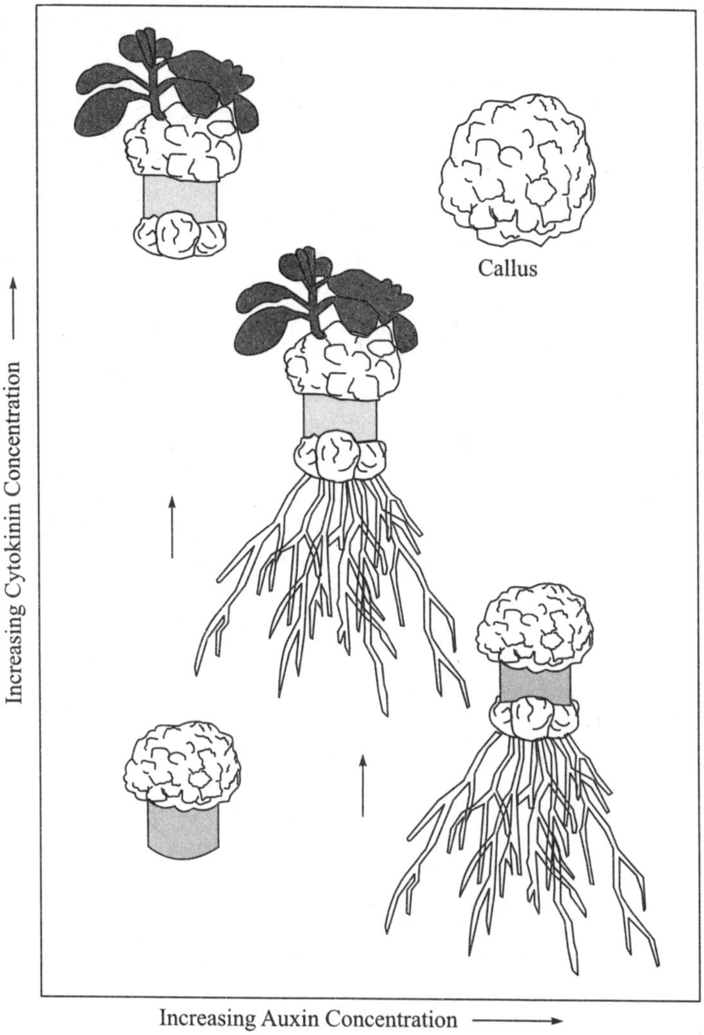

Figure 23.1 Multiple root formation after addition of two-harmones—auxin and cytokinin in the medium in meristem culture.

The clones could be developed from the flower meristems, if transferred to a sterile culture dish, where cell multiplication takes place to form a callus—a mass of undifferentiated cells. At the surface of callus, embryo-like structure is formed, called somatic embryo. These somatic embryos develop into shoots and later form roots to develop into a plant (Figure 23.1). This method has been utilized for micropropagation of ornamental plants and straberrys; plantation crops like oil palm, banana, medicinal plants, aromatic plants and trees. In the 1960s, a multinational company 'Unilever Corporation', owner of oil palm plantation in Malaysia and in other tropical countries, used this method to multiply oil palm seedlings about 3 millions every year to replace aging trees. In addition to high yields and high oil content, (30 per cent on average), the new trees are dwarf in height making mechanical harvesting easier and economically cheaper.

The *first* advantage of micropropagation is that the genetic make will not be changed due to quick multiplication of the meristem.

The *second* advantage of this method is that the plants obtained from the meristem culture are disease-free. Because, before the meristems are transferred to the tissue culture medium, these are thoroughly surface sterilized to kill the bacteria and fungus spores lying on the outside of the tissue. It is also advantageous to produce virus-free trees. In fact, the meristems are free from virus infection and plants obtained from micropropagation through meristem culture are also virus-free. Potato, strawberry and sugarcane propagation could be done from micropropagation.

Generation of plants from a single cell

The embryonic region of plants, the root tips and shoots apexes, can be cultured in the laboratory, to generate plants, when small sections of leaves, stem or root are cut, sterilized and put in the culture medium. Such plant pieces form calluses—masses of undifferentiated cells. The cells in such calluses divide, in large and divide again provided the medium contains not only minerals and sucrose but also one or more plant hormones (auxin and cytokinin). The calluses are grown on the nutrient medium solidified with agar. The healthy growing calluses can also be cut into small pieces that could be shaken in a liquid nutrient medium without agar, in a rotary shaker. The single cell or small clumps of cells will separate from the surface of the callus pieces, and swirl around in the nutrient medium. The small clumps break apart and the cells continue divide and enlarge. This is called *cell culture* (Figure 23.2).

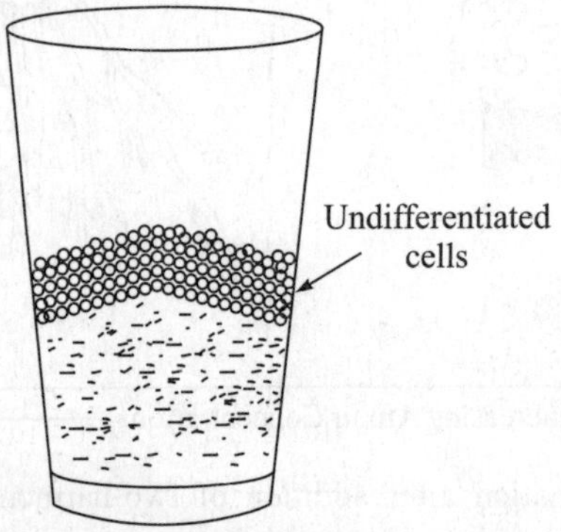

Figure 23.2 Multiple undifferentiated cell arising in a sterile tissue culture medium.

The ability of a single mature plant cell to give rise to an entire organism is called *totipotency*. Earlier, it was believed that mature plant cells could not be utilized for the tissue culture. However, when mature cells are used for callus formation, the cell walls are digested with enzyme and then protoplasts (naked cells) are isolated which are quite fragile, but these cells too could be cultured to regenerate into plants. When protoplasts are cultured in a nutrient medium, these cells form clumps. When small clumps are transferred to solid medium, it grows into callus that can form shoots. These shoots could be cut and transplanted into medium that induces root formation and a plant is regenerated in this way.

In some plant species, the small clumps can develop into embryos resembling the embryos developed from the fertilized egg cells. This is called *somatic cell embryos*. The plants obtained by tissue culture are genetically alike. But sometime mutation occurs in one out of million cells. When somatic embryos with altered genetic constitutions develop into a plant, this phenomenon is called *somaclonal variation*.

Some people feel that cells of a mature plant do not have same genetic constitution as mutations may occur and cause somaclonal variation. This is very rare as micromutation (molecular) is not visible. In author's view, the plant generated from a single cell, meristem or protoplast must be 99 per cent genetically identical and such plants will not be much different from their parents or donors.

Embryo Culture

Regeneration of whole plant from an embryo in the culture medium is known as embryo culture. This technique is used when there is disharmony between the embryo and endosperm. This technique is used in case of distant hybridization such as interspecific and intergeneric crosses to make the hybridization successful for obtaining haploids, etc. The production of haploids through embryo culture is called *embryogenesis*.

Anther Culture

Haploids produced through anther culture is done by placing anthers into a suitable culture medium. Care should be taken to avoid any type of injury otherwise it will reduce the formation of callus from anther walls. The production of haploids through anther culture is called *androgenesis*.

Ovule Culture

Whole plant (haploid) can be regenerated from unfertilized ovules.

Cell Suspension Culture

Since very long time, bacterial and fungal cultures have been widely used in the biotechnology industry for producing industrial feeds (chemicals) and pharmaceuticals (Penicillin) and enzymes.

In addition, many plants produce chemicals, which are being used as drugs, cosmetics, flavouring agents and agriculture chemicals. So far, these chemicals have been extracted from the plants grown in the fields and forests (Table 23.1, M.W. Flower, 1983). Now, it is possible to grow these cells that can produce the desired chemicals in the '*cell suspension medium*' containing nutrients and hormones; as microorganism have been domesticated to produce the industrial products, similarly the embryonic region of plants such as root tips, shoot appex can be cultured in laboratory in sterile culture. When small sections of leaves are cut out, sterilized and put in culture, they form calluses (masses of relatively undifferentiated cells). The cells in such calluses divide, enlarge and divide again provided medium contains not only mineral and sucrose but auxin and cytokinin. The vigorously growing calluses can be cut into small pieces that can be shaken into liquid nutrient medium (without agar) in a rotary shaker. The single cell or small clumps of cells will separate from the surface of the callus pieces and swirl around the nutrient medium. These cells continue to divide and enlarge and small clumps continuously break apart. This is called *suspension culture.*

TABLE 23.1 The plant products of industrial uses

Industry	*Plant product*	*Plant species*	*Industrial uses*
Pharmaceuticals	Codein (alkaloid)	*Papaver somniferum*	Analgesic
	Diosgenin (steroid)	*Dioscorea deltoidea*	Antifertility agents
	Quinine (alkaloid)	*Cinchona ledgeriana*	Antimalarial
	Digoxin (cardiac glycoside)	*Digitalis lanata*	Cardiatonic
	Scopolamine (alkaloids)	*Datura stramonium*	Antihypertensive
	Vincristine (alkaloid)	*Catharanthus roseus*	Antileukemic
Agrochemicals	Pyrethrin	*Chrysanthemum cinerariaefolium*	Insecticide
Food and drinks	Quinine (alkaloids)	*Cinchona ledgeriana*	Bittering agent
Sweetner	Thaumatin (protein)	*Thaumatococcus danielli*	Nonnutritine sweetner
Cosmetics	Shikonin	*Lithosperum erythrorhizon*	Dye

Application of tissue culture technique in crop improvement

Tissue culture technique is useful in crop improvement in the following ways.

1. Generation of variability such as somaclonal, gametoclonal or protoconal. This is mythical in the sense that tissue culture techniques do not change the nuclear or plasma genes composition except in rare cases if mutation can occur.
2. Development of haploids is possible by anther culture, pollen culture and ovule culture.
3. Embryo rescue in distant crosses is possible in tissue culture.
4. Somatic hybridization is possible by protoplast fusion.
5. Selection of disease-resistant lines by infecting the culture with disease spores is possible.
6. Selection of salt tolerance: Millions of cells are subjected to saline solution in the flask but the cells having genes for resistance to salt can survive.
7. Selection for drought: In the culture medium drought is induced by using chemical PEG (polyethylene glycol).
8. Micropropagation is possible by rapid mass multiplication of plant materials belonging to horticultural and plantation crops.

CULTURE MEDIUM

Culture medium is defined as, the *in vitro* nutritional requirements of the plant cells in a culture medium which are similar to those nutrients that are required by the plants in natural condition. Culture medium differs from crop species to species. Culture medium of Murashige and Skoog (1962) which is also known as MS culture medium. Culture medium consists of the following plant nutrients:

Macronutrient	*Formula*	*Quantity* (mg/l)
(i) Ammonium nitrate	NH_4NO_3	1650
(ii) Potassium nitrate	KNO_3	1900
(iii) Calcium chloride dihydrate	$CaCl_2 \cdot 2H_2O$	440
(iv) Magnesium sulphate heptahydrate	$MgSO_4 \cdot 7H_2O$	370
(v) Potassium dihydrogen phosphate	KH_2PO_4	170
Micronutrient	*Formula*	*Quantity* (mg/l)
(i) Potassium iodide	KI	0.830
(ii) Boric acid	H_3BO_3	6.200
(iii) Manganese sulphate tetrahydrate	$MnSO_4 \cdot 4H_2O$	22.300
(iv) Zinc sulphate heptahydrate	$ZnSO_4 \cdot 7H_2O$	8.600
(v) Sodium molybdate dehydrate	$Na_2MoO_4 \cdot 2H_2O$	2.250
(vi) Copper sulphate pentahydrate	$CuSO_4 \cdot 5H_2O$	2.250
(vii) Cobalt chloride hexahydrate	$COCl_2 \cdot 6H_2O$	0.250
(viii) Sodium salt of EDTA	Na_2 EDTA	37.300
(ix) Ferrous sulphate heptahydrate	$FeSO_4 \cdot 7H_2O$	27.800
Vitamin		*Quantity* (mg/l)
(i) Inositon	–	100.00
(ii) Glycerine	–	2.000
(iii) Thiamine hydrochloride	Thiamine HCl	0.100
Others		*Quantity* (mg/l)
(i) IBA	–	0.010
(ii) GA	–	0.100
(iii) Sucrose	–	20.000
(iv) pH	–	5.5

ALIEN GENE TRANSFER BY PROMISCUS BACTERIA

Transfer of a segment of DNA between the different species could be called promiscuous. Genetic engineering involves transfer of genes between the unrelated species. However, biotechnology has very limited scope such as transfer of genes from alien sources to incorporate resistance to plant pathogen, insect-pests and quality characteristic, etc. using soil bacterium *Agrobacterium tumefaens,* i.e. which causes tumour crown gall formation in many plants. When soil bacterium infects a stem at wound site near the ground, infection disturbs the wound healing process. Instead of making wound callus that heals the wound, it causes cell proliferation into cancerous growth. The cause of cancerous growth is the bacterium which is attached with the plant cell. These bacterial genes get integrated with the chromosomes of the plant cell which encode production of auxin and cytokinin. The abnormal high level of the harmones produced is the cause of cancerous growth in the plants.

Bottlenecks in Alien Gene Transfer

1. Many economically useful genes have not yet been identified. Some specific viral genes have been identified that when transferred, check the virus from multiplying. Similarly, bacterium, *Bacillus thuringiensis* carries a gene that makes an insecticidal protein. When the bacterial gene is transferred in to the plants, the plant cells make the insecticidal protein and when the insect larvae eat the leaves or root, they die.
2. Regeneration of plants from the cells into which alien gene has been transferred from the unrelated species is not very simple. The claim so far is that a number of crops have been transformed and regenerated: These are:
   - Cereals—rice, wheat, corn, barley and rye
   - Fibre—cotton
   - Food legumes—soya bean, pea, bean, canola and flax
   - Horticultural plants—carrot, cauliflower, celery, cucumber and lettuce, potato, sugar beet, petunia, tomato, tobacco
   - Pasture—Alfalfa, white clover, orchard grasses
   - Trees—Poplar, apple and walnut

IMPORTANCE OF BIOTECHNOLOGY IN PLANT BREEDING

Inspite of all tall claims and propaganda, Biotechnology will be an aid, not an alternative of conventional breeding methods. It will aid in two ways:

(i) Aid in hybrid seed production by genetically engineered male sterility into any crop species that can be transformed to produce male sterile plants or to control pollination.
(ii) Molecular markers—small pieces of DNA called RAPDs (Random Amplified Polymorphic DNAs) can be used to mark a plant's entire genome (total of all its genes) by the plant breeders where it is very difficult to change the components which are the multigenic characters like yield components by the conventional breeding.

Biotechnology finds its importance in the following ways:

1. *Production of biopesticides:* One of the major objectives of the biotechnology is the production of biopesticides to reduce the environmental pollution by chemical pesticides.
2. *Mycoinsecticide:* Not only the plants, but also the insect-pests like caterpillars, aphids, grasshoppers and insect larvae are attacked by fungi. There are some mycoinsecticides (fungi), which are being utilized for controlling insect-pests in the Third World.

3. *Nematode–bacterial symbiosis:* There are a number of nematode (round worms) species that harbour symbiotic bacteria. The newly emerging nematodes (juvenile) enter the natural openings or muscles of insects which then release into the bloodstream of the insect the bacteria. This type of bacterial infection kills the insects and allows the nematode to multiply in the dead body of insects. These nematodes are insect-specific and do not harm the plants and the mammalians. The production of nematodes in the laboratory was not commercially possible but some American company like Biosys is able to grow nematodes on nutrient broth and by this 3.5 trillian nematodes could be produced in a week and about 5 million nematodes are needed to control soil dwelling pests in one hectare. Biosys is working on *Steinernema* species of nematode that does not harm plants.
4. *Insect-killing bacteria:* *Bacillus thuringiensis* produces toxic proteins that is poisonous and kills the insect larva when bacterium reaches the guts of insects after feeding the plants, though the bacterial strains are insect-specific but six strains of *B. thuringensis* are being commercially produced in the large fermenters and after production these could be stored for two years.
5. *Insect virus:* The insect-infecting viruses are species-specific and infect a few species of insects. There are seven such viruses. They do not cause harm to the plants and animals. It has been noticed that application of *Bacalo* viruses could reduce the use of chemical by 60–70 per cent but they are not very popular due to narrow range of hosts.
6. *Mycoherbicides:* The way fungi can cause crop diseases, they can also cause weed diseases. But production of such fungi spores and to store to prolong their self-life is yet to be worked out. In Canada 'Biomal' active against smooth leaved mallow has been got registered by a biotechnology company, Philom-Bios (1992).

Biotechnology is being forcefully popularized as 'gold mine' by private firms industries as if, this science is going to bring not only the economic prosperity but also the freedom from hunger too to man in 21st century, but I have doubt due to the following limitations.

LIMITATIONS OF BIOTECHNOLOGY (FOR ALIEN GENE TRANSFER)

1. Every living organism/species from bacteria to man possesses a specific genomic size, i.e. fixed number of chromosomes and also fixed number of nucleotides in the chromosome which are arranged in a linear order on the DNA helix which determines the specific characters of a particular living species. Any deletion or addition of nucleotide(s) which affect directly the species character is always fatal.
2. The genes/gene complexes present in each species are in a 'well-punctuated genetic equilibrium', i.e. they are well integrated in the DNA molecule, so that they resist any change because such a "*well-integrated coadapted gene complex constitutes an evolutionary unit and provides 'happy home' for genes to function inspite of its intrinsic variability*. Any disharmonious alien gene/gene combination, which attempts to become incorporated in such a gene complex will be discriminated by selection (Mayr, 1954 and Ram, 2005).

3. It is presumed that genetic information is written in a triplet code wherein the nucleotides are arranged in the triplet form on the *Reading frame* on a DNA double helix. The correct reading of the nucleotides sequence of a gene is assured by beginning the translation into amino acid sequences from a fix starting point and proceed triplet by triplet (Codon to codon) to the end. Hence, if the triplet in the *reading frame* happens to be altered by addition or deletion of one nucleotide, then the *entire reading frame* (all the subsequent triplet code) get altered in the DNA chain, and a totally incorrect translation is done from point on-ward which adversely affect the protein (enzyme) synthesis (Figure 23.3).

Normal cistron	CAT CAT CAT CAT CAT CAT CAT
G^+ base added	CAT CAT GCA TCA TCA TCA TCA
	A
A^- base removed	CAT CAT CAT CTC ATC ATC ATC ATC
	A
Base added and base removed	CAT CAT GCA TCA TAT CAT CAT CAT

Figure 23.3 Effect of addition and deletion of a nucleotide on the transcription of genetic message.

4. The severity of the 'shift' of nucleotides in the 'reading frame' by addition or deletion could be understood from the fact that the 'sickle-cell anaemia' (a dreaded human disease in which the red blood cells (RBC) afflicted with this disease carry low oxygen pressure which may lead to serious breathing difficulties), is caused due to the replacement of a single amino acid, glutamic acid by valine in the haemoglobin (protein) and a 'shift' of a single nucleotide from A→U in the codon GAG or GAA coding for glutamic acid to codon GUG or GUA coding for valine, is responsible for the shift in the *reading frame* for occurrence of such a deadly heritable disease in the man. Sickle-cell mutant gene is recessive, hence heterozygous individual escape the disease (Figure 23.4).

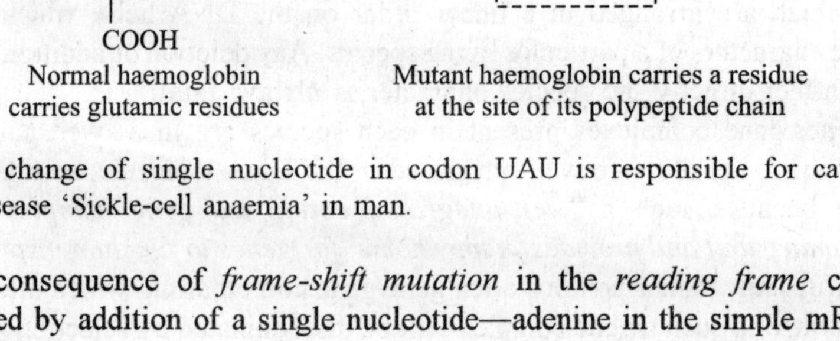

Figure 23.4 A change of single nucleotide in codon UAU is responsible for causing a deadly disease 'Sickle-cell anaemia' in man.

The consequence of *frame-shift mutation* in the *reading frame* can be further illustrated by addition of a single nucleotide—adenine in the simple mRNA sequence (Figure 23.5).

Leu Leu Leu Leu

CUG CUG CUG CUG

CUG CAG GCU GCU G

Leu Histidine Alanine

Figure 23.5 Consequences of frame-shift mutation in amino acid synthesis.

Not the least, a single base change in codon in UAU coding for tyrosine can produce nine codons which may code independently for different amino acids, viz.

UAU → Tyrosine → UAU → Aspertic acid → AAU → Aspirin → CAU → Histidine → UGU → Cystein → UUU → Phenylalanine → UCU → Tyrosine UAA stop → UAG stop

According to the new definition, a gene is a DNA fragment consisting of hundred to thousands of nucleotides bases. Now, the impact of insertion and deletion of an alien gene consisting of hundred/thousands of nucleotides can be very well understood by any person, how much complexity and disruption in the biosynthetic pathway it can bring in the host is 'unimaginable'.

5. The genes of eukaryotes (higher life forms) do not show expression in the prokaryotic genetic backgrounds, e.g. bacteria *Escherichia coli* and Viruses; it may happen vice versa too because of the differences in the size of regulatory gene mechanism, and the differences in the size of ribosomes (70s in prokaryotes and 80s in higher form), which are called 'workshop of protein synthesis'. Let me explain what is a gene?

   "*A gene is a more complex (much longer really) than just a protein-coding string of nucleotides*".

   First of all the protein-coding region, in eukaryotic DNA may be interrupted by short stretches (introns) of DNA whose function is not yet clear. These short stretches (introns) are also transcribed when mRNA is first transcribed in the nucleus. These short non-coding stretches are called 'introns' and the coding regions of DNA are called 'exons'. Before translation of genetic message starts, the *introns* are sliced away through the help of enzyme during mRNA processing in the cytoplasm called gene 'splicing', as a result only 10–20 per cent of hn mRNA remains functional in eukaryotes. In contrast, bacterial and viral mRNAs do not undergo processing. Eukaryotic mRNA is monocistronic while bacterial and viral mRNAs are polycistronic. These are the reasons that prokaryotic genes can not express their effect in eukaryotic genetic back ground and vice-versa.

6. A gene directly does not take part in protein synthesis. There are 'regulatory' genes (DNA segments) on each side of the protein-coding region (structural genes) called inducer (I), promoter (p) and operator (o) genes, that specify when gene is to be activated during development; to which *stimuli* (hormonal or environmental) the gene should respond; in which cell it should become active; how much mRNA should be made and when gene is to be turned 'on and off.'. The regulatory regions are of course of great importance for the orderly development of a cell and for gene to function in an organism. For example, a disease resistance gene may encode a protein that allows the plant to detect the invader quickly so that the plant can turn 'on' its defenses. The gene

that encodes the detector protein responds to a *stimuli* from the pathogen, perhaps a chemical or metabolite made by the pathogen and this response involves *several other genes*. When a plant breeder transfers a resistance gene from a wild or alien source to a cultivated crop, the entire regulatory cascade, i.e. promoters and enhancers must be working correctly (Figure 23.6). If the new combination carries the resistance gene but not the regulatory genes, the resistance may not be expressed correctly and the new variety whether transgenic or conventional would not be resistant to the pests or pathogen, for which the entire exercise was done. The genetic engineers and biotechnologists who are involved in the development of transgenic crop plants must know that the gene they want to transfer must not have only structural gene (protein-coding region) but must be equipped with its regulatory region also. Of course, they can equip the gene with a new regulatory region that should ensure correct expression. But size of regulatory gene complex in bacteria and viruses differ from higher organisms (plants and animals).

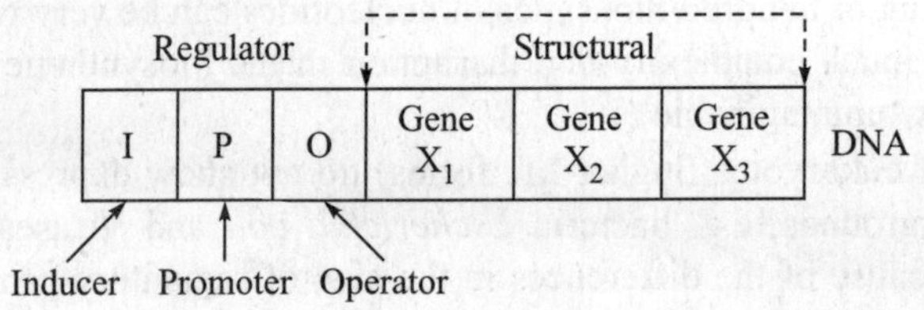

Figure 23.6 Structure of regulatory and structural gene in the living organisms.

7. The regulatory regions of bacterial gene generally do not work in plants because transcriptional factors (proteins) are different from those of plants. Similarly, a regulatory region of the genes from a mono-cotyledonous plant may not work in a dicotyledonous crop (bean), therefore, a full understanding of regulatory gene expression is of great importance to the genetic engineers and the molecular biologists, who are advocates of the alien gene transfer and development of transgenics.
8. Mutation is a natural phenomenon. Mutation in regulatory genes can make the structural genes ineffective. However, mutation can turn the individual into wild type. In chapter dealing with 'mutation' mention has been made that when NP9 rice variety was treated with 30 kr gamma rays. Three types of wild genotypes appeared—grassy mutant, spready plant habit and grains with shattering tendency (Ram, 1974). Biotechnology can convert cultivated crop plants again to wild-type crops in 100 to 1000 years in future, if bacterial and viral genes continue to be transferred in cultivated types.
9. Not all the regions of DNA are functional; hardly 10 to 20 per cent of DNA molecules consisting of unique DNA (exon) are functional; rest are inert due to repetitive DNA sequences or (intron). Therefore, finding a gene of interest is like searching a needle in the wheat-hay stack.
10. Similarly, an alien gene must be transferred in the non-repetitive DNA sequences, i.e. unique DNA sequences otherwise the alien gene may or may not express its type effect, provided replaced unique DNA sequences or 'exon' are not carrying important functions.

11. Through alien gene substitution, we can transfer, a single gene-controlled character like 'race-specific' resistance, while resistance to most of the pests and pathogens is multigenic. Not the least, crop varieties require multiple resistance together for a number of diseases and pests which is not possible through alien gene transfer.
12. The alien gene should not be placed in a genetic background which determines the edible parts like grain, oil, protein and sucrose, etc. It may cause health hazard particularly if mutation takes place in negative direction in the regulator or structural genes.
13. There are some overenthusiastic scientists who say that through 'Biotechnology' dream of second green revolution can be achieved. It is their misconception because they do not know that *yield has no genes*. Yield is governed by a constellation of yield contributing characters such as plant population per unit area (tillers plant), grains per spike and grain weight, etc. All the yield contributing characters are polygenically controlled and thus it is beyond the scope of biotechnology to improve the yield by alien gene transfer.
14. While transferring the alien gene (segment of DNA), the segment must be placed in a proper direction. If placed in reverse direction, it may cause antisensing of gene.
15. Before transfer of alien gene from 'plasmid' to host cell, the alien DNA segment symbolising gene must be properly purified, otherwise a small piece of plasmid DNA can be hazardous to health of the organism where it is placed.

In the last, I am of the view that almost all molecular biologists are biochemists, they know little about genetics. Similarly, almost all modern day genetists are biometricians, they know little about biochemistry and molecular level of genes. Hence there is a great gap in the understanding of both about exact nature and function of gene at molecular level. Moreover, I have already written earlier that DNA structure, genetic code, split gene, mRNA precessing and tRNA/adopter RNA concepts are still hypothetical and unsolved problems. Untill these issues which are basics for molecular genetics are solved, the concept of "*alien-gene transfer and development of transgenic plants and animals*" will remain a myth.

REVIEW QUESTIONS

1. What is meristem culture? Give the advantages of meristem culture.
2. What is single-cell culture? How it helps in preparing 'suspension culture'?
3. Give some plant products of industrial uses—pharmaceuticals, agrochemicals and cosmetic, etc. which can be obtained by 'suspension culture'.
4. What is promiscus bacterium? How is it helpful to man in genetic transformation of crop plants? Give its bottleneck.
5. Give importance of biotechnology in agriculture.
6. Can biotechnology become alternative of conventional breeding methods? Give its limitations.
7. What do you mean by transgenic plants? Explain briefly giving some examples?
8. How did the idea of transgenic come into the mind of man? Write a short note.
9. What are the methods of transfer of alien genes into cultivated types? Describe briefly.

10. Write in detail the method of *Agrobacterium*-mediated transfer of alien gene.
11. What is Ti-plasmid? Does Ti-plasmid contains genes? How does it transfer its gene to the plant cell?
12. In which area, the concept of transgenic could be successfully utilized.
13. Is theory of transgenic a myth? If so, give the reasons.

REFERENCES

Chrispeels, M.J. and D.E. Sadava (1994): *Plants, Genes and Agriculture*, Jones and Bartlett, London.

Cocking, E.C. (1986): The tissue culture revolution, *in: Plant Tissue Culture and Its Agriculture Applications*, Anderson L. and P.G. (Eds.).

Jensen, M.M. and W.L. Collins (1985): Hydroponic vegetable production, *Horticulture Review*, 7, 483–538.

Marx, J.E. (1989): *A Revolution in Biotechnology*, Cambridge University Press, Cambridge.

Moss, J.P. (1992): *Biotechnology and Crop Improvement in Asia*, International Crop Research Institute for Semiarid Tropics, Hyderabad.

CHAPTER

24

Plant Genetic Engineering

Genetic Engineering is a new branch of technology for transferring a gene (s) from one organism to another unrelated organism without involvement of sexual fusion of gametes. This technique can breakdown the sexual barriers so that any gene from any organism can be introduced in a desired genetic background of crops or animals. In Chapter 25, 'gene cloning and organism cloning' and their advantages and disadvantages will be explained. In this chapter, bacterial-mediated gene transfer in the cultivated crops have been discussed. For the successful transfer of gene to meet the desired objective, the gene to be transferred must be correctly identified; the gene must be correctly expressed at right time and in the right organ (cell or tissue), and the protein that gene encodes must have right function.

The goal of plant genetic engineering is to isolate one or more specific genes from unrelated species and introduce these into the commercial crop plants. The improvement of a crop plant can often be achieved by introducing a single gene, and genes can now be transferred to the plants by using the natural gene transfer system of a pathogenic soil bacterium called *Agrobacterium tumefaciens*.

The soil bacterium *Agrobacterium tumefaciens* causes tumour (crown gall) in many plants. When these bacteria infect a wound site, specially on the stem close to the ground, the infection disturbs the wound-healing process; instead of making of wound callus that covers the wound, the cells proliferate into cancerous growth. The cells from this tumour can be grown in tissue culture and unlike normal plant cells, they continue to grow like cancerous growth without addition of any growth harmone in the culture medium.

The microbiologist found that after bacteria attach itself to the wound site, they transfer some of their genes to the plant cell. These genes carry information necessary to code for *auxins* and *cytokinin*. Abnormally, high levels of these harmones produced as a result of this gene transfer prompt the cancerous growth. After genes have been transferred, from a single bacteria to a single plant cell and intergrated into a chromosome of the plant cell, the cell which is transformed with bacterial gene begins to proliferate. This way bacterial gene will be present in the every cell derived from the transformed cells. This transfer of DNA segment between unrelated species could be called 'promiscuous' as in normal course, DNA is transferred between the related species through the sexual reproduction.

The moment the molecular biologists realized that if they substitute some other gene(s) of their interest for the genes that bacterium transferred to the plant cell that formed the callus, they could obtain transformed plant cells that carry any gene they desire to introduce into the crop plants, provided they could regenerate the transformed cell into a plant. This is beginning of 'Genetic Engineering'.

In fact, *Agrobacterium tumefaciens* is now being used as one of the media to transfer the foreign genes (from unrelated species) into the plant cells. The genes transferred from bacterium to the plant are carried on a bacterial 'Plasmid', a circular DNA molecule. Like any other bacteria, *Agrobacterium* too has a single circular chromosome but in addition, it has some plasmids on which genes are located. The plasmid of *A. tumefaciens* is called 'Ti-plasmid' (Tumour-inducing plasmids) because genes found on plasmids induce tumours. The transfer of DNA occurs only after bacteria have attached themselves to the cell walls of wounded plant cells. The substances that leak out of the wounded site activate bacterial genes that encode enzyme necessary for cutting DNA out of the bacterial plasmids and reintegrate it into the plant DNA.

There is no restriction which gene could be transferred, so one can look any where for the useful genes and pathogens can often be thwarted by using their own genes. Specific viral genes have been identified that when introduced into the plant, keep the virus off from multiplying or contain their spread from cell to cell and thus contain the damage to be caused by Virus. However, the gene(s) to be transferred must be correctly identified and located on the plasmids/chromosomes; the transferred gene must be correctly expressed at the right time and in the right region of the cell tissues, and the protein that gene encodes must have the right function as stated earlier.

METHODS OF GENE TRANSFER

There are four ways to transfer the genes of unrelated species.

1. *Agrobacterium*-mediated gene transfer
2. Microprojectile gene transfer
3. Electroporation
4. Germ line gene transfer in animal (see Chapter 25)

Agrobacterium-mediated Gene Transfer

The *Agrobacterium*-mediated gene transfer is the most convenient way to create transformed plants. The small pieces of plant tissues (leaves, roots, stems are cut into pieces about 2–5 mm long) are cultivated with the bacterium *Agrobacterium* that carries the interesting genes on its 'Ti-plasmid'. This allows the bacterium to transfer then plasmid DNA to the individual plant cells at the cut surface of the plant (Figure 24.1). These transformed plant cells will conveniently give rise to a transformed plant under the correct conditions of the tissue culture. Thus, the transformation has three phases:

(a) Creating a strain of bacterium that has a 'Ti-plasmid' with the gene. Scientists want to transfer this to the plant.
(b) Culturing the bacteria with pieces of plant tissues so that DNA transfer can occur.
(c) Regenerating a plant from the transformed cells; at the same time the non-transformed cells are being killed by antibiotics.

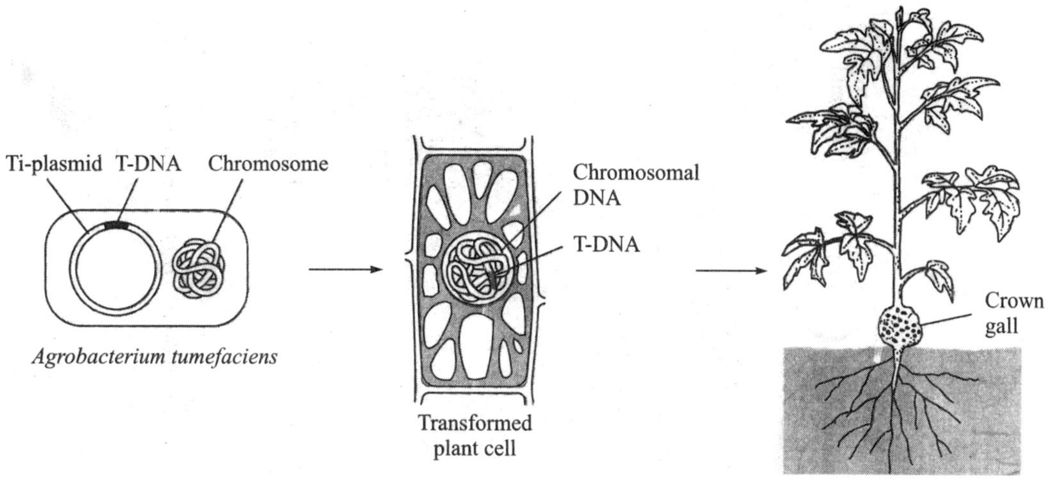

Figure 24.1 Transforming of Ti-plasmid of *Agrobacterium tumefaciens* to the plant cell which forms the gall.

The first stage involves molecular manipulation of DNA. The genes in the 'Ti-plasmid' (about seven in all) that cause the tumorous growth of the plant must be removed and be replaced by the genes some one wants to transfer to the plant. These genes occur in Ti-plasmid in one continuous DNA segment. The restriction enzyme is used to remove the genes, but care should be taken to leave the border (regulatory gene) intact as they are crucial for DNA transfer to the plant and integration into the plant DNA (Figure 24.2, after Chrispal and Sadava, 1994).

Second, introduce the target genes equipped with a suitable regulatory region and an antibiotic resistance gene, also with its own regulatory region into a 'Ti-plasmid' of a vector. The product of the antibiotic resistance gene usually an enzyme that inactivates a specific antibiotics which will let the cells that have this gene (all the cells that have been transformed) survive in a culture medium that contains the antibiotics. The newly constructed 'Ti-plasmid' must then be reintroduced into a suitable strains of *Agrobacterium tumefaciens*.

In the third phase, the *Agrobacterium tumefaciens* bacterial strains are co-cultivated with small pieces of leaves, stems or roots. Some specific biochemical fluids leak out of the wound sites of the plant materials when the tissues have been cut, making DNA transfer possible by activating certain genes present in the bacteria. The transfer involves cutting one strand of the double-stranded DNA molecule, first at right border and later at the left border, with the simultaneous copying of the DNA into the small single-stranded molecules that contain only the genes contained between the two borders. These DNA molecules which are coated with protein to protect it against degradation enters the nucleus of the plant cell to which the *Agrobacterium* is attached and are then integrated into the plant DNA. Integration is followed by the synthesis of second strand and repair of the break. The plant DNA is 'as good as new' but carries some new genes that will be faithfully duplicated each time the DNA replicates as cell divides. The pieces of the plant tissues are then transferred to a culture medium with two antibiotics. One to kill the bacteria (they are not needed any more) and other to kill the non-transformed plant tissues. The transformed cells now generate new cells and form calluses, if the necessary hormones are put in the culture medium to which the pieces of leaves, stem and roots were transferred.

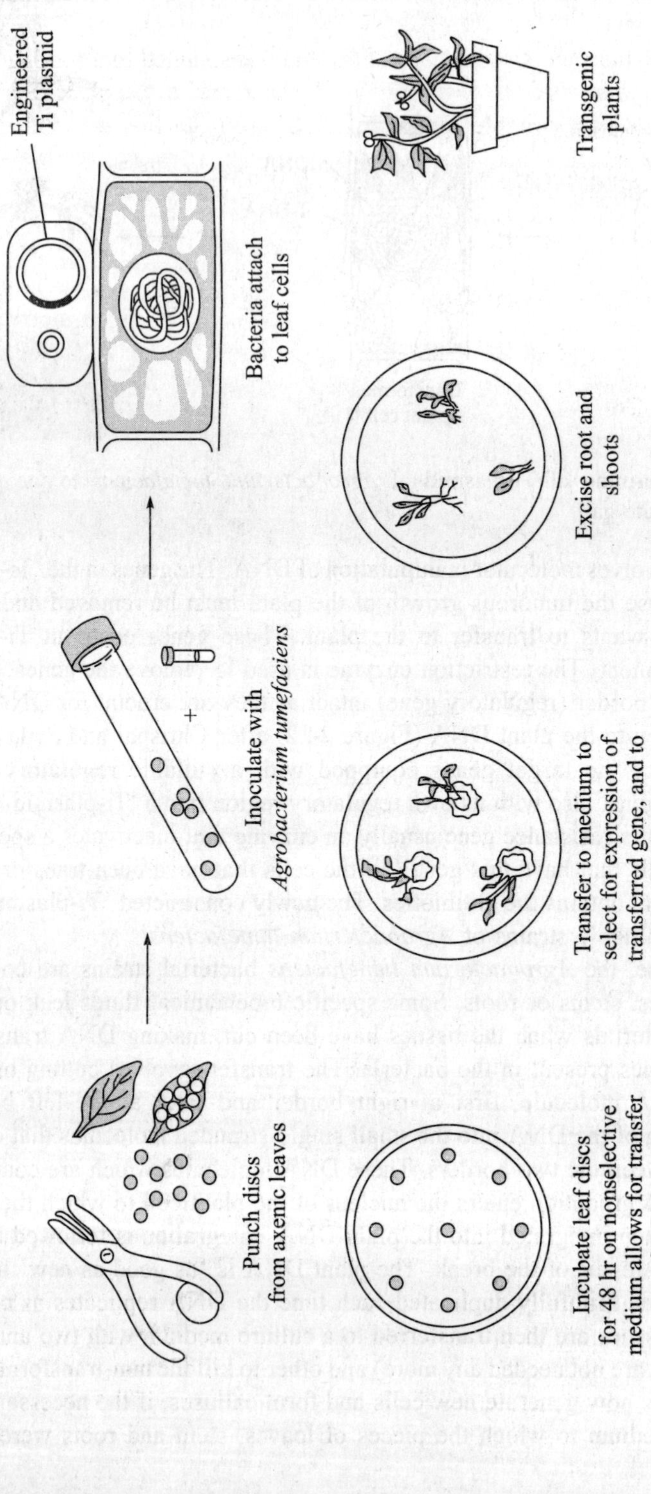

Figure 24.2 The method of transformation of plants using Ti-plasmid of *Agrobacterium tumefaciens*.

In the forth phase, the small callus pieces are transferred to a medium that induces shoots (at higher cytokinin level).

In the fifth phase, the shoots should be cut off and transplanted to a rooting medium. In this system, any gene placed between the two borders is transferred to the plant DNA. But the genes should be accompanied by a suitable control region or promoter that determines when the gene will be active, and in which tissue, etc.

However, this method is useful only in those plants which are susceptible to *Agrobacterium tumefaciens*.

Microprojectile Gene Transfer

It is done by the introduction of DNA into the cell by bombardment of DNA particles, known as (*ballistic impregnation*) (Figure 24.3).

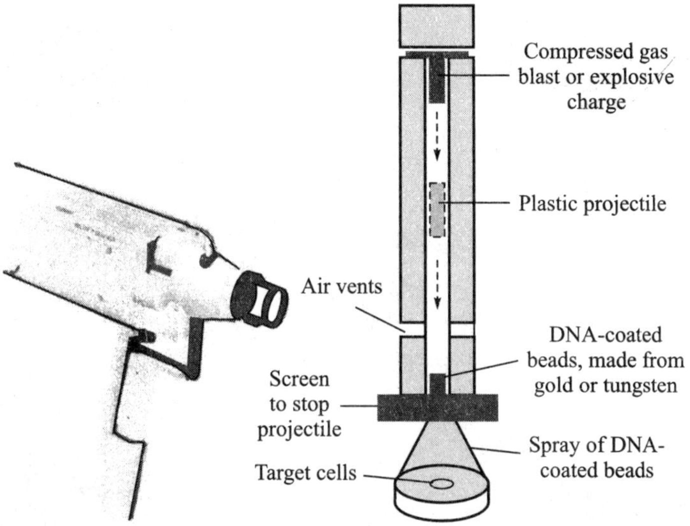

Figure 24.3 Particle-bombardment method to ballistically impregnant alien DNA fragment in the target cell.

There are many plant species especially the grasses and cereals which are not susceptible to infection by *Agrobacterium*. Similarly, some legume species such as soybean, the transformed cells do not regenerate in tissue culture from calluses. Thus, most important crops like rice, corn, wheat, sorghum, sugar cane, beans and soybean do not respond to the *Agrobacterium*-mediated gene transfer, though it is possible to introduce DNA easily into the protoplast of such species but it is difficult to regenerate fertile plants from the protoplast.

J.C. Sanford and his colleagues at Cornell University, USA, developed a microprojectile gun to deliver DNA directly into the plant cells by shooting it through the cell wall and the cell membrane. They coated small (4 mm in diameter) tungsten particles with DNA and constructed a gun that could accelerate the particles with enough speed that the particles would enter the first cell layer of a tissue. After lodging in the cells, DNA was transcribed into RNA and RNA was translated into proteins. This way introduced DNA was genetically active.

In Figure 24.4 comparison of *Agrobacterium* and particle gun methods is shown to create transgenic plants. (Gasser, C.S. and R.T. Fraley, 1992).

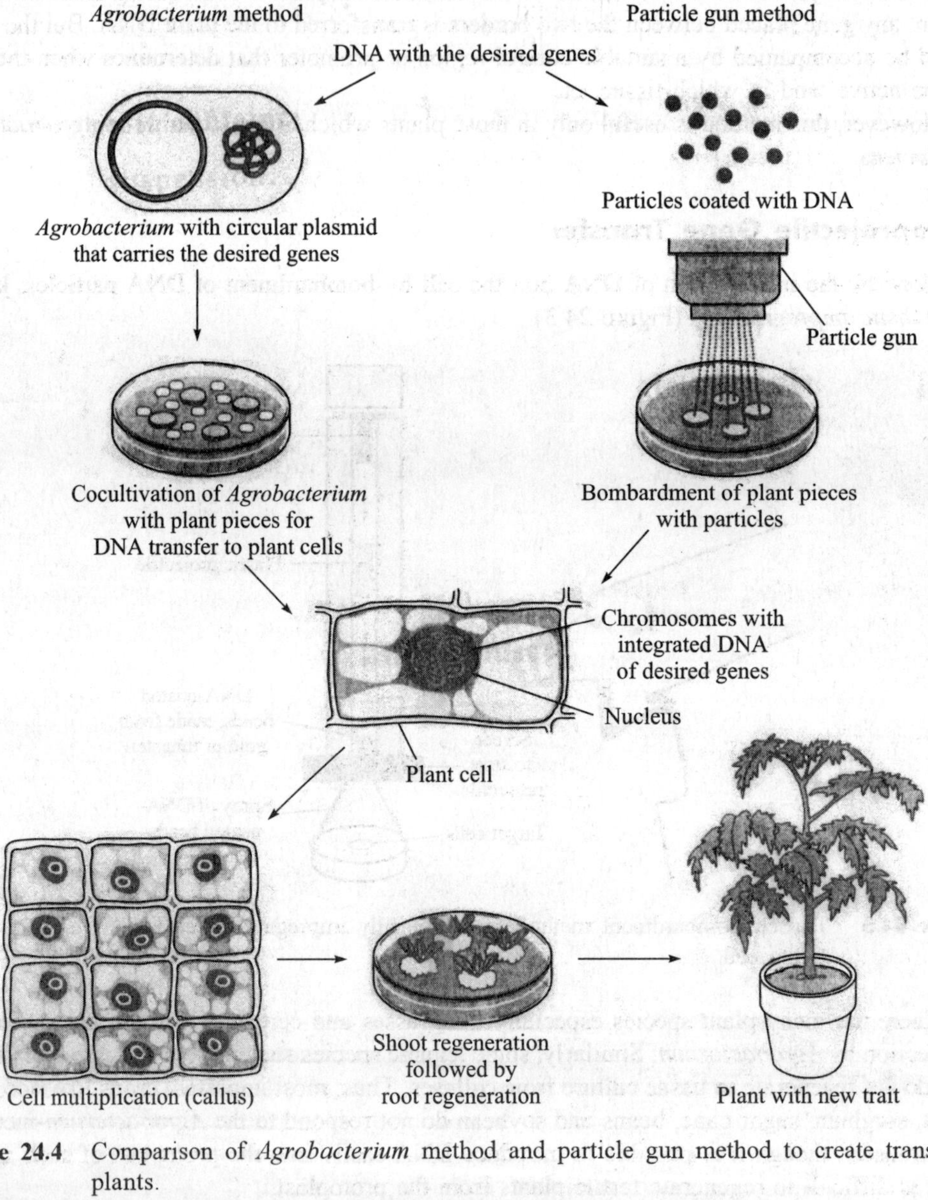

Figure 24.4 Comparison of *Agrobacterium* method and particle gun method to create transgenic plants.

Electroporation

Electroporation is totally different way of transferring foreign DNA into cells. It is applied to the protoplasts (cells whose cell walls have been removed by enzymes) but has been found successful for introducing DNA into the cells of apical meristen that are partially digested with enzymes that degrade cell walls. Electroportion involves a sudden electrical discharge in a very small container that holds plant protoplasts and DNA molecules in a liquid suspension. The discharge opens the plant plasma membrane simultaneously in many different places, allowing the DNA molecules to go into cell. The next step is to culture the protoplasts and try to regenerate plants from them. There is some difficulty in this system as many plant species do not readily regenerate from a single protoplasts.

But there are two bottlenecks for improving the crop plants by foreign gene transfer.

(a) Useful genes should be first precisely identified.
(b) Regeneration of plants from the cells into which the new gene(s) has/have been transferred.

The major difference between medical biotechnology and agriculture technology is that former aims to produce high-value products that will be used in small amounts (pharmaceuticals, vaccines, kits to diagnose diseases and so on) whereas later seeks to produce a large amount of low-value products (seeds of plants transformed with one or more genes from different species). Such plants are called transgenic plants.

ECONOMIC USES OF GENETIC ENGINEERING

As given above, the Biotechnology Companies are generating the transgenic crop plant varieties for various purposes such as insect-resistant and herbicide-tolerant cotton, slow-ripening tomatoes, high-starch potatoes, nutritious (protein, lysine, methionine rich) crop plants (maize, soya bean) and low volume, high pharmaceuticals. In the genetically-engineered plants, the new genes biosynthesize new enzymes which help in the production of new products of economic utility. But by genetic engineering technique, production of specific enzyme could be suppressed by using 'Antisense' genes.

Antisensing of Gene (Technology)

The sequence of bases along only one strand of the DNA double helix directs the production of proteins. The strand is called *sense* strand. *Antisense* RNA is an RNA sequence transcribed from the wrong (*antisense*) strand of DNA. When *antisense* and *sense* mRNA molecules meet they combine and prevent the production of protein encoded by the normal mRNA.

The process which occurs naturally in the control of bacterial genes, can be used by the biologists to switch off selected genes. This is done by inserting an appropriate piece of inverted DNA fragment into *sense* strand of the cell's DNA. *Antisense* mRNA molecules are then formed from the inverted DNA section when DNA is transcribed. This technique has been used to delay the ripening of tomato in USA.

Softening in 'tomato' occurs by cell wall degradation through enzyme—pectin methyl esterase and polygalacturonase. Biologists from Calgene Davis, California checked the enzyme polygalacturonase formation using antisensing gene technology as a result tomatoes could be left on vines for longer time for marketing than normal ripening time.

Production of Nutritious Crops as Food and Feed

We require eight amino acids that must be met through the diet. These are isoleucine, leucine, lysine, methionine, phenylalanine, threonine, tryptophan and valine. If even one is missing or if not enough is supplied of it, human body can not make protein it needs for proper functioning. The methionine-rich beans or lysine-rich corn could provide proteins with complete amino acid profile for the poorest and most poorly fed people who live on a single staple food grain such as maize, rice or wheat. But it is always beneficial to take diversified diet that is adequately nutritive even if plant-based. So far alien genes have been transferred to 60 crops to make them market-friendly and 20 are under cultivation in USA. Some of these are tomato with fish gene; potato with chicken gene; banana gene to cure cholera; rice of pea gene. In addition, antifreeze flaunder gene from fish to produce frost-resistant tomato; human gene in pig to produce leaner animals; and genes from bovines have been transferred in chickens to hasten their growth (45 days for slaughter). Earlier, it was taking 80 days to reach the table. More interestingly, genes of tropical fishes are being transferred to arctic fishes making them able to move to the warmer water to increase the fish catch for human consumption.

Animals grow faster on protein-rich diet, therefore, to capture this market, some seed companies in USA (Larry-Beach and Pioneer) are transferring genes from sunflower and Brazil nut to encode methionine-rich protein in soybean to feed animals the soybean meal.

The protein-rich sulphur-containing amino acids are being expressed in the leaves of alfalfa and clover to increase the sulphur amino acid level of feed consumed by sheep that produces wool. Wool is very rich in sulphur-based amino acids (wool is a protein). By eating a diet richer in sulphur-based amino acids, the same number of sheep will produce more wool.

Molecular Farming

Plants are like factories which need water, sunlight, minerals and right combination of genes which work like 'labour' in the biological machine (plant) to produce exactly what industries want. The incorporation of right genes, plant could produce the modified starches, proteins, valuable industrial oils, plastics, pharmaceutical drugs or enzymes for use in the food industries.

Production of mammalian neuropeptide

One such example of molecular farming is the production of mammalian neuropeptide 'leu-enkephaline' in the seed of canola (*Brassica napus*) by plant biotechnology in Belgium. 'Leu-enkephaline' is a short peptide that displays opiate activities and regulates brain function. The biotechnologists modified the gene of seed protein in such a way that mutant protein contained five amino acids of 'leu-enkephaline' in the right sequence and the 'leu-enkephaline' peptide could be conveniently cleaved (cut apart) from the purified seed protein. This modified gene was introduced in canola and seeds of transferred plants yielded about one milligram of 'leu-enkephaline' per 50 grams of seeds.

Production of industrial oils

The edible oils and many non-edible oils all have rather similar fatty acids with 18 carbon atoms (C18) and one, two or three unsaturated fatty acid bonds. The oils found in different crops are largely interchangeable, the exceptions are olive oil and canola oil which are valued for low levels of saturated fatty acids and cocoa butter, which is valued for fatty acid that confers physical properties desirable for confectionary and cosmetic industries.

A survey of 10,000 plant species shows that 95 per cent of all plant oils are made from six fatty acids, though a thousand different fatty acid structures have been identified mostly in undomesticated oil plants.

The approach is not to cultivate the undomesticated plants that produce these fatty acids but to transfer the genes encoding these fatty acids to the oil seed crops that are already under cultivation. Canola (rapeseed), soy bean, flax, cotton, sunflower and safflower produce oil seeds, and these have been transformed with foreign genes. The task ahead is to isolate the necessary genes and transfer in the cultivated crop plants. For example, castor seed contains four plant defense chemicals: a toxic lectin, a potent allergen, a toxic alkaloid and an oil that contains 90 per cent ricinoleic acid well known for its cathartic effect. The industrial use of castor oil is the synthesis of nylon and manufacture of lubricants, hydraulic fluids, plastics, cosmetics and other materials. The research findings indicated that a single enzyme in castor seed converts a normal fatty acid 'oleic acid' into ricinoleic acid. Because oleic acid is synthesized by other oil seed crop plants, the introduction of this enzyme into the seeds of a cultivated oil seed crop would result in the synthesis of ricinoleic acid for industrial uses. Short chain fatty acids such as lauric acid are used very widely to make detergents and are now obtained primarily from tropical oils such as coconut oil.

Production of enzymes for industries

There are a number of enzymes used in the food industries and their commercial production is quite profitable. About 20 enzymes are already being marketed. Many of them are active at 55°C and 90°C. They are produced by microorganisms which have been selected over in the years. Such strain of microorganisms do not occur in nature. The genes for these enzymes are leaf-specific, seed-specific, and can be introduced into the plants that will cause the enzymes to accumulate in these organs. These enzymes are purified and separated from other plant proteins and substances before they are used in food industries. Amylase, and glucoamylase are produced by fungus, *Aspergillus* which completely hydrolyzes starch to produce glucose. Pectinase is produced by a fungi; breakdown complex carbohydrates of cell wall and used to 'Clear' fruit extract so that they can be filtered. Complex carbohydrates clog the filter.

- Glucose is produced by fungi; breaking down the complex carbohydrates of cell walls of the cereals and used to make beer filterable.
- Glucose isomers are produced by bacteria or fungi, which convert glucose to fructose, the main ingradient of corn syrup. Corn syrup is mostly used as sweetner.
- Lipaxynase is produced by plants (legume seeds) and used to whiten wheat flour.
- Rennet extract is produced in the abdomen of calf and cow; contain several enzymes of which chymosin is the most important which coagulates milk proteins so that whey and curds can be separated.
- Lactase is produced by yeast and fungi; converts lactose (milk sugar) into galactose and glucose; used to prepare milk products such as lactose for people who have lactase deficiency and cannot tolerate lactose.

- Protease is produced by bacteria, fungi and plant proteases, find wide application in the cheese industry. It is suitable of rennet and ripening of cheese and also as meat tenderizes.
- Phytase is produced in the plants and microorganisms; breaksdown phytase in seeds to release phosphate; can be used to treat animal feed so that more of the phosphate can be absorbed by animals.

The fermentation technology for production of above said enzymes of industrial uses cheaper and need very little enzyme purification because enzymes are secreted into the fungal culture medium where they accumulate at very high level. The use of plant materials that contain an enzyme for modifying animal feed is quite useful. Because phytase produced by fermentation is added to the animal feed to help in digestion of the phytic acid. It would be simple to produce phytase in tobacco leaves or potato tuber by transferring the alien genes and mix these with phytic acid containing feeds. Cellulase is an other example of an enzyme that could be produced in transgenic plants and can be used as feed additive which helps to digest cellulose, the plant materials that contain cellulose should be finally dispersed as powder or in form of a slurry to release cellulase from the tissue.

Production of special starches

The starch is a source of energy but as a raw material, it is used as food stuffs (sweetner and thickners) or industrial products (adhesives, products for paper sizing) can be made. Fermentation process can convert starch into other chemicals and 10 per cent of total starch produced is utilized in this way.

Starch is a mixture of two large complex carbohydrates—amylase and amylopectin. Starch is insoluble in water but when a suspension of starch is heated, the starch grains swell and a gelatinous mass is formed. The clarity, viscosity and taste of paste differ from different sources of starches such as potato, corn, wheat and tapioca. For thickening the sauce, potato starch is used, not the wheat base.

The starch synthesis is controlled by three enzymes and their relative level determines the relative amount of amylose and amylopectin. If any enzyme is lacking, it will stop synthesis of amylose.

In 1991, scientists from Calgene transferred a bacterial gene to potato that encodes an enzyme which synthesizes small cyclic starch molecule rings that have seven or eight glucose molecules. These molecules are called cyclodextrins. Dextrins are used in the pharmaceutical industries for drug delivery and to remove unwanted compounds such as caffeine from foods. Cytodextrins are quite expensive but production of cytodextrin from genetically-engineered potato may be less expensive to the industries and also consumers. Potatoes are rich in starch. Starch, a polymer of glucose, is made by transferring glucose molecules from a glucose donor called ADP-glucose to a glucose acceptor; the growing starch molecules. The enzymes involved this transfer is not abundant in potato. The gene for this enzyme was cloned and isolated from a bacterium and transferred to potato that synthesise ADP-glucose more rapidly and produced more starch at Michigan state university.

Production of plastics by plants

Polyhydroxylbutyrate (PHB) is a polyster that accumulates as a natural storage material in many species of bacteria, in the same way as plants accumulate oil, starch or sucrose as storage compounds. The polyhydroxylbutyrate produced by bacteria in large fermentors is presently used as a renewable source for manufacturing biodegradable plastics. However, production of polyhydroxylbutyrate by fermentation is much more expensive than their production by plants. The synthesis of PHB requires three enzymes and the first enzyme (ketothiolase) in the pathway is already present in all the plants (Figure 24.5).

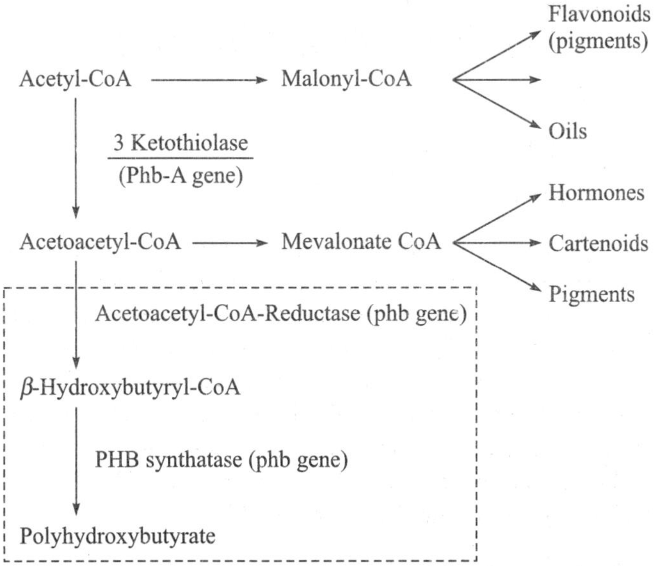

Figure 24.5 Biosynthesis of plastic by plants.

Chris Somerville and his colleagues at Michigan State University introduced the genes encoding the other two enzymes (reductase and PHB synthase) into *Arabidopsis thaliana* plants and found that the transgenic plants synthesized PHB (Polyhydroxyl butyrate) and that accumulates plastic as small granules (0.2–0.5 p.m.) in the cytoplasm, nucleus and vacuoles of the plant cells. But plant growth remains stunted possibly as a result of draining the important intermediate product 'acetoacetyl-CoA' from the other pathways for which it is needed.

Improvement of post harvest

Senescence is a natural phenomenon in the plants that occurs in the end of life cycle of the organ. For example, leaf senescence in the decideous trees occurs in autumn and leaves turn yellow. In the fruits, senescence is called ripening and this is associated with desirable changes in the fruits such as conversion of starch to sugar, loss of chlorophy11, synthesis of red yellow pigments, synthesis of aromatic compounds and finally softening of the fruit mass. All these processes are controlled by hormones. In the leaves, senescence does not occur as long as levels of auxin and cytokinins are high. But soon their levels drops, ripening of fruits is accelerated by the hormone 'ethylene'. Banana industry has taken advantage of the information of hormonal control of fruit ripening. Bananas are picked up green, transported half-way around the world and then gassed with ethylene to start the ripening process.

Tomatoes are also picked up green, transported and gassed with ethylene. In fact regulation of fruit ripening and slow senescence in flowers and vegetable are challenge to the biotechnology industries because handling and storage results a lot of losses. Genetic engineering to alter the ripening of tomatoes has been undertaken by several laboratories and approach involves suppression of ethylene synthesis by tomato fruits itself.

Biosynthesis of ethylene involves following two steps:

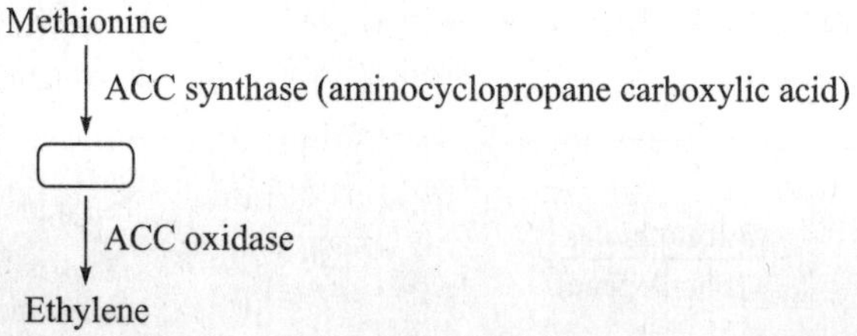

The ethylene synthesis can be suppressed with 'antisense' technology using the genes that encode either one of these enzymes in reverse orientation and hooked to the control region that directs gene expression in the entire plants. Scientists of Monsanto seed company used different approaches to get the same results. They introduced a bacterial gene into tomatoes that encodes an enzyme that degrades ACC (aminocyclopropane carboxylic acid) as soon as it is formed. The net result is again to suppress the ethylene synthesis and delay in the fruit ripening. In this way, it will be possible to leave the fruits on the vine (plants) for longer period of time. Those fruits seen turning red could be picked up and sent to the market. The fruits will also not need refrigeration during shipping to the other parts of world.

Production of herbicide tolerant plants

Just as transgenic plants may reduce the use of chemicals, they may also increase the use of herbicides in agriculture. The companies that market herbicides have started producing genetically-engineered crop plants that tolerate herbicides. For example, Monsanto's glyphosphate (Round up) Danpont's sulfonylureas and Hoechst's glufosinate are probably good herbicides. Mansanto has crop plants resistant to 'Round Up'; Dunpont has plant registrant to Glean and Hoechst has plant resistant to Basta. However, plant resistant to herbicides are also non-biodegradable and one indeed quite toxic, so environmentally unfriendly.

Insect-resistant plants

The insects damage not only the yield but also the quality of products. There are several species of bacteria identified so far which produce specific proteins that kill insect larvae which ingest these bacteria with their food. The most widely studied of these bacteria is *Bacillus thuringiensis* (Bt) that is found all over the world and there are a large number of subspecies and strains within these. When these bacteria form spores, they also form a large crystal-like structure in the bacterial cytoplasm, that is made out of protein. One of the proteins in this crystal structure is called the Bt protoxin. When insect larvae eat the bacterial cells, the spores and crystal-like structure containing protoxin which when released in the larvae gut, when digestive enzymes cleave the protoxin, produces an active toxin which kills the larvae. The scientist working in Mansanto seed company in St. Louis, Missouri introduced a gene from a strain of *Bacillus thuringiensis* into tomato against tomato fruits worm and tomato pinkworm as well as in cotton against cotton pink boll worm. Bt. cotton has been released in many countries including China. India has released 'Bt. cotton', only in the year, 2002 for commerical cultivation.

The scientists from CIBA Research in Triangle Park, N.C. have succeeded in transferring the corn plant with a Bt. toxin gene that makes plant resistant to European corn borer. Mansanto and Pioneer seed companies are also attempting to produce corn that is resistant to corn root worm.

Production of crop plants resistant to fungi and bacterial diseases

The approach to evolve crop plants resistant to fungal and bacterial diseases involves introduction of genes in the plants that encode enzymes that degrade cell walls of fungi and bacteria. These cell walls are completely different from the plant cell walls and enzymes that degrade the fungal and bacterial cell walls and do not damage the plant cell walls.

The fungal cell walls consist of two polymers—glucan and chitin that can be broken down by glucanases and chitinases respectively. A particular chitinase may be effective against one species of fungus but not against another. Therefore, effective use of chitinase gene for genetic engineering again may depend on the exact matching of the genes with the target pathogen just as with Bt. toxin or the digestive enzyme.

The bacterial cell walls are made up of unusual polymers called 'peptoglycans' which possess both proteins and carbohydrate components, which can be digested by enzymes lysozyme. Lysozyme is found in chicken eggs and in cow stomach when they digest the cellulose and cell walls of plant tissues. The cow lysozyme can degrade cell walls of plant pathogens and bacteria, therefore, can be used for biotechnological approaches to control bacterial diseases. The scientists at SIBIA Biotechnology Company in San Diego, California, have introduced genes that encode cow lysozyme into a yeast *Pichia pastoris*. When these transgenic yeast cells are grown in large fermentors, they produce substantial quantities of lyzyme which are secreted into culture medium. By treating the seeds with very dilute solutions of (25–100 part per million) of lysozyme, bacteria on the seed could be killed.

Production of biopesticides

1. There are some fungi which attack the insect-pest, caterpillar and larvae, aphids and grasshoppers, etc. Laboratory production of mycoinsecticides is in progress in USA and other countries. But they are very slow in their effectiveness.
2. There are only a few species of bacteria which have microbial insecticidal properties. But potential of *Bacillus thuringiensis*, the only one now is used commercially. *Bacillus thuringiensis* produces toxic protein that acts as poison in the gut when larvae feed upon the leaves. Several hundred natural strains (isolates) of *Bacillus thuringiensis* have considerable specificity toward various groups of insects such as the lepidopetra (butterflies and moths), coleoptera (beetles) or diptera (mosquitoes). At least, six strains are being commercially developed. Commercial production of *Bacillus thuringiensis* is done in large fermentors and after formulation, the bacteria can be stored for one or two years. In addition, the gene that encodes toxic proteins can be transferred to other bacteria such as *Pseudomonas* that are more easily grown in mass cultures.

Microherbicides

Just as fungi can cause crop diseases, they can also cause weed diseases. But problem appears to be mostly technical not biological, how to grow fungi economically in laboratory; how to store fungus and spores and prolong their life and have to apply to the needs because fungi grow in a wet environment. In Canada, one fungal mycoherbicide 'BioMal' active against smooth-leaved mallow has been produced by a Canadian Biotech company Philon-Bios. in Saskatoon.

Insect viruses

Seven major groups of viruses infect insects and could be developed as biological control agents. The known insect viruses are usually highly species-specific and may infect only one insect or a few species of insects. These are harmless to plants and animals, mammals: They can reduce the use of chemicals in control of pests (Chrispal and Sadava, 1994).

CONTROVERSY ON GENETICALLY ENGINEERED FOOD GRAINS

In the recent years a lot of controversy have arisen throughout the world as well as in India too, whether genetically modified plants will be safe to eat, because world market getting flooded with genetically-modified vegetables and fruits. The consumers eat about 1,00,000 genes daily and this DNA is efficiently brokendown in the stomach. The genetically engineered plants will have additional gene/gene complex to which they are not accustomed to eat, may eat by eating genetically engineered food products—fruits, vegetable, milk and meat, etc.

SAFETY OF BACTERIAL ENGINEERED PLANT AS FOOD (GENETICALLY MODIFIED FOOD)

The genetically engineered plants contain bacterial gene (s) that make the plant cells resistant to antibiotics. The antibiotic resistance of the plant cells is an important step in the plant transformation process. It is feared that bacterial gene may pose a health hazard. But new transformation technique by Du Pon and also by US Department of Agriculture to remove the bacterial gene after transformation has been achieved. The bacterial antibiotic resistance gene is cut out of the host plant DNA using restriction nuclease enzyme produced within the cell itself.

In addition, the new protein that has been encoded by the gene may not be present in the plant part which we eat because presence or absence of a protein in a particular organ depends on the gene control region used to make transgenic plant. For example, if a bacterial toxin gene (Bt) is expressed with a root-specific control region, then protein will be present only in the roots. However, such gene will provide insect resistance to roots only. In this case, root should not be eaten. Similarly, where leaves are eaten or seeds are eaten, leaf and seed region-specific bacterial genes should not be transferred. For example, if gene, for 2-amylase inhibitor found in the beans is transferred to other legumes to inhibit development of bruchid beetles, this inhibitor gene inhibits human 2-amylase as well as insect 2-amylase. But this protein is eaten in large quantities all over the world who eat the legume pulses. Where food is cooked and then eaten, it is safe to eat.

In some cases, food technologists have identified the specific proteins that are allergenic, therefore, when a gene is transferred from a plant known to have allergenic proteins, there is a possibility that the gene may encode such a protein. In that case, the genetically engineered food will have to be tested and the company that wants to market such new food will have to test and ascertain that the food product will not be allergenic. In fact, the company has to label the genetically engineered food grains, vegetables and fruits for the information of the consumers. It is up to the consumers to buy the transgenic tomato or tomato soup or transgenic chicken or chicken soup, etc.

The countries or the group of people or individuals are opposing not only to the pharmaceutical products such as interferon or insulin, but also to the genetically engineered plants and animals which are directly used for food. In 1980, many research centres and laboratories were destroyed by the people in USA, UK and Netherland, etc. In 1990 there was a consumer boycotts.

TECHNIQUES TO DETECT FOREIGN PROTEINS IN GENETICALLY MODIFIED (GM) PLANTS

There are two methods to detect foreign proteins in the GM plants:

1. ELISA (enzyme-linked immunosorbent assay) technique
2. PCR (polymerase chain reaction) technique

ELISA Technique

Detecting the foreign protein in a GM plant product is mainly done by employing a technique called ELISA (enzyme-linked immunosorbent assay). For this, the foreign protein produced in various parts of a GM plant is first extracted and bound to a plastic plate. An antibody molecule which specifically binds to the extracted foreign protein is then added. The antibody molecule is actually labelled with an enzyme which catalyzes a reaction that gives a coloured product. The intensity of the colour produced is proportional to the amount of foreign protein detected. However, highly processed foods derived from GM plants usually contain the foreign protein in a degraded state that cannot be recognized by the antibody. Therefore, ELISA is useful in detecting the products of a GM plant when samples are fresh, raw food taken at early stages of processing. Another drawback of the foreign protein detection in GM plants is the fact that some proteins are not expressed at the same level in all parts of the plant. For example, in transgenic maize containing the foreign Bt. toxin gene, the expression of the Bt. protein in the kernels, the part that is used in foods, is about 1,000 times lower than in the leaves. So detecting the Bt. protein in kernels by ELISA is not feasible.

PCR Technique

The most promising target for detecting a GM plant is to identify the inserted foreign gene. For this a technique called PCR is put to use. By this highly sensitive DNA amplification technique, a tiny stretch of DNA could be selectively multiplied. A major requirement for carrying out PCR is the availability of short pieces of DNA flanking or present on either side of the region to be amplified. Known as 'primers', these segments of DNA bind to either side of the GM DNA extracted from a food sample. To enable the binding of primers with the extracted foreign DNA, the latter is first subjected to heat treatment for separating the two strands of DNA. Now in presence of an enzyme called Tag polymerase, the two single strands of foreign DNA act as templates for the synthesis of two new strands, thus giving rise to two copies of the parent foreign DNA. PCR involves repeated cycles of DNA replication. In about 25 rounds of PCR, more than a million copies of target DNA are produced that can be identified by running the amplified DNA on agarose gel across which electric charge is applied. An automated PCR machine where temperature of the sample mixture can be controlled is nowadays used for amplifying DNA. Known as the DNA thermal cycler, it is indeed a DNA photocopier (Chawla, 2000).

To improve the specificity of PCR further, nested PCR is done. This technique involves an initial amplification of a longer target fragment by PCR, followed by a second PCR where primers are used that bind to a section within the previously amplified DNA. A shorter segment of DNA is thus amplified which is highly specific to the target DNA sequence. So nested PCR basically involves an additional PCR step for amplifying an internal fragment of the foreign DNA for its sensitive analysis. For identifying a GM plant, DNA sequences specific to the inserted foreign gene are amplified.

The DNA detection method is, however, not applicable if in the course of food processing, plant DNA is completely separated or destroyed such as in starch products, sugars and refined oils. Some processed foods from which inserted foreign DNA has been shown to be amplified by PCR are tomato soup, tomato purèe, tinned fish, chips, animal feeds and noodles among others. DNA can even be extracted from raw pressed oils but not heavily refined oils. DNA degradation in highly processed foods is mainly due to the prolonged heat treatment, low pH of the food, and prolonged storage of fresh foods.

REASONS FOR CONSUMER'S BOYCOTT

Safety Consideration

The genetically engineered food grains, vegetables and fruits may not be safe to eat. There is all possibility that genetically engineered foods contain novel allergense.

Anticorporate Argument

It is true that there are some multinational seeds companies/corporations who want to make money and they often twist the truth to serve their own profit-making purposes. The consumers should not believe in their propaganda that they want to promote human welfare.

High Input Agriculture

The high hope that biotechnology will raise the yield per se or make the agriculture production more sustainable is a mirage. The transgenic plants may increase the cost of production by increasing the use of chemicals (herbicide-tolerant plants) or decrease their use (pest-tolerant plants) but any way it is going to create more problems by way of unforeseen ecological consequences than upgrading the yield by raising transgenic crops.

Ethical Considerations

The genome of an individual is highly interactive coadapted gene complex and have passed through the million of years of natural selection processes. Such a well-integrated, coadapted gene complex constitute an evolutionary unit. Any disharmonious gene (foreign gene) or gene complex, which attempt to become incorporated in such a well-integrated gene complex by human intervension will either be eliminated by selection or if it finds a place in the genome of host, it may show adverse effect if it has pleiotropic gene effect.

It is true that plant breeders too transfer the gene(s) governing resistance, etc. to diseases and pests by conventional breeding techniques from wild (species and genera) in the cultivated varieties. But these are from related species and genera which are much safer and ethical than the transferring genes from bacteria or virus vectors which are parasitic in nature.

Philosophical Considerations

There is general opinion that population is growing day by day. Today population is 6 billion, within a decade that may grow to 10 billion. And we have to feed them from earth, therefore, earth's atmosphere and productivity should not be disturbed by our exploitative mode of action.

BIOTECHNOLOGY IS A MYTH

The expectation from biotechnology is more myth than reality. It is an old wine put in a new bottle by making sheer propaganda by the multinational seed companies and being sold in the developing countries taking advantage of liberalization the New Economic Reform Policy enforced by advance west over east to have hold over the agriculture of the developing countries. Bt cotton, Bt maize, Banana with vaccine, rice with vitamin A, genes, etc. are new marketing techniques coined by the multinational seed companies to create seed market in the developing countries because people here specially these in power could be easily bribed and brain-washed by white skin businessmen. I fear that in long run these multinational seed companies may cause irrepairable damage to our indigenous germplasm materials and expertise turning our research institutions like I.C.A.R. and State Agriculture Universities as cowsheds. I.C.A.R.'s best institutions like I.A.R.I., New Delhi, NDRI, Karnal and I.V.R.I., Izzatnagar have already turned as cowsheds by Indian politicians by appointing their men as Directors, Deputy Director Generals and Director Generals, who enjoyed the system as Princes of Medieval Periods of India. The system is already on rot.

It is well known that biotechnology is a baby of the developed world—Western Europe, Japan and North America. It is also but natural that further advancement will be attained in these countries. Biotechnology researches have been so far handled by private multinational seed companies and government institutions are relatively minor players in biotechnology. The herbicide-tolerant plants, high starch potatoes, oil and plastic-producing plants and enzymes produced for the food industries are their main 'mythical' goals. The beneficiaries of these new plants are the industrialists of the developed world and not the farmers of the developing countries except 5–10 per cent who are rich and able to afford high cost of seeds. The benefit of green revolution was also squarely enjoyed by the rich farmers of the developing countries, same way, the benefit if any of genetic engineering, will be enjoyed by the rich only.

In India, there has been a lot of opposition against the introduction of genetically-modified seeds in the agriculture. But inspite of the opposition 'Bt cotton' was introduced and released for cultivation. In case of cotton we do use cotton fibre for cloth making but seed is also fed to the milking animals. If Bt gene affects the cotton seed protein, then it would be dangerous because it will certainly adversely affect the quality of milk protein and other milk products such as cheese, etc. However, if Bt gene does not affect the seed protein of cotton seed, then there is no harm.

Introduction of the high-yielding varieties of crops in mid-sixties did increase the yield but they caused a great genetic erosion of existing germplasm materials of crop plants, wild types which were useful sources of diseases and pest resistance; animals due to canversion of vast grasslands and village barren lands including ponds into rice and wheat fields. Mechanization of agriculture replaced drought force and created vast unemployment in rural areas. In addition, injudicious uses of chemical fertilizers, herbicides and pesticides have been found biggest sources of environmental pollution. Thus, introduction of transgenic crop varieties will cause further damage to the existing gene pool, environment, health hazard and crop plant sterility and will convert cultivated crop plants into grasses.

However, biotechnology is not an alternative of conventional plant breeding, of course through biotechnology we can produce new genetic variants which could be utilized in advancing yield, adaptability, quality and resistance to the diseases and pests by traditional breeding techniques provided 'alien' gene (s) under transfer are safe, beneficial and able to recombine to the genes of the recipient plants.

We can incorporate resistance to major diseases and pests of crop plants from wild germplasm sources as well as various grain quality characteristics, through the traditional plant breeding techniques.

REVIEW QUESTIONS

1. What do mean by plant genetic engineering? How is it different from mechanical engineering?
2. What are the most prevalent methods of genetic engineering? Describe briefly one of them.
3. How alien genes could be transferred from one species to distantly related species/genera? Name some crop plants where alien genes have been transferred.
4. What do you mean by transgenic crop plants/animals? Give advantages and disadvantages of transgenic crop plants.
5. Write economic uses of genetic engineering and its limitations.
6. What is antisensing of gene? How is it useful in agriculture? Give a brief description.
7. Why there is a controversy about use of genetically modified (GM) food. Give reasons of boycotts by the consumers?

8. How can you detect foreign proteins in the genetically modified food? Describe briefly.
9. What is PCR technique? For what purpose it is used?
10. Do you agree with this that biotechnology is a more mythical science than reality? If so why? If not, how?

REFERENCES

Chawla, P. (2000): Food for thought, *Science Reporter*, **2,** 25–26.

Chrispeels, M.J. and D.E. Sadava (1994): *Plants, Genes and Agriculture*, Joes and Bartlett, Bostona and London.

Cocking, E.C. (1983): Application of protoplast technology to agriculture: *Experientia*, **46,** 123.

Corbin, D.R. and H.J. Klee (1991): *Agrobacterium tumefaciens* mediated plant transformation, *Current Opinion in Biotechnology*, **2,** 147–152.

Dasilva, E.J. (Ed.) (1992): *Biotechnology: Economic and Social Aspects*, Cambridge University Press, Cambridge.

Gasser, C.S. and R.T. Fraley (1992): Transgenic crops, *Scientific American*, **266** (6), 62–67.

Hiatl, A. (Ed.) (1992): *Transgenic Plants*, Dekker, New York.

Ram, M. (2005): *Unfolding Reality of Life*—The *DNA Word*, Dominant Publications, New Delhi.

CHAPTER

25

Concept of Gene Cloning and Procedure for Reconstruction of Recombinant DNA

There was no proper knowledge of molecular nature of gene until 1940 when Avery, McCleod and McCarthy (1944) and Harsey and Chase (1952) came out with their experimental results that DNA is the genetic material. Before that, it was thought that genes were made of proteins. From 1952 to 1966, structure and function of DNA was worked out; the genetic code was discovered and decoded and processes of transcription and translation of the genetic information were deciphered. In the years 1971–1973 the new era of 'recombinant DNA technology' (genetic engineering and gene cloning technology) dawned which pushed the science of genetics ahead of the science of atomic theory in physics.

Main skill needed to carry out gene cloning experiments are:

1. Clear concept of gene cloning.
2. Types of gene cloning.
3. Tools (vehicle) for gene cloning.
4. Methods of extraction of DNA and its quantification.
5. Procedure for preparation of pure DNA samples (Follow T.A. Brown, *Gene Cloning*, 1990).
6. Cutting down DNA molecules.
7. Introduction of DNA into host cell.
8. Joining of DNA molecules together.
9. Identification of cell containing DNA fragment.
10. Utilization of recombinant DNA.

WHAT IS GENE CLONING

The development of a population of identical cells containing identical recombinant DNA molecules is called *gene cloning*.

TYPES OF GENE CLONING

There are two types of gene cloning:

1. Genomic cloning
2. Complementary DNA cloning

Genomic Cloning

In genomic cloning, a chromosomal sequence of DAN is inserted into a vector 'plasmid' or a phage chromosome that can replicate automatically in a bacterial cell.

Complementary DNA Cloning

In complementary DNA cloning (cDNA clone), a sequence of DNA that is faithfully complementary copy of a RNA or that contains part of a RNA sequence as it exist in the cell is inserted in the plasmids through the use of restriction endonucleases.

There are some basic differences between genomic and complementary cloning of gene. A genomic clone whether from prokaryotes or eukaryotes contains sequences that flank the structural gene. In the eukaryotic genomic clones, sequences from both 'introns' and 'exons' are present while a cDNA (complementary) clone contains only sequences corresponding to the messenger RNA (mRNA) transcriptional unit. It will lack 'intron' sequences if it is formed from a mature mRNA.

These two types of clones give different information. For example, a genomic clone gives information about structure of chromosomal DNA and the cDNA clone tells the structure of mature mRNA (processed RNA) (Figure 25.1).

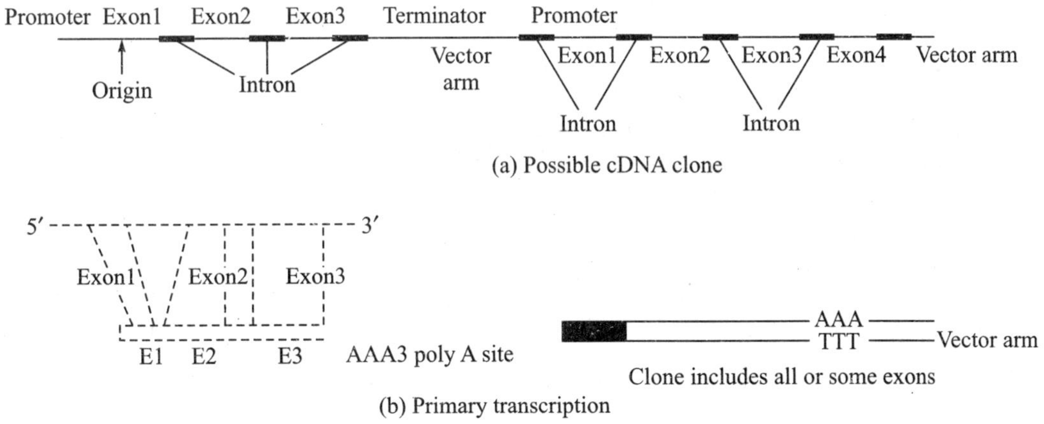

Figure 25.1 The differences between genomic clones and cDNA clones.

The main steps involved in the gene cloning are:

1. A fragment of DNA containing the particular gene to be cloned is inserted in the circular DNA molecule called 'vector' to produce a chimera or recombinant DNA molecule.

2. A vector is also called a 'vehicle', which transports the gene into a host cell which is usually a bacterium, though other types of living cells can also be used.
3. Within the host cell, the vector multiplies and produces numerous identical copies not only of its own but also of the foreign genes it carried.
4. When host cell divides, copies of the recombinant DNA molecules passed on to the progeny and further vector multiplication takes place.
5. After a large number of cell divisions, a colony of 'clone' of identical host cells is produced. Each cell is the clone which contains one or more copies of the recombinant DAN molecules. The gene carried by the recombinant DNA molecule is now called the clone.

TOOLS OF GENE CLONING

Vector as the Vehicle

Gene cloning requires some specialized tools and procedures. For example, gene cloning requires 'vehicle' through which the gene to be cloned need to be transported into the host cell. The cloning vehicle is a DNA molecule. It must be capable to enter the host cell as that inside the host cell, it can produce multiple copies of its own. There are two types of DNA occurring molecules which are being used as a vehicle. These vehicles are also called vector.

1. Plasmids
2. Bacteriophage DNA (virus chromosomal DNA)

The cloning vehicle should be small, ideally less than 10 kilobase (kb) in size because large molecules breakdown during purification and are difficult to manipulate.

PLASMIDS

These are small circular DNA molecules found in bacteria and other organism. The plasmids replicate independent of bacterial chromosome (DNA).

The basic feature of the plasmid DNA molecule is that it is circular that leads independent existence in the bacterial cell. Plasmids always carry one or more genes. Some time these genes are responsible to provide resistance to bacterium against toxic concentrations of antibiotics such as *ampicillin* due to presence of antibiotic resistance gene in the plasmids. The antibiotic resistance is often used as a 'selectable marker' to ensure that bacteria in a culture contains a particular plasmid (Figure 25.2).

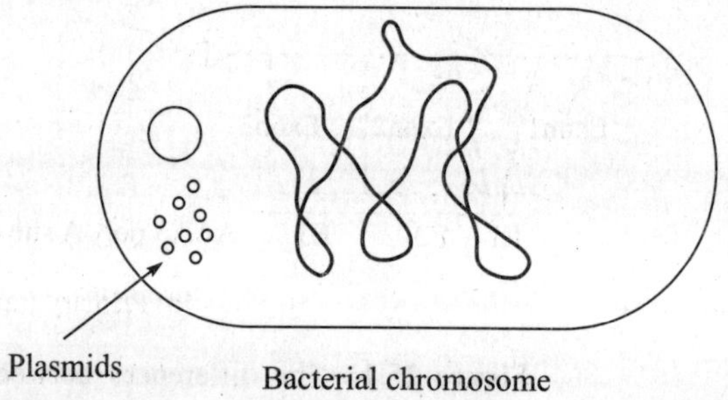

Figure 25.2 Plasmids are independent of bacterial genetic elements.

All plasmids contain at least one DNA sequence that can help in origin of replication, so that they can multiply within the cell independent of the bacterial chromosome. The smaller plasmids make use of host cells' own DNA replication enzyme to make copies of themselves, while the larger ones carry genes that code for special enzyme that are needed for plasmid replication.

Some plasmids get integrated with the bacterial chromosomes and they are maintained in this form through numerous cell divisions but some, exist as independent elements (Figure 25.3).

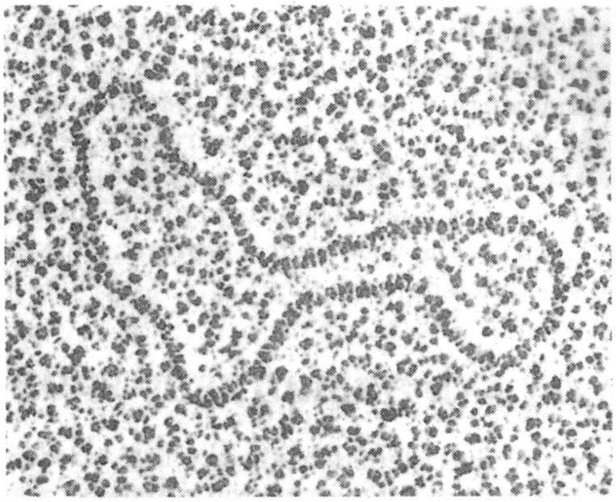

Figure 25.3 Plasmids in *E. coli*.

The size of plasmids varies from 1.0 kb (smallest) to over 250 kb for the largest plasmids. In Table 25.1, plasmid size found in different organisms is given.

TABLE 25.1 Plasmid size in different organisms

Plasmid	*Size nucleotide length (kb)*	*Molecular wt. (MDa)*	*Organism*
PBR_{345}	0.7	0.46	*E. coli*
PBR_{322}	4.362	2.9	*E. coli*
Col E_1	6.36	4.2	*E. coli*
RP_4	54	36	*Pseudomonas* and others
F	95	63	*E. coli*
Tol	117	78	*Pseudomonas putida*
PT_1Ach_5	213	142	*Agrobacterium tumefaciens*

Note: 1kb (1 kilobase) = 1000 nucleotides

Types of Plasmids

Plasmids are of two types: *conjugative* and *non-conjugative*. Conjugative plasmids promote sexual conjugation between the bacterial cells. It is a process that results into spread of conjugative plasmid from one cell to all other cells in a bacterial culture. Conjugation and plasmid transfer are controlled by a set of transfer or 'tra, gene', which are present in conjugative plasmids but absent in non-conjugative plasmids.

Non-integrative plasmids do not integrate with bacterial chromosome at the time of cell division.

Integrative plasmids (*episomes*) integrate with bacterial chromosome at the time of cell division (Figure 25.4).

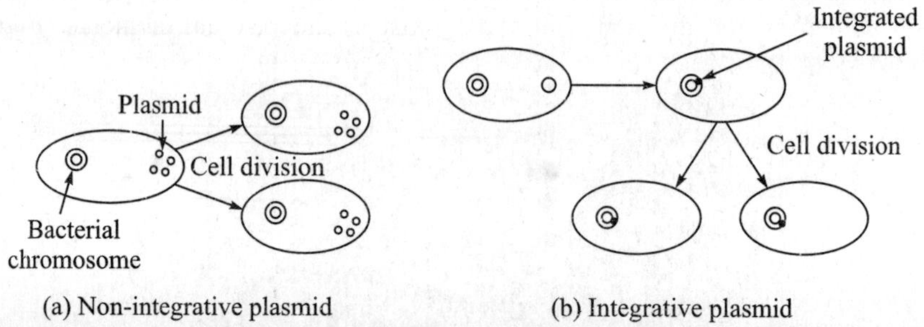

Figure 25.4 Replication of non-integrative and integrative plasmids (episomes).

Many different types of plasmids may be found in a single cell including more than one conjugative plasmid at any one time; some of the *E. coli* cells contain up to seven types of different plasmids which coexist due to compatible nature; those which are not compatible are eliminated from the cell. Such plasmids are called incompatible. Conjugative bacteria transfer the plasmids through pilus to other bacteria (Figure 25.5).

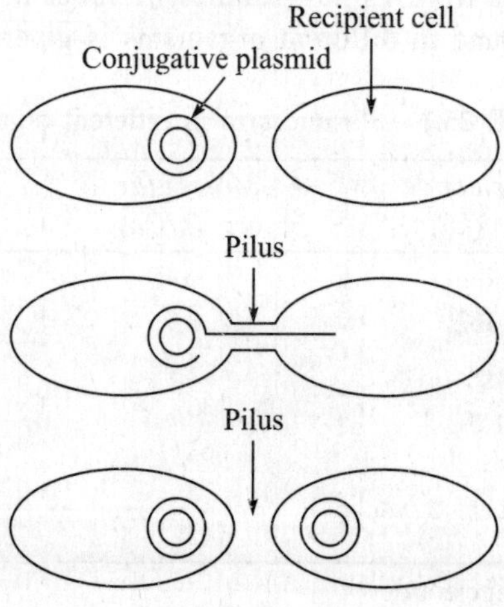

Figure 25.5 Transfer of plasmid through conjugation between bacterial cells.

Plasmids in Other Organisms than Bacteria

Among the eukaryotes, plasmid in the '2 μm circle' is found in many strains of *Saccharomyces cerevisiae.* The discovery of '2 μm' plasmid was very fortuitous as it has allowed the construction of vectors for cloning genes because *S. cerevisiae* is an industrial organism as a host. Search for plasmids in fungi, plants and animals has been found futile.

BACTERIOPHAGES

Bacteriophages are viruses which infect bacteria. During infection, the bacteriophages inject their DNA into the host cell where it undergoes replication.

The DNA molecules which serves as vehicle for gene cloning has to display several features. The most important is, it must be able to replicate within host cell, so that numerous copies of the recombinant DNA molecules can be produced and passed on to the daughter cell.

Phages are also viruses. They specifically infect bacteria like all viruses. Phages are simple in structure consisting of a DNA (occasionally RNA) molecules carrying a number of genes enclosed by a protective coat or capsid made up of protein molecules (Figure 25.6).

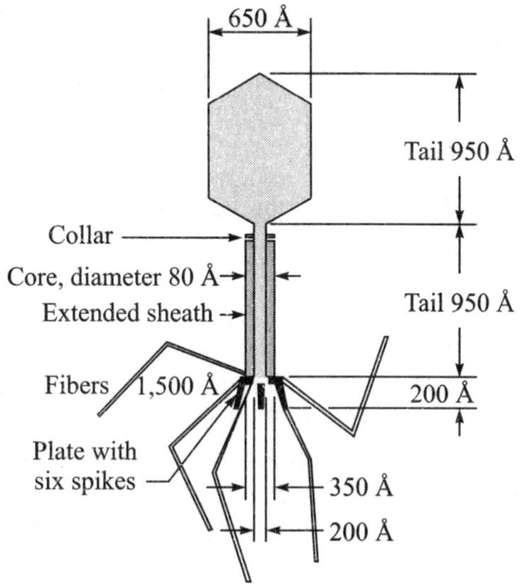

Figure 25.6 Structure of bacteriophage.

Infection

1. The phage particle attaches to the outside the bacterium and injects its DNA into the cell.
2. The phage DNA replicates by specific phage enzyme coded by gene located on the phage chromosome.
3. Other phage genes direct the synthesis of the *capsid* protein. New phage particles are produced and get released after lysis of the bacterium.

The entire infection cycle is completed within 20 minutes. This is also called lytic cycle because new phage particles are released after lysis of the bacterial cell.

Soon after replication of phage DNA, synthesis of capsid proteins takes place. This way phage DNA is never maintained in a stable form.

The bacteriophages are of two types—lytic and lysogenic the lytic bacteriophage do not integrate their DNA content with host. They multiply in the cell and get released after causing lysis of the host cell (Figure 25.7).

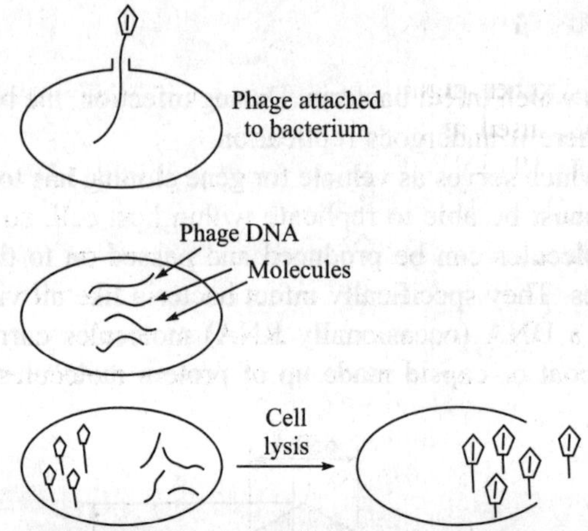

Figure 25.7 Lysis of the infected bacterial cells.

Lysogenic Phages

In contrast to lytic cycle, lysogenic infection is characterized by retention of the phage DNA molecule in the host bacterium for the cycle of cell division. The *lysogenic phage DNA is inserted into the bacterial genome as shown in Figure* 25.8 *in episomal manner*. The integrated form of the phage DNA (called the prophage) is *quiescent* and a bacterium (lysogen) which carry 'prophage' is usually physiologically indistinguishable from the normal (uninfected) bacteria. After some time, the prophage is released from the host genome and as soon as the phage reverts to the lytic cycle, it lyses the cell.

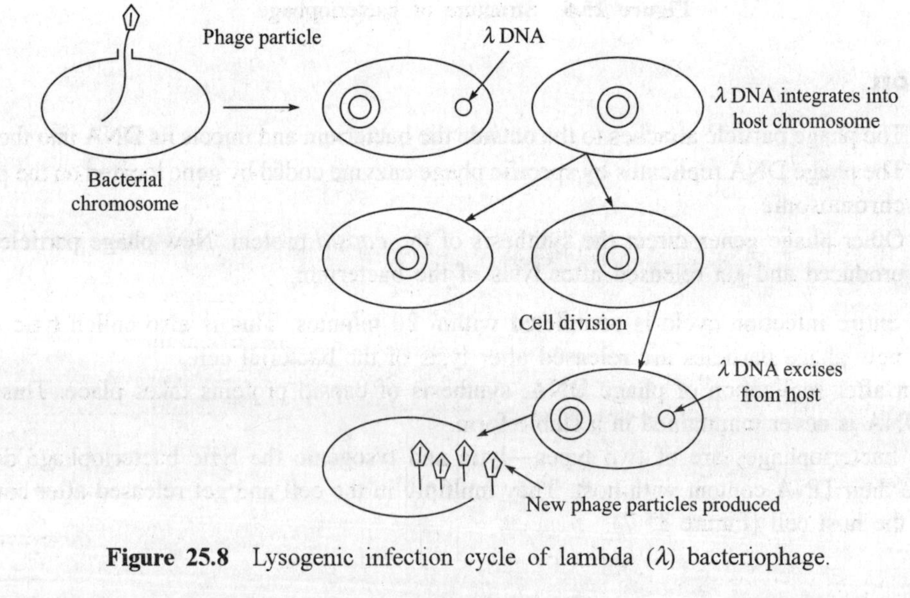

Figure 25.8 Lysogenic infection cycle of lambda (λ) bacteriophage.

VIRUSES AS CLONING VECTOR

Plasmids are found not in other organisms except bacteria and yeast whereas viruses have considerable potential to be used as cloning vehicle for both plants and animals. Mammalian viruses such as *simian virus* 40 (SV 40) and *cauliflower mosaic virus* (CaMV) are some viruses that have received present day attention.

METHODS FOR EXTRACTION OF DNA

Methods of extraction of DNA from plants and plasmid DNA from bacteria and quantification of DNA in a solution are given at the end of this chapter for practical purpose to the students.

PROCEDURES FOR PREPARATION OF DNA FOR GENE CLONING

The procedures described by T.A. Brown in 1990 should be followed.

Preparation of DNA

Genetic engineering needs three types of DNA at different occasions:

(i) Total cell DNA from where desired gene has to be cloned. This may be from bacteria or plant cell or animal cell.
(ii) Pure plasmid DNA: The procedure for preparation is the same as above except at some stage, plasmid DNA must be separated from bulk of the chromosome present in the cell.
(iii) Phage DNA is needed if phage is used as cloning vehicle. Phage DNA is prepared from bacterial phage particles, not from infected cells. So that there should be no contamination from the bacterial DNA. Special technique is needed to remove phage *capsid* (head). M13 double-stranded replicative DNA is prepared from *E. coli* cell. This is an exception because these are bacterial plasmid.

Preparation of Total DNA for Purification

The bacterial cell can be divided into four stages:

(i) A culture of bacteria is grown in a liquid medium with balanced nutrients and then *harvested.*
(ii) The bacterial cells are brokendown to get released their contents.
(iii) The extract is treated to remove the entire component except DNA.
(iv) Thus, obtained DNA solution is then concentrated by centrifugation.

The bacteria are grown in a liquid culture medium consisting of balanced nutrient mixture with concentration on which bacteria can grow. There are two types of media: first is M9, in which all nutrients are known and their concentration are also known and second is exa. M9, which consists of inorganic nutrients like Na_2HPO_4 (6.0), KH_2PO_4 (3.0), NaCl (0.5), NH_4Cl (1.0), $MgSO_4$ (0.5), $CaCl_2$ (0.01) and glucose (2.0) gram/litre. Some trace elements and vitamins are also added in the medium.

Second medium is not well defined because precise identity and quantity is not known. This is called L.B. method which is quite complex because two ingradients—tryptone and yeast extracts are complex mixture with unknown identity. The tryptone supplies amino acids and small peptides. *Yeast* provides nitrogen, sugar and inorganic/organic nutrients.

For the preparation of cell extract, small quantity of bacteria are obtained. Harvesting is done by spinning the culture in a centrifuge (Figure 25.9) and at low centrifugation speed, the bacterial pellete is deposited at the bottom of centrifuge tube allowing culture medium to be poured off.

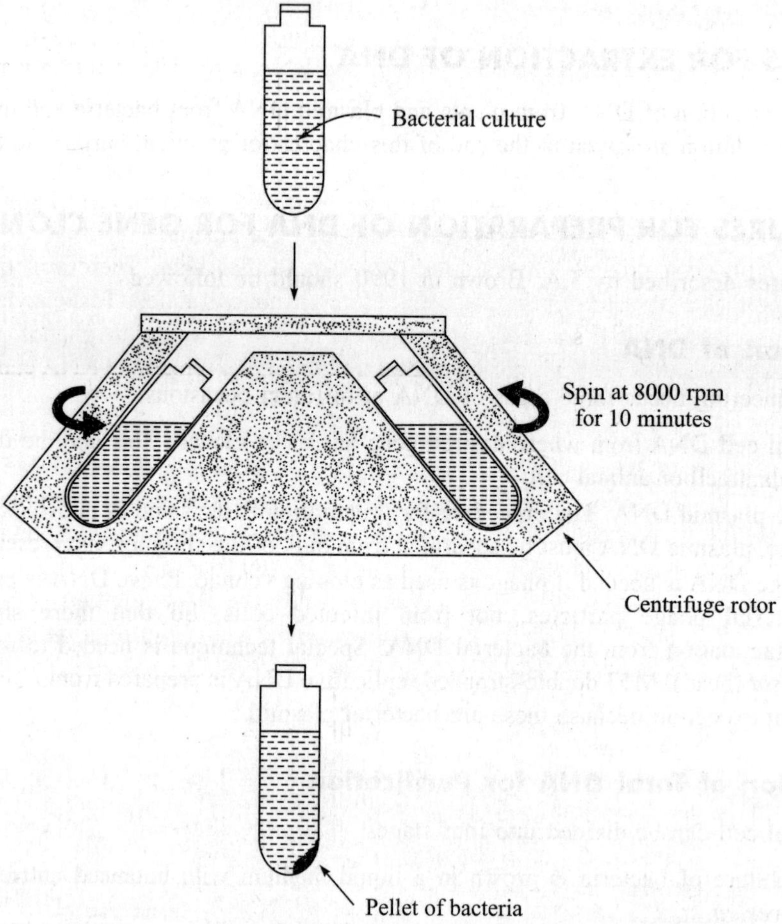

Figure 25.9 Harvesting bacteria by centrifugation method.

The bacterial cell is encased in a cytoplasmic membrane and surrounded by a rigid cell wall. The cytoplasmic membrane and cell wall are broken to release the cell components. The cell wall can be broken by mechanical force and also by chemical methods to lyse the cell. Chemical methods are used if the objective is DNA preparation.

Chemical lysis involves one agent for cell breaking and another for membrane breaking. The chemicals used for breaking the cell wall depends upon the species involved but for *E. coli* and related species lysozyme, ethylenediamine or combinations of both are applied. Lysozyme is an enzyme.

After cell lyses, the insoluble cell debris are removed by centrifugation leaving the cell extract consisting of DNA, RNA and protein. The lysozyme digests the polymeric compounds while EDTA removes the cell envelope and also inhibits cellular enzyme to degrade DNA. A detergent such as sodium dodecyl sulphate (SDS) is also added to cause the lysis by removing lipid molecule, thereby causing membrane breaking.

Purification of DNA

In addition to DNA, cell extracts possess protein and RNA also. The standard way to remove the protein is to add phenol or a 1:1 mixture of phenol and chloroform. These organic solvents precipitate proteins but leave the nucleic acid (DNA and RNA) in aqueous solution. The cell extract is then mixed with solvent and lays then separated by centrifugation. The white coagulate protein mass are left at the interface between the acquous and organic layers (Figure 24.9). The acquous solution of nucleic acid can be then removed by a pipette.

The centrifugation process breaks some DNA molecules also. Therefore, treat the cell extract with protease (pronase or proteinase K) before phenol extraction. This enzyme will break the protein (polypeptide) into smaller units, which are easily removed by phenol.

Some RNA molecules are removed by phenol treatment; the remaining RNA can be removed with enzyme ribonuclease by degrading into smaller units.

Concentration of DNA for making thick solution of DNA is needed for successful preparation. The most frequently used method for the concentration of DNA is *Ethanol* precipitation. In the presence of Na^+ salt, absolute ethanol efficiently precipitate polymeric nucleic acid at 20°C temperature with thick solution of DNA; the *ethanol* can be layered on the top of the sample causing molecules to precipitate at the interface.

A glass rod could be pushed through the ethanol into the DNA solution. When the rod is removed, the DNA molecule will adhere and will be pulled out of the solution in the form of a long fibre (Figure 25.10).

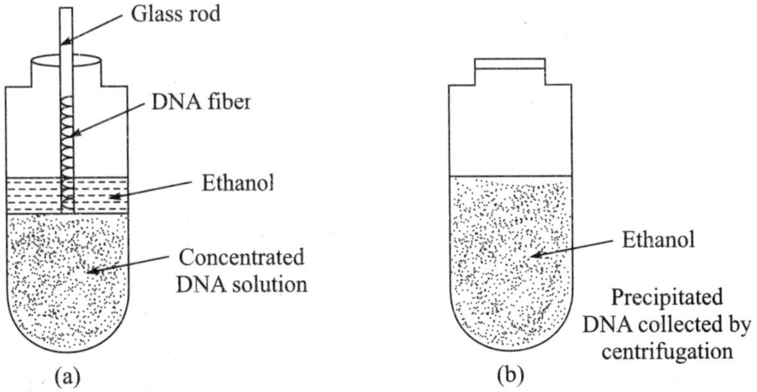

Figure 25.10 Collection of DNA by ethanol centrifugation. (a) Concentrated solution of DNA fibre can be withdrawn with a glass rod. (b) Precipitated DNA can be collected by centrifugation.

There is another method for DNA concentration. If ethanol is mixed with a dilute DNA solution, the precipitate can be collected by centrifugation and then redissolved in an appropriate volume of water. Ethanol precipitation leaves in solution short chain and monomeric nucleic acids components.

The DNA concentration can be accurately measured by *ultraviolet observance spectrophotometry* in that amount of ultraviolet radiation absorbed by the solution of DNA is directly proportional to the amount of DNA in the sample.

Preparation of Total Cell DNA from Other Organism

Other than bacteria, cell DNA is also required in the genetic cloning such as plants and animal, etc. if the objective is to clone genes from these organisms. The basic steps in DNA purification are the same but in case of eukaryotic cell, some changes are needed because cell walls of plants are stronger than bacteria. The chemical used for breaking the cell wall of bacteria and also lysozyme will not be effective in the plant cell. Some degradative enzymes are known, but grinding of frozen materials with a mortar and pestle is also effective. Since animal cells possess no cell wall, therefore, they can be lysed by treating with some detergent only.

Purification of Plasmid DNA

Purification of plasmids of bacteria is done in the same way as purification of total cell DNA. The bacterial cells are grown and harvested as well as cell extract is taken out has been detailed earlier. The extract is deproteinized and RNA is also removed and a cell extract is prepared. The DNA is concentrated by ethanol precipitation technique. Thereafter plasmid DNA is separated from the bacterial DNA present in the cell extract (Figure 25.11). Separation of plasmid DNA from bacterial DNA is difficult but essential because presence of slightest fragment of bacterial DNA would be dangerous in the gene cloning purposes. Only pure plasmid DNA should be used

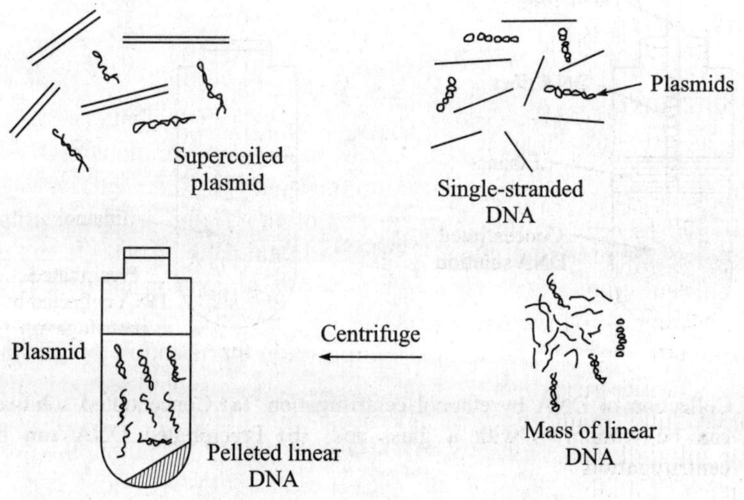

Figure 25.11 Plasmid purification.

as cloning vehicle. There are two techniques mostly used to separate them, i.e. on the basis of size of DNA and on the basis of confirmation, i.e. linear or circular. There are differences between the plasmid DNA and chromosomal DNA. Plasmid DNA are hardly 8 per cent of the size of bacterial (*E. coli*) chromosome. In addition, plasmid and bacterial DNA differ in the conformation, i.e. overall spatial configuration of the two molecules. The plasmid DNA are circular while in the cell extract, the bacterial chromosome on breaking becomes linear although in the normal cell both are circular. Thus, pure plasmid DNA could be obtained by separating circular from linear molecules.

Plasmid Amplification

The objective of the amplification is to increase the number of plasmids by checking the replication of bacterial chromosome. Some plasmids multiply in the absence of protein synthesis while bacterial chromosome cannot. This characteristic could made use during growth of a bacterial culture for plasmid purification. After a satisfactory growth has reached, a protein synthesis inhibitor is added (chloramphenicol) and the culture is incubated for 12 hours. During this period, plasmid molecules continue to multiply while chromosomal replication and cell division is inhibited, as a result, plasmid copy number reaches to several thousands. This plasmid amplification is needed for increasing its yield (Figure 25.12).

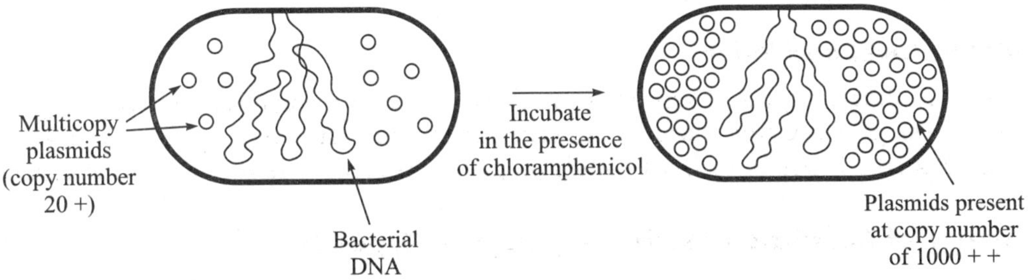

Figure 25.12 Amplification of bacterial plasmids.

Purification of Bacteriophage DNA

There is a big difference between phage DNA purification and preparation of cell DNA or plasmid DNA. The phage is the starting material with no cell extract. The bacteriophage particles are obtained in large number from extracellular medium of an infected bacterial culture. When such an infected bacterial culture is centrifuged, the bacteria are pelleted out, leaving phage particles in suspension. The phage particles are then collected from suspension and their DNA is extracted by a single deproteinization step removing the phage capsid (protein).

The procedure is straightforward and simple than purification of plasmid. The main difficulty, specially with λ is growing on infected culture in such a manner and care that the extracellular phage 'titre' (number of phage particles per millilitre of culture) is reasonably high. That is, 10^{10} per ml; will yield 500 ng of DNA. Infact 1000–2000 ml culture are needed to get substantial quantity of λ DNA.

RESTRICTION ENDONUCLEASE

The main step involved in the gene cloning is the production of defined DNA fragment. For this purpose, the DNA molecules should be cut in a very precise and reproducible manner needed for gene cloning.

The DNA fragments are obtained by treatment of DNA by restriction enzymes. These restriction enzymes are 'nucleases' that were discovered by the nobel laureate W. Arber, H. Smith and D. Nathan (1978). First discovery of restriction endonuclease was made in 1950, when it was found that some bacterial strains are immune to the bacterial phases called 'host-controlled' restrictions. Restriction occurs because some bacterium produces an enzyme that degrades the phase DNA before it replicates and develops new phases. The bacterium's own DNA carries additional methyl group that blocks the degradative enzyme to act. These degradative enzymes are called restriction nucleases and are synthesized in about 1200 species.

The DNA of bacterium contains a particular short sequence of nucleotides matching the restriction site of the enzyme. The most restriction sites consist of 4–6 nucleotides within which restriction enzyme makes two breaks in a single DNA strand; one in each strand generating '3-OH' and 5′-P′ groups at each position. The sequence recognized by restriction enzymes are 'palindrome' that means sequence is identical in both strands which gives same meaning forward and backward from central axis. Several hundred restriction enzymes with different restriction sites have been isolated from microorganisms.

MADAM, Level and Refer

So for, three different types of restrictions nucleases have been identified with differing mode of action. Out of these, type, I and III are very complex and have very limited role to play in cloning. But type II possesses DNA cutting properties, hence it is important in gene cloning.

Property of Restriction Nuclease (Enzyme) Types II

The basic property of the restriction nuclease (enzyme) is that each enzyme cut DNA molecule at a specific DNA recognition sequence point. For instance, restriction nuclease Pvul isolated from *Proteus vulgaris* cut DNA at the hexanucleotide "CG AT CG", in contrast, another enzyme Pvul II obtained from same source cuts DNA at the hexanucleotides CA GC TG.

There are nucleases which recognize only hexanucleotide sites, but there are some nucleases which cut at four, five as well as eight nucleotide sequence sites, e.g. *Sau3A* from *Staphylococcus aureus* strain 3A cut at four (GATC), *Alul* from *Arthrobacter luteus* cut at AGCT. Some more examples are cited in Table 25.2 which are generally used in cutting the DNA in gene cloning experiments.

There are examples where restriction endonuclease recognizes one site, but cut nucleotide sequences at related sites, e.g. Hin F_1 (Haemophilus influenzae strain F1) recognizes GANTC, but cut at related sites—GAATC, GATTC, GAGTC and GACTC. Thus, number and sizes of DNA fragments depend upon the exact position of recognition sequences for the endonucleases in the original molecules.

TABLE 25.2 Some restriction nucleases used in gene cloning experiments

Restriction enzyme	*Source*	*Recognition nucleotide sequence*	*Nature of cuts*
Eco R1	*Escherichia coli*	GAATTC	Sticky
Bam HI	*Bacillus amyloliquefaciens*	GGATCC	Sticky
Pvul II	*Proteus vulgaris*	CAGCTG	Blunt
HinF1	*Haemophilus influenzae*	GANTC	Sticky
Alul	*Arthrobacter luteus*	AGCT	Blunt
Not I	*Nocardia otitids caviarum*	GCGGCCGC	Sticky
HinF1	*Haemophilus influenzae R1*	GAATC	Sticky

Nature of cuts

The restriction nucleases make two types of cuts—'Sticky end' and 'Blunt end' (flush end). For example, Bam HI recognizes hexanucleotide sequence GGATCC. Sau3 A recognizes tetranucleotide sequence GATC both produce sticky ends.

1. ***Sticky-end cut:*** In the 'sticky end' cut, the cleavage is staggered, usually by two or four nucleotides, as a result DNA fragments have short single-stranded over hangs at each end and can be joined to each other because each fragment carries a complementary sticky end.

   N-N-G-A-A-T-T-C-N-N Eco R_1 N-N-G   A-A-T-T-C-N-N

   N-N-C-A-A-T-T-C-N-N   N-N-CTTAA   G-N-N

   Sticky end

   N = A, G, C, T

2. ***Blunt-end cut:*** On the other hand, many restriction enzymes make cut in the both strands (double) in the middle. It is called blunt end (flush end). Pvu II and Alu I are the blunt-end cutter enzymes.

   N-N-N-G C T N N AluI NNNG¦CTNN

   N-N-N-G G A N N → NNTC¦GANN

   Blunt ends

One important point to note is that almost every recognition sequence is Palindrome, i.e. they read the same from in each direction. DNA AND MADAM. LEVEL REFER.

Thus, there are three features of restriction enzymes:

(a) Restriction enzyme makes breaks in palindromic sequences.
(b) The breaks need not be directly opposite one another in two DNA strands.

(c) Enzyme that cleaves the DNA strands asymmetrically generates DNA fragments with complementary ends.

EcoR1 isolated from *E.coli* cuts the strand between G and A. The fragments produced by many restrictions could spontaneously form circles, which should be made linear on heating. The $EcoR_1$ restriction sites are:

- 5′ GAATTC 3′
- 3′ CTTAAG 5′

Cloning DNA with Complementary Single-Stranded Ends (Sticky end)

The single-stranded ends of EcoR1 fragments (as fragment produced by digestion of DNA with EcoR1 restriction enzyme) will be complementary regardless of the source of DNA. Thus, EcoR1 fragment cut from mouse chromosome, chicken chromosomes and so on can be incorporated into phage lambada chromosomes or plasmids that have been similarly cleaved with 'EcoR1' restriction enzyme. Once a gene or restriction enzyme fragments is inserted into a 'plasmid' the recombinant plasmid is introduced into a host cell, such as *E.coli*, by transformation (uptake of naked DNA by cells). When *E.coli* is used, the cell must be first made permeable to DNA by treatment with a calcium salt. An important advantage of the procedure over some other cloning procedure (Figure 25.13 A) is that after cloning, the cloned sequence can be precisely cut out of the recombinant chromosomes by using the same restriction enzyme that was used in the original cloning process (the restriction sites will remain at the same position).

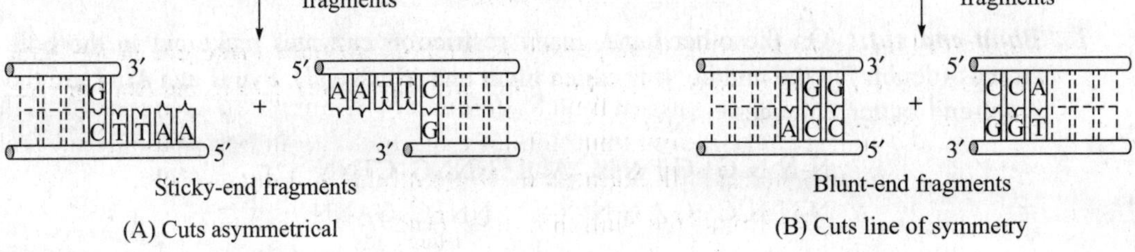

Figure 25.13 Two types of cuts made by restriction enzymes. The arrows indicated the cleavage sites. (A) Cuts made in each strand at an equal distance from the centre of symmetry of the restriction site. In this example with the enzyme EcoR1, the resulting molecules have complementary 5′ overhangs. Other restriction enzymes produce fragments with complementary 3′ overhangs. (B) Cuts made in each strand at the centre of symmetry of the restriction site. The products have blunt ends. The specific enzyme in this example is Ball.

Cloning DNA with 'Blunt' Ends by Adding Complementary Traits

A second procedure for the '*in vitro*' construction of recombinant DNA molecules involves the synthesis of complementary single-stranded traits on the 3′ ends of strands of DNA by using the enzyme '*terminal transferase*'. The procedure has an advantage over the restriction enzyme produced in that it can be used to join DNA molecules that terminate with blunt ends produced by cleavage with restriction enzymes that cut both strands of DNA at the same point (Figure 25.13 (B)). Once, the recombinant DNA molecules are produced using the *terminal transferase* procedure, they are cloned in the same way as those produced using complementary ends generated by restriction enzyme that makes stagger cuts. Both these methods are original techniques used to construct recombinant DNA molecules. Several modifications as well as new methods have been developed now.

Construction of Recombinant DNA

As stated earlier, preparation of pure DNA vector is the first step; and construction of DNA recombinant is the second. For these purposes, the DNA to be cloned is cut at specific points and then rejoined together. The cutting and rejoining of DNA fragment is carried out by special enzymes known as restriction endonuclease for cutting and ligases for joining of DNA fragments. There are a range of such enzymes available so far. These are:

1. ***Nucleases:*** These are enzymes which cut, degrade and shorten the DNA molecules. The nucleases degrade DNA molecules by breaking the phosphodiester bonds that link one nucleotide to the other in a DNA strand. Nucleases are of two types:

   (i) *Exonucleases:* Remove nucleotides—one at a time at the end of DNA strand (Figure 25.14).

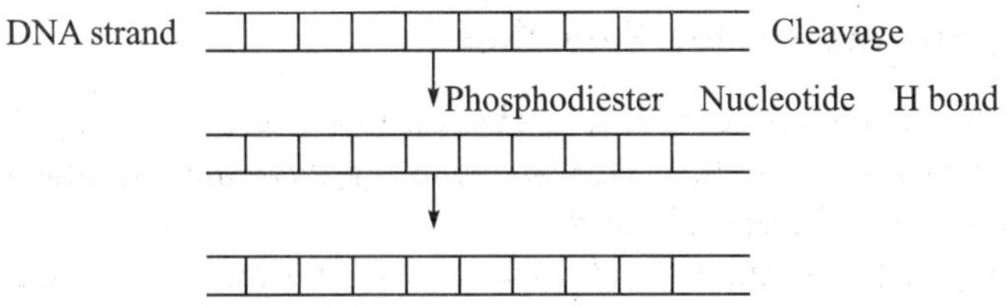

Figure 25.14 Removal of nucleotide from end of DNA strand.

   The enzyme Bal 31 purified from bacterium *Alteromonas espejiana* is an example of exonuclease that removes nucleotides from both the strands and enzyme such as *E. coli* exonuclease III degrades only one strand of a double helix.

   (ii) *Endonucleases:* Break the phosphodiester bond from inside of the DNA strand (Figure 25.15).

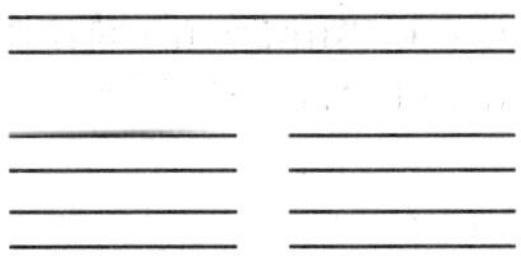

Figure 25.15 Breaking of phosphodiester bond from inside of the DNA strand.

Similarly, S1 endonuclease from the fungus *Aspergillus oryzae* cleaves only one strand whereas deoxyribonuclease (DNAse I) obtained from cow pancreas cuts both the strands. It is non-specific in its action and cuts phosphodiester bond at any internal part of the DNA (Figure 25.16(A)).

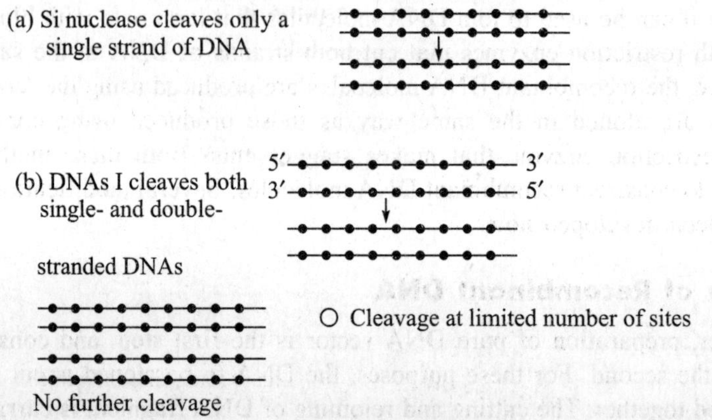

Figure 25.16(A) Reactions catalyzed by different types of endonucleases.

2. ***Ligases:*** Join DNA fragments together.
3. ***Polymerases:*** Help in making copies of DNA molecules.
4. ***Modifying enzymes:*** Remove or add chemical group.
5. ***Topoisomerases:*** Introduce or remove supercoils from covalently closed circular DNA.

Legation of DNA Molecules Together

The final step in the construction of a recombinant DNA is the joining of the vector molecule and the DNA to be cloned together. This is called *legation* and the enzyme that catalyzes the reaction is called *legase*. (Figure 25.16(B)).

All the living cells produce DNA ligases which carry important role to repair any discontinuities arising on account of the breaking of cells, DNA molecules due to replication or recombination in one of the DNA strands or due to any reason, a phosphodiester bond between the adjacent nucleotides is missing. In the test tubes, purified DNA ligases join the single-strand or two ends of the same DNA molecule. But for genetic engineering purposes, generally, enzyme used is purified from *E. coli* bacteria that have been infected from T4 phage.

The legation of complementary *sticky ends* is much easier and efficient than *blunt end* ligation because the ligase enzyme is unable to catch hold of the molecule to be legated. However, 'blunt' end legation can be done at high DNA concentration, which increases the possibility of ends coming together in correct way. But if the phosphodiester bonds are not synthesized quickly, in that case the sticky ends will fall apart (Figure 25.17).

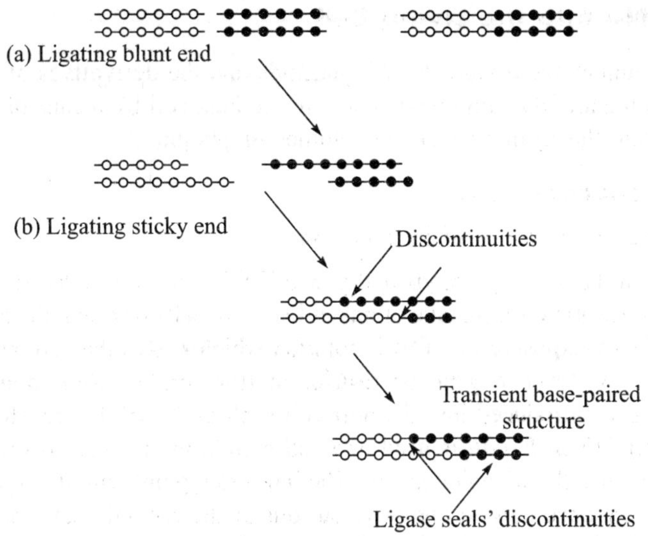

Figure 25.16(B) Ligating (joining) the cut DNA molecules together.

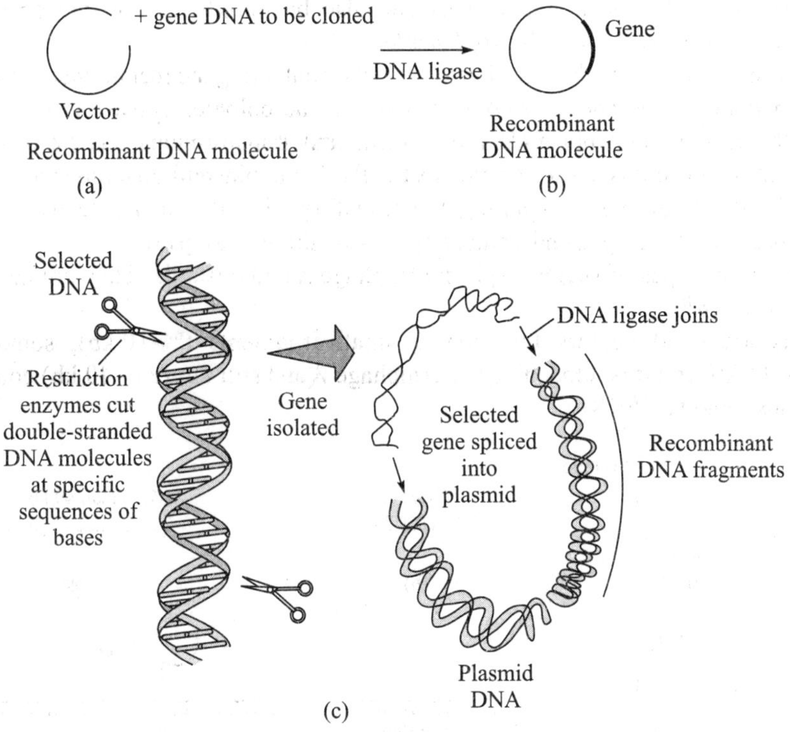

Figure 25.17 Construction of recombinant DNA molecule.

Insertion of DNA into the Living Cell

At present most common vectors are *E. coli* plasmids and the derivatives of the bacteriophages λ and M13. Recombinant DNA can be depicted in the host cell by means of genetic features or markers evident from the formation of the colonies or plaque.

Methods of introduction

There are following two methods of introduction:

Transformation: In this case, plasmids and phage DNA can be introduced in the bacterial cell such as *E. coli* by transformation (uptake of naked DNA by cell) wherein, the cells gain the ability to take up free DNA by exposure to a $CaCl_2$ solution which makes the cell membrane permeable for the entry of DNA. Once a gene or restriction fragment is inserted into a plasmid, the recombinant plasmid is introduced into the host cell such as *E. coli* by transformation (uptake of naked DNA by cell). When *E. coli* is used, the cell membrane is made permeable for the entry of DNA by treating the cell with calcium salt. The important point with the procedure is that after cloning, the clone sequence can be precisely cut out of the recombinant chromosome by same restriction enzyme that was used in the original cloning process because restriction site will remain at the same position.

Electrophoration: The recombinant DNA can also be introduced into the bacterial cell by electrophoretic procedure called *electrophoration*.

After the introduction of DNA, the bacterial cells, containing the recombinant DNA are plated on a solid medium. If the added DNA is a plasmid, the colonies consisting of bacterial cells containing the recombinant plasmids are formed, and transformants can be detected by the phenotype that the plasmids confer on the host cells. If the plasmid contains one or more genes for resistance to antibiotics, on plating the transformed cells on a selective medium with antibiotics, prevent all but plasmid containing cell is allowed to grow.

There are three types of vectors—plasmids, phage λ and cosmids. They are commonly used for cloning in *E. coli*.

Plasmids are good vectors for cloning small fragments (%-10 kb); somewhat larger fragments (– 15 kb) could be cloned in bacteriophage λ and still larger (– 40 kb) could be cloned in the cosmids (Figure 25.18).

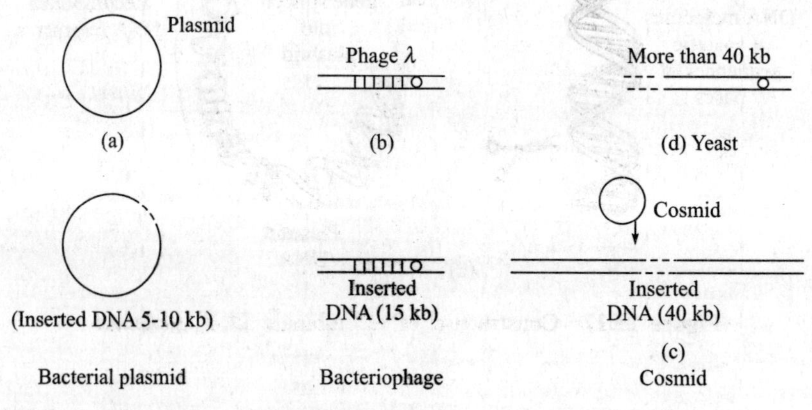

Figure 25.18 Common cloning vectors for use with *E. coli*.

How DNA fragments are inserted in plasmids

This is important to know how the foreign DNA fragments are inserted in the plasmids. The insertion of DNA fragments in the plasmid is done in two ways:

1. The circular plasmids are cut opened or 'linearized' with a restriction endonuclease so that a piece of foreign DNA can be inserted.
2. The foreign DNA fragments are obtained by digestion with a restriction enzyme and the cut ends of the plasmids have complementary single-stranded region that allow them to pair, and this way the foreign DNA fragments get inserted into the plasmids.

It has been already mentioned that restriction nucleases have unique characteristics of cleaving DNA sites at *Palindromes* such as Hind III restriction endonuclease cleaves DNA fragment at sequence 5 $AAGCTT_3$ which is complementary to

- 3 TTCGAA5′
- 5′AAGCTT3′
- 3 TTCGAA5′

The number of restriction sites in a plasmid varies depending upon the restriction nuclease. For instance, a small 9200 base pair plasmid 'pSC 101' has single site for Hind III.

There are two ways to insert the foreign DNA fragment into the cleaved plasmids to regain again a circular structure. The most easy way is to cleave the foreign DNA with same restriction endonucleage that has been used to cut the plasmid.

The plasmid with inserted plasmid reinfects the bacteria and there produces a large number of foreign insert through autonomous replication.

If we digest human DNA with Hind III, 7,50,000 fragments with different sequences each on average about 4000 base pairs could be produced. No doubt, we can do cleaving of both 'plasmid' and the foreign DNA fragment, but there is some limitation.

(i) The inserted foreign fragment would be small while for constructing sequence library; we need large pieces of foreign DNA.

(ii) It may also be possible that restriction sites, in the foreign genome may be absent or poorly developed as a result cleavage size between these two sites must be large and may exceed even the upper limit of 10 kb. While foreign DNA fragment must not exceed in the size in order to make any problem to the plasmid to enter the bacterial cell (Figure 25.19).

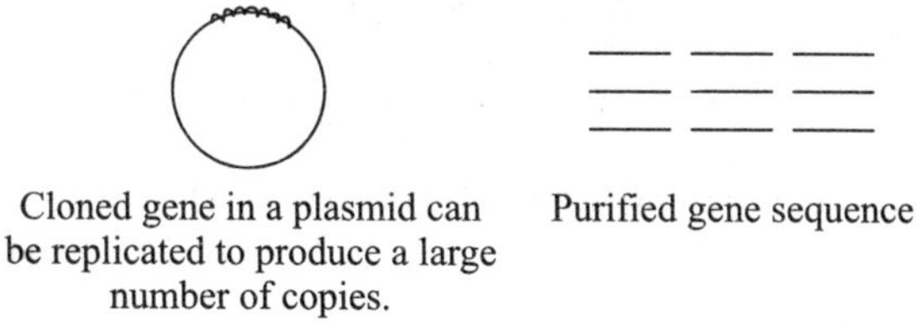

Figure 25.19 How to produce a large number of desired genes.

However, these two problems could be solved by mechanical shearing of the DNA in a 'high-speed blender' to produce the average size of the fragment. But they may be of blunt ends, not of single-stranded complementary ends, as cleaved by Hind III cleaved plasmid. The restriction endonucleases have the property of cleaving DNA at a particular specific site usually at *Palindromes*. As may be seen below, the restriction endonuclease Hind III cleaves DNA between two as in the sequence of 5′ AAGCTT3′. The two complementary strands of *Palindrome* are:

- 5′AAGCTT3′
- 3′TTCGAA5′

Bacterial cell with plasmids containing foreign gene could be recovered in the presence of calcium ions. In the calcium solution, about one to ten *E. coli* cells out of one million will be picked up from the medium in which plasmids are carrying foreign genes (DNAs). These rare cells that contain the plasmids are identified and recovered by using selective technique. For instance, 'pSc101' plasmid contains a gene called 'tet' that confers resistance to the antibiotic 'tetracycline'. We should choose bacterial cells that are sensitive to tetracycline (tet) and genetically transform them by infecting them with plasmid carrying the 'tet' gene. The bacterial cells containing the 'tet' pSc101 plasmids are resistant to the antibiotic and can be identified in a large population of bacterial cloning that lack resistance. Since plasmid replicates autonomously, the plasmid inserted with foreign gene will be present in each bacterial cell of a colony developed from the transformed cell. Figure 25.20 shows an example of cloning plasmid. A fragment of DNA from any organism has been placed in the plasmid. The recombinant DNA then used to transform a bacterial cell so that foreign DNA should multiply in the bacterium many more times.

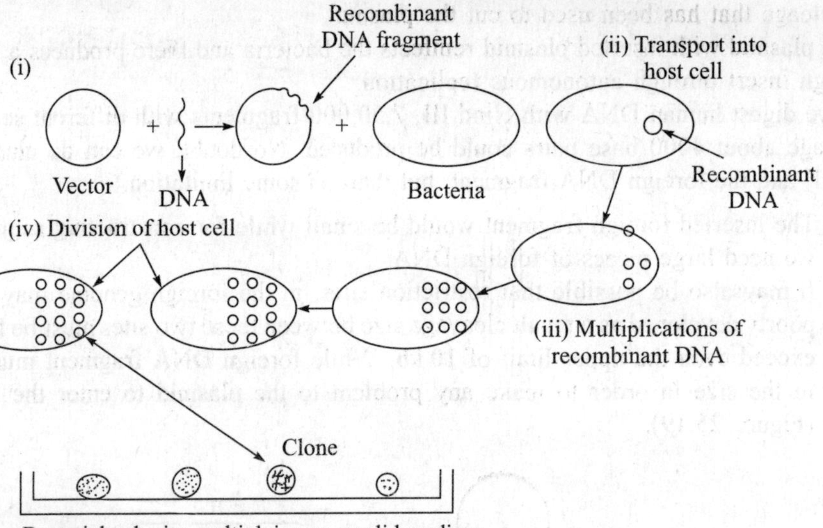

Figure 25.20 Main steps involved in gene cloning that is construction of a recombinant DNA molecule.

If the length of phage DNA is longer, viz. 50 kb, then the non-essential (intron) portion of DNA could be removed and replaced with donor DNA. After the donor DNA has been legated in phage, the recombinant DNA is packaged into mature phage *in vitro* and the phage is used to infect bacterial cell.

Joining the sticky ends does not ensure that a functional gene has been obtained. Suppose original DNA has been cleaved to produce four fragments A, B, C, D. whose original sequence was A, B, C, D; the reassembly of fragments can yield, A, C, B, D and B, A, D, C fragments in the rearrangement instead of A, B, C, D because B and C have the same pair of sticky ends.

The DNA ligase made by *E. coli* phage T4 can join both sticky as well as blunt ends.

COMPLEMENTARY (cDNA) GENE CLONING

Theoretically, a gene is easily detected by its expression, i.e. selection on the basis of phenotypic expression confirms that a gene has been cloned. But not all prokaryotic genes produce such detectable phenotypes. Similarly, an eukaryotic gene that has been inserted into a bacterial plasmid (vector) may fail to express its phenotypic effect (formation of functional protein). However, this problem can be solved by direct cloning of coding mRNA in pure form. For example, chicken gene for albumin (egg white protein) may be cloned into a bacterial plasmid. Such a plasmid could be used as a source of ovalbumin gene for DNA sequencing. *A recombinant plasmid could be formed by treating both chicken cellular DNA and E. coli plasmid DNA with same restriction enzyme mixing the fragments, annealing and finally transforming E. coli with DNA mixture, DNA containing recombinant plasmid.* But a bacterial colony containing this plasmid could not be identified by test for the presence of ovalbumin because bacterial cell containing avalbumin gene may not be able to synthesize ovalbumin because of presence of *introns* (non-coding fragments) present in the genes, because bacteria lack the enzyme to remove the *introns* in the RNA processing. The reason is that a heterogenous mRNA (hnRNA) is synthesized in the cell with a large fraction of total mRNA while for synthesis of ovalbumin, a single mRNA species—in this case ovalbumin mRNA is needed. In cloning genes whose products are the major cellular proteins, the purified (processed) mNRA containing coding sequences of the gene are needed.

The method depends on the use of an unusual polymerase '*reverse transctiptase*' which can be used as a single-stranded RNA molecule, viz. mRNA as template to synthesize a double-stranded DNA copy called *cDNA* (complementary DNA). Figure 25.21 shows how a *template* RNA molecule synthesizes an mRNA whose *introns* have been removed from the primary script? If the purpose is to identify recombinant DNA or to synthesize gene products in bacterial cell, the cDNA formed from the processed mRNA is the material of the choice for cloning.

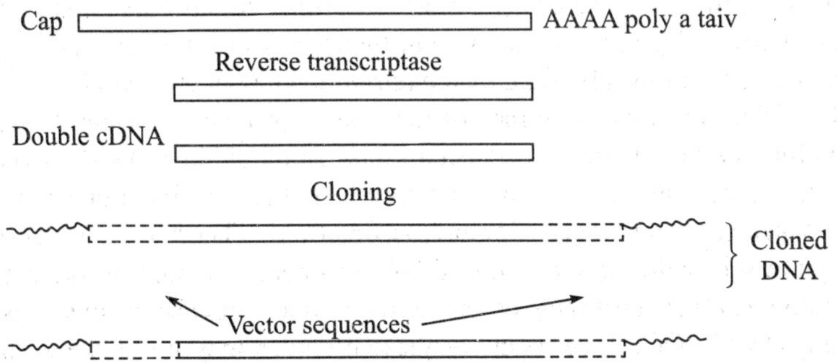

Figure 25.21 Cloning of cDNA molecules derived from mRNA by enzyme reverse *transcriptase*. *Introns* are removed from the primary transcript in RNA processing.

Identification of Recombinant DNA

After cleaving a vector DNA (plasmid) by restriction enzyme and when it is renatured in presence of a number of restriction fragments of DNA from an organism (donor), many types of molecules are obtained. *Some self-joined vector DNA (plasmids) with acquired DNA fragments, some vector which contains more than one DNA fragment and some DNA molecules having joined many fragments together.* While our objective is to isolate (i) the vector that indeed possesses an 'inserted' DNA segment and (ii) that segment is of our interest. There are some procedures to detect recombinant vector:

- The first step is to isolate bacteria that contains plasmids.
- The common procedure is to use a plasmid possessing a gene marker for 'antibiotic-resistant' gene.
- To grow transformed bacteria on a medium containing the antibiotics; in which only the bacterial cell containing plasmids having antibiotic-resistant gene will form the colony. A good example is the plasmid PBR322 (4362 base-pair) which contains two genes resistant to different antibiotics—tetracycline (tet-r) and ampicillin (amp-r). Therefore, the transformed bacterial cell that contains PBR 322 will be easily isolated by growing cell on the medium containing either tetracyclin or ampicillin. Another good point with PBR322 is that it contains only one cleavage site, for each of seven restriction enzymes. When donor DNA is inserted into any of these sites, the exact position of insertion is known (Figure 25.22).

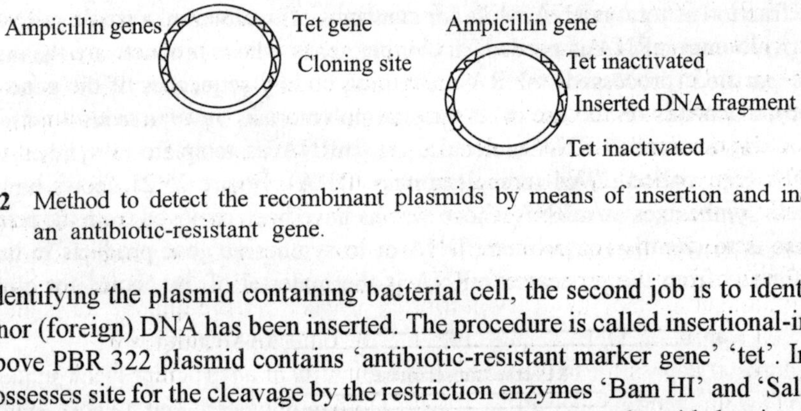

Figure 25.22 Method to detect the recombinant plasmids by means of insertion and inactivation of an antibiotic-resistant gene.

After identifying the plasmid containing bacterial cell, the second job is to identify plasmid in which donor (foreign) DNA has been inserted. The procedure is called insertional-inactivation. As stated above PBR 322 plasmid contains 'antibiotic-resistant marker gene' 'tet'. In PBR 322, 'tet' gene possesses site for the cleavage by the restriction enzymes 'Bam HI' and 'SalI'. Insertion of donor (alien) DNA fragment at either of these sites produce a plasmid that is not effective against tetracycline; in other words 'tet-s' because insertion of alien DNA fragment interrupts the 'tet' gene activity as a result its function is inactivated but recombinant plasmid remains resistant to ampicillin because resistance gene 'amp-r' is functional. Therefore, if transformed bacterial cell containing plasmid with inserted alien DNA fragment is plated on the medium containing ampicillin antibiotics; all the surviving colony will be 'amp-r' and hence must possess the plasmid with recombinant DNA. Of course, some of these colonies may be of 'tet-s' and 'tet-r', which could be identified by replica plating on tetracyclin containing medium. Thus, the bacterial cell that contains 'amp-r, tet-s' must be having 'recombinant' PBR 322 plasmid.

Screening of a Particular Clone Gene (Recombinant DNA Fragment)

The most simple and common procedure is to detect a cloned gene (leu+ gene) by 'complementation' in a leu-host grown on a medium lacking leucine. Other procedure is 'colony hybridization'. This method permits detection of the presence of a gene for which DNA or RNA is labelled with radioactivity or some other mean is available. *The colonies to be tested are transferred (lifted) from solid medium onto a intracellulose or nylon filter by gently pressing the filter onto the surface of the solid.*

The filter is treated with sodium hydroxide (NaOH), which simultaneously breaks open the cell and denatures the DNA. The filter is then saturated with labelled DNA or RNA complementary to the gene being sought and the cellular DNA is renatured. The labelled nucleic acid used in hybridization, functions as 'probe'. After washing the unbound "probe", the position of the bound probe identify the desired colonies. The desired colonies are identified by auto-radiography due to presence of radioactively-labelled probe.

DNA Library

DNA libraries of foreign DNA sequences can be established in bacteria by inserting DNA sequence into the plasmids. The bacterial plasmids are the primary tools for molecular genetical studies and also in the alien gene cloning mechanism because these are modified to use as a vehicle for carrying foreign DNA fragments. These plasmids are first purified from bacterial chromosome by centrifugation so that pieces of DNA fragments from any organism including plants, animals and human beings could be inserted into the plasmids and they are returned to the bacterial cells. In the bacterial cells, the plasmid replicates automatically and hence the inserted; alien DNA segment also replicates. In this way, a large number of DNA fragments could be manufactured and a sequence library could be built up in the bacteria of entire foreign genome so that a copy of each and every sequence of foreign genome is present in the plasmid and a colony of such genetically-modified bacteria could be established. All the bacteria and their descendents will be carrying a plasmid with the same sequence of alien DNA.

Polymerase Chain Reaction (PCR)

Through recombinant technology, it is possible to produce a large quantity of cloned DNA that can be obtained for use in DNA sequencing and for other application.

In some cases, it is possible to produce a large quantity of a particular DNA sequence without any cloning. This method is called PCR. It uses DNA polymerase and a short, synthetic *oligonucleotides* may be about 20 nucleotides, as primer that should be complementary in sequence to the ends of any DNA fragment of interest. Starting with a mixture having as little as one molecule of the fragment of interest; repeated round of replication of DNA increases the number of molecules exponently. Suppose, we have started with a single molecule; thirty times of DNA replication will result $2^{30} = 10^9$ molecules. This is also called *amplification*. This amplified DNA without purification can be used for DNA sequencing as probe in DNA-hybridization or for cloning purposes.

The procedure for DNA sequence amplification is given in Figure 25.23. The solid line shows sequence to be amplified and the dotted line shows oligonucleotide sequences. The oligonucleotides are also called "*primer*" sequence, because, they anneal to the ends of sequence to be amplified and are used as primers for chain formation by DNA polymerase process.

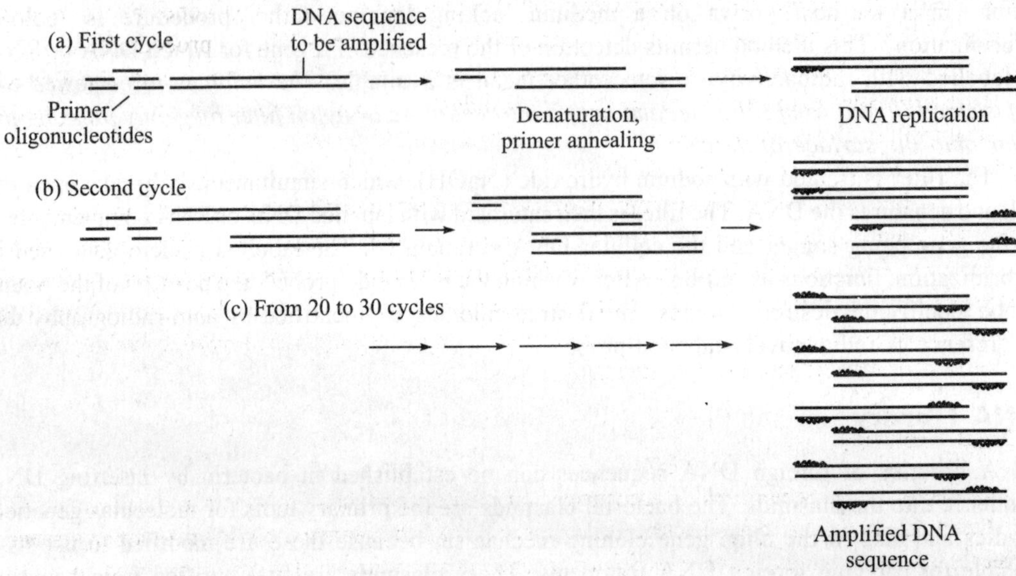

Figure 25.23 PCR technique for amplification of DNA sequences.

In the first cycle of PCR amplification, the DNA is denatured by heating or by use of alkali to separate the strands and then renatured on cooling in the presence of a vast number of oligonucleotides. Thereafter, DNA polymerase is added. Elongation of the primer produces double-stranded molecules. The first cycle in PCR produces two copies of each molecule possessing sequences complementary to the *primer*; the second cycle of PCR is similar to the first in which DNA is denatured and renatured in the presence of an excess of *primer* oligonucleotides and DNA polymerase is added to elongate the *primers*. After this cycle, there are four copies of each DNA molecule in the original mixture. These steps of denaturation, renaturation and replication by addition of DNA polymerase are repeated from twenty to thirty times and each time (cycle) number of molecules of the amplified DNA sequence is doubled.

The *primers* are designed to anneal to opposite DNA strands at the extreme ends of the region to be amplified. The *primers* are oriented with their 3′ ends facing the sequence to be amplified so that in DNA synthesis *primers* grow toward each other but along opposite template strand. Each newly produced strand terminates in a sequence that can anneal with complementary *primer*, therefore, it can be used for further amplification. PCR amplification is very useful in producing large quantities of specific DNA sequences without need for cloning.

Limitations

1. The DNA sequences at the ends of the region to be amplified must be known so that *primer* oligonucleotides can be synthesized.
2. The sequence to be amplified must be smaller than about 3000 nucleotides to allow the replication steps to occur efficiently.

In Figure 25.24, *primer* sequences used in PCR amplication are shown along with newly synthesized DNA shown in dotted chambers. After first round of replication, each newly synthesized DNA molecule functions as a new template. About 3 kb sequence could be amplified.

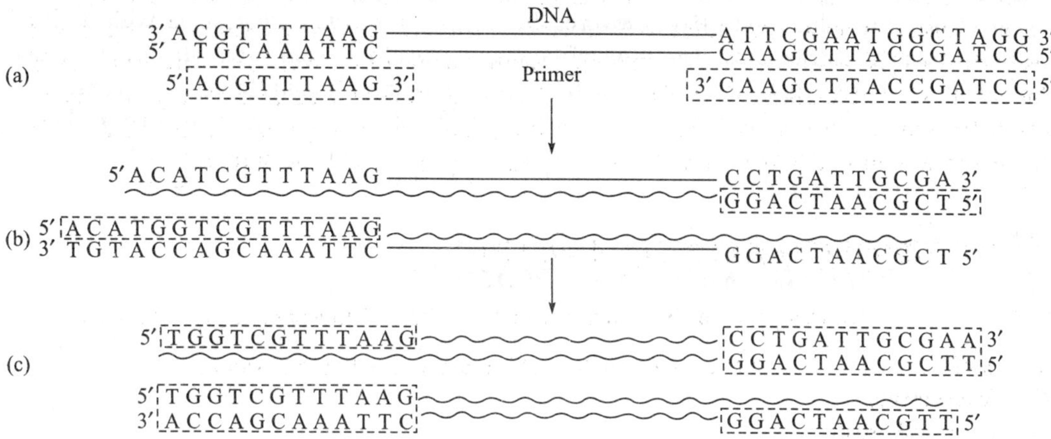

Figure 25.24 Primers used in PCR amplification of DNA. Newly synthesized DNA sequences are shown in dotted boxes. After first round of replication, each newly synthesized DNA form a new template and this way number of copies go on increasing.

REVERSE GENETICS

Mutation is considered as a source of origin of new genes. The recombinant DNA technology has enabled the man to clone the wild-type gene and to mutate it in a specific way, and to reintroduce the gene back into the organism to facilitate the study of its phenotypic effect. Because mutations are predetermined, a very fine level of resolution is possible in defining promoter and enhancer sequences that are necessary for transcription and the sequences necessary for RNA splicing, particularly amino acids that are essential for protein function. This is called *reverse genetics*.

CLONING IN YEAST: ARTIFICIAL CHROMOSOME

A large DNA molecule can be cloned intact in a yeast cell with the use of special vector by creating yeast artificial chromosome (YAC). The YAC vector possesses four kinds of genetic elements:

1. A cloning site.
2. A yeast centromere and genetic markers that are selectable in yeast.
3. An *E. coli* origin of replication and genetic marker that is selectable in *E. coli*.
4. A pair of telomere sequences from tetrahymena. This way, a YAc vector is a *shuttle vector* that can replicate and can be selected in yeast as well as in *E. coli*.

During cloning vector is cut with restriction enzymes A and B to free the chromosome arms. These arms are ligated with DNA fragment and yeast cells are *transformed*.

The circular vector (plasmid) is isolated after growth in *E. coli* and cleaned with two different restriction enzymes one that cuts at A site and the other at B site.

The DNA to be cloned is very gently isolated and broken into large fragments either by physical means or by restriction enzyme. The isolated DNA fragments are sandwiched between two arms of YAC. This product is recovered because it is a true chromosome with a single centromere and two telomeres. The product containing two YAC left arms are decentric and those with two YAC right arms are acentric and hence they are unstable. The yeast cells containing YAC, with inserted DNA fragment can be identified in a number of ways including colony-hybridization. The greatest advantage of using YAC for cloning vector is that there is no theoretical size limit to a YAC insert. An average size could be taken as 300 kb.

GENETIC TRANSFORMATION IN PLANTS

There are three procedures used in the genetic transformation of the plant cells. These are:

(i) Soil bacterium (*Agrobacterium tumefaciens*) plasmid-mediated system.
(ii) Microprojectile.
(iii) Electrophoresis.

These procedures are discussed in detail in Chapter 24 dealing with plant genetic engineering.

GENETIC TRANSFORMATION IN ANIMALS

Here too there are two methods.

Germ-line Transformation

Germ-line transformation in animals can be carried out in the following ways.

The DNA is directly injected into the germ line of the host. The injected DNA is often maintained in germ-line nuclei and some time spontaneously become incorporated into the chromosome.

However, the most common is direct injection of vector DNA into the nucleus of a fertilized egg, which is then transferred to uterus of a foster mother for development. The vector is usually a modified *retrovirus*. The retroviruses have RNA as their genetic material and code for reverse transcriptase that converts the retrovirus genome into double DNA that is inserted into the genome in the injected cells. Genetically-engineered *ritroviruses* containing inserted DNA (gene) undergo same process. Such animals are called *transgenic* (Hartle, 1991).

Embryonic Stem Cell

This method of transformation is used in the mammals. The embryonic stem cells are present in the blastocyst, that give rise to embryo. The embryonic cells are very soft. They can be isolated, then grown and manipulated in the culture, where the recombinant DNA vector can be introduced. These stem cells can be then injected into the cavity of another blastocyst, where they are incorporated into the developing embryo. If any mutation takes place into the group embryotic stem cells, it is incorporated into the germ line of the living animals.

APPLICATION OF RECOMBINANT DNA TECHNOLOGY

The recombinant DNA technology is considered as a revolution in the biological sciences because of the following uses:

1. Production of useful proteins (Figure 25.25).
2. Creation of cells capable of synthesizing economically useful molecules.
3. Generation of DNA and RNA sequences as resource tools or in medieval diagnosis.
4. Manipulation of the genotypic constitutions of plants and animals.
5. Correction of genetic defects; gene therapy.
6. Production of drugs and medicines according to the genetic constitution of a man.

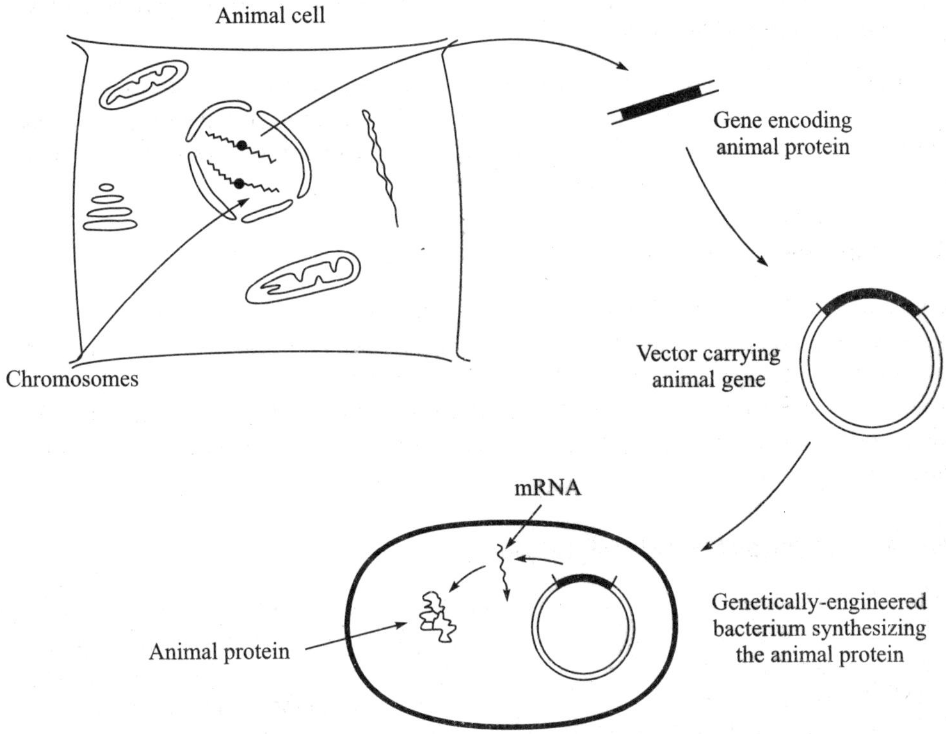

Figure 25.25 Procedure for the production of an animal protein by a bacterium through gene cloning.

Practical Utility of Recombinant DNA Technology

1. Production of genetically-transformed marine bacteria that can clean up oil spills in the ocean by metabolizing petroleum.

2. Designing bacteria that can synthesize industrial chemicals or degrade the industrial waste.
3. Designing bacteria that are able to compost waste more efficiently and fix nitrogen.
4. Efforts are being made to produce bacteria that can convert biological waste into alcohol.
5. Human insulin and growth hormones are being produced in bacteria.
6. Efforts are being made to alter the genotype of the plants by means of recombinant DNA technology by transforming genes from unrelated species to enhance the yield, hardiness and resistance to diseases and pests.
7. Glyphosphate is a weed killer that has been manufactured as biopesticide which inhibits formation of essential enzymes in plants. But glyphosphate is not a selective herbicides. It kills other crop plants also. However, target gene of glyphosphate is present in the bacterium *Salmonella typhimurium*. A resistant form of the gene was obtained by mutagenesis and was cloned in *Salmonella* in the presence of glyphosphate. This gene was cloned in *E. coli* and recloned in the T-DNA of *Agrobacterium*. Transformation of plants with the glyphosphate-resistant gene yielded varieties of maize, cotton, and tobacco that are resistant to herbicides.

Bacillus thurigiensis produces a protein that is lethal to black cut worm and cotton boll worm. Gene coding for lethal proteins has been introduced into the soil bacterium *Pseudomonas fluorescens* which lives in association with plant roots. Inoculation of soil with the engineered *P. fluorescens* helps to control the black entworm.

Uses in Scientific Researches

Recombinant DNA technology is being used as a research tool in molecular genetics. The *reverse genetics* makes it possible to isolate and alter a gene at will and reintroduce it back into a living cell or even in the germ line. This saves time and labour. *Reverse genetics* enables in the construction of mutants that cannot be produced by any other way. Greatest effect of recombinant DNA technology has been on basic studies in eukaryotic gene regulation and development.

Production of Specific Useful Protein

Recombinant DNA technology is useful in production of some valuable proteins in large quantities. A DNA sequence coding for the desired protein is cloned in a vector adjacent to a bacterial promoter. This step is usually done with cDNA because cDNA has all the coding sequences spliced together in a right order. Using a vector with high copy number ensures that many copies of the coding sequences will be present in each bacterial cell, which can result in the synthesis of gene product at a concentrated range from 1–5 per cent of the total cellular protein.

Protein that do not fold or become modified properly in bacterial cell can be produced in genetically-engineered cells of yeast or any other suitable eukaryote. If yeast is taken as an example, a genetically-engineered plasmid called a *shuttle vector* containing sequences from both *E. coli* and yeast and capable of undergoing replication and segregation in both the organisms. First, cloning is done in *E. coli* because it is easier; than plasmid is isolated and transferred in yeast background for extrusions of cloned gene.

Identification of Hereditary Diseases

Probes derived by recombinant DNA methods are widely used in prenatal detection of diseases. Such as detecting cystic fibrosis, Huntington chorea, sickle-cell anaemia and many other genetic disorders. In some cases probes derived from the gene itself are used and in other cases, restriction fragment polymorphism genetically linked to disease gene are employed. If the disease gene itself or a region close to it in the chromosome differ from normal chromosome, in the position of one a more cleavage sites for restriction enzymes, these differences can be detected by *southern blot* with use of cloned DNA form the region as the probe.

DNA FINGERPRINTING

If a large number of genes in a sexually-reproducing species continue to recombine, the number of genotypes in a few generations would be so distinct that no two persons would be genetically alike. This way it is theorised that each and every person must be carrying his/her own genetic identity called 'fingerprint'. However, it is not possible to score the individual in a population for so many genes though some genes have been scored such as genes conferring different blood group antigens. These blood group antigens have been used to decide paternity disputes and tracking criminals of rape, murder and theft cases by judiciary.

Another method with decisive precision has been developed by forensic laboratories is the *fingerprinting* technique. Here molecular marker RFLP (Restriction Fragment Length Polymorphism) is used to identify the individual with accuracy because, very large number of loci can be scored. For example, about 1000 odd microsatellites are present in the human genomes, most of which are highly polymorphic. If a restriction enzyme cleaves a specific site which it recognizes on either side of each of the two alleles of a microsatellite, the fragment of varying length will be produced.

If alleles at a large number of different loci are scored using different restriction enzymes, the observed variability in the population of fragments generated would be so large that no two individuals are expected to show same pattern of variation. Therefore, DNA available from any crime scene (source) is compared with the DNA of a suspected individual to determine whether or not the two match. But in India where democracy is the concubine of the criminals *raj* (criminal corrupt, mafioso politicians) this technique has no validity because if the laboratories fail to abide them, they will blast it.

Methods for Extraction of DNA

Extracting DNA from plants (Onion)

This is a crude but effective method for *isolating DNA and RNA from onion.* First, the tissue is broken up mechanically. Household detergent is used to degrade both the cell membranes and those surrounding the nuclei. Cell fragments are separated by filtration; the nucleic acids and soluble proteins remain. Protease is used to partly degrade soluble proteins, then the nucleic acids are precipitated into ice cold ethanol. The onion DNA can be run on an electrophoresis gel, if it is first dried then dissolved in TE buffer.

Preparation: The ethanol used must be ice cold. Place it in a plastic bottle in a freezer at least 24 hours before we should attempt this activity.

Materials and equipment:

- A small onion.
- Washing-up liquid, 10 cm^3 (use a watery type, not the thicker and concentrated variety).
- Table salt, 3 g.
- Distilled water, 90 cm^3.
- Very cold ethanol, about 6 cm^3, straight from the freezer (industrial methylated spirit (IMS) is suitable, but please see the safety note given below).
- *Novozymes neutrase*$^{(R)}$ (a protease), 2–3 drops.
- Ice, in a jug with cold water.
- Coffee filter paper (do not use laboratory filter paper, as liquid takes too long to pass through it).
- Plastic syringe of 1 cm^3 (without a needle).
- Sharp vegetable knife and chopping board.
- Large plastic funnel.
- Two beakers of 250 cm^3
- Boiling tube.
- Glass rod or spoon for stirring the mixture.
- Waterbath, maintained at 60°C.

Procedure:

1. Dissolve the salt in 90 cm^3 of distilled water. Add the washing-up liquid and mix gently.
2. Chop the onion into small pieces; roughly 5 mm × 5 mm. Add the chopped onion to a beaker with the salty washing-up liquid solution.
3. Stand the beaker in a waterbath at 60°C for exactly 15 minutes.
4. Cool the mixture by placing the beaker in an ice waterbath for 5 minutes, stirring frequently.
5. Pour the mixture into a blender and blend it for not more than 5 seconds.
6. Filter the mixture into a second beaker. Ensure that any foam on top of the liquid does not contaminate the filtrate.
7. Add 2–3 drops of protease to about 10 cm^3 of the onion extract in a boiling tube and mix well.
8. Very carefully pour the ice cold ethanol down the side of the boiling tube to form a layer on top of the onion extract.
9. Leave the tube undisturbed for a few minutes. Nucleic acids (DNA and RNA) will precipitate into the upper (ethanol) layer.

Safety measures while using ethanol in freezers: Most freezers are not spark-proof. There have been instances of explosions when flammable vapours, e.g. from ethanol have accumulated within a freezer and been ignited by a spark. If you do not have a spark-proof freezer, you must ensure that any ethanol is placed in a sealed, vapour-tight plastic container. If you do not have access to a freezer at all, put some ice in an insulated container, then stand the bottle of ethanol in it for several hours before use.

Quantification of DNA in a solution

A standard method of measuring the amount of DNA in a solution is to measure its absorbance of ultraviolet light. Most schools do not have an UV spectrophotometer, however, so the following method, which uses a colorimeter capable of measuring absorbance at 660 nm, can be used instead. For this, denatured DNA (strictly deoxypentose) is reacted with acidified diphenylamine to produce a dark blue-coloured compound.

Preparation: The diphenylamine reagent can be prepared in advance, but once diluted, it must be used within 24 hours.

Materials and equipment:

- Standard DNA solution (containing 3 mg per cm^3), e.g. salmon sperm DNA in saline citrate buffer.
- Diphenylamine reagent, 3 cm^3 per sample (see recipe, below).
- DNA sample in saline citrate buffer.
- Saline citrate buffer (used as a 'blank' control).
- Test tubes.
- Plastic syringes of 10 cm^3 (without needles) for dispersing liquids.
- Glass marbles, to cover the test tubes.
- Antibumping granules (e.g. broken porcelain).
- Glass rod, for stirring liquids.
- Maker pen, fine-tipped, permanent.
- Safety spectacles.
- Spectrophotometer or colorimeter and cuvettes to measure absorbance at 660 nm. for the quantitative method.

Procedure:

1. Dispense 3 cm^3 of the DNA solution prepared in the preceding exercise (onion) into a test tube. Label the tube.
2. Place 3 cm^3 of a standard DNA solution in a second tube and 3 cm^3 of saline citrate buffer in a third tube. Label all of the tubes appropriately. These will provide a 'standard' and a 'control' respectively with which you may compare your test sample.
3. Wear eye protection. Add 3 cm^3 of diphenylamine reagent to each tube.

Safety measures while using diphenylamine reagent and boiling liquids: Eye protection should be worn. Diphenylamine is harmful, if ingested or inhaled, and may irritate skin or eyes, if it comes into contact with them.

Extraction of plasmid DNA from bacteria

This is the 'classical' method of isolating plasmid DNA; with it we can see the purpose of each step. Quicker, easier methods, using small resin filters that nest within a microcentrifuge tube are also available as kits from suppliers of materials for molecular biology, but these tend to be too costly for everyday teaching purposes.

Preparation: An overnight culture of *E. coli* containing a suitable (i.e. self-cloned) plasmid is required. Sterile tubes and tips are not needed.

Timing: This activity takes about 60 minutes, including a period of 30 minutes when the materials are being chilled.

Materials and equipment:

- One cm^3 liquid culture of *E. coli* containing plasmid DNA, grown overnight at 37°C. It is best to use a plasmid with a high-copy number (i.e., where numerous copies exist within each bacterial cell) such as pUC18 or pUC19. Antibiotic is needed in the medium to maintain the plasmid (e.g., ampicillin for pUC18, pUC19 or pBLU).
- Glucose/EDTA/tris (GET) buffer, 100 µl.
- SDS/sodium hydroxide (SDS/NaOH), 200 µl.
- Potassium acetate/ethanoic acid (KOAc), 150 µl.
- Tris/EDTA (TE) buffer, 15 µl.
- Ice cold ethanol (95%), 400 µl.
- Two microcentrifuge tubes of 1.5 cm^3
- Micropipettes or similar dispensing devices.
- Tips for micropipette, 8.
- Crushed i.e., in an expanded polystyrene cup.
- Microcentrifuge for 1.5 cm^3 microcentrifuge tubes, capable of producing at least 8000 × g. With a slower centrifuge (e.g. 2000–2003 xg) the times stated here should be doubled.
- Hair drier.

Procedure:

1. Spin down 1 cm^3 of bacterial culture in a 1.5 cm^3 microcentrifuge tube for 1 minute or until a mass of cells some 2–3 mm in diameter is visible.
2. Decant the supernatant (liquid) into disinfectant in a waste container.
3. Ensure that all the liquid is removed, using a micropipette to draw off any that remains in the tube.
4. Add 100 µl of GET buffer to the tube. Resuspend the cells by closing the tube, then flicking the side repeatedly with a finger. EDTA in the buffer chelates Mg^{2+} and Ca^{2+} ions, which are necessary for DNAse action and cell membrane stability.
5. Add 200 µl of SDS/NaOH solution. Mix gently by inverting the closed microcentrifuge tube. Leave to stand for 5 minutes at room temperature. SDS is an ionic detergent, which dissolves the phospholipids and proteins of the cell membrane. This lyses the cells releasing the DNA. The sodium hydroxide splits both the plasmid and chromosomal DNA into single-stranded molecule, but each denatured plasmid remains stuck as two interwined rings.
6. Add 150 µl of KOAc and mix gently by inverting the closed microcentrifuge tube. Leave to stand on ice for 5 minutes. Treatment with KOAc restores the pH of the solution to neutral—the DNA consequently renatures. KOAc also precipitates the SDS, phospholipids, proteins and large DNA molecules.
7. Centrifuge the tube for 5 minutes to spin down the cell debris and chromosomal DNA.

8. Very carefully transfer 400 μl of the supernatant (liquid) to a clean microcentrifuge tube. Ensure that none of the cell debris is carried over. The supernatant contains the plasmid DNA.
9. Add 400 μl of ice cold ethanol. Mix gently.
10. Leave the tube in a deep freeze (–18°C – 20°C) for 10–30 minutes, then centrifuge for 10 minutes at high speed. You should now have a very small but just visible pellet.
11. Decant the supernatant into a waste container, taking care not to discard the pellet of plasmid DNA.
12. With a micropipette, remove the last drops of liquid. If convenient, leave the uncapped tube to stand at room temperature for 15 minutes or use a hair drier so that all the alcohol evaporates.
13. Dissolve the pellet in 15 μl of TE buffer; vigorous mixing is needed to ensure that this happens. Store the plasmid solution in a freezer or run it on a gel.

Procedure:

1. Mix 15 μl of plasmid DNA solution with 4 μl of loading dye.
2. Very carefully load the sample into a well in a 0.8 per cent agarose gel.
3. Run the gel until the loading dye has moved 30–40 mm.
4. Pour off the buffer, leave the gel in the tank and pour on 0.04 per cent.
5. Wash the gel surface with water.
6. The plasmid bands will appear as the stain moves down through the gel. Three bands, corresponding to three physical forms of the plasmid should be seen.

Safety measures while handling microorganisms and ethanol in freezers: Good microbiological practice should be observed.

Precaution: Do observe safety measures while handling chemicals as well as microorganisms.

REVIEW QUESTIONS

1. In what way the introduction of recombinant DNA molecules into the host cells is similar to mutation and in what way it is different.
2. What is the function of intron or intervening sequences located between coding sequences within gene in eukaryotic organisms?
3. What is value of recombinant DNA and gene cloning technology to geneticists?
4. What determines the sites at which DNA molecules will be cleaved by a restriction endonucleases?
5. Restriction nucleases are invaluable tools for biologists. However, gene encoding restriction enzymes obviously did not resolve to provide tools for scientists. What is value of restriction endonucleases to the microorganism that produce them.
6. Why is the DNA of a microorganism not degraded by a restriction endonuclease that it produces even though its DNA contains recognition sequences normally cleaved by the endonucleases.

7. What major advantages does the polymerase chain reaction (PCR) technique have overall other methods for analyzing nucleic acid structure and function?
8. Almost all of the sophisticated cloning vectors—phages, plasmids or phage—plasmid hybrids in use today have one component (in addition to the same type of origin of replication) in common. What is the component and what is its function?
9. What advantages have the transcription vectors over other types of cloning vectors?
10. While comparing the nucleotide base-pair sequences of genomic DNA clones with the cDNA clones of specific genes of higher plants and animals, what are the most common differences observed?

REFERENCES

Abelson, J. and E. Butz (1980): Recombinant DNA, *Science*, 209, 1317.

Brown, D.D. (1973): The isolation of gene, *Scientific American* (August).

Brown, J.K. (1996): Agarose gel DNA quantification (sic). Access excellence activities exchange, www access excellence org/AE/AEC/EF brown dnahtml.

Brown, T.A. (1990): *Gene Cloning*, Chapman and Hull, London.

Burstonk (1956): A study of condition and mechanism of the diphenyl ribonucleic acid amine reaction for the calorimetric estimation of deoxy *Biochemical Journal*, **62**(2), 314–322.

Chilton, M.D. (1983): A vector for introducing new genes into plants, *Scientific American* 248 (June), 50-9 The Ti. Plasmid.

Cohen, S.N. (1987): The manipulation of genes, *Scientific American* (July).

Cooke, H. (1987): Cloning in yeast, an appropriate scale for mammalian genomes. *Trends in Genetics*, **3:**173.

Drlica, K. (1984): *Understanding of Gene Cloning*, Wiley.

Emery, A.E.H. (1984): *An Introduction to Recombinant DNA*, Wiley.

Hardy, K. (1986): *Bacterial Plasmid*, 2nd ed., Van Nostrand Reinhold, Wokingham.

Hartl, Daniel L. (1991): *Basic Genetics*, Jones and Bartlett, Boston.

Lin, E.C.C. (1984): *Bacteria, Plasmid and Phages*, Harvard University Press.

Luria, S.E. (1970): The recognition of DNA in bacteria, *Scientific American Journal*, August.

Marcur, J. (1961): A procedure for the isolation DNA from microorganism, *Journal of Molecular Biology*, **3**(2), 208–218.

Miller, R. (1996): DNA from bacterial cell, Richardson (Ed.), *SSERC Bulletin*, **189,** 8.

Pouwels, P.H. et al. (1985): *Cloning Vectors*, Elsevier, Amsterdom.

Smith, D.H. (1979): Nucleotide sequence specificity of restriction enzymes, *Science*, **205,** 455.

Smith, H.O. and K.W. Wilcox (1970): A restriction enzyme from *Haemophilus influenzae*, *Journal of Molecular Biology*, **51,** 379–391.

Part IV

Quantitative Genetics

CHAPTER

26 Concept of Probability

The laws of probability are of fundamental importance (a) in predicting the likelihood of certain results from a given cross, (b) in assessing how well the phenotypic ratio of a progeny fits in a particular hereditary postulate, and (c) in understanding the operation of genetic mechanism. One of the fundament rules of probabilities is that the chance (probability) of the simultaneous occurrence of two or more independent events is equal to the product of the probabilities of each event. Burn and Brunitto (1989) have explained the concept of probability to be used in the genetical researches, very nicely.

PROBABILITY OF TWO INDEPENDENT NONGENETIC EVENTS

The laws of probability can be applied to any random event such as tossing a coin into air to decide in a game like cricket who should bat first out of two parties on the basis of head or tail provided the highly improbable chance of falling of the coin on edge is neglected. There is always a probability that coin may fall either on head or tail side, i.e. chance of a head or tail is 1 in 2 or 1/2(50:50). But if the same coin is tossed four times, a ratio of 2 heads: 2 tails may or may not happen. There is always possibility of occurrence of 4 heads: 0 tail; 3 heads: 1 tail; 3 tails: 1 head and 4 tail: 0 head. However, if the number of tossing are made more than 4, i.e. 10, 20, 30, 40, 50 and so on; the expected results would be near to 1:1 ratio because successive tosses would be presumably independent of each other.

Two Coin Tosses

Now when two coins are taken and tossed simultaneously by two persons suppose 50 times, what would be the outcome? Let us say, outcome of any single toss of two coins together would be hh (both heads) or ht (one head and one tail) or tt (both tails) provided the coins are to be shaked in the closed cupped hands before letting them fall on the table or ground. Suppose, the results of the 50 times tossing of the two coins was 12 hh: 27 ht: 11 tt. Now the question is whether these results are according to the expectations based on the assumption that each coin has an equal chance of landing either head or tail. Similarly, in a genetical study after making crosses, one can study what possible genetic mechanism may be operating in order to produce expected results. In either a coin tossing or genetical studies one has to formulate certain hypothesis to predict or explain the results whether in the term of hypothesis, deviation from the predicted result is within the limits set by chance alone, if the hypothesis could be used to predict the results of other experiments. If not, the results were according to hypothesis; the hypothesis would have to be altered. However, in order to make any definite conclusion, one has to take care of certain assumptions in case of tossing coins or genetical experiments. First, whether coins tossed were unbiased that is the coins were tossed in such a way that each coin has equal chance of landing either head or tail? Second, whether falling on head or tail of one coin has any side effect on the nature of the falling of other coin? That is, does each has an equal chance of falling on head or tail side? It means behaviours of the two coins are independent of each other. In case of genetical studies behavior of two genes in segregation are independent of each other. Third, that successive tosses (events) are also independent of each other. That is, toss resulting two heads will not affect the outcome of the next toss.

Based on the above said assumptions, it is concluded that the coins or genetical events are unbiased; independent of each other and outcome of several tosses or crosses are likewise independent of each other. If so, what result could be expected?

Each coin has one chance in two or a probability of 1/2 as head; and a probability of 1/2 of landing as tail.

By product rule of probability, chances of two coins falling on head side is $1/2 \times 1/2 = 1/4$ and on tail side also it is the $1/2 \times 1/2 = 1/4$.

Let us see a third situation where it is possible that by tossing two coins, one head (H) and one tail (T) may be obtained. If one landing on head (1/2) and one landing on tail 1/2, then there is a probability of $1/2 \times 1/2 = 1/4$. On this assumption total array of possibilities may be (l/4hh + 1/4tt + 1/4ht) = 3/4; it may be devoid of 1/4 possibilities unaccounted. However, the rule of additive probability is that the chance of one head plus one tail is really the sum of the probability of coin I being head and coin II being tail plus the probability of coin I being tail and coin II being head. That is, the ht category is really ht plus th or 2ht. Therefore, the probability of one head and one tail of two coins is $2(1/2 \times 1/2) = 1/2$.

The general formula for two independent events falling together is:

$$(a + b)^2 = a^2 + 2ab + b^2$$

where

a = probability of head

b = probability of tail

2 = number of times the events occur

The observations and expectations of the two coins tossing can be summarized as:

Tossing of coin I

Class	*Observed*	*Expected*
hh	12	12.5
ht	27	25.0
tt	11	12.5
Total =	50	50.0

where

h = head

t = tail

Tossing of coin II

Class	*Observed*	*Expected*
hh	10	12.5
ht	33	25.0
tt	7	12.5
Total =	50	50.0

We can examine these observations applying χ^2 (chi-square) formula to judge their goodness of fit.

Four Coin Tosses

Let us see an experiment or game involving four coin tossed together 100 times. The possible results of heads and tails combinations obtained from 100 tosses are as under:

Class	*Observed*
hhhh	9
hhht	32
hhtt	29
httt	25
tttt	5
Total =	100

Like, the ideal results of two coin tosses 50 times have been determined from the expansion of $(a + b)^2$, the binomial $(a + b)^4$, can be expanded to determine the expected results

$$a = \text{probability of a head for any coin} = \frac{1}{2}$$

$$b = \text{probability of a tail for any coin} = \frac{1}{2}$$

Expansion of binomial

$$(a + b)^4 = a^4 + 4a^3b + 6a^2b^2 + 4ab^3 + b^4$$

On substitution of numerical values of a and b

$$\left(\frac{1}{2}\right)^4 + 4\left[\left(\frac{1}{2}\right)^3 \cdot \left(\frac{1}{2}\right)\right]^3 + 6\left[\left(\frac{1}{2}\right)^2 \cdot \left(\frac{1}{2}\right)^2\right]^2 + 4\left[\left(\frac{1}{2}\right) \cdot \left(\frac{1}{2}\right)^3\right]^3 + \left(\frac{1}{2}\right)^4$$

$$= 1$$

$$= \frac{1}{16} + \frac{4}{16} + \frac{6}{16} + \frac{4}{16} + \frac{1}{16}$$

$$= 1$$

Explanation

(a) The first term of expression $(1/2)^4$, indicates the probability of all four coins coming up head simultaneously.
(b) The second term indicates the probability of three heads (a^3) and one tail (b).
(c) The third term indicates that the probability of head (a^2) and tail (b^2) together (a^2b^2) is 2 + 2.
(d) The fourth term indicates the probability of coming up one head (a) and three tail (3b) together simultaneously.
(e) The fifth term indicates the probability of all the four coins coming up tail simultaneously.

The observed results can be compared with the expected results in a four coin toss together based on the rules that the events are independents (Burns and Brunitto, 1989).

Class	*Observed*	*Expected*
hhhh	9	6.25
hhht	39	25.00
hhtt	29	37.50
httt	25	25.00
tttt	5	6.25
Total =	100	100

The above said observed and expected deviations could be determined to know whether the above said differences are due to chance deviation, if tossing of four coins together are unbiased.

Expression of trinomial

The solution to many genetic problems can be provided by binomial approach. But it is necessary to know the methods of binomial expansion and the expression of the terms.

1. The power of binomial chosen, that is the value of n is determined by the number of coins, individual, etc. For the two coins, expression $(a + b)^2$ is used even the coin is tossed 10, 20 or 50 times; for four coins tossed simultaneously $(a + b)^4$ is used and so on. The power of binomial remains constant for a given number of events, irrespective of whether the number of tosses going up or down.
2. The number of terms in the expansion is n + 1. Thus, $(a + b)^2$ expands to $a^2 + 2ab + b^2$ have three terms and $(a + b)^4$ have five terms.
3. Every term of expansion contains both a + b. The power of a in the first term is equal to n.

Expansion of binomial

$(a + b) = a + b$
$(a + b)^2 = a^2 + 2ab + b^2$
$(a + b)^3 = a^3 + 3a^2b + 3ab^2 + b^3$
$(a + b)^4 = a^4 + 4a^3b + 6a^2b^2 + 4ab^3 + b^4$
$(a + b)^5 = a^5 + 5a^4b + 10a^3b^2 + 10a^2b^3 + 5ab^4 + b^5$
$(a + b)^6 = a^6 + 6a^5b + 15a^4b^2 + 20a^3b^3 + 15a^2b^4 + 6ab^2 + b^6$

If a and b are each equal to 1/2, the symmetrical probability curve may be shown as given in Figure (26.1).

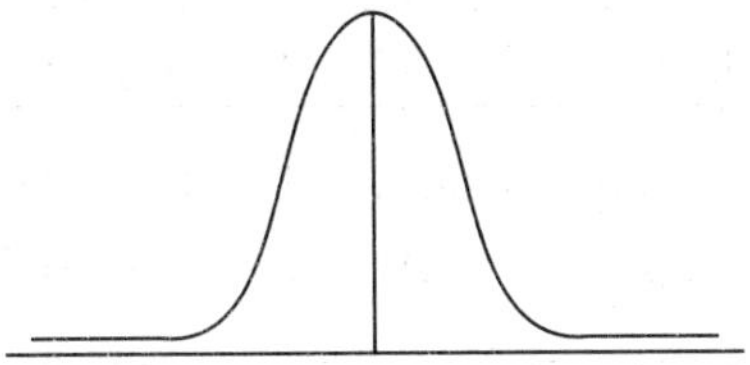

Figure 26.1 Probability curve drawn on the basis of binomial expansion where a = b = 1/2.

Uses of Binomial in Genetic Experiments

Different combinations in groups of given size representing a particular ratio can be calculated by binomial expansion of $(P + Q)^2$, where P and Q represent binomial of occurrence of two alternative events, e.g. P presents probability of a boy and Q represents the probability of a girl. This could also be symbolized as ♂, the symbol represents the shield and spear of man, the Roman war god ♀ is the female symbol which represents the mirror of Venus [Roman golden of love] and n represents the size of group involved. How many boys and how many girls would be expected in the randomly selected from families of 2, 3, 4, 5, or more. The combinations of families of a given size may be calculated by the binomial expansion of $(p + q)^n$ for 2 children in a family, the chances for all the boys; all the girls and 1 boy and 1 girl can be calculated as:

$$(p + q)^2 = p^2 + 2pq + q^2$$
$$p = q = 1/2$$

where

p^2 = probability of two boys = 1/4
q^2 = probability of two girls = 1/4
$2pq$ = probability of 1 boy and 1 girl = 1/2

For 3 children of a family, the probability of all the boys; all the three girls; 2 boys + 1 girl and 2 girls + 1 boy would be

$$(p + q)^3 = p^3 + 3p^2q + 3pq^3 + q^3$$
$$p^3 = 3 \text{ boys} = (1/2)^3 = 1/8$$
$$q^3 = 3 \text{ girls} = (1/2) = 1/8$$

Probability of boys and girls in families

Number of children	$(p + q)^n$, *where* p = ♂ *and* q = ♀
1.	$(1/2 + 1/2)^1$ = 1/2(1 ♂) + 1/2(1 ♀)
2.	$(1/2 + 1/2)^2$ = 1/4(2 ♂) + 1/2(1 ♂ + 1 ♀) + 1/4(1 ♀)
3.	$(1/2 + 1/2)^3$ = 1/8(3 ♂) + 3/8(2 ♂ : 1 ♀) + 3/8(1 ♂ : 2 ♀) + 4/16 (1♂ + 3 ♀)

4.	$(1/2 + 1/2)^4$ = 1/16(4 ♂) + 4/16(3 ♂ + 1 ♀) + 6/16(2 ♂ + 2 ♀) + 11/16(1 ♂ + 3 ♀) + 1/16(4♀)
5.	$(1/2 + 1/2)^5$ = 1/32(5 ♂) + 5/32(4 ♂ + 1 ♀) + 10/32(3 ♂ + 2 ♀) + 10/32(2 ♂ + 3 ♀) + 5/32(1 ♂ + 1 ♀) + 1/32(5 ♀)

$3p^2q$ = 2 boys + 1 girl = $3(112)^2$ (1/2) = 3/8

$3pq^2$ = 2 girls + 1 boy = $3(1/2)^2$ (1/2) = 3/8

In a family of 5 children, the probability of boys would be as follows:

1/32 = all boys = $(1/2)^5$
5/32 = 4 boys + 1 girl
10/32 = 3 boys + 2 girls
10/32 = 3 girls + 2 boys
5/32 = 1 boy + 4 girls, and
1/32 = all girls = $(1/2)^5$

When the probability of only a certain combination in a given size group is required, the factorial may be used. These are products of factors derived from functions by successfully increasing or decreasing by a constant usually '1'. For example, factorial 4 (4!) is the product of 4 × 3 × 2 × 1 or 24 [Factorial 0(0!) = 1 and ++ Factorial (1!) = 1 by definition].

The probability for a particular combination could be calculated from a formula based on appropriate term in bionomial expansion:

$$(p + q)^n = p = n!$$
$$x!\ (n - x)!\ p^X\ q(n - x)$$

where n is the product of integer B making up the total size of the group (n = total number in the group).

x! is the product of integers for the class with the probability p and q and (n – x)! is the product of the integers for the class with the probability q.

The symbol p represents the probability for a boy and symbol q represents the probability for a girl.

If 6 babies are born in a hospital on the same day; what is the probability that 2 will be boys and 4 may be girls.

$$p = q = \frac{1}{2}$$

$$p = \frac{n!}{x!(n-x)!\,p^*\,q(n-x)}$$

$$= 6\frac{1}{2!4!}(1/2)2\ (1/2)4$$

$$= \left(6\times5\times4\times3\times2\times\frac{1}{2}\times1\right)\ (4\times3\times2\times1)\ \left(\frac{1}{4}\right)\left(\frac{1}{16}\right)$$

$$= 15\times\frac{1}{4}\times\frac{1}{16}$$

$$= \frac{15}{64}$$

This 15/64 is the probability of 2 boys and 4 girls in a group of 6.

Example

If the trait being considered is in a human family and the parents are known to be heterozygous (Aa), the probability for a normal pigmented child (p) would be 3/4 and the probability for an albino child (q) would be 1/4. In the families of 4, what is the probability that 2 will be normal pigmented and 2 will be albino. Substituting the formula:

$$p = \frac{n!}{x!(n-x)!\,p^{*}q(n-x)}$$

$$= \frac{4!}{2!(2)!}\left(\frac{3}{4}\right)^2\left(\frac{1}{4}\right)^2$$

$$= 4 \times 3 \times 2 \times \frac{1}{2} \times 1(2 \times 1)\left(\frac{3}{4} \times \frac{3}{4}\right)\left(\frac{1}{4} \times \frac{1}{4}\right)$$

$$= \frac{54}{256}$$

$$= \frac{27}{128}$$

Thus, probability for 2 normal pigmented and 2 albino children in families of 4 children from heterozygous (Aa) parents, would be 27/128.

PROBABILITY OF SEPARATE OCCURRENCE OF INDEPENDENT EVENTS

The application of probability approach can be taken up in determining the frequency of an allele in a population for genetical studies 'say' allelism in human population. The studies till conducted show that 1 in 10,000 babies born in some Scandinavian countries is albino. Albino character in human and plant both are governed by a single recessive gene pair found on autosomes. The albino babies are born with the union of two independent gametes carrying recessive gene alleles coming from two independent sources with equal frequency. The frequency in this case is 0.0001 (1 in 10,000) and the probability that any given gamete in the population carries the allele for albinism is equal to 0.0001 or 0.01.

Since this allele is a member of a pair of alleles whose total frequency must be 1, the frequency of dominant alleles should be 1 – 0.01 = 0.99

After getting these two values, the frequencies of three possible genotypes can be determined:

a = frequency of allele of pigmentation (A) = 0.99
b = frequency of alleles of albino (a) = 0.01

Substitute the binomial expansion $(a + b)^2 = 1$, which gives genotypic frequency.

Binomial Expansion

$$
\begin{aligned}
& a^2 + 2ab + b^2 \\
\text{Genotype frequency} &= AA + Aa + aa \\
&= (0.99)^2 + 2(0.99 \times 0.01) + (0.01)^2 \\
&= 0.9801 + 0.0198 + 0.0001
\end{aligned}
$$

Goodness of Fit

One does not expect always to get the result exactly to one's calculation and the question arise how much deviations from the expected results one can accept as a chance error. Mendel too had faced the same problem while calculating the ratios. Therefore, some mathematical tool is needed to determine 'goodness of fit' of such ratios and chi-square test is one such tool. The formula for calculating chi-square is

$$\chi^2 = \Sigma(O - E)^2/E$$

REVIEW QUESTIONS

1. In tossing three coins simultaneously, what is the probability in one toss of (a) three heads, (b) two heads, and one tail?
2. A couple has two boys and expecting a third child. What is the probability that the third child would be a girl?
3. A family has eight children, all girls. What is the probability of their ninth child would be a boy?
4. How often in the families of eight children will all eight be boys?
5. In crossing two heterozygous deep *Coleus* plants, what is the probability of the occurrence in the F_1 of (a) deep, and (b) shallow?
6. (a) Give the third term in the expansion of (a + b).
   (b) If $a = b = \frac{1}{2}$, what is the numerical value of this term?
   (c) If a = 1 and b = 1, what is the numerical value of this term?
7. For a given recessive allele in a particular population, four persons in 100 are homozygous. What is the probability that any given gamete in that population carries the recessive alleles?
8. In an Australian, population of 10,000 individuals have cystic fibrosis, an inherited disease of pancreas that is fatal to the children homozygous for the recessive allele, same is the case in USA. What is the probability that any given gamete in these population carries this recessive allele?
9. In Australian and American population given as above, what is calculated frequency of person heterozygous for cystic fibrosis?

10. A man and his wife both have free ear lobes, but both are heterozygous. They possess four children. What is the probability that these children include one girl, with free ear lobes, one boy with attached ear lobes and two girls with attached lobes?

REFERENCES

Burns, G.W. and P.J. Brunitto (1989): *Science of Genetics*, Macmillan, New York.

Chandrasekharan, S.N., S.V. Parthasarthy and K. Ramiah (1953): *Cytogenetics and Plant Breeding,* Vardhanachary and Co., Madras.

Gardener, E.J., M.J. Simmons and D.P. Snustad (1984): *Principles of Genetics*, John Wiley and Sons, Inc., New York.

Poehlman, J.M. and D.A. Sleper (1995): *Breeding Field Crops*, Panima Publication, New Delhi.

Snedecor, G.W. (1956): *Statistical Methods*, 5th ed., Ames. Iowa State College Press.

CHAPTER

27 Multiple Factor Hypothesis and Polygenic Inheritance

The seven pairs of characters studied by Mendel were luckly governed by the qualitative genes, also called *oligogenes* or *major genes*, each with full type effect. But all the characters do not lend for accurate qualitative description. For instance, height, grain size, grain weight, duration of flowering, flower size, and chemical composition of some oil content in corn have to be measured quantitatively in some appropriable units of measurements (centimetre or gram). Such characters are also called metric characters. These characters show continuous variation while major genes (oligogenes) show discontinuous variations in distinct classes. Thus, inheritance is governed by two types of variations: discontinuous (major genes effect) and continuous variation caused by quantitative (minor) genes.

It is believed that Mendel as well as rediscoverers of Mendel's laws of inheritance (1900) might have deliberately neglected the continuous variation in their research presumably because they felt it would only confuse their objectives. This sort of variability had, however, been main object of an active research by Galton and his students in England in the later part of the nineteenth century. They were unable to know the mode of transmission of continuous variation from parent to progeny. They were analyzing their data by applying statistical methods—frequency distribution, standard errors and coefficient of variables, etc. Because characters related to continuous variation can be specified accurately only in the metric terms such as length, time, weight, etc. they are referred as *quantitative characters*.

The classic study of quantitative inheritance such as colour of wheat was done by Swedish geneticist and plant breeder, Nilsson-Ehle, H. (1908) who crossed two varieties of wheat one with very dark red grain and the other with white grain. The F_1 produced 'intermediate colour' grains and in F_2, the colours of grain ranged from deep red to white.

$R_1R_1\ R_2R_2$	↓	$r_1r_1r_2r_2$	Parents
(Deep red)	×	(White grain)	
F_1	$R_1r_1\ R_2r_2$		
	(Medium red)		

He selfed the F_1 generation. In F_1 generation there was no dominance. In F_2 generation, five classes of grain colours from deep red to white appeared in the ratio of 1:4:6:4:1 as shown in Table 27.1.

TABLE 27.1 Number of phenotypes and genotypes appeared in F_2 generation

F_2 progeny (Genotype)	*Colour*	*Number of phenotype*	*Number of dominant gene*
$R_1R_1\ R_2R_2$	Very deep red	1	4
$R_1R_1\ R_2r_2$ $R_1r_2\ R_2r_2$ $R_1R_1\ r_2r_2$	Deep red	4	3
$R_1r_1R_2r_2$ $r_1r_1R_2R_2$ $R_1r_1r_2r_2$ $r_1r_1R_2r_2$	Pink	6	2
$R_1r_1\ r_2r_2$	Light red	4	1
$r_1r_1r_2r_2$	White	1	0

The distribution of the F_2 plants into five colour classes exhibited continuous variation. Here only two genes were involved. If the number of genes are large, a huge population has to be raised to have a good chance of recovering the parental types.

TRANSGRESSIVE SEGREGATION

Sometimes plants with traits arise outside the range of parental type by segregation. This is called *transgressive segregation.* Transgressive segregation occurs when the parents are intermediate to the extremes of the segregating population.

	$R_1R_1\ r_2r_2$ (Medium red)	×	$R_2R_2\ r_1r_1$ (Medium red)	Parents
	↓			
F_1		$R_1r_1\ R_2r_2$ (Medium red)		

In F_1 generation, there was no dominance because colour of grain was intermediate (medium red).

In F_2 generation, a number of genotypes with distinct seed coat colour appeared, which were not available in parental population shown in Table 27.2 (after Falconer, D.S. and Makay, 1996).

Although the multiple factor hypothesis viewed the genotypes for a given quantitative character as simply the sum of the effects of some definite number of genes. Differences did not visualise coordinated genetic systems of the type now regarded as basic to quantitative inheritance. It was one of the most important contributions to the genetic thoughts.

Here genotype ($R_1R_1\ R_2R_2$) of very deep red grain colour is obtained as a result of *transgressive segregation.* This principle is very useful in plant breeding to obtain rare segregates superior to the parental types. Such types of *transgressive variants* can be obtained in F_3 and also in $F_4 - F_6$ generations, when a character is governed by many genes and linkage is absent.

TABLE 27.2 Quantitative gene effect in the expression of seed coat colour in F_2

F_2 generation	Genotype	Colour	Phenotype	Number of dominant gene
1	$R_1R_1\ R_2R_2$	Very deep red	1	4
2	$R_1R_1\ R_2r_2$	Deep red		
2	$R_1r_2\ R_2R_2$		4	3
1	$R_1R_1\ r_2r_2$	Medium red		
1	$r_1r_1\ R_2R_2$		6	2
4	$R_1r_1\ R_2r_2$			
2	$R_1r_1\ r_2r_2$	Light red		
2	$r_1r_1\ R_2r_2$		4	1
1	$r_1r_1\ r_2r_2$	White	1	0

HUMAN EYE COLOUR

In human beings, eye colour differs a lot because of variation in the melanin pigmentation in the iris. Except in albinos, almost every individual possesses some eye pigmentation. Studies conducted indicated that a number of genes alleles are involved in the expression of eye colours. If we consider that the number of phenotypic classes are more than twice the number of pairs of polygenes (2n + 1), nine classes would result from the action of four pairs of alleles. According to this hypothesis, the human eye colours must have the following alleles in genotypes:

Number of genotype	Contributing allele in eye colour
0	Light colour
1	Medium blue
2	Dark blue
3	Gray
4	Green
5	Hazed
6	Light brown
7	Medium brown
8	Dark brown

It is clear that eye colour is a polygenic character. Similarly, in human being, skin colour is also polygenically-controlled character.

In cattle, animals with solid (uniform) and spotted colours are frequently observed. Solid is due to dominant allele S and spotting is due to recessive alleles. Studies conducted indicate that the degree of spotting in ss individual depends upon a rather large series of polygenes.

The quantitative inheritance cited here is based on 'multiple genes' which produce small but equal and are cumulative in their effect. They do not exhibit dominance between alleles. Quantitative genes (polygenes) are highly Influenced by the environmental factors. Sometime, environmental factors mask the expression of these polygenes.

MULTIPLE FACTOR HYPOTHESIS

East (1916) gave conclusive evidence that inheritances of quantitative characters are governed by Mendelian genes. East studied variation of corolla length in tobacco by crossing true breeding parents and he studied variation from F_2 to F_3 generation using statistical tools. Based on his studies, he gave the concept of multiple factor hypothesis through which he postulated that a definite but large number of genes (factors) each with small but similar and cumulative effect determine the inheritance of quantitative characters.

TYPES OF GENE ACTION

With multiple genes four types of gene action have been recognized in quantitative traits. They are additive, dominance, epistasis and overdominance. The nature of their action are explained below.

Additive

In this type of gene action, each gene enhances the phenotypic expression of the trait. In the expression of a quantitative trait (height for example) the effect of a gene adds one increment in the phenotypic expression of the trait; two genes will add two units and so on, whether they are alleles at one locus or independent genes at several locus, where $aabb = 0$, $Aabb = 2$, $AAbb = 3$, $AABB = 4$.

Dominance

Dominant gene effect is deviation from additivity so that heterozygote is more like one parent than other. With complete dominance, heterozygote and one of the parent will have equal effect, e.g. $aa = 0$; $Aa = 2$ and $AA = 2$.

Overdominance

In this case, each allele contributes a separate effect and the combined alleles contribute an effect greater than that of either allele separately. If the effect of each allele is one as above, therefore, $aa = 1$, $AA = 1$ and $Aa = 2$.

Epistasis

Epistasis is the results of nonallelic gene interactions, i.e. the interaction of genes located at different loci, the two genes may have no effect individually on phenotypes, but when they come together, the homozygote (*AABB*) and heterozygote (*AaBb*) show effect on the phenotype. For example,

$$AAbb = 0$$
$$aaBB = 0$$

but

$$AB = 4$$

For two genes *AA* and *BB,* the possible combined effects have been studied by Comstock (1964) which is given in Table 27.3 (Comstock, 1964). He has given five levels of gene effects (0–4) with two pairs of genes, where *A* = *B*. But this does not hold true for all the genes, because two genes may effect the expression of trait in different way. However, for the development of varieties, additive gene effect is the best, for heterosis breeding, dominance, overdominance and epistasis too are the best proposition.

TABLE 27.3 Gene models for two loci show effects of gene action involving alleles of two genes at different loci assuming that each positive effect adds an increment of one to the expression of four quantitative characters

Model I: *Additive gene effect*			
Genotype	*AA*	*Aa*	*aa*
BB	+4	+3	+2
Bb	+3	+2	+1
bb	+2	+1	0
Model II: *Dominant gene effect*			
Genotype	*AA*	*Aa*	*aa*
BB	+4	+4	+2
Bb	+4	+4	+2
bb	+2	+2	0
Model III: *Epistatic gene effect (complementary)*			
Genotype	*AA*	*Aa*	*aa*
BB	+4	+4	0
Bb	+4	+4	0
bb	0	0	0
Model IV: *Overdominant gene effect (complex)*			
Genotype	*AA*	*Aa*	*aa*
BB	+2	+3	+1
Bb	+3	+4	+2
bb	+1	+2	0

Effect of gene action involving alleles of two genes at different loci assuming that each adds one + increment in the expression of a quantitative character could be ascertained by gene models as given in Table 27.3.

QUANTITATIVE CHARACTERS

Some quantitative characters such as yield component characters (yield has no gene) spike length, grain per spike, grain weight and tillers per plant and oil contents are influenced by many genes and these characters are equally influenced by the environment also.

Johanson (1909), Nilsson-Ehle, H. (1908) and East (1916) demonstrated that quantitative characters are inherited according to Mendelian genetic principles. Since then it became clear that variation in quantitative characters arise from joint action of genotype and environment.

If P is the phenotype, it can be represented by:

$$P = \mu + g + e + (ge)$$

where

μ is the general population mean
g shows the genotypic effect
e shows the environmental effect
ge shows an interaction effect of genotype × environment

Any genotype or environment adds or subtracts from the mean value of the phenotype depending on whether its effect on character (yield in case of crop plant) is above or below the general population mean. The interaction (G × E) will be zero only when all the genotypes behave consistently in all the environments equally that is, there is no G × E (genotype × environment).

The phenotype is the interaction product of G × E and the genotype is the particular assemblage of the genes possessed by the individual and the environment in all the nongenetic circumstances that influence the phenotypic value. The two components of value associated with genotype and environment are genotypic value and environmental deviations. The genotype confers a certain value (a) on the individual and the environment causes a deviation from this (a) in one (+a) or other direction (–a) could be symbolically written as

$$P = G + E$$

where

P is the phenotypic value
G is the genotypic value, and
E is the environmental deviation

Genotype	A_2A_2		A_1A_1	A_1A_1
Genotype value	–a	0	d	+a

The genotypic value d, of the heterozyote A_1A_2 depends on the degree of dominance, if there is no dominance d = 0. However, if A_1 is dominant over A_2, then d is positive and if A_2 is dominant over A_1, then d is negative. If the dominance is complete, then d is equal to $+a$ or $-a$ and if there is overdominance d is greater than $+a$ or lesser than $-a$. The degree of dominance may be expressed as d/a.

THEORY OF POLYGENES

The multiple factor hypothesis of Nelson-Ehle and East was extended to polygenes theory by Mather (1943) and his colleagues. The polygenes possess two important characteristics. First, their individual contribution to the expression of a character is very small .as a result their identification by segregation, recombination and mutation test is difficult because effect is nonvisible. Second, a large number of genes are involved to express the characters they control which are widely distributed throughout the genome. These genes too have positive (+) and negative (–) effects and hence the phenotype is product of total of positive and negative effects of polygenes. According to Mather, these genes form a group of closely linked genes which tend to inherit as a unit because of the cross-over difficulty between the genes to yield out recombination.

LINKAGES OF QUANTITATIVE AND QUALITATIVE GENES

The quantitative genes also show linkages with qualitative genes. One of such example of genetic linkage between qualitative and quantitative characters was observed by Sax (1950) in dry bean *Phaseolus vulgaris* in which pigmented and white-seeded strains differed considerably in their seed weight. He made crosses between pigmented plants with larger seed weight. In F_2 generation, the pigmented plants showed larger seed weight, than white seed types. He concluded that gene for pigment production was linked with gene governing larger seed weight. Castle (1932) observed an association of coat colour with body size in mice, rats and rabbits.

QUANTITATIVE TRAIT LOCI (QTL)

From the recent studies, the idea of effective factors has received support for quantitative trait loci. The discovery of molecular markers along the length of chromosomes in the genome have greatly facilitated the linkage studies. Using different restriction enzymes, it is possible to map markers on each of the chromosomes in a genome. A number of quantitative traits in both plants and animals have been found to be linked with these markers.

Edward et al. 1992 in F_2 population of maize observed the characters like ear length, ear number, days to silk, grain weight, kernel row number and tasseel branch number, all quantitative traits are found to be linked with one or other molecular markers. They observed that molecular marker facilitated investigations are effective means of elucidating factors influencing quantitative inheritance.

Presently, QTL are best described as identifiable chromosome regions, which can be picked up in segregating population through their linkage with molecular markers. They are complex loci carrying a number of lightly linked genes so that they are inherited as unit.

REVIEW QUESTIONS

1. What is multiple factor hypothesis? Explain giving an example the types of gene action involved.
2. What are polygenes? How they are inherited?
3. What is transgressive segregation? Explain its importance in improving the crop plants.
4. What do you mean by quantitative traits? Explain the factors responsible for variation in quantitative traits and methods of measurements.
5. Write short notes on:
   (a) Transgressive segregation
   (b) Additive genes action
   (c) Epistasis
   (d) Quantitative trait loci.

REFERENCES

Comstock, R.E. (1964): Proc. of 19th Annual Hybrid Corn, Indus. Research Conference.
East, E.M. (1916): Studies of the seed size inheritance in *Nicotiana, Genetics,* **1,** 164–176.
Jain, H.K. (2000): *Genetics: Principles, Concept and Implications,* Oxford and IBH, Delhi.
Mather, K. (1943): Polygenic inheritance and natural selection, *Biol. Rev.*, **18,** 32–64.
Nilsson-Ehle, H. (1908): Lunds University Arskrift, N.F. Adv. 2: Bd.5.

CHAPTER

28 Genetic Components of Variances

The genetics of quantitative (metric) characters centres around the study of variation. The basic idea is the study of variation and its partitioning into the components and these components are attributable to the different causes. The relative magnitude determines the genotypic properties of the population in particular the degree of resemblance between relatives.

The amount of variation in the quantitative characters is measured and expressed as variance. When the values are expressed as deviation from the population mean, the variance is simply the mean of the sum of square of deviations

$$\frac{\Sigma d^2}{N}$$

The genotypic variance is the variance of genotypic values and the environmental variance is the variance of environmental deviations. The total variance is the phenotypic variance or the variance of the phenotypic values which is sum of the separate components. The components of the total variance and the values whose variances are measured are given in Table 28.1.

TABLE 28.1 Components of variances

Component	*Symbol*	*Value whose variances are measured*
1. Phenotypic	V_P	Phenotypic value
2. Genotypic	V_G	Genotypic value
(i) Additive	V_A	Breeding value
(ii) Dominance	V_D	Dominance deviation
(iii) Interaction	V_I	Interaction deviation
3. Environmental	V_E	Environmental deviation

The variance of phenotypic values (total variance)

$$V_P = V_G + V_E$$
$$= V_A + V_D + V_I + V_E$$

where

V_P = phenotypic variance
V_G = genotypic variance

V_E = environmental variance
V_A = variance due to additive gene effect
V_D = variance due to dominance gene effect
V_I = variance due to genetic interaction

The partitioning into genotypic and environmental variances does not give proper understanding of the genetic properties of a population and also does not reveal the cause of resemblance, therefore, it is further divided according to the division of the genotypic value into breeding value (additive gene effect), dominance deviation and interaction deviation so,

Values $G = A + D + I$

Variance components $V_G = V_A + V_D + V_I$

Genotypic components = Additive + Dominance + Interaction

ADDITIVE VARIANCE

This is also called variance of breeding value. It is an important component because it is the main cause of the resemblance between the relatives. Therefore, main cause of observable genetic properties of the population and the response of the population to selection. The additive genetic variance is fixable and important for true breeding homozygous lines/strains specially in the self-pollinated crops where out-crossing is limited. Such genetic variability is useful to the varietal breeding and improvement through pure line selection. Therefore, the important portion of genotypic variances are:

(a) Additive genetic variance
(b) Non-additive genetic variance
(c) Environmental variance

This partitioning would also be expressed as a ratio of additive genetic variances to the total phenotypic variance called as heritability.

$$\text{Heritability} = \frac{V_A}{V_P}$$

The procedure for estimation of additive genetic variance rest on the observation of the degree of resemblance between the relative and will be given later in this chapter. Our immediate concern here is to show how the genetic components of variances are influenced by gene frequency and the assigned genotypic value.

DOMINANCE

When a single locus is only under consideration, the difference between the genotypic value G and the breeding value A of a particular genotype is known as dominance.

$$G = A + D$$

The dominance deviation arises from the property of dominance among the alleles (at a locus). In the absence of dominance, the breeding values and the genotypic values coincide. The dominance deviations are the result of interaction between alleles within same locus (intraallelic). Since dominance variance is nonfixable, hence it could not be utilized in breeding varieties but for heterosis breeding.

INTERACTION VARIANCE

The properties of interaction variance have been worked out by Cockerham (1954), Kemtharne (1953(a), (b), 1954) and little is yet known about its importance in relation to the other components. It seems however, that the amount of variance contributed by it, is usually rather small and that neglect of it, is therefore not likely to lead serious error. The description of the properties of interaction variance rests on further subdivision into different subcomponents.

It is first divided according to the number of loci involved; two-factor interaction arises from the interaction of two loci; three-factor from three loci, etc. Interactions involving larger number of loci contribute so little variances that they can be ignored.

The next subdivision of the interaction variance is according to whether the interaction involves breeding value (additive values) or dominance deviations. There are thus three sorts of two-factor interactions; interaction between two breeding values gives rise to additive × additive variance (V_{AA}); interaction between breeding value of one lacus and the dominance deviation of other gives rise to additive × dominance variance (V_{AD}) and interaction between the two dominance deviations gives rise to dominance × dominance variance (V_{DD}). Therefore, the interaction variance can be subdivided into:

$$V_I = V_{AA} + V_{AD} + V_{DD} + \cdots$$

The partitioning we have made here is essentially theoretical one, and before we pass on, we should consider how much of it could be made actually in practice. When observations of resemblance between relatives are available, we can estimate the additive variance and so make the partition, V_A: ($V_D + V_I + V_E$) and if inbred lines are available we can estimate this environmental variance and so make the partition V_G: V_E. If both these partitions are made, we can separate the additive genetic variance from the rest of the genetic variance and so make the three fold partition into additive genetic, nonadditive genetic and environmental variance (Table 28.2).

TABLE 28.2 Partitioning of variance of four characters in *Drosophila melanogaster*: components as percentage of total phenotypic variance

Component		*Character*		
		Bristle/100	Thorax/100	Eggs/100
Phenotypic	V_P			
Additive, genetic	V_A	52	43	18
Nonadditive genetic	$V_D + V_I$	19	6	44
Environmental	V_E	39	51	38

V_A: ($V_D + V_I$): V_E, the dominance and the interaction components are summed together as nonadditive genetic variance.

ENVIRONMENTAL VARIANCE

Environmental variance by definition embraces all the variations of nongenetic origin, can have great variety of causes and its nature depends very much on the character and the organism studied. Generally speaking, the environmental variation is a source of error that reduces precision in genetical studies and the aim of the experiment is to reduce it as much as possible by careful management or by proper designing of the experiments by increasing the number of replications, plot size, etc. The nutritional, climatic factors are the commonest external causes of environmental variation. They are partly under control. Maternal effects form another source of environmental variation that is some time important in mammals, but less susceptible to control.

Error in measurement is another source of variation, though it is quite trivial. Some character, however, cannot be strictly measured and hence have to be graded by eye judgment into classes, viz. Carcass quantities of livestock. In addition, to the variation arising from recognizable causes, such as those mentioned above, there is usually a substantial amount of non-genetic variation whose cause is unknown and which cannot be eliminated by experimental designs. This is generally called *intangible* variation. Some of the intangible variations may be called 'environmental'. Circumstances in the common meaning of the word that is by circumstances external to the individual even though their nature is not known. Some however may arise from 'developmental' variation that is, which cannot be attributed to the external circumstances, but is attributed in the ignorance of its exact nature to 'accidents' or 'errors' of development as a general cause. The characters more susceptible to the influence of external/environment in contrast, are those connected with the metabolic processes such as growth, fertility and lactation.

Example

Human birthweight provides an example of a character, subject to much environmental variation, whose nature has been analyzed in detail (Penrose, 1954; Robson, 1953). Partitioning of the phenotypic variance given in Table 28.3 indicates the relative importance of all the identified sources of variation. Birthweight being regarded as a character of the child. All the environmental variation is maternal in the sense that it is connected to the parental environment, but several distinct components of the maternal environment are distinguished. Maternal genotype which accounts for 20 per cent of the total phenotypic variance reflects genetic variance (chiefly additive) between mothers in the birthweight of their child, i.e. birthweight is regarded as the character of mother. Mother's environment generally accounts for another 18 per cent; reflects non-genetic variation between mothers in the same way. *Maternal environment immediately causes attributable to mother but* differing in successive pregnancies. Two causes of the same nature, age of mother and parity (i.e., whether child is first, second, etc.) are separately identifiable. Finally, the intangible variation is all the remainder of which causes cannot be identified.

TABLE 28.3 Partitioning of variance of human birthweight components as percentage of the total phenotypic variance

Causes of variation	*Total (%)*
Genetic: additive	15
Nonadditive (approximate)	1
Sex	2
Total genotypic	18
Maternal genotypic	20
Maternal environment, general	18
Maternal environment immediate	6
Age of mother	1
Parity	7
Intangible	30
Total environmental	82

GENOTYPE–ENVIRONMENT INTERACTION

A phenotype is considered as an interaction product of the genotype and the environment. A specific difference in the environment may have greater effect on some genotype than others or there may be a change in order of the productivity of a series of genotypes when evaluated under different environments. That is to say genotype x may be superior to genotype y in A_1 environment but inferior in environment A_2.

For example, two strains of mice between 3–6 weeks age were reared at two levels of nutrition:

	Good (A_1) nutrition	Bad (A_2) nutrition
Strain x	17.2 g	12.6 g
Strain y	16.6 g	13.3 g

Strain x grows better (A_1) than y in good nutritional supply than bad nutritional (A_2) supply.

An interaction between genotype and environment gives rise to an additional component of the variance and this interaction variance can be minimized by replicating the genotype by the use of inbreeds and F_1. In an experiment, which removes the genotypic variance by the use of all identical individuals (genotypes) with no genetic variance the result of partition.

Example

Partition of phenotypic variance into genotypic and environmental components has been done for several characters in *Drosophila melanogaster* (length of thorax)

Population	*Component*	*Observed variance*
Mixed	$V_G + V_E$	0.366
Uniform	V_E	0.186
Difference	V_G	0.180

$$\frac{V_G}{V_P}$$

$$= 0.180 \times \frac{100}{0.366}$$

$$= 49\%$$

That is, 49 per cent variation in the thorax length in the genetically-mixed population of *Drosophila melanogaster* is attributed to genetic differences between the individuals and 51 per cent due to nongenetic differences.

GENOTYPE–ENVIRONMENT CORRELATION (ASSOCIATION)

Correlation between genotype and environment is not of so much importance because by randomization, the environmental effect could be minimized. However, there are some circumstances in which correlation or associations do exist. In dairy cattle, yield is correlated with the quantity of feeds provided by the dairymen. The better genotypes are given feed grains according to the milk yield per (kg). This is also true in case of nitrogen (doses) given to the wheat crop varieties (Figure 28.1).

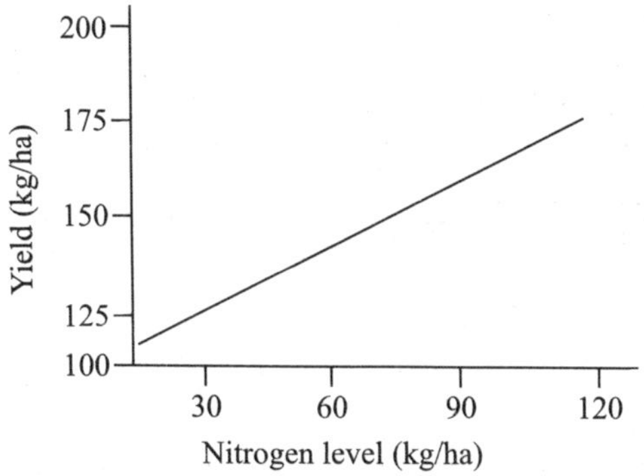

Figure 28.1 Effect of nitrogen level in wheat crop.

If inbred lines or F_1 are used in the experiment, the interaction variance remains with the environment in the phenotypic variance measured in the genotypically-uniform individuals. In the normal circumstances, therefore, this variance due to G × E interaction since it cannot be separately measured, is best known as part of the environmental variance under the large differences in the environment such as between different habitats, are under consideration. The presence of G × E interaction becomes important in connection with the adoption of animal breeds, crop varieties to local conditions.

GENOTYPE AND ENVIRONMENTAL VARIANCE

The relative importance of genotype and environment interaction components determines the phenotypic value P = G + E. The relative importance of a cause of variance means the amount of variation that it gives rise to as a proportion of total. Therefore, the relative importance of genotype as determinant 'backbone' of the phenotypic value can be given by the ratio of genotypic to the phenotypic variance

$$\frac{V_G}{V_P}$$

The genotypic and environmental components can be estimated in the experimental population. The environmental variance can not be separated because it includes all the nongenetic variance much beyond the experimental control but genotypic variance can be separated by use of inbreds, F_1 of a cross of two inbred lines with no genotypic variance. If a group of such individuals is raised under the normal range of environmental circumstances, their phenotypic variance provides an estimate of environmental variance V_E. Subtraction of this from the phenotypic variance of a genetically-mixed population then gives to rise an estimate of genotypic variance of this population.

The phenotypic variance was measured first in a genotypically-mixed, i.e. a random bred population and there is a genetically-uniform population consisting of F_1 of the inbreds lines (F.W. Robertson, 1957(b)) as follow:

Variation in F_2 and Backcrosses

The variations in the true breeding parental lines and their F_1 are exclusively nonheritable. But variation in F_2 and onward is always heritable because of the segregation of the heritable components (alleles). In the absence of differences in the fertilization and viability, the constitution of F_2 in respect of the segregating pairs of alleles A–a will be 1/4AA : 1/2Aa : 1/4aa and these genes will contribute increments of da+, ha: – da to the measurements of the individuals in the three classes respectively; the derivation from the mid-parent may be 1/2ha; and taking all genes into account, the F_2 mean will be

$$\frac{1}{2S(ha)} = \frac{1}{2[h]}, \quad \text{i.e. half of the } F_1 \text{ mean}$$

Considering only Aa locus of our model, the values associated with the genotypes are:

Genotypes	*Values*
AA	da
Aa	ha
aa	–da

Since all the deviations are measured from AA—aa mid-parent as the origin, the difference between the AA and aa homozygotes has the value 2da. It can be shown like

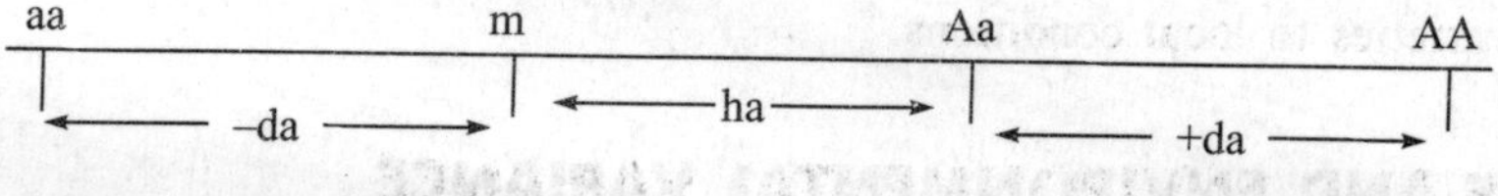

The capital letter symbol refers to the allele with a plus effect on the character and does not denote either dominance or recessiveness. When character is governed by many genes the phenotype of a true-breeding strain will be

$$\Sigma(+d) + \Sigma(-d) + - \ldots$$

where

$$\Sigma = \text{summation (total sum)}$$

The F_1 of a hybrid AA × aa will be phenotypically ha

- If ha = 0, no dominance
- If ha = da, full dominance
- If ha > da, overdominance

If F_1 deviates from mid-parental value, some dominance obviously must exist at one or more loci. Therefore, the F_1 between two homozygous strains differing in a number of genes will be Σ(h).

If all (+) alleles by which two strains differ are concentrated in one parent (AA) and all (–) alleles in other; the difference between two strains must be of Σ(d). However; in practice the difference between two strains will usually behave a lesser value than Σ(d), because isodirectional distribution of (+) and (–) alleles is not very likely. The contribution of F_2 generation as regard the gene A--a is expected to be

Genotypes:	$\frac{1}{4}$AA	: $\frac{1}{2}$Aa	: $\frac{1}{4}$ aa
Phenotypes:	da	ha	–da

The mean of the F_2 generation is expected to be

$$\frac{1}{4}(da) + \frac{1}{4}(-d)$$

$$a + \frac{1}{2}(ha) = \frac{1}{2}ha \quad \text{— Generation mean}$$

The contribution of this gene to the total sum square (sum of the squares of deviation from the mid-parent) will be

$$\frac{1}{4}(da)^2 + \frac{1}{2}(ha)^2 + \frac{1}{4}(-da)^2 = \frac{1}{2}(da)^2 + \frac{1}{2}(ha)^2$$

The contribution to the sum of squares of deviation from F_2 mean will be

$$\frac{1}{2}(da)^2 + \frac{1}{2}(ha)^2 - \frac{1}{2}(ha)^2 = \frac{1}{2}(da)^2 + \frac{1}{4}(ha)^2$$

Since the generation mean is $\frac{1}{2}$ha, the variance of F_2 is

$$\frac{1}{2}(da)^2 + \frac{1}{2}(ha)^2 - \frac{1}{4}(ha)^2 = \frac{1}{2}(da)^2 + \frac{1}{4}(ha)^2$$

This is the only genetic portion of the variance; provided that there is no nonallelic gene interaction and linkage does not exist; an environmental portion will occur in the actual experimental results.

Let

$$D = \Sigma(d)^2$$
$$H = \Sigma(h)^2$$

The total heritable variance exerted by k genes in F_2 will be

$$V_{F_2} = \frac{1}{2}D + \frac{1}{4}H + E \tag{28.1}$$

where

$$D = \Sigma(da)^2 \text{ and}$$
$$H = \Sigma(ha)^2$$

The heritable variation in F_2 is, therefore, divided into two portions:

D = additive portion and
H = dominance portion

By similar methods the summed variance of backcrosses to the parents is found to be:

$$V_{B_1} + V_{B_2} = \frac{1}{2}D + \frac{1}{2}H + 2E \tag{28.2}$$

Let me cite here an experimental result to illustrate the general feature of the estimation procedure.

Example

In an experiment heading date was determined for the wheat varieties Ramona (P_1) and Baart (P_2) and for F_1, F_2 and first backcross generation with both parents. The mean and variance for these crosses are given in Table 28.4 (after Allard, R.W.).

TABLE 28.4 Data of heading of homozygous wheat varieties, Ramona (P_1), Baart (P_2) and hybrid generations derived from cross $P_1 \times P_2$

Generation	*Number*	*Mean* ($\bar{X}$) *(data of heading)*	*Variance* (δ_2)
P_1	159	12.99	11.036
B_1	329	15.63	17.352
F_1	171	18.45	5.237
F_2	552	21.20	40.352
B_2	314	23.38	34.288
P_2	148	27.61	10.320

$$V_{F_2} = \frac{1}{2}D + \frac{1}{4}H + E = 40.352 \tag{28.3}$$

$$V_{B_1} + V_{B_2} = \frac{1}{2}D + \frac{1}{2}H + 2E = 51.640\ (17.352 + 34.288) \tag{28.4}$$

Hence,

$$2F_2 = D + \frac{1}{2}H + 2E = 80.70\ (40.352 \times 2)$$

$$V_{B_1} + V_{B_2} = \frac{1}{2}D + \frac{1}{2}H + 2E = 51.640$$

$$\frac{1}{2}D = 29.060$$

By using the mean variance of nonsegregating generation, we estimate the value of E. Therefore,

$$E = \frac{V_{P_1} + V_{P_2} + V_{F_1}}{3}$$

$$= \frac{11.036 + 10.320 + 5.237}{3}$$

$$= \frac{26.593}{3}$$

$$E = 8.864$$

By substituting the estimate of D + E in Eq. (28.1) or (28.2), we can estimate the value of H.

$$\frac{1}{2}D - \frac{1}{2}H + 2E = 51.640$$

$$\frac{1}{2}H = 51.640 - 29.060 - 2(8.864)$$

$$\frac{1}{2}H = 4.792$$

$$H = 9.584$$

$$\frac{1}{2}D = 29.060$$

In F_2 total variance due to environment is

$$\frac{8.864}{40.35} = 0.22 \times 100$$

$$= 22\%$$

Therefore, variance due to genetic cause is 100 – 22 = 78%
Thus, 78% is the heritability in broad sense.
Heritability in narrow sense = 29.060/40.352 = 72.0%
Genetic advance at 5% selection pressure

$$(G) = (K)\ (\delta A)\ (\delta a_2/\delta A_2) = (2.06)\ (40.352)\ (0.720)$$
$$= 9.42 \text{ days}$$

Mean of the F_2 population was 21.20 days hence the mean of the F_2 progeny of the selected individuals is expected to be 21.20 – 9.42 = 11.87 days.

BACKCROSSING

In backcrossing of the F_1 to either of its true breeding homozygous parents AA or aa, half the progeny will be homozygous, and half heterozygous in respect of each gene pairs, by which parents differ, therefore, the contribution of A–a to the mean magnitude of the two backcrosses as measured from the mid-parent will be

$$B_1 = \frac{1}{2}(da + ha)$$

and
$$B_2 = \frac{1}{2}(ha - da)$$

The contributions to the variance of the two backcrosses can be easily shown as

$$B_1 = \frac{1}{4}(da + ha)^2$$

and
$$B_2 = \frac{1}{4}(da + ha)^2$$

where B_1 was backcross to AA, and B_2 was backcross to aa, parent. The contribution of d and h to these variances are not separable as they stand, but if two variances are summed, A–a contributes $1/2(da^2 + ha^2)$ to the total contribution of d and h now being separable.

If the genes are independent in their expression and transmission; the heritable components of the two variances and their sum will be

$$V_{B_1} = \frac{1}{4}\Sigma(da + ha)^2$$

$$V_{B_2} = \frac{1}{4}\Sigma(da + ha)^2$$

$$V_{B_1} + V_{B_2} = \frac{1}{2}\Sigma(d^2) + \frac{1}{2}\Sigma(h^2)$$

$$= \frac{1}{2}D + \frac{1}{2}H$$

The V_{B_1} and V_{B_2} variances will also be including nonheritable component E. The full expressions of their sum will be

$$V_{B_1} + V_{B_2} = \frac{1}{2}D + \frac{1}{2}H + 2E$$

If $E > 0$, the $V_{B_1} + V_{B_2}$ will exceed

$$VF_2 = \frac{1}{2}D + \frac{1}{2}H + E,$$ even though dominance may be absent.

Examples

1. In a tomato cross between Denmark × red currant, the parental means were 1.672 and −0.053 and the mid-parental value was 0.809 (fruit wt. expressed on a log scale). The

F_1 mean was 0.707 suggesting preponderance of the dominance of genes for smaller parent 'red currant'. Powers in 1942 estimated heritable portions of the variance (HV) as HV_{BRC} (backcross to red currant) = 0.1353, HV_{BD} (backcross to Denmark) = 0.01907 and V_1F_2 = 0.03192. Thus, $HV_{BRC} + HV_{BD}$ exceeds HV_1F_2, by 0.00068. So, giving H = 0.00272, it follows that D = 0.06248 and S(dh) = 0.00554.

Dominance though clear but not marked S(dh). This indicates that genes for smaller fruit size parents are preponderantly dominant.

2. In another case, corolla length in *Nicotiana*, dominance appears to be more marked, in an interspecific cross:

   Nicotiana longsdorfii × *N. sanderae* (Smith, 1937)

   The parental means were:

   S = 1292

   Mid-parental value = 664.5 L = 37

   F_1 mean = 742

   Average variance of the reciprocal backcrosses to *N. sanderae* = 85.5

   N. longsdorfii = 98.5

   S(dh) is thus in direction with excess of F_1 over mid-parental value.

The nonheritable items included in V_{BL}, V_{BS} and V_{F2} may be estimated from the average parental value and F_1 mean as 742. Deduction of this leaves 43.5 and 56.5 as the heritable components of two average backcross variance and 88.5 as that the average F_2 variance. Then,

$$V_{B1} + B_2 = \frac{1}{2}D + \frac{1}{2}H$$

$$= 43.5 + 56.5 = 100$$

$$V_{F2} = \frac{1}{2}D + \frac{1}{4}H = 88.5$$

The degree of dominance = $\sqrt{\dfrac{H}{D}}$

$\sqrt{\dfrac{H}{D}}$ provides an estimate of the average dominance of k genes.

Similarly, contribution of gene A–a to F_3, the variance of F_3 mean will be

$$\frac{1}{4}d^2a + \frac{1}{2}\left(\frac{1}{2}ha\right)^2 + \frac{1}{4}(-da)^2 - \left(\frac{1}{4}ha\right)^2 = \frac{1}{2}d^2a + \frac{1}{16}h^2a$$

The overall mean of F_3 will be

$$\frac{1}{4}(h)$$

and total heritable variance of F_3 means will be

$$\frac{1}{4}S(d^2) + \frac{1}{16}S(h^2) = \frac{1}{2}D + \frac{1}{16}H$$

where S is the sum of total deviation of square.

COMPONENTS OF QUANTITATIVE GENETIC VARIATION (V_G)

It has been already explained in the beginning that the genetic variance is composed of three major components:

(a) Additive genetic variance (V_A)
(b) Dominance component variance (V_D)
(c) Nonallelic gene interaction or *epistasis* (V_I)

The genetic variation (V_G) it can be shown as

$$V_G = V_A + V_D + V_I$$

where additive genetic variance is the variance contributed by gene alleles having linear quantitative genetic effect. The resemblance between parent and offspring is determined by the additive genetic effects. Dominance components variance represents the deviation of the heterozygote (F_1) from the mid-parental value or the average of the homozygous parents ($P_1 + P_2/2$). The dominance variance may be smaller than the additive variance but unstable than the additive variance, and nonallelic gene interaction or epistasis is the result of:

(i) Additive × additive gene components
(ii) Additive × dominance gene component
(iii) Dominance × dominance gene component

The magnitudes and role of nonallelic gene interaction are smaller than the additive and dominance variations which are very difficult to evaluate.

Similarly, the covariance of F_3 mean with its F_2 parent's measurement will be

$$\frac{1}{2}S(d^2) + \frac{1}{8}S(h^2)$$

$$= \frac{1}{2}D + \frac{1}{8}H$$

and mean variance of F_3 families will be

$$\frac{1}{4}S(d^2) + \frac{1}{8}S(h^2)$$

$$= \frac{1}{2}D + \frac{1}{8}H \text{ (Fisher, Immer and Tendeis, 1932)}$$

GENETIC PARAMETER FOR ESTIMATION OF NONALLELIC INTERACTION

If sufficient generation means are available, one can detect the effects of nonallelic interaction on these means and can also estimate their magnitudes. Where if two back-crossed generations B_1, B_2 are available, the formula for six parameters given by Jink and Jones (1958) can be applied:

$$m = \frac{1}{2}P_1 + \frac{1}{2}P_2 + 4F_2 - 2B_1 - 2B_2$$

$$d = \frac{1}{2}P_1 - \frac{1}{2}P_2$$
$$h = \sigma B_1 + \sigma B_2 - 8F_2 - F_1 - 1\frac{1}{2}P_1 + 1\frac{1}{2}P_2$$
$$i = 2B_1 + 2B_2 - 4F_2$$
$$j = 2B_1 - P_1 - 2B_2 + P_2$$
$$l = P_1 + P_2 + 2F_1 + 4B_1 - 4F_2$$

where

m = mean
d = measures additive variance
h = dominance variance
i = additive × additive interaction
j = dominance × dominance interaction
l = additive × dominance interaction

The standard error of these estimates can be obtained by

$$V(d) = \frac{1}{4}V_{P1} + \frac{1}{4}V_{P2}$$
$$S(d) = \sqrt{V(d)}$$

and significance of d can be tested by

$$t = \frac{d}{s(d)}$$

If back-cross generations B_1 and B_2 are not available, then only five-parameter model described by (Hayman, 1955) can be applied. In five-parameter model, L (additive × dominance) parameter is absent.

EPISTASIS

The epistasis is (nonallelic interactions) of two types:

1. Complementary (C)
2. Duplicate (D)

Those in which 'h' and 'l' have same sign + + or – – we will refer it as complementary type. And those in which 'h' and 'l' have opposite sign + and – it can be concluded that interactions are on balance mainly of the duplicate type; dominant epistasis or recessive suppresser type.

CONCEPT OF HERITABILITY

The question whether a character is hereditary or environmental has no meaning unless the genes have the proper environment to cause a character to develop and conversely no amount of manipulation of the environment will cause a character to develop unless necessary genes are present. Nevertheless, one must recognize that the variability observed in some character is caused primarily by differences in the genes carried by different individual and that the variability in other characters is primarily due to the differences in the environment to which individuals have been exposed. In order to have a quantitative statement one must have a clear idea of the effect of environment on variability and also a clear idea of the effect of genes on variability.

Heritability of a metric character is one of the most important properties in the genetical studies. Heritability could be defined as the ratio of additive genetic variance to phenotypic variance.

$$h^2 = \frac{V_A}{V_P} \quad \text{(Narrow Sense)}$$

It can also be defined as the regression of breeding value on phenotypic value too. It measures the extent to which the variability in the yield or in any character in the population is the result of genetic factors which is thus transmitted to the progeny of the selected plants/individuals and how much variability in the yield is due to environmental factors in which it has grown.

$$h^2 = b_{AP}$$

Heritability specifies the properties of total variability that is due to the genetic causes or the ratio of the genetic variance to the total (phenotypic) variance:

$$H = \frac{\delta G^2}{\delta G^2 + \delta E^2} \quad \text{(Broad Sense)}$$

where δG^2 is the genetic variance; δE^2 is the environmental variance, and V_E (δ) is

$$= \frac{(V_{P_1} + V_{P_2})}{2}$$

or

$$V_E = V_{F_1} + V_{P_1} + \frac{V_{P_2}}{3}$$

$$2V_G = V_A + V_D + V_E$$

AVERAGE EFFECT

Average effect of a gene is the mean deviation from the population mean of individuals which received that gene from one parent, the gene received from the other parent having come at random from the population. This can be explained in the following ways:

> *"Let a number of gametes all carrying A_1 alleles unite at random with the gametes from the population; then the mean deviation from the population mean of the genotypes so produced is equal to the average effect of the gene A_1. The concept of average effect is easier to grasp in the form of the average effect of a gene substitution which can be more conveniently be used when only two alleles at a locus are under consideration. If we could change A_2 gene into A_1 at random in the population and could then note the resulting change of value, this would be average effect of the gene substitution. It is equal to difference between the average effects of the two genes involved in the substitution."*

The average effect of a gene or gene substitution depends on the gene frequency. The average effect is, therefore, a property of the population as well as of the gene. Consider the substitution of a recessive gene a, for its dominant allele A. The substitution will change the value only when the individual already carries one recessive allele, i.e. heterozygous Aa,. Changing AA to Aa will not affect the value but changing Aa to aa will. When the frequency of recessive allele a is low, there will be many AA individuals as a result substitution will not affect; but when frequency of recessive allele is high, there will be very few AA individuals, and the substituted aa individuals will give the effect.

Let a locus with two alleles A_1 and A_2 of frequency P and q respectively and take first the average effect of the gene A_1 for which the symbol α is given. If gametes carrying A_1 unite at random with gametes from the population, the frequency of the genotypes produced will be p for A_1A_1 and q for A_1A_2 and then mean would be pa + qd. The difference between this mean value and the population mean is the average effect of gene A_1 (Table 28.5).

TABLE 28.5 Average effect of gene

Types of gamete	*Values and frequencies of genotypes produced*			*Mean value of genotype produced*	*Population mean to be deducted*	*Average effect of gene*
Genotypic	A_1A_1	A_1A_2	A_2A_2			
value	a	d	−a			
A_1	P	q		pa + qd	−[a(p − q) + 2dpq	q(a + d) (q − p)
A_2		p	q	−qa + pd	−[a(p − q) + 2dpq	−p(a + d) (q − p)

Let us see how the average effect is related to the genotypic value a and d, consider a locus is with two alleles A_1 and A_2 with frequency p and q. First take average effect of gene A_1 using symbol α. If gametes carrying A_1 unite at random with gametes in the population. The frequency of genotypes produced will be P of A_1A_1 and q of A_1A_2, the genotypic value of A_1A_1 is +a and that of A_1A_2 is d, mean of these taking into account of proportion while they occur is pa + qd. The difference between this mean value and the population mean is the average effect of the gene A_1. Taking into account the value of the population mean from equation.

$$m = a(p - q) + 2dpq$$
$$\alpha_1 = pa + qd - [a(p - q) + 2dpq]$$
$$q = [a + d\ (q - p)]$$

Similarly, the average effect of the gene A_2 is

$$\alpha_2 = -p[a + d(q - p)]$$

If A_1 is substituted for A_2, a proportion of p will be found in A_1A_2 and a proportion q in A_2A_2 genotypes. In the former, the substitution will change the value from d to +a and in latter from −a to d. The average change is, therefore,

$$p(a - d)\ + q(d + a)$$

or

$$a + d\ (q - p)$$

which on rearrangement → becomes

$$a + d (q - p)$$

then, average effect of gene substitution would be

$$\alpha = a + d(q - p)$$

where α is the average effect of gene substitution.

Example

Consider the pygmy gene and its effect on bodyweight for which a = 4 g and d = 2 g. If frequency of pg gene is q = 0.1, the average effect of substitution for pq would be by equation $\alpha = a + d(q - p) = 4 + 2 \times 0.8 = 2.4$ g. If frequency is q = 0.4, the average effect of gene substitution would be $\alpha = 4 + 2 \times -0.2 = 3.6$ g. This way average effect would be greater, if the gene frequency is greater. The average effect of genes separately by equation $\alpha_1 = q\alpha$ and $\alpha_2 = p\alpha$ are

Average effect of +

$$\alpha_1 = \frac{q = 0.1}{+0.24} \quad \frac{q = 0.4}{+1.44}$$

Average effect of pg

$$\alpha_2 = -2.16 \quad -2.16$$

BREEDING VALUE

The concept of average effect indicates that parents pass on their genes, not their genotypes to the next generation. It is, therefore, average effect of parent's genes that determine the mean genotypic value of its progeny. The values of an individual 'suppose' a bull is judged by the mean value its progeny. If an individual male is mated to a number of females taken at random from the population, then its breeding value is twice the mean deviation of the progeny from the population mean. The deviation has to be doubled because the parent in question transmit only half the gene (haploid) to the progeny, the other half come at randomally chosen individuals from the populations mean. The breeding value should be expressed in absolute units but conveniently in form of deviation from the population mean. Just as the average effect is a property of the gene and the population; so the breeding value is a property of the individual and the population from which its mates are drawn.

DOMINANCE DEVIATION

Dominance is the product of intraallelic gene action and when a single locus is under consideration, the difference between genotypic value G and the breeding value A of a particular genotype is called dominance deviation (D)

$$G = A + D$$

Dominance deviation arises due to an intraallelic interaction at a locus and in absence of dominance, the genotypic value is equivalent to breeding values

$$G = A + O$$

Because average effects of genes and the breeding values of the genotypes depend on the gene frequency in the population; dominance deviations also dependent on gene frequency. The differences between the breeding values and the genotypic values are dominance deviation.

INTERACTION DEVIATION

When a single locus is under consideration, the genotypic value is made up of the breeding value and the dominant deviation only; when the genotype refers to more than one locus, the genotypic value may contain an additional deviation due to nonadditive combination.

Suppose GA be the genotypic value of an individual attributable to one locus and G the aggregate genotypic value attributable to both loci together, then

$$G = G_A + G_B + I_{AB}$$

where I_{AB} is the deviation from additive combination of these genotypic values. If I is not zero for any combination of the genes at different loci, those genes are said to be 'interact' and deviation is called interactive deviation a epistatic depiction.

If the interactive deviation is zero, the genes concerned are said to 'act additively' between loci. So, for all loci together we should say

$$G = A + D + I$$

where A is the sum of breeding value, attributable to separate loci and D is the sum of dominance deviations.

REFERENCES

Allard, R.W. (1960): *Principles of Plant Breeding*.

Falconer, D.S. (1960): *Introduction to Quantitative Genetics*, Longman, London.

Fisher, R.A. (1918): The correlation between relatives on the supposition of Mendelian inheritance 1941: Average excess and average defect of gene substitution. Trans. *Roy. Soc. Edin*, **52:** 399–433.

Hayes, A.K., Immer, F.R. and Smith, D.C. (1955): *Methods of Plant Breeding*, 2nd ed., McGraw-Hill, New York.

Hayman, B.I. (1955): Description and analysis of gene action and interaction, *Cold Spring Harb. Symp. an Quant. Biol.*, **20,** 79–84.

Hayman, B.I. and K. Mather (1953): The progress of inbreeding when homozygotes are at disadvantage, *Heredity*, **7,** 165–183.

Lic, C. (1955a): *Population Genetics*, Chicago University Press, London.

Lush, J.L. (1945): *Animal Breeding Plants*, 3rd ed., State Agriculture College Press, Iowa.

Mather, K. (1949): *Biometrical Genetics*, Mithuen and Co., London.

Race, R.R. and R. Sanger (1954): *Blood Groups in Man*, 2nd ed., Blackwell, Oxford.

CHAPTER

29 Estimation of Genotypic Variability in Varying Environments

Johannsen (1909) demonstrated the differences between genotype and environment by his famous experiment on lima bean Nilsson-Ehle, H. and East provided the evidences that quantitative characters are inherited according to the Mendelian laws. This made clear that variation in population arises from joint action of genotype and environment. If the yield is denoted by A, the phenotypic expression could be represented as

$$A = \mu + a + e + ae$$

where μ stands for mean, a for genotypic effect, e for environment and ae for G × E interaction.

Any particular genotype (or particular environment) adds or subtracts from the mean depending on whether its effect on yield is above or below the general population mean. The interaction (ae) will be zero, if genotypes behave constantly in all environments. Which means there are no genotype–environment interactions. To identify the effect of environmental factors affecting the performance of a genotype(s), soil fertility, rainfall pattern, temperature, sunshine, etc. are taken into consideration. To reduce the environmental effect, the genotypes are tested at a number of locations, years and also in the replicated randomized trials and the analysis is conducted to estimate the variances of above said factors influencing the phenotypic expression say yield (A), then

$$A = \mu + a + r + l + y + al + ay + ly + aly + e$$

where

μ = population mean
a = genotypic mean
r = replication
l = location
y = year

al, ay, ly and aly = genotype × location (al), genotype × year (ay), location × year (ly), and genotype × location × year (aly) interactions
e = error of measurement

PROCEDURE FOR ANALYZING GENOTYPE × ENVIRONMENTAL VARIABILITY

This is done by analyzing the variance for data on the lines tested in a replicated trial at a number of locations for two or more years (Table 29.1).

TABLE 29.1 Analysis of variance of data on the lines compared in replicated trials for two or more years and locations

Source of variance	*d. f.*	*Expected mean square*
Lines (genotype)	F-1	σ^2e + R σ^2aly + Ry σ^2al + RL σ^2ay + RLY σ^2a
Lines × years	(F-1) (y-1)	σ^2e + R σ^2aly + RL σ^2ay
Lines × locations	(F-1) (L-1)	σ^2e + R σ^2aly + Ry σ^2ay
Lines × year × locations	(F-1) (y-1) (L-1)	σ^2e + R σ^2aly
Error	(R-1) (FYL-1)	σ^2e

where

σ^2a = genotypic component of differences arising among the lines
σ^2al = a component arising from interaction between lines × locations
σ^2ay = line × year variance component
σ^2aly = line × location × year variance component
σ^2e = error variance

This type of analysis has been carried out by Hansen, Robinson and Comstock (1956) with Korean bean *Lespedeza*. They chose 284 lines (families) from 3 different hybrids in F_3 and F_4 generations, i.e. 136 lines from I cross; 65 lines from II cross and 83 lines from III cross. Each line (family) had been raised from seed of single randomly chosen F_2 plant. The individuals within an F_3 lines were bulked for F_4 tests. The trial was conducted for two years in two locations.

ESTIMATION OF COMPONENTS OF VARIANCE FOR TWO CHARACTERS IN KOREAN LESPEDAZA

Character	*Population*	σ^2a	σ^2al	σ^2ay	σ^2aly	σ^2e
Total yield	1	2480	0000*	2057	3099	16,578
	2	3354	0000*	858	2070	17,432
	3	2084	1777	1745	1461	15,071
Seed yield	1	317	27	101	259	1061
	2	206	9	123	322	933
	3	122	34	59	117	403

* = Negative estimates

$$\sigma A^2 = \sigma a^2 + \frac{\sigma al^2}{L} + \frac{\sigma av^2}{Y} + \frac{\sigma aly}{LY} + \frac{\sigma e^2}{RLY}$$

which is

$$\sigma A^2 = 2480 + \frac{0000}{2} + \frac{2057}{2} + \frac{3099}{2} + \frac{16,578}{8} = 6355.48$$

Thus,

$$H = \frac{2480}{6355.48}$$
$$= 0.3902$$
$$= 39.0\%$$

$$G = (k)(6A)\ \frac{(6a^2)}{(6A^2)}$$
$$= (2.06)\ \sqrt{(6355.48)}\ (0.3902)$$
$$G_a = 64 \text{ g/plot}$$

GENETIC ADVANCE

The important aspects of most breeding programmes are:

(a) Selection within a base population of genetically variable individuals or families
(b) Utilization of selected materials for the creation of new population to be used as commercial varieties or use as base population for new cycle of selection.

However, genetic advance under selection in new population as against the base population will depend upon:

(a) The amount of genetic variability, that is the magnitude of differences (in a newly selected plants) among different individuals (or families) in the base population, and
(b) The magnitude of masking effect of the environmental and interaction components of the variability on the genetic variability that is (e) and (ae), locations and year.
(c) A third factor, the intensity of selection will also influence the rate of genetic advance under selection.

Under the situation explained above, suppose, that selection has to be done among a large number of homozygous lines of a self-pollinated species. Furthermore, suppose that this selection is based on the information obtained from testing of each of these lines for yielding ability in replication yield trials conducted at several locations for several years. From the data obtained, it is possible to find mean yield denoted by A for each line. Presuming that this value of A is normally distributed with S.D. (6a), the frequency distribution of the mean value can be represented by Figure 29.1.

We can select the lines falling for in q portion of the total n lines. That is, all lines having yield a greater than $\overline{X}$ (mean yield) and some fixed values, $A^1(\overline{X})$. The expected genetic advance of the selected lines will be

$$Ga = (K)\ (6a)\ (H)$$

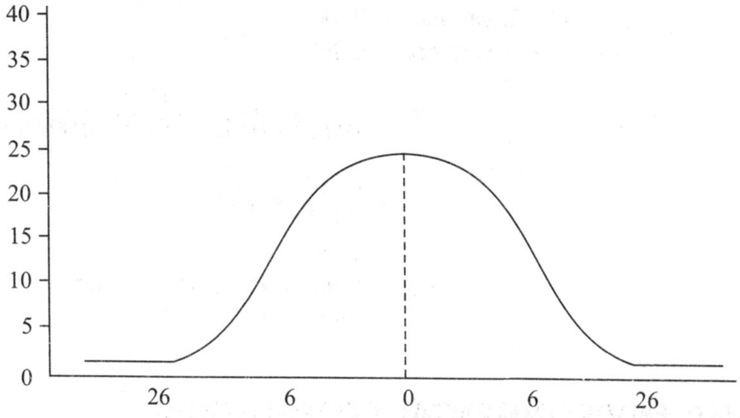

Figure 29.1 Theoretical distribution of mean yield of lines compared in the replicated yield trial.

Ga measures the difference between mean genotypic values of the q selected lines $\overline{X}_1$ and mean genotypic value of the n original line $\overline{X}_2$, 6A is the phenotypic standard deviation of the mean yield of the n lines. H is the heritability coefficient, K is the selection differential which takes into account the mean phenotypic value of q selected families (A), the mean phenotypic value of all n lines tested, the phenotypic standard deviation ($6A^2$) and the intensity of selection q/n.

Selection (%)	Intensity (i)
1	2.665
5	2.063
10	1.755
20	1.400

If a breeder is interested in calculating genetic advance per year of selection over the number of years required to complete the cycle of selection, If so, the gain per year Gy would be

$$\text{Ga} = \frac{K \times i \times V_p \times h^2}{Y}$$

where

i = 1, when both the parents are taken into consideration

y = number of years required to complete the cycle of selection. If selection is skewed to one, this value will be 0.5

Example

Taking magnesium (Mg) concentration in random sample of an outcrossing species, tall. The variance components are:

Additive variance = 0.114
Year variance = 0.022
G × Y variance = 0.015 for Mg concentration

$$G \times L \text{ variance} = 0.011$$
$$G \times L \times Y \text{ variance} = 0.009$$

$$h^2 = \frac{0.114}{(0.114 + 0.015 + 0.011 + 0.009 + 0.022)} \times 100$$
$$= \frac{0.114}{0.171} \times 100$$
$$= 67\ \%$$
$$Ga = 2.063 \times 1 \times 0.171 \times 0.67$$
$$= 0.57$$

GENETIC AND ENVIRONMENTAL CORRELATION

The association between two characters that can be directly observed is the correlation of phenotypic values or phenotypic correlation. Similarly, there is genetic correlation, which is the correlation of breeding values and environmental correlation. Thus a correlation, whatever its nature may be, is the ratio of the appropriate covariance to the product of two standard deviations, viz. phenotypic correlation coefficient is

$$rp = \frac{Covp}{\sigma P \times \sigma PY}$$

or

$$rp1.2 = \frac{Covp\,1.2}{\sqrt{(\sigma^2 p^1)(\sigma^2 p^2)}}$$

whereas phenotypic variance is the sum of genetic and environmental covariances. So, we can write phenotypic correlation as

$$rp = \frac{Cova + Cove}{\sigma P \times \sigma PY}$$
$$rp = h \times hyrA + ex\ ey\ rE$$

The method by which genetic correlation can be estimated is given below.

Estimate of Genotypic Correlation Coefficient

Here only principle is given, not procedure. Instead of computing the components of variance of one character from analysis of variance, we compute the components of covariance of two characters from an analysis of covariance, which takes exactly the same form as the analysis of variance. Instead of starting from the squares of the individual values and partitioning the sum of squares of the individuals' values and partitioning the sum of square according to the source of variation, we start from the product of the values of the two characters in each individual and partitioning the sum of products character to the source of variations. This leads to the estimates of the observational components of covariance. Therefore, genetic correlation is obtained as

$$rA = \frac{\text{Cov xy}}{\sqrt{\text{Var x, Var y}}}$$

$$rg = \frac{\text{Cov g 1.2}}{\sqrt{(\sigma^2 g^1)(\sigma^2 g^2)}}$$

Environmental correlation coefficient

$$re\ 1.2 = \frac{\text{Cove 1.2}}{\sqrt{(\sigma^2 e_1)(\sigma^2 e_2)}}$$

where

Covp 1.2 = phenotypic covariance for a pair of traits 1 and 2 (x x y)
Covg 1.2 = genotypic covariance for a pair of traits 1 and 2 (x x y)
Cove 1.2 = environmental covariance for a pair of traits 1 and 2 (x x y)
$\sigma^2 g_1 \sigma^2 p_1$ = genotypic and phenotypic variance for trait first (x)
$\sigma^2 g_2 \sigma^2 p_2$ = genotypic and phenotypic variance for trait second (y)

where covariances of offspring and parents for each characters are needed, then genetic correlation = rA

$$= \frac{\text{Cov xy}}{\sqrt{\text{Var xx, Var yy}}}$$

where Cov xy is called cross covariance and Cov xx and Cov yy are the offspring–parent correlation.

REFERENCES

Allard, R.W. (1960): *Principles of Plant Breeding*, John Wiley and Sons, New York.
Falconer, D.S. (1960): *Introduction to Quantitative Genetics*, Longman, London.
Mather, K. (1949): *Biometrical Genetics*, Mithuen & Co., London.
Panse and Sukhatme (1967): *Statistical Methods for Research Workers*, Indian Council of Agriculture Research, New Delhi.

CHAPTER

30 Population Genetics

Population genetics deals with the natural population consisting of interbreeding individuals, which shared a common gene pool. The gene alleles interact with each other in the pool with the environment and thus a selection pressure is exerted on these gene alleles. In 1908, G.H. Hardy a British mathematician and a German physician W. Weinberg independently developed a mathematical concept called *Hardy–Weinberg law* to determine the genetic equilibrium in a free mating population. This principle is the foundation of population genetics.

HARDY–WEINBERG LAW

According to their proposition, if a gene is represented in an indefinitely large random mating population by the neutral alleles, A and a in the ratio of qA: (1 – q) a, the frequencies of these alleles will remain constant in a specific genotypic proportions, viz. q2AA + 2q (1 – q) Aa + (1 – q)2 aa = 1 unless the changes are brought in the composition of population by selection, mutation, migration and non-random mating. This is very important proposition in the population genetics. To illustrate this proposition, suppose there are N individuals in a population and each individual has freedom to mate with any other individual. Furthermore, suppose in a random mating population D individuals are homozygous, H are heterozygous and R are homozygous recessive. Since AA individuals carry A alleles, A a individuals carry one A allele and other a, therefore, total number of A alleles in the population would be 2D + H, i.e. the proportion of A alleles would be 2D + H/2N or D + 1/2H/N. This proportion is called gene frequency of A. Then gene frequency of a may be H + 2R/2N or 1/2H + R/N. So, sum of A + a = unity.

The frequencies of A and a are recorded as q and (1 – q) or P and q, where q + (1 – q) = 1 or P + q = 1.

Example

If a sample of 100 individuals drawn from a population where composition of AA: A a and a alleles are in proportion of 25AA, 10Aa and 65aa. It means D = 25, H = 10 and R:65, then gene frequencies of A may be

$$Aq = \frac{25+5}{100}$$

$$= 0.30$$

$$(1 - q) = \frac{5+65}{100}$$

$$= 0.70$$

This proportion can be presented in other way, like

$$D + H + R = 1$$

$$q = D + \frac{1}{2H}$$

$$1 - q = R + \frac{1}{2H}$$

If AA = 0.25, Aa = 0.10 and aa = 0.65, then

$$q = 0.25 + \frac{1}{2}(0.10)$$

$$= 0.30$$

$$1 - q = 0.65 + \frac{1}{2}(0.10)$$

$$= 0.65 + 0.05$$

$$= 0.70$$

This is similar to as given above.

$$R = aa$$

HARDY–WEINBERG EQUILIBRIUM

In a large random mating population, both the gene frequencies and genotype frequencies remain constant from generation to generation in absence of forces like migration, mutation and selection and the genotypes frequencies are determined by the gene frequencies. This law can be used to determine the frequencies of each allele of a pair or of a series of genes as well as frequencies of homozygotes and heterozygotes in the population. Such a population is called Hardy–Weinberg equilibrium.

The deduction of the Hardy–Weinberg law involves three steps:

1. From parents to progeny, the gametes they produce.
2. From union of gametes to genotypes in zygotes they produce.
3. From genotypes of the zygotes to gene frequencies in the progeny generation.

1. The parents may have gene and genotype frequencies as follows:

Genotype:	A_1	A_2	AA_1	A_1A_2	A_2A_2
Frequency:	p	q	P	H	Q

Two shorts of gametes are produced: A_1 and A_2

The frequencies of these gametes are the same as the gene frequencies p and q in the generation producing them.

Reason: Since A_1A_1 individuals produce only A_1 gametes and A_1A_2 individuals produce two types of gametes in equal number of gametes bearing gene A_1 and A_2 (if no anomaly). So, the frequency of A_1 gametes produced by whole population would be P + 1/2H, which is the gene frequency of A_1.

2. Random mating between individual is equivalent to random union among their gametes:

	♂ gametes and their frequencies: A_1 p	A_2 q
♀ gametes and their frequencies: A_1 p	A_1A_1 p^2	A_1A_2 pq
A_2 q	A_1A_2 pq	A_2A_2 q^2

So, the genotype frequencies would be

$$\begin{array}{ccccc} 1A_1A_1 & : & 2A_1A_2 & : & 1A_2A_2 \\ p^2 & : & 2pq & : & q^2 \end{array} \qquad (30.1)$$

It is important to note that genotype frequencies depend only on the gene frequencies in parents and not on the parental genotype frequencies, if the parents mate at random.

3. Lastly, we use these genotype frequencies to determine the gene frequency in the offspring generation. Applying Eq. (30.1), we find the gene frequency of A_1 as $p^2 + 1/2\ (2pq) = p(p + q) = p$, which is same as in parent generation.

The properties of a population with respect to a single locus as expressed in the Hardy–Weinberg law and showed above are:

(a) In the absence of forces like migration, mutation and selection, a large random mating population is stable with respect to gene and genotype frequency.

(b) The genotype frequencies in the progeny produced by random mating are determined entirely by gene frequencies among the parents. Therefore:

(i) A population in Hardy–Weinberg equilibrium has relationship as expressed in Eq. (30.1), between the gene and genotype frequency in any one generation.

(ii) These Hardy–Weinberg genotype frequencies are established by one generation of random mating in respect of genotype frequencies of the parents.

Mating frequencies are another proof of Hardy–Weinberg law. Mating frequencies of the parents are good proof of Hardy–Weinberg law though little cumbersome. Let us again consider a locus with two alleles and also let the frequencies of genes and genotypes in the parents as before.

	Gene		*Genotype*		
	A_1	A_2	A_1A_1	A_1A_2	A_2A_2
Frequency	p	q	P	H	Q

The genotypes may be having 9 types of mating
(Genotypes and frequencies of male)

♂ gametes and their frequencies:

♀ gametes and their frequencies:	A_1A_1 p	A_1A_2 H	A_2A_2 q
A_1A_1, P	P^2	PH	PQ
A_1A_2, H	PH	H^2	HQ
A_2A_2, Q	PQ	QH	Q^2

$A_1A_1 \times A_1A_1$ produce only A_1A_1 offsprings. So, a proportion P^2 is A_1A_1 genotypes. Similarly one quarter. 1/4 of offspring of $A_1A_2 \times A_1A_2$ will be A_1A_1. So, this type of mating which has frequency of H^2 which contributes a proportion of 1/4 H^2 of total A_1A_1 progeny (Table 30.1).

TABLE 30.1 Genotype and frequency of progeny

Mating type	*Frequency*	A_1A_1	A_1A_2	A_2A
$A_1A_1 \times A_1A_1$	P^2	P^2		—
$A_1A_1 \times A_1A_2$	2PH	PH	PH	—
$A_1A_1 \times A_2A_2$	2PQ	—	2PQ	—
$A_1A_2 \times A_1A_2$	H^2	$1/4H^2$	$1/2H^2$	$1/4H^2$
$A_1A_2 \times A_2A_2$	2HQ	—	HQ	HQ
$A_2A_2 \times A_2A_2$	Q	—	—	Q^2
	Sum	$(P + 1/2H)^2$	$2(P + 1/2H)$ $(Q + 1/2H)$	$(Q + 1/2H)^2$
		P^2	2Pq	q^2

Thus, sum after simplification gave the same as given in Eq. (30.1) to be equal to $P^2 + 2pq + q^2$. This is called Hardy–Weinberg equilibrium.

To determine the genetic compositions of a group of individuals, we have to specify their genotypes and their number. Take for example, a single locus (single gene). Suppose at A locus, there are two different alleles A_1 and A_2 then there would be three possible genotypes, A_1A_1, A_1A_2 and A_2A_2. If one group of individual A_1A_1 are 1/4 of the total, the frequency of this genotype would be 0.25 or 25 per cent. Similarly, frequency of A_2A_2 would be 25 per cent and A_1A_2 would be 50 per cent, if the frequencies of all the genotypes together added up to unity or 100 per cent.

SINGLE LOCUS

Example

The M–N blood groups in man are determined by alleles at a locus. The three genotypes with three blood groups would be M, MN, N.

Mourant (1954) studied the blood group frequencies among Eskimoes of East Greenland and also among Icelanders, which are presented as

	Blood group			Number of individuals
	M	MN	N	
Greenland	83.5	15.6	0.9	569
Frequency %				
Iceland	31.2	51.5	17.3	747

	Gene		Genotype		
	A_1	A_2	A_1A_1	A_1A_2	A_2A_2
Frequencies	p	q	P	H	Q
Therefore,	p + q = 1	P +	H +	Q = 1	

Since each individual contains two genes (A_1A_2), the frequency of A_1 gene would be 1/2 (2P + H) and the relationship between gene frequency and genotype frequency among the individuals counted would be

$$p = P + \frac{1}{2}H$$

$$q = q + \frac{1}{2}H \tag{30.2}$$

The frequency of M gene in Greenland would be = 0.835 + 1/2 (0.156) = 0.913.
The frequency of N gene would be = 0.009 + 1/2 (0.156) = 0.087.
Therefore, gene frequencies in two population as expressed in percentage would be

	Gene	
	M	N
Greenland	91.3	8.7
Iceland	57.0	43.0

The genetic properties of a population are influenced in the process of transmission of genes from generation to generation by a number of factors. Some of these are: (i) differences in fertility and viability, (ii) migration and mutation, and (iii) mating system (random or promixia).

The Hardy–Weinberg law can be used to calculate the gene frequency under a number of conditions including completely dominant, codominant genes, multiple genes and sex-linked genes.

Single Locus Dominance

An interesting character with not known selective value is the ability or inability to taste the chemical *phenylthiocarbamide* (PTC) also called *phenylthiourea*. This test is very simple and can be easily performed by any genetics class. The testing ability depends on a completely dominant allele designated as taster (TT or Tt) and nontaster (tt).

Example

In our study, out of 280 students, 198 were taster and 82 were nontaster. From this data frequency of gene T and t in the sample can be calculated.

The 82 nontaster (29.29%) are persons with genotypes tt which according to the Hardy–Weinberg principle can be represented as q^2. Therefore,

$$q^2 = 0.2929$$

$$q = \sqrt{0.2929}$$

$$= 0.5412$$

This is the frequency of gene, because only a pair of allele is involved. Frequencies of two genes must be again total 1; p + q = 1.

Hence

$$p = 1 - q$$

$$p = 1 - 0.5412$$

$$= 0.4588$$

This is the frequency of gene allele T.

Now frequencies of homozygotes and heterozygotes can be calculated. The binomial expansion $p^2 + 2pq + q^2$ shows the distribution of three possible phenotypes:

$$p^2 = TT$$

$$= (0.4588)^2 = 0.2105$$

$$2pq = Tt$$

$$= 2(0.4588 \times 0.5412)$$

$$= 0.4966$$

$$q = tt$$

$$= (0.5412)^2$$

$$= 0.2929$$

Thus, by testing representative samples of different population, frequencies of gene alleles T and t as well as frequencies of homozygotes and heterozygotes can be calculated.

Single Locus Codominance

The M–N blood group represents a good example of a series of phenotypes due to a pair of codominant allele. The frequencies of two alleles in a population can be calculated from samples of two different groups.

Example

In a study consisting of 6129 white Americans living in New York City Boston and Columbus were found in the genotypic ratio of 1:2:1. These individuals can be grouped according to MN blood system as

Genotype	*Number*
MM	1787
MN	3039
NN	1303
	Total = 6129

For calculating the frequencies of two alleles M and N, one thing must be kept in mind that 6129 persons possess, a total of 6129 × 2 = 12,258 alleles. Wherein, the number of M alleles, for example is 1787 + 1787 + 3039. Thus, calculation of M and N frequencies may be done as follows:

$$M = \frac{1787 + 1787 + 3039}{12,258}$$

$$= \frac{6613}{12,258}$$

$$= 0.5395$$

$$N = \frac{1303 + 1303 + 3039}{12,258}$$

$$= \frac{5645}{12,258}$$

$$= 0.4605$$

Thus, frequencies of two alleles in this sample are almost equal and these are closer to the 1:2:1 genotypic ratio.

The frequencies of three genotypes in the population can be expressed as follows:

Genotype	*Phenotype*	*Phenotypic frequency*
MM	M	0.5395 × 0.5395 = 0.2911
MN	MN	2(0.5395 × 0.4605) = 0.4968
NN	N	(0.4605 × 0.4605) = 0.2121

These genotypic frequencies follow binomial distribution. If we say p represents frequency of M and q represents of N, the total arrays of probabilities of three genotypes (MM, MN, NN) may be

$$(p + q)^2 = P^2 + 2Pq + q^2$$

where $p + q = 1$, so $(p + q)^2 = 1$

Value of P = 0.5395 is already given. Similarly, the probability of MN (heterozygote) is given by 2pq and of N by q^2.

The frequencies of M and N can also be calculated by another approach:

$$q^2 \text{ has value } \frac{1303}{6129} = 0.2126$$

$$q = \sqrt{0.2126}$$

$$= 0.46, \text{ because } p + q = 1.$$

$$p = 1 - q$$

or $1 = 0.46 + 0.54$ – gene frequency

Example

A sample of Nawaho Indian from Mexico is as follows:

Phenotype	*Number*
MM	305
MN	52
NN	4
	Total = 361

Total number of alleles = 2 × 361 = 722

This sample does not represent 1:2:1 genotypic ratio. This indicates that frequency of alleles is not equal in the population. The data indicate M is more frequent than N.

$$\text{Let P is the frequency of M allele} = 305 + 305 + \frac{52}{722}$$
$$= 0.9169$$

$$\text{q is the frequency of N} = 52 + 4 + \frac{4}{722}$$
$$= 0.0831$$
$$p + q = 0.9169 + 0.0831$$
$$= 1.00$$

On the basis of these gene frequencies, the sample produces three types of genotypic frequencies:

Genotype	*Gene frequency*	*Genotype probability*	*Genotypes in sample*	
			Expected	*Observed*
MM	$P^2 = (0.9169)^2$	0.8407	303.5	305
MN	$2\ pq = 2\ (0.9169 \times 0.0831)$	0.1524	55.0	52
NN	$q^2 = (0.0831)^2$	0.0069	2.5	4
			361.0	361.0

$x^2 = 1.071$

MULTIPLE ALLELES

If we want to cover more than two alleles, no new principle is needed. Treat the situation as two allele system by considering each allele in turn and lumping the other together. In general, if q_1 and q_2 are the frequencies of any two alleles A_1 and A_2 of a multiple series, then the genotype frequencies under Hardy–Weinberg law would be

Genotype	A_1A_1	A_1A_2	A_2A_2
Frequency	q_1^2	$2q_1q_2$	q_2^2

These frequencies are also attained by one generation of random mating.

Example

The ABO blood groups in man are determined by a series of allelic genes. Let the frequencies of A, B and O genes be p, q and r, so that $p + q + r = 1$. Table 30.2 shows the (i) genotype,

(ii) blood groups (phenotypes) corresponding to different genotypes, and (iii) expected frequencies of blood groups in terms p, q, and r on assumption of Hardy–Weinberg equilibrium and (q) observed frequencies of blood groups (Race and Sanger, 1954) of 190177 airmen in UK.

TABLE 30.2 Genotypic, phenotypic (blood groups) and expected frequencies of blood groups

(i) Genotype	AA, AO	BB, BO	OO	AB
(ii) Blood group	A	B	O	AB
(iii) Frequency (%)				
Expected:	$p^2 + 2pr$	$q^2 + 2qr$	r^2	2pq
Observed:	41.716	8.560	46.684	3.040

The frequency of O gene is square root of the frequency of O group (Ceppellini, et al., 1955). Next, sum of the frequencies of the B and O groups is

$$q^2 + 2qr + r^2 = (q + r)^2 = (1 - p)^2$$

So

$$p = 1 - \sqrt{\bar{B} + \bar{O}}$$

where $\bar{B}$ and $\bar{O}$ are the frequencies of blood groups B and O

In the same way $q = 1 - \sqrt{\bar{A} + \bar{O}}$ and we have seen that

$$r = \sqrt{\bar{O}}$$

This method gives the following gene frequencies in the sample:

A gene:	p = 0.2567
B gene	q = 0.0598
O gene	r = 0.6833
	Total = 0.988

0.002 is sampling error to reach unity 1 = 0.

The expected frequency of AB from estimate of p and q is 3.070 per cent which is in good agreement with observed frequency of 3.040 per cent. ($x^2 = 0.7$, with i.d.f. calculated by method given by Race and Sanger).

Consider the following sample of 23,787 persons:

Phenotype	*Number*	*Frequency*
A	9,943	0.418
B	2,379	0.100
AB	904	0.038
O	10,561	0.444
Total =	23,787	1.000

Frequency of each allele can be calculated from these data, where p, q and r represent the frequencies of genes I^A, I^B and i respectively. The value of r, that is the frequency of gene r, can be calculated

$$r^2 = 0.444$$

Hence, $r = \sqrt{0.444} = 0.6663$—this is the frequency of i

Sum of the A + O phenotypes is given by

$$(p + r)^2 = 0.418 + 0.444 = 0.862$$

Therefore,

$$p + r = \sqrt{0.862} = 0.9284$$

So,

$$p = (p + r) - r = 0.9284 - 0.6663$$

$= 0.2621$—this is the frequency of I^A, because $p + q + r = 1$.

$$q = 1 - (p + r) = 1 - 0.9284$$

$= 0.0716$—this is the frequency of I^B.

The genotypic frequencies for a sample of 23,787 persons can now be calculated as:

Phenotype	*Genotype*	*Genotypic frequency*	*Population based on sample*	*Probability*
O	ii	r^2	0.4440	0.4440
A	$I^A I^A$	p^2	0.0687	0.4180
	$I^A i$	2 pr	0.3493	
B	$I^B I^B$	q^2	0.0051	
			0.0954	0.1005
AB	IA IB	2pq	0.0375	0.0375

SEX LINKAGE

In human beings, male possesses only one X chromosome while female two XX. Therefore, males cannot reflect the bionomial distribution for random combination of pairs of sex-linked genes, as females do. Hence, equilibrium distribution of genotypes for a sex-linked trait, when $p + q = 1$ could be given by

$$♂—p + q$$
$$♀—p^2 + 2pq + q^2$$

The red green colour blindness is controlled by sex-linked recessive gene (r). About 8 per cent of all males suffer from this deutan colour blindness. This shows that q, the frequency of r is 0.08 and p which represent the frequency of R (normal allele) is 0.92. Therefore, the frequency of colour blind female may be $q^2 = 0.0064$.

The sex-linked dominants can be dealt in the same way:

- In case of normal colour vision $p = 0.92$.
- The incidence of normal women $p^2 + pq = 0.9936$.

CHANGE IN GENE FREQUENCY

A large random mating population is though stable, but migration, mutation and selection, these three systematic processes change the gene frequency in the population. Figure 30.1 shows the impact of migration on the composition of new populations.

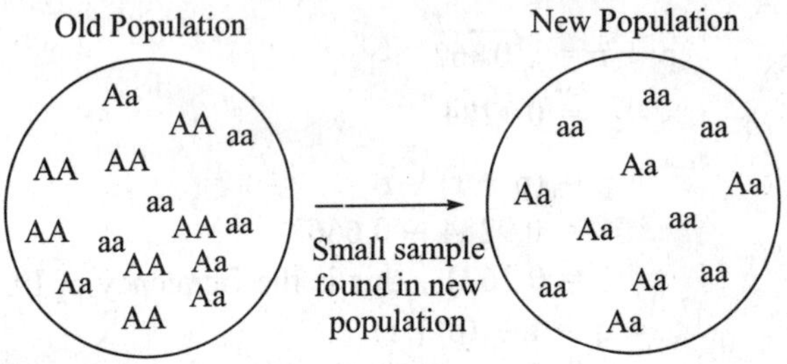

Figure 30.1 How migration of gene changes the genetic composition of new population.

MIGRATION

Let us suppose that a large population consists of a population m of new immigrants in each generation, the remainder 1 – m is the native. If the frequency of a gene being qm among the immigrants and q0 among the natives, the frequency of the gene in the mixed population q1 will be

$$
\begin{aligned}
q1 &= mqm + (1 - m)\, q0 \\
&= m\,(qm - q0) + q0
\end{aligned}
$$

The change in the gene frequency brought in one generation of the immigrants is the difference between the frequency before immigration and gene frequency after immigration

$$
\begin{aligned}
Q &= q1 - q0 \\
&= m\,(qm - q0)
\end{aligned}
$$

MUTATION

Mutation is of two types:

1. Nonrecurrent mutation
2. Recurrent mutation

Nonrecurrent Mutation

If mutation is taking in a single gene allele A → a, there will be one individual in a population of AA. Therefore, frequency of mutated gene is quite low. With a gene at a very low frequency, may be lost from the population. Therefore, unique mutation without selective advantage cannot produce permanent change in the population.

Recurrent Mutation

In this case, each mutational event recurs regularly with characteristic frequency and in a large population frequency of a mutant gene is never so low that complete loss may occur from sampling. Suppose gene A_1 mutates to A_1 and A_2 with a frequency u per generation (u is the proportion of all A_1 genes that mutate to A_2 between one generation and the next). If the frequency of X_1 in one generation is P_0 and the frequency of newly mutated X_2 genes in the next generation may be up_0. Therefore, frequency of new gene X_2 may be P_0—up_0 and the change of gene frequency may be up_0.

However, gene in question mutates in both the directions. Suppose, there are only two alleles X_1 and X_2 with frequencies P and Q; X_1 mutates to X_2 @ u per generation and X_2 mutates to X_1 @ v per generation, then mutation rate would be

$$X_1 \underset{v}{\overset{u}{\rightleftharpoons}} X_2$$

When initial gene frequency is P_0 and q_0, then change of gene frequency in one generation would be:

$$Q = up - uq$$

At equilibrium

$$Pu = qv$$

or

$$\frac{p}{q} = \frac{u}{v}, \quad \text{if Q is zero}$$

SELECTION

Selection causes changes in gene frequency and consequently in genotype frequency.

FITNESS AND SELECTION

Fitness of a genotype refers to its relative reproductive success and is measured by the proportionate contribution of the offspring to the next generation. Fitness indicates the adoptive selective value that is, fitness alone determines whether or not a genotype will be selected for the perpetuation of its genes in the next generation.

Change in the gene frequency due to selection depends on the degree of dominance with respect to the fitness. Various mathematical models have been constructed to explain how selection affects the genes frequently in a population.

GENETIC DRIFT

In a randomly mating population, one of the assumptions of Hardy–Weinburg law is that population is in the equilibrium state. However, genetic drift caused by sampling error affects the gene frequency in the population. This is because each population starts with a new gene frequency, and there is a continuous drift in the gene frequency from that of original population due to sampling error. The magnitude of drift that occurs in each generation is an independent event and depends largely on the size of the population in that generation.

In addition, drift in each generation is a random event, therefore, direction of change cannot be predicted. Continued process of transmission of genes from one generation to the other results in either fixation or elimination of some alleles from the population. Subdivision of a large population into small local population also creates greater opportunities for genetic drift. Extreme form of genetic drift occurs when a few individuals break off from a larger population to form a new colony. This is called *founder effect*.

REFERENCES

Ceppellini, R., C.D.M Siniscal and C.A.B Smith (1953): Estimation of gene frequencies in a random mating population, *Ann. Human. Genet.*, London, **20,** 97–115.

Comstock, R.E. and H.F. Robinson (1952): Estimation of average dominance of genes *in*: *Heterosis*, J.W. Gowen (Ed.), Ames Iowa State College Press.

Falconer, D.S. (1960): *Introduction to Quantitative Genetics*, Longman, London.

Hayes, K.K., F.R. Immer, and Smith (1955): *Methods of Plant Breeding*, McGraw-Hill, New York.

Murant, A.E. (1954): *In Distribution of Human Blood Groups*, Blackwell, Oxford.

Race, R.R. and R. Sanger (1954): *Blood Groups in Man*, 2nd ed., Blackwell, Oxford.

Sncdecor, G.W., *Statistical Methods*, 5th., Ames Iowa State College Press.

CHAPTER

31 Measurement of Quantitative Characters

If we have to classify individuals in a population according to their height, we cannot put them into group labels 'tall' and 'short' because there are all degrees of height and a division into classes will be basically arbitrary. The variation without natural discontinuities is called continuous variation and the characters which exhibit continuities are called *quantitative characters* or *metric characters* because their study depends upon measurement in metric units (cm, mm, mg, g, ml, etc.).

STATISTICAL PROCEDURES FOR ANALYSIS OF QUANTITATIVE CHARACTERS

A number of statistical methods have been evolved time and again to analyse the quantitative variations to make predictions about resemblance between the relatives or changes occurring in the population due to sampling or selection. The most useful statistical parameters to test the significance between population mean and sample mean are given in Table 31.1 with their notations:

TABLE 31.1 Statistical parameters to test the significance

Characteristics	*Sample*	*Population*
Number of variates	n	N
Mean	$\bar{x}$	m or μ
Variance	V	σ^2
Standard deviation	S.D.	σ
Correlation coefficients	r	ρ
Regression coefficients	b	β

Before I proceed to explain the application of above statistical parameters, it is important to make clear the differences between sample and population, the statistics and parameters used because these expressions represent the ways the statistical hypotheses are to be tested. The term *population* refers the infinitely large hypothetical population and sample represents a part or all of a population. The population is characterized and described by certain metric quantities called 'parameters' and function of observations on a random sample, which provide information about parameter is called statistics. Two most useful statistics for analysing the quantitative characters are *mean* (m or $\bar{x}$) and standard deviation (σ) variance (σ^2).

If the measurements are taken from two different characteristics say x, and y, of each series of the individuals, the variations of two variables are measured by covariance which is used to find out correlation and regression measure of the central tendency. Arithmetic mean provides the measure of central tendency of a population. *Mean* or average is the some of the observations divided by number. *Mode* is the class of greatest frequency in a series of central tendency of a population.

It can be also defined as "mode is that value of the variate which occurs most frequently".

$$\bar{X} = X_1 + X_2 + X_3 + X_4 + X_5 \ldots + X_n$$

$$= \Sigma \frac{X_i}{n}$$

where

X_1, X_2 = values of the observations

n = total number of individuals whose values were recorded

Σ = summation of all x values

$i = 1 \ldots n$

$\bar{X}$ = arithmetic mean of samples which is estimate of population mean

μ = population mean

Median

It is the value which is located in the middle of a series when observations are arranged in order of magnitude and it divides the series into two equal halves, half lying above it and half number of observations below.

Measure of Deviation (Dispersion)

The mean gives a satisfactory measure of central tendency of the population but it does not tell, how the individual observations are distributed around their mean. Two different samples may have the same mean value but in one case the individual observations may be closer to mean, while in other case, they may show a wider spread around it indicating presence of greater variability.

STANDARD DEVIATION (S.D.)

It is good measure of variation among different members of samples. Standard deviation is square root of the variance. Variance is the sum of squares of deviations (d) of individual readings from mean divided by degree of freedom (D.F. = n – 1) where n represents the number of readings.

Let us find out the S.D. value from data given in Table 31.2.

TABLE 31.2 Data to calculate S.D. and S.E.

Cross-value	*Frequency*	*Assumed mean (A.M.)*	*Deviation from assumed mean (A.M.)*	*fd*	fd_2
24	4	–	–6	–24	–144
26	8	–	–4	–32	128
28	10	–	–2	–20	40
30	4	30	0	0	0
32	6	–	+2	+12	24
	32			–64	336

$$\text{Actual mean} = \text{A.M.} + \frac{fd}{n}$$

$$= 32 + \left(\frac{-64}{32}\right)$$

$$\text{M} = 30$$

$$\text{Mean deviation} = \Sigma\ \frac{f.d.}{N}$$

$$\text{Standard deviation (S.D.)} = \frac{\sqrt{fd^2\ \frac{(\Sigma fd)^2}{n}}}{n-1} \quad \text{or S.D.} = \sqrt{\frac{\Sigma d^2}{N}} \quad \text{(Panse and Sukhatmi 1957)}$$

$$= \frac{\sqrt{336 - \frac{-(64)^2}{31}}}{31}$$

$$= 5.60$$

When S.D. is expressed as percentage of mean, it is termed as coefficient of variability (C.V.)

$$\text{C.V.} = \text{S.D.} \times \frac{100}{M}$$

$$= 5.60 \times \frac{100}{28}$$

$$= 20\%$$

STANDARD ERROR (S.E.)

Instead of a single sample, if more samples are drawn form a population and the standard deviation for each mean is calculated, the different values of S.D. will be different of statistical constant which measures such kind of dispersion called standard error (S.E.).

$$\text{S.E.} = \frac{\text{S.D.}}{\sqrt{n}}$$

$$= \frac{5.60}{\sqrt{32}}$$

$$= 1.8$$

Variance is sum of squares of standard deviation or S.E. *Coefficient of variability* is a measure of variability expressed in percentage of mean making it possible to compare the relative variability of two populations with widely different means.

Calculation

For the calculation of these statistics, it is better to produce some data. Hayes and Smith (1955) have worked out different statistical constants to measure variance by taking an example given in Table 31.3.

TABLE 31.3 Frequency of plots with yield grouped on 0.20 lb interval

Yield	*Class centre value* (*x*)	*No. of plots* (*f*)	*fx*	fx^2
2.8	4	↓	11.2	31.36
3.0	15	↓	45.0	135.00
3.2	20	↓	64.0	204.80
3.4	47	↓	159.8	543.32
3.6	63	↓	226.8	816.48
3.8	78	↓	296.4	1126.32
4.0	88	↓	352.0	1408.00
4.2	69	↓	289.8	1217.16
4.4	59	↓	259.6	1142.24
4.6	34	↓	156.6	719.44
4.8	11	↓	52.8	253.44
5.0	8	↓	40.0	200.00
5.2	4	↓	20.8	108.16
	500		1774.6	7905.72

where

s = means summation
f = frequency or number of plot
X = class center value, and
N = total number of plots

The formula for the calculation of m, x is

$$= \frac{s(fx)}{N}$$

In the present problem

$$S(X) = S\,[(4 \times 2.8) + (15 \times .30) \ldots + (4 \times 5.2)]$$

$$= 1974.6$$

To obtain mean yield, this value is divided by N (N = total number of plots (500). Numerically, this will be $\frac{1974.6}{500}$

$$m = 3.9492 \text{ lb per plot}$$

Mean is 4.0 lb (approximate)
Common measures of variation are *standard error* and *variance.*
Variance is the square of standard error (σ^2)

$$\text{Variable } (Vx) = \frac{\Sigma(fi(xi - x))^2}{\Sigma(fi) - 1}$$

where

Σ = summation of overall classes
fi = frequency of class fi
xi = value of class i
Σfi = summation of the number of observations overall classes
$\Sigma fi, xi$ = summation of produce
fi and xi = overall classes

$$\text{Standard error } (\sigma) = \frac{\sqrt{s(fx^2) - s(fx)\,x}}{N - 1}$$

$$= \sqrt{\frac{7905.72 - 1974.72\ (3.9492)}{499}}$$

$$= 0.464 \text{ lb}$$

The above formula could also be expressed as

$$\frac{\sqrt{s[f\,(x - x)]}}{N - 1}$$

This is the S.E. of a single determination.

$$\text{Correction factor} = \frac{s(fx)^2}{N}$$

The standard error of a mean is given by

$$\frac{S}{\sqrt{N}}$$

where

S is S.E. of a single determination.
N is the number of observations from which mean is determined.

Coefficient of variability (C.V.) is the relative measure of variability expressed in percentage. This can be also called S.D. expressed as a percentage of mean.

$$\text{C.V.} = \frac{\text{S.D.}}{\bar{X}} \times 100$$

S = standard error

$\overline{X}$ = mean of the sample

The standard error of difference (S as diff) could be calculated as

$$\sqrt{S^2a + S^2b - 2r\,ab.Sa.Sb}$$

where *a* and *b* represent the two treatments being compared and *r* is the correlation between separate measurements of the quantities.

When $r = 0$, the formula becomes

$$\sqrt{S^2a + S^2b}$$

where $Sa = Sb$ and $r = 0$, so S diff becomes $S = \sqrt{2}$

In this and in other similar problems, the significance of a difference is determined by comparing it with the S.E. of the difference.

COEFFICIENT OF CORRELATION

Coefficient of correlation (r) is used as a measure of the degree of association between two characters. Perfect positive correlation is +1 and perfect negative correlation is −1. No correlation is given by $r = 0$; the intermediate values denote degree of association. A convenient working formula is

$$rxy = \frac{S(xy) - S(x)\,S(y)/N}{\sqrt{\dfrac{s(x^2) - [S(x)]^2}{N}}\sqrt{\dfrac{s(y^2) - [S(y)]^2}{N}}}$$

where *x* and *y* are the two variables.

LINEAR REGRESSION

It measures the relationship between two variables and can be expressed by means of regression coefficient. This gives rate of change in one variable (the dependent variable) per unit rate of change in another (independent) variable.

Regression coefficient (*byx*)

$$= \frac{S(x - \overline{x})(y - \overline{y})}{S(x - \overline{x})^2}$$

$$= \frac{\dfrac{S(xy - S(x)\,S(y)}{N}}{\dfrac{S(x^2) - [S(x)]^2}{N}}$$

$$byx = \frac{\text{Sum of products of } xy}{\text{Sum of square of } x}$$

$$\text{Regression equation} = y = y + b(x - x)$$

where

Y = predicted yield
$\bar{y}$ = observed mean yield
x = number of kernel per spikelet

COVARIANCE

A new variance component known as covariance (cov) of X and Y is calculated by the following formula:

$$\text{Cov }(X \cdot Y) = \frac{\Sigma(Xi - X)(Yi - Y)}{n - 1}$$

Covariance measures how much the value of two traits, X and Y vary from one another about their respective mean values.

Correlations coefficient (r) of two traits in a sample is calculated as

$$r = \frac{\text{Cov}(xy)}{\sqrt{Vx \cdot Vy}}$$

Simple regression analysis deals with estimations and test of significance of two parameters α and β in the equation $y = \alpha + \beta x$; where α is the intercept of the line on y axis; β is the linear regression coefficient in the slope of the line, and y is the population mean corresponding to any x. Linear regression coefficient determines the extent of change in y with each unit change in x. If a and b denote α and β estimates, the regression line is determined by $Y = a + bx$.

The formula for calculation of a and b is

$$a = Y - bx$$

and

$$\frac{by}{x} = \frac{\text{Cov.}(xy)}{Vx}$$

where X and Y are mean values of X and Y traits and b(or $\frac{by}{x}$) is the regression of Y and X.

REFERENCES

Burns, P.W. and P.J. Brunitto (1989): *Science of Genetics*, Macmillan, New York.

Hayes, K.K., F.R. Immer and Smith (1955): *Methods of Plant Breeding*, McGraw-Hill, New York.

Panse and Sukhatmi (1967): *Statistical Methods for Agriculture Research Workers*, Indian Council of Agricultural Research, New Delhi (India).

Snedecor, G.W. (1956): *Statistical Methods*, 5th ed., Ames Iowa State College Press.

Glossary

Acentric chromosome: Chromosome segment lacking centromere.
Achiasmata: Meiosis without chiasma formation and crossing over.
Acrocentric: A chromosome or chromatid that has its centromere near the end.
Adapter RNA: It is also called transfer RNA or soluble RNA (sRNA). It is carries amino acids to RNA template in protein synthesis.
Adenine: A purine base found in RNA and DNA.
Adenosine: When adenine is attached with ribose (sugar).
Agglutinin: An antibody in blood plasma that brings about clumping (agglutination) of blood cells.
Albinism: Having no pigment in the skin, hair and eyes in animals/human beings and having no chlorophyll) in plants.
Alkaptonuria: An inherited metabolic disorder, where the patient excretes excessive amount of homogentisic acid (alkapton) in urine.
Allele: An alternative form of gene is called allele. When more than two alleles are situated on a locus, they are called multiple alleles.
Allotetraploid: An organism with four genomes derived from hybridization of different species.
Amino acid: Amino acids are building blocks of protein-containing amino (NH_2) group and carboxyl (–COOH) group.
Aminocentesis: A procedure for obtaining aminiotic fluid from pregnant woman.
Amorph: A mutation that obliterates gene function: anull mutation.
Amphidiploid: An interspecific hybrid derived from doubling the chromosomes in F_1 generation.
Anaphase: A stage of mitosis and meiosis during which the daughter chromosome passes from the equatorial plane to opposite pole of the cell.
Androgenetic haploid: Generation of haploid organism from male gamete or pollen in plants.
Aneuploid (heteroploid): An organism or cell having a chromosome number that is not an exact multiple of the monopod (n), e.g. 2n + 1 or 2n – 1.
Annealing: Pairing of complementary DNA strands to form a double helix. This is a technique to identify homologous region in two DNA strands.
Antibody: Substance in a tissue or fluid of the body that acts in antagonism to a foreign substance (antigen).

Anticodon: Three bases in a transfer RNA molecule that are complementary to the three bases of a specific codon in messenger RNA.

Antigen: A substance, usually a protein introduced into a living organism that elicit antibody formation.

ATP: Adenosine triphosphate, an energy-rich compound that promotes certain activities in the cell.

Attenuation: A mechanism for controlling gene expression in prokaryotes that involves premature termination of transcription.

Attenuator: A nucleotide sequence in the 5 region of a prokaryotic gene (or in RNA) that causes premature termination of transcription probably by forming secondary structure.

Auxotroph: A mutant microorganism (bacterium or yeast) that will not grow on minimal medium but require addition of semi-compound such as an amino acid or a vitamin.

Bacteriophage: Virus that attacks bacterial cells.

Bacteriocin: Toxic proteinaceous substance produced by bacteria that is toxic to sensitive bacterial strains. *E. coli* produces colicines.

Base analogue: Unnatural purine or pyrimidine base, differing slightly from normal base that can be incorporated into nucleic acid.

Balbioniring: A large ring-shaped, puffed configuration of a polytene chromosome.

Biotypes: Physiologic races of fungal diseases with similar genetic make-up.

Breeding value: Part of deviation from the population mean that is due to additive gene action.

Broad sense heritability: The proportion of the total phenotypic variance that is due to genetic variance.

Buoyant density: This is measure of ability of a substance to float in some standard fluid, viz. cesium chloride.

Bouquet configuration: Orientation of chromosome ends in contact with one region of nuclear membrane during leptotene stage of meiosis.

Cap: A structure formed at terminal base of an eukaryotic mRNA (5' GTP) by addition of methylated G after transcription.

Carcinogen: A substance capable of inducing cancer in an organism.

Centimoragan: Crossover unit.

Centriole: An organelle in the animal cell that is involved in the formation spindle during mitosis.

Centromere: Spindle fibre attachment region of a chromosomes.

Chiasma (Pl, chiasmata): A visible exchange of partners or a cross-over between two of four chromatids of two homologous chromosomes during first phase (pachytene) of meiosis.

Chimera (animal): Individual derived from two embryos.

Chimera (plants): A group of plant tissues composed of genetically-different cells.

Chloroplastid: Green structure in the plant cytoplasm that contains chlorophyll.

Chromatid: Identical strands of a chromosome after duplication.

Chromocentric: Body produced by the fusion of heterochromatic region of the chromosome (repetitive sequences) in the polytene tissues.

Chromonema (Pl. chromonemata): Single, thread-like chromosome.

Chromosome banding: Repetitive sequences forms bands along the length of chromosomes that take light dark stains.

Chromosome: Darkly stained, thread-like nucleoprotein body that serves as packing maternal of DNA.

Chromosome bridge: Dicentric chromosome forms bridge at anaphase of mitosis and meiosis.

Cistron: A unit of function in DNA.

C-mitosis (colchicine mitosis): Mitosis in which spindle formation is inactivated leading to increased chromosome contraction, as a result, normal distribution of chromatids is disturbed at anaphase and chromosome got doubled.

Codon: A set of three adjacent nucleotides in an mRNA molecule that specifies the amino acid incorporation in a polypeptide synthesis.

Constitutive heterochromatin: The inert state of permanently non-expressed DNA sequences—satellite DNA.

Coincidence: A ratio of observed frequency of double crossovers to the expected frequency.

Cointegrate: A DNA molecule formed by the fusion of two different DNA molecules usually mediated by a transposable element.

Colchicine: An alkaloid derived from *Autumn crocus* that arrests spindle formation.

Colinearity: A relationship in which the units in one molecule occur in the same sequence as the units in another molecule that they specify.

Complementary test (cis-trans test): An introduction of two recessive mutations into the same cell but on different chromosomes to determine whether they are alleles of the same genes and affect the same genetic function.

Constitution enzyme: An enzyme that is produced in fixed a quantity regardless of need.

Conversion: A process of nonreciprocal recombination, as a result, meiotic products occur in 3A: 1a or 1A: 3a ratio instead of 2A: 2a.

Coordinate regression: Correlated regulation of the structural genes in an operon by a molecule that interacts with the operator sequence.

Copolymers: The polymers of two kinds of monomers (organic bases) such as uracil and cytosine (poly, etc.) have been combined by genetic coding studies.

Coupling (cis-arrangement): The condition in which a double heterozyote has received two-linked mutations from a parent and their wild-type alleles from the other parent, e.g. ab/ab × → ab/++ (Cl-repulsion).

Covariance: A measure of the statistical association between two variables.

Crossing over: A process in which chromosomes exchange genetic material through breakage and reunion of their DNA molecules.

Cytoplasmic inheritance: Hereditary transmission of nonnuclear genetic materials through cytoplasm, e.g. plastids and mitochondria.

Decriminalization: Repelling movement of chiasmata towards the end of bivalent.

Deficiency (deletion): Loss of a segment of chromosome.

Degeneracy: Subjected to genetic code in which a particular amino acid is coded by more than one codon.

Deme: A local population of organism.

Denaturation: Loss of native configuration of micromolecule.

Dominance: A condition in which one member of an allelic pair is manifested to exclusion of other allele.

DNase: Any enzyme that hydrolyses DNA.

Dose compensation: A phenomenon in which the activity of a gene is increased or decreased according to the number of copies of that gene in the cell.

Ecotype: A population or strain of an organism that is adapted to a particular habitat.
Enchancer: A substance or object that increase a chemical activity or a physiological process.
Endomitosis: Duplication of chromosome without division of nucleus.
Endonuclease: An enzyme that breaks strands of DNA at internal position.
Endoplasmic reticulum: Network of membrane in the cytoplasm to which ribosome adheres.
Enzyme: A protein molecule that accelerates biochemical process in the cell.
Epigenetic: A term used for the nongenetic causes of a phenotype.
Eugenic: Application of genetic principles for improvement of human race.
Eukaryotes: The group of organism having nuclei enclosed in the membrane in their cells.
Euploid: An organism or cell having chromosome number that is exact multiple of the monopioid (n) or haploid.
Exon: Coding region of DNA.
Exonuclease: An enzyme that digests DNA or RNA beginning at the ends of strands.
Extrachromosomal: Nonchromosomal or cytoplasmic inheritance.
Flagellum: A whip-like organelle of locomotion in certain cell.
Frame-shift: Mutation that upsets reading frame of mRNA.
Gene: A fragment of DNA that controls specific biological function.
Gene conversion: A process in which one allele is replicated at the expense of another leading to nonmendelian segregation ratios in whole tetrad.
Gene flow: Spread of genes from one breeding population to another by migration.
Gene pool: The sum total of all different alleles in the breeding population.
Genetic load: The proportion by which the fitness of the optimum genotype is decreased by deleterious genes.
Genome: A complete set of chromosomes inherited as a unit by an organism.
Genotype: The genetic constitution of an organism.
Germplasm: The hereditary materials transmitted to the offsprings through germ cells.
Guanine: A purine base found in DNA and RNA.
Haploid (monoploid): An organism or cell having n number of chromosomes.
Helix: Any structure with spiral shape.
Haemoglobin: Conjugated protein molecule containing iron.
Heritability: Degree to which a given trait is controlled by genes.
Heterochromatin: Chromatin staining darkly during interphase.
Heterogametic sex: Producing two types of gametes with regard to sex chromosomes.
Histones: A group of proteins rich in basic amino acids.
Hfr: High frequency recombination strain of bacteria (*E. coli*).
Holandric gene: A gene carried on the Y chromosome and, therefore, transmitted from father to son.
Homeobox: A DNA sequence found in several genes that are involved in the specification of organs in the different body part of the animals.
Homeofic mutation: A mutation that causes a body part to develop in an appropriate position in an organism.
Homogametic sex: Producing all gametes alike.
Hypersensitive site: Regions of DNA that are highly susceptible to digestion with endonuclease.
Hypomorph: A mutation that reduces but does not completely abolish gene expression.

Inducer: A substance of low-molecular weight that inactivates a repressor by combining with it thereby stimulating gene expression.
***In situ*:** A Latin word meaning in natural condition.
Interference: Crossing over at one point reduces the chances of another crossing over to take place in its vicinity.
Interphase: The stage in the cell cycle when the cell is not yet dividing but DNA replication occurs.
Intersex: The phenotype which shows intermediate character of male and female.
Intron: Intervening sequences of DNA bases which do not code for any genetic information.
Inversion: Rearrangement of gene order on the chromosome.
***In vitro*:** A Latin word, meaning within glass; biologically it means raising plant species/ microorganism within glass under controlled conditions of temperature and humidity.
Isogenic stock: Genetically-uniform homozygous (similar) lines.
Kappa particle: *Paramoecium aurelio* which controls synthesis of paramicin, a toxic substance.
Karyotype: The chromosome constitution of a cell arranged according to length.
Lampbrush chromosome: Large diplotene chromosome present in the oocyte nuclei particularly found in amphibians.
Lethal allele: An allele that is fatal to the cell/organism.
Ligase: An enzyme that joins the ends of two broken DNA strands.
Linkage: When two or more genes are located on the same chromosome.
Linkage map: A linear or circular diagram that shows the relative position of gene on a chromosome as determined by genetic analysis.
Locus: Location of a gene or its allele on the chromosome.
LTR (Long Terminal Repeat): A DNA sequence present at each end of a retrotransposon.
Lysis: Bursting of cell following infection by virus.
Lysogenic bacteria: Bacteria harbouring temperate bacteriophages.
Map unit: Denotes crossover unit.
Mendelian population: A natural interbreeding sexually-reproducing plants or animals with common gene pool.
Messenger RNA (mRNA): RNA that carries information necessary for protein synthesis as dictated by DNA.
Microtubules: Hollow filaments in the cytoplasm, composed of mitotic spindle and make part of locomotor apparatus of motile cells.
Mitochondria: Organelles in the cytoplasm of plant and animal cells where oxidative phosphorelation takes place.
Modifiers: Genes that affect the expression of other genes.
Monomers: A single-molecular entity that may combine with others to form more complex structure.
Monosomic: A diploid organism lacking one chromosome of its genomic complex (2n – 1).
Mutation pressure: Repeated occurrence of mutation in a population.
Mutagen: Chemical or physical mutagenic agents.
Narrow-sense heritability: The proportion of genotypic variance that is due to additive genes.
Nondisjunction: Failure of separation of homologous chromosomes in mitotic/meiotic anaphase.
Nucleic acid: A macromolecule composed of DNA and RNA.

Nucleolar organizer: A chromosomal segment containing some genes located at secondary constriction of some chromosome that controls the synthesis of ribosomal RNA.

Nucleolus: An RNA-rich spherical sac in the nucleus of metabolic cell associated with nucleolar organizer.

Nucleosome: A spherical subunit of eukaryotic chromatin that is composed of histones and nucleotide pair.

Nucleotide: An unit of DNA/RNA molecule consisting of a nitrogen base, pentose sugar and phosphoric acid.

Nullisomic: A diploid cell/organism lacking a pair of chromosome (2n – 2).

Null-mutation: A mutation that abolishes the expression of a gene.

Octaploid: An organism having eight sets of a genome.

Oncogene: A gene that causes tumor formation in animals.

Oocyte: The egg mother cell that undergoes two meiotic divisions to form female gametes (egg cells).

Oogenesis: Formation of egg cell, in an ovum in animals.

Operon: A group of regulatory genes that control the function of structural genes.

Operator: A part of an operon that controls the activity of one or more structural genes by binding a regulatory protein.

Pair-rule gene: A gene that influences the formation of body segments in *Drosophila*.

Palindrome: A segment of DNA in which base-pair sequence reads the same in both the directions from a point of symmetry.

Panmictic population: A population in which mating occurs randomly.

Paracentric inversion: An inversion that is entirely within one arm of a chromosome and does not include centromere.

Penetrance: Percentage of individuals that show full phenotypic expression.

Peptide: A chain of amino acids dictated by DNA to form a particular protein.

Pericentric inversion: An inversion including centromere.

Phage: Bacteriophage (viruse).

Phenocopy: An organism whose external appearance has been changed due to environment to resemble other organism but not the genotype.

Phenotype: External appearance of an organism.

Phenylketonuria: Metabolic disorder controlled by recessive genes causing mental retardation.

Plaque: A clear area on a culture plate where bacteria have been killed by virus.

Plasmid: An extrachromosomal hereditary determinant that exists in an autonomous state in the bacterial cell.

Plastid: A cytoplasmic body found in the plant cell.

Pleiotropy: A single gene that controls more than one character.

Polydactyly: Occurrence of more than the usual number of fingers or toes.

Polymer: A compound composed of many smaller subunits which result from the process of polymerization.

Polymerase chain reaction (PCR): A procedure involving multiple cycles of denaturation, renaturation and polynucleotide synthesis that amplifies a particular DNA.

Polymerization: Chemical union of two or more molecules of the same kind to form a polymer of DNA/RNA.

Polymorphism: Two or more kinds of individuals maintained in a breeding population.

Polynucleotides: A linear sequence of nucleotides of DNA/RNA.
Polypeptide: A linear sequence of amino acid.
Polyploids: An organism with more than two sets of chromosomes (genomes).
Polytene chromosome: Giant chromosome produced due to interphase replication without division, as a result, many identical chromatides are arranged side by side.
Position effect: A difference in the phenotypic expression of a genotype due to change in the location of gene.
Primer: A short nucleotide sequence with a reactive 3-OH that can initiate DNA synthesis.
Promotor: A nucleotide sequence to which RNA polymerase binds and initiates transcription.
Prototroph: An organism (bacterium) that can grow on the minimal medium.
Provirus: A viral chromosome that has integrated into a host gene.
Pseudogenes: An inactive but stable component of a genome resembling a gene derived by mutation from active genes.
Quantitative inheritance: Inheritance of measurable traits.
Random genetic drift: Changes in the allele frequencies in a small breeding population.
Renaturation: Restoration of a molecule to its native form in nucleic acid biochemistry.
Restitution nucleus: A nucleus with unreduced or doubled chromosome number that results from failure of a meiotic or mitotic division.
Restriction enzyme: An endonuclease that recognizes a specific short sequence in the DNA and cleaves the DNA molecule.
Restriction map: A linear or circulate site of DNA that is cleaned by different restriction enzymes.
Restrotransposon: A transposable element that resembles the integrated form of retrovirus.
Reverse transcriptase: An enzyme that catalyzes the synthesis of DNA using an RNA template.
RFLP (Restriction Fragment Length Polymorphism): Genetic difference between individuals that is detected by comparing DNA fragment released by digestion with one or more restriction enzymes.
Rh^- factor: Antigen in the red blood corpuscles of certain pedicle that is, if transferred to a pregnant woman with Rh^-, the child may be aborted due to haemolytic disorder.
Satellite DNA: A short highly repetitive sequence that can be isolated from rest of the DNA by centrifugation.
Segment polarity gene: A gene that function to define the anterior and posterior components of body segments in *Drosophila*.
Selection pressure: Effectiveness of differential survival.
Selection differential: Difference between the mean of the individuals selected to be parents and the overall mean of the population.
Selection response: Difference between the mean of the individuals to be parents and the mean of their offspring.
Sex factor: A bacterial episome (e.g. F. plasmid in *E. coli*) that enables the donor cell to transmit its genetic material in the cytoplasm of recipient cell.
Silencer: A DNA sequence that helps to reduce or shut-off the expression of a nearby gene.
Spermatogenesis: The process by which maturation of gamctcs (sperms) of male takes place.
Splicity: The mRNA processing by which exons join together to form mature mRNA by eliminating the introns.

Structural genes: The genes that specify the synthesis of a polypeptide.
Synopsis: Pairing of homologous chromosomes.
Temperate phage: A phage (virus) that invades but does not destroy (lyse) the host (bacteria).
Template: One of the DNA strands, which dictates transcription of genetic information in messenger RNA.
Tetraploid: An organism/plant having four sets of same chromosome/genome.
Tetrasomic: An organism or nucleus possessing four numbers of one of its chromosomes (2n + 2).
Tetrad: Four cells arising out of second meiotic division in plants, the pollen grains and ascospores in fungi.
Totipotent cells: Undifferentiated cells (nuclei).
Transduction: Transfer of a segment of DNA from one bacteria (donor) to other (recepient) through bacteriophages which lead genetic recombinations in recepient bacteria.
Transfection: The uptake of DNA by a eukaryotic cell.
Transgenic: A term applied to an organism/plant having introduced alien gene in its genome.
Transposase: An enzyme that catalyzes movement of a DNA sequence.
Transposon: The DNA element that can move from one position to other in DNA and causes mutation.
Transversion: A mutation caused by substitution of a purine base for a pyrimidine base and vice versa in DNA/RNA.
Unequal crossing over: Crossing over between retreated DNA sequences that creates duplicated and deficient products.
Univalent: Unpaired chromosomes of meiosis I phase.
Vector: A plasmid or viral chromosome that can be used to construct recombinant DNA molecule for introduction into living cells.
Western blot: Transfer of proteins from an electrophoretic gel to a cellulose or nylon.
Wobble hypothesis: Explains how one tRNA recognizes two codons. The first two bases of the mRNA codon and anticodon pair properly, but third base in the anticodon has some play (or Wobble) that permits it to pair with more than one base.
Xenia: Genetic influence of the pollen parent on the endosperm of the plant.
Y chromosome: Male-determining chromosome in human being and in many animals.
Zygote: The cells produced by the union of male and female gametes.

Index